One-Dimensional Polymeric Nanocomposites

One-dimensional nanomaterials are emerging as promising materials for their many unique characteristics. This book covers their synthesis and applications in batteries, supercapacitors, fuel cells, solar cells, green energy production, flexible electronics, electrochemical sensors, and biomedicine.

Progress in nanotechnology offers an opportunity to synthesize materials with unique properties. The properties of nanomaterials can be further improved by growing them in one-dimension structures with variations in their architecture. One-dimensional polymeric nanocomposites offer various advantages such as nano dimensions, high surface area, structural stability, and the ability to tune their electrochemical, electronic, and optical properties. The book covers basic concepts, chemistries, properties, and the importance of one-dimensional nanomaterials, along with their wide applications and state-of-the-art progress in the energy, flexible electronics, sensor, and biomedical fields. The fundamentals of electrochemical behavior and our understanding of them for various applications are also discussed in detail.

This book will provide new direction to scientists, researchers, and students to better understand the chemistry, technologies, and applications of one-dimensional polymeric nanocomposites.

One-Dimensional Polymeric Nanocomposites

Synthesis to Emerging Applications

Edited by

Ram K. Gupta, Ph.D.

Associate Professor at Pittsburg State University

Tuan Anh Nguyen, Ph.D.

Head of Microanalysis Department at the Institute for Tropical Technology,
Vietnam Academy of Science and Technology

CRC Press is an imprint of the
Taylor & Francis Group, an **informa** business

First edition published 2023
by CRC Press
6000 Broken Sound Parkway NW, Suite 300, Boca Raton, FL 33487-2742

and by CRC Press
4 Park Square, Milton Park, Abingdon, Oxon, OX14 4RN
CRC Press is an imprint of Taylor & Francis Group, LLC

Library of Congress Cataloging-in-Publication Data
Names: Gupta, Ram K., editor. | Nguyen, Tuan Anh (Chemist), editor.
Title: One dimensional polymeric nanocomposites : synthesis to emerging applications / edited by Ram K. Gupta, Tuan Anh Nguyen.
Description: Boca Raton : CRC Press, 2023. | Includes bibliographical references and index.
Identifiers: LCCN 2022033927 (print) | LCCN 2022033928 (ebook) | ISBN 9781032116211 (hardback) | ISBN 9781032122502 (paperback) | ISBN 9781003223764 (ebook)
Subjects: LCSH: Nanocomposites (Materials) | Polymeric composites.
Classification: LCC TA418.9.N35 O3875 2023 (print) | LCC TA418.9.N35 (ebook) | DDC 620.1/92--dc23/eng/20221110
LC record available at https://lccn.loc.gov/2022033927
LC ebook record available at https://lccn.loc.gov/2022033928

ISBN: 9781032116211 (hbk)
ISBN: 9781032122502 (pbk)
ISBN: 9781003223764 (ebk)

DOI: 10.1201/9781003223764

Typeset in Palatino
by Deanta Global Publishing Services, Chennai, India

Contents

Contributors

Parisa Pour Shahid Saeed Abadi
Michigan Technological University
Houghton, MI

Nazlı Albayrak
School of Medicine
Acibadem M. A. Aydınlar University
Istanbul, Turkey

Joseph Sahaya Anand
Universiti Teknikal Malaysia Melaka (UTeM)
Melaka, Malaysia

Mohammad Arjmand
Nanomaterials and Polymer Nanocomposites Laboratory
School of Engineering
University of British Columbia
Kelowna, BC

Priya Arumugam
Alagappa University
Karaikudi, India

Mukesh Kumar Awasthi
College of Natural Resources and Environment
Northwest A & F University
Xianyang, China

Sonia Bahrani
Health Policy Research Center
Health Institute
Shiraz University of Medical Sciences
Shiraz, Iran

Trishna Bal
Birla Institute of Technology
Mesra, India

Anasuya Bandyopadhyay
Indian Institute of Technology Roorkee
Saharanpur, India

Sanjib Banerjee
Indian Institute of Technology
Bhilai, India

Anna Basa
University of Bialystok
Bialystok, Poland

Parameswaran Binod
Microbial Processes and Technology Division
National Institute for Interdisciplinary Science and Technology (CSIR-NIIST)
Kerala, India

Trinath Biswal
Veer Surendra Sai University of Technology
Odisha, India

Xian-Wu Cao
National Engineering Research Center of Novel Equipment for Polymer Processing
Key Laboratory of Polymer Processing Engineering of Ministry of Education
South China University of Technology
Guangzhou, China

Pooja Chawla
ISF College of Pharmacy
Moga, India

Vedhi Chinnapiyan
V.O. Chidambaram College
Thoothukudi, India

Angelica Corpuz
College of Engineering and Architecture
Cagayan State University
Tuguegarao City, Philippines

Itishree Jogamaya Das
Birla Institute of Technology
Mesra, India

Ubong Eduok
College of Engineering
University of Saskatchewan
Saskatoon, SK

Zari Fallah
University of Mazandaran
Babolsar, Iran

Richika Ganjoo
Lovely Professional University
Phagwara, India

Jose Garcia-Torres
Biomaterials, Biomechanics and Tissue Engineering Group
Research Center for Biomedical Engineering (CREB)
Universitat Politècnica de Catalunya (UPC)
Barcelona, Spain
and
Barcelona Research Center in Multiscale Science and Engineering
Universitat Politècnica de Catalunya
Barcelona, Spain
and
Institut de Recerca Sant Joan de Déu
Esplugues de Llobregat, Spain

Roger Gonçalves
Federal University of Sao Carlos
São Carlos, Brazil

Ram K. Gupta
National Institute for Materials Advancement
Pittsburg State University
Pittsburg, KS

Seyyed Alireza Hashemi
Nanomaterials and Polymer Nanocomposites Laboratory
School of Engineering
University of British Columbia
Kelowna, BC

Golnaz Heidari
University of Guilan
Rasht, Iran

Merve İlgar
Istanbul University-Cerrahpasa
Istanbul, Turkey

Divya Jayoti
Centre for Nano and Soft Matter Sciences, Shivanapura
Bengaluru, India
and
Indian Institute of Technology Madras
Chennai, India

Mizan İbrahim Kahyaoğlu
Ondokuz Mayis University
Samsun, Turkey

Karthikeyan Kandaswamy
Alagappa University
Karaikudi, India

Selcan Karakuş
Istanbul University-Cerrahpasa
Istanbul, Turkey

Masoud Kasraie
Michigan Technological University
Houghton, MI

Ashish Kumar
NCE
Department of Science and Technology
Government of Bihar
India

Devendra Kumar
Indian Institute of Technology
Bhilai, India

Sanjay Kumar
School of Biosciences and Bioengineering
DY Patil International University
Akurdi, India

Filiz Kuralay
Hacettepe University
Ankara, Turkey

Sevinc Kurbanoglu
Ankara University
Ankara, Turkey

Erlantz Lizundia
Life Cycle Thinking Group
Department of Graphic Design and Engineering Projects
Faculty of Engineering in Bilbao
University of the Basque Country (UPV/EHU)
Bilbao, Spain
and
Basque Center for Materials, Applications and Nanostructures
University of the Basque Country (UPV/EHU), Science Park
Leioa, Spain

Aravind Madhavan
Rajiv Gandhi Center for Biotechnology
Kerala, India

Sunil Kumar Baburao Mane
Khaja Bandanawaz University, Kalaburagi
Karnataka, India

G. Manjunatha
Shri Siddhartha Institute of Technology
Tumkur, Karnataka, India

Sk Arif Mohammad
Indian Institute of Technology
Bhilai, India

Seyyed Mojtaba Mousavi
National Taiwan University of Science and Technology
Taipei, Taiwan

Hari Prasath Nagaiah
Alagappa University
Karaikudi, India

Loc Thai Nguyen
School of Environment, Resources and Development
Asian Institute of Technology
Klong Luang, Thailand

Rafia Nimal
Quaid-i-Azam University
Islamabad, Pakistan
and
Ankara University
Ankara, Turkey

Zeynep Orhan
Ataturk University
Erzurum, Turkey

Cemal Özeroğlu
Istanbul University-Cerrahpasa
Istanbul, Turkey

Ashok Pandey
Centre for Innovation and Translational Research
Indian Institute for Toxicology Research (CSIR-IITR)
Lucknow, India

Tressia A.P. Paulose
Bishop Moore College
Kerala, India

A.A.P.R. Perera
National Institute for Materials Advancement
Pittsburg State University
Pittsburg, KS

Suguna Perumal
School of Chemical Engineering
Yeungnam University
Gyeongsan, Republic of Korea

Karapagavinayagam Petchimuthu
Environment, DCW Ltd
Thoothukudi, India

S. Krishna Prasad
Centre for Nano and Soft Matter Sciences, Shivanapura
Bengaluru, India

Parteek Prasher
University of Petroleum & Energy Studies
Dehradun, India

B.T. Punchihewa
University of Missouri–Kansas City
Kansas City, MO

Baby Suneetha Ragupathy
V.O. Chidambaram College
Thoothukudi, India

Reshmy Rajasekharan
Bishop Moore College
Kerala, India
and
Providence College of Engineering
Kerala, India

Aditya Dev Rajora
Birla Institute of Technology
Mesra, India

Seeram Ramakrishna
Center for Nanofibers and Nanotechnology
National University of Singapore
Singapore

Alimorad Rashidi
Nanotechnology Research Center
Research Institute of Petroleum Industry (RIPI)
Tehran, Iran

Sindhu Raveendran
Microbial Processes and Technology Division
National Institute for Interdisciplinary Science and Technology (CSIR-NIIST)
Kerala, India

Manuel Salado
Basque Center for Materials, Applications and Nanostructures
University of the Basque Country (UPV/EHU), Science Park
Leioa, Spain

Pragnya Satapathy
Centre for Nano and Soft Matter Sciences, Shivanapura
Bengaluru, India

Afzal Shah
Quaid-i-Azam University
Islamabad, Pakistan

Naghma Shaishta
Khaja Bandanawaz University, Kalaburagi
Karnataka, India

Uday Shankar
Indian Institute of Technology Roorkee
Saharanpur, India

Soheila Sharafinia
Shahid Chamran University of Ahvaz
Ahvaz, Iran

Mousmee Sharma
Uttaranchal University
Arcadia Grant, India

Shreya Sharma
Birla Institute of Technology
Mesra, India

Shveta Sharma
Lovely Professional University
Phagwara, India

Karutha Pandian Shunmugiah
Alagappa University
Karaikudi, India

Muhammad Siddiq
Quaid-i-Azam University
Islamabad, Pakistan

Shashi Singh
School of Biosciences and Bioengineering
DY Patil International University
Akurdi, India

Surabhi Sonam
School of Biosciences and Bioengineering
DY Patil International University
Akurdi, India

Felipe de Souza
National Institute for Materials Advancement
Pittsburg State University
Pittsburg, KS

Feiyan Su
College of Materials Science and Chemical Engineering
Harbin Engineering University
Harbin, China

Ozge Surucu
Ege University
Izmir, Turkey

Ezgi Tan
Istanbul University-Cerrahpasa
Istanbul, Turkey

Abhinay Thakur
Lovely Professional University
Phagwara, India

Yi-Zhang Tong
National Engineering Research Center of Novel Equipment for Polymer Processing
Key Laboratory of Polymer Processing Engineering of Ministry of Education
South China University of Technology
Guangzhou, China

Kaique Afonso Tozzi
Federal University of Sao Carlos
São Carlos, Brazil

Fahad Uddin
Birla Institute of Technology
Mesra, India

Didem Nur Unal
Ankara University
Ankara, Turkey

Bengi Uslu
Ankara University
Ankara, Turkey

Xiaodong Wang
College of Materials Science and Chemical Engineering
Harbin Engineering University
Harbin, China

Krzysztof Winkler
University of Bialystok
Bialystok, Poland

Wei Wu
City University of Hong Kong
Kowloon, Hong Kong SAR, China
and
National Engineering Research Center of Novel Equipment for Polymer Processing
Key Laboratory of Polymer Processing Engineering of Ministry of Education
South China University of Technology
Guangzhou, China

Anthony C.Y. Yuen
School of Mechanical and Manufacturing Engineering
University of New South Wales
Sydney, Australia

Ehsan Nazarzadeh Zare
School of Chemistry
Damghan University
Damghan, Iran

Hui Zhao
Guangxi Key Laboratory of Clean Pulp & Papermaking and Pollution Control
School of Light Industry and Food Engineering
Guangxi University
Nanning, China

Ting Zheng
College of Materials Science and Chemical Engineering
Harbin Engineering University
Harbin, China

Editors

Ram K. Gupta is an Associate Professor at Pittsburg State University, Kansas. Dr. Gupta's research focuses on conducting polymers and composites, green energy production and storage using biowastes and nanomaterials, optoelectronic and photovoltaic devices, organic-inorganic heterojunctions for sensors, bio-based polymers, flame-retardant polymers, bio-compatible nanofibers for tissue regeneration, scaffold and antibacterial applications, corrosion-inhibiting coatings, and bio-degradable metallic implants. Dr. Gupta has published over 250 peer-reviewed articles, made over 350 national, international, and regional presentations, chaired many sessions at national/international meetings, edited many books, and written several book chapters. He has received several million dollars for research and educational activities from many funding agencies. He is currently serving as Editor-in-Chief, Associate Editor, and Editorial Board Member of numerous journals.

Tuan Anh Nguyen completed his BSc in Physics from Hanoi University, Vietnam, in 1992, and his Ph.D. in Chemistry from Paris Diderot University, France, in 2003. He was a Visiting Scientist at Seoul National University, South Korea, in 2004 and the University of Wollongong, Australia, in 2005. He then worked as a Postdoctoral Research Associate and Research Scientist at Montana State University from 2006 to 2009. In 2012, he was appointed as Head of the Microanalysis Department at the Institute for Tropical Technology at the Vietnam Academy of Science and Technology. He has overseen four Ph.D. theses as thesis director and three more are in progress. He is Editor-in-Chief of *Kenkyu Journal of Nanotechnology & Nanoscience* and Founding Co-Editor-in-Chief of *Current Nanotoxicity & Prevention*. He is the author of four books in Vietnamese and editor of 32 Elsevier books in the *Micro & Nano Technologies Series*.

1

One-Dimensional Polymeric Nanocomposites: An Introduction

Sunil Kumar Baburao Mane, Naghma Shaishta, and G. Manjunatha

CONTENTS

1.1 Introduction

Polymer nanocomposites (PNCs) are filler-reinforced polymers in which the filler size is of the order of nanometers (less than 100 nm) in at least one dimension. According to the number of dimensions beyond the nanometric range, nanofillers can be classed as zero-dimensional (spherical), one-dimensional (layered), or two-dimensional (fibrous and tubular) nanofillers [1]. PNCs have attracted the interest of both academia and industry

DOI: 10.1201/9781003223764-1

since their first appearance in the 1990s [2]. Since then, they have been shown to offer better qualities in terms of low cost, barrier resistance, thermal stability, and low weight, as well as excellent optical, mechanical, and electrical qualities [3]. Organic or inorganic nanofillers improve the qualities of current polymers and contribute new qualities to the hybrid material, broadening the range of possible applications while maintaining the polymer's simplicity of production and processing [4]. The nano-size of the filler is responsible for the distinctive features of PNCs. Nanofillers, unlike micro- or macroscopic fillers, have a very high volume-to-surface ratio, which increases filler–filler and filler–matrix communications [5]. These materials have found commercial success in a variety of applications, such as flame retardants, stimulus-responsive materials, and automobile components [6]. In general, composite systems include three major matrixes: polymer, metal, and ceramic, with additional additives like flakes, particles, and lamina.

Nanocomposites are defined as composites having one constituent on a nanometric scale. Among the common matrices, polymers are far more complicated than metals or ceramics but they are also cheaper and easier to process. Polymers also have a reduced modulus and strength when used at lower temperatures [7]. As a result, these materials must be improved in terms of these latter qualities, depending on the requirements of the material being developed. Nanofillers are used as filler components in PNCs and are classified according to their physical structure, particle morphologies, and chemical origins. Based on their morphologies and nano dimensions, these particles are classed as one-dimensional linear (carbon nanotubes), two-dimensional powder (montmorillonite), or three-dimensional powder (silver nanoparticles) [8]. Over the past two decades, efforts have been made to exploit the properties of PNCs in the creation of materials with advanced applications in the field of batteries, fuel cells, supercapacitors, biomedicines, drug delivery, smart materials, etc. The roadway map of the past two decades (2001 to 2021) for research publications with the term "PNC/one-dimensional", based on the subject area, article type, and publication title from ScienceDirect, is shown in Figures 1.1 and 1.2.

1.2 Nanoscale Filler Classifications

In response to different requirements, nanoparticles with consistent characteristics can be designed into the polymer system to achieve a predictable effect. According to their diverse dimensions, those nanofillers appropriate for PNC applications may be broadly classified as one-, two-, or three-dimensional materials (Table 1.1).

1.2.1 One-Dimensional Nanofillers

These are plate-like bodies less than 100 nm in any one dimension, a few nanometers thick, and forming rather long sheets. Most one-dimensional nanofillers have distinct structural possessions, like a nano-wall, a nano-plate, a nano-disk, etc., which are vital in operative nano-devices. Montmorillonite clays, nanographene platelets, and ZnO nanosheets are among the one-dimensional materials that have received a lot of attention. Fe_3O_4 nanosheets are frequently employed in the domains of micro/nano electronics, biosensors, and chemical engineering, due to their exceptional optical, electrical, and magnetic characteristics [9].

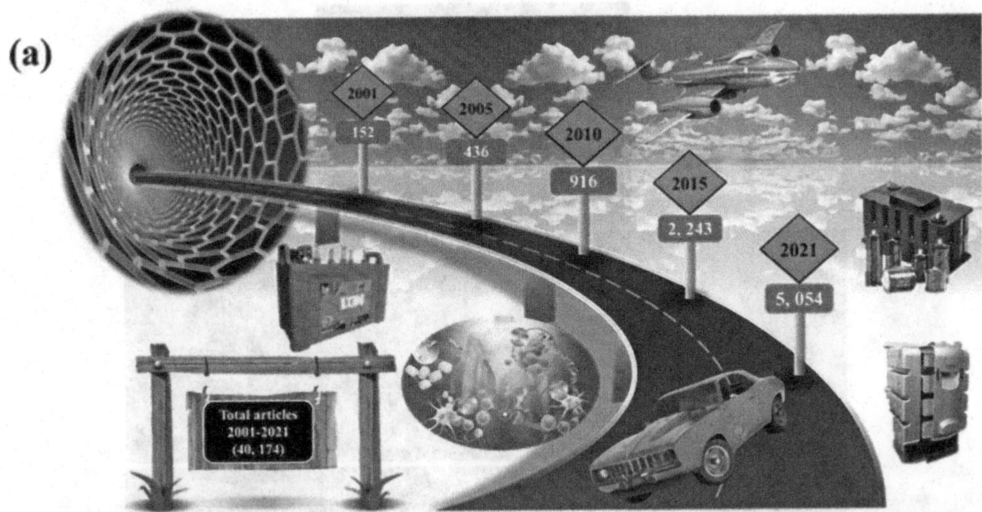

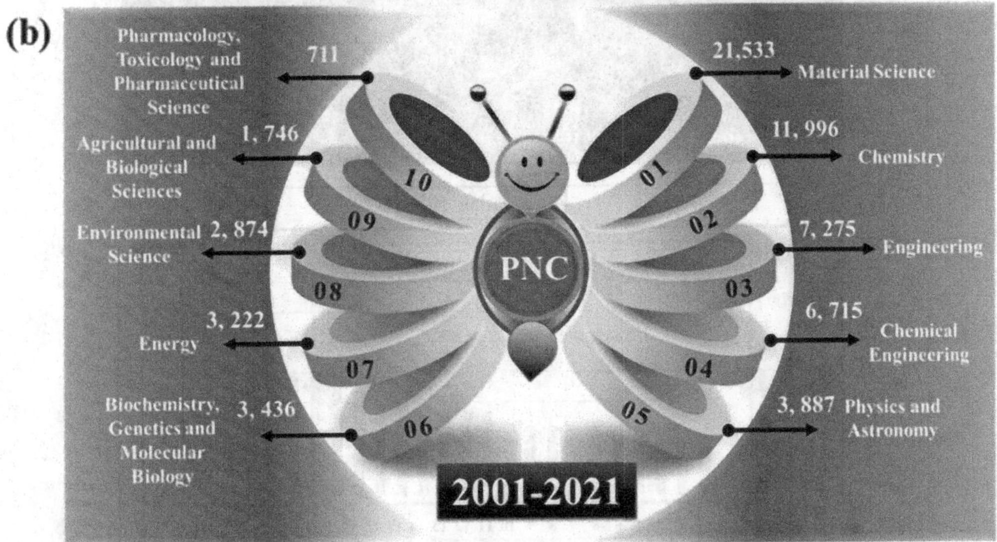

FIGURE 1.1
(a) Roadway map of two decades of publications and (b) number of publications on PNCs/one-dimensional based on the subject area from ScienceDirect (2001 to 2021).

1.2.2 Two-Dimensional Nanofillers

These are materials having two dimensions smaller than 100 nm, often in the form of rods. Carbon nanofibers, carbon nanotubes, halloysite nanotubes, nickel nano strands, and aluminum oxide nanofibers are examples of two-dimensional nanomaterials. Furthermore, nanotubes, plant fibers, nanowires, carbon fibers, oxides, graphene, molybdenum disulfide, and hexagon boron nitride are the most widely used two-dimensional nanofillers in PNCs. These fillers offer greater flame retardancy and striped features than one- and three-dimensional fillers, addressing an extensive number of requirements in the fields of catalysis, electronics, optics, sensing, and energy [10].

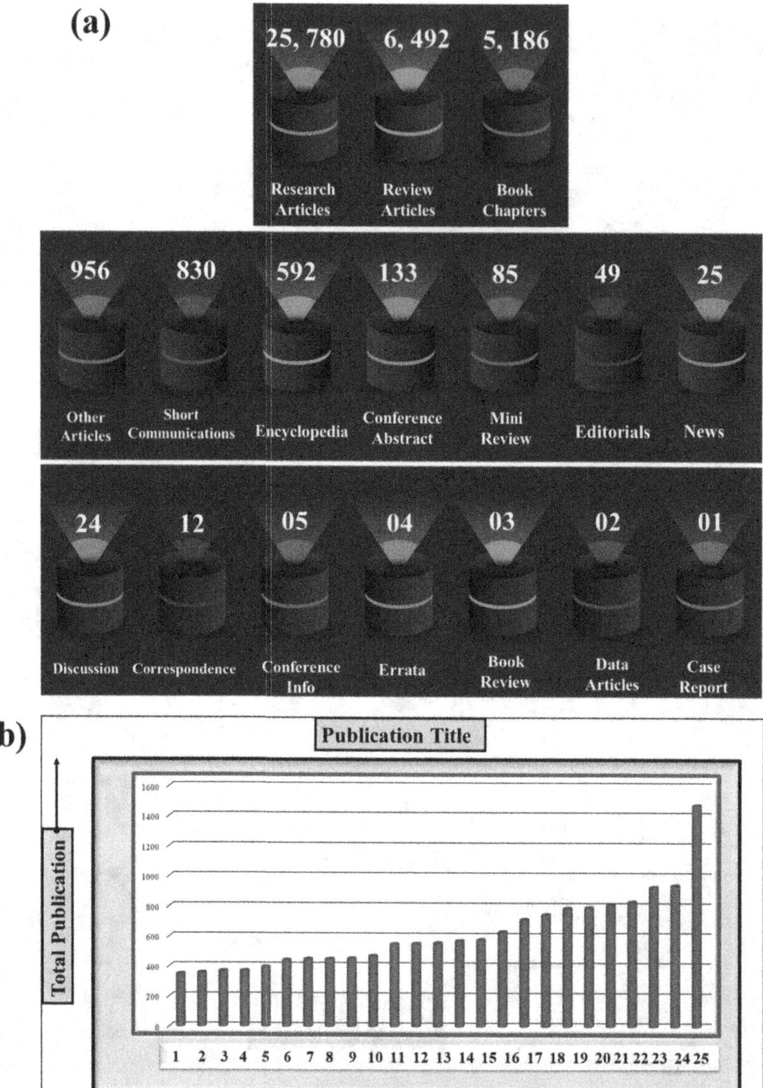

FIGURE 1.2
The number of publications of PNCs/one-dimensional (a) based on article type and (b) based on publication title from ScienceDirect (2001 to 2021). [Inset: 1: Int. J. Hydrog. Energy, 2: Colloids Surf A Physicochem Eng Asp, 3: Appl. Clay Sci, 4: Mater. Today: Proc, 5: Ceramics Int., 6: J. Membr. Sci, 7: Biosens. Bioelectron, 8: J. Power Sources, 9: Nano Energy, 10: Mater. Chem. Phys, 11: Mater. Sci. Eng. C, 12: Carbon, 13: Compos. Part A Appl. Sci. Manuf, 14: J. Colloid Interface Sci, 15: Eur. Polym. J, 16: J. Alloys Compd, 17: Int. J. Biol. Macromol, 18: Sens. Actuator B Chem, 19: Compos. Part B Eng, 20: Appl. Surf. Sci. 21: Electrochim. Acta, 22: Chem. Eng. J, 23: Carbohydr. Polym, 24: Compos. Sci. Technol, 25: Polymer.]

1.2.3 Three-Dimensional Nanofillers

These are nanomaterials with three dimensions measured on the nanometer scale; as a result, they are generally spherical or cube-shaped, and are also known as zero-dimensional particles. Polyhedral oligomeric silsesquioxane, nanosilicon, nano metal particles,

TABLE 1.1

A List of Nanomaterials Categorized According to Their Nanoscale Dimensions

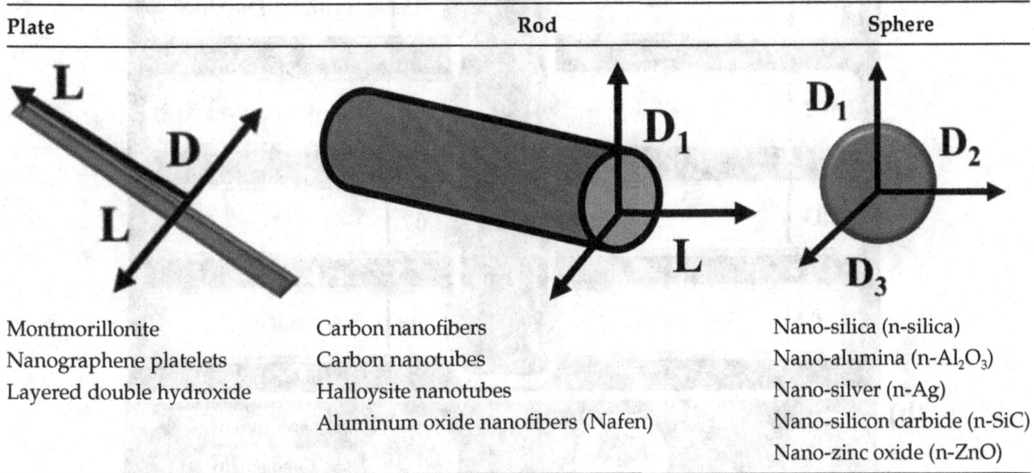

Plate	Rod	Sphere
Montmorillonite	Carbon nanofibers	Nano-silica (n-silica)
Nanographene platelets	Carbon nanotubes	Nano-alumina (n-Al_2O_3)
Layered double hydroxide	Halloysite nanotubes	Nano-silver (n-Ag)
	Aluminum oxide nanofibers (Nafen)	Nano-silicon carbide (n-SiC)
		Nano-zinc oxide (n-ZnO)

metal oxides, and quantum dots are the most frequent 3D nanomaterials. Metals and metal oxide nanoparticles, in particular, have the benefits of high stability, high catalytic activity, and ease of manufacture, and they are frequently used in the purification, coating, catalysis, and biological sectors, together with diverse polymers [11].

1.3 The Properties of Polymer Nanocomposites

One-dimensional (1D) nanostructured materials such as nanowires, nanofibers, and nanotubes (Figure 1.3a) have caught the attention of researchers due to their interesting properties and wide range of potential applications. A significant amount of research has been conducted on innovative techniques for the controlled synthesis of 1D nanomaterials. The distinguishing characteristics may lead to applications in fields like electronics, magnetism, optics, and catalysis. As a result, researchers in 1D nanomaterials are always seeking to produce novel basic science as well as prospective practical applications. As a result, the need for a platform that allows active researchers in this field to share their latest developments in a timely and efficient manner is critical.

Numerous features of the unique polymers and new qualities coming from the inclusion of nanoparticles can be considerably improved in PNCs. The key features of PNCs are described in Figure 1.3b, and they encompass physical, chemical, and biological aspects. The size, loading capacity, dispersion uniformity, aspect ratio, and interface interaction of the nanofiller with the polymer matrix affect the extent to which nanocomposite properties can be improved. Most polymers, with the exception of a few conducting polymers, lack conductivity because of covalent bonding and a lack of electron pathways. Surprisingly, novel PNCs produced by combining conductive nanofillers with insulating polymers demonstrate a wide range of electrical characteristics.

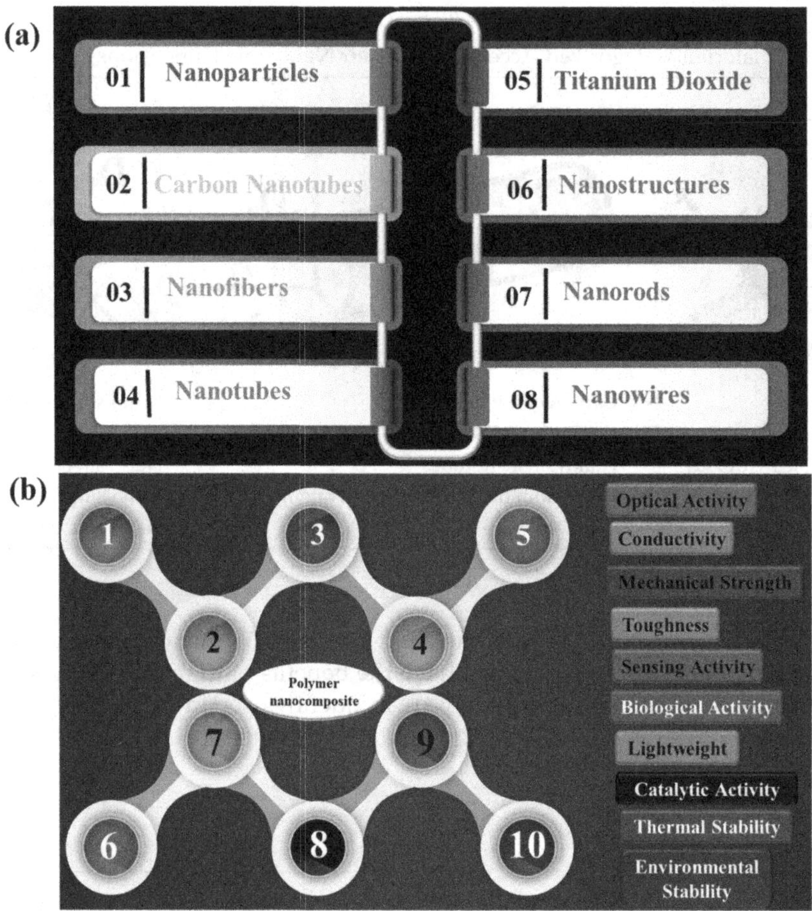

FIGURE 1.3
(a) Examples of PNCs and (b) important properties of PNCs.

1.4 Polymer Nanocomposite Design

The design of nanocomposites needs meticulous planning, material selection, preparation, and production. It involves deciding on a polymer in terms of the matrix and the filler, identifying the best production technique, considering end-use requirements, and then designing and constructing the nanocomposite. In response to real-world demands, the designer can select from a wide range of presently available polymers, like thermoplastics, thermoset, and elastomers, as well as fillers.

When selecting the filler and the polymer, a designer should consider the compatibility of the bonds and comforts for the integration and dispersion of the filler into the polymer; the selection of an appropriate, compatible combination of polymer and filler is a relatively difficult operation for the design of the nanocomposite polymer. The designer must evaluate many aspects, including the required properties, applications, operation, costs, environmental effects, and dispensations to achieve a successful nanocomposite polymer (Figure 1.4a). The use of these restrictions needs a rationale, specific features, and practical applications in order to develop nanocomposite polymers.

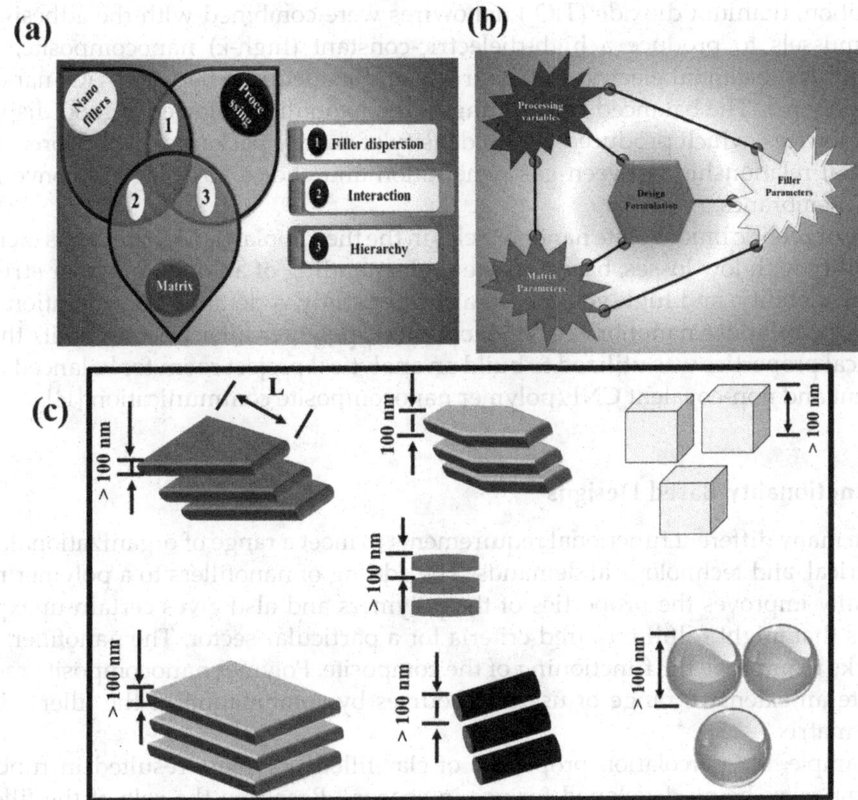

FIGURE 1.4
(a). Design tactic of PNCs, (b) parameters that affect the design of PNCs and (c) numerous shapes of nanofiller.

1.4.1 Designing with Rationality in Mind

In the engineering sector, it is usually a priority to have systems to achieve accurate measurements. The use of nanofillers in nanocomposites generates substantial chemical, macromolecular, and crystal-like compound transformations, which influence the properties of the bulk material. By anticipating how these assemblies affect composite behavior, the designer may produce new materials with specified functionality. For instance, it is significant that the electro-thermo-polarization (ETP) of polymer nanocomponents is reliant on the subsequent macroscopic conduction rather than the arrangement of the nanofiller, and this motivates nanofiller choice, the amount of nanofiller added, and the processing techniques, which govern morphology.

This sort of design necessitates a thorough grasp of the physical and chemical characteristics, as well as their interactions, among crystalline, chemical, and macromolecular structures. The essence of a sensible strategy links these assemblies with the major features. This requires a detailed understanding of the chemistry and physics of fillers and polymer components. The understanding of the essential importance of edge modes in graph/elastomer nanocomposites has been developed as a novel technique for durable covalent arcs. The rubber composites were strengthened with quinone-modified graphene. The alteration has been shown to effectively limit nanosheet repackaging and to result in even distribution and strong interface contact [12].

In addition, titanium dioxide (TiO_2) nanowires were combined with the adhesive protein in mussels to produce a high-dielectric-constant (high-κ) nanocomposite, which responded to substantial electrical incongruity and inadequacy between TiO_2 nanowires and the polymer. The balanced functioning of the nanofillers allowed a good dispersion of the nanowires, which produced a high density of energy packaging [13]. Moreover, the commercial relationship between gas penetration and choice has plagued conventional polymer membranes.

The use of zeolitic imidazolate nanoparticles in the thermoplastic Matrimid has increased permeability with low losses, based on the understanding of an open polymer structure, molecular mobility, and high volume. In a further study, a detailed consideration of the effects of the cellulose nanofibril (CNF) biomimetic polymer filler fraction on its thermo-mechanical properties was utilized to build an analytical project room for balanced design of covalent and non-covalent CNF/polymer nanocomposite communication [14].

1.4.2 Functionality-Based Designs

There are many different functional requirements to meet a range of organizational, medical, electrical, and technological demands. The adding of nanofillers to a polymer matrix additionally improves the properties of the polymers and also gives certain unexpected attributes that might fulfill required criteria for a particular sector. The nanofiller, therefore, works to improve the functioning of the composite. Polymer nanocomposite research can create an extensive range of useful resources by combination of the filler with the polymer matrix.

For example, the percolation properties of clay-filled polymers resulted in functional nano-composites being developed for use in sensors. Based on the role of the filler, the designer can develop a nanocomposite material. Some nanocomposites with functional features are described later in this subsection to highlight functionality-based designs. An example of the production of antibacterial functional polymer nanocomposites has been achieved by reactions between chitosan and metal salts to form metallic nanoparticles. A functional nanocomposite with a potential sensing capacity was developed using chitosan and Au/Ag nanoparticles. The introduction of Ag nanoparticles allowed Cu (II) ions to be detected.

The same nanocomposite can be changed by substituting the Au nanoparticle with nanoclay for NO_3^- detection. The combination of graphene oxide and poly(acryloyl-6-aminocaproic acid) was also developed in another study [15]. The additional work has resulted in the development of self-healing composites with sensing capacity, including reduced graphene oxide and polyborosiloxane [16]. Using carbon nanotubes and conducting polyurethanes from memory foam, functional nanocomposites were produced for actuator applications. For precise temperature control, a thin spray-coatable composite film, combining natural rubber latex, carbon fibers, and exfoliated graphite, has been developed [17].

To produce polarization-based absorbing characteristics appropriate for smart, flexible packaging films, Bernab et al. [18] utilized polyvinyl acetate (PVA) and silver nanoparticles. ZnSe nanowires in poly[2-methoxy-5-((2-ethylhexy)oxy)-1,4-phenylenevinyl] have been included at the p2-n heterojunction of an electroluminescence device to address electrical shortcomings [19]. CdSe/ZnS quantum dots (QDs) are employed as the blue emission layer in a polymer light-emitting diode in an alternating triarylamine functionalized fluorine-based copolymer [20].

1.4.3 Custom-Made Property-Based Designs

When employed to improve the polymer matrix, the nanoscale dimensions of the filler in these polymer nanocomposites generate a quantic effect, resulting in an extraordinary combination of features. The role and impact of the chemical filler is out of proportion to its size. In micro-sized, particulate-packaged composites, these effects are not achieved. For example, the temperature, module, and strength of the thermal distortion of clay were significantly improved without sacrificing impact strength or optical clarity when nanofiller was applied to nylon-6 [21].

Many desirable material characteristics, such as physical, chemical, optical and magnetic features, can be achieved simultaneously by adding nanoscale fillers into one polymer material [22]. This idea enables the fabrication of polymer nanocomposites with unique characteristics. Several types of chemical nature, particle form, particle size, and process parameters can be used to customize particular physical and chemical characteristics in a material [23]. Various examples can demonstrate significant achievements in this area. Research studies show that the processing and filler characteristics can be changed to produce a nanocomposite with regulated light dispersion capabilities. The use of a polymethacrylate (PMMA) matrix with TiO_2 nanofillers produced one of the most remarkable effects in this field, generating a 'blue sky look' [24].

To create large-scale nanocomposite panels with adjustable optical diffusion, researchers adapted an industrial method to synthesize PMMA and to dope the matrix with TiO_2 nanofillers. In addition, researchers were able to construct individual solar cells with built-in photovoltaic-utilizing QDs matched to PMMA and Stokes shift cadmium sulfate (CdS) or cadmium selenide (CdSe). The CdS core diameter ratio-to-shell-thickness (zinc sulfide) was altered to provide a varied absorption and emission range in the concentrator. The points were subsequently integrated into the PMMA, using a modified, manufacturing technique [25]. Another study generated hydraulic polymers at specific TiO_2 concentrations with customized refractive indexes within the 1.52–1.65 range. At a 44% TiO_2 concentration, the highly transparent nanocomposite has an organic molecule (2-ethyl-2-oxazoline) that partially covers its surface. The nanocomposite was shown to exhibit high visible light transparency and to absorb UV radiation but with a specific refractive index [26].

In addition, other researchers have added silica-coated Fe_3O_4 nanoparticles to polyvinyl-phosphonic acid to produce a drug delivery agent and absorbents to remove organic pollutants Niobium carbide (NbC) nanocomposite thin films developed on Si and polystyrene substrate also exhibit useful mechanical and electrical conductivity properties. The adjustment of the Nb flow to alter the nanocrystal size and the free carbon phase of films produced the required features [27]. To obtain exceptional gas barrier properties, researchers have developed an interconnected network of natural graphene oxide platelets. Organically modified nanoparticles of montmorillonite clay were combined with the epoxy-based negative photoresist SU-8 to build microcantilevers with tunable sensor sensitivity.

Appropriate modulus and thermal stability are provided by the resulting nanocomposites. All of them have benefited from talcum powder-filled polypropylene with customized elastic, thermal, and electrical characteristics, as well as infrared shielding, agricultural foils, or packaging compounds. A highly compatible, customized, mechanical feature has been produced using hybrid polypropylene (PP) strengthened with hexagonal boron hydride (h-BN) and nanohydroxyapatite [28]. Other remarkable, customized designs of properties include simulcasting mechanical and thermal properties through the incorporation of poly(tetrafluoroethylene) in tubular natural halloysite [29]. Mechanical and thermal distortions coupled with nylon-6 filled with ε-caprolactam montmorillonite resulted in better mechanical and thermal acetate characteristics and increased polylactic acetate mechanical and thermal acetate properties [30].

1.4.4 Design Parameters

The design of polymer nanocomposites, as indicated in Section 1.2.1, depends on several factors. Figure 1.4b illustrates the parameters controlling the design of polymer nanocomposites. The designer must take these aspects into account when designing a unique polymer nano-composition material, irrespective of the design. Process variables include, among other things, route, temperature, stress, and time. Filler factors include filler form, size and type; surface nature of filler; volume of filler or proportion by weight; and treatment. The matrix characteristics include polymer type, surface type, polymer chemistry, volume, and weight fraction and structure. There are three parameters: aspect ratio (AR), interface area, and orientation.

1.4.4.1 Aspect Ratio

A nanofiller is incorporated into one of three types of nanocomposite: 1D, 2D, or 3D. Wires, fibers, rods, tubes, disks, flakes, platforms, spheroids, and ellipsoids occur in many shapes and sizes. Nanofillers are classed as 2D materials despite significant variations in form and structure.

The nanofillers with these different sizes and shapes are described in Figure 1.4c. These measurements describe the geometrical characteristics of 2D nanofillers that are important to the manufacture of the surface and aspect ratio of nanocomposite polymers. The area in which nanofillers are distinguished from microfillers is one of the features; this is the field that is responsible for the matrix and the participation of the filler, for which the surface of nanofillers is too big.

AR is generally used to achieve easy determination of the geometrical measure of nanofillers, which is the ratio between the filler's greatest length and its smallest dimension. Measurements of all imaginable forms can be reduced to AR. The AR has a major impact on the interface area per filler volume [31]. Filler AR affects a wide variety of polymer nanocomposite properties. Yao and colleagues examined the influence of AR on dielectric and percolation properties of polyvinylidene fluoride plates [32].

Growth may be attributed to the increasing total surface area of the nanotubes. Likewise, the PMMA modulus for microelectromechanical systems, loaded with carbon nanotubes (CNTs), of the AR-dependent Young's modulus, has been found. The AR has been shown to have a critical value above which it is not impacted by the modulus [33]. The AR filler also influences thermal properties. In studies of clay-reinforced nylon-6 nanocomposites, the modulus, aspect ratio, and transition temperature of nanocomposites were shown to decrease as the AR fell [34]. The greatest electrical conductivity has been attained with a relatively high AR. Every treatment that lowered the AR level was also related to an increase in the percolation threshold (functionalization, ultrasonics). Another study utilizing epoxy and multi-wall CNTs has revealed that enhanced electrical conductivity is linked to higher AR. As far as AR is concerned, scientists have also shown that the percolation threshold is conductive after epoxy treatment [35]. The effect of ARs on the flammability of polystyrene nanocomposites by multiwall CNTs has been researched, finding that higher ARs reduce the flammability more effectively [36].

Nanoparticle AR has been shown to have an impact on the electric conductivity of graphene/polyurethane, polyester, or exfoliated graphite nanocomposites. Colonna and collaborators identified the improvements in electrical and thermal conductivity as a result of increased AR of graphene nanoplatelets (GNP) poly(butylene terephthalate) nanocomposite loading [37]. According to Liang and Pearson, AR has no impact on some polymer nanocomposites [38]. The oscillatory and nonmonotonic connections between the percolation threshold and the CNT AR have been found in thermoplastic elastomers by Dang and colleagues [39].

1.4.4.2 Interface

The interplay between nanofillers and the polymer matrix influences the mechanical, thermal, and electrical characteristics. For example, nanocomposite strength and modulation are controlled by the efficiency of stress transfer between matrix and filler, which, in turn, is influenced by interface area. When fillers are dispersed in a polymer matrix, the filler surface area creates a 'communication region' that changes the polymer's behavior, morphology, spatial cargo distribution, and bond splitting.

These changes demonstrate that the interface ratio of the nanocomposite is greater than that of microparticle-filled composites for the same volume of fillers. A large contact zone with a high interface area implies that the polymer's behavior and structure vary considerably. As a result of the structural property linkage of the components, the nanocomposite will have substantial differences in properties from those of microcomposites. This highlights the importance of interfacial contact in polymer nanocomposites. The degree of the impact of the filler presence is dependent on how the filler and the matrix interact. Communication between the filler and the polymer is a multifaceted process, including numerous processes of interaction, which are specific to the filler and the matrix. The nature of the communication and the parameters involved are depicted in Figures 1.5a and b.

These contact procedures differ, depending on the type, source, size, shape, and preparation of the polymers and fillers used. Weighting relates to a tendency of polymers to flow through the filler, cover the surface entirely and remove any air present; this is a measure of the polymer viscosity and the surface critical energy. The roughness of the surface affects the mechanical locking and has a major impact on shear strength. The interaction between loads is called the electrostatic interaction at the molecular level. It is highly active at relatively modest distances. In nanocomposites, the distance between fillers is very small. As such, an essential phenomenon is the electrostatic interaction at the filler–matrix border. Between some polymers and fillers, covalent bonding can occur. In contrast, certain polymers can penetrate the surface of the filler. These interactive properties are responsible for the broad variety of properties and phenomena shown by the nanocomposites when compared to the parent polymers. The cumulative interplay arising from these features determines the nature of strong or weak adhesion.

These representations are usually imprecise since they do not take into consideration all the properties of the fiber–matrix interface. Two methods, namely adhesion strength and adherence, are commonly used for experimental assessment of the nature of this interactive communication between polymer and fillers [40].

The contact angle is often employed as a quantitative measure of adherence. Using the Young equation, the surface and interface energy assessments may subsequently be utilized for calculation of adherence [41]. On the other hand, adhesion strength of scattered fibers is typically assessed using a single-fiber fragmentation test. Such an approach

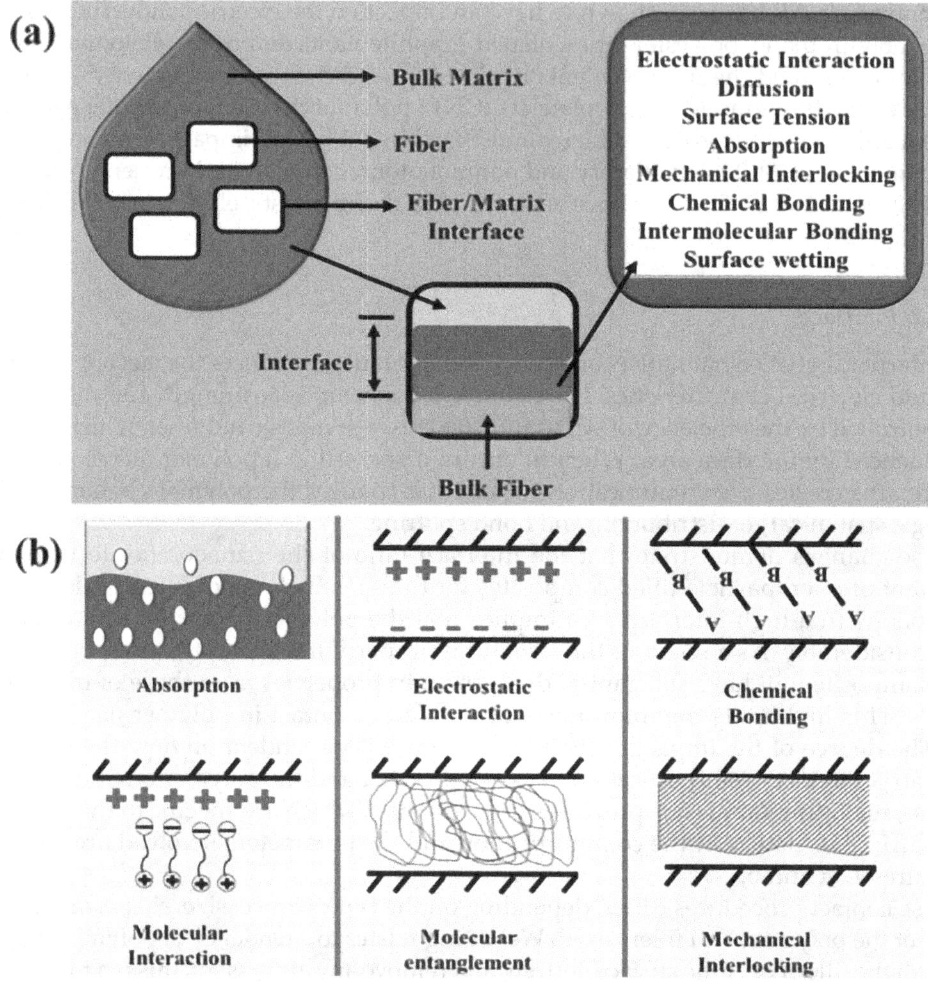

FIGURE 1.5
(a) Fiber–matrix interface interaction and (b) mechanism of interaction at the fiber–matrix interface.

includes placing the composite along the axis of a single fiber in a polymer matric and measuring tensile strength. The fragmentation of the fiber into separate lengths is subsequently associated with interfacial shear force.

Silica particle research has proven this to be true in packaging nylon-6 nanocomposite polymers. Researchers found that the treatment of aminobutyric acid silica nanoparticles for filler/matrix adhesion has led to a considerable increase in the tensile force of the nanocomposite. In addition, adapted composites showed particle dispersion, relative to the original composites. Comparable outcomes were reported with polypropylene filled with silica and nylon-6 filled with kaolin. According to Thio and colleagues [42], the tensile strength of the glazed polypropylene glass beads is enhanced by amplified interface adhesion. Greater interfacial strength has been found to offer increased yield strength for barium sulfate-filled polypropylene.

The fracture strength of silica particles included in nylon-6 polymer nanocomposites depends upon the interface strength of the fiber–matrix contact. This was also seen in the case of PP/$CaCO_3$ nanocomposites. Since most of the nanofillers are chemically inert, the polymer matrix is weak. CNTs, for instance, show little interfacial contact and are difficult to distribute in a polymer matrix due to their high electrostatic attraction [43]. There have been additional reports of chemically inert graphite and a few reports on hydrophobic graphite. This was also found in graphene and graphene oxide. On the other hand, natural fillers are inherently hydrophilic and do not properly adhere to naturally hydrophobic thermoplastic polymers. Fillers and polymers are often altered (functionalized) to improve interfacial adhesion. The two different forms of functionalization are covalent (chemical) and noncovalent (physical) (Figure 1.6a).

In covalent functionality [44], functional groups are bonded to the filler surface. This can be done in one of two ways. The first is the direct method, which means that the functional groups are directly connected to the filler walls. Bonding them directly onto the filler surface introduces oxygen-containing groups like carbonyl and carboxylic groups. The reaction of the filler with specific, chemically reactive molecules is typically achieved.

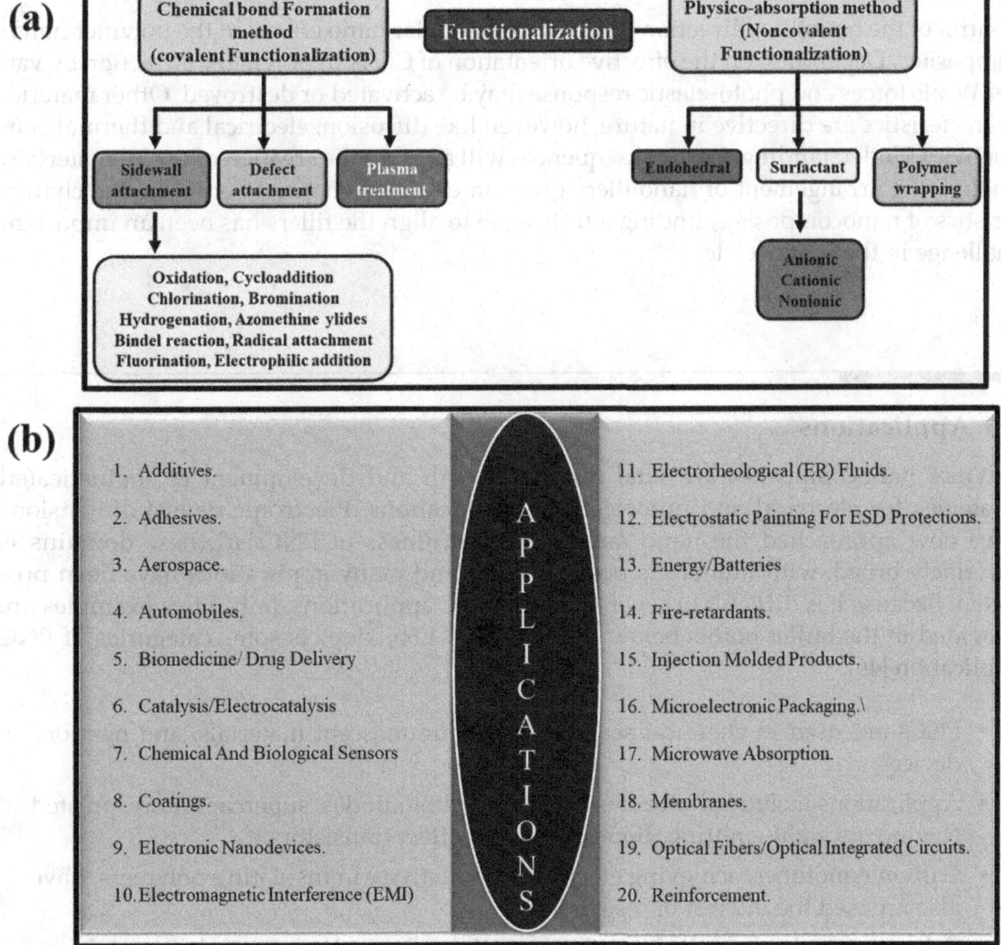

FIGURE 1.6
(a) Functionalization approach of PNCs and (b) various applications of PNCs.

The second technique is indirect functionalization, where feature groups are attached to the filler surface faults. Faults are produced intentionally by the use of a strong acid or by oxidation using an oxidant, stabilized and then linked to functional groups [45]. The development of a considerable number of such defects and the breakup of the fillers into smaller lengths are two drawbacks of covalent functionalization.

In non-covalent functionalization, plasma treatments are employed. Fluorine is covalently attached to the filler surface, before being exposed to an initiator of polymerization. The noncovalent functionalization uses interactions of hydrophobic stacking, electrostatic attraction, and polymer wrapping to functionalize the surface of the fillers with useful molecules or vigorous types like polymers and surfactants as assembly mediators. The mediators generated then act as anchors for other molecules. It is known that filler dispersion is hindered by noncovalent ways of operation. The capillary effect is used to specifically functionalize hollow tubular fillers, which consist of molecules stored in the interior cavity of the filler.

1.4.4.3 Orientation

Control of the nanofiller direction can lead to particular nano effects in the polymer nanocomposites. Depending on the effective orientation of CNTs in polymers, attraction by van der Waals forces and photo-elastic response may be activated or destroyed. Other material characteristics are directive in nature, however, like diffusion, electrical and thermal conductivity. Understanding these consequences will assist in the creation of useful materials. Though the arrangement of nanofillers gives an excellent chance to enhance the characteristics of nanocomposites, finding a technique to align the fillers has been an important challenge in the past decade.

1.5 Applications

Polymer nanocomposites are vital for the growth and development of sophisticated strategies for electrical and optoelectronic applications. Electronic device dimensions have now approached the nano range. The usefulness of PNCs in these domains is extremely broad, with numerous possible uses, and many applications have been projected. Because it is difficult to cover all of the PNC applications, only a few examples are provided in the bullet points below, while Figure 1.6b, depicts some categories of PNC application [46].

- PNCs are used in chemical sensors, electroluminescent materials, and memory devices;
- Applications include photovoltaic cells and photodiodes, supercapacitors, printed conductors, light-emitting diodes, and field-effect transistors;
- Carbon nanotubes, achieving electrical conductivity in insulating polymers, have also aroused the interest of researchers;
- Polymer-based solar cells have the potential to be used to create low-cost, large, flexible panels. The sole disadvantage is that the efficiency is significantly lower than that of commercial solar cells.

1.6 Synthesis of Polymer Nanocomposites

To achieve the purposes and advantages of fillers, it is important to evenly distribute the fillers within the matrix during the synthesis of PNCs. Because fillers are nanoscale, their distribution differs significantly from that of microscale fillers, as seen by the following characteristics. First, if the filling process is done by volume fraction, considerably more nanometer fillers are needed than micro fillers at the same volume percentage. As a result, the nanoparticles in the matrix are extremely tightly packed, resulting in increased van der Waals forces and electrostatic communications among them, making even distribution impossible. Second, anisotropic nanofillers have a high aspect ratio, making them more likely to clump together. Monolayer graphene, for example, has an aspect ratio of around 10^4, thus it tends to lower its surface energy by π–π stacking.

Third, a significant number of nanofillers with a large superficial area are injected into the polymer matrix, resulting in a very large interface area, altering the PNC's overall performance. As a result, even dispersion of nanofillers in the polymer matrix is the most important element in the production of PNCs. The common approaches for dispersing nanofillers and preventing nanoparticle agglomeration by employing peripheral energy are listed in Table 1.2.

1.6.1 Ultrasonication-Assisted Solution Mixing

This is the method most frequently used for producing PNCs [47]. The nanofillers and polymer are first dissolved in a solution. Then, the nanofillers are uniformly dispersed throughout the matrix with the use of ultrasound. The PNCs are then produced by evaporating the solvent. The ultrasonic energy, which is greater than the energy of communication among

TABLE 1.2

Detailed General Techniques Used for the Synthesis of PNCs

Methods	Appropriate Filler	Appropriate Matrix	Solvent	Regulatory Aspects
Ultrasonication-assisted solution mixing	All types	Liquid or various monomers of thermosets	Required	Time and sonication power
Shear mixing	Nanosheets	Liquid or various monomers or oligomers of thermosets	Required	Shapes of the rotor blades, rotating speed and time
Three-roll milling	Nanosheets and nanotubes	Liquid / various monomers or oligomers of thermosets	Not required	Speed of roller, gap between adjacent rollers,
Ball milling	All types	Liquid or solid thermoplastic and thermosets	Not required	time and mixing Ball size, rotating speed, ball or nanofiller ratio
Double-screw extrusion	All types	Solid thermoplastic	Not required	Processing temperature, screw configuration, rotation speed
In-situ synthesis	All types	Liquid or various monomers or oligomers of thermosets	Required	Chemical reaction conditions, temperature, condensation rate

the nanomaterials in the aggregates, separates the nanoparticles from their aggregation state into smaller units.

The aggregates of nanofillers are fragmented into smaller ones as the ultrasonic duration increases, and eventually they form individual nanoparticles that are self-governing with extra nanoparticles in the polymer. This technique has been widely employed in the study of novel nanocomposites due to its ease of use and consistent performance. However, because ultrasonic has a limited impact in high-viscosity solutions, most polymers must be kept at a low concentration and dissolved in a solvent with a high boiling point, which will impair the solvent elimination procedure and in turn, reduce the performance of the PNCs.

1.6.2 Shear Mixing

This is a far more widespread and straightforward approach than ultrasonic-assisted mixing, requiring simply stirring and having the potential for industrial-scale mass output. The shear force created by the stirrer spinning is utilized to separate the collection of nanofillers during the stirring operation. The nanoparticles will be separated by stirring and then collected later; owing to the low strength of the shear force, it is usually essential to raise the agitator speed to maximize this separation. Because this technique does not damage the structure of the nanofillers, it may be used to separate weakly linked nanoaggregates as well as to peel off certain stacked nanosheets.

1.6.3 Three-Roll Milling

This is a technique for distributing nanofillers within high-viscosity matrixes, like ink, adhesive coatings, and so on, using the shearing force between the rollers. Three cylindrical rollers move at various speeds in a three-roll milling machine, with neighboring rollers revolving in the opposite direction. Because the speeds of the rollers and the spacing among them can be adjusted, the particle size and the homogeneity of the packing can be precisely regulated. Furthermore, because the shearing force created among the rollers is greater than that produced by stirring, the technique may be used to process high-viscosity resources with little or no sludge.

As a result, CNTs, graphene nanosheets, nanoclays, and other anisotropic nanofillers are often dispersed using this approach [48], though, because the distance between adjacent rollers must be at least 1 μm, the dispersion effect of nanospheres with 3D dimensions of less than 100 nm is poor. Nanosphere aggregates can only be fragmented into smaller units, not into separate particles. The spinning of the roller, on the other hand, necessitates the use of adhesive resources, and nanofillers can only be distributed in a thermosetting matrix, not in a thermoplastic matrix.

1.6.4 Ball Milling

In the metallurgy and mineral dispensing industries, ball milling is frequently employed. It works on the basis of grinding and mixing powders in a compact area by utilizing the massive shear and compression forces created by severe ball collisions. This technique can distribute carbon nanotubes, graphene nanoparticles, Si nanoparticles, and BNs into thermoplastic and thermosetting polymers for the production of PNCs [49]. Ball milling produces a strong shear force that can tear off some 2D nanostructures like graphene, MoS_2, and BNs, but it cannot separate the interlayer structure linked by ionic bonding. The ball milling technique is also appropriate for solvent-free conditions, allowing nanofillers to be directly distributed in solid thermoplastic matrixes like polyphenylene sulfide, polyethylene, and polymethyl methacrylate.

1.6.5 Double-Screw Extrusion

The enormous shear force produced by the high-speed spinning of the double screw at high temperature during double-screw extrusion scatters nanofillers within the thermoplastic matrix [50]. Because of the advantages of solvent-free and environmentally sustainable technologies, this approach is extensively utilized in industry. The fillers may be distributed into the polymer at a high concentration to produce well-controlled performance, and this technique can be used with a variety of nanoparticle sizes, including graphene sheets, carbon nanotubes, and silicon dioxide.

This technique necessitates a higher temperature, which is beneficial in reducing polymer viscosity and for loading more nanofillers, but it also risks disintegrating polymers and nanofillers. The reason for this is that the resources include functional groups with limited heat stability. When the temperature is too high, the PNCs will fracture, causing their performance to deteriorate. Furthermore, the space between the screws is too big to uniformly distribute certain nanofiller aggregates, preventing consistent monodispersal of such nanofillers. As a result, additional technologies must be combined to improve performance.

1.6.6 In-Situ Synthesis

In addition to the above-mentioned ways of distributing organized nanofillers into polymers, *in-situ* synthesis, which directly produces nanoparticles in polymers from molecular precursors, is an additional, significant option. Physical and chemical strategies are the two types of synthesis, with chemical methods like the hydrothermal technique and the sol-gel technique. Physical *in-situ* synthesis involves converting a gas-phase precursor into inorganic nanoparticles by plasma activation, then condensing organic molecules over the surface of the inorganic nanoparticles to cover the polymer shell and create PNCs.

1.7 Conclusion and Prospects

Throughout this chapter, the fundamental concepts, characteristics, and synthesis processes of PCNs are all thoroughly covered. The construction and presentation of the composites enable them to be used in a variety of applications. The characteristics of PNCs are influenced by the orientation, shape, particle size, dispersion, and volume dispersion of the nanofillers. The interface contact among the matrix and filler regulates the majority of the mechanical, physical, and chemical properties of PNCs. As a result, the most essential characteristic in the synthesis of PNCs is the homogeneous dispersion of nanofillers.

The design space is not always a precise model, and it must be built with an awareness of the proper impacts of matrix, filler, and process factors. A critical condition in the formation of innovative PNCs with specified functionality is the ability to predict how chemical and macromolecular alterations, as well as changes in polymer crystalline structures, would disturb the overall performance of composites. To achieve improved qualities needs a deep grasp of the engagement process. Although these disciplines are still in their infancy, research into PNCs has lately developed much up-to-date expertise.

Acknowledgment

The first author, Sunil Kumar Baburao Mane, is thankful to University Grants Commission (UGC) Major Research Project (MRP) [F. no. 37-171/2009(SR)] New Delhi, India for financial support.

References

1. J. J. Joshua, P. S. Venkatanarayanan, D. S. Lohchab, R. Karandikar. *Mate. Today: Proceed.* 2021, DOI: 10.1016/j.matpr.2020.12.971.
2. Y. Kojima, A. Usuki, M. Kawasumi, A. Okada, Y. Fukushima, T. Kurauchi, O. Kamigaito. *J. Mater. Res.* 1993, 8(5), 1185–1189.
3. S. Fu, Z. Sun, P. Huang, Y. Li, N. Hu. *Nano Mater. Sci.* 2019, 1(1), 2–30.
4. D. Ponnamma, J.-J. Cabibihan, M. Rajan, S. S. Pethaiah, K. Deshmukh, J. P. Gogoi, S. K. Pasha, M. B. Ahamed, J. Krishnegowda, B. Chandrashekar. *Mater. Sci. Eng. C* 2019, 98, 1210–1240.
5. S. Ghasemi, E. N. Amini, M. Tajvidi, A. Kiziltas, D. F. Mielewski, D. J. Gardner. *J. Appl. Polym. Sci.* 2020, 138, 49983.
6. V. Kavimani, B. Stalin, P. M. Gopal, M. Ravichandran, A. Karthick, M. Bharani. *Adv. Polym. Technol.* 2021, 2021, DOI: 10.1155/2021/6627743.
7. I. I. Salakhov, N. M. Shaidullin, A. E. Chalykh, M. A. Matsko, A. V. Shapagin, A. Z. Batyrshin, G. A. Shandryuk, I. E. Nifant'ev. *Polymers* 2021, 13(11), 1821.
8. J. Barton, A. Niemczyk, K. Czaja, L. Korach. *Chemik* 2014, 68, 280–287.
9. T. Li, G. Ding, S. T. Han, Y. Zhou. *Polymer Nanocomposite Materials: Applications in Integrated Electronic Devices.* WILEY-VCH GmbH, 2021.
10. K. S. Rao, J. Senthilnathan, J. M. Ting, M. Yoshimura. *Nanoscale* 2014, 6(21), 12758–12768.
11. A. C. Matos, C. F. Marques, R. V. Pinto, I. A. C. Ribeiro, L. M. Gonçalves, M. A. Vaz, J. M. F. Ferreira, A. J. Almeida, A. F. Bettencourt. *Int. J. Pharm.* 2015, 490(1–2), 200–208.
12. Z. Yang, J. Liu, R. Liao, G. Yang, X. Wu, Z. Tang, B. Guo, L. Zhang, Y. Ma, Q. Nie, F. Wang. *Compos. Sci. Technol.* 2016, 132, 68–75.
13. G. Wang, X. Huang, F. Liu, J. Zhang, P. Jiang, *IEEE International Conference on dielEctrics (ICD),* Montpellier, 2016, 864–867.
14. A. J. Benítez, F. Lossada, B. Zhu, T. Rudolph, A. Walther. *Biomacromolecules* 2016, 17(7), 2417–2426.
15. H. P. Cong, P. Wang, S. H. Yu. *Chem. Mater.* 2013, 25(16), 3357–3362.
16. E. D'Elia, S. Barg, N. Ni, V. G. Rocha, E. Saiz. *Adv. Mater.* 2015, 27(32), 4788–4794.
17. E. Sauerbrunn, Y. Chen, J. Didion, M. Yu, E. Smela, H. A. Bruck. *Phys. Status Solid* 2015, 212, 2239–2245.
18. M. Bernabo, A. Pucci, H. H. Ramanitra, G. Ruggeri. *Mat. (Basel)* 2010, 3, 1461–1477.
19. G. H. Chen, S. J. Ho, H. S. Chen. *J. Phys. Chem. C* 2014, 118(44), 25816–25822.
20. S. P. Mucur, D. Lenkeit, I. Kanelidis, S. E. San, E. Holder, E. Tekin. *J. Polym. Sci. Part B Polym. Phys.* 2014, 52(2), 147–156.
21. E. I. Akpan, X. Shen, B. Wetzel, K. Friedrich, *Elsevier* 2019, 47–83.
22. A. Noll, K. Friedrich, T. Burkhart, U. Breuer. *Polym. Compos.* 2013, 34(9), 1405–1412.
23. T. Hanemann, D. V. Szabo. *Mat.* 2010, 3, 3468–3517.

24. A. Colombo, F. Tassone, F. Santolini, N. Contiello, A. Gambirasio, R. Simonutti. *J. Mater. Chem. C* 2013, 1(16), 2927–2934.
25. F. Meinardi, A. Colombo, K. A. Velizhanin, R. Simonutti, M. Lorenzon, L. Beverina, R. Viswanatha, V. I. Klimov, S. Brovelli. *Nat. Photonics* 2014, 8(5), 392–399.
26. A. Colombo, F. Tassone, M. Mauri, D. Salerno, J. K. Delaney, M. R. Palmer, R. D. L. Rie, R. Simonutti. *RSC Adv.* 2012, 2(16), 6628–6636.
27. L. Yate, L. E. Coy, G. Wang, M. Beltran, E. Diaz-Barriga, E. M. Saucedo, M. A. Ceniceros, K. Załęski, I. Llarena, M. Möllera, R. F. Ziolo. *RSC Adv.* 2014, 4(106), 61355–61362.
28. K. W. Chan, H. M. Wong, K. W. K. Yeung, S. C. Tjong. *Mat. (Basel)* 2015, 8, 992–1008.
29. S. Gamini, V. Vasu, S. Bose. *Mater. Res. Express* 2017, 4(4), 45301.
30. D. Notta-Cuvier, M. Murariu, J. Odent, R. Delille, A. Bouzouita, J. M. Raquez, F. Lauro, P. Dubois. *Macromol. Mater. Eng.* 2015, 300(7), 684–698.
31. R. Gulotty, M. Castellino, P. Jagdale, A. Tagliaferro, A. A. Balandin. *ACS Nano.* 2013, 7(6), 5114–5121.
32. S. H. Yao, Z. M. Dang, M. J. Jiang, H. P. Xu, J. Bai. *Appl. Phys. Lett.* 2007, 91, 072912.
33. B. Ashrafi, P. Hubert, S. Vengallatore. *Nanotechnology* 2006, 17(19), 4895–4903.
34. F. H. Gojny, M. H. G. Wichmann, B. Fiedler, I. A. Kinloch, W. Bauhofer, A. H. Windle, K. Schulte. *Polymer (Guildf).* 2006, 47(6), 2036–2045.
35. M. R. Ayatollahi, S. Shadlou, M. M. Shokrieh, M. Chitsazzadeh. *Polym. Test.* 2011, 30(5), 548–556.
36. B. H. Cipiriano, T. Kashiwagi, S. R. Raghavan, Y. Yang, E. A. Grulke, K. Yamamoto, J. R. Shields, J. F. Douglas. *Polymer (Guildf).* 2007, 48(20), 6086–6096.
37. S. Colonna, M. M. Bernal, G. Gavoci, J. Gomez, C. Novara, G. Saracco, A. Fina. *Mater. Des.* 2017, 119, 124–132.
38. Y. L. Liang, R. A. Pearson. *Polymer (Guildf).* 2010, 51(21), 4880–4890.
39. Z. M. Dang, K. Shehzad, J. W. Zha, T. Hussain, N. Jun, J. Bai. *Jpn. J. Appl. Phys.* 2011, 50(8R), 080214.
40. S. Y. Fu, X. Q. Feng, B. Lauke, Y. M. Mai. *Compos. Part B Eng.* 2008, 39(6), 933–961.
41. D. E. Packham. *Int. J. Adhes. Adhes.* 1996, 16(2), 121–128.
42. Y. S. Thio, A. S. Argon, R. E. Cohen. *Polymer (Guildf).* 2004, 45(10), 3139–3147.
43. S. J. Nikkhah, M. R. Moghbeli, S. M. Hashemianzadeh. *J. Mol. Model.* 2015, 21(1), 1–12.
44. M. Karimi, N. Solati, M. Amiri, H. Mirshekari, E. Mohamed, M. Taheri, M. Hashemkhani, A. Saeidi, M. A. Estiar, P. Kiani, A. Ghasemi, S. M. M. Basri, A. R. Aref, M. R. Hamblin. *Expert Opin. Drug Deliv.* 2015, 12(7), 1071–1087.
45. S. C. Wang, K. S. Chang, C. J. Yuan. *Electrochim. Acta* 2009, 54, 4937–4943.
46. M. Tyagi, D. Tyagi. *Int. J. Electron. Electr. Eng.* 2014, 7, 603–608.
47. X. Shen, Z. Wang, Y. Wu, X. Liu, Y. B. He, J. K. Kim. *Nano Lett.* 2016, 16(6), 3585–3593.
48. Y. Li, H. Zhang, E. Bilotti, T. Peijs. *MRS Adv.* 2016, 1(19), 1389–1394.
49. J. Gu, Y. Guo, X. Yang, C. Liang, W. Geng, L. Tang, N. Li, Q. Zhang. *Compos. Part A: Appl. Sci. Manuf.* 2017, 95, 267–273.
50. J. Fawaz, V. Mittal. *Synthesis of Polymer Nanocomposites: Review of Various Techniques.* Wiley-VCH, 2014.

2

One-Dimensional Polymeric Nanocomposites: Current State-of-the-Art

Sk Arif Mohammad, Devendra Kumar, and Sanjib Banerjee

CONTENTS

DOI: 10.1201/9781003223764-2

2.1 Introduction

2.1.1 Carbon Nanotube-Strengthened Polymer Composites

Carbon nanotubes (CNTs) possess fascinating electrical conductivity and mechanical properties. In 1991, Iijima [1] first reported CNTs. There are two kinds of CNT: single-walled carbon nanotubes (SWCNTs) and multi-walled carbon nanotubes (MWCNTs). Ajayan *et al.* [2] reported on CNT/epoxy nanocomposites with improved electrical and mechanical properties. The electrical/mechanical properties of these composites are affected by purity, integrity, concentration, dispersion of CNTs, and polymer types. In the following sections, the synthesis of various CNT-polymeric nanocomposites (PNCs) and their properties will be discussed.

2.1.1.1 Thermoset Polymer Composites

CNTs tend to self-aggregate, diminishing the full potential of the CNT/PNCs. Hence, there has been a lot of research into improving the dispersibility of CNT/PNCs, using commonly used surfactants like sodium dodecyl sulfate (SDS), poly(phenylene ethynylene)s (PPEs), polystyrene sulfate (PSS), and sodium dodecylbenzene sulfonate (SDBS) (Figure 2.1), as

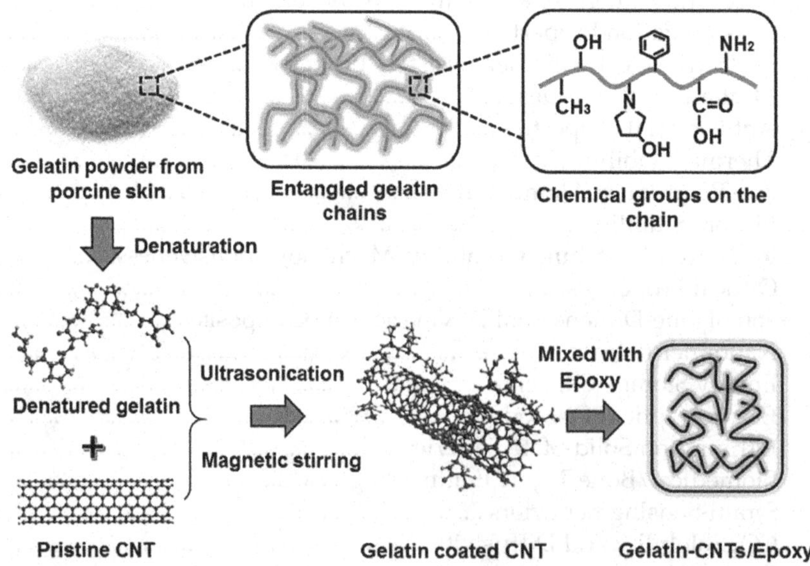

FIGURE 2.1
The creation of a gelatin-CNT/epoxy nanocomposite is depicted. Adapted with permission from Reference [4], Copyright (2018), Elsevier.

well as *via* attachment of appropriate chemical functionalities. Mei *et al.* [3] synthesized a CNT/epoxy composite with moderate conductivity *via* an ice templating approach.

2.1.1.1.1 Polyurethane-Based Matrix

Thermoplastic polyurethane (TPU) is composed of a hard segment (composed of aromatic/aliphatic isocyanates) and a soft segment (composed of polyether/polyester polyol). Because of this wide range of physical properties, which are tunable based on the hard/soft segment ratio, TPU is the desired candidate for preparing CNT/PNCs with unique features. To this end, Yakovlev *et al.* [5] achieved the preparation of a CNT/TPU composite with a much-reduced percolation threshold (Figure 2.2).

2.1.1.1.2 Matrix Based on Polycarbonate or Polyamide-6

Polycarbonate (PC) and polyamide-6 (PA-6) are extensively employed as thermoplastic polymers due to their high chemical, mechanical, and abrasion resistance, along with their thermal, optical, and dimensional stabilities. However, CNT/PC and CNT/PA-6 nanocomposites suffer from insufficient CNT dispersibility. To improve the performance of such nanocomposites, surface modification of CNT is a viable option. Babal group [6] synthesized MWCNT/PC composites in the presence of a low concentration of filler with a twin-screw extruder with a reverse flow channel.

2.1.1.1.3 Polyethylene-Based Matrix

Polyethylene (PE) is one of the widely used polymers. Based on the density and branching, PE is classified into linear low-density polyethylene (LLDPE), high-density polyethylene (HDPE), and ultrahigh molecular weight polyethylene (UHMWPE). Due to its excellent moisture resistance, mechanical, chemical, and tribological qualities, UHMWPE is an excellent choice for application in the nanocomposite field. Such CNT/PE nanocomposites are prepared via dry mixing, ball milling combined with hot pressing, sintering, and blown film extrusion.

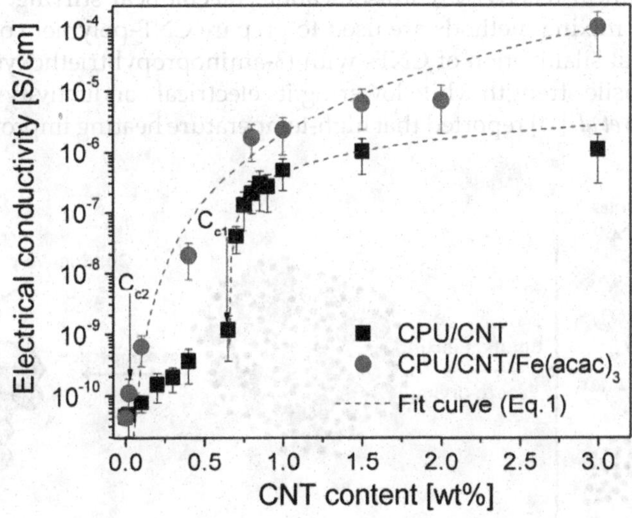

FIGURE 2.2
Electrical conductivity of complex polyurethane (CPU)/CNT and CPU/CNT/Fe(acac)$_3$ composites *vs.* filler weight fraction (a) and a log-log plot of σDC *vs.* C-Cc for the same composites (b). Adapted with permission from Reference [5], Copyright (2017), Elsevier.

2.1.1.1.4 *Polypropylene-Based Matrix*

Unique thermal, chemical, mechanical, dielectric, and optical properties of polypropylene (PP) make it suitable for application in the automotive, medical, construction, and toy industries. Extrusion/injection molding and hot pressing are the most typical methods for fabricating CNT/PP composites, which exhibit appreciable electrical conductivity (Figure 2.3) [7].

2.1.1.1.5 *Polystyrene-Based Matrix*

Because of its low production cost, excellent mechanical properties, and high chemical stability, polystyrene (PS) is commonly used in a wide range of consumer products, such as commercial packaging, bags, cups, and dishes. Injection molding, hot pressing, casting, or spin coating are typical manufacturing processes for CNT/PS composites. However, most of these techniques require sonication, which might affect the physicochemical properties of such composites.

2.1.1.1.6 *Polymethacrylate-Based Matrix*

Poly(methylmethacrylate) (PMMA) is a good alternative to polycarbonate and glass because of its transparency, excellent mechanical properties, and low weight. As with other thermoplastic polymer-based nanocomposites, the most important hurdle to overcome is how to produce a nanofiller dispersion in a suitable matrix. Apart from the well-known latex method, the functionalization of such nanocomposites may greatly improve the electrical and mechanical performance of the composite. Koysuren group [9] utilized a chemical vapor deposition (CVD) approach to coat CNTs with non-covalent poly(glycidyl methacrylate) (PGMA), considerably improving the nanofiller–matrix interaction.

2.1.1.1.7 *Polymer Composites Strengthened by Carbon Nanofibers*

Carbon nanofiber (CNF) is another one-dimensional (1D) carbon nanofiller that is commonly employed. CNF's active carbon atoms are visible on the materials' exterior surface, increasing its chemical reactivity. Ultrasonication, mechanical stirring, solvent casting, and other solution mixing methods are used to prepare CNF-polymer composites. Nie *et al.* [10] reported that silanization of CNFs with (3-aminopropyl)triethoxysilane increased the composite's tensile strength while lowering its electrical conductivity by several magnitudes. Guadagno *et al.* [11] reported that high-temperature heating improved the stiffness

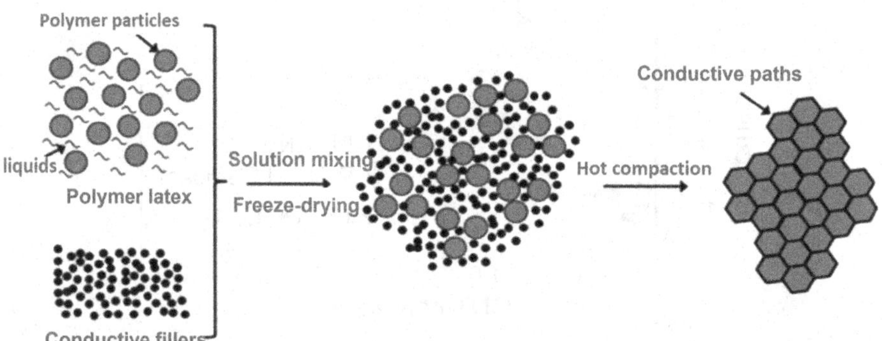

FIGURE 2.3
Fabrication of segregated conductive nanocomposites using the latex technique. Adapted with permission from Reference [8], Copyright (2014), Elsevier.

and surface smoothness of the material, increasing electrical conductivity and lowering the percolation threshold.

2.2 Synthesis of One-Dimensional Polymeric Nanocomposites

2.2.1 Synthesis Via Electrospinning Technique

Zhang *et al.* [12] prepared a series of unaligned $0.5Ba(Zr_{0.2}Ti_{0.8})O_3$–$0.5(Ba_{0.7}Ca_{0.3})TiO_3$-poly(vinylidene fluoride) (BZCT-PVDF) and aligned BZCT-PVDF and BZCT@SiO_2-PVDF composites *via* a combination of electrospinning and hot-pressing. Finally, the annealed composites were quenched in ice water after being annealed at 200 °C for 8 minutes.

2.2.2 Solution-Casting Method

Isobe *et al.* [13] prepared a polymer-containing organo-gel by immersing the polymer into a cellulose gel, with the solution being shaken moderately for over five hours. The dry casting of solutions on glass plates was also used to create pure polymer films. Yin *et al.* [14] prepared functionalized boron nitride nanosheets-1D copper nanowire-poly(vinyl alcohol) [f-BNNS-CuNWs/PVA] film with different concentrations of nanofillers by varying the amounts of functionalized boron nitride nanosheets (f-BNNS) and copper nanowires (CuNWs).

2.2.3 Hot-Stretching Process

Song and co-workers prepared a ternary nanocomposite containing poly(vinylidene fluoride) (PVDF) and a fixed $BaTiO_3$ (BT) content of 10%, with a volume percentage of Bi_2S_3 nanorods (fBi_2S_3) ranging from 0% to 120% [15]. Stretched nanocomposite films were then cut into pieces of 10×10 mm. The unstretched films of the nanocomposite were then cut into small pieces and stacked in the hot-press mold for random composites.

2.2.4 Melt Compounding

Melt compounding is a method of producing a polymer nanocomposite by combining nanofiller and a polymer melt. This process requires a high temperature under solvent-free conditions.

2.2.5 Method of In-Situ Polymerization

In-situ polymerization for the preparation of nanocomposites involves mixing nanofillers directly with monomer solutions and dispersing them as the polymerization takes place, shortening the time it takes to fabricate polymer nanocomposites.

2.2.6 In-Situ Polymer Composite Synthesis

Roy group employed an *in-situ* PNC synthesis approach [16]. For this, Al^{2+}: Nd_2O_3 nanowires were added to an aniline hydrochloride solution in varying proportions by weight (5, 10, 15, 20, and 25 wt%). Then, at a temperature of 0 to 5 °C, 0.25 M ammonium persulphate was gently added to the mixture with steady stirring, and it was left to react for 12 hours, yielding a greenish-black precipitate. Structural and morphological characterizations and electrical measurements of the nanocomposites were carried out.

2.2.7 Template-Based Method

In this method, oxidation of pyrrole (to polypyrrole, PPy) was carried out with $FeCl_3$/ methyl orange, using a template-based technique to produce PPy-based nanocomposites. Using a digital caliper, the polymer film thickness was measured to be roughly 150–170 mm. In recent work, the authors employed atomic force microscopy to explore the surface topography of polymer films. The mean surface roughness of the polymer sheets was discovered to be 250 nm, which would represent a simple sensor component [17].

2.2.8 Sonication

Polypropylene fumarate (PPF) nanocomposites with a 0.2 wt% loading of boron nitride nanotubes (BNNTs) and boron nitride nanoplatelets (BNNPs) were produced *via* sonication. After the addition of the nanomaterial dispersions to a 1:1 mixture of PPF/N-vinyl pyrrolidone (NVP) crosslinker, the mixture was sonicated for 15 minutes. Finally, cylindrical specimens were fabricated by pouring the nanocomposite mixture into Teflon molds and curing overnight at 60 °C [18].

2.2.9 Radical Polymerization

By altering the quantity of the crosslinking agent or *N*-isopropyl acrylamide (NIPAM), Ma *et al.* [19] produced thermochromic 1D photonic crystal films *via* radical polymerization. The samples were then stored in water at ambient temperature for later use.

2.2.10 Melt Intercalation Method

Daniel *et al.* [20] used a volumetric-type feeder, the Cloisite-30B (C30B) (1, 2, 3, 5 wt%), PP granules, and Elvaloy-AC3427 (EA), and processed them in a 40:1 length-to-diameter (L/D) ratio with an intermeshing twin-screw extruder. A pelletizer was used to pelletize the water- and air-cooled PP/C30B/EA nanocomposite.

2.3 Properties of One-Dimensional-Polymeric Nanocomposites

1D-polymeric nanocomposites have many properties, such as crystallization, mechanical, dielectric, antibacterial, thermal, flammability, biocompatibility, resistance-switching capability, and optical properties based on their composition. Some of these valuable properties are discussed below.

2.3.1 Crystallization Properties

Some 1D-polymeric nanocomposites show crystallization properties. Hao *et al.* [21] synthesized a 1D-polymeric nanocomposite which is used to enhance the crystallization property of the polymer. They looked at the components of the crystallization cycle of polymers grafted onto 1D nanorods, as well as the impact of polymer nanorod cooperation and grafting density on comparison of crystallization results from dynamic Monte Carlo simulations. With the increase in the grafting density, some other parameters increase, such as the number of crystallites, segment density close to the interfacial areas, and the mean square radius of gyration ($<R^2_g>$), which are helpful in the improvement of crystallizability. Idumah *et al.* [22] reported that halloysite nanotubes (HNTs) as another 1D naturally occurring nanomaterial showing a very unique collection of high-aspect-ratio, tube-like structures, normally accessible, with decent biocompatibility, extraordinary functionality, and generally more noteworthy mechanical performance, greater thermal stability, and simplicity of accessibility. Nanocomposites loaded up with HNTs display greatly improved tensile/ductile properties, toughness and elastic moduli, and flexural strength. Changes in the crystallization behavior of the polymer generated by the integration of HNTs are linked to the primary constituents of HNTs and interfacial interactions in nanocomposites. Therefore, HNT-polymer nanocomposites (HNT-PNCs) have developed a new crystallization property. HNT-PNCs exhibited excellent medication-exemplifying and delivering affinities because of their cylindrical biocompatibility and microstructure, making them ideal for use as medication delivery agents and scaffolding for tissue engineering. The various HNT preparation methods are depicted in Figure 2.4.

2.3.2 Dielectric/Electrical/Piezoresistive Properties

Some 1D-polymeric nanocomposites exhibit dielectric/electrical/piezoresistive properties based on their composition or polymer matrix. The dielectric properties of PNCs is important because they express the potential for dissipation/energy storage when subjected to an electric field. Pierini *et al.* [23] prepared 1D-conducting PNCs to check their electrical properties by using electrospinning methods. They synthesized nanofibers based on geometric chrysotile nanotubes and polyaniline/poly(ethylene oxide)-doped blends. Electrospun fibers are generally produced from pristine polymers with identical blend-loaded materials mixed with the carbon nanotubes that are also utilized as the reference

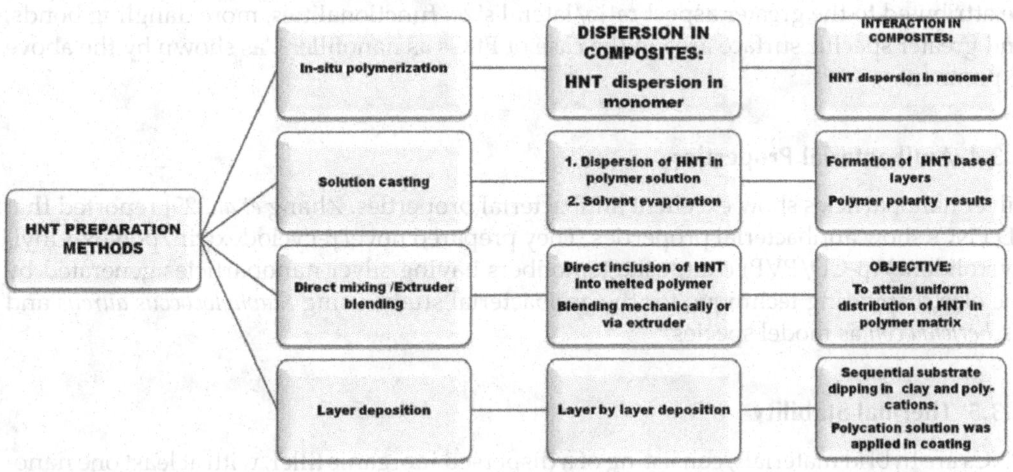

FIGURE 2.4
Preparation methods of HNT-polymer composites. Adapted with permission from Reference [22], Copyright (2019), Taylor & Francis.

materials to investigate the electrical and structural properties of such materials. The expansion of materials with greater conductivity altered the electrical characteristics of the materials for the most part. These electrospun fibers that are fabricated with a conventional cover like chrysotile tend to show greater electrical conduction than do the pristine materials. Khurram *et al.* [24] demonstrated the correlation of the dielectric response of the nanocomposites in the microwave range. Avilés *et al.* [25] investigated the piezoresistive responses, electrical conductivity, and mechanical properties of 1D-, 2D-, and 3D-CNT/polymer composites, few layer thermally reduced graphene oxide (FLG), and cubic-shaped few-layer graphene shells (CGSs). Composites incorporating a 1D/2D hybrid mix of CNTs and FLGs or CNTs alone have greater piezoresistive sensitivity. The increased piezoresistive sensitivity of the material is attributed to their higher excluded volume. Chu *et al.* [26] demonstrated that the larger aspect ratio led to improved dielectric properties of polymer nanocomposites. 1D nanofibers (TiO_2/C nanofibers, NFs) are practical examples of it. Chu *et al.* also synthesized a TiO_2/CNFs /poly(vinylidene fluoride-hexafluoropropylene) (TiO_2/CNFs/P(VDF-HFP)) nanocomposite to improve dielectric properties.

2.3.3 Mechanical Properties

Traditional metals have been extensively replaced by PNCs mainly because of the latter's unique properties in various engineering applications. The enhanced mechanical properties, like increased wear/fatigue resistance, also allow the use of these materials in military and aerospace applications [27]. Pradhan *et al.* [28] investigated the effect of filler dimensionality on the mechanical properties and morphology of the nanocomposites by utilizing different kinds of nanofillers like layered silicate (2D filler), multi-walled carbon nanotubes (1D filler), and boehmite (3D filler). The nanofillers were first dispersed in the matrix of a polyolefin-based elastomer known as ethylene-1-octene copolymer (EOC). Based on the results of the microhardness measurements and tensile tests of the samples, it appeared that the nanofiller's reinforcing action occurred in the following order: 1D filler > 2D filler > 3D filler. Avilés *et al.* [25] reported that the findings of the preceding experiments showed that composites containing a 1D-2D hybrid blend of CNTs and FLGs or CNTs alone had better mechanical properties. This improved performance can be attributed to the greater aspect ratio/lateral size, functionalities, more dangling bonds, and greater specific surface area in the case of FLGs as nanofillers, as shown by the above experiment.

2.3.4 Antibacterial Properties

Silver nanoparticles show excellent antibacterial properties. Zhang *et al.* [29] reported that 1D PNCs show antibacterial properties. They prepared novel β-cyclodextrin/poly(*N*-vinylpyrrolidone) (β-CD/PVP) composite nanofibers having silver nanoparticles generated by the electrospinning technique for the antibacterial study, using *Staphylococcus aureus* and *Escherichia coli* as model species.

2.3.5 Thermal Stability

PNCs are hybrid materials, consisting of a dispersed inorganic filler, with at least one nanometer-scale dimension, in an organic polymer matrix. Adding a small amount of nanofiller changes the polymer's macroscopic characteristics dramatically. In addition, the fillers have a relatively high aspect ratio. In comparison to plain polymer materials, nanocomposites offer much improved features, such as improved heat stability and mechanical characteristics. HNTs have a wide range of possible uses in PNCs, owing to their superior heat stability. HNTs are naturally occurring 1D nanomaterials made up of mainly hollow tubular nanostructures with high aspect ratios. As a result, they have excellent mechanical strength, biocompatibility, thermal stability, and availability. The HNT-PNCs show improved potential for the design and growth of high-performance compounds for environmental safety, biomaterials, and the aircraft/automobile industries. These materials have been utilized to thermally stabilize polyamide-6 (PA6), polypropylene (PP), linear low-density polyethylene (LLDPE), polyethylene terephthalate (PET), and polybutylene terephthalate (PBT). These materials are used as environmentally friendly inclusions in polymer materials due to their inherent barrier effect against mass movement and heat at the time of the nanocomposite's combustion. When iron was added to HNTs, the thermal stability of the nanocomposites improved further.

2.3.6 Fire Retardance/Flammability

Based on the intended thermal application of the materials, the addition of the nanofillers can be designed for one of two purposes, namely increased thermal conductivity or reduced flammability. The heat release rate, which is a critical factor in assessing a product's flammability and fire hazard, can be used to determine the flammability of any material. In recent years, the scientific community has shown a strong interest in polymeric materials with unique qualities such as fire resistance, greater mechanical strength, and largely decreased or increased temperatures. Problems caused by fires include pollution of the environment by the production of large volumes of smoke, combustion products, and thermal decomposition, while the excessive heat that is produced during the results in significant material losses. Combining polymers with a variety of fire retarding agents to generate composites with particular component characteristics can significantly increase polymer fire resistance. HNTs can be used as a very efficient fire retardant material for incorporation into polymers because of their tubular structure and thermal stability. Due to their surface acidity, HNTs can also be catalyzed, resulting in a very protective polymeric coat-like char that decreases the heat release rate (HRR). Passive fire protection materials (PFP) have been closely examined quantitatively through constant heat-flux propane-burner tests, with PFPs fabricated on a steel substrate. The findings of the burner tests show that the materials can produce a large, continuous heat flux, similar to what is seen in a strong fire. Furthermore, the results demonstrate that the HNT materials included created exceptional fire resistance, in contrast to other nano-materials that only showed a minor improvement. Dumazert *et al.* [30] studied the poly (1, 4-butanediol succinate) (PBS) nanocomposites, having two halloysite and sepiolite materials to investigate and demonstrate its thermal stability and fire retardancy properties *via* thermogravimetric analysis (TGA). Zhang *et al.* [31] developed a novel hybrid (GHNT) consisting of HNTs and graphene oxide (GO), as well as composites based on cyanate-ester (CE) resins. The effects of GHNT on fire retardancy, smoke suppression, and heat resistance in GHNT/CE composites were examined.

2.3.7 Biocompatibility

Biocompatibility is another important property of 1D PNCs. Farshid *et al.* [32] synthesized PNCs by utilizing decomposable polymers such as poly(propylene fumarate), *N*-vinyl pyrrolidone (NVP) as the crosslinking agent, and a nanomaterial (such as SWCNT/MWCNT, graphene oxide nanoplatelets, tungsten disulfide nanotubes, or molybdenum disulfide nanoplatelets). The authors also demonstrated *in-vitro* cytocompatibility of such nanocomposites (Figure 2.5).

Armentano *et al.* [33] investigated the biocompatibility of HNTs. This property helps HNTs to gain the position as an essential nanomaterial for increasing numbers of biomedical applications. Recently, HNT-PNCs are practically utilized in drug delivery, tissue engineering, bone implants, cancer therapy, and cosmetics. HNT-based materials have also been fabricated into biodegradable polymers materials, like polylactide (PLA), polyvinyl alcohol (PVA), chitosan, and poly(lactic-co-glycolic) acid (PLGA) for the synthesis and design of scaffolds for tissue engineering.

2.3.8 Resistance-Switching Capability/Mechanoresponsiveness

The option for flash memory or other dynamic random-access memory (DRAM), such as magnetoresistive RAM (MRAM), resistance-switching RAM (RRAM), ferroelectric RAM (FeRAM), and phase-change RAM (PCRAM), has been recommended. RRAM based on metal–insulator–metal (MIM) structures has recently been thoroughly investigated as one of the most promising next-generation memory candidates due to its low power consumption, non-volatility, high operation speed, similarity to conventional Si-based process, and simple structure. Jeon *et al.* [34] prepared 1D conductive nanomaterials, such as CNF or CNT coated with TiO_2. In repeated switching investigations, the carbon nanotube-TiO_2 and CNF-TiO_2 mixed composites show a constant resistance-switching action with a high on/off ratio, as well as increased switching stability.

2.3.9 Optical Properties

Photonic crystals (PCs) have two components in a periodic array with different refractive indices, such as colloidal crystalline arrays (CCA) and Bragg stacks. In comparison with

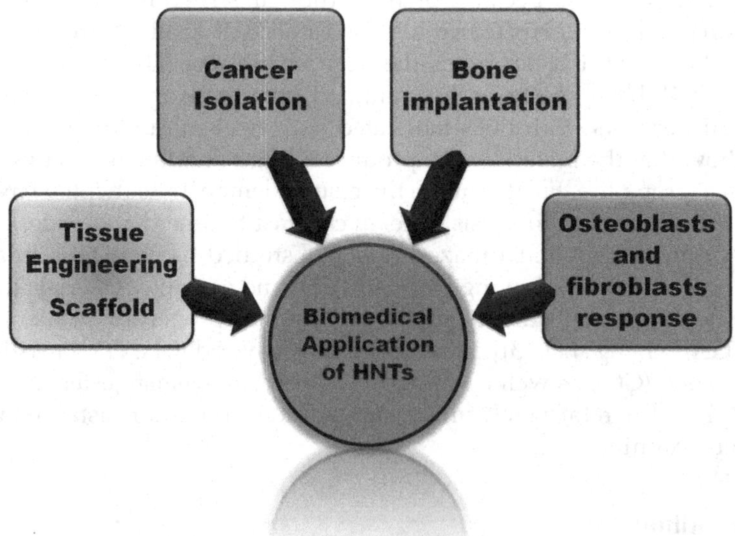

FIGURE 2.5
Various biomedical uses of HNT-PNCs. Adapted with permission from Reference [22], Copyright (2019), Taylor & Francis.

other photonic crystals (PCs), the responsive photonic crystals are generally able to make a change in the refractive index contrast or periodic lattice spacing. Therefore, they can modulate the color induced by different stimuli like magnetic fields, ionic concentrations, electric fields, pH, temperature, and stress. As a result, responsive photonic crystals are a prospective choice for structural displays and color printing, bioassays, optical devices, outdoor signage, sensors, and anti-counterfeiting devices, among other applications. Ma *et al.* [19] demonstrated the optical properties of 1D polymer nanocomposite materials. They used a sterically stabilized magnetically responsive photonic crystal (MRPC) for producing flexible thermochromic films on a centimeter scale.

2.4 Application of One-Dimensional Polymeric Nanocomposites

2.4.1 Nanofiller Release

Polymer nanocomposites (PNCs) are used for various commercial purposes, such as food packaging, building infrastructure, biomedical devices, and implants. Duncan *et al.* [35] explained that the nanofiller (such as engineered nanomaterials (ENMs)) could be released from the PNCs into the liquid medium. There are four possible mechanisms (desorption, dissolution, diffusion, and degradation of the matrix) to release the ENM from PNC materials (Figure 2.6).

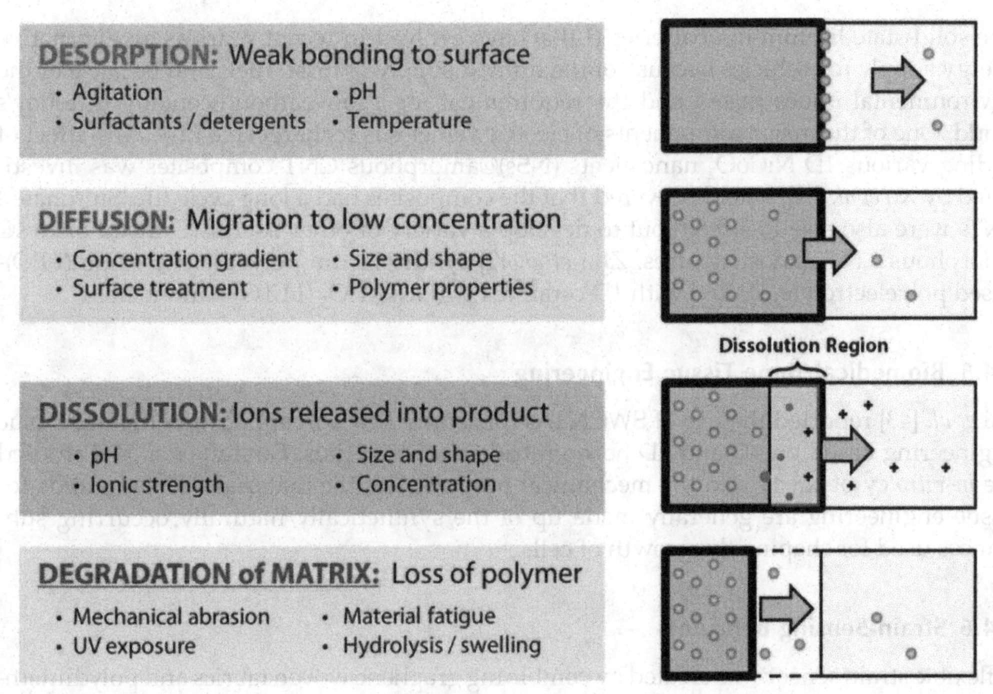

FIGURE 2.6
Schematic diagram of four possible mechanisms of ENM release from PNC materials. Adapted with permission from Reference [36], Copyright (2014), Wiley.

2.4.2 Energy Storage Capacitor

The design and development of compact, affordable, and super-functioning electric energy storage devices have been challenging for researchers around the world. The polymer-based nanocomposite film capacitors are generally considered to be superior to other materials because of their high-speed charging/discharging, high power density, long cycle life, and high operation voltage for energy storage devices. Pan group [37] prepared 1D lead-free $0.5(Ba_{0.7}Ca_{0.3})TiO_3$–$0.5Ba(Zr_{0.2}Ti_{0.8})O_3$ nanofibers (BCZT NFs) through the electrospinning method. The same research group [38] subsequently reported the synthesis of artificial PNCs, using a new hybrid structure that includes 1D Ag@$BaTiO_3$@polydopamine@Ag nanofibers (1D ABPAs NFs) as fillers and P(VDF-HFP) as a matrix. The material developed demonstrated remarkable inclusive performance. Lin *et al.* [39] looked into sandwich structure ceramic/polymer composites, using 1D PVDF/$K_{0.5}Na_{0.5}NbO_3$ nanofibers composites as the intermediate layer and PVDF as the outer layer for energy storage applications.

2.4.3 Electro-Optic (EO) Modulator

Inou *et al.* [40] designed and tested an electro-optic (EO) modulator made up of 1D photonic crystal nanobeam waveguides, composed of polymer/silicon hybrids. Results revealed that the EO polymer/silicon hybrid photonic crystal nanobeam waveguides produce an efficient and superior platform.

2.4.4 Lithium-Ion Solid-State Batteries

The solid-state lithium-ion batteries (LIBs) have evolved in recent years as an alternative energy supply for vehicles because of the limited supply of fossil fuels and because of the environmental issues raised and the requirement for a low-carbon economy in today's world. One of the major components of electric vehicles is rechargeable LIBs. The effect of adding various 1D $NiCoO_2$ nanosheets (NSs)@amorphous CNT composites was investigated by Xu *et al.* [41], who discovered that the composites had a long cycle life. Sulfonated PNTs were also able to stretch out to develop a variety of other useful 1D metal oxides@ amorphous CNT nanostructures. Zhu *et al.*[42] synthesized a polyethylene oxide (PEO)-based polyelectrolyte, loaded with 1D ceramic $Li_{0.33}La_{0.557}TiO_3$ (LLTO) nanofibers.

2.4.5 Biomedical/Bone Tissue Engineering

Shi *et al.* [43] reported the use of SWCNT/eco-friendly PNCs in application to bone tissue engineering based on 1D and 2D boron nitride nanomaterials. Farshid *et al.* [44] studied the *in-vitro* cytotoxicity and the mechanical properties of the materials. The scaffolds for tissue engineering are generally made up of the synthetically/naturally occurring substances used for shaping the growth of cells.

2.4.6 Strain-Sensing Behaviors

A flexible strain sensor was created by combining graphene woven fabrics and polydimethylsiloxane (PDMS). This wearable sensor can also identify human beings in action, which also include expression change, pulse, hand clenching, phonation, blinking, and breathing. Li *et al.* [45] developed a wearable wrist bracelet composed of a carbon sponge (CS)/

PDMS composite. Chen *et al.* [46] demonstrated the effect of filler dimensionality on the strain-sensing behavior of isoprene rubber (IR)-constructed conductive composites on the surface of a computer.

2.4.7 CO_2 Solubility and Diffusivity

Several investigations reported swelling of polymer/CO_2 systems and changes in the CO_2 diffusion coefficient in various pure polymers. Hu *et al.* [47] studied the effects of polymer aggregation states (melt and solid phases) and layered filler-containing 1D nanostructures on CO_2 diffusivity and solubility (nano-MMT).

2.4.8 Thermoelectric Composites

Design and development of flexible advanced thermoelectrics depend upon the mechanical flexibility and superior thermoelectric (TE) performance, and the sustainability of essential elements. Xu *et al* [48] studied a flexible thermoelectric composed of PVDF/Ta_4SiTe_4 organic-inorganic composites. The authors demonstrated a proof-of-concept investigation with flexible thermoelectric modules, explaining the efficacy of the dimensionality/morphology-matching technique.

2.4.9 Flexible Piezoresistive Tactile Sensors and Actuators

The different industrial processes in the agriculture field, including fertilizer production, and the food technology industry, require the sensitive and accurate detection of ammonia due to the toxicity of ammonia gas. In addition, increased inhalation of ammonia may lead to many harmful, clinical conditions. The ammonia gas-sensing action of 1D polypyrrole-metal oxide nanocomposites was demonstrated by Joulazadeh *et. al.* [49], using zinc oxide (ZnO) or tin dioxide (SnO_2) nanoparticles/polypyrrole (PPy) nanocomposites.

2.4.10 Tensioning Cables

Recently, researchers have focused on the use of 1D polymeric nanocomposites in the tensioning cable application, along with various other applications as discussed above. CNT strands have a larger surface area than carbon fiber due to their small size, which may allow for more developed load transfer between the polymer and the filler. The mechanical parameters of the CNT/bismaleimide (BMI) composite fiber were the most notable, with modulus and tensile strength of 232–315 GPa and 4.5–6.94 GPa, respectively. CNT/epoxy nanocomposites have been employed in high-tension applications, including pressure vessels (COPVs), which are used to keep a spacecraft's cryogenic tank at a consistent pressure.

2.4.11 Shielding against Electromagnetic Interference

Electromagnetic interference (EMI) shielding can affect the signal/response of electrical devices and is also harmful to human health. As a result, EMI should be protected against electromagnetic radiation. In recent years, shielding devices have decreased in size, and shielding with a thin- profile and low-weight device has become popular. To scatter, reflect, or absorb electromagnetic radiation, the EMI shielding materials must be conductive and/or magnetic. Metal, on the other hand, has quite high effectiveness in terms of electromagnetic interference shielding (EMI SE). The polymer-based nanocomposite materials tend to attract more attention as they show desirable properties such as being lightweight and non-susceptible to corrosion, with high processability.

2.4.12 Thermal Barriers

Nanofillers are mainly used to reduce the flammability of material and to enhance its thermal conductivity. Heat discharge rate is an extremely significant characteristic for determining a material's flammability and fire hazard; the higher the value, the faster the heat transfers across a material. The nacre-inspired ceramic-based nanosheet materials can be considered for thermal barrier applications as they have extremely high thermal resistance and are a very favorable candidate for thermal barrier applications. Several nacre composite materials utilize flammable polymers like polyacrylic acid or polyvinyl alcohol. The nacre-based materials' flame resistance can be encouraged by selecting a fire-resistant polymer, such as the micron-thick montmorillonite/polyanion composite, which has much higher flame retardancy than montmorillonite (MTM)/polyvinyl alcohol.

2.5 Summary and Perspectives

In conclusion, one-dimensional polymeric nanocomposite materials, such as carbon nanotubes, graphene, and fullerene, provide a unique chance to combine the features of polymers and nanocomposite materials. These combinations open up new possibilities for engineering, high-tech enterprises, and medicinal applications. However, successful formation of polymeric nanocomposites for industrial application with suitable viscosity, thermal, and conductive properties will require more research on the polymerization protocols/processes. Once accomplished, this will lead to a major shift in the potential applications of such materials in everyday life. Several strategies have been developed to produce polymer nanocomposites from a solution-casting method, hot stretching process, sonication, melt intercalation, and electrospinning technique. In this chapter, we have looked at the engineering and biomedical properties of various one-dimensional polymer nanocomposites to explore their potential applications.

Acknowledgments

The authors acknowledge the financial support from the Science and Engineering Research Board, India for the Ramanujan Fellowship Award (SB/S2/RJN-113/2016), the Early-Career Research Award (ECR/2018/001990), and IIT Bhilai for the Research Initiation Grant. DK thanks CSIR, Government of India for the fellowship.

References

1. Iijima, S., Helical microtubules of graphitic carbon. *Nature*, 1991. 354(6348): p. 56–58.
2. Ajayan, P.M., Stephan, O., Colliex, C. and Trauth, D., Aligned carbon nanotube arrays formed by cutting a polymer resin—Nanotube composite. *Science*, 1994. 265(5176): p. 1212–1214.

3. Mei, H., Xia, J., Han, D., Xiao, S., Deng, J. and Cheng, L., Dramatic increase in electrical conductivity in epoxy composites with uni-directionally oriented laminae of carbon nanotubes. *Chem. Eng. J.*, 2016. 304: p. 970–976.
4. Li, Y., Li, R., Fu, X., Wang, Y. and Zhong, W.H., A bio-surfactant for defect control: Multifunctional gelatin coated MWCNTs for conductive epoxy nanocomposites. *Compos. Sci. Technol.*, 2018. 159: p. 216–224.
5. Yakovlev, Y.V., Gagolkina, Z.O., Lobko, E.V., Khalakhan, I. and Klepko, V.V., The effect of catalyst addition on the structure, electrical and mechanical properties of the cross-linked polyurethane/carbon nanotube composites. *Compos. Sci. Technol.*, 2017. 144: p. 208–214.
6. Babal, A.S., Gupta, R., Singh, B.P., Singh, V.N., Dhakate, S.R. and Mathur, R.B., Mechanical and electrical properties of high performance MWCNT/polycarbonate composites prepared by an industrial viable twin screw extruder with back flow channel. *RSC Adv.*, 2014. 4(110): p. 64649–64658.
7. Tjong, S.C., Liang, G.D. and Bao, S.P., Electrical behavior of polypropylene/multiwalled carbon nanotube nanocomposites with low percolation threshold. *Scr. Mater.*, 2007. 57(6): p. 461–464.
8. Pang, H., Xu, L., Yan, D.X. and Li, Z.M., Conductive polymer composites with segregated structures. *Prog. Polym. Sci.*, 2014. 39(11): p. 1908–1933.
9. Koysuren, O., Karaman, M. and Ozyurt, D., Effect of noncovalent chemical modification on the electrical conductivity and tensile properties of poly (methyl methacrylate)/carbon nanotube composites. *J. Appl. Polym. Sci.*, 2013. 127(6): p. 4557–4563.
10. Nie, Y. and Hübert, T., Effect of carbon nanofiber (CNF) silanization on the properties of CNF/epoxy nanocomposites. *Polym. Int.*, 2011. 60(11): p. 1574–1580.
11. Guadagno, L., Raimondo, M., Vittoria, V., Vertuccio, L., Lafdi, K., De Vivo, B., Lamberti, P., Spinelli, G. and Tucci, V., The role of carbon nanofiber defects on the electrical and mechanical properties of CNF-based resins. *Nanotechnology*, 2013. 24(30): p. 305704.
12. Zhang, Y., Zhang, C., Feng, Y., Zhang, T., Chen, Q., Chi, Q., Liu, L., Li, G., Cui, Y., Wang, X., Dang, Z. and Lei, Q., Excellent energy storage performance and thermal property of polymer-based composite induced by multifunctional one-dimensional nanofibers oriented in-plane direction. *Nano Energy*, 2019. 56: p. 138–150.
13. Isobe, N., Sekine, M., Kimura, S., Wada, M. and Kuga, S., Anomalous reinforcing effects in cellulose gel-based polymeric nanocomposites. *Cellulose*, 2011. 18(2): p. 327–333.
14. Yin, C.G., Liu, Z.J., Mo, R., Fan, J.C., Shi, P.H., Xu, Q.J. and Min, Y.L., Copper nanowires embedded in boron nitride nanosheet-polymer composites with enhanced thermal conductivities for thermal management. *Polym.*, 2020. 195: p. 122455.
15. Song, S., Wang, Y., Luo, Y., He, D., Abella, A. and Deng, Y., One-dimensional oriented microcapacitors in ternary polymer nanocomposites: Toward high breakdown strength and suppressed loss. *Mater.*, 2018. 140: p. 114–122.
16. Roy, A.S., Antistatic and dielectric properties of one-dimensional Al^{2+}:Nd_2O_3 nanowire doped polyaniline nanocomposites for electronic application. *Sens. Actuator A Phys.*, 2018. 280: p. 1–7.
17. Joulazadeh, M. and Navarchian, A.H., Ammonia detection of one-dimensional nano-structured polypyrrole/metal oxide nanocomposites sensors. *Synth. Met.*, 2015. 210: p. 404–411.
18. Farshid, B., Lalwani, G., Shir Mohammadi, M., Simonsen, J. and Sitharaman, B., Boron nitride nanotubes and nanoplatelets as reinforcing agents of polymeric matrices for bone tissue engineering. *J. Biomed.*, 2017. 105(2): p. 406–419.
19. Ma, H., Zhu, M., Luo, W., Li, W., Fang, K., Mou, F. and Guan, J., Free-standing, flexible thermochromic films based on one-dimensional magnetic photonic crystals. *J. Mater. Chem. C*, 2015. 3(12): p. 2848–2855.
20. Daniel, D.J. and Panneerselvam, K., Abrasive wear of polypropylene/Cloisite 30B/Elvaloy AC 3427 nanocomposites. *J. Compos. Mater.*, 2018. 52(13): p. 1833–1843.
21. Hao, T., Ming, Y., Zhang, S., Xu, D., Zhou, Z. and Nie, Y., The effect of grafting density on the crystallization behaviors of polymer chains grafted onto one-dimensional nanorod. *Adv. Polym. Technol.*, 2019: p. 6491532.

22. Idumah, C.I., Hassan, A., Ogbu, J., Ndem, J.U. and Nwuzor, I.C., Recently emerging advancements in halloysite nanotubes polymer nanocomposites. *Compos. Interfaces*, 2019. 26(9): p. 751–824.
23. Pierini, F., Lanzi, M., Lesci, I.G. and Roveri, N., Comparison between inorganic geomimetic chrysotile and multiwalled carbon nanotubes in the preparation of one-dimensional conducting polymer nanocomposites. *Fibers Polym.*, 2015. 16(2): p. 426–433.
24. Khurram, A.A., Rakha, S.A., Ali, N., Munir, A., Zhou, P. and Raza, M.A., Comparison of the dielectric response of hybrid polymer composites filled with one-dimensional and two-dimensional carbonaceous materials in the microwave range. *Adv. Polym. Technol.*, 2018. 37(3): p. 890–897.
25. Avilés, F., May-Pat, A., López-Manchado, M.A., Verdejo, R., Bachmatiuk, A. and Rümmeli, M.H., A comparative study on the mechanical, electrical and piezoresistive properties of polymer composites using carbon nanostructures of different topology. *Eur. Polym. J.*, 2018. 99: p. 394–402.
26. Chu, H., Fu, C., Xu, J., Li, W., Qian, J., Nie, W. and Ran, X., Carbon-doped inorganic nanoassemblies as fillers to tailor the dielectric and energy storage properties in polymer-based nanocomposites. *Mater. Des.*, 2020. 188: p. 108486.
27. Banakar, P., Shivanand, H.K. and Niranjan, H.B., Mechanical properties of angle ply laminated composites-A review. *Int. J. Pure Appl. Sci. Technol.*, 2012. 9(2): p. 127.
28. Pradhan, S., Lach, R., Le, H.H., Grellmann, W., Radusch, H.J. and Adhikari, R., Effect of filler dimensionality on mechanical properties of nanofiller reinforced polyolefin elastomers. *ISRN Polym. Sci.*, 2013. 2013: 284504, 1–9.
29. Zhang, L., Li, C., Zhang, X. and Wang, S., Synthesis, characterization and antibacterial properties of novel one-dimensional composite nanofibers. *Asian J. Chem.*, 2013. 25(13): p. 7145.
30. Dumazert, L., Rasselet, D., Pang, B., Gallard, B., Kennouche, S. and Lopez-Cuesta, J.M., Thermal stability and fire reaction of poly (butylene succinate) nanocomposites using natural clays and FR additives. *Polym. Adv. Technol.*, 2018. 29(1): p. 69–83.
31. Zhang, Z., Xu, W., Yuan, L., Guan, Q., Liang, G. and Gu, A., Flame-retardant cyanate ester resin with suppressed toxic volatiles based on environmentally friendly halloysite nanotube/graphene oxide hybrid. *J. Appl. Polym. Sci.*, 2018. 135(31): p. 46587.
32. Farshid, B., Lalwani, G. and Sitharaman, B., In vitro cytocompatibility of one-dimensional and two-dimensional nanostructure-reinforced biodegradable polymeric nanocomposites. *J. Biomed. Mater. Res. A*, 2015. 103(7): p. 2309–2321.
33. Armentano, I., Tarpani, L., Morena, F., Martino, S., Latterini, L. and Torre, L., Nanostructured biopolymer-based materials for regenerative medicine applications. *Curr. Org. Chem.*, 2018. 22(12): p. 1193–1204.
34. Jeon, W., Kim, Y. and Lee, S.S., Resistance switching capable polymer nanocomposites employing networks of one-dimensional nanocarbon wrapped by TiO_2 conformal layer. *IEEE Trans. Nanotechnol.*, 2018. 17(3): p. 567–573.
35. Duncan, T.V., Release of engineered nanomaterials from polymer nanocomposites: The effect of matrix degradation. *ACS Appl. Mater. Interfaces*, 2015. 7(1): p. 20–39.
36. Noonan, G.O., Whelton, A.J., Carlander, D. and Duncan, T.V., Measurement methods to evaluate engineered nanomaterial release from food contact materials. *Compr. Rev. Food Sci. Food Saf.*, 2014. 13(4): p. 679–692.
37. Pan, Z., Yao, L., Zhai, J., Wang, H. and Shen, B., Ultrafast discharge and enhanced energy density of polymer nanocomposites loaded with $0.5(Ba_{0.7}Ca_{0.3})TiO_3–0.5Ba(Zr_{0.2}Ti_{0.8})O_3$ one-dimensional nanofibers. *ACS Appl. Mater. Interfaces*, 2017. 9(16): p. 14337–14346.
38. Pan, Z., Xing, S., Jiang, H., Liu, J., Huang, S. and Zhai, J., Highly enhanced discharged energy density of polymer nanocomposites via a novel hybrid structure as fillers. *J. Mater. Chem. A*, 2019. 7(25): p. 15347–15355.
39. Lin, Y., Sun, C., Zhan, S., Zhang, Y. and Yuan, Q., Ultrahigh discharge efficiency and high energy density in sandwich structure $K_{0.5}Na_{0.5}NbO_3$ nanofibers/poly (vinylidene fluoride) composites. *ACS Appl. Mater. Interfaces*, 2020. 7(9): p. 2000033.

40. Inoue, S.I. and Otomo, A., Electro-optic polymer/silicon hybrid slow light modulator based on one-dimensional photonic crystal waveguides. *Appl. Phys. Lett.*, 2013. 103(17): p. 171101.
41. Xu, X., Dong, B., Ding, S., Xiao, C. and Yu, D., Hierarchical $NiCoO_2$ nanosheets supported on amorphous carbon nanotubes for high-capacity lithium-ion batteries with a long cycle life. *J. Mater. Chem. A*, 2014. 2(32): p. 13069–13074.
42. Zhu, P., Yan, C., Dirican, M., Zhu, J., Zang, J., Selvan, R.K., Chung, C.C., Jia, H., Li, Y., Kiyak, Y., Wu, N. and Zhang, X., $Li_{0.33}La_{0.557}TiO_3$ ceramic nanofiber-enhanced polyethylene oxide-based composite polymer electrolytes for all-solid-state lithium batteries. *J. Mater. Chem. A*, 2018. 6(10): p. 4279–4285.
43. Sitharaman, B., Shi, X., Walboomers, X.F., Liao, H., Cuijpers, V., Wilson, L.J., Mikos, A.G. and Jansen, J.A., In vivo biocompatibility of ultra-short single-walled carbon nanotube/biodegradable polymer nanocomposites for bone tissue engineering. *Bone*, 2008. 43(2): p. 362–370.
44. Farshid, B., Lalwani, G., Shir Mohammadi, M., Simonsen, J. and Sitharaman, B., Boron nitride nanotubes and nanoplatelets as reinforcing agents of polymeric matrices for bone tissue engineering. *J. Biomed. Mater. Res. Part B Appl. Biomater.*, 2017. 105(2): p. 406–419.
45. Li, Y.Q., Zhu, W.B., Yu, X.G., Huang, P., Fu, S.Y., Hu, N. and Liao, K., Multifunctional wearable device based on flexible and conductive carbon sponge/polydimethylsiloxane composite. *ACS Appl. Mater. Interfaces*, 2016. 8(48): p. 33189–33196.
46. Chen, J., Li, H., Yu, Q., Hu, Y., Cui, X., Zhu, Y. and Jiang, W., Strain sensing behaviors of stretchable conductive polymer composites loaded with different dimensional conductive fillers. *Compos. Sci. Technol.*, 2018. 168: p. 388–396.
47. Hu, D., Chen, J., Sun, S., Liu, T. and Zhao, L., Solubility and diffusivity of CO_2 in isotactic polypropylene/nanomontmorillonite composites in melt and solid states. *Ind. Eng. Chem. Res.*, 2014. 53(7): p. 2673–2683.
48. Xu, Q., Qu, S., Ming, C., Qiu, P., Yao, Q., Zhu, C., Wei, T.R., He, J., Shi, X. and Chen, L., Conformal organic–inorganic semiconductor composites for flexible thermoelectrics. *Energy Environ. Sci.*, 2020. 13(2): p. 511–518.
49. Joulazadeh, M. and Navarchian, A.H., Ammonia detection of one-dimensional nano-structured polypyrrole/metal oxide nanocomposites sensors. *Synth. Met.*, 2015. 210: p. 404–411.

3

Methods for Preparation of One-Dimensional Polymeric Nanocomposites

Tressia A.P. Paulose, Reshmy Rajasekharan, Aravind Madhavan, Sindhu Raveendran, Parameswaran Binod, Mukesh Kumar Awasthi, and Ashok Pandey

CONTENTS

3.1 Introduction

Composites are made up of more than one material and have qualities that are better than those of the individual materials. To improve the mechanical and physical properties, fillers of various sizes are commonly used to strengthen polymers. The filler, which can be macro, micro, or nano in size, usually determines the properties of the polymer composites [1]. When nanofillers are dispersed in a polymer matrix they are known as polymer nanocomposites. The addition of modest concentrations of nanofillers to specific polymers has been shown to enhance their physical, mechanical, thermal, and flammability properties without altering their processability [2].

Nanofillers provide far superior reinforcement (nano-effect) compared to macro- and microfillers at the equivalent filler loading because of their greater degree of interaction between the nanofiller and the polymer. Several aspects contribute to the reinforcing action of the filler, including polymer matrix characteristics, nanofiller nature, type and concentration, aspect ratio, size, orientation, and distribution of particles [3]. A variety of synthetic polymers, natural polymers, elastomers, and biopolymers have been reinforced with various nanofillers for different applications [4].

Polymer nanocomposites may be classified into three categories based on the size of the distributed nanofillers as (a) nanocomposites with three-dimensional (3D) nanoscale

DOI: 10.1201/9781003223764-3

fillers, including spherical silica, semiconductor nanoclusters, and quantum dots, (b) nanocomposites with two-dimensional (2D), namely graphene or layered silicate present in polymeric matrices as sheets with a thickness of a few nanometers and a length of hundreds to thousands of nanometers, and (c) nanocomposites with one-dimensional (1D) structures, such as nanotubes (NTs), nanofibers (NFs), nanofilaments, and nanowires (NWs). Uniform dispersion of nanofillers is the most difficult challenge to overcome to maximize the potential of the composites [5]. 1D nanofillers have a lower specific surface area, which is beneficial for lowering surface energy and preventing nanofiller aggregation in polymeric matrices, resulting in homogeneous nanofiller dispersions and powerful interfacial interactions, allowing for efficient load transferring from the polymeric matrix to the nanofillers.

3.2 Preparation of 1D Polymeric Nanocomposites

The physical and chemical variabilities between different types of polymers make it hard to develop a single universal method for manufacturing all sorts of polymeric nanocomposites. Depending upon the processing efficiency, nanofillers, and desired attributes, each polymer system requires unique processing conditions. The quality of the polymer nanocomposites depends significantly on the method used for preparing the polymer nanocomposites. Melt intercalation, solution mixing, *in-situ* polymerization, and electrospinning are some of the typical processing techniques employed for the manufacture of polymer nanocomposites.

3.2.1 Melt Intercalation

Melt intercalation is the most common method for making nanocomposites out of polymers that are insoluble in solvents, such as thermoplastic polymers. This process entails heating the polymer matrix to high temperatures to create a viscous liquid into which the nanofillers are introduced and mixed, using commercialized industrial technologies such as single- or twin-screw extruders/compounders and injection molding [1]. Because of the high temperature, the polymers melt quickly, while the shear force in the extruder aids in the dispersion of the filler. The shear forces created inside the extruder are mostly determined by the viscosity of the melt and the speed with which the mixing parts rotate. The strain that the polymer imposes on the filler during mixing in the melt state is governed by its molecular weight and weight distribution. A large amount of shear stress breaks down large agglomerates into smaller ones, thereby enhancing their dispersion in the polymer matrix. Composites that have been processed at a high shear rate show higher mechanical characteristics than those processed at a low shear rate because of better dispersion of the filler in the former. The compounded polymer-nanofiller composites can be further processed by techniques such as blow molding, extrusion, and injection molding. The melt compounding technique has the advantage of being able to use the already-existing melt processing method to produce nanocomposites. Good dispersion of the nanofiller improves heat stability, while improved mechanical properties are some of the other advantages of the melt compounding approach. It is also a more environmentally friendly and cost-effective process than solution intercalation as it does not need any solvents. However, for surface-modified nanofillers, using high temperatures can harm the modified surface. Degradation can be avoided by operating at lower temperatures or employing more thermally stable changes. The processing parameters, such as time, temperature, screw speed, residence time, shear stress, and filler surface modification have to be fine-tuned as they are all important factors in how effective dispersion can be achieved. Melt blending has been utilized to make polymer nanocomposites from several matrices, including polypropylene (PP), polylactic acid (PLA), poly(methyl methacrylate) (PMMA), polycarbonate PC), polyamide 6 (PA6), polyurethane (PU), and polyvinyl chloride (PVC).

Masterbatch dilution is a very common method used for better dispersion of the nanofillers to construct composites *via* melt compounding. In this technique, composites with highly loaded filler concentrations, known as masterbatches, are initially produced in the extruder. The masterbatches are then diluted with neat polymer in a subsequent melt mixing process to attain lower concentrations of filler nanocomposites. The nanofillers become well dispersed upon wetting of the agglomerates with polymer in the course of the masterbatch technique. Multi-walled carbon nanotubes (MWCNT)-reinforced PC composites have been fabricated by diluting a masterbatch of 15 wt% MWCNT in PC with neat PC, using melt mixing at 260°C [1]. High shear stress during melt mixing is caused by high CNT concentrations in the masterbatches, which results in high melt viscosity. Fabricators will also find it much easier to work with polymer-bound CNTs. MWCNTs dispersed in PLA have been prepared using a twin-screw extruder with a modular assembly which can differ in the aspect ratio of the screw, the screw profile, and the feeding positions [2]. MWCNT-PU nanocomposite fibers have been produced by blending acid-treated/oxidized MWCNTs using a double-screw extruder over 15 min at 180°C and 60 rpm, followed by extrusion to fabricate fibers with a diameter of 200 μm [3]. Good dispersions are obtained for MWCNT loading of up to 9.32 wt%, even as localized CNT aggregations are observed for the composites containing higher CNT concentrations. Carbon nanofiber- (CNF) reinforced PCs have been developed by melt mixing using micro compounder at 260–300°C, and screw speeds of 50 and 150 rpm applied for 5–30 min, whereas CNF-reinforced PP composites are prepared using a twin-screw kneader [4].

Ethylene-vinyl acetate (EVA) copolymer, reinforced with sepiolite composites, has been prepared *via* melt blending using a Brabender twin-screw extruder [5]. The compounding temperature is 155°C at the feed portion and 170°C at the die head, and the materials are extruded at a screw speed of about 50 rpm. The pelletized composites are dried and injection molded to obtain rectangular sheets. Sepiolite-filled nanocomposites have also been prepared with two different commercial polymeric matrices, poly (butylene terephthalate) (PBT) and PP [6]. As pristine sepiolite is hydrophilic, modified sepiolites are employed to increase the affinity for the selected polymers. PP-based composites are prepared with surface-modified apolar sepiolite modified using dimethyl hydrogenated tallow ammonia (2MHT) and a slightly polar sepiolite modified with $PrSH(OEt)_3$-Si, whereas PBT composites, based on a polar ester resin, are synthesized using the highly polar sepiolite surface modified using bis-2-hydroxyethyl methyl tallow quaternary ammonium (MT2EtOH) and the polar aromatic sepiolite modified using dimethyl benzyl hydrogenated tallow quaternary ammonium (2MBHT). The surface modifiers are depicted in Figure 3.1. Nanocomposites with micronized pristine sepiolite are also prepared to investigate the effects of sepiolite modification on the desired properties of sepiolite nanocomposites. Surface modification of the sepiolite results in better dispersion which causes an efficient shielding from oxygen, thereby reducing the catalytic effects of the sepiolite on PP thermal degradation. In addition, greater stability toward thermoxidation is attained. Grafting using mercaptosilane on catalytically active hydroxyl groups of sepiolite significantly reduces the available

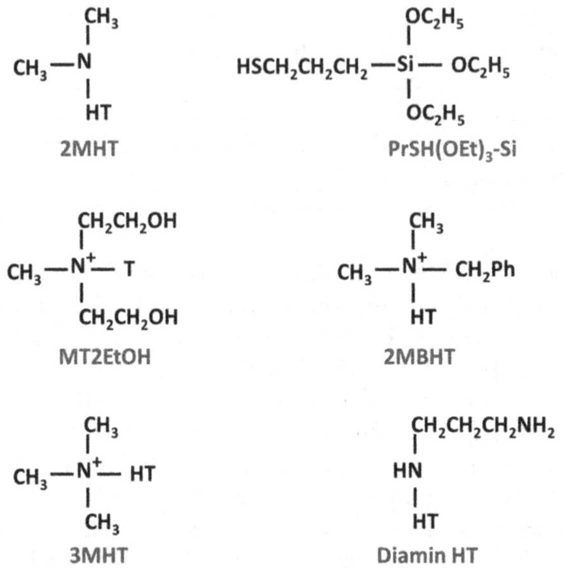

2MHT

$PrSH(OEt)_3$-Si

MT2EtOH

2MBHT

3MHT

Diamin HT

FIGURE 3.1
Sepiolite surface modifiers.

catalytic sites to reduce its degradation effect on PP, compared with the other ammonium or amine surface modifiers.

Sepiolite organo-modified with different types of modifiers, such as 2MBHT, *N*-hydrogenated tallow1,3-diaminopropane (DIAMIN HT), and trimethyl hydrogenated tallow quaternary ammonium (3MHT), have also been used to prepare PA6 composites by melt extrusion [7]. The extruded material is injection molded (at 240–250°C) into dumbbell shapes using an injection molder after being dried at 80°C for 24 hours. The effect of organofiller concentration on the mechanical properties and nanostructure of the PA6/sepiolite nanocomposites at organofiller concentrations of 6, 9, and 12 wt% are investigated. The catalytic effect of the sepiolite on the polymer matrix decreases as the modifier concentration increased [8]. The nature of the organic modifier (size and chemical structure) and the amount of nanofiller affect the mechanical properties of the PA6/organo-sepiolite nanocomposites.

A dual approach has been adopted to improve the interfacial bonding due to the incompatibility and poor dispersion of nanofillers in the PP matrix [9]. Firstly, surface-modified nanofillers are used to reinforce the aforementioned polymer matrix. Both surface-modified sepiolite and CNTs have been tested as nanofillers. CNTs are oxidized *via* an HNO_3 treatment. The surface modifications of sepiolite are achieved by treating the micronized sepiolite with a cationic surfactant, cetyltrimethylammonium bromide (CTABr), which can exchange Mg^{2+} ions of the sepiolite for positively charged amino groups, and, at the same time, introduce alkyl chains from the cetyl group that can enhance the compatibility of the sepiolite with PP. Secondly, due to the non-polar and incompatible nature of PP with silicate surfaces, a compatibilizer is used to facilitate the interaction between the modified clays and the polymer matrix. A commonly used compatibilizer for improving the dispersion efficiency of sepiolite in PP is maleic anhydride grafted polypropylene (PP-g-MAH). Composites are prepared by manually mixing pelletized PP with pelletized PP-g-MAH and the modified nanofiller (sepiolite or CNTs).

PA6/HNTs composites have been synthesized by melt extrusion of PA6 and HNTs in a twin-screw extruder [10]. As the silanized fillers do not readily react with PP under the processing conditions, PP chains are grafted onto the surface of HNTs [11]. To prepare the PP chain-grafted nanofiller, the HNTs are added to an ethanol-water solution of (3-amino-propyl)triethoxysilane and refluxed at 80°C for 3 hours. The precipitated HNTs are washed with an ethanol-water solution and dried. This modified halloysite is mixed with PP-g-MAH solution in xylene at 120°C and the precipitated PP-grafted halloysite is extracted with xylene and dried. The schematic for the preparation of PP -halloysite is shown in Figure 3.2. PP and the PP-grafted halloysite is blended using a twin-screw extruder and then injection molded. Recently, PU/CNT composites have been prepared by melt compounding PU and CNT at 210°C and a rotating speed of 100 rpm for 15 min [12]. Details of different polymer nanocomposites prepared by melt intercalation are depicted in Table 3.1.

3.2.2 Solution Intercalation

The solution intercalation method generally consists of the following steps: (i) dispersing the filler in the polymeric solution through agitation; (ii) uniformly mixing the filler and polymeric solution by further agitation; and (iii) controlling evaporation of solvents and/

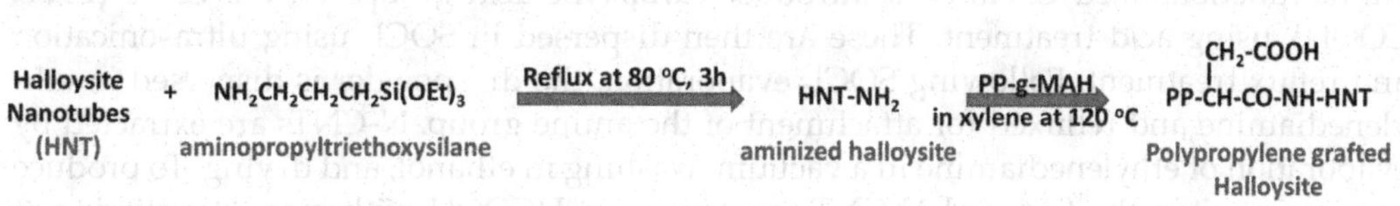

Halloysite Nanotubes (HNT) + $NH_2CH_2CH_2CH_2Si(OEt)_3$ aminopropyltriethoxysilane → Reflux at 80 °C, 3h → HNT-NH_2 aminized halloysite → PP-g-MAH in xylene at 120 °C → PP-CH(CH$_2$-COOH)-CO-NH-HNT Polypropylene grafted Halloysite

FIGURE 3.2
Schematic representation of the preparation of polypropylene grafted halloysite nanotubes.

TABLE 3.1
1D Polymer Nanocomposites Prepared *via* Melt Intercalation

Matrix	Filler	Temp (°C)	Screw Speed (rpm)	Surface Modification of Filler/Polymer	Ref
PC	MWCNT	260	150	—	[1]
PLA	MWCNT	180–220	100/500	—	[2]
PU	MWCNT	180	60	Acid-treated CNTs	[3]
PP	CNF	190–230	50/150	—	[4]
PC	CNF	260–310	50/150	—	[4]
EVA	Sepiolite	155–170	50	—	[5]
PP	Sepiolite	190	60	Filler modified with trimethyl hydrogenated tallow ammonia (2MHT). Filler modified with PrSH(OEt)$_3$-Si.	[6]
PBT	Sepiolite	235	60	Filler functionalized using methylbis(2-hydroxyethyl) tallow quaternary ammonium (MT2EtOH)	[6]
PA 6	Sepiolite	240–250	200	Filler modification using benzyl(hydrogenated tallow alkyl)dimethyl quaternary ammonium (2MBHT)	[7]
PA 6	HNT	200–240		—	[10]
PP	HNT	180–200	100	PP chain grafted onto surface of filler	[11]
PU	CNT	210	100	—	[12]

or precipitation of nanocomposites. Agitation is usually achieved by magnetic stirring, refluxing, ultrasonication, or shear mixing. After mixing, the dispersion holding the filler in the polymeric solution can be injected into the spray dryer that converts the material into powder form. Unlike melt intercalation, this method is not environmentally benign since the method uses large volumes of solvents. Solution intercalation is an excellent method if the solvent employed is not particularly toxic, so this technique is widely utilized for water-soluble polymers to produce intercalated composites based on polymers, such as poly(ethylene oxide), poly(vinyl alcohol), poly(acrylic acid), or poly(vinylpyrrolidone).

Aromatic polyamide/MWCNT nanocomposite membranes have been prepared by a polymer grafting technique using *N,N*-dimethylacetamide (DMA) as a solvent for CNTs and the polyamide (PA) polymer [13]. MWCNTs are dispersed *via* ultrasonication in DMA solvent prepared with lithium chloride salts. Benzoyl peroxide initiator is added to form free radicals on both CNTs and PA, which gives polymer-grafted nanotubes that are better dispersed. After adding the initiator, the mixture is heated with stirring at 80°C, followed by evacuating the solution to remove the dissolved gas and then cast onto a dried, clean glass Petri dish by spreading with the help of a glass rod to form a uniform, thin film which is then placed in an oven to evaporate the solvent. PA6/CNT composites have also been prepared with amine-functionalized CNTs (N-CNTs) [14]. The first step to produce amine-functionalized CNTs is to introduce carboxylic acid groups on the CNTs (CNT-COOH), using acid treatment. These are then dispersed in $SOCl_2$ using ultrasonication and reflux treatment. Following $SOCl_2$ evaporation, the dry powder is dispersed in ethylenediamine and refluxed for attachment of the amine group. N-CNTs are extracted by evaporation of ethylenediamine in a vacuum, washing in ethanol, and drying. To produce nanocomposites, the PA6 and N-CNTs are mixed in HCOOH with magnetic stirring at 100°C for 8 hours followed by evaporation of the solvent. Thin composite films are prepared from a hot press at 235°C and 200 bar pressure.

CNT/polyvinyl-alcohol composite films have been prepared for mechanical characterization [15]. Acid-treated CNTs are used to produce an electrostatically stabilized dispersion of NTs in water. Nanocomposite films with a broad spectrum of NT loadings are made by careful mixing of aqueous poly(vinyl alcohol) $[CH_2CH(OH)]_n$ (PVOH) solutions with CNT dispersions, while stirring continually at 480 rpm, followed by subsequent casting and controlled water evaporation. CNT-reinforced polyacrylonitrile (PAN) composite fibers have been developed by utilizing the solution impregnation route [16]. The composite fibers produced are placed in a vacuum oven to evaporate all the residual solvent. A solution-mixing technique is adopted to develop the MWCNT/PMMA nanocomposite since it leads to high-quality dispersions of MWCNTs and does not cause any damage to the MWCNTs [17]. Acid-treated MWCNTs are dispersed in a homogeneous solution of PMMA in chloroform, using high-power sonication. The polymer/MWCNTs suspension is cast in flat pans and then extruded using a mixing extruder at 240°C to remove any remaining solvent. The precipitation method has also been used whereby the polymer/MWCNT suspension is added drop-by-drop into a large volume of methanol. The solution is continuously stirred vigorously using a mechanical stirrer and then the precipitated solids are filtered and dried. In yet another modified process, the MWCNTs are floated and sonicated in chloroform for extended periods of time to obtain a stable suspension. PMMA solution in chloroform is subsequently added to the MWCNT suspension and sonicated for about 20 min. The suspension is finally added drop-by-drop into a large amount of methanol to precipitate the solids, which are then dried and extruded.

Carbon nanofiber (CNF)-reinforced polybenzimidazole (PBI) nanocomposite films have been prepared by directly mixing CNFs in a dilute solution of PBI in DMA with mechanical stirring at 300 rpm for 30 min [18]. Alternatively, the CNFs can be mixed with DMA and dispersed thoroughly by ultrasonication for 30 min at 60°C. After ultrasonication, CNFs dispersed in DMA are mixed with the PBI solution under both ultrasonication and magnetic stirring. The mixture is then used to cast on the glass plate to produce 60- to 80-µm-thick films. The film is dried in the vacuum oven at 80°C for 2 hours and cured at 200°C overnight. The addition of CNFs improves the storage modulus, thermal stability, and tensile strength of the PBI film.

Chitosan is an exceptional dispersing agent of single-walled CNTs (SWCNTs). A SWCNT-reinforced chitosan polymer nanocomposite is prepared by sonicating SWCNTs in a solution of chitosan dispersed in distilled water in the presence of acetic acid [19]. The mixture is then cast into a glass dish and dried to obtain a uniform film. The SWCNT-chitosan nanocomposite showed better heat barrier properties than the pure chitosan films. Four different surface-modified nanowires (NWs) (TiO_2, $Na_2Ti_3O_7$, $SrTiO_3$, and $BaTiO_3$) have been used to fabricate high-dielectric-constant ferroelectric polymer nanocomposites with poly(vinylidene fluoride-co-hexafluoropropylene) (P(VDF-HFP)) as the matrix [20]. Surface modification of NWs is carried out by dispersing the NWs in an aqueous solution of dopamine hydrochloride with stirring for 24 hours at 60°C. To prepare the P(VDF-HFP)-based nanocomposites films, the surface-functionalized NWs are dispersed in DMF by ultrasonication and then mixed with a homogeneous solution of P(VDF-HFP) in DMF. The mixture is agitated and ultrasonicated with piezoelectric vibrator before casting onto a glass plate. Finally, the nanocomposites are compressed into thin films of around 50 µm thickness by hot-pressing at 190°C with a pressure of about 20 MPa. Details of the preparation of 1D polymer nanocomposites through solution intercalation are provided in Table 3.2.

3.2.3 In-Situ Polymerization

The *in-situ* polymerization method is specific in terms of the polymer synthesized in the presence of the nanomaterial. Nanofillers are normally mixed simultaneously with

TABLE 3.2

1D Polymer Nanocomposites Prepared *via* Solution Intercalation

Matrix	Filler	Solvent	Surface Modification	Dispersion	Ref
Aromatic PA	MWCNT	DMA	Benzoyl peroxide initiator for free radicals to form polymer-grafted nanotubes	Ultrasonication	[13]
PA 6	CNT	Formic acid	Amine-functionalized CNTs	Stirring	[14]
PVA	CNT	Water	Acid-treated CNTs	Stirring	[15]
PAN	CNT	DMF	—		[16]
PMMA	MWCNT	Chloroform/ Methanol	—	Ultrasonication	[17]
PBI	CNF	DMF	—	Stirring/ ultrasonication (60°C)	[18]
Chitosan	SWCNT	Distilled water with acetic acid	—	Sonication	[19]
$Na_2Ti_3O/TiO_2/$ $BaTiO_3/SrTiO_3$	P(VDF-HFP)	DMF	Polydopamine film on the surface of NWs	Ultrasonication	[20]

monomer solutions and distributed while polymerization takes place, shortening the time it takes to generate polymer nanomaterials. Even before the polymerization method begins, the nanofiller should be thoroughly distributed within monomer solution, allowing the polymer to form between both the nanomaterials. Several approaches can be used to start polymerization (thermal, radiation, or catalyst). The *in-situ* method of polymerization is an alternative for producing insoluble or thermally fragile polymers that cannot be handled with solution or melt intercalation. Organic modifiers have been used to aid the dispersion of nanoparticles and to take part in polymerization. This method allows for increased nanofiller concentration without agglomeration, improved end product efficiency, expansion to a solvent-free state, covalent bonding between nanoparticle functional groups and the polymer matrix, and the use of both thermoplastic and thermoset polymers. However, nanofillers can sometimes hinder polymerization. For polymerization without nanoparticles, the equipment employed can be the same.

In one of the earliest methods, *in-situ* polymerization has been used in preparing MWCNT-reinforced polymer composite film. The carbon-arc discharge process is used to produce MWCNTs, which is then sonicated in ethanol, dried, and distributed on a glass plate [21]. A blade is used to carefully distribute an aqueous polymer combination (diacrylate/urethane oligomer) onto the dried NT-containing graphite powder, bringing the liquid film down to a width of approximately 200 μm. A mini-conveyor UV-curing unit is then used to polymerize the films for 2 min. MWCNT-reinforced polyurethane composites have also been prepared by mixing a poly(ethylene adipate) dispersion in acetone with MWCNTs dispersed in acetone, using high-intensity ultrasound [22]. The mixture is dried in a vacuum reactor for 2 hours at a temperature of 120°C under a pressure 2–5 hPa, while being intensely stirred. To synthesize the polyurethane 4,4′-diphenylmethane diisocyanate, glycol and glycerine are added and the reaction is kept at 120°C for 16 hours. PU/MWCNT elastomeric nanocomposites have been prepared using amide-functionalized CNTs in the presence of sonication [23]. The CNTs are amide functionalized *via* a stepwise procedure in which carboxylic acid groups are first introduced onto the CNTs (CNT-COOH), using acid treatment. These are then dispersed in $SOCl_2$ by ultrasonication and reflux treatment and then filtered to give CNT–COCl powder which is cleaned, using anhydrous tetrahydrofuran, and then dried. The CNT–COCl powder is then stirred with excessive ethylenediamine (EDA), filtered, cleaned using anhydrous tetrahydrofuran, and dried to obtain the CNT-COEDA. The grafted amide groups on the CNTs easily react with the toluene diisocyanate (TDI) and polyoxytetramethylene glycol (PTMO) to make the PU composites.

SWCNT/polyimide nanocomposites have been prepared using the monomers 1,3-bis(3-aminophenoxy) benzene as the diamine and 2,2-bis(3,4-anhydrodicarboxyphenyl) hexafluoropropane as the dianhydride to synthesize the aromatic polyimide [24]. A dilute, sonicated dispersion of SWCNT in DMF is mixed with the diamine and the dianhydride in an inert atmosphere. The whole reaction process is carried out in an ultrasonic bath in flasks until the viscosity of solution had increased and stabilized. To obtain SWCNT–poly(amic acid) solution, the combination is agitated overnight without sonication. To chemically imidize the SWCNT–poly(amic acid), acetic anhydride and pyridine are utilized as catalysts.

HCl-doped poly(anthranilic acid) (PAA) has been reinforced with carboxylic groups, including MWCNTs by *in-situ* chemical oxidative polymerization of PAA [25]. The carboxylic functionalization of the CNTs (c-CNTs) is carried out by H_2SO_4-HNO_3 treatment. PAA/c-MWCNT composites are formed by adsorbing anthranilic acid monomers on the top of MWCNTs and polymerizing them. Scanning electron microscopy (SEM) analysis revealed a tubular layer of coated copolymer film on the surface of c-MWCNTs due to the strong contact between both the c-MWCNTs and the co-monomer.

Recently, MWCNTs have been used as a core template to direct the self-assembly of polypyrrole (PPy) on the sidewalls of the CNTs [26]. MWCNT/PPy composites are prepared by adding the CNTs to a solution of the surfactant sodium dodecyl sulfate (SDS) in absolute ethanol, followed by the pyrrole monomer under ultrasonication. The initiator, $FeCl_3$, is added drop-by-drop to the resulting dispersion and stirred, filtered, washed using distilled water and ethanol, and dried. Perfect coatings of PPy on the sidewalls of MWCNTs are obtained as 1D core-shell structures. Polyaniline/multi-walled CNT (PANI/MWCNT) composites have been made through an *in-situ* method of polymerization of aniline in the presence of highly purified MWCNTs [27]. To a dispersion of aniline monomer and MWCNTs in HCl solution, an ammonium persulfate initiator is added and stirred. The oxidative polymerization is carried out at room temperature. The resulting mixture is filtered and cleaned with deionized water. To produce a core-shell structure, the CNTs are entirely covered within the PANI matrix. The interaction between aromatic molecules and PANI/MWCNT composites is physical or by non-covalent interactions. Sulphonated MWCNTs have been used as the self-assembled template for the formation of PANI nanostructures [28]. The hydrogen sulfated MWCNTs (MWCNT-$(OSO_3H)_n$) are prepared by directly sulfating the hydroxyl groups on the nitric acid-oxidized MWCNTs in the presence of concentrated H_2SO_4. The composites are prepared by adding the initiator ammonium persulphate to an ultrasonicated suspension of the monomer aniline and MWCNT-$(OSO_3H)_n$.

To make polymer-grafted NTs, *in-situ* polymerization is normally employed. The free radical addition method is employed for grafting styrene polymers onto the surfaces of MWCNTs [29]. The material is precipitated and vacuum dried after being heated at 90°C for 11 hours with MWCNTs, styrene monomers, toluene, and azobisisobutyronitrile initiator. Melt-mixing and masterbatch dilution are used to efficiently create MWCNT-reinforced thermoset phthalocyanine (Pc/CNTs) nanocomposites [30]. The Pc monomer 4,4'-bis(3,4-dicyanophenoxy) biphenyl (BPH) is melted at 250°C and mixed with CNTs to form the BPH/CNTs masterbatch as a dark green solid, which is then shattered for 5 min in a high-speed disintegrator to generate the BPH/CNTs masterbatch powder. Pc/CNT nanocomposites with various CNT loadings are made by diluting a 5 wt% BPH/CNTs masterbatch with weight-measured BPH. Finally, the mixture is placed into a warmed (250°C) mold and cured for 4 hours.

MWCNT epoxy composites are prepared by dispersing the CNTs in an epoxy resin using ultrasonication, followed by curing with a triethylenetetraamine thickener by gelling overnight at room temperature and curing at 100°C for 2 hours [31]. The nanotube distribution in the matrix is improved in composites with a minimal nanotube loading (5 wt%). The sonicated combination of the resin and MWCNTs in chloroform can be spin-coated onto a clean, conventional 4-inch silicon wafer to generate thin films of these epoxy composites [32]. After the solution has set, a thin film composite can be made by prebaking at 95°C for 15 min, then exposing to UV radiation for 15 seconds, and then post-baking at 75°C for 25 min. Because more solvent is supplied during ultrasonication and the mixture is less viscous, the composite films are thinner than the corresponding pure epoxy films. Because of the high viscosity of the slurry, a higher CNT concentration complicates the spin-coating process. MWCNT-reinforced liquid crystalline epoxide resin has been prepared *via* ultrasonication-assisted solution-mixing using the liquid crystalline epoxy resin, diglycidyl ether of 4,4′- dihydroxy-α-methyl stilbene as the matrix, and sulphanilamide as the curing agent [33]. The CNTs surfaces are chemically oxidized by refluxing with aqueous HNO_3. The resin, curing agent, and filler are ultrasonicated for 1 h in acetone, then the solvent is evaporated. In a vacuum oven, the combinations were cured at 150°C for 4 hours and then post-cured at 175°C for 1 h. The interfacial adhesion of CNTs with the resin is improved by surface chemical oxidation.

SWCNT/epoxy nanocomposites with NT loadings of 0.01 and 0.05 wt% were produced by combining the epoxy resin and SWCNTs under high-shear and heat processing the mixture before adding the hardener [34]. SWCNTs are purified using HCl, which causes flaws in the NTs by introducing a few carboxylic acid groups (–COOH) at defect sites. Because processing SWCNT/epoxy composites can be challenging as a result of the increased viscosity, the epoxy resin diglycidyl ether of bisphenol F is utilized because it is less viscous than the more commonly used bisphenol A epoxy resin. It also offers greater functionality, allowing for a greater cross-linking density. Another benefit is that it needs higher-functioning amine-based curing agents with lower viscosities. A cycloaliphatic diamine (4, 4′-methylenebiscylohexanamine) is used as the hardener/curing agent. The epoxy resin is ultrasonicated for 1 h with a sonicated suspension of CNTs in DMF. Sonication is used to distribute the epoxy resin in a sonicated suspension of CNTs in DMF. The DMF is dried by heating the SWCNT/resin suspension at 100°C for 24 hours. Another mixture of SWCNT/resin is heated at 150°C for 24 hours under vacuum in a comparable way. A twin-screw batch mixer is used to mix both batches and subject the solution to high-shear mixing. After that, the curing agent is added to the homogeneous mixture and compounded for another 30 min. To finish the cross-linking process, the mixture is placed in a Teflon pan, degassed, and cured for 2 hours at 80°C under vacuum, accompanied by another 2 hours at 150°C. The schematic for the preparation of SWCNT/epoxy polymer nanocomposites is shown in Figure 3.3. After grafting epoxy resin to SWCNTs *via* an esterification reaction, good dispersion of the NTs is obtained. The covalent bonds between both the NTs and the epoxy resin are formed when the SWCNT/resin combination is heated before adding the hardener. Preheating the SWCNT/epoxy resin mixture to eliminate the solvent, on the other hand, increases the viscosity and makes processing at greater SWCNT loadings unfeasible.

A conventional calendaring procedure is used to create epoxy composites made up of catalytically generated double-wall carbon NTs (DWCNTs) and an epoxy matrix [35]. The three cylindrical rollers rotate at varying speeds in a three-roll machining process, with

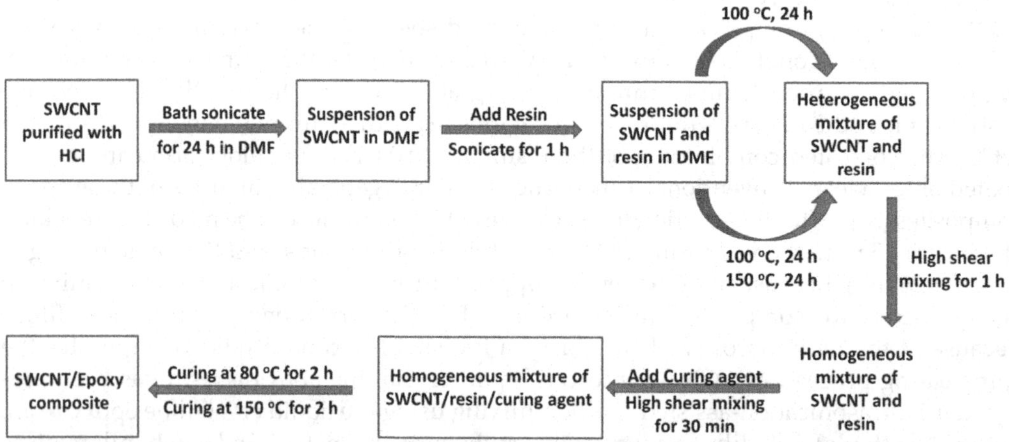

FIGURE 3.3
Scheme for the preparation of SWCNT/epoxy polymer nanocomposites.

neighboring rollers revolving in opposite directions. The resin is manually mixed with 0.1 wt% amino-functionalized DWCNTs, accompanied by a three-roll calendar dispersion stage. The suspension is accumulated, combined with the hardener for 10 min while stirring, cured at room temperature for 24 hours, and then post-cured at 60°C for another 24 hours. Similar composites are also prepared with non-functionalized DWCNTs. Because the shearing force produced between the parallel rollers is greater than that produced by stirring, the process can be used to operate high-viscosity materials with little or no solvent. The use of calendars to exfoliate and scatter CNTs in epoxies appears to be a suitable and safe process. Only a minimal number of CNTs is required to increase mechanical characteristics. The qualitative observation of increased dispersibility and interface of amino-functionalized NTs in an epoxy matrix, as well as the mechanical stabilities acquired, show the significance of CNT functionalization. Using a three-roll milling machine, with parallel rollers running at 200 rpm, with a 5-mm gap size between the rollers, CVD-produced MWCNT is completely dispersed into the PU combination of the two components of PU (PF-359 and E-145) [36].

Epoxy nanocomposites reinforced with sepiolite have been prepared by blending sepiolite with the epoxy resin in a Haake mixer at 130°C, accompanied by 1 hour of ultrasonication [37]. The hardener diglycidyl ether of bisphenol is then incorporated into the mixture (epoxy: hardener 1: 0.875). After degasification, the solution is immediately put into a warmed steel mold. HNT-epoxy resin hybrids have been prepared by adding hardener bisphenol A dicyanate ester into a dispersion of HNT in the epoxy resin diglycidyl ether of bisphenol A. The resulting mixture is degassed and poured into steel molds with Teflon coating and then cured [38].

Alkoxysilanes have been used as coupling agents to disperse SWCNTs throughout the polymer matrix. Siloxane bonds between the cross-linking agent and the SWCNTs improve the adherence of the organic interface to the inorganic interface. The condensation of silanol groups on the cross-linking agent with hydroxyl or carboxyl functionalities on the SWCNTs forms this bond. At defect sites and tube ends, cleansing of SWCNTs with acid removes remaining catalysts and amorphous carbon, introducing functionality such as carboxylic acid and other oxygen-containing groups (e.g., hydroxyl groups). The reaction of 4,4'-hexafluoroisopropylidiene diphthalic anhydride (6-FDA) with 1,3-bis(3-aminophenoxy)benzene (APB) yields an alkoxysilane-terminated amide acid (ASTAA) polymer that is end-capped with aminophenyltrimethoxysilane (APTS). An ultrasonicated mixture of SWCNTs in DMA is added to a preweighed ASTAA solution to make nanocomposite mixtures. The solutions are being used to cast unoriented thin films by placing them on clean, dry glass plates and drying them after 16 hours of continuous stirring in a nitrogen environment. The films on glass are step dried for 1 hour at 100, 200, or 300°C in a forced-air oven [39]. Table 3.3 depicts the main parameters of preparation of 1D polymer nanocomposites through the *in-situ* polymerization method.

3.2.4 Electrospinning

The electrospinning technique facilitates the development of continuous polymer NFs from polymeric or melts in high electric fields. A thin polymer jet is released when the electric force on the charges produced on the polymer liquid overcomes surface tension. Electrospinning composites can be used to build fibrous polymeric sheets with fiber sizes ranging from m to nm. To achieve greater control over fiber alignment, one of several collector topologies, namely a flat plate or a spinning drum, can be used. Electrospun NFs can be oriented to create specific functional nanostructures, such as NWs, thin nonwoven

TABLE 3.3

1D Polymer Nanocomposites Prepared *via In-Situ* Polymerization

Matrix	Filler	Solvent	Surface Modification	Dispersion	Initiator/Curing Agent	Ref
Urethane diacrylate oligomer	MWCNTs	Ethanol			UV	[21]
PU	MWCNTs	Acetone		Stirring at 120°C		[22]
PU	MWCNTs		Amide-functionalized CNTs	Sonication		[23]
Polyimide	SWCNTs	DMF		Ultrasonication	Acetic acid and pyridine	[24]
PAA	MWCNTs	HCl solution	Carboxylic acid- functionalized CNTs	Ultrasonication	Ammonium persulphate	[25]
PPy	MWCNTs	Absolute ethanol	Sodium dodecyl sulfate as surfactant	Ultrasonication	$FeCl_3$	[26]
PANI	MWCNTs	HCl solution		Stirring	Ammonium persulphate	[27]
PANI	MWCNTs		Sulphonated MWCNTs	Ultrasonication	Ammonium persulphate	[28]
PS	MWCNTs	Toluene		Heating at 90°C		[29]
Phthalocyanine	MWCNTs			Melt mixing at 250°C		[30]
Epoxy	MWCNTs			Ultrasonication,	Triethylenetetraamine	[31]
Epoxy	MWCNTs	Chloroform		Ultrasonication	UV	[32]
Epoxy	MWCNTs	Acetone	Oxidation of CNTs with HNO_3	Ultrasonication	Sulphanilamide	[33]
Epoxy	SWCNTs	DMF	Carboxylic acid-functionalization of SWCNTs using HCl	Ultrasonication, high-shear mixing and heat treatment	4, 4′-methylenebiscylohexanamine	[34]
Epoxy	DWCNTs		Amino- functionalized DWCNTs	Three-roll milling	Amine-curing agent H 137i	[35]
PU	MWCNTs	Toluene		Three-roll milling	Diglycidyl ether of bisphenol A	[36]
Epoxy	Sepiolite			Mixing at 130°C followed by ultrasonication	Bisphenol A dicyanate ester	[37]
Epoxy	HNTs					[38]
Alkoxysilane-terminated amide acid polymer	SWCNTs	DMA	Acid-treated SWCNTs	Sonication		[39]

mats, NTs, and aligned nanocomposites. The polymer NFs with a broad range of qualities, including low porosity, high mechanical strength, high electrical conductivity, and low weight, as well as specific surface functions, can be produced based on the polymeric material. Because of these characteristics, polymer NFs are a better material for a wide range of applications.

PVA-reinforced MWCNT NFs have been fabricated *via* such an electrospinning process [40]. A water-soluble polymeric system is employed to avoid the use of surfactants. MWCNTs are acid processed and then electrostatic interactions spun to generate filler-enclosed NFs with a diameter of 100–200 nm in deionized water with PVA. The MWCNTs/PVA solution will be added to a syringe. The solution is transferred at a flow rate of 0.1–5 ml/hour through to the syringe needle. The electrospun fibers are collected on a rotating metal drum with a diameter of 9 cm and a speed of rotation of 25 rpm. The syringe needle has an inner diameter of 0.2 mm, and the distance between the needle and the rotating metal drum is kept at 10–15 cm. The potential difference between the needle filled with MWCNT/PVA solution and the metal drum is 10–20 kV. The color of the fiber sheets changed with the filler concentration in the composite. The distortion of the fiber sheets increases with increasing NT concentration The texture and surface morphology of spun NFs are shown to be affected by the MWNT/PVA composite components as well as the operational conditions. Electrospinning is used to make MWCNT-containing polyvinyl alcohol (PVA) and polyethylene oxide (PEO) NFs [41]. It produces random fiber fabrics or continuous yarns with fiber diameters in the nanoscale range by elongating a CNT–polymer mixture. The modification of the polymer matrices and, most likely, the MWCNTs may be expected because this approach uses a very high electric voltage to stretch the fibers. Ultrasound disperses acid-modified MWCNTs and sodium dodecyl sulfate in water for 30 min. To prepare a viscoelastic spinnable mixture, combine this mixture with PEO/ethanol/water solution or PVA/water solution. The positive electrode is 20 cm away from the grounded metal collector. For the collecting of the NFs samples, a copper grid layered with a holey carbon layer is pasted on the grounded metal collector.

Using the electrospinning technique, lignosulfonic acid sodium salt (LSA) is employed to successfully disperse virgin CNTs in water, resulting in a PVA/LSA/CNT hybrid fiber with a steadily increasing CNT load [42]. PVA, LSA, and MWCNTs are dissolved in a water-ethanol solution (75% deionized water, 25% ethanol) and ultrasonicated for 1 hour before electrospinning. PVA/LSA/CNT hybrid fibers are electrospun over a distance of 24 cm between both the syringe needle and the grounded collector with an electric field of 18 kV. The solution is delivered at a rate of 0.12 ml/hour. Individual, isolated electrospun fibers are collected at the collector using a transmission electron microscope (TEM) grid on a silicon substrate. The bundling and non-uniform distribution of CNTs occur in each fiber despite ultrasonication and the usage of LSA, resulting in varying CNT loadings. The overall elastic modulus of PVA-LSA-CNT fibers is much higher than that of PVA-LSA fibers, although CNT loading does not affect the elastic modulus. Table 3.4 illustrates the conditions for the preparation of 1D polymer nanocomposites *via* an electrospinning method.

3.2.5 Non-Traditional Methods

3.2.5.1 Using a Magnetic Field

Inorganic polymer nanocomposites consisting of 1D magnetic NWs are produced by the assembling of iron oxide (γ-Fe_2O_3) colloidal nanoparticles in polymer matrices under the

TABLE 3.4

1D Polymer Nanocomposites Prepared *via* Electrospinning

Matrix	Filler	Solvent	Filler Modification	Dispersion	Ref
PVA	MWCNT	Distilled water	Acid treated MWCNTs	—	[40]
PVA	MWCNT	Water	Acid treated MWCNTs	Ultrasonication; Sodium dodecyl sulfate as surfactant	[41]
PEO	MWCNT	Ethanol/Water	Acid treated MWCNTs	Ultrasonication; Sodium dodecyl sulfate as surfactant	[41]
PVA	MWCNT	Ethanol/Water		Ultrasonication; Lignosulphonic acid sodium salt as dispersant	[42]

application of a magnetic field (MF) [43]. The NWs are formed *in-situ*, during the evaporation of a drop-casted copolymer solution of polyethylmethacrylate-co-methylacrylate mixed with -Fe_2O_3 nanoparticles in chloroform, by applying a weak homogeneous MF produced by two permanent magnets parallel to the substrate during the deposition and evaporation stages. The external MF causes the assembly of nanoparticles along the path of the MF lines, the formation of magnetic NWs.

3.2.5.2 Supercritical CO_2-Assisted Mixing

Supercritical CO_2 ($scCO_2$) is nontoxic, nonflammable, inexpensive, and abundant. When $scCO_2$ is added to the polymer phase, it acts as a lubricant, reducing chain–chain interactions while increasing free volume. As a result, $scCO_2$ has a significant plasticization effect on the polymer, and the physical characteristics of the polymer are dramatically altered, including a decrease in Tg, interfacial tension, and melt viscosity of the polymer. CO_2 dissolution in a polymer promotes monomer and initiator diffusion as well as reactive component dispersion inside the polymer matrix. Supercritical CO_2 has been used for the preparation of sepiolite-reinforced polypropylene nanocomposites [44]. PP and sepiolite, together with the compatibilizer PP-g-MAH, are placed in an autoclave. The ratio of PP-g-MAH in the nanocomposite is the same as the proportion of sepiolite in the nanocomposites (1:1 weight ratio). After being refrigerated to –6°C, CO_2 is fed into the autoclave using an Isco model 260D syringe pump. The autoclave is kept at 15 MPa and 200°C for 30 min, while stirring with a pitched bladed turbine. It is subsequently cooled to room temperature in water and the CO_2 is released. The bulk materials acquired from the autoclave were freeze-dried for a few seconds before being split into bits and hot-pressed into dumbbell-shaped specimens at 200°C. When $scCO_2$ is used instead of typical melt combining, better sepiolite dispersion and less breaking of sepiolite fibers are produced. Even without the use of PP-g-MAH as a compatibilizer, the nanocomposites treated in $scCO_2$ have a relatively homogeneous sepiolite distribution throughout the PP matrix.

3.2.5.3 Bucky Paper Composites

MWCNT/PU composites are prepared by bucky paper infiltration, which involves filtering an ultrasonicated dispersion of acid-treated MWCNTs in DMF through a polytetrafluoroethylene membrane with an average pore size of 0.2 μ, under a conditioned vacuum pressure of 200 ± 20 mbar, followed by infiltration of the polymer solution in THF [45]. When the infiltration time has reached roughly 12 hours, the infiltration pressure is dropped to 50 ± 10 mbar to accelerate the process. The filter membranes are pulled away to reveal uniform high-MWCNT-loaded nanocomposite films with the polymer uniformly covering the CNTs.

3.3 Conclusions

Although many methods have been utilized for the synthesis of 1D polymeric nanocomposites, melt intercalation, solution intercalation, *in-situ* polymerization, and electrospinning are the most commonly used methods. Homogeneous dispersion of the nanofillers is very important to enhance the properties of the composites. Several physical, mechanical, and chemical routes have been adopted for improving the dispersion of 1D nanofillers, such as carbon nanotubes, CNFs, and sepiolite. On the basis of processing efficiency, types of nanofiller, and desired end product quality, each polymer system requires unique processing conditions.

Acknowledgments

Reshmy Rajasekharan and Sindhu Raveendran acknowledge the Department of Science and Technology, New Delhi for funding projects under the DST WOS-B scheme.

References

1. P. Potschke, A. R. Bhattacharyya, and A. Janke, "Melt mixing of polycarbonate with multi-walled carbon nanotubes: Microscopic studies on the state of dispersion," *Eur. Polym. J.*, vol. 40, pp. 137–148, 2004, doi: 10.1016/j.eurpolymj.2003.08.008.
2. T. Villmow, P. Potschke, S. Pegel, L. Haussler, and B. Kretzschmar, "Influence of twin-screw extrusion conditions on the dispersion of multi-walled carbon nanotubes in a poly(lactic acid) matrix," *Polymer*, vol. 49, pp. 3500–3509, 2008.
3. W. Chen, X. Tao, and Y. Liu, "Carbon nanotube-reinforced polyurethane composite fibers," *Compos. Sci. Technol.*, vol. 66, pp. 3029–3034, 2006.
4. E. Hammel, X. Tang, M. Trampert, T. Schmitt, K. Mauthner, A. Eder, and P. Potschke, "Carbon nanofibers for composite applications," *Carbon*, vol. 42, no. 5–6, pp. 1153–1158, 2004.
5. H. C. Bidsorkhi, M. Soheilmoghaddam, R. H. Pour, H. Adelnia, and Z. Mohamad, "Mechanical, thermal and flammability properties of ethylene-vinyl acetate (EVA)/sepiolite nanocomposites," *Polym. Test.*, vol. 37, pp. 117–122, 2014.
6. G. Tartaglione, D. Tabuani, G. Camino, and M. Moisio, "PP and PBT composites filled with sepiolite: Morphology and thermal behaviour," *Compos. Sci. Technol.*, vol. 68, no. 2), pp. 451–460, 2008.
7. D. García-López, J. F. Fernández, J. C. Merino, and J. M. Pastor, "Influence of organic modifier characteristic on the mechanical properties of polyamide 6/organosepiolite nanocomposites," *Compos. Part B*, vol. 45, no. 1, pp. 459–465, 2013.
8. D. García-López, J. F. Fernández, J. C. Merino, J. Santarén, and J. M. Pastor, "Effect of organic modification of sepiolite for PA 6 polymer/organoclay nanocomposites," *Compos. Sci. Technol.*, vol. 70, no. 10, pp. 1429–1436, 2010.

9. V. H. Orozco, A. F. Vargas, W. Brostow, T. Datashvili, B. L. Lopez, K. Mei, and L. Su, "Tribological properties of polypropylene composites with carbon nanotubes and sepiolite," *J. Nanosci. Nanotechnol.*, vol. 14, no. 7, pp. 4918–4929, 2014.
10. B. Guo, Q. Zou, Y. Lei, M. Du, M. Liu, and D. Jia, "Crystallization behavior of polyamide 6/halloysite nanotubes nanocomposites," *Thermochim. Acta*, vol. 484, no. 1–2, pp. 48–56, 2009.
11. M. Du, B. Guo, M. Liu, and D. Jia, "Preparation and characterization of polypropylene grafted halloysite and their compatibility effect to polypropylene/halloysite composite," *Polym. J.*, vol. 38, no. 11, pp. 1198–1204, 2006.
12. M. Pourmohammadi-Mahunaki, V. Haddadi-Asl, H. Roghani-Mamaqani, M. Koosha, and M. Yazdi, "Preparation of polyurethane composites reinforced with halloysite and carbon nanotubes," *Polym. Compos.*, vol. 42, no. 1, pp. 450–461, 2021.
13. H. A. Shawky, S. R. Chae, S. Lin, and M. R. Wiesner, "Synthesis and characterization of a carbon nanotube/polymer nanocomposite membrane for water treatment," *Desalination*, vol. 272, no. 1–3, pp. 46–50, 2011.
14. N. Mahmood, M. Islam, A. Hameed, and S. Saeed, "Polyamide 6/multiwalled carbon nanotubes nanocomposites with modified morphology and thermal properties," *Polymers (Basel)*, vol. 5, no. 4, pp. 1380–1391, 2013.
15. M. S. P. Shaffer and A. H. Windle, "Fabrication and characterization of carbon nanotube/poly(vinyl alcohol) composites," *Adv. Mater.*, vol. 11, no. 11, pp. 937–941, 1999.
16. M. Li, Y. Song, C. Zhang, Z. Yong, J. Qiao, D. Hu, Z. Zhang, H. Wei, J. Di, and Q. Li, "Robust carbon nanotube composite fibers: Strong resistivities to protonation, oxidation, and ultrasonication," *Carbon*, vol. 146, pp. 627–635, 2019.
17. C. Zeng, N. Hossieny, C. Zhang, and B. Wang, "Synthesis and processing of PMMA carbon nanotube nanocomposite foams," *Polymer*, vol. 51, no. 3, pp. 655–664, 2010.
18. H. M. S. Iqbal, S. Bhowmik, R. Benedictus, J. B. Moon, C. G. Kim, and A. H. I. Mourad, "Processing and characterization of space-durable high-performance polymeric nanocomposite," *J. Thermophys. Heat Transf.*, vol. 25, no. 1, pp. 87–95, 2011.
19. G. Venugopal, J. C. Veetil, N. Raghavan, V. Singh, A. Kumar, and A. Mukkannan, "Nanodynamic mechanical and thermal responses of single-walled carbon nanotubes reinforced polymer nanocomposite thin-films," *J. Alloys Compd.*, vol. 688, pp. 454–459, 2016.
20. G. Wang, X. Huang, and P. Jiang, "Tailoring dielectric properties and energy density of ferroelectric polymer nanocomposites by high-k NWs," *ACS Appl. Mater. Interfaces*, vol. 7, no. 32, pp. 18017–18027, 2015.
21. H. D. Wagner, O. Lourie, Y. Feldman, and R. Tenne, "Stress-induced fragmentation of multiwall carbon nanotubes in a polymer matrix," *Appl. Phys. Lett.*, vol. 72, no. 2, pp. 188–190, 1998.
22. J. Ryszkowska, M. Jurczyk-Kowalska, T. Szymborski, and K. J. Kurzydłowski, "Dispersion of carbon nanotubes in polyurethane matrix," *Phys. E Low-dimensional Syst. Nanostructures*, vol. 39, no. 1, pp. 124–127, 2007.
23. J. Xiong, Z. Zheng, X. Qin, M. Li, H. Li, and X. Wang, "The thermal and mechanical properties of a polyurethane/multi-walled carbon nanotube composite," *Carbon*, vol. 44, no. 13, pp. 2701–2707, 2006.
24. C. Park, Z. Ounaies, K. A. Watson, R. E. Crooks, J. Smith Jr., S. E. Lowther, J. W. Connell, E. J. Siochi, J. S. Harrison, and T. L. St. Clair, "Dispersion of single wall carbon nanotubes by in situ polymerization under sonication," *Chem. Phys. Lett.*, vol. 364, no. 3–4, pp. 303–308, 2002.
25. M. P. Dash, M. Tripathy, A. Sasmal, G. C. Mohanty, and P. L. Nayak, "Poly(anthranilic acid)/multi-walled carbon nanotube composites: Spectral, morphological, and electrical properties," *J. Mater. Sci.*, vol. 45, pp. 3858–3865, 2010.
26. N. Baghdadi, M. S. Zoromba, M. H. Abdel-Aziz, A. F. Al-Hossainy, M. Bassyouni, and N. Salah, "One-dimensional nanocomposites based on polypyrrole-carbon nanotubes and their thermoelectric performance," *Polymers (Basel)*, vol. 13, pp. 278–293, 2021.
27. W. Li and D. Kim, "Polyaniline/multiwall carbon nanotube nanocomposite for detecting aromatic hydrocarbon vapors," *J. Mater. Sci.*, vol. 46, pp. 1857–1861, 2011.
28. Z. Wei, M. Wan, T. Lin, and L. Dai, "Polyaniline nanotubes doped with sulfonated carbon nanotubes made via a self-assembly process," *Adv. Mater.*, vol. 15, no. 2, pp. 136–139, 2003.

29. X. Wu and P. Liu, "Polymer grafted multiwalled carbon nanotubes via facile in situ solution radical polymerisation," *J. Exp. Nanosci.*, vol. 5, no. 5, pp. 383–389, 2010.
30. Z. Wang, X. Yang, J. Wei, M. Xu, L. Tong, R. Zhao, and X. Liu, "Morphological, electrical, thermal and mechanical properties of phthalocyanine/multi-wall carbon nanotubes nanocomposites prepared by masterbatch dilution," *J. Polym. Res.*, vol. 19, no. 9, pp. 9969–9976, 2012.
31. L. S. Schadler, S. C. Giannaris, and P. M. Ajayan, "Load transfer in carbon nanotube epoxy composites," *Appl. Phys. Lett.*, vol. 73, no. 26, pp. 3842–3844, 1998.
32. X. Xu, M. M. Thwe, C. Shearwood, and K. Liao, "Mechanical properties and interfacial characteristics of carbon-nanotube-reinforced epoxy thin films," *Appl. Phys. Lett.*, vol. 81, no. 15, pp. 2833–2835, 2002.
33. J. Jang, J. Bae, and S. Yoon, "A study on the effect of surface treatment of carbon nanotubes for liquid crystalline epoxide – Carbon nanotube composites," *J. Mater. Chem.*, vol. 13, pp. 676–681, 2003.
34. M. Moniruzzaman, F. Du, N. Romero, and K. I. Winey, "Increased flexural modulus and strength in SWNT/epoxy composites by a new fabrication method," *Polymer*, vol. 47, pp. 293–298, 2006.
35. F. H. Gojny, M. H. G. Wichmann, U. Kopke, B. Fiedler, and K. Schulte, "Carbon nanotube-reinforced epoxy-composites: Enhanced stiffness and fracture toughness at low nanotube content," *Compos. Sci. Technol.*, vol. 64, pp. 2363–2371, 2004.
36. H. Souri, I. W. Nam, and H. K. Lee, "Electrical properties and piezoresistive evaluation of polyurethane-based composites with carbon nano-materials," *Compos. Sci. Technol.*, vol. 121, pp. 41–48, 2015.
37. Y. Zheng and Y. Zheng, "Study on sepiolite-reinforced polymeric nanocomposites," *J. Appl. Polym. Sci.*, vol. 99, no. 5, pp. 2163–2166, 2006.
38. M. Liu, B. Guo, M. Du, X. Cai, and D. Jia, "Properties of halloysite nanotube-epoxy resin hybrids and the interfacial reactions in the systems," *Nanotechnology*, vol. 18, no. 45, p. 455703, 2007.
39. J. G. Smith Jr., J. W. Connell, D. M. Delozier, P. T. Lillehei, K. A. Watson, Y. Lin, B. Zhou, and Y.-P.Sun, "Space durable polymer/carbon nanotube films for electrostatic charge mitigation," *Polymer*, vol. 45, pp. 825–836, 2004.
40. J. S. Jeong, J. S. Moon, S. Y. Jeon, J. H. Park, P. S. Alegaonkar, and J. B. Yoo, "Mechanical properties of electrospun PVA/MWNTs composite nanofibers," *Thin Solid Films*, vol. 515, no. 12, pp. 5136–5141, 2007.
41. W. Zhou, Y. Wu, F. Wei, G. Luo, and W. Qian, "Elastic deformation of multiwalled carbon nanotubes in electrospun MWCNTs-PEO and MWCNTs-PVA nanofibers," *Polymer*, vol. 46, no. 26, pp. 12689–12695, 2005.
42. K. K. H. Wong, M. Zinke-Allmang, J. L. Hutter, S. Hrapovic, J. H. T. Luong, and W. Wan, "The effect of carbon nanotube aspect ratio and loading on the elastic modulus of electrospun poly(vinyl alcohol)-carbon nanotube hybrid fibers," *Carbon*, vol. 47, no. 11, pp. 2571–2578, 2009.
43. D. Fragouli, B. Torre, G. Bertoni, R. Buonsanti, R. Cingolani, and A. Athanassiou, "Formation and microscopic investigation of iron oxide aligned NWs into polymeric nanocomposite films," *Microsc. Res. Tech.*, vol. 73, no. 10, pp. 952–958, 2010.
44. J. Ma, E. Bilotti, T. Peijs, and J. A. Darr, "Preparation of polypropylene/sepiolite nanocomposites using supercritical CO_2 assisted mixing," *Eur. Polym. J.*, vol. 43, no. 12, pp. 4931–4939, 2007.
45. B. Fernández-d'Arlas, U. Khan, L. Rueda, L. Martin, J. A. Ramos, J. N. Coleman, M. L. Gonzalez, A. Valea, I. Mondragon, M. A. Corcuera, and A. Eceiza, "Study of the mechanical, electrical and morphological properties of PU/MWCNT composites obtained by two different processing routes," *Compos. Sci. Technol.*, vol. 72, no. 2, pp. 235–242, 2012.

4

Architectural Aspects of One-Dimensional Nanocomposites and Various Applications

Trinath Biswal

CONTENTS

DOI: 10.1201/9781003223764-4

4.1 Introduction

One-dimensional (1D) nanostructured materials are being developed in various forms such as fibers, rods, wires, whiskers, belts, rings, and tubes, which have attracted considerable attention over the past three decades owing to their improved novel characteristics and fascinating applications in several vital fields [1, 2]. 1D nanostructured materials are capable of improving the production of energy processes because of their unique properties. Different techniques have been developed for synthesizing materials with a 1D nanostructure with appropriate chemical composition and morphology. Among the various methods, the method of electrospinning is the most useful technique for synthesizing the various forms of 1D nanostructure materials, such as nanofibers, nanorods, nanowires, etc. As compared with the method of mechanical spinning for producing microfibers, the process of electrospinning normally utilizes the electrostatic repulsions among the surface charges, leading to the decrease in the diameter of glassy filaments or viscoelasticity of the jets. Significant benefits in the process of electrospinning include low cost and ease of fabrication of a large number of various types of nanofiber. Other benefits of the method of electrospinning are its capability to control the diameter of the fibers, to increase the surface-to-volume ratio, and to achieve an acceptable aspect ratio and adequate pore size with non-woven fabrics [3, 4]. Another benefit of fabricating 1D composite nanomaterials by electrospinning is the multifunctional properties of such materials which make them suitable for many applications. Usually, four categories of 1D composite nanomaterial can be synthesized, namely organic polymer/organic polymer, inorganic/inorganic, polymer/inorganic, and composites. In this chapter, the synthesis, architectural aspects, classification, and applications of 1D nanostructured materials are explained [5, 6].

4.2 Types of Nanomaterials

Based on the size of the particles, the nanomaterials (NMs) are classified into the following four different categories. Figure 4.1 describes the different types of nanomaterial:

1. Zero-dimensional nanomaterial;
2. One-dimensional nanomaterial;
3. Two-dimensional nanomaterial;
4. Three-dimensional nanomaterial [7].

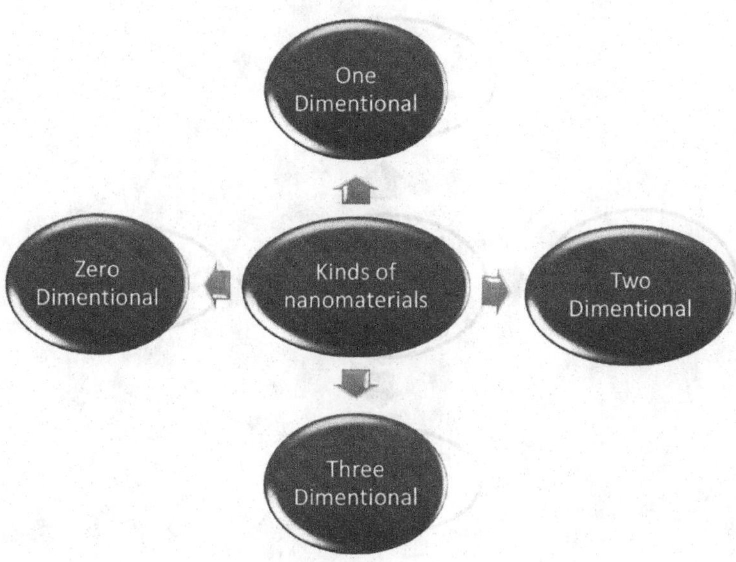

FIGURE 4.1
Classification of nanomaterials.

4.3 Synthesis of 1D Nanostructured Materials

The nanomaterials can be synthesized by two different approaches;

- The smaller-sized materials in the nanoscale can be obtained by reducing the bulk materials;
- By synthesizing the materials to nanoscale by some other techniques.

The first approach is known as 'top-down', whereas the second one is called 'bottom-up'. The top-down method is more popularly used in the microelectronic industry, which follows the reduction of circuits and various components with spatially arranged structures, with a tolerance of about a few nanometres. The most exciting feature of this process of synthesis is the characteristics and properties of some bulk substances which are preserved in the treated materials.

One of the significant demerits of this process is low yield. The top-down process normally includes different methodologies like ion etching, metal-assisted chemical etching, or anodic oxidation. Contrary to the different top-down processes, the bottom-up method normally follows the chemical self-assembly and molecular recognition of the molecules, which permits getting the structures with sizes and can vary from the range of a few nanometers to several microns. This methodology includes various methodologies, the most significant of which are vapor-phase growth, liquid-phase growth, template-assisted etching, and electrospinning [8, 9, 10]. The various methodologies of synthesis are shown in Figure 4.2.

4.3.1 Vapour Phase or Gas-Phase Synthesis of 1D Nanomaterials

Gas-phase synthesis is amazing because it uses a well-designed pathway to control and regulate the process parameters for producing 1D nanomaterial of appropriate shape,

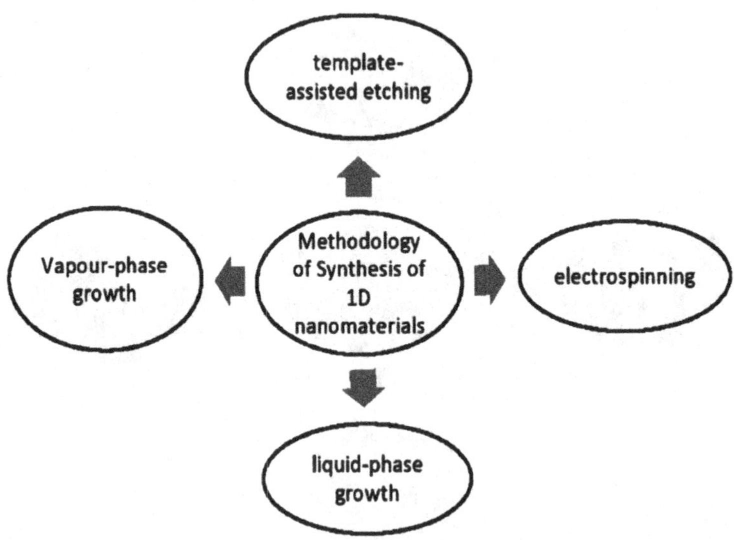

FIGURE 4.2
Methodology of synthesis of ID nanomaterials.

size, and chemical composition. In the traditional method of synthesis, using chemical vapor deposition (CVD), some gaseous substances generated are permitted to react either heterogeneously or homogeneously according to specific applications. In the process of homogeneous CVD, the particles produced in the vapor phase diffuse toward the surfaces with lower temperature because of the thermophoretic stresses and may either be put down on the surface of the substrate, producing a substance termed particulate films, or scuffled from the surfaces of lower-temperature forms of powder in the nanoscale. In the heterogeneous synthesis of CVD, the solid phase is produced on the surface of the substrate, which acts as a catalyst in the reaction, resulting in the formation of a dense film. The 1D nanomaterials of high-quality purity can be synthesized by using various routes. The vapor phase technique is the condensation of the vapors into nanosized particles. Condensation happens when the vapors are in supersaturated form. In this method, the homogeneous state of nucleation in the vapor phase is utilized, producing suitable nanoparticles [11, 12].

4.3.2 Mechanical Grinding

The method of mechanical abrasion is a top-down process of synthesizing 1D nanomaterials. This is a popular and widely applied method for synthesizing 1D nanomaterials because of its simple procedure, low-cost equipment, and applicability to the production of every class of substances. The most important benefit is the probability of simply scaling up to the amounts of the materials, which can be applied to various fields. But the major problems related to this method are as follows [13, 14]:

- Pollution from the atmosphere and/or milling media;
- Consolidation of the product materials without toughening the nanocrystalline microstructure.

4.3.3 Sol-Gel Technique

The technique of sol-gel comprises the development of inorganic networks by the formation of a suspension containing colloidal particles (sol) and gelation of the suspension to form a system in a continuous liquid phase, termed a gel. The components used for the synthesis of the colloidal form of particles normally consist of an element of the metalloid or metal type surrounded by different kinds of sensitive ligands. The preliminary raw material is treated to produce the dispersible oxide and also to produce a sol in interaction with dilute acid or water. If a liquid is removed from the sol, then the material obtained is termed a gel, where the transition between gel and sol regulates the shape and size of the particles. During the process of calcination, the gel forms oxide materials. The processing of sol-gel is normally carried out *via* condensation and hydrolysis of the alkoxide-based materials such as 4-(tetraethyl orthosilicate), Si $(OC_2H_5)_4$. During the process, the reaction which occurs based on condensation and hydrolysis of the metal alkoxides (MOR) can be represented as follows:

$$\text{MOR} + \text{H}_2\text{O} \rightarrow \text{MOH} + \text{ROH} \ (\text{During the process of hydrolysis})$$

$$\text{MOH} + \text{ROM} \rightarrow \text{M}-\text{O}-\text{M} + \text{ROH} \ (\text{During the process of condensation})$$

The technique of sol-gel for the formation of NMs is widespread among researchers and was widely used for producing oxide-based materials [15, 16].

4.3.4 Ultrasonic Spray Pyrolysis

In this technique, the precursors are perfectly nebulized, after which the undesired components undergo combustion and the essential material remains, such as ZrO_2 when using the precursor of $Zr(C_3H_7O)_4$. Flame hydrolysis, usually a variant of this method, is applied for the synthesis of fused silica. In this technique, $SiCl_4$ is heated in an oxy-hydrogen combustible flame, which forms silica particles, which are widely dispersed. A white-colored fine powder is produced, which contains spherical particles of a diameter of 7–40 nm. The flame used for production is a gas mixture containing H_2 and O_2 or C_2H_2 and O_2, which provides the source of energy. If the pressure of the gas during the time of reaction is too high, then extremely agglomerated powders are formed, which is a disadvantage for the subsequent processing [17, 18].

4.3.5 Gas Condensation Processing (GPC)

In this method, an inorganic material based on a metal, such as a sub-oxide, is normally vaporized. The resources necessary for thermal evaporation include devices related to electron beam evaporation, crucibles, and an inert gas such as Ar, Kr, or Ne. A cluster will be formed within the area surrounding the source through homogenous nucleation within the gas phase, which again undergoes growth by incorporation and coalescence of atoms present in the gas phase. The size of the particle or cluster depends upon the residence time of the synthesized particle within the growth system and is affected by the pressure of the gas, the nature of the inert gas used, and the rate of evaporation or vapor pressure generated from the material to be evaporated. With increasing vapor pressure, gas pressure, and weight or volume of the inert gas used, the mean dimension of the nanoparticles

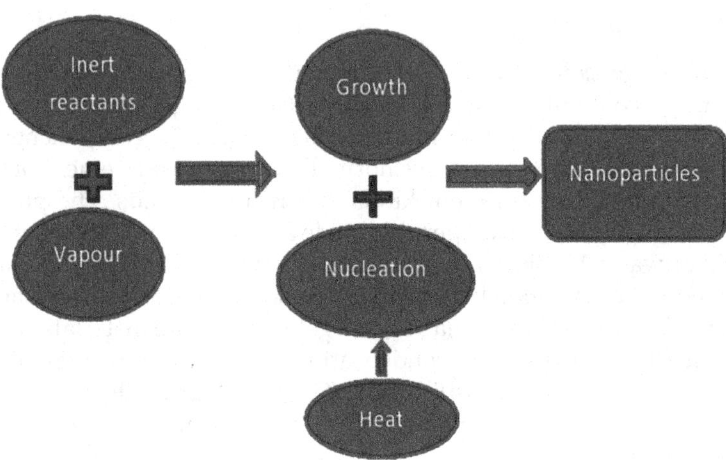

FIGURE 4.3
Steps for the synthesis of 1D nanomaterials in a vapor phase.

synthesized also increases [19, 20]. Figure 4.3 shows a simple procedure for the production of 1D nanomaterial in a vapor phase

The production of crystalline nanoscale pure metal is quite straightforward, requiring evaporation. Those metals which have the capability of reacting with the crucible material or which have extremely high melting points, such as Zr and W, are usually evaporated by sputtering, electron beam evaporation, or a laser technique. The synthesis of the nanoparticles of intermetallic substances or alloys through the method of thermal evaporation can occur only in extraordinary cases when the vapor pressures of the elements are found to be similar. The different composite materials, like W/Ga or Cu/Bi, can be synthesized by evaporating these elements in two distinct crucibles simultaneously on a rotating system of the collection device, where intermixing of the various-sized particles are observed. However, the composition of the elements involved in the process is difficult to control and the level of productivity is extremely low. The powdered forms of the nanocrystalline oxide are synthesized through the post-oxidation of basic nanoparticles of suitable pure metal like Ti to TiO_2 or a sub-oxide such as ZrO to ZrO_2 in a controlled manner. Although the technique of gas condensation has been widely used to synthesize many different kinds of ceramic and metallic nanomaterials, this process is restricted to the laboratory scale only [21, 22].

4.3.6 Chemical Vapor Condensation (CVC)

In this technique, the source of evaporation applied in GPC is substituted by a reactor having a hot wall. The residence time of the precursor used regulates whether particles or films are produced. At some specific residence times, the formation of both film and particles can be achieved. By regulating the residence time of the molecules present in the precursor by adjusting the rate of gas flow, a difference in pressure occurs between the key chamber and delivery system. Figure 4.4 represents the various physical techniques for the synthesis of 1D nanomaterials.

The temperature created in the hot wall reactor causes the effective creation of particles of ceramics in the nano range and metals as films for processing in CVD. Normally, the precursors of metal-organic are added to the high-temperature zone of the reactor by utilizing a controller based on mass flow. This technique may be extended to the synthesis of

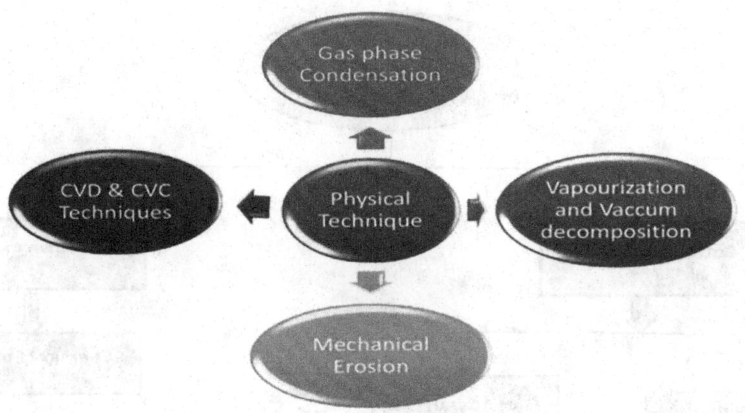

FIGURE 4.4
Physical technique of synthesis of 1D nanomaterials.

nanoparticles, which need the determination of an improved parameter field to promote the formation of the particles rather than that of the film [23, 24].

4.3.7 Sputtered Plasma Processing

This technique is usually a version of the process of gas-phase condensation, excluding the detail that the source of the materials is an effective sputtering target, which is sputtered by consuming inert gases, along with the components obtained which are allowed to gather with the production of the nanomaterial. Both radiofrequency and direct current sputtering have been effectively utilized for producing particles in the nanoscale range. The multitarget sputtering or reactive sputtering can also be utilized to produce oxides or alloys, nitrides, and carbides of materials. This process is particularly effective for the synthesis of ultrapure metal nanoparticles without agglomeration [25].

4.3.8 Microwave Plasma Processing

This process of forming 1D nanomaterials is near-analogous to the CVC technique, but here plasma rather than elevated temperature is used for the degradation of the metal-organic parts. The foremost benefit of using plasma-assisted pyrolysis in comparison to thermal activation is the requirement for lower temperatures in the reaction, leading to a decrease in the risk of agglomerating the primary particles. Furthermore, it was observed that, by using additional precursors into the tubular reactor (second reaction zone) *via* extruding the guide tubes of the microwave, the particles produced can be coated over the second phase [26]. For example, nanoparticles of ZrO_2 can be coated using Al_2O_3 where the inner core of ZrO_2 is crystalline whereas the coated material Al_2O_3 material is amorphous. The synthesis of the primary particles is generally carried out through homogeneous nucleation, which can be easily and effectively measured by application of the kinetics of the gas reaction. The metallic nanoparticles synthesized using different techniques are shown in Figure 4.5 [27, 28].

4.3.9 Particle-Precipitation-Assisted CVD

In this technique, the colloidal clusters of substances are normally utilized to synthesize the nanoparticles. This process is a modified form of the conventional CVD technique,

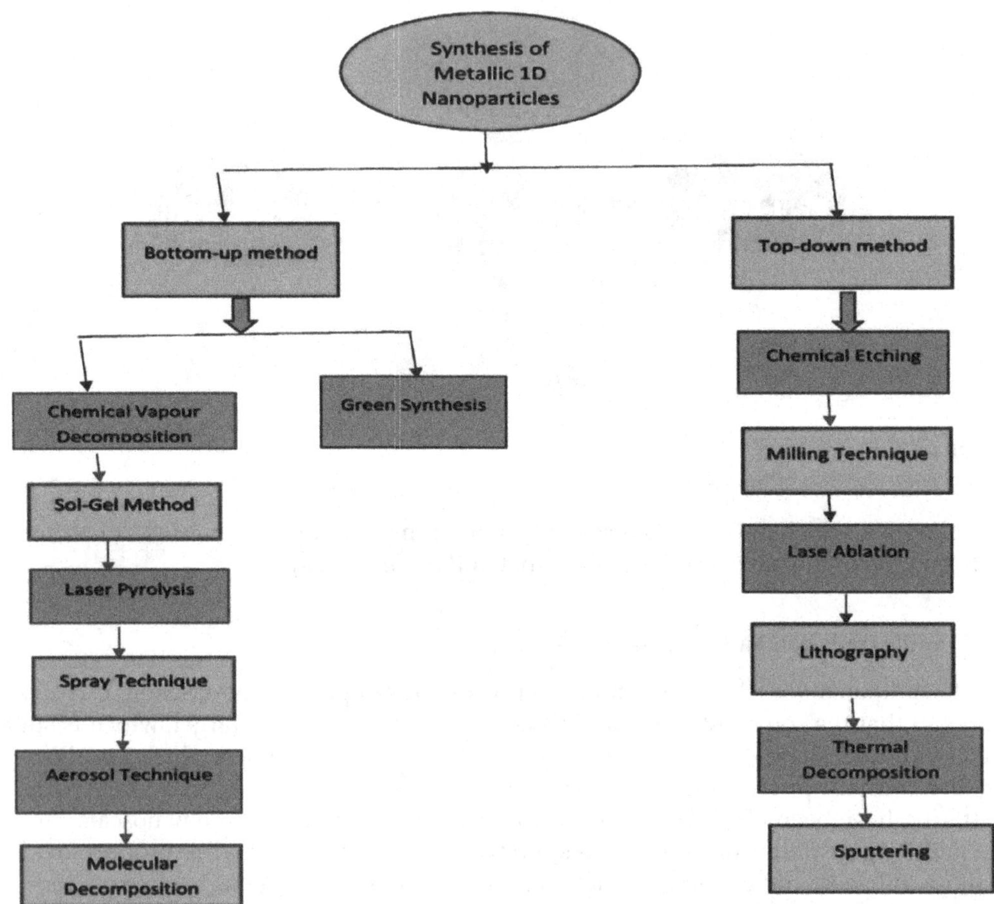

FIGURE 4.5
Various methods of synthesis of 1D nanoparticles.

where the aerosol particles are normally produced in the gas phase at higher temperatures and the synthesized particles are deposited on a substrate at lower temperatures. In this method, the reaction conditions are set up in such a manner that the particles are synthesized by condensation over the gas phase and then collected on the surface of the substrate, which is usually carried out under varying conditions, permitting heterogeneous nucleation. By adopting this technique, both particulate films and nanoparticles can be synthesized. For example, the nanoparticles of SnO_2 can be synthesized effectively by using this technique [29, 30, 31].

4.3.10 Laser Ablation

This is an advanced technique widely applied for the synthesis of 1D films of particulates and nanoparticles. In this method, the laser beam of light is utilized as a source for the ablation of solid materials. When a beam of laser is irradiated on the surfaces of the materials, the materials vaporize quickly. The collisions between the evaporated species lead to the excitation of the electronic state, which is coupled with the source of light emission, and produces ions and electrons, resulting in the formation of a laser-induced plasma plume. The spectrum of emission and the size of the plasma plume normally depends upon the nature and type of the targeted substance, ambient medium such as gas or liquid, ambient pressure, and conditions of the laser light. In the liquid medium, laser ablation is normally confined within the plasma plume in a small area and disperse the nanoparticles in the liquid state directly. The laser spark atomizer has been used to synthesize extensively mesoporous dense films; the porosity can again be changed by controlling the rate of flow of the carrier gas. The thick nanoparticulate films of ZrO_2 and SnO_2 with quite a similar microstructure were produced effectively by applying this technique. The formation of nanoparticles of other materials, like lithium manganate, carbon, and silicon, was also successfully conducted by this technique [3, 32].

4.3.11 Electrospinning Method of Synthesis

The electrospinning technique, in combination with carbonation or calcination of nanofibers of ceramic materials or other inorganic nanofibers, is successfully used for the synthesis of 1D materials. 1D composite nanomaterials with good stability and with structures exhibiting excellent useful properties can be synthesized using electrospinning in combination with various approaches, listed below [33].

4.3.11.1 Direct-Dispersed Electrospinning

To introduce a nanocomponent of inorganic material into a fiber made of polymeric materials, the most basic pathway is the dispersion of the nanocomponent inorganic material into the solution form of the polymer, followed by electrospinning. For example, quantum dots (QDs) of CdTe are added into the solid nanofibers of polyvinylpyrrolidone (PVP) through the method of direct-dispersed electrospinning, for the synthesis of 1D nanomaterials. In addition to nanoparticles of semiconductors or metals, carbon nanotubes (CNTs) serve as a candidate material in combination with electrospun nanofibers to enhance the mechanical properties and electronic and thermal conductivities of polymer fibers [34–36].

4.3.11.2 Gas-Solid Reaction

The inorganic nanoparticles can accumulate and do not distribute perfectly in the matrix of the polymer fiber during the process of electrospinning. Incorporation of the nanostructures of semiconductors into polymer nanofibers can be done in the gas-solid reaction using the electrospinning method. The approach to the synthesis of 1D nanomaterials is as follows:

- The co-dissolution of the polymer and metal salt into a solvent to form a homogeneous solution;
- Electrospinning this prepared solution to obtain the nanofiber composite of the metal/polymer salt.

The exposure of the nanofibers of the polymer/metal salt composite synthesized by electrospinning in H_2S gas at normal temperature produces the NPs of PbS in the process of *in-situ* polymerization. Cadmium acetate can be used to synthesize nanofibers of PVP/CdS composite in place of lead acetate. The synthesized semiconductor nanoparticles obtained by adopting this technique were observed to be present inside as well as outside the polymer nanofibers [37, 38].

4.3.11.3 *In-Situ* Photochemical Reduction Technique

The method of photochemical reduction is a potential technique for synthesizing metal colloids. UV irradiation in the aqueous medium of $AgNO_3$ stimulates photo-oxidation of H_2O through excited Ag^+ ions, leading to the formation of Ag atoms, H^+, and OH^- ions. Then, the Ag atoms will gather together and form colloidal Ag particles. For the synthesis of a 1D polymer/ Ag composite, the $AgNO_3$ can be uniformly distributed into the matrix phase of the polymer, then the composite formed is exposed to UV radiation to carry out photo-reduction. Based on this technique, the nanofiber composite of polymer reinforced with Ag can be effectively made by using the method of electrospinning. The ultrafine, antimicrobial cellulose acetate (CA) fibers can be synthesized by the addition of Ag nanoparticles. The nanofibers of CA/$AgNO_3$ composite can be synthesized by using the solution of CA containing $AgNO_3$ in the process of electrospinning. After photoreduction using UV light, the color of the CA fibers slowly transform into light yellow, confirming the synthesis of Ag nanoparticles. Based on this method, the nanofiber composite of TiO_2 reinforced with Au can also be produced through photoreduction by UV irradiation of $HAuCl_4$ in the presence of organic stabilizers and TiO_2 nanofibers [39, 40].

4.3.11.4 Electrospinning Sol-Gel Technique

In combination with the sol-gel technique, many different varieties of nanofiber made from composite materials can also be produced by electrospinning, which includes metal/ polymer and polymer/sulfide polymer/metal oxide. Again, by using the process of calcination, the polymer nanofibers based on metal oxides can be changed into nanofibers of ceramic composite/metal oxide [41, 42].

4.3.11.5 Emulsion Electrospinning Method

The technique of emulsion electrospinning is now predicted to be an invaluable technique for future applications for controlled drug delivery. Hydrophobic drugs can be encapsulated into electrospun poly(l-lactic acid) (PLLA) fibers, which exhibit approximately zero-order kinetics at the time of drug release. Water-soluble drugs can also be electrospun into water-soluble polymeric material fibers, although it is challenging to control drug delivery. The method of emulsion electrospinning can provide a solution to this issue. In this process, drugs with hydrophilic characteristics are uniformly dispersed within the fibers, which restricts the burst of release. An aqueous solution of bovine serum albumin (BSA) can appropriately encapsulate in the poly(ethylene-co-vinyl acetate) fibers. The 1D PEO/ PEG-PLA of the di-block copolymer core of nanofibers can be produced by the method of electrospinning in the emulsion of water-in-oil [43, 44].

4.3.11.6 Co-Evaporation Method

Discharge of the polymer solution into the jet results in elongation and instability, permitting the jet and the solution to become long and thin. The nanofibers in a solid state can be synthesized at the time of solvent evaporation. By co-electrospinning, some other constituents can dissolve in the polymer solution to form composite nanofibers after the evaporation of the solvent [45, 46].

4.3.11.7 Coaxial Electrospinning Technique

Electrospinning generally contains a single capillary, which can synthesize solid fibers. If a single capillary is replaced by coaxial capillaries, then the sheath/core nanofibers can be obtained. Hence, the method of coaxial electrospinning provides us with an opportunity to fabricate functional composite nanomaterials. The composite nanofibers synthesized by using the coaxial electrospinning method include polymer/polymer, inorganic/inorganic, and polymer/inorganic composites. Recently, the sheath/core nanofibers of poly(dodecylthiophene) (PDT)/PEO and polysulfone (PSU)/PEO were synthesized using coaxial electrospinning. The fibers of core/sheath $LiCoO_2$/MgO have also been fabricated by coaxial electrospinning in coupling with a sol-gel technique [47, 48].

4.4 Architectural Aspects and Their Effects on the Properties of 1D Polymeric Materials

Many 1D polymeric materials have unique architectural aspects, which impact their properties. Some examples are given below.

4.4.1 Titanate (TNTs) and Halloysite Nanotubes (HNTs)

This material was first recognized in the 1990s and exhibits outstanding performance and characteristics. The nanotubes of TiO_2 show improved catalytic activity compared to regular TiO_2, greater stability, greater surface area, strong interactions with metal, and excellent performances in both acidic and alkaline media. The 1D structure of the nanotube form of TiO_2 shows bundled tubes having lengths 1–220 nm and diameters in the range 30–80 nm. The TNT morphology mainly depends upon the process of synthesis such as electrochemical anodization, sol-gel method, template-assisted technique, etc. The 1D TNT can be used as filler material in the synthesis of fuel cell membranes. A 1D nanomaterial can be synthesized by using halloysite nanotubes with natural minerals of two-layered aluminosilicate clay, which was quaternized with imidazolium groups having various functionalities, such as decyl, benzyl, or butyl moieties. It was found that the addition of 5 wt% of HNTs increased Young's modulus and tensile strength up to 1100 MPa and 52 MPa, respectively [49]

4.4.2 Carbon Nanotubes (CNTs)

There are two main types of CNT:

- Single-walled carbon nanotubes (SWCNT), which contain a single graphitic sheet wound upon itself;
- Multi-walled carbon nanotubes (MWCNT), which are synthesized by the combination of many coaxial sheets wound one on top of the other.

Both the SWCNTs and MWCNTs show extremely promising electrical properties. They are characterized by high strength and flexibility, and are therefore appropriate for the application of reinforcing fiber materials in the formation of 1D composites [50].

4.4.3 Poly(vinyl Alcohol) (PVA)

Poly(vinyl alcohol) is normally an inert polymer and is soluble in warm water, exhibiting ionic conductivity following the addition of some alkaline solutions such as NaOH or KOH. A promising 1D composite material was synthesized using PVA as the matrix and MWCNTs as reinforcing material, and it can be used in the manufacture of direct methanol alkaline fuel cells (DMAFCs). It was reported that the electrical conductivity increases in response to an increasing concentration of MWCNTs. The MWCNTs were mixed with PVA solution, stirred continuously at a temperature of 80 °C for 3 hours to graft PVA onto the MWCNTs. The dried PVA/MWCNT films were immersed in a solution of KOH. It was found that PVA containing 0.05% of functionalized MWCNTs exhibited maximum power density. This is because the functionalized MWCNTs, with reinforced nanoparticles of Fe_3O_4, were introduced into the PVA film. To enhance the chemical compatibility between PVA and CNTs, the CNTs were functionalized with chain segments of PVA [51].

4.4.4 Chitosan (CS)

One of the new composites was synthesized by using the biopolymer, chitosan, as the matrix, with CNTs as the reinforcing material. CNTs were functionalized and chitosan was quaternized to improve the mechanical properties of the composite prepared. The nanoscale form of CNTs was functionalized by quaternizing the silica derived through *in situ* sol-gel technique. During the preparation of the composite, the addition of 5 wt% of functionalized CNTs increased the conductivity of the material up to 43 mS/cm at 80 °C, which is almost double that of quantized chitosan alone [52].

4.4.5 Aromatic Polymers

The imidazolium group (Im) is found to be suitable for synthesizing most of the 1D composites consisting of aromatic polymers. The quaternized poly(p-phenylene oxide) (PPO) was blended with methyl-Im(PPO–MIm) in functionalized MWCNTs by polymeric ionic having Im moieties (PIL(BF4)–MWCNTs). The conductivity of the composite material was improved by the addition of 95% as filler to the blend (PPO–MIm). The 1D long-range structure showed good ion-conducting power, and therefore exhibited higher conductivity (135 mS/cm at 70 °C). The material Im-PEEK (Im- Polyether-ether-ketone) serves as a hosting matrix phase in the case of poly(vinyl imidazole) with functionalized carbon nanotubes (PVI@CNTs). The imidazolium groups present in the filler materials on reaction with Im-PEEK form the structural cross-linking attributes and nanotubes. The membrane of nanohybrid features, having around 15 wt% of reinforcing materials (Im-PEEK/PVI@CNT-15), exhibits a conductivity of 121 mS/cm at 70 °C. The increase in ionic conductivity was caused by the extra anion transport pathways formed at the interface between the matrix phase of polymer and the PVI@CNTs, which resulted in the improvement of ionic connectivity and ionic movement [53].

4.5 Applications of 1D Nanomaterials

1D nanomaterials can be applied to various fields, and some important fields in which they can be applied are presented below.

4.5.1 Applications in Optical and Electronic Nanodevices

The electro-optical and optical properties of nanowires, based on some specific semiconductors and metals, have been extensively studied because of their applications in the manufacture of nanoscale optoelectronic and electronic devices. 1D composite nanomaterials were then identified as suitable materials for electro-optical and electronic devices. The 1D nanocomposites of Ag-NP/TCNQ/PVP show reasonable electrical properties. The 1D TiO_2, on the coating with a layer of conducting polymeric materials, show very good optical properties, and therefore are used for the manufacture of optoelectronic devices [54].

4.5.2 Applications in Sensing

One of the vital applications of 1D nanomaterial is sensing in the security checking, medical, and environmental fields. This is due to the exceptional surface-to-volume ratio exhibited by these nanostructured materials, therefore developing electrical properties associated with the species adsorbed onto the surface. A 1D nanomaterial which acts as a humidity sensor was synthesized by doping LiCl onto the TiO_2 nanofibers by applying the technique of electrospinning. The nanofibers of TiO_2 doped with LiCl showed much greater sensitivity than do nanofibers of pure TiO_2. A similar sensing property was observed by doping KCl over the nanofibers of TiO_2. The thin film of nanofibers of poly(acrylic acid)-poly(pyrene methanol) (PAA-PM) cross-linked with PU was synthesized by electrospinning, which served as an optical sensor. The reduction in the fluorescence intensity of the composite fiber film in response to an increase in the concentration of analytes was caused by the effective quenching as a result of the analytes being deficient in electrons. The nanofiber of PANI/PEO composite exhibited extensive electrical conductivity. The 1D CNTs (both MWCNTs and SWCNTs) served as electrochemical sensors. The nanofibers of the PANI/PEO composite were used in gas, optical, and humidity sensors. The urease-immobilized PVP fibers were synthesized by using an ethanol solution in the process of electrospinning, and served as a urea biosensor [55, 56].

4.5.3 Applications in Catalysis

The nanofibers of ceramic materials, such as nanofibers of ZnO and TiO_2, are used as photocatalysts. The stable ceramic or polymer fibers function as very good catalysts because of their combination of high porosity with a huge surface area. The nanoparticles of Pd act as an interesting catalyst in PAN-AA nanofibers. The insertion of Pd nanoparticles in the nanofibers of the PAN/AA copolymer shows 4.5 times greater catalytic power than the Pd/Al_2O_3 catalyst. The nanofibers of the PAN/Pd composite show greater catalytic power, recovery, reusability, and resistance to leaching. Again, the polymeric core fibers having Pd, Pt, Ru, or any bimetallic nanoparticles, in coupling with the technique of electrospinning, show better catalytic hydrogen reduction. The nanofibers of the silica/Ag composite show good catalytic activity toward methylene blue dye reduction by using $NaBH_4$ as the reducing agent. It was observed that the nanofiber composite exhibits an excellent property of catalytic activity without any poisoning or deactivation. Nanowires of Pt having a length of about 125 nm were coated on the nanofibers of the TiO_2 surface. The fiber mats showed good catalytic properties for hydrogenation of the dye methyl red. The results indicated that the ability of the catalyst toward the mat decorated with Pt nanowires was much greater than that of the mat coated with the Pt nanoparticles, which could be due to the huge active site at the surface of the nanowire based on Pt-related fibers [57, 58].

4.5.4 Applications in the Environmental Field

Water pollution caused by heavy metals and air pollution due to volatile organic compounds (VOCs) are some of the major environmental problems. The nanofiber mat synthesized by the electrospinning method is a vital candidate material for decontaminating the environment because of its huge surface area, low cost of synthesis, and porous structure. The inorganic nanoparticles in combination with the electrospun nanofibers, in particular, have been widely developed for cleaning water owing to their property of protecting the discharge of NPs into the ecosystem and the decrease in the cost of their synthesis. The nanoparticles of boehmite can also be incorporated into both hydrophobic PCL polymers and hydrophilic polyamides. The electrospun polymer composite fibers can be effectively used as an adsorbent for the separation of heavy metal ions. Excellent composite fibers were developed by using wool keratose in combination with silk fibroin (SF) for the removal of heavy metals and ions from water and soil through adsorption. In addition to toxic heavy metals, volatile organic compounds and toxic poisonous gases are also removed by electrospun nanofibers. The ultrafine porous carbon fibers (UPCF), which are synthesized through the electrospinning process and modified by using nitrogen-based compounds, showed high performance in the removal of SO_2. The 1D composite composed of PVC and a catalyst (used for decontamination of nerve agents) can be synthesized by using *o*-iodosobenzoic acid and β-cyclodextrin. The composite is highly effective at removing paraoxon, which serves as a nerve agent stimulant. The composite nanomaterials synthesized by the electrospinning process are widely used in environmental decontamination through various types of filtration [59, 60].

4.5.5 Energy Applications

The energy crisis is recognized to be a critical global issue faced by the world in the twenty-first century. The fossil fuel resources, such as coal, natural gas, and petroleum oil are the major sources of energy production and use. However, these are stock and limited (non-renewable) resources. Also, the combustion of fossil fuel sources causes an increase in greenhouse gas emissions, causing global warming and environmental issues. Hence, the development of sustainable devices and technologies for clean energy and effective management is vital. In the case of dye-sensitized solar cells (DSSCs), the 1D structure of metal oxide nanostructures shows improved conduction of charges to enhance the efficiency of energy conversion. The nanofibers or nanorods prepared by the electrospinning method are widely used as an electrode for DSSCs owing to their very large surface area and a huge number of pores, which facilitates the adsorption of dye sensitizers. To enhance adhesion between the conductive substrates and 1D TiO_2 nanomaterials, many methods have been proposed. The photovoltaics based on organic materials can be achieved by conjugating polymer matrix and blending metal oxide semiconductors, between which the charge transfer will be carried out. The 1D composite nanofiber of poly(3-hexylthiophene)/ZnO synthesized by the electrospinning technique is highly effective for photovoltaic devices. Now, researchers are focusing on the development of fuel cells because of their eco-friendly nature. Among the various kinds of fuel cells, direct methanol fuel cells (DMFCs) are very popular due to their high energy density and room-temperature operation. In DMFCs, supporting material is needed to load the nanoparticles, such as Pt or its alloy as the catalyst. The supporting material exhibits a substantial impact on the activity of the catalyst. To improve the performance of the catalyst, the subsidiary materials must possess adequate stability, high surface area, and a better electron conductor than the catalyst. Many carbon materials including CNTs, mesocarbon, and carbon fibers, have been used extensively as supports or stimulators for the catalytic activity of DMFCs. The nanoporous carbon fibers and nanoscale carbon fibers synthesized by the electrospinning method represent good candidate materials for the catalyst supports. The Pt-cluster reinforced with carbon fibrous mats (CFMs/Pt) through the process of electrodepositing and electrospinning was successfully synthesized for DMFCs. These materials find their widest application in electrochemical energy storage devices [61, 62].

4.5.6 Biomedical Applications

1D nanomaterials are found to be interesting materials with biomedical applications. Such materials are used for drug delivery, immobilized enzymes, artificial blood vessels, wound healing, and many others [63].

4.5.6.1 Applications in Drug Delivery

Nanofibers possess many benefits, such as reduced influence on the application of drug activity, ease of implementation, and control of drug release. In many cases, drugs which are soluble in water can be readily inserted into water-soluble polymeric materials for drug delivery applications. 1D composite nanofibers of PLGA/PEG-*b*-PLA/PLA synthesized by the electrospinning technique exhibit a sustained-release outline after the first burst compared with only the scaffold of PLGA. To control its release, the drug was incorporated in nanofibers, which was challenging. The coaxial electrospinning method can effectively encapsulate drugs within the PCL fibers and form the structure of the core/sheath. The degradation or disintegration of the composite nanofibers and rate of release of the drug are associated with the hydrophilic characteristics of the drugs. In this case, the release of the drug was comparatively smooth and better than with the PLGA scaffold and no release occurred due to burst. Conducting polymeric materials, including poly(3,4-ethylenedioxythiophene) (PEDOT) and polypyrrole (PPy), can also be used as effective materials since the conducting polymers are able to actuate under the voltage applied, so that it may be used to control and regulate the rate of drug release [64, 65].

4.5.6.2 Applications in Tissue Engineering

Since the nanofiber scaffolds possess high porosity and high surface area, they serve as good candidate materials which are applicable to tissue engineering. There are many biodegradable and biocompatible polymer scaffolds, which show improved biological functionality and physical properties. The composite fibers of PCL/gelatin can be synthesized through the electrospinning method and can serve as an auspicious scaffold for the culture of the bone marrow stromal cells (BMSCs). It has been recognized to be favorable for migration and growth of BMSCs as compared with only the PCL scaffold. This is mainly owing to the property of better cellular affinity, hydrophilicity, and dissolution of the composite nanofibers of PCL/gelatin at the time of cell culture, deformation, and elongation behavior of the composite scaffold.

Conducting polymeric materials containing gelatin fibers are very suitable for tissue engineering applications. The nanofibers were synthesized using a PLLA/laminin composite using three different processes, namely blended electrospinning, covalent binding, and physical adsorption. The collagen-coated PCL nanofibers synthesized by the coaxial electrospinning method showed preferential cell proliferation. The human dermal fibroblast (HDF) density on coupling with the nanofibers of PCL/collagen composite had improved structures of the core/sheath. The nanofibers of PCL coated with collagen showed stimulated cell migration within the scaffolds, representing outstanding integration between the scaffolds and cells. The nanofibers aligned uniaxially were always more successful as a potential scaffold than were arbitrary electrospun nanofibers, because the former nanofibers could control the growth of neurons. The nanofiber composite of PCL/collagen exhibited improved cell attachment, axonal regeneration, neurite orientation, and cell migration, and were therefore considered to be an excellent scaffold material for artificial implants. Nanomaterials of aligned 1D PCL/gelatin composite synthesized from 30:70 of gelatin and PCL in the nanofiber structure showed improved biodegradation and better mechanical properties for cell culture. The nanofibers of these composites exhibited improved proliferation and nerve differentiation than normal PCL nanofibers [66, 67].

4.5.6.3 Wound Dressing Applications

The nanofibers of polymer/Ag composites have been extensively studied for the use as wound dressings owing to the antimicrobial activities of Ag. Nanofibers of PVA/Ag composite were synthesized from an aqueous solution of PVA/$AgNO_3$ through an electrospinning process followed by UV irradiation or heat treatment. During the process, Ag nanoparticles were produced and distributed on the surface. The composite fiber obtained by the heat treatment process normally achieved greater stability than was obtained by UV irradiation because, at elevated temperatures, crosslinking between hydroxyl groups (-OH groups) can happen, which decreases the Ag dissolution in a high-moisture-content environment. The nanofibers of an electrospun PVA/Ag composite exhibited outstanding antibacterial properties, but the major demerit was the creation of a gray-blue color on the skin. To prevent discoloration, the nanoparticles of Ag-loaded-Zr are suggested for wound dressings. The 1D composite material obtained by encapsulation of an Ag-imidazole cyclophane gem-diol complex into the hydrophilic Tecophilic nanofibers exhibited considerable anti-bactericidal activity. These types of scaffolds of composite fibers are normally effective against pathogenic microorganisms such as *Escherichia coli, Saccharomyces cerevisiae, Pseudomonas aeruginosa, Aspergillus niger, Staphylococcus aureus,* and *Candida albicans* [68, 69].

4.6 Conclusion

The nanomaterials with 1D morphology include rods, tubes, rods, belts, etc. There are many methods available for the synthesis of 1D nanomaterials such as vapor phase or gas phase synthesis, mechanical grinding, sol-gel technique, ultrasonic spray pyrolysis, gas condensation processing, particle precipitation-assisted chemical vapor deposition, laser ablation, electrospinning technique, etc. Among the various architectural aspects of the 1D nanocomposite, synthesis by the electrospinning method is particularly significant. The 1D nanomaterials show excellent properties such as ease of control, chemical composition, good surface properties, adequate crystal structure, and phase purity. The technique of electrospinning is a versatile and easy method for the synthesis of thin fibers with improved mechanical properties, and enormous suface area. The 1D nanomaterials are widely and effectively used in different applications, such as optical and electronic devices, biological and chemical sensors, catalysis, energy storage, environmental, and various biomedical applications, including wound healing, drug delivery, and tissue engineering.

References

1. Erik, G., Liqiang, M., & Peidong, Y. (2019). Introduction: 1D Nanomaterials/Nanowires. *Chemical Reviews*, 119(15), 8955–8957.
2. Jaison, J., Ahmed, B., Yen, S. C., Dufresne, A., & Danquah, M. K. (2018). Review on Nanoparticles and Nanostructured Materials: History, Sources, Toxicity and Regulations. *Beilstein Journal of Nanotechnology*, 9, 1050–1074.
3. Kumar, P. J., Das, G., Leonardo, F. F., et. al. (2018). Nano Based Drug Delivery Systems: Recent Developments and Future Prospects. *Journal of Nanobiotechnology*, 16(1), 71–103.
4. Machín, A., Fontánez, K., Arango, J. C., Ortiz, D., De León, J., Pinilla, S., Nicolosi, V., Petrescu, F. I., Morant, C., & Márquez, F. (2021). One-Dimensional (1D) Nanostructured Materials for Energy Applications. *Materials*, 14(10), 2609–2655.
5. Xia, Y., Yang, P., Sun, Y., Wu, Y., Mayers, B., Gates, B., Yin, Y., Kim, F., & Yan, H. (2003). One Dimensional Nanostructures: Synthesis, Characterization and Applications. *Advanced Materials*, 15(5), 353–389.
6. Lu, X., Wang, C., & Wei, Y. (2009). One-Dimensional Composite Nanomaterials: Synthesis by Electrospinning and their Applications. *Small*, 5(21), 2349–2370.
7. Yeo, M.-K., & Nam, D.-H. (2013). Influence of Different Types of Nanomaterials on their Bioaccumulation in a Paddy Microcosm: A Comparison of TiO_2 Nanoparticles and Nanotubes. *Environmental Pollution*, 178, 166–172.
8. Zhao, Y. S., Fu, H., Peng, A., Ma, Y., Liao, Q., & Yao, J. (2010). Construction and Optoelectronic Properties of Organic One-Dimensional Nanostructures. *Accounts of Chemical Research*, 43(3), 409–418.
9. Han, N., & Ho, J. C. (2014). *One-Dimensional Nanomaterials for Energy Applications, Nanocrystalline Materials* (Second Edition). Elsevier, 75–120.
10. Weng, B., Liu, S., Tang, Z., & Xu, Yi-J. (2014). One-Dimensional Nanostructure Based Materials for Versatile Photocatalytic Applications. *RSC Advances*, 4(25), 12685–12700.
11. Lu, X., Zhang, W., Wang, C., ChinWen, T., & Wei, Y. (2011). One-Dimensional Conducting Polymer Nanocomposites: Synthesis, Properties and Applications. *Progress in Polymer Science*, 36(5), 671–712.
12. Vucaj, N., Quinn, M. D. J., Baechler, C., Notley, S. M., Cottis, P., Hojati-Talemi, P., Evans, D. R., Wallace, G. G., Murphy, P. J., & Evans, D. R. (2014). Vapor Phase Synthesis of Conducting Polymer Nanocomposites Incorporating 2D Nanoparticles. *Chemistry of Materials*, 26(14), 4207–4213.
13. Yadav, T. P., Yadav, R. M., & Singh, D. P. (2012). Mechanical Milling: A Top Down Approach for the Synthesis of Nanomaterials and Nanocomposites. *Nanoscience and Nanotechnology*, 2(3), 22–48.
14. Tulinski, M., & Jurczyk, M. (2017). Nanomaterials Synthesis Methods. *Metrology and Standardization of Nanotechnology*, 75–98. doi:10.1002/9783527800308.ch4.
15. Parashar, M., Shukla, V. K., & Singh, R. (2020). Metal Oxides Nanoparticles via Sol–Gel Method: A Review on Synthesis, Characterization and Applications. *Journal of Materials Science: Materials in Electronics*, 31(5), 3729–3749.
16. Bhagwat, V. R., Humbe, A. V., More, S. D., & Jadhav, K. M. (2019). Sol-Gel Auto Combustion Synthesis and Characterizations of Cobalt Ferrite Nanoparticles: Different Fuels Approach. *Materials Science and Engineering: B*, 248, 1–5.

17. Rahemi Ardekani, S., Sabour Rouh Aghdam, A., Nazari, M., Bayat, A., Yazdani, E., & Saievar-Iranizad, E. (2019). A Comprehensive Review on Ultrasonic Spray Pyrolysis Technique: Mechanism, Main Parameters and Applications in Condensed Matter. *Journal of Analytical and Applied Pyrolysis*, 141, 1–19.
18. Solero, G. (2017). Synthesis of Nanoparticles through Flame Spray Pyrolysis: Experimental Apparatus and Preliminary Results. *Nanoscience and Nanotechnology*, 7(1), 21–25.
19. Gao, J., Qian, H., Xu, S., & Han, M. (2013). Investigation of 1D Silver Nanoparticle Arrays for Use as Molecule Concentration-Specific SERS Substrates. *Journal of Nanomaterials*, 2013, 1–8. doi:10.1155/2013/989803.
20. Suryanarayana, C., & Prabhu, B. (2007). Synthesis of Nanostructured Materials by Inert-Gas Condensation Methods. *Nanostructured Materials*, 47–90. doi:10.1016/b978-081551534-0.50004-x.
21. Sivaprahasam, D., Sriramamurthy, A. M., Vijayakumar, M., Sundararajan, G., & Chattopadhyay, K. (2010). Synthesis of FeCu Nanopowder by Levitational Gas Condensation Process. *Metallurgical and Materials Transactions B*, 41(4), 841–856.
22. Zhu, X., ten Brink, G. H., de Graaf, S., Kooi, B. J., & Palasantzas, G. (2020). Gas-Phase Synthesis of Tunable Size Germanium Nanocrystals by Inert Gas Condensation. *Chemistry of Materials*, 32(4), 1627–1635.
23. Vasilyeva, E., Nasibulin, A., Tolochko, O., Rudskoy, A., Sachdev, A., & Xiao, X. (2015). Application of WSe_2 Nanoparticles Synthesized by Chemical Vapor Condensation Method for Li-Ion Battery Anodes. *Zeitschrift Für Physikalische Chemie*, 229(9), 1429–1437.
24. Chang, W., Skandan, G., Hahn, H., Danforth, S. C., & Kear, B. H. (1994). Chemical Vapor Condensation of Nanostructured Ceramic Powders. *Nanostructured Materials*, 4(3), 345–351.
25. Saito, G., & Akiyama, T. (2015). Nanomaterial Synthesis Using Plasma Generation in Liquid. *Journal of Nanomaterials*, 2015, 1–21. doi:10.1155/2015/1236.
26. Lu, Q., Gao, F., & Komarneni, S. (2004). Microwave-Assisted Synthesis of One-Dimensional Nanostructures. *Journal of Materials Research*, 19(6), 1649–1655.
27. Ouyang, B., Zhang, Y., Xia, X., Rawat, R. S., & Fan, H. J. (2018). A Brief Review on Plasma for Synthesis and Processing of Electrode Materials. *Materials Today Nano*, 3, 28–47.
28. Wojnarowicz, J., Chudoba, T., & Lojkowski, W. (2020). A Review of Microwave Synthesis of Zinc Oxide Nanomaterials: Reactants, Process Parameters and Morphologies. *Nanomaterials*, 10(6), 1086–1235.
29. Kolahalam, L. A., Kasi Viswanath, I. V., Diwakar, B. S., Govindh, B., Reddy, V., & Murthy, Y. L. N. (2019). Review on Nanomaterials: Synthesis and Applications. *Materials Today: Proceedings*, 18, 2182–2190.
30. Manawi, Y., Ihsanullah, Samara, A., Al-Ansari, T., & Atieh, M. (2018). A Review of Carbon Nanomaterials' Synthesis via the Chemical Vapor Deposition (CVD) Method. *Materials*, 11(5), 822–852.
31. Kim, M., Osone, S., Kim, T., Higashi, H., & Seto, T. (2017). Synthesis of Nanoparticles by Laser Ablation: A Review. *KONA Powder and Particle Journal*, 34, 80–90.
32. Rybaltovsky, A., Epifanov, E., Khmelenin, D., Shubny, A., Zavorotny, Y., Yusupov, V., & Minaev, N. (2021). Two Approaches to the Laser-Induced Formation of Au/Ag Bimetallic Nanoparticles in Supercritical Carbon Dioxide. *Nanomaterials*, 11(6), 1553–1564.
33. Mani, K. P., George, V., Ramakrishnan, B. P., Joseph, C., Viswambharan, U. N., & Abraham, I. M. (2015). Synthesis and Photoluminescence Studies of One Dimensional Sm_2MoO_6 Nanofibers Derived from Electrospinning Process. *Journal of Materials Research and Technology*, 4(2), 224–227.
34. Wang, L., Yang, G., Peng, S., Wang, J., Yan, W., & Ramakrishna, S. (2019). One-Dimensional Nanomaterials Toward Electrochemical Sodium-Ion Storage Applications via Electrospinning. *Energy Storage Materials*, 25, 443–476.
35. Zhu, W., Cheng, Y., Wang, C., Pinna, N., & Lu, X. (2021). Transition Metal Sulphides Meet Electrospinning: Versatile Synthesis, Distinct Properties and Prospective Applications. *Nanoscale*, 13(20), 9112–9146.
36. Kuchibhatla, S. V. N. T., Karakoti, A. S., Bera, D., & Seal, S. (2007). One Dimensional Nanostructured Materials. *Progress in Materials Science*, 52(5), 699–913.

37. Ghorbani, H. (2014). A Review of Methods for Synthesis of Al Nanoparticles. *Oriental Journal of Chemistry*, 30(4), 1941–1949.
38. Leonés, A., Lieblich, M., Benavente, R., Gonzalez, J. L., & Peponi, L. (2020). Potential Applications of Magnesium-Based Polymeric Nanocomposites Obtained by Electrospinning Technique. *Nanomaterials*, 10(8), 1524–1556.
39. Mutlu, S., Metin, E., Aydin Yuksel, S., Bayrak, U., Nuhoglu, C., & Arsu, N. (2021). In-Situ Photochemical Synthesis and Dielectric Properties of Nanocomposite Thin Films Containing Au, Ag and MnO Nanoparticles. *European Polymer Journal*, 144, 1–13.
40. Wang, S.-T., Lin, Y., Nielsen, M. H., Song, C. Y., Thomas, M. R., Spicer, C. D., Stevens, M. M., Ercius, P., Aloni, S., & Stevens, M. M. (2019). Shape-Controlled Synthesis and In Situ Characterisation of Anisotropic Au Nanomaterials Using Liquid Cell Transmission Electron Microscopy. *Nanoscale*, 11(36), 16801–16809.
41. Soto-Nieto, F., Farías, R., & Reyes-López, S. Y. (2020). Sol–Gel and Electrospinning Synthesis of Silica–Hydroxyapatite–Silver Nanofibers for SEIRAS and SERS. *Coatings*, 10(10), 910–937.
42. Malwal, D., & Packirisamy, G. (2018). Recent Advances in the Synthesis of Metal Oxide (MO) Nanostructures. *Synthesis of Inorganic Nanomaterials*, 255–281. doi:10.1016/b978-0-08-101975-7.00010-5.
43. Zhang, C.-L., & Yu, S.-H. (2014). Nanoparticles Meet Electrospinning: Recent Advances and Future Prospects. *Chemical Society Reviews*, 43(13), 4423–4448.
44. Xue, J., Wu, T., Dai, Y., & Xia, Y. (2019). Electrospinning and Electrospun Nanofibers: Methods, Materials, and Applications. *Chemical Reviews*, 119(8), 5298–5415.
45. Kulkarni, S. K. (2015). Synthesis of Nanomaterials—I (Physical Methods). In: *Nanotechnology: Principles and Practices*. Springer, Cham. doi:10.1007/978-3-319-09171-6_3.
46. Manawi, Y., Ihsanullah, Samara, A., Al-Ansari, T., & Atieh, M. (2018). A Review of Carbon Nanomaterials' Synthesis via the Chemical Vapor Deposition (CVD) Method. *Materials*, 11(5), 822–857.
47. Buzgo, M., Mickova, A., Rampichova, M., & Doupnik, M. (2018). Blend Electrospinning, Coaxial Electrospinning, and Emulsion Electrospinning Techniques. *Core-Shell Nanostructures for Drug Delivery and Theranostics*, 325–347. doi:10.1016/b978-0-08-102198-9.00011-9.
48. Ma, Q., Wang, J., Dong, X., Yu, W., & Liu, G. (2013). Electrospinning Fabrication of High-Performance magnetic@photoluminescent Bifunctional Coaxial Nanocables. *Chemical Engineering Journal*, 222, 16–22.
49. Grylewicz, A., Szymański, K., Darowna, D., & Mozia, S. (2021). Influence of Polymer Solvents on the Properties of Halloysite-Modified Polyethersulfone Membranes Prepared by Wet Phase Inversion. *Molecules*, 26(9), 2768–2785.
50. Eatemadi, A., Daraee, H., Karimkhanloo, H., Kouhi, M., Zarghami, N., Akbarzadeh, A., Abasi, M., Hanifehpour, Y., & Joo, S. W. (2014). Carbon nanotubes: Properties, synthesis, purification, and medical applications. *Nanoscale Research Letters*, 9(1), 393–412.
51. Liu, Z., Lanier, O. L., & Chauhan, A. (2020). Poly (Vinyl Alcohol) Assisted Synthesis and Anti-solvent Precipitation of Gold Nanoparticles. *Nanomaterials*, 10(12), 2359–2375.
52. Li, J., Cai, C., Li, J., Li, J., Li, J., Sun, T., Wang, L., Wu, H., & Yu, G., (2018). Chitosan-Based Nanomaterials for Drug Delivery. *Molecules*, 23(10), 2661–2686.
53. Zhou, D., Zhang, Y., Zhu, J., Yu, J., Wang, Y., & Hu, Z. (2019). Tailoring the Architecture of Aromatic Polymers for Highly Efficient Dispersion of Carbon Nanomaterials and their High-Performance Composites. *Carbon*, 148, 297–306.
54. Zhang, C., Yan, Y., Sheng Zhao, Y., & Yao, J. (2013). Synthesis and Applications of Organic Nanorods, Nanowires and Nanotubes. *Annual Reports Section "C" (Physical Chemistry)*, 109, 211–239.
55. Hu, K., Wang, F., Shen, Z., Yan, Y., & Liu, H. (2021). Enhancement Methods of Hydrogen Sensing for One-Dimensional Nanomaterials: A Review. *International Journal of Hydrogen Energy*, 46(38), 20119–20138.
56. Liyanage, T., Qamar, A. Z., & Slaughter, G. (2021). Application of Nanomaterials for Chemical and Biological Sensors: A Review. *IEEE Sensors Journal*, 21(11), 12407–12425.

57. Ashik, U. P. M., Viswan, A., Kudo, S., & Hayashi, J. (2018). Nanomaterials as Catalysts. *Applications of Nanomaterials*, 45–82. doi:10.1016/b978-0-08-101971-9.00003-x.
58. Wang, J., & Gu, H. (2015). Novel Metal Nanomaterials and their Catalytic Applications. *Molecules*, 20(9), 17070–17092.
59. Zhu, Y., Feng, Y., Chen, S., Ding, M., & Yao, J. (2020). Carbon Nitride Nanotubes-Based Materials for Energy and Environmental Applications: A Review of Recent Progress. *Journal of Materials Chemistry A*, 8(48), 25626–25648.
60. Yuan, B., & Cademartiri, L. (2015). Flexible One-Dimensional Nanostructures: A Review. *Journal of Materials Science and Technology*, 31(6), 607–615.
61. Shi, X., Zhou, W., Ma, D., Ma, Q., Bridges, D., Ma, Y., & Hu, A. (2015). Electrospinning of Nanofibers and their Applications for Energy Devices. *Journal of Nanomaterials*, 2015, 1–20. doi:10.1155/2015/140716.
62. Wei, Q., Xiong, F., Tan, S., Huang, L., Lan, E. H., Dunn, B., & Mai, L. (2017). Porous One-Dimensional Nanomaterials: Design, Fabrication and Applications in Electrochemical Energy Storage. *Advanced Materials*, 29(20), 1–39.
63. Shahriar, S., Mondal, J., Hasan, M., Revuri, V., Lee, D., & Lee, Y.-K. (2019). Electrospinning Nanofibers for Therapeutics Delivery. *Nanomaterials*, 9(4), 532.
64. de Jong, W. H., & Borm, P. J. (2008). Drug Delivery and Nanoparticles: Applications and Hazards. *International Journal of Nanomedicine*, 3(2), 133–149.
65. Hassan, S., Prakash, G., Bal Ozturk, A., Saghazadeh, S., Farhan Sohail, M., Seo, J., Khademhosseini, A., & Zhang, Y. S. (2017). Evolution and Clinical Translation of Drug Delivery Nanomaterials. *Nano Today*, 15, 91–106.
66. Zheng, X., Zhang, P., Fu, Z., Meng, S., Dai, L., & Yang, H. (2021). Applications of Nanomaterials in Tissue Engineering. *RSC Advances*, 11(31), 19041–19058.
67. Hasan, A., Morshed, M., Memic, A., Hassan, S., Webster, T., & Marei, H. (2018). Nanoparticles in Tissue Engineering: Applications, Challenges and Prospects. *International Journal of Nanomedicine*, 13, 5637–5655.
68. Pormohammad, A., Monych, N. K., Ghosh, S., Turner, D. L., & Turner, R. J. (2021). Nanomaterials in Wound Healing and Infection Control. *Antibiotics*, 10(5), 473–491.
69. Barroso, A., Mestre, H., Ascenso, A., Simões, S., & Reis, C. (2021). Nanomaterials in Wound Healing: From Material Sciences to Wound Healing Applications. *Nano Select*, 10(5), 473–490.

5

Understanding Interfacial Influence on the Properties of One-Dimensional Nanocomposites

Hui Zhao, Wei Wu, Yi-Zhang Tong, Xian-Wu Cao, Anthony C.Y. Yuen

CONTENTS

5.1 Introduction

With the fast development of material science and technology, polymer nanocomposites (PNCs) have attracted considerable attention from both academia and industry. PNCs are a class of mixtures consisting of polymers and nanomaterials, with one component material in the nanometer range of 1 nm to 100 nm. The presence of nanofillers in polymer matrices can lead to significant improvements in the physical and chemical properties of PNCs, such as thermal stability, flame retardancy, gas barrier performance, and mechanical properties. The fillers can be generally classified into four types: zero-dimensional (0D), one-dimensional (1D), two-dimensional (2D), and three-dimensional (3D), depending on the dimensional features of the nanofillers [1]. Among them, the 1D nanofillers are usually in the form of nanotubes, nanowires, and nanofibers, which have one dimension of less than 100 nm. Different from the traditional bulk materials, the 1D nanofillers have a high specific surface area, resulting in more interfacial interaction with the polymers. Therefore, the interfacial interaction between the nanofillers and polymer matrices is the key challenge to realizing the full potential of PNC property enhancement. This chapter will cover the basic concept of interfacial interaction. The effects of surface functionalization, intrinsic properties, and microstructure of 1D nanofillers, as well as the synergistic effects with other fillers, on the properties of polymer/1D nanofiller nanocomposites, are discussed.

DOI: 10.1201/9781003223764-5

5.2 Theory of Interfacial Interaction

Interfacial interaction in PNCs can be categorized into filler–polymer interaction and filler–filler interaction. These interactions can significantly affect the physical and chemical properties of PNCs, especially their mechanical properties. Considering the rigidity of the 1D nanofiller, interfacial bonding plays an important role in transferring the interfacial stress from the surrounding polymer matrix to the 1D nanofiller, resulting in significant effects on the mechanical properties of PNCs. The enhancement of the mechanical properties of PNCs is attributable not only to the addition of the nanoparticles themselves but also to the interaction between particles and polymers. On the one hand, polymer chains can form an adsorption layer on the surface of nanoparticles. On the other hand, nanoparticles can form a network structure through the bridging of polymer chains, changing the mechanical properties of the system. Generally, the interactions between the nanofiller and the polymer matrix can be categorized into either non-covalent or covalent interactions [2].

5.2.1 Non-Covalent Interaction

The non-covalent interactions in PNCs mainly involve van der Waals (VDW) forces, electrostatic forces, and hydrogen bonding. Interaction can be improved through various mechanical interaction improvement strategies, such as bridging, wrapping, and increment of specific interfacial areas. With the simultaneous incorporation of two or more types of reinforcing nanofillers, bridging can be established between polymer chains and nanofillers. The possible bridging phenomenon depends on the ratio of the gyration radius of the polymer chain to the average distance between the nearest reinforcing filler. Thus, it is possible to increase the bridging interaction by increasing the nanofiller loading and/or increasing the molecular weight of the polymer. Wrapping nanofillers will not only increase the interfacial interaction but also contribute to the homogeneous dispersion of nanofiller in the polymer matrix. The formation of π-π stacking, the presence of hydrophobicity, and van der Waals interactions all contribute to the wrapping mechanism. Soft polymer chains, such as poly(methyl methacrylate) (PMMA) and polystyrene (PS), have been widely utilized to wrap around 1D nanofillers; the chemical composition of wrapped polymer chains plays an important role in the wrapping mechanism. In addition, high-molecular-weight polymers usually have long molecular chains that will be more likely to wrap. Specific surface area, which is defined as the total surface area of a material per unit of mass, is another important factor that affects non-covalent interactions in PNCs. It has been reported that the reinforcement of PNCs has a linear relationship with the overall interfacial area of nanofillers [3]. Thus, the greater the specific surface area, the greater the interfacial interaction area, which will contribute to increasing the possibility of non-covalent interfacial interactions.

5.2.2 Covalent Interaction

Covalent interaction means there are some chemical bonds between the nanofillers and the polymer chains. A large number of studies have demonstrated that the establishment of covalent interactions is a simple and effective strategy by which to improve the performance of PNCs. The surface treatments of 1D nanofillers by oxygen plasma and/or UV-ozone can bring in active polar groups, including hydroxy, carbonyl, and carboxyl groups, onto the surface, which can enhance the interaction with polymer chains. In-situ polymerization is another useful method for introducing monomers to react with each other and the fillers to form chemical bonds. The monomer of the polymer will be polymerized in the presence of dispersed nanofillers so that the growth of polymer chains will take place on the surface of the nanofillers, thereby providing a strong interfacial interaction between the nanofillers and the polymer matrix. In addition, the in-situ polymerization also helps the nanofillers disperse homogeneously in the polymer matrix. Surface functionalization is also a simple and effective way to create covalent interactions. For example, the use of silane coupling agents can bring in various types of reactive groups on the surface of 1D nanofillers. However, to achieve high-efficiency interfacial interaction, the choice of the surface modifier with different substituents needs to be considered in detail.

5.2.3 Characterization of Interfacial Interaction

Contact angle and surface tension are simple and effective methods to evaluate the wettability of the fillers in order to quantitatively characterize the interfacial interaction of PNCs. The surface tension (γ_s) can be calculated by the following Equations (5.1) and (5.2) [4]:

$$\gamma_L(1+\cos\theta) = 2\sqrt{\gamma_s^D \gamma_L^D} + 2\sqrt{\gamma_s^P \gamma_L^P} \tag{5.1}$$

$$\gamma_s = \gamma_s^D + \gamma_s^P \tag{5.2}$$

where γ_L refers to the surface tension of the liquid, γ_L^D and γ_L^P represent the contribution of the dispersion force and the polar force to the surface tension γ_L of the liquid used, respectively, and γ_S^D and γ_S^P represent the contribution of the dispersive component and the polar component to the γ_s of the solid, respectively. If the contact angle of liquids in the equilibrium approaches 0°, it means that a liquid with a surface tension less than the critical value of a given solid will completely wet the solid surface. If the contact angle is in the range of 0°–90°, it will partially wet the solid. Liquid drops will not wet the surface when the contact angle is greater than 90°.

In addition to the macroscopic wetting experiments, atomic force microscopy (AFM) has been utilized for microscopic wetting measurements. A carbon nanotube is attached to an AFM tip, then immersed in the polymer melt. A downward force on the CNT can be calculated by the cantilever deflection, reflecting the surface tension of the polymer melt. Based on the Wilhelmy model [5], the tension force can be calculated by the following Equation (5.3):

$$F_r = \gamma_L \times \pi \times \left(d_{\text{out}} \times \cos\theta_{\text{out}} + d_{\text{in}} \times \cos\theta_{\text{in}}\right) \tag{5.3}$$

where γ_L represents the surface tension of the polymer melt (N/m), θ is the measured contact angle, θ_{in} and θ_{out} stand for the inside and outside contact angles of the nanotube, respectively, and d_{in} and d_{out} are the inside and outside diameters of the nanotube, respectively.

5.3 Effects of Surface Functionalization of 1D Nanofiller

As previously stated, the surface functionalization of 1D nanofiller can enhance the interfacial tension as well as improving the dispersion of 1D nanofiller in polymer matrices, resulting in an improvement in properties of PNCs. However, the interface between the filler and the polymer needs to be well designed to achieve these goals.

The surface treatment of 1D nanofillers with coupling agents is an effective way to improve the adhesion characteristics. For instance, Wu et al. modified the surface of halloysite nanotubes with gamma-glycidoxy-propyltrimethoxysilane to improve the dispersion [6]. The results revealed that the functionalized halloysite nanotubes could disperse uniformly in the poly(butylene succinate) (PBS) matrix with high compatibility. With the addition of 7 wt% modified halloysite nanotubes (HNTs), the PBS composites exhibited increases of 13.4%, 16.2%, and 26.0% in the flexure modulus, tensile modulus, and impact strength, respectively. Similarly, CNTs modified with the silane coupling agent endowed the poly-dimethylsiloxane (PDMS) with excellent composite flexibity. Zhao et al. demonstrated that CNTs containing reactive epoxide groups and long poly(methyl methacrylate) (PMMA) tails could serve as a thermodynamic compatibilizer for immiscible poly(vinylidene fluoride)/polylactide (PVDF/PLLA) blends [7]. The presence of CNTs could bridge the two immiscible phases through physical entanglement and chemical reaction. The tensile strength and percentage elongation at break of PVDF/PLLA blends were improved simultaneously. Shi et al. used mussel protein-inspired chemistry and reversible addition fragmentation chain transfer (RAFT) polymerization to graft biocompatible monomer poly(ethylene glycol) monomethyl ether methacrylate (PEGMA) onto the surface of CNTs [8]. The PEGMA-functionalized CNTs (CNT@PEGMA) exhibited good dispersibility in various organic solvents. In addition, the modified CNTs also had good biocompatibility with HeLa cells and A549 cells. Wu et al. prepared core-shell-structured silicon carbide nanowires with continuous polypyrrole (PPy) shells [9]. The SiC nanowire was used as "nanofiber seeding" to synthesize the PPy shell layer. With the addition of 5 wt% of PPy@SiC nanowires, the composites could accurately achieve effective electromagnetic absorption in the K_u and X bands. Cha et al. modified carbon nanotubes with melamine (M-CNTs) [10]. Compared with the pristine CNTs, the M-CNTs could disperse more uniformly in the polymer matrix and exhibited a 95% increase in elongation at break and a 22% increase in tensile strength. The presence of M-CNTs in the epoxy composite could effectively construct bridging structures to dissipate more energy during fracture.

5.4 Effects of the Intrinsic Property of 1D Nanofiller

1D nanofillers usually present excellent mechanical properties, with high elastic modulus and high tensile strength. The addition of 1D nanofillers will achieve significant improvements in the mechanical properties of polymers while preserving the inherent flexibility of the polymer substrate. Tsentalovich et al. [11] found that the diameter, number of walls, graphitic degree (determined by Raman G/D ratio), and purity of CNTs would affect the quality of the CNT fibers. The higher aspect ratio of CNTs could improve the strength and conductivity of the polymer simultaneously. Erpek et al. compared the reinforcement effect of CNTs with that of the natural tubular clay HNTs [12]. It was observed that the addition of CNTs could improve the tensile strength and elongation at break whereas the HNTs had negative effects on these properties. When the thermoplastic polyurethane (TPU) was utilized as a compatibilizer agent, the CNTs exhibited a better toughening effect on the properties of polylactide than those of HNTs due to the short and straight structure of HNTs.

The introduction of 1D nanofiller into the polymer can not only improve the mechanical properties of the polymer, but also enhance its thermal conductivity. Zakaria et al. added multi-walled carbon nanotubes (MWCNTs) to epoxy resin to prepare an improved composite material [13]; it was found that the bending strength, tensile strength, and thermal conductivity of epoxy resin were increased by 29%, 26%, and 60%, respectively. The epoxy composites containing MWCNTs exhibited greater mechanical properties than those of epoxy/graphene composites. The authors believed that the main reason for this phenomenon was the strong interfacial interaction between the 1D filler and the polymer substrate. In another, similar study, Zakaria et al. [14] also compared the effects of single-layer graphene (SLG) and single-walled carbon nanotubes (SWCNTs) on the thermal conductivity and mechanical properties of epoxy. The results demonstrated that the epoxy/SWCNT nanocomposites had better mechanical properties and slightly lower thermal properties as compared with the epoxy/SLG nanocomposites. Due to the good interaction between SWCNTs and the epoxy matrix, the bending strength, tensile strength, and thermal conductivity of the epoxy composites were increased by 21%, 10%, and 26%, respectively.

Based on the strong quantum transport effect, 1D nanofillers have unique electrical properties that make them among the preferred electronic fillers. Initial studies showed that the addition of CNTs could endow the insulating polymer substrate with increased electrical conductivity, even if the concentration of CNTs was very low. The composite materials developed by combining CNTs with excellent electrical conductivity and flexible polymers possess wide application prospects in integrated electronic products. In recent years, CNTs grown on oxide silicon wafers have been transferred to polymer substrates, and high-performance CNT/polymer circuits have been fabricated by the technology of deposition and photolithography [15]. Bi et al. prepared a novel type of core-shell-structured silicon carbide/silicon dioxide whisker (SiC/SiO_2-W) to mix with poly(vinylidene fluoride) (PVDF) [16]. The results revealed that the SiC/SiO_2-W/PVDF composite had better mechanical performance and drastically suppressed dielectric loss and conductivity, as compared with the SiC-W/PVDF composite. Control of SiO_2 shell layer thickness can adjust the dielectric permittivity, dielectric loss, and AC conductivity of the SiC/SiO_2-W/PVDF composite because the SiO_2 is an insulating material that can reduce the interfacial polarization and electric leakage in the investigated frequency range. In another study, He et al. used a modified spinning method to incorporate core-shell SiC@SiO_2 whiskers into a PVDF matrix [17]. They found that the highly insulating SiO_2 shell could help to suppress the dielectric loss. With the addition of 20 wt% well-aligned SiC@SiO_2 whiskers, the PVDF composite had a dielectric constant of 854 at 1000 Hz, which was 95 times larger than that of pure PVDF [17].

The addition of 1D nanofillers can also change the surface wettability of the polymer. For example, the introduction of SiC into the polymer could lead to an increase in the surface roughness of the composite film, while the introduction of these 1D materials directly leads to the appearance of more airbag structures on the surface of the composite film, thus changing its surface wettability [18]. Qian et al. investigated the effects of different types of carbon nanofillers, including CNTs, carbon nanofibers (CNFs), graphene nanoplates (GnPs), and carbon black (CB), on the hydrophobicity of PVDF/fluorinated ethylene propylene/SiO_2 composites [19]. The authors found that the addition of different carbon fillers affected the surface roughness and surface energy of the composite coating. The superhydrophobic composite coating containing CNTs had the highest antiscaling performance, which was attributed to the homogeneously dispersed CNTs in the composites and the tackling effect of CNTs, that could prevent the adhesion of $CaCO_3$ to the surface of the coating. Mirjalili et al. [20] compared the fracture behaviors of the epoxy nanocomposites containing different dimensional fillers (i.e., 1D, 2D, and 3D). It was shown that the addition of CNTs or clay platelets to epoxy matrices could modestly delay crack growth. However, the presence of ternary component rubber had a positive hybrid toughening effect with CNTs to significantly improve the nano-rubber/CNT/epoxy composites.

The incorporation of 1D nanofillers can also impart electromagnetic shielding properties to the polymer. Chiu et al. introduced SiC into the resin matrix through the carbothermic reduction method [21]. The results suggest that the electromagnetic wave absorption was significantly enhanced as the proportion of SiC increased, as shown in Figure 5.1. When the concentration of SiC nanowires was increased by up to 25 wt%, the minimum reflection loss of the composite was 32.4 dB. This may be due to the high dielectric loss of SiC networks. In another study, it was found that when the SiC mass ratio was increased by up to 35%, the composite materials showed dual-frequency electromagnetic wave absorption of 31.7 dB and 9.8 dB at 8.3 GHz and 27 GHz, respectively [22]. Such phenomena were caused by the formation of a dielectric network structure and the electric polarization of SiC nanowires in the matrix. Moreover, SiC nanowires exhibited higher electromagnetic shielding efficiency as compared with bulk SiC, which was because the 1D structure of SiC nanowires with a large specific surface area and high polarization could consume more microwaves.

In addition to the above effects, the intrinsic properties of 1D nanofillers will directly affect the rheological behavior of composites. Banerjee et al. prepared PP/MWCNT composites containing two different types of MWCNTs *via* melt compounding [23]. Rheological analysis revealed that the MWCNTs with a smaller size would achieve a lower rheological percolation threshold than would larger MWCNTs. Furthermore, the storage modulus of the PP/MWCNT composite with smaller-sized MWCNTs was larger than that of the PP/MWCNT composite with larger-sized MWCNTs at the same concentration, revealing the intrinsically lower values associated with the aggregation of smaller-sized MWCNTs. In addition, a higher level of interfacial interaction was observed in PP composites with smaller-sized MWCNTs, as was evident from extensive rheological analysis. Chiu et al.

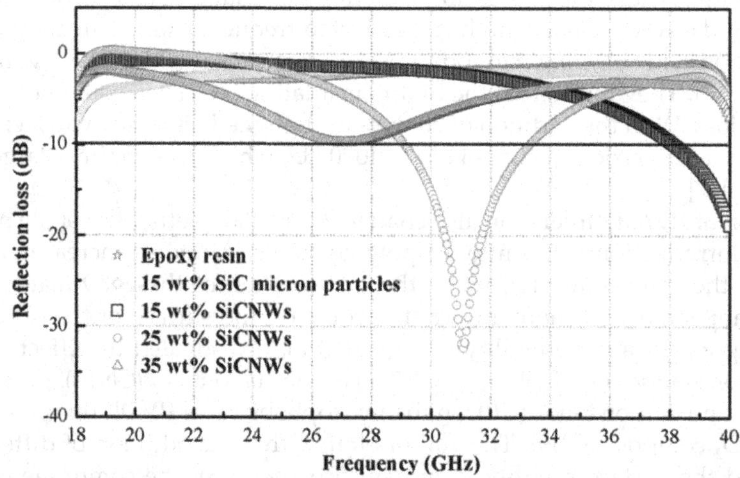

FIGURE 5.1
Electromagnetic shielding performance test results of epoxy composites. Adapted with permission from [21]. Copyright (2010) American Chemical Society.

compared the rheological behavior of polystyrene composites filled with GnPs and CNTs [24]. The authors demonstrated that the CNTs could be wrapped by the neighboring PS chains to form the hybrid network in PS/CNT composites, whereas the GnPs could absorb only the PS chains on the surface to form a PS/GnPs hybrid network. But the scaling exponent value in the percolation scaling law for the PS/GnPs composites is higher than that for the PS/CNT composites, suggesting that GnPs can strengthen the melt viscosity and solid electrical conductivity more effectively than CNTs.

5.5 Effects of the Microstructure of 1D Nanofiller

The microstructure of 1D nanofillers can also affect the properties of composites. It is well known that the microstructure characteristics of 1D nanofillers have a high aspect ratio, which can affect the surface morphology, mechanics, thermodynamics, electrical conductivity, and other properties of composite materials when they are introduced into polymer substrates. Published studies have shown that the introduction of CNTs into other materials can transfer the internal stress of the composites more effectively than common fillers, thus improving the mechanical properties. This is due to the long rod-like microstructure of CNTs with a high aspect ratio [25]. However, it has been reported that the smooth microstructure of CNTs will weaken their improvement effect on the mechanical properties of composites, especially when CNT aggregates are in the composites. Such a negative effect on the composites is more serious than most. In this regard, various methods to reduce the agglomeration of CNT should be utilized to overcome these adverse phenomena. Khan et al. reported that PMMA composites with oriented CNTs had better mechanical properties than composites with randomly oriented CNTs [26]. With the addition of 0.3 wt% CNTs, the well-aligned CNTs could achieve the maximum improvements in elastic modulus and fracture toughness of the composites of approximately 40% and 50%, respectively. In addition, the electrical percolation threshold of randomly oriented CNTs was about 0.034 vol%, which was more than one order of magnitude higher than the 0.0031 vol% value for the well-aligned CNTs. However, it should be noted that, if the CNTs concentration is too high, the toughness of the composite will deteriorate due to the agglomeration of such fillers.

The introduction of CNTs provides not only a significant improvement in the electrical conductivity at low CNT concentrations but also benefits the enhancement of the heat transfer of PNCs. Linzundia et al. investigated the effects of MWCNTs on the microstructure and thermal conductivities of polylactide (PLA) composites [27]. With the addition of 5 wt% MWCNTs, the thermal conductivity of PLA/MWCNTs composites was raised to 0.345 W/m K. The experiments revealed that the thermal conductivity of composites was much lower than at an existing theoretical mode, an effect which was attributed to the large thermal resistance (R_k) of $1.8 \pm 0.3 \times 10^{-8}$ m^{2}K/W at the interface of MWCNTs and the PLA matrix. In addition, the formation of an interconnected conductive nanotube network for thermal diffusion in PLA/MWCNT composites demonstrated that it was an efficient strategy to successfully increase the thermal conductivity of insulator polymers by the addition of CNTs. Unlike randomly dispersed CNTs, Liao et al. found that the aligned CNT/polyethylene (CNT/PE) composites could achieve high thermal conductivity by non-equilibrium molecular dynamic simulations [28]. The analyses indicated that the combination of the intrinsic high thermal conductivity of SWCNTs, the well-aligned SWCNTs with

PE chains, and the non-bonded interactions between SWCNTs and PE chains could help to significantly improve the thermal conductivity of PE. Zhou et al. demonstrated that the thermal conductivity of PLA/CNT composites was affected by the alignment, dispersion, volume fraction, and size of CNTs as well as the polymer size [29]. The thermal conductivity of the random PLA/CNTs composites could be increased by increasing the volume fraction of CNTs. Meanwhile, the thermal conductivity along the alignment of CNTs can be obviously improved to 16 times that of pure PLA by increasing the volume fraction, dispersion degree, and length of CNTs. Interestingly, the alignment of the CNTs was the most important factor among these. Shen and his colleague integrated vertically aligned 3D thermally conductive anisotropic silicon carbide microwire (SiCMW) networks into the epoxy matrix [30]. The epoxy/SiCMWs composites obtained showed a significant increase in thermal conductivity at an ultralow loading of SiCMWs. With the addition of 1.32 vol% SiCMWs, the epoxy composite achieved a thermal conductivity value of 0.62 W/m·K, which was 2.3 times that of the epoxy composite containing randomly aligned SiCMWs. Wang et al. prepared thermally conductive PVDF-based dielectric composites *via* surface functionalization and orientation of SiC nanowires [31]. The presence of well-aligned, amino-functionalized SiC (f-SiC) nanowires contributed to the suppression of dielectric loss as well as the enhancement of thermal conductivity of the PVDF composites. With the incorporation of 13.8 vol% f-SiC nanowires, the PVDF/f-SiC composite showed a thermal conductivity that was twice that of the pure PVDF while maintaining a low dielectric loss of 0.048 at the frequency of 1000 Hz. Vu et al. developed a carbon nanotube-poly(methylmethacrylate) hybrid (CNT@PMMA) for thermo-conductive epoxy composites [32]. The non-conductive PMMA shell layer contributed to achieving the uniform dispersion of CNTs in the epoxy matrix and improving the interfacial interaction. The thermal conductivity of the epoxy composites containing 1 wt% of CNTs in CNT@pPMMA beads increased from 0.19 W/m·K to 0.96 W/m·K as compared with pure epoxy. Similarly, Wang et al. reported [33] a type of epoxy-based composite by intercalating a low loading of core-shell silicon carbide nanowire-silica-amino (termed SiCnw-SiO_2-NH_2) hybrids, which exhibited simultaneously exhibited high permittivity and thermal conductivity while maintaining rather low dielectric loss. The cobweb-structured SiCnw-SiO_2-NH_2 hybrids exhibited high thermal conductivity for the epoxy matrix at low filler loading because of their space microstructures and hydrogen bond interactions.

The previous study proposed that the high aspect ratio of 1D materials also contributes to the increase in electrical conductivity of the composites [4]. For instance, the conductivity of the composite material conferred by the introduction of 1D materials is better controlled or improved than that of other materials such as reduced graphene oxide (rGO) or carbon black (CB) [34]. Furthermore, it has been reported that the rod-like structure of CNTs can increase their contact area, and that a greater contact area can be generated by controlling the orientation of CNTs in the substrate, thereby improving the electrical properties of CNTs and their corresponding composites [35]. Specifically, the research found that, when the conductivity of the composites with randomly oriented CNTs reached semiconductor level, the conductivity of composites with oriented CNTs was nearly one thousand times that of the former.

The large aspect ratio of 1D materials results in a very low percolation threshold in a thermosetting matrix. In addition, the inherent rod-shaped morphology of ID material will lead to gaps in the conductive network formed when the CNTs are randomly oriented. Thus, a large amount of impermeable space will be generated between adjacent 1D materials, which will affect the electrical conductivity of the composite [36]. This viewpoint can also explain the research results and phenomena mentioned above.

Other studies have shown that the aspect ratio of 1D nanofillers will also affect the dielectric constant of composites. For example, Cai et al. analyzed the dielectric composites with enhanced dielectric constant and dielectric breakdown strength by the construction of the ceramic nanofillers through a finite element method and a phase-field model [37]. The results revealed that the orientation of ceramic nanofibers had a significant effect on the dielectric constant of nanocomposites. Cao et al. developed core-shell-structured polydopamine-functionalized silicon carbide (SiC@PDA) nanowhiskers. The presence of polydopamine (PDA) could effectively reduce the dielectric loss of PVDF/SiC@PDA composites because it could reduce the connection between SiC@PDA fillers and inhibit the interfacial polarity. In addition, the PDA shell layer enhanced the interfacial interaction between the SiC and the PVDF matrix. The insulating PVDF/SiC@PDA composite with 30 wt% of SiC@PDA achieved a dielectric constant of 35.1 at 1 KHz, although the dielectric loss was still maintained at as little as 0.037. Wu et al. fabricated TiO_2-modified MWCNTs (MWCNTs@TiO_2) in aqueous solution *via* a sol-gel method. The decorated, insulating TiO_2 nanoparticles could help to inhibit the formation of the conductive network of MWCNTs, suppressing the dielectric loss of the PLA composites. With the addition of 5 wt% MWCNTs@TiO_2, the PLA nanocomposite had a dielectric constant of 26.6 at 1 KHz, which was 8.3 times higher than that of the pure PLA (3.2). Guo et al. decorated CNTs with amorphous carbon (CNTs@C) via a facile hydrothermal method, then mixed them with PVDF to prepare dielectric composites [38]. It was found that the CNTs@C could serve as numerous microcapacitors in the PVDF matrix before the formation of a conductive network, resulting in a significant enhancement of the dielectric constant. The dielectric constant of the PVDF/CNTs@C composites reached a maximum dielectric constant of 5910 (at 1 kHz), which was considerably better than that of the PVDF/CNTs composites.

5.6 Synergistic Effect of 1D Filler and Other Fillers

Because of the particularity of the 1D material structure, it is easy to agglomerate under the influence of van der Waals forces, which greatly affects the performance of 1D material/polymer composites. To reduce the related adverse effects, it has become commonplace to add other fillers into 1D material/polymer composites (namely the construction of hybrid materials) [39]. In general, hybrid materials have the following advantages: compared with 1D materials systems, the dispersion is improved; due to the overlap/connection between fillers, the percolation threshold of materials is decreased, and most importantly, the synergy of hybrid systems can greatly improve mechanical properties, thermal conductivity, and electrical conductivity [39]. However, it should be noted that the performance of the hybrid materials is not a simple sum of the performance of each filler but also involves the interactions between filler and filler, and between filler and matrix [40].

It has been reported that the mechanical properties of CNT/graphene nanosheets (GnPS)/polymer composites are better when the mass ratio of CNT to GnPS is 3:16. The internal reason was that the glass transition temperature (T_g) and porosity affected the mechanical properties of the materials, so that the increase in T_g and the low porosity would lead to the improvement of the tensile strength of the composites [41]. The formation of a CNTs-GnPs 3D structure could dramatically increase the mechanical properties as well as the thermal and electrical conductivities of the hybrid composites since both of the fillers have superior, intrinsic, physical properties. Furthermore, Paszkiewicz et al. also investigated how different shapes (1D, 2D) and particle sizes of carbon nanoadditives affect the barrier and mechanical properties of poly(ethylene terephthalate) (PET) [42]. With the incorporation of expanded graphite (EG) and SWCNTs, the PET ternary composites exhibited a higher Young's modulus and an increase in gas barrier performance, which demonstrated the positive synergistic effect of 1D material with other fillers. In particular, the presence of the SWCNTs/EG hybrid might impede the diffusion and rearrangement of the long polymer chains due to the interaction between both the fillers and the PET matrix, thus affecting the overall crystallization process of the PET composites.

In terms of thermal conductivity, it has been reported that aluminum nitride (AlN) particles can be combined with MWCNTs to form a hybrid system, which can be introduced into the immiscible polycarbonate/polyamide 66 (PC/PA66) composites to improve their thermal conductivity [43]. In addition, a non-covalent functionalization approach was utilized to improve the interfacial interaction between the boron nitride (BN) filler and the epoxy matrix without damaging the intrinsic structures of BN by using 1-pyrenebutyric acid (PBA) molecules. The epoxy composites with 2 wt% filler loading generated a three-dimensional thermal network consisting of BN nanotubes (1D) and BN nanosheets (2D) that had a thermal conductivity of 0.47 W/m·K [44]. The enhanced thermal conductivity of epoxy composites could be attributed to the positive synergistic effects between BN nanotubes and BN nanosheets.

Previous studies had indicated that, compared with the single CNT system, the CNT/rGO mixed filler system can bring a lower seepage threshold and a greater conductivity to the composite material under the same filler concentration [45]. The ultrahigh-molecular-weight polyethylene (UHMWPE) composites filled with GnPs or MWCNTs exhibited low percolation thresholds (0.25 and 0.20 wt%, respectively). With an equivalent concentration of GnPs and MWCNTs, the ternary composites exhibited a lower percolation threshold of 0.10 wt% with a higher conductivity behavior (10^{-2} S/m). Similarly, some studies found that the penetration threshold of CNT/CB/epoxy resin was lower than that of CNT/epoxy resin systems and CB/epoxy resin systems, as shown in Figure 5.2. Further studies pointed out that, in this hybrid system, 1D materials would prevent the re-accumulation of other

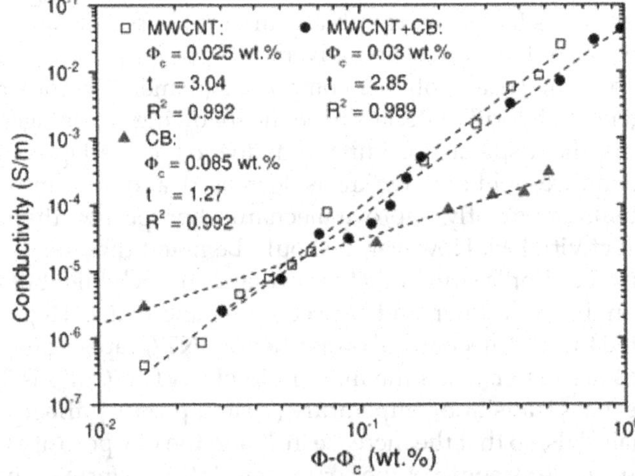

FIGURE 5.2
The relationship between electrical conductivity and filler content of epoxy nanocomposites with different filler types and quantities. Adapted with permission from [47]. Copyright (2009) Springer Nature.

fillers, and other fillers would form a network of isolation layers on the polymer surface to prevent CNT from diffusing into the polymer, thus effectively optimizing the conductive network [46]. In addition, a similar epoxy nanocomposite involving the MWCNT/CB hybrid system has been reported. Due to the synergistic effect of CNT network formation and charge transfer, the conductivity of a composite constructed by a hybrid system is higher than that achieved by a single type of filler. Sumfleth et al. built a relationship between electrical conductivity and filler content of epoxy nanocomposites with different filler types and quantities [47]. It was observed that the threshold of electrical seepage of the MWCNT/CB/epoxy resin system was about 0.025 wt%, whereas the threshold in the MWCNT/epoxy system was about 0.03 wt%. Researchers believed that the reason for this result was that, when CB particles were doped, they would fill the space between CNTs, resulting in a reduction in the average distance between conductive particles.

In addition to the binary hybrid system mentioned above, the multivariate hybrid system is also of interest to researchers. For MWCNT/BN/GNP hybrid composites, there was a synergistic effect not only between MWCNT and both GNP and BN fillers, but also between GNP and BN. The fine particle structure of BN contributed to better dispersion of GNP in an epoxy resin matrix, which helped to improve the mechanical and electrical properties of composites [48].

5.7 Summary

In this chapter, we have reviewed the basic concepts of interfacial interaction in PNCs and summarized several important factors affecting the interfacial interaction between 1D nanofillers and polymer matrices. It requires a well-designed interface between 1D nanofillers and polymer chains to achieve the full potential of property enhancement of PNCs. However, there are still many challenges to overcome to prepare high-performance PNCs containing 1D nanofillers. The biggest issue is the dispersion of 1D nanofillers in the polymer matrices. The high specific surface area will result in easy aggregation. The combination of surface functionalization of 1D nanofillers and the introduction of the third type of nanofiller may be a simple and effective way to alleviate aggregation at high filler loading. On the other hand, we can take advantage of the anisotropy of 1D materials and utilize special processing methods to obtain an oriented structure of 1D material in the polymer matrix, thereby reducing agglomeration to some extent and improving the performance of PNCs. Although the successful integration of 1D nanofillers into PNCs to achieve homogeneous dispersion is still a challenge, many important achievements in fabricating nanocomposites with 1D nanofillers have been made. The fundamental understanding of the interfacial interactions in PNCs will lead to optimized interfacial interactions, therefore improving the properties of PNCs.

References

1. Kuchibhatla SVNT, Karakoti AS, Bera D, Seal S. One dimensional nanostructured materials. *Progress in Materials Science* 2007;52(5):699–913.

2. Kashfipour MA, Mehra N, Zhu JH. A review on the role of interface in mechanical, thermal, and electrical properties of polymer composites. *Advanced Composites and Hybrid Materials* 2018;1(3):415–439.
3. Cadek M, Coleman JN, Ryan KP, Nicolosi V, Bister G, Fonseca A, Nagy JB, Szostak K, Beguin F, Blau WJ. Reinforcement of polymers with carbon nanotubes: The role of nanotube surface area. *Nano Letters* 2004;4(2):353–356.
4. Bauhofer WG, Kovacs JZ. A review and analysis of electrical percolation in carbon nanotube polymer composites. *Composites Science and Technology* 2009;69(10):1486–1498.
5. Hoecker F, KargerKocsis J. Surface energetics of carbon fibers and its effects on the mechanical performance of CF/EP composites. *Journal of Applied Polymer Science* 1996;59(1):139–153.
6. Wu W, Cao XW, Luo J, He GJ, Zhang YJ. Morphology, thermal, and mechanical properties of poly(butylene succinate) reinforced with halloysite nanotube. *Polymer Composites* 2014;35(5):847–855.
7. Zhao XW, Wang HT, Fu Z, Li YJ. Enhanced interfacial adhesion by reactive carbon nanotubes: New route to high-performance immiscible polymer blend nanocomposites with simultaneously enhanced toughness, tensile strength, and electrical conductivity. *ACS Applied Materials and Interfaces* 2018;10(10):8411–8416.
8. Shi YG, Zeng GJ, Xu DZ, Liu MY, Wang K, Li Z, Fu LH, Zhang QS, Zhang XY, Wei Y. Biomimetic pegylation of carbon nanotubes through surface-initiated RAFT polymerization. *Materials Science and Engineering C-Materials for Biological Applications* 2017;80:404–410.
9. Wu F, Sun MX, Chen CC, Zhou T, Xia YL, Xie A, Shang YF. Controllable coating of polypyrrole on silicon carbide nanowires as a core shell nanostructure: A facile method to enhance attenuation characteristics against electromagnetic radiation. *ACS Sustainable Chemistry and Engineering* 2019;7(2):2100–2106.
10. Cha JM, Kim JH, Ryu SW, Hong SH. Comparison to mechanical properties of epoxy nanocomposites reinforced by functionalized carbon nanotubes and graphene nanoplatelets. *Composites Part B Engineering* 2019;162:283–288.
11. Tsentalovich DE, Headrick RJ, Mirri F, Hao JL, Behabtu N, Young CC, Pasquali M. Influence of carbon nanotube characteristics on macroscopic fiber properties. *ACS Applied Materials and Interfaces* 2017;9(41):36189–36198.
12. Erpek CEY, Ozkoc G, Yilmazer U. Comparison of natural halloysite with synthetic carbon nanotubes in poly(lactic acid) based composites. *Polymer Composites* 2017;38(11):2337–2346.
13. Zakaria MR, Abdul Kudus MH, Md. Akil H, Mohd MZ. Comparative study of graphene nanoparticle and multiwall carbon nanotube filled epoxy nanocomposites based on mechanical, thermal and dielectric properties. *Composites Part B Engineering* 2017;119:57–66.
14. Zakaria MR, Abdul Kudus MH, Md Akil H, Thirmizir MZM, Abdul Malik MFI, Othman MBH, Ullah F, Javed F. Comparative study of single-layer graphene and single-walled carbon nanotube-filled epoxy nanocomposites based on mechanical and thermal properties. *Polymer Composites* 2018;40(S2):E1840–E1849.
15. Dodabalapur A. Organic and polymer transistors for electronics. *Materials Today* 2006;9(4):24–30.
16. Bi JY, Gu YZ, Zhang ZC, Wang SK, Li M, Zhang ZG. Core-shell SiC/SiO_2 whisker reinforced polymer composite with high dielectric permittivity and low dielectric loss. *Materials and Design* 2016;89:933–940.
17. He DL, Wang Y, Song SL, Liu S, Deng Y. Significantly enhanced dielectric performances and high thermal conductivity in poly(vinylidene fluoride)-based composites enabled by SiC@SiO_2 core-shell whiskers alignment. *ACS Applied Materials and Interfaces* 2017;9(51):44839–44846.
18. Yu HL, Zhu JQ, Yang L, Dai B, Baraban L, Cuniberti G, Han JC. Superhydrophobic carbon nanotube/silicon carbide nanowire nanocomposites. *Materials and Design* 2015;87:198–204.
19. Qian HJ, Zhu ML, Song H, Wang HY, Liu ZJ, Wang CJ. Anti-scaling of superhydrophobic poly(vinylidene fluoride) composite coating: Tackling effect of carbon nanotubes. *Progress in Organic Coatings* 2020;142:105566.
20. Marouf BT, Mai YW, Bagheri R, Pearson RA. Toughening of epoxy nanocomposites: Nano and hybrid effects. *Polymer Reviews* 2016;56(1):70–112.

21. Chiu SC, Yu HC, Li YY. High electromagnetic wave absorption performance of silicon carbide nanowires in the gigahertz range. *The Journal of Physical Chemistry C* 2010;114(4):1947–1952.
22. Shen ZZ, Chen JH, Li B, Li GQ, Zhang ZJ, Hou XM. Recent progress in SiC nanowires as electromagnetic microwaves absorbing materials. *Journal of Alloys and Compounds* 2020;815:152388.
23. Banerjee J, Kummara S, Panwar AS, Mukhopadhyay K, Saxena AK, Bhattacharyya AR. Influence of carbon nanotube type and novel modification on dispersion, melt-rheology and electrical conductivity of polypropylene/carbon nanotube composites. *Polymer Composites* 2021;42(1):236–252.
24. Chiu YC, Huang CL, Wang C. Rheological and conductivity percolations of syndiotactic polystyrene composites filled with graphene nanosheets and carbon nanotubes: A comparative study. *Composites Science and Technology* 2016;134:153–160.
25. Martone A, Faiella G, Antonucci V, Giordano M, Zarrelli M. The effect of the aspect ratio of carbon nanotubes on their effective reinforcement modulus in an epoxy matrix. *Composites Science and Technology* 2011;71(8):1117–1123.
26. Khan SU, Pothnis JR, Kim JK. Effects of carbon nanotube alignment on electrical and mechanical properties of epoxy nanocomposites. *Composites – Part A: Applied Science and Manufacturing* 2013;49:26–34.
27. Lizundia E, Oleaga A, Salazar A, Sarasua JR. Nano- and microstructural effects on thermal properties of poly (L-lactide)/multi-wall carbon nanotube composites. *Polymer* 2012;53(12):2412–2421.
28. Liao QW, Liu ZC, Liu W, Deng CC, Yang N. Extremely high thermal conductivity of aligned carbon nanotube-polyethylene composites. *Scientific Reports* 2015;5:16543.
29. Zhou B, Luo W, Yang JQ, Duan XB, Wen YW, Zhou HM, Chen R, Shan B. Thermal conductivity of aligned CNT/polymer composites using mesoscopic simulation. *Composites – Part A-Applied Science and Manufacturing* 2016;90:410–416.
30. Shen ZM, Feng JC. Achieving vertically aligned SiC microwires networks in a uniform cold environment for polymer composites with high through-plane thermal conductivity enhancement. *Composites Science and Technology* 2019;170:135–140.
31. Wang B, Yin XH, Peng D, Lv RH, Na B, Liu HS, Gu XB, Wu W, Zhou JL, Zhang Y. Achieving thermally conductive low loss PVDF-based dielectric composites via surface functionalization and orientation of SiC nanowires. *Express Polymer Letters* 2020;14(1):2–11.
32. Vu MC, Bae YH, Yu MJ, Islam MA, Kim SR. Core-shell structured carbon nanotube-poly(methylmethacrylate) beads as thermo-conductive filler in epoxy composites. *Composites – Part A-Applied Science and Manufacturing* 2018;109:55–62.
33. Wang ZD, Wang XZ, Zhao NN, He JY, Wang SL, Wu GL, Cheng YH. The desirable dielectric properties and high thermal conductivity of epoxy composites with the cobweb-structured SiCnw-SiO_2-NH_2 hybrids. *Journal of Materials Science: Materials in Electronics* 2021;32(16):20973–20984.
34. Avilés F, May-Pat A, López-Manchado MA, Verdejo R, Bachmatiuk A, Rümmeli MH. A comparative study on the mechanical, electrical and piezoresistive properties of polymer composites using carbon nanostructures of different topology. *European Polymer Journal* 2018;99:394–402.
35. Peng HS, Sun XM. Highly aligned carbon nanotube/polymer composites with much improved electrical conductivities. *Chemical Physics Letters* 2009;471(1–3):103–105.
36. Philipse AP. The random contact equation and its implications for (Colloidal) rods in packings, suspensions, and anisotropic powders. *Langmuir* 1996;12(5):1127–1133.
37. Cai ZM, Wang XH, Luo BC, Hong W, Wu LW, Li LT. Nanocomposites with enhanced dielectric permittivity and breakdown strength by microstructure design of nanofillers. *Composites Science and Technology* 2017;151:109–114.
38. Guo QK, Xue QZ, Sun J, Dong MD, Xia FJ, Zhang ZY. Gigantic enhancement in the dielectric properties of polymer-based composites using core/shell MWCNT/amorphous carbon nanohybrids. *Nanoscale* 2015;7(8):3660–3667.
39. Ke K, Yue L, Shao HQ, Yang MB, Yang W, Manas-Zloczower I. Boosting electrical and piezoresistive properties of polymer nanocomposites via hybrid carbon fillers: A review. *Carbon* 2021;173:1020–1040.

40. Szeluga U, Kumanek B, Trzebicka B. Synergy in hybrid polymer/nanocarbon composites: A review. *Composites – Part A: Applied Science and Manufacturing* 2015;73:204–231.
41. Shachar-Michaely G, Pinsk N, Cullari LL, Nadiv R, Regev O. Mixed dimensionality: Highly robust and multifunctional carbon-based composites. *Carbon* 2021;176:339–348.
42. Paszkiewicz S, Kwiatkowska M, Rosłaniec Z, Szymczyk A, Jotko M, Lisiecki S. The influence of different shaped nanofillers (1D, 2D) on barrier and mechanical properties of polymer hybrid nanocomposites based on PET prepared by in situ polymerization. *Polymer Composites* 2016;37(7):1949–1959.
43. Xiao C, Leng XY, Zhang X, Zheng K, Tian XY. Improved thermal properties by controlling selective distribution of AlN and MWCNT in immiscible polycarbonate (PC)/Polyamide 66 (PA66) composites. *Composites – Part A: Applied Science and Manufacturing* 2018;110:133–141.
44. Su JL, Xiao Y, Ren M. Enhanced thermal conductivity in epoxy nanocomposites with hybrid boron nitride nanotubes and nanosheets. *Physica Status Solidi (A)* 2013;210(12):2699–2705.
45. Ren PG, Di YY, Zhang Q, Li L, Pang H, Li ZM. Composites of ultrahigh-molecular-weight polyethylene with graphene sheets and/or MWCNTs with segregated network structure: Preparation and properties. *Macromolecular Materials and Engineering* 2012;297(5):437–443.
46. Tang C, Long G, Hu X, Wong KW, Lau WM, Fan M, Mei J, Xu T, Wang B, Hui D. Conductive polymer nanocomposites with hierarchical multi-scale structures via self-assembly of carbon-nanotubes on graphene on polymer-microspheres. *Nanoscale* 2014;6(14):7877–7888.
47. Sumfleth J, Adroher XC, Schulte K. Synergistic effects in network formation and electrical properties of hybrid epoxy nanocomposites containing multi-wall carbon nanotubes and carbon black. *Journal of Materials Science* 2009;44(12):3241–3247.
48. Vovchenko L, Lazarenko O, Matzui L, Perets Y, Zhuravkov A, Fedorets V, Normand FL. Mechanical and electrical properties of the epoxy composites with graphite nanoplatelets and carbon nanotubes. *Physica Status Solidi (A)* 2014;211(2):336–341.

6

Liquid Crystals in One-Dimensional Polymeric Nanonetworks: Physics and Applications

Pragnya Satapathy, Divya Jayoti, and S. Krishna Prasad

CONTENTS

6.1 Introduction

The self-assembly of polymers is an emerging new field within material sciences, offering many potential applications. Crystalline polymers are good examples of spontaneous self-assembly, presenting nanostructures in the form of crystalline lamellae, while engineering self-assembly of polymers by realizing designed structures is currently much pursued [1]. Of specific relevance to the present chapter are polymeric architectures in which liquid crystallinity is the guiding force [2, 3]. Although many liquid crystal (LC) structures are available for the purpose, here we concentrate on cases where the LC has only nematic character; a nematic is simply an orientationally ordered fluid. In principle, the co-presence of LCs and polymers with clearly demarcated regions of occupation are possible in two ways [4], as schematically represented in Figures 6.1a–b. Having the polymer as the major component, PDLCs comprise random dispersions of LC droplets in a polymer matrix (Figure 6.1a). Generally, between droplets, no correlation exists for the intra-droplet director orientation. While allowing the LC molecules to continue having the required electro-optical character, PDLCs are robust, exhibiting the mechanical advantages with polymers. However, the high polymer content also increases the threshold voltage and, consequently, the operating voltage of the device. The second route for achieving the co-presence of LCs and polymer is through polymer-stabilized liquid crystals, where the polymer is in the minority. The realized polymer networks are dispersed in the LC matrix (Figure 6.1b). Since this allows the structure of the LCs present during the polymerization to be transferred to the polymer, the networks can be anisotropic. Typically [5–7], the network formed is of the fibrous type (Figure 6.1c). Inclusion of inorganic nanostructures,

DOI: 10.1201/9781003223764-6

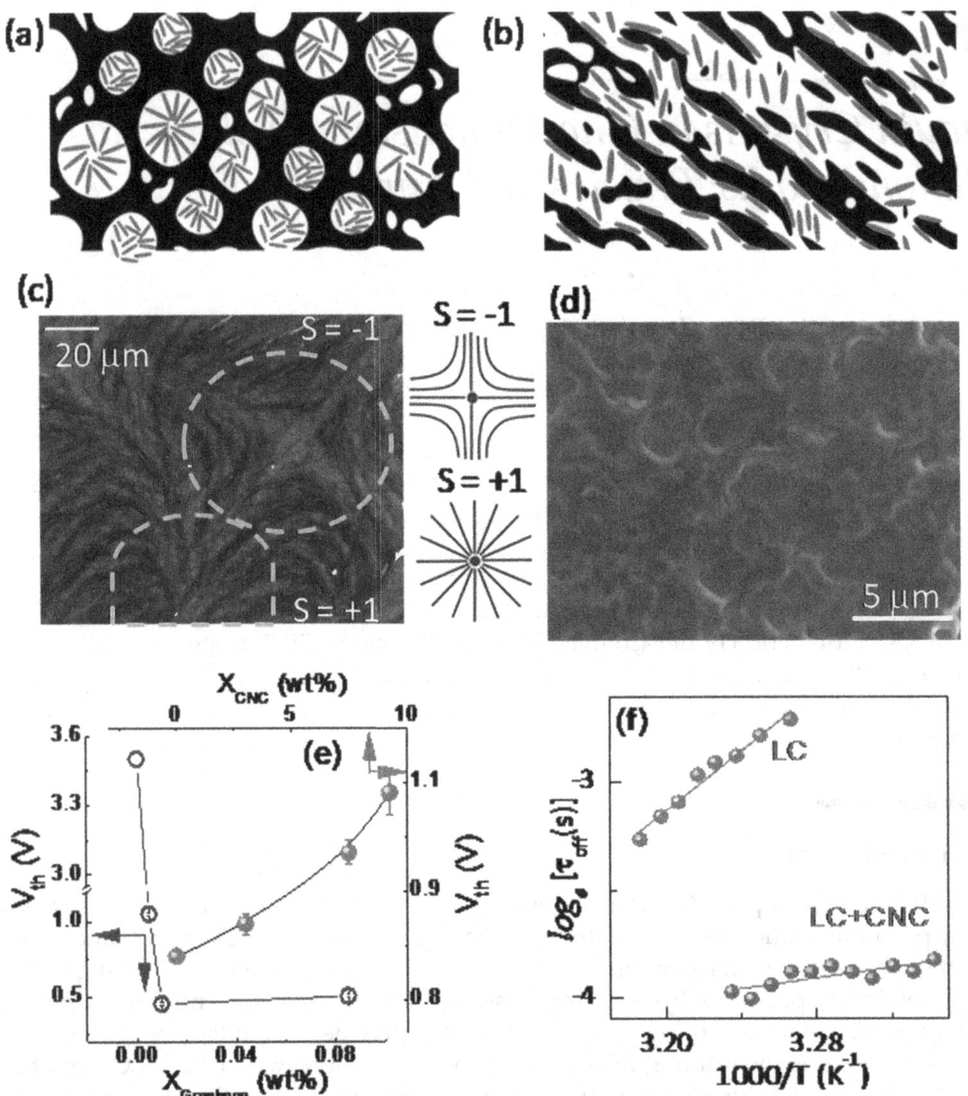

FIGURE 6.1
Schematics of (a) PDLC and (b) PSLC architectures showing LC molecules (shaded ellipses) and polymer regions (dark areas). In PDLCs, the LC molecules are contained as droplets in a polymer matrix, whereas, in PSLCs, they form a continuum interspersed with polymer fibers. FESEM images for PSLCs (c) formed by a nematogenic (a solid that forms a nematic liquid crystal when heated), polymerizable monomer showing fibrous architecture, and (d) formed by a biopolymer comprising cellulose nanocrystals (CNC) presenting the Swiss-cheese morphology. The regions enclosed by dashed lines in the image (c) bring out the advantage of PSLCs in mapping the topological defects associated with the spatial variation of the director as shown schematically adjoining the image; S = ± 1 indicate the sign and strength of such defects. (e) The concentration dependence of the Frédericksz threshold voltage (V_{th}) in two cases: PSLCs in which a carbon nanoform (graphene) is included as an additional component, and a biopolymer which itself is made up of cellulose nanostructures. (f) The thermal variation of the director relaxation time subsequent to the removal of the voltage applied to the liquid crystal cell for cases of no-polymer (LC) and the one having the cellulose polymer (LC+CNC).

such as carbon nanotubes (CNTs) or graphene [8, 9], which can get into polymer bundles, acting as reinforcement, can result in a Swiss-cheese morphology, a feature that we found recently [10] even when the network is formed by nanocrystals of a biopolymer, cellulose (Figure 6.1d). The transfer of the LC structure to the polymer morphological aspect can be very well exploited to map the spatial variation of the nematic director using scanning electron microscopy. This serves as one of the most convenient ways to identify the sign and strength of the surface defects [11, 12], as is seen in Figure 6.1c, which depicts a pair of strength 1 defects having opposite signs. We are currently performing studies on other phases, such as the twist-bend nematic phase, to map out the director patterns in these more esoteric structures. A final point that we want to make about the morphologies is that the strand diameters of the polymer fibrils are much smaller when the nanostructures are present. In the case of carbon nanostructures as inclusions, a possible reason for the thinner fibers is the strong van der Waals interaction between the carbon nanoforms, which tends to make the polymer bundles curvy, as seen, for example, in the case of PLA/PEG blends containing MWCNTs [13]. But, having observed such curvy or Swiss-cheese architecture in the cellulose nanocrystal case as well, we suspect that mechanical/elastic features govern the behavior. Investigations of mechanical properties of the formed PSLC networks with and without the nanostructures are expected to throw more light on this feature.

6.2 Electro-Optical Properties

We now look at two examples of electro-optical properties of the liquid crystal which are significantly influenced by a polymer network. Both these examples were experimentally obtained through Frédericksz transformation of the nematic director. Under equilibrium conditions, this involves having the nematic molecules oriented in the plane of the substrates. Since the LC material exhibits a positive dielectric anisotropy (= $\varepsilon_{||}$ $\varepsilon_{\perp}$ 0, $||$ and $\perp$ representing values parallel and perpendicular to the director, respectively), an AC electric field (1 kHz sine wave) applied normal to the substrates can reorient the director. The threshold voltage (V_{th}) at which such a reorientation is initiated has a significant thermal variation (increasing or decreasing the temperature) in both the no-polymer and with-polymer (PSLC) systems. In stark contrast, incorporating CNTs, graphene, or cellulose nanocrystals makes V_{th} essentially independent of temperature over the entire thermal range of the nematic phase. The Swiss-cheese morphology of the network formed when these nanostructures are present appears to cause this novel effect, which is highly desirable for simplifying the device driver circuits. However, the V_{th} dependence on the nanostructure concentration differentiates between the cases of carbon nanoforms or cellulose nanocrystals. For example, V_{th} is lowered by a factor of ~7 on adding about 0.01 wt% of graphene but increases by ~15% upon adding 10% cellulose nanocrystals (CNC) (Figure 6.1e). We should, of course, note one difference between the two systems: in the former, the nanostructure is an external feature of the polymer fiber but the cellulose itself is present in the nanoform. Whether the difference in the V_{th} behavior is due to the variation is a topic of current investigations. The dynamics of switching the device between the equilibrium bright state and the field-driven dark state is not much influenced by the presence of the nanostructures. But the magnitude and the temperature dependence of the switch-off dynamics are substantially diminished when the nanostructures are present. These features are exemplified in Figure 6.1f for the CNC case: Whereas the values decrease by a factor of 3, the temperature dependence, which follows the Arrhenius equation, shows that for the nanostructure case, the activation energy decreases by a factor of 6 in comparison to the host (no-polymer) situation. The temperature-independent dynamics again help in simplifying the device driver circuitry.

6.3 Application of 1D Polymer Networks

From the beginning of the concept, PSLCs have been considered potential candidates for various applications, with smart privacy windows that switch between transparent and opaque states being among the favorites [14–16]. We begin this section by looking at the influence of nanostructures on the PSLC-based smart windows. This is followed by the effect of the polymer network on photoluminescence and its anisotropy. A brief description of liquid crystalline elastomeric actuators concludes the section.

6.3.1 Smart Windows

The category of smart windows which we consider is the electrically operated devices that switch between opaque and transparent states by exploiting the refractive index mismatch between the two constituents. Here, we restrict the discussion to devices that are scattering under equilibrium conditions (no electrical voltage) and which become transparent when a sufficient voltage is applied. The LC- and network-forming polymers are chosen such that the ordinary refractive index of the LCs (n_o) matches that of the polymer (n_p). The positive birefringence for the chosen LCs means that its extraordinary index (n_e) is mismatched with n_p, creating the required scattering in the field-off state. It is desirable to have a high contrast, quantified in terms of a parameter referred to as haze. Being a measure of the amount of impinging light scattered by the sample, the goal is to have high (low) haze values in the off (on) states. Figures 6.2a–d schematically represent the protocol to determine the haze (H) values by employing an integrating sphere in conjunction with a double-beam spectrophotometer (Perkin-Elmer model Lambda 750), and using the ASTM 1003-92 expression,

$$H = \left(\frac{T_4}{T_2} - \frac{T_3}{T_1} \right) \times 100 \qquad (6.1)$$

With this definition (Eqn. 6.1), H is 0% for a perfectly transparent sample and 100% for a white diffuser. Furthermore, an attractive feature is to have H weakly dependent on the wavelength of light. Figure 6.2e shows the haze spectrum in the scattering state of one of the PSLCs with cellulose nanocrystals. The high H values as well as a mere 15% variation over the entire visible region are quite appealing. The presence of nanostructures improves the high haze values achieved as evident from the bar diagram (Figure 6.2f) representing the data for a PSLC without any nanostructure (PMMA-b) and two systems with different types of nanostructures (CNTs and CNCs). As is desirable in all cases, the H value in the field-driven transparent state was found to be only a few %, resulting in high contrast of the device.

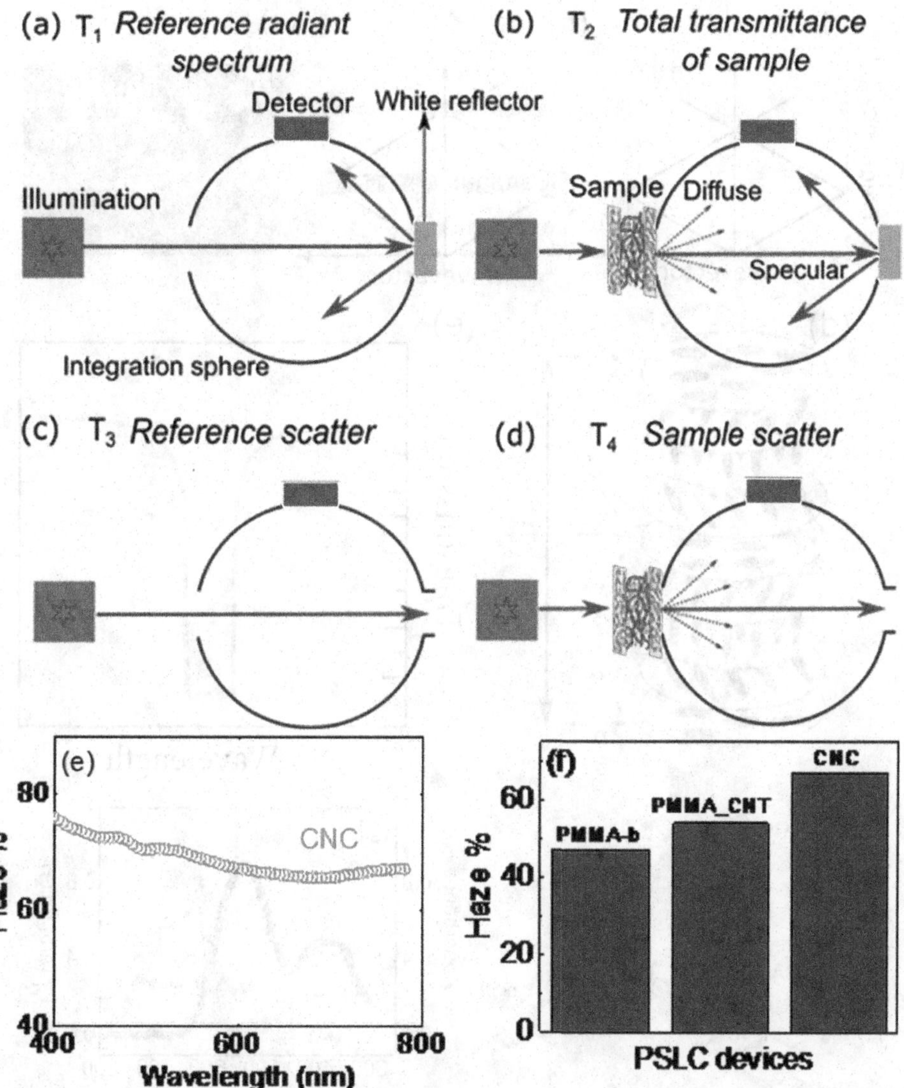

FIGURE 6.2
(a)–(d) Schematic representations of the optical geometries required to calculate the haze factor. (e) The haze spectrum showing very weak dependence over the entire visible wavelength region for the LC+CNC system. (f) Comparative bar diagram of the realized haze at 540 nm for three different PSLC devices. Evidently, the device formed by the CNC network provides the highest haze value.

6.3.2 Enhancement of Photoluminescence and Its Anisotropy by Polymer Networks

In much the same way as the electron transport works, the frequency *vs.* wavevector relationship for a photon in a vacuum has no gaps (Figure 6.3a). However, if the passage is through a periodic dielectric medium (Figure 6.3c), gaps – referred to as photonic bandgaps (PBG) – open up, representing frequency regions (Figure 6.3b) where propagating photons are not allowed. Cholesteric or chiral nematic structures, having a helical modulation of the local nematic director (Figure 6.3d), are among the finest examples of a self-assembled

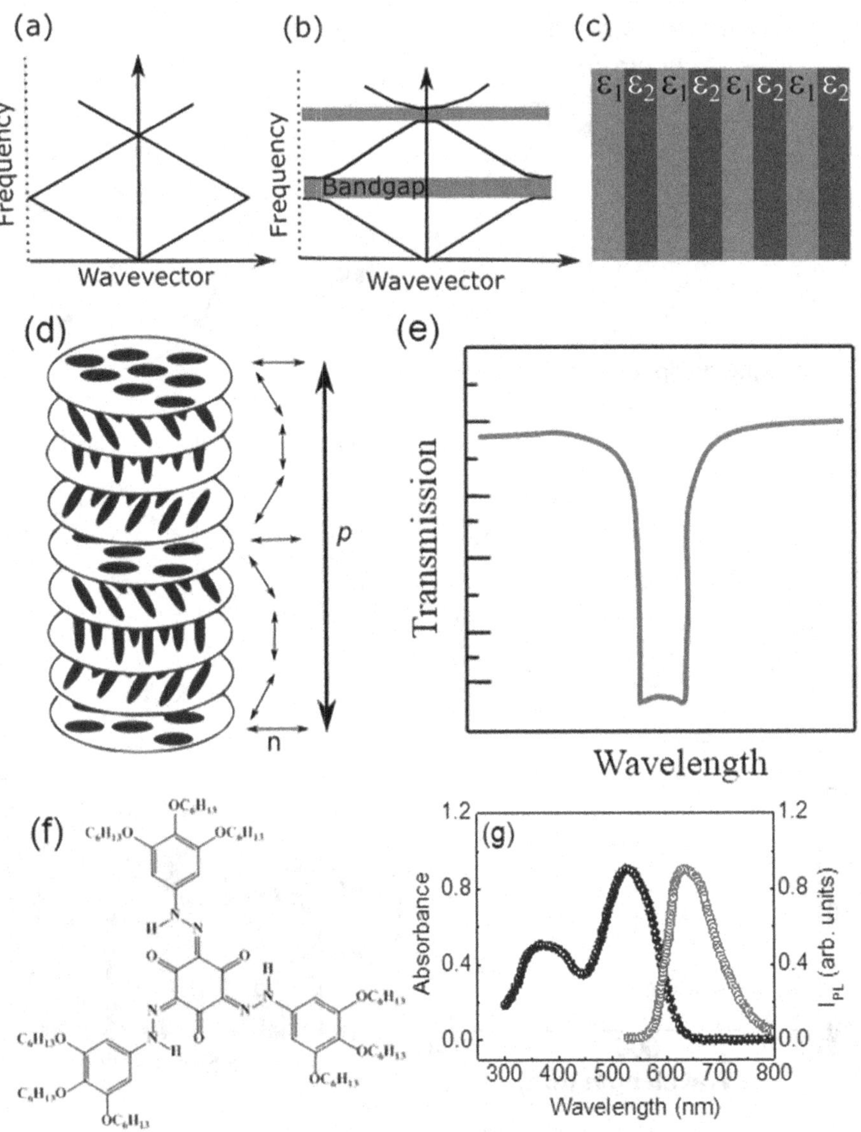

FIGURE 6.3
Band diagram of (a) a uniform one-dimensional medium allowing continuous propagation of photons, and (b) a medium with periodic dielectric (ε_1, ε_2) variation as schematically shown in (c), having non-propagating photons over certain frequency regions, and thus allowing a photonic bandgap (PBG) to open up. (d) Schematic representation of the molecular arrangement in the chiral nematic phase giving rise to a helical structure leading to photonic bandgap character; n represents the local director direction, and p the pitch of the helix. (e) Transmission spectra from a chiral nematic phase showing a minimum in the PBG region. The molecular structure (f) and the absorption/emission spectra (g) of the employed fluorophore taken in an organic solvent.

system exhibiting such a photonic bandgap. In such materials, the light spectrum presents a range of wavelengths, characteristic of PBG, over which the transmission drops to a minimum (Figure 6.3e). We compare the influence of imposed polymer networks on the photoluminescence (PL) in organic emitting systems where PBG is present or absent. We look at a similar influence where the emitters are quantum nanostructures.

The molecular structures of the organic fluorophore employed in these studies are shown in Figure 6.3f, and their absorption and photoluminescence characteristics in the pristine form are presented in Figure 6.3g. This fluorophore is a weak emitter and was chosen precisely for that reason to emphasize the influence of the presence of the network and PBG. The effect of the polymer network on the PL magnitude when the fluorophore is present at a low concentration (5 wt%) in a nematic LC is shown in Figure 6.4a. The network was created when the constituent LC and the fluorophore were mixed with a photopolymerisable monomer, which is also a nematic-forming mesogen (RM82, Merck), and polymerized under optimized conditions. The enhancement of PL is certainly seen when the network is present, although its dependence on the polymer concentration is non-monotonic. This

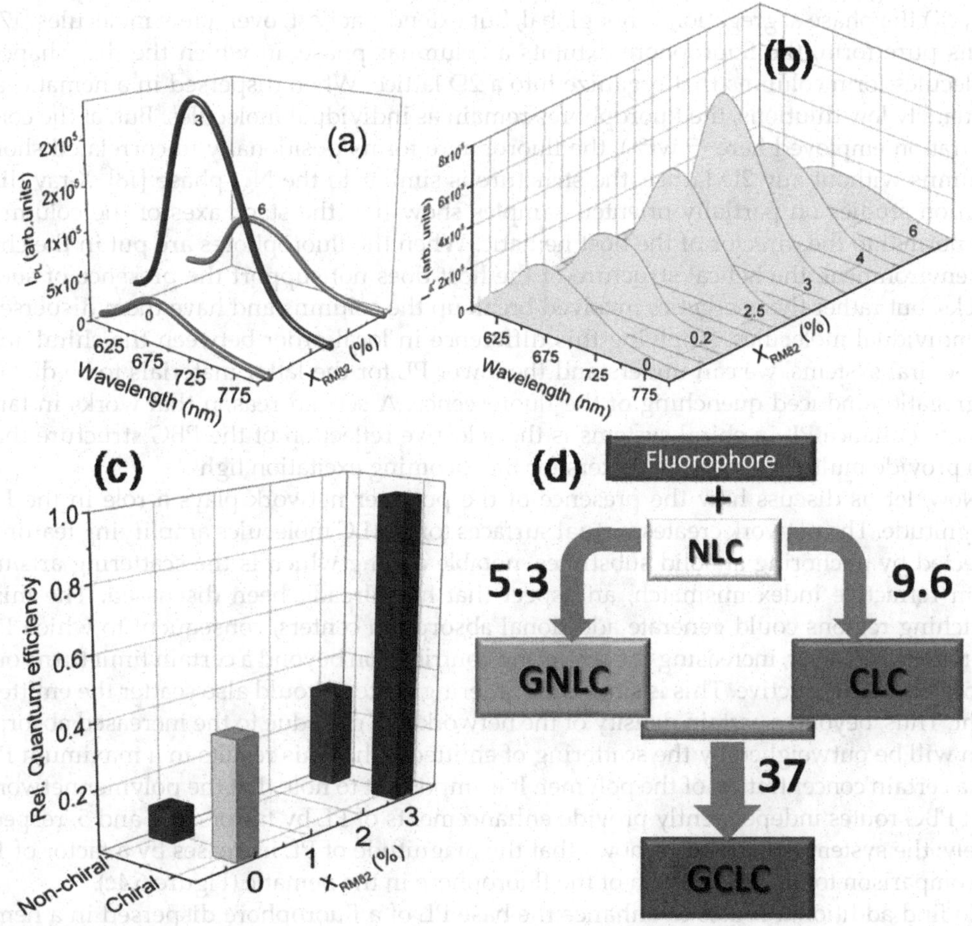

FIGURE 6.4
PL spectra for (a) the non-chiral and (b) the chiral PSLC systems for polymer (RM82) concentrations indicated against each profile. Specifically, to be noted is that the polymer concentration has a non-monotonic influence on the magnitude of emission. (c) Bar diagram indicating the peak emission values for the four cases of non-chiral, no-polymer; chiral, no-polymer; polymer, non-chiral; and polymer, chiral system. The polymer cases are with 3% RM82. (d) Schematic diagram to illustrate the combination-approach to enhance PL with the synergistic influence of plasmonic nanoparticles and the chiral system; the values next to the arrows represent the enhancement factors (NLC: host nematic LC, CLC: NLC with a chiral component, GNLC: NLC with gold nanoparticles, and GCLC: CLC with gold nanoparticles).

feature becomes even more emphatic (Figure 6.4b) when the LC has a chiral component such that the length scale of the photonic bandgap matches the wavelength of the light employed to excite the fluorophore. It may be recalled that we had recently shown that this length-scale matching, especially where the pitch of the chiral nematic helix is comparable to the excitation, rather than the emission, wavelength, is a very efficient protocol to enhance the base PL of the fluorophore.

Before looking at the reasons for the non-monotonic dependence of the PL magnitude on the concentration of the polymer (or equivalently the density of the polymer network), let us discuss why the chiral systems have a higher PL than their non-chiral counterparts. Firstly, we need to identify that the nematic host is calamitic, i.e., the constituents are rod-like molecules. In contrast, the fluorophores are disc-shaped, and thus are strongly prone to phase segregation with the host system. However, since their concentration is quite low (5 wt%) the phase segregation is not global, but extends, at best, over a few molecules [17]. In its pure form, the fluorophore exhibits a columnar phase, in which the disc-shaped molecules form columns that organize into a 2D lattice. When dispersed in a nematic, at extremely low dilutions, the fluorophores remain as individual molecules. But, at the concentration employed here (5 wt%), the fluorophore forms positionally uncorrelated short columns without any 2D lattice; the structure is similar to the N_{col} phase [18]. X-ray diffraction studies on partially oriented samples show that the stack axes of the columns are normal to the director of the host nematic. When the fluorophores are put in the chiral environment, the helical structure of the host does not support the presence of such stacks, but rather the energetics involved break up the columns and have them dispersed as individual molecules. Applying this difference in local order between the chiral and non-chiral systems, we can understand the lower PL for the latter materials to be due to aggregation-induced quenching of the fluorescence. A second reason that works in tandem to enhance PL in chiral systems is the selective reflection of the PBG structure that can provide multiple excitation centers for the incoming excitation light.

Now, let us discuss how the presence of the polymer network plays a role in the PL magnitude. The network creates virtual surfaces for the LC molecules amplifying features affected by anchoring at solid substrates, notable among which is the scattering arising from refractive index mismatch, an aspect that has already been discussed. The mismatching regions could generate additional absorption centers, consequent to which PL increases. However, increasing the scattering contribution beyond a certain limit turns out to be counterproductive. This is since the scattering centers would also scatter the emitted light. Thus, beyond a certain density of the network, the gain due to the increased absorption will be outweighed by the scattering of emitted light. This results in a maximum PL for a certain concentration of the polymer. It is important to note that the polymer network and PBG routes independently provide enhancements of PL by factors of 2 and 5, respectively; the system having both shows that the magnitude of PL increases by a factor of 16 in comparison to the base system of the fluorophore in the nematic (Figure 6.4c).

To find additional routes to enhance the base PL of a fluorophore dispersed in a nematic, we incorporated plasmonic nanoparticles and observed metal-enhanced fluorescence. Again, by combining this aspect with an imposed PBG character, it was found that the base fluorescence increased nearly 40-fold, a factor that was much higher than a mere addition of the enhancement that could be achieved by only the plasmonic or PBG routes; this scheme is schematically represented in Figure 6.4d. Possible explanations for this conjunctive enhancement were provided as summarised in the following. Efficient non-radiative energy transfer between the donor and acceptor requires that they are within the range of 10 nm [19]. Possibilities of fluorescence quenching are higher if the fluorophore is in direct contact with the surface of the gold nanoparticle. The dimensions of the nanoparticles and the fluorophores are comparable (a few nm). Although this promotes a stronger Förster Resonance Energy Transfer (FRET) interaction, the liquid-like nematic (or chiral nematic) environment lessens the chances of a constant direct contact helping enhancement of PL. The Purcell effect, involving interaction between a metal nanoparticle and the fluorophore, separated by greater distances (~50 nm), could also be contributing [20]. To conclude this section, we mention two further features of these systems. The first is that, by appealing to the creation of an intermediate-field focal conic state and a high-field homeotropic state, a 3-level PL memory device could be fabricated. The second feature is the large dissymmetry factor, circularly polarised luminescence (CPL), which is the quantification of how the chirality of the material can relate to the chiral light emission. CPL spectroscopy measures the difference between the left (I_{LCPL}) and the right (I_{RCPL}) circularly polarised luminescence, providing g_{lum} (eq 6.2) the dissymmetry factor defined as, with limiting values of 0 and ± 2. In view of the fact that simple chiral organic molecules exhibit quite low values of g_{lum} (~ 10^{-2}) [21], the value of 0.33 ± 0.01 observed with the systems described above emphasizes the role of self-assembly and macroscopic chiral effects of these materials.

$$g_{lum} = 2\frac{\left(I_{RCPL} - I_{LCPL}\right)}{\left(I_{RCPL} + I_{LCPL}\right)} \tag{6.2}$$

This section describes investigations on composites of nematic with quantum emitters. Our programme on this topic has been initiated to achieve liquid crystal display (LCD) devices that are emissive. As is already evident from the discussions above, LCs are non-emissive and, therefore, for their operation, LCDs require backlighting, a function that has been met conventionally by compact fluorescent tubes, but these are progressively being replaced by LEDs and more recently by quantum dots. Traditionally, quantum dots (QDs) are formed by the II–VI group of elements such as CdS and CdSe but are being supplanted by the two classes of nanoparticles, gradient-controlled QDs and all-inorganic halide perovskite, good examples of which have been employed in the studies to be described below: CdSeS and $CsPbBr_3$ QDs. The latter are especially popular owing to their high PL quantum yield and narrow emission line widths. The emphasis in the present manuscript is on the anisotropy of PL and its modulation by applied electric or actinic light fields.

Let us begin with the studies on a CsLC composite comprising $CsPbBr_3$ nanoparticles dispersed in a nematic liquid crystal. The transmission electron microscopy (TEM) images of even the as-synthesized nanoparticles, exhibiting cuboid shapes, a feature that was supported by the orthorhombic structure as evidenced by X-ray diffraction, already indicated the system's inherent tendency to support anisotropic structures. The TEM images of the drop cast sample of the CsLC composite presenting organization of these quantum cuboids (QCs) along their body diagonal provided further credence to this aspect. The attractive feature is that the organization along the diagonals (Figure 6.5a) turns out to be dictated by the liquid crystal along with the nematic director. Presumably, this organization is possible due to the interaction between the passivating entities (oleylamine and oleic acid) decorating the QCs and the LC molecules. Polarising optical microscopy images provide unequivocal evidence that this occurs not just at the microscopic level but on macroscopic scales as well (Figure 6.5b). That the linear assembly of dark regions in this image is indeed from the QCs is clearly established by their emissions obtained when excited by a 404 nm laser light (Figure 6.5c). As we shall see a little later, such a superstructure, realized simply by fixing the nematic director direction, provides a powerful handle with which

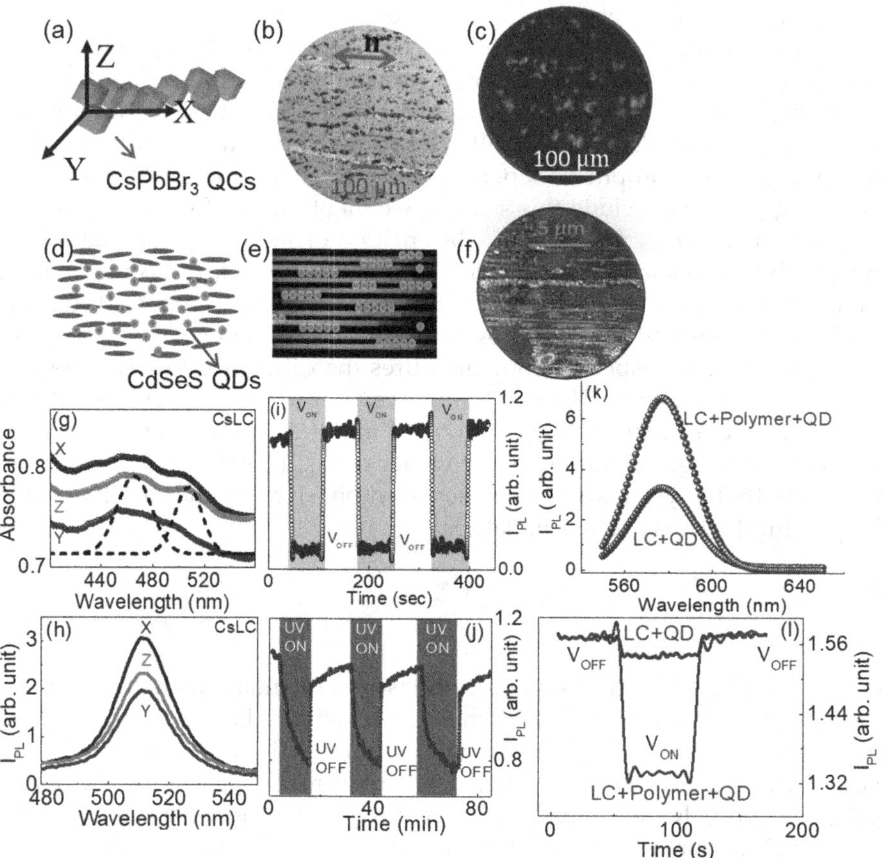

FIGURE 6.5
(a) Schematic illustration of the linear self-assembly of CsPbBr3 quantum cuboids (QCs) along the nematic director direction (n) that coincides with direction X. The self-assembly is evident from the polarising microscope image obtained with (b) white light and (c) 404 nm laser excitation. The QCs appear as black and grey regions in the two images, respectively. The self-assembly that is absent when spherical CdSeS quantum dots (QDs) are incorporated into a nematic (d) is favored by an imposed, oriented polymer network (e), as indeed seen from the FESEM image shown in (f). Absorption (g) and emission (h) spectra for the nematic+QC composite (CsLC) bring out the dual-anisotropic character arising out of the linear assembly of QCs. The individual absorption bands determined after a background subtraction are shown as dashed lines for the Z configuration. Modulation of the PL that is achievable by an AC electric field or an actinic light (365 nm) in a composite of LC+CdSeS QDs is presented in (i) and (j), respectively. As in the case of organic emitters, imposition of a polymer network on the emitting system enhances the magnitude of the base PL as well as its modulation, as shown in panels (k) and (l), respectively.

to achieve the anisotropic photophysical properties of QCs. The non-isotropic shape of the QCs seems to be certainly a necessary ingredient since the spherical gradient-controlled CdSeS do not readily give rise to any linear assembly when incorporated into an oriented nematic (Figure 6.5d). Oriented polymer networks, obtained in the manner already described, are effective at achieving linear assemblies from these spherical QDs, as suggested in Figure 6.5e. A composite consisting of the nematic, CdSeS QDs, and a thermally polymerizable monomer, when polymerized under suitable conditions generates these assemblies. A field-emission scanning electron microscopy (FESEM) image (Figure 6.5f) obtained by leaching out the LC after polymerization demonstrates the creation of the linear assemblies, with EDX measurements lending the necessary structural proof.

For the description of the photophysical properties, we consider the reference frame indicated in Figure 6.5a for a cell fabricated using indium tin-oxide (ITO)-coated glass substrates pre-treated with a polyimide layer and containing the LC+QD composites. Here, under equilibrium conditions, the nematic director is oriented along the X direction lying in the plane of the substrates, and the other in-plane direction is labeled Y. An electric field when applied to reorient the LC director is applied normal to the substrate, a direction coincident with Z. The absorption spectra for the three directions, X, Y, and Z, using the CsLC composite, are shown in Figure 6.5g, after subtracting the background. Photoselection to get the data for the in-plane directions was done by using a polariser before the sample which allowed selective excitation of the QC assembly aligned along that direction; the data for the Z-direction were obtained by applying an electric field which was well above the Frédericksz threshold such that the LC molecules are essentially in the homeotropic geometry (director normal to the substrate). Although the profiles are similar, the peak values (A) are indeed different and satisfy the relation $A_X > A_Z > A_Y$, bringing out the existence of the double anisotropy in absorption. These two anisotropies can be expressed as $\Delta A_{XY} = (A_X - A_Y)/(A_X + 2A_Y)$ and $\Delta A_{XZ} = (A_X - A_Z)/(A_X + 2A_Z)$; here, the denominators indicate the anisotropic average of the absorptions involved. Two interesting features were observed: ΔA_{XY} is an order of magnitude higher than ΔA_{XZ}, and the two anisotropies have opposite temperature dependences. These features have been explained using the refractive indices involved and the inherent absorptions [22]. Figure 6.5h presents the PL spectra obtained with 404 nm excitation in the X, Y, and Z configurations, and, as to be expected from the behavior of absorption, the peak value of PL (I_{PL}) magnitude followed the relationship $I_X > I_Z > I_Y$. This feature permits double anisotropies to be determined for PL as well. It was observed that ΔI_{XY} was always higher than ΔI_{XZ} and that their ratio increased from a value of 4 at room temperature to about 8 near the transition to the isotropic phase.

The next aspect of the photophysical properties of the LC+QD composite that we describe is the modulation of PL by two external controls, an electric field and actinic light. As indicated in the previous paragraphs, the electric field influence is through the Frédericksz transformation of the nematic, changing the PL value from I_X to I_Z. Of course, this is a completely reversible change with the value reverting to I_X when the field is switched off. It is well-known that, in the Frédericksz transformation, the switch-off process is quite slow, being governed by viscous forces. Thus, the return to the equilibrium I_X value from I_Z is a slow process. To hasten this, we specifically chose a nematic that exhibits a dual-frequency character, in which both X-Z as well as Z-X transformations are obtained in the presence of an electric field, with the former using a low-frequency field and the latter a high-frequency field. This protocol accelerates the recovery of the equilibrium situation by a factor of 40, permitting real-time large-magnitude modulation of PL that is highly reproducible and repeatable over many cycles (Figure 6.5i). PL modulation can also be obtained by a simultaneous appeal to another photo-driven effect, viz., photoisomerization of the constituent molecules. For this purpose, we incorporated a small amount (5%) of an azobenzene derivative into a LC-CdSeS QD composite. In the presence of actinic UV light (365 nm wavelength), the azobenzene compound undergoes a conformational change from the *trans* to the *cis* form. The associated shape change from rod to a bent form of this minority constituent is sufficient to bring about an isothermal transition from the nematic to the isotropic phase in the presence of UV light. Switching the light off, the system returns to the nematic phase spontaneously but more slowly, but this can be accelerated by applying a DC bias. The temporal profile of this all-optical, highly reproducible and fatigue-free PL modulation is shown in Figure 6.5j.

The final aspect that we present with which to conclude this section is the influence of the polymer network. Our preliminary work on a system that employs a composite comprising a nematic and CdSeS QDs (2 wt%) shows that the base PL of the LC-QD composite is substantially enhanced (Figure 6.5k) when the material is contained in a network made of a thermally polymerized system, just as in the case of organic emitters previously described. The network also significantly amplifies the modulation that is driven by an electric field (Figure 6.5l).

6.3.3 Orientation Order-Coupled Rubber Elasticity: LCE Actuators

In this section, we discuss a slightly different type of liquid crystalline phase in which, in addition to the long-range orientation order of the usual nematic LCs, rubber elasticity is also present [23, 24]. In these liquid crystalline polymers, the rod-like LC subunits form a part of the main chain or the side chain of the polymer network [25]. To realize large, macroscopic, mechanical effects driven by external optical or thermal stimulus, these polymer chains are cross-linked together to form liquid crystal networks (LCNs) [26, 27]. Thus, the degree of crosslinking is an important, tunable parameter in the LCNs.

In the studies to be described here, LCPs were mixed with small quantities (0.1–3.0 wt%) of photo-thermal agents that absorb the incoming radiation. This absorption causes a thermal gradient through the thickness of the LCP film, which results in a macroscopic shape deformation. The networks, consisting of an acrylate monomer, crosslinker, photoinitiator, and a photothermal agent, are prepared by the cross-linking of LC monomers via either a single- or a two-step crosslinking process. Since the interest was to make the anisotropic, structural character of the LC also play a role, the fabrication of the LCP film involved preparing it so that the two faces had an antagonistic orientation of the LC molecules. This was achieved by filling the prepared composite in a glass cell (of nominal thickness $d \sim 50\ \mu$) with pre-treated substrates, each having a specific but different influence on the orientation direction: one of them promoting planar and the other homeotropic direction. As opposed to a cell with a uniform planar orientation (Figure 6.6a), this resulted in a hybrid situation wherein, across the cell thickness, a splay-bend deformation of the director existed, shown schematically in Figure 6.6b. After UV curing of the sample, the film was peeled off and used for the photothermal actuation experiments.

The mechanical deformation experiments were performed by having the films vertically suspended from supports and illuminating them with a NIR laser (wavelength 808 nm) beam having a footprint of 3 mm × 4 mm. The photothermal stimulus causes the film to heat up such that it transforms from the nematic to the isotropic phase. The consequent loss in the orientational order from the nematic (Figure 6.6c) to the isotropic state (Figure 6.6d) leads to a macroscopic contraction in the longitudinal direction with respect to the LC director and concomitant expansion in the transverse direction as depicted in Figures 6.6c and d. It should be noted that the deformation is reversed when the sample is subsequently cooled to the nematic phase. To study these macroscopic deformations, we shone an NIR laser on the films kept in a hanging configuration with one end tethered to the support which can be seen in Figure 6.6e. Since the film was tethered from a solid support the top, the spot that was illuminated by the laser controls a spatially dependent magnitude and shape of actuation as seen in the movie frames shown in Figure 6.6e; the maximum deformation was observed in the case where the laser spot is closest to the tethered end of the film. Different cases of the actuation of an untethered film laser illuminated from

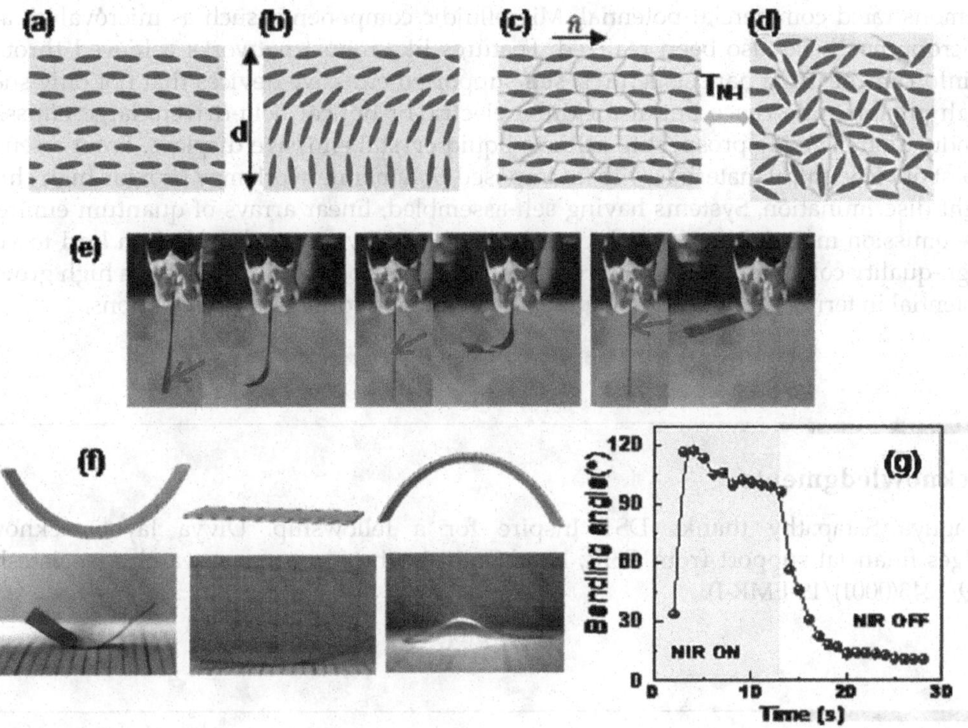

FIGURE 6.6
Schematic showing (a) the planar and (b) the hybrid director orientation along the LCN film thickness d. Representation of a planar LCN film in the (c) nematic and (d) isotropic phase obtained by NIR illumination. (e) Photographs of actuations of a suspended LCN film (size: 14 mm × 4 mm) illuminated at different vertical spots (indicated by the arrow). (f) Schematic (top) and actual photograph (bottom) of orientation-side dependent deformation of the unilluminated film (middle) and NIR-illuminated film on the planar side (left) and homeotropic side (right). (g) Temporal variation of the actuation and subsequent relaxation (Note that the film is never at 0° due to residual internal crosslinking stresses developed during film formation).

the top are shown in Figure 6.6f in the form of schematic representations as well as actual photographs. The concave (left) and convex (right) shaped deformations are obtained by illuminating the planar and homeotropic sides of the hybrid film, respectively. This illumination-side-dependent shape change is due to the fact that the absorption cross-section and consequently the photothermal conversion rates are high on the planar side. The temporal monitoring of the process shows that the actuation is quite fast and achieves a maximum bending angle of 120° (Figure 6.6g). All these features suggest that these LCNs have potential for complex actions, including bending, stretching, and curling, studies on which are in progress.

6.4 Outlook

Applications such as smart windows and electrically switchable THz devices, based on polymer-modified liquid crystals similar to those discussed in this chapter, have already demonstrated commercial potential. Microfluidic components, such as microvalves and micropumps have also been realized. Features like robust networks achieved through reinforcement using nanostructures, self-supported emissive devices that not only show high photoluminescence but also exhibit electric or optical field-driven, large emission modulation have the prospect to achieve liquid crystal emissive displays. From a sensor point of view, chiral materials with an imposed polymer network may provide high chiral light discrimination. Systems having self-assembled, linear arrays of quantum emitters, the emission magnitude of which can also be tuned by external fields, can lead to very high-quality color-tunable polarisers. Thus, polymer-modified LCs present a high growth potential in terms of both fundamental and technology-oriented investigations.

Acknowledgments

Pragnya Satapathy thanks DST Inspire for a fellowship. Divya Jayoti acknowledges financial support from CSIR, New Delhi in the form of a research associateship (09/1243(0001)/19-EMR-1).

References

1. Dahman, Y. Chapter 10 - Self-Assembling Nanostructures. By Yaser Dahman, Gregory Caruso, Astrid Eleosida, and Syed Tabish Hasnain. In: *Micro and Nano Technologies*; Dahman, Y. B. T.-N. and F. M. for E., Ed.; Elsevier, 2017; pp 207–228. https://doi.org/10.1016/B978-0-323-51256-5.00010-1.
2. Thakur, V. K.; Kessler, M. R. *Liquid Crystalline Polymers Volume 1–Structure and Chemistry*, 1st ed.; Thakur, V. K., Kessler, M. R., Eds.; Springer, Cham, 2016.
3. Drzaic, P. S. *Liquid Crystal Dispersions*; World Scientific, 1995; Vol. 1. https://doi.org/doi:10.1142/2337.
4. Nagaraj, M. Polymer Modified Liquid Crystals. *Liq. Cryst. Today*, 2019, *28*(3), 68–69. https://doi.org/10.1080/1358314X.2019.1693101.
5. Fung, Y. K.; Yang, D.-K.; Ying, S.; Chien, L.-C.; Zumer, S.; Doane, J. W. Polymer Networks Formed in Liquid Crystals. *Liq. Cryst.*, 1995, *19*(6), 797–801. https://doi.org/10.1080/02678299508031102.
6. Dierking, I. Polymer Network–Stabilized Liquid Crystals. *Adv. Mater.*, 2000, *12*(3), 167–181. https://doi.org/10.1002/(SICI)1521-4095(200002)12:3<167::AID-ADMA167>3.0.CO;2-I.
7. Madhuri, P. L.; Hiremath, U. S.; Yelamaggad, C. V.; Madhuri, K. P.; Prasad, S. K. Influence of Virtual Surfaces on Frank Elastic Constants in a Polymer-Stabilized Bent-Core Nematic Liquid Crystal. *Phys. Rev. E*, 2016, *93*(4), 42706. https://doi.org/10.1103/PhysRevE.93.042706.
8. Krishna Prasad, S.; Baral, M.; Murali, A.; Jaisankar, S. N. Carbon Nanotube Reinforced Polymer-Stabilized Liquid Crystal Device: Lowered and Thermally Invariant Threshold with Accelerated Dynamics. *ACS Appl. Mater. Interfaces*, 2017, *9*(31), 26622–26629. https://doi.org/10.1021/acsami.7b08825.
9. Baral, M.; Bramhaiah, K.; John, N. S.; Krishna Prasad, S. Graphene-Augmented Polymer Stabilization: Drastically Reduced and Temperature-Independent Threshold and Improved Contrast Liquid Crystal Device. *ACS Omega*; Centre for Nano and Soft Matter Sciences, Jalahalli, Bengaluru 560013; India, 2019, 403–411. https://doi.org/10.1021/acsomega.8b03026.

10. Satapathy, P.; Parthasarathi, S.; Rao, D. S. S.; Bano, S.; Negi, Y. S.; Prasad, S. K. Switchable Smart Windows Using a Biopolymer Network of Cellulose Nanocrystals Imposed on a Nematic Liquid Crystal. *Appl. Phys. Lett.*, 2020, *117*(10), 103702. https://doi.org/10.1063/5.0020982.
11. Dierking, I.; Archer, P. Imaging Liquid Crystal Defects. *RSC Adv.*, 2013, *3*(48), 26433–26437. https://doi.org/10.1039/C3RA45390A.
12. Dierking, I. Recent Developments in Polymer Stabilised Liquid Crystals. *Polym. Chem.*, 2010, *1*(8), 1153–1159. https://doi.org/10.1039/C0PY00087F.
13. Ahmad, A. F.; Aziz, S. A.; Obaiys, S. J.; Zaid, M. H.; Matori, K. A.; Samikannu, K.; Aliyu, U. S. Biodegradable Poly (Lactic Acid)/Poly (Ethylene Glycol) Reinforced Multi-walled Carbon Nanotube Nanocomposite Fabrication, Characterization, Properties, and Applications. *Polymers*, 2020. https://doi.org/10.3390/polym12020427.
14. William Doane, J.; Wu, B.; Erdmann, J. H.; William Doane, J.; Shutters, P. D. L. C. Where has this Technology Gone? *Liq. Cryst.*, 2006, *33*(11–12), 1313–1322. https://doi.org/10.1080/02678290601119708.
15. Mondal, I.; Kiruthika S.; Ganesha, M. K.; Baral, M.; Kumar, A.; Vimala, S.; Madhuri, P. L.; Nair, G. G.; Prasad, S. K.; Singh, A. K.; et al. ITO-Free Large Area PDLC Smart Windows: A Cost-Effective Fabrication Using Spray Coated SnO_2 on an Invisible Al Mesh. *J. Mater. Chem. A*, 2021, *9*(40), 23157–23168. https://doi.org/10.1039/D1TA05820G.
16. Hu, X.; Zhang, X.; Yang, W.; Jiang, X.-F.; Jiang, X.; de Haan, L. T.; Yuan, D.; Zhao, W.; Zheng, N.; Jin, M.; et al. Stable and Scalable Smart Window Based on Polymer Stabilized Liquid Crystals. *J. Appl. Polym. Sci.*, 2020, *137*(30), 48917. https://doi.org/10.1002/app.48917.
17. Madhuri, P. L.; Rao, D. S. S.; Yelamaggad, C. V.; Achalkumar, A. S.; Prasad, S. K. Fast Photoluminescence Switching in the Nematic Phase of Calamitic–Discotic Composites. *Adv. Opt. Mater.*, 2015, *3*(8), 1116–1124. https://doi.org/10.1002/adom.201500126.
18. Kouwer, P. H. J.; Mehl, G. H. Full Miscibility of Disk- and Rod-Shaped Mesogens in the Nematic Phase. *J. Am. Chem. Soc.*, 2003, *125*(37), 11172–11173. https://doi.org/10.1021/ja037075y.
19. Jares-Erijman, E. A.; Jovin, T. M. FRET Imaging. *Nat. Biotechnol.*, 2003, *21*(11), 1387–1395. https://doi.org/10.1038/nbt896.
20. Li, M.; Cushing, S. K.; Wu, N. Plasmon-Enhanced Optical Sensors: A Review. *Analyst*, 2015, *140*(2), 386–406. https://doi.org/10.1039/C4AN01079E.
21. Sánchez-Carnerero, E. M.; Agarrabeitia, A. R.; Moreno, F.; Maroto, B. L.; Muller, G.; Ortiz, M. J.; de la Moya, S. Circularly Polarized Luminescence from Simple Organic Molecules. *Chem. – A Eur. J.*, 2015, *21*(39), 13488–13500. https://doi.org/https. https://doi.org/10.1002/chem.201501178.
22. Satapathy, P.; Santra, P. K.; Prasad, S. K. Synergistic Path for Dual Anisotropic and Electrically Switchable Emission from a Nanocomposite of CsPbBr3 Quantum Cuboids and Nematic Liquid Crystal. *Crystals*, 2019, *9*(8), 378. https://doi.org/10.3390/cryst9080378.
23. Kato, T.; Uchida, J.; Ichikawa, T.; Soberats, B. Functional Liquid-Crystalline Polymers and Supramolecular Liquid Crystals. *Polym. J.*, 2018, *50*(1), 149–166. https://doi.org/10.1038/pj.2017.55.
24. Finkelmann, H. Liquid Crystalline Polymers. *Angew. Chemie. Int. Ed. Engl.*, 1987, *26*(9), 816–824. https://doi.org/10.1002/anie.198708161.
25. Küpfer, J.; Finkelmann, H. Nematic Liquid Single Crystal Elastomers. *Makromol. Chem. Rapid Commun.*, 1991, *12*(12), 717–726. https://doi.org/10.1002/marc.1991.030121211.
26. Mistry, D.; Gleeson, H. F. Mechanical Deformations of a Liquid Crystal Elastomer at Director Angles between 0° and 90°: Deducing an Empirical Model Encompassing Anisotropic Nonlinearity. *J. Polym. Sci. Part B Polym. Phys.*, 2019, *57*(20), 1367–1377. https://doi.org/10.1002/polb.24879.
27. Finkelmann, H.; Nishikawa, E.; Pereira, G. G.; Warner, M. A New Opto-Mechanical Effect in Solids. *Phys. Rev. Lett.*, 2001, *87*(1), 15501. https://doi.org/10.1103/PhysRevLett.87.015501.

7

Toxicity and Risk Assessments of One-Dimensional Nanocomposites

Ubong Eduok

CONTENTS

7.1 Introduction

In their nanostates, nanomaterials possess properties that are uniquely different from the same material in their bulk forms [1]. In fact, the physicomechanical properties of newly built nano-devices qualify them with potential for use in modern drug delivery, space intergalactic exploration, and even energy generation. The insatiable knowledge behind nanotechnology has advanced miniaturization, with most devices developed to fit our pockets and capable of housing enormous histories within tiny spaces. Whereas some devices are also fabricated to mimic living systems, others are made of specially designed molecules capable

DOI: 10.1201/9781003223764-7

of executing complex functions at the nanoscopic levels (e.g., nanorobots in medicine). The applications of nanotechnology are as vast as the numerous nanomaterials utilized in their various functions. This makes nanotechnology synonymous with a modern innovation tool with sustainable promise [1]. The promising aspect of this field is the creation of functional materials with flexible and novel properties at the nanoscale. Their unique size, reduced weight, and high sensitivity are some of the properties required for sensor designs towards enhanced functions in nanomedicine. Most nanorobots are capable of tissue repairs and drug dispensaries. Most nanosensors of various forms (e.g., nanowires) are also used in devices for security and defense in the military for detecting chemicals and biological weapons. Although tremendous successes have been recorded with the placement of nanosensors within the blood cells of astronauts to help detect radiation exposure, there are also some developments in deploying drug-loaded nanoshells for the detection and treatment of tumor cells without secondary damage around healthy tissues [2].

The current global enthusiasm for transistor miniaturization has also revolutionized modern computing. New transistors are now faster as they are smaller in size. Downsizing by size is now also leading the frontiers of modern nanolithography. Fabrication of nanostructures by patterning is gaining momentum in this field as newer materials, systems, and devices will be shrunk to reduced size and cost while also increasing processing speed. The innovations behind molecular switchable concepts may become building blocks for future computers capable of replacing today's etched patterning in silicon boards. Between manufacturing lines, science, technology, agriculture, and medicine, applicable examples of nanotechnology are enormous. Science and technology, at the nanoscale, are blooming research areas, with so many scientists and engineers working on various aspects to solve human and environmental problems [1, 2].

Between the late '90s and early 2000s, so much has been invested worldwide by governmental and industrial agencies in R&D involvements in nanotechnology. Even when most data within the literature are not completely dependable, however, it can be observed that a lot of financial investments have been invested in research in nanotechnology. Lux Research [3] predicted USD 4.1 billion from investments by global governments in 2005 and USD 1 trillion was reported for 2015 from products incorporating nanotechnology and from sponsored projects from government and industrial regimes. Products of engineered nanomaterials, systems, and devices are now everywhere. From routine domestic life to industrial experiences, these materials are relevantly used in many industries: medicine (e.g., special therapeutics), beauty (e.g., cosmetics), food (e.g., packaging films), and solar (e.g., biosensors), etc. So many commercial products, like wound-dressing ointments, laundry detergents, and antimicrobial surface paints, also contain various nanomaterials (particles and composites) in varying concentrations. They were introduced for defined applications to serve renewed purposes while improving quality of life. The emergence of nanotechnology is now responsible for the promotion and introduction of about three to four consumer products per week in modern markets [3].

For industrial use, a US-based National Nanotechnology Initiative [4] has reported several tons of nanomaterials (e.g., alumina) utilized for making precision polishing substances. A typical example of these materials is the abrasive particles in silicon wafers. The production and use of fullerenes and metal oxides could also be measured at 1500–2000 tons in Japan alone, not including single-walled and double-walled carbon nanotubes. The 2008 Summer Olympics had 803 regulated products containing nanoparticles and nanocomposites, according to the Project on Emerging Nanotechnologies [5]. Stable in their current states, these nanomaterials are difficult to degrade. Within the environments where they are dispersed, these nanomaterials continue to possess significant threats to human life [5].

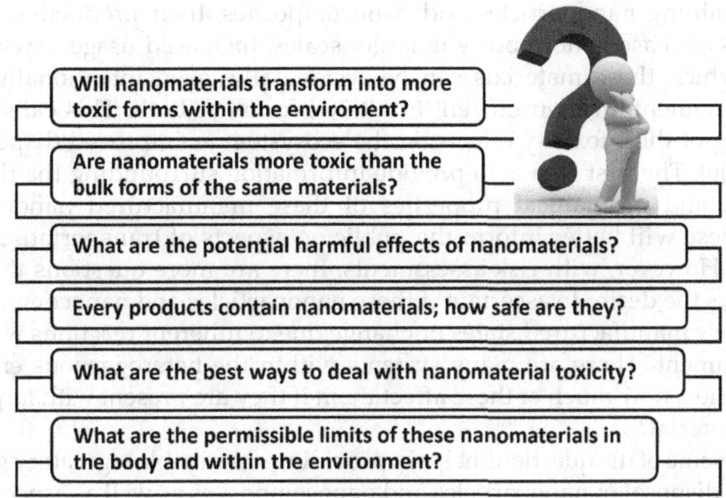

FIGURE 7.1
Common toxicity and risk assessment questions about nanoparticles and nanocomposites.

As their fields of applications and modernization increase, the risk of environmental exposure increases exponentially. So many daily questions have been raised over the years on the impacts of nanomaterials on our daily lives. These questions come as a caution to the revolutionizing of nanomaterials through mass production in processes of various kinds. One outstanding issue surrounding these materials is their potential toxicity to humans and their environment. Modern debates have addressed man's susceptibility to any adverse biological effects they might pose but much is expected from regulatory bodies towards usage control.

To understand their effects on lives is to probe their biodegradation processes within the cellular structures of these living systems. When considering the biochemistry of cells, it is pertinent to ask questions about cellular responses as the accumulation of these nanomaterials may lead to more than physiological changes, mutation, and gene alterations, if biological disruptions (catalyzed by this biodegradation mechanism) in organelle integrity persist. Figure 7.1 depicts common toxicity and risk assessment questions about nanoparticles and their nanocomposites. Normally, the approval of nanomaterials for everyday products should be accompanied by sufficient information about their nanotoxicology. Part of this involves the creation of unsafe backgrounds for latent undesirable properties and their biological impacts on the environment. Modern research suggests complete avoidance of outright toxic ones, whereas the quests for more green nanomaterial alternatives are also recommended. This chapter will address the potential toxicity impacts of emerging nanomaterials on life and the environment. It also covers discussions that center on recent trends in risk assessments on nanomaterials that are composites of two or more principal constituents.

7.2 Dispersion and Fate of Nanocomposites during Transport

It is widely accepted that nanotechnology is leading the greatest innovations in science and engineering in this century. With the increasing number of everyday commercial products containing nanoparticles and nanocomposites, their production and utilization have also increased enormously to larger scales. Increased usage now means more avenues by which these materials can be accidentally (even intentionally) dispersed into the environment, in their current or altered forms or both. This calls for a better understanding of their toxicity effects on the ecosystem as improved disposal systems are designed [6]. The first step is to provide information surrounding the thermophysical, chemical, and mechanical properties of these manufactured nanomaterials, so that maybe these will better inform the public on aspects of transportation within the environment. However, with risk assessments, there are more questions than answers. There is always the desire to ascertain if these nanoparticles and nanocomposites can be retained in their manufactured states or change due to inherent reactions with elements of the environment. There are many systems within the heterogeneous environments where they reside; how much of these affect them if they are present with larger particles of the same material?

Addressing some of the questions helps in providing sustainable regulatory guidelines in managing the disposal of nanoparticles and nanocomposites as well as aspects surrounding the protection of ecosystems from indiscriminate disposal in the quest to enhance the advantages of nanotechnology for everyone. Nanomaterials can be introduced into the environment by atmospheric emissions and even through waste streams from production plants. Industrial accidents associated with wear and erosion may also lead to disposal of nanomaterial into the environments rather than from general use. From surface coatings, wearable fabrics, electronic products, and self-care products, these nanoparticles, and nanocomposites enter the environment in varying proportions. Here, whether solid or fluidic states, depending on the type, nanoparticles and nanocomposites are capable of equal impacts due to their infinitesimally minute sizes. Heavier ones and even smaller nanomatters aggregate and sink into the seas while others are capable of being transported around as they float on surfaces. Others are ultimately emitted from fumes and effluents and are deposited on land and solid surfaces. There, they contaminate soils and surface structures, alter their qualities, and are later transported *via* surface drainages, rainwater runoffs, and spills into bodies of water and groundwater aquifers. The cycle continues until they get back to man through edible food and fruits, drinking water, and even dermal adsorption [7]. The environment is more impacted by direct particulate discharges in waste than by accidents though the populace still needs to be efficiently educated with regard to all pollution sources.

In the air, manufactured nanoparticles and nanocomposites can be directly inhaled due to their nanoscale sizes (others are smaller than micrometers). It is also worthy of note to mention that agglomerated particles may possess properties uncharacteristic of individual particles, and this enhances their persistency during indoor and outdoor inhalation [8]. The high surface area and reduced diameters of these particles make their deposition for adsorption within respiratory organs very possible. They may diffuse within finely divided blood vessels and interact at high reactivity rates after uptake, depending on their chemistries and surface morphologies. It is also possible that most of them can degrade into products with an increased tendency for biological toxicity within the lungs. This may even be severe in patients with respiratory diseases (e.g., asthma). Nanoparticles in the lungs also increase the tendencies toward chronic inflammatory diseases related to respiratory obstruction [9]. The probability of developing these diseases decreases significantly in healthy subjects since they already have improved clearance ability; only one-quarter of particles ranging in size between 50 and 100 nm are eliminated after the first day of inhalation [9]. On alveolar surfaces, aspirated nanoparticles initiate chemotactic signals and macrophage responses, depending on size-specific factors. These also affect the tendency for fast and firm adhesion within capillary ends of alveolar membranes where gas exchanges occur.

The skin is the largest organ in the body and participates in the adsorption of a significant percentage of nanoparticles and nanocomposites into the body through unintended means. During the manufacture of engineered nanomaterials, workers may be exposed to substances with anthropomorphized characters; this may count as an unintentional source. However, dermal exposure could also be intentional when the skin is in contact with moisturizers, detergents, and lubricants carrying nanoparticles and nanocomposites. The usage of water-repellent titania in sunscreen is a typical example. The hospitals readily use some pairs of socks and wound-dressing ointments containing antibacterial AgNP agents. It has been opined in some quarters that dermal adsorption and penetration of nanomaterials have greater effects than when adsorbed through other organs. This is partly linked with how the body metabolizes these nanoparticles and nanocomposites, especially those that have been photoactivated upon accumulation through the skin. Changes in the forms of nanomaterials also affect their modes of action, especially if their hosts have also undergone significant changes during their conditions of use (e.g., stress). For instance, some fabrics bearing nanoparticles and nanocomposites that are regularly laundered have the probability of nanomaterial release since their nanofibers are repeatedly overstretched. This physical process impacts like the combined action of mechanical scrubbing (with detergent) and heat (during drying and iron-pressing) during the few minutes of the laundry activity [9, 10].

7.3 Suitability of Parameters When Studying Toxicity

The key to developmental prospects with nanoparticles and nanocomposites should involve the management of risks associated with their manufacture, usage, and disposal. To protect factory workers, product end-users, the general public, and the ecosystem, educational and regulatory information should be made available to all and sundry, especially the risks involved with the use of all products containing these nanoparticles and nanocomposites as presented in Table 7.1. It is generally accepted that the toxicity of nanomaterials depends on their chemistry, size, and morphology. It is also pertinent to note that every parameter is suitable when studying the toxicity of nanomaterials. Most authors have utilized a variety of strategies (from varying cell lines, culture conditions, and incubation times). However, there is always the question of the physiological relevancy of certain cytotoxicity observations across various reported studies (knowing that the nature of nanomaterials utilized during toxicity tests may vary significantly). The living conditions of various test subjects (cells, plants, animals, etc.) for nanotoxicological trials are not the same across the board. 1D nanocomposites are indeed present in a wide range of products that can be divided into classes, including matrix nanocomposite materials that are ceramic, metallic, stimulus-responsive nanoparticles and nanocomposites. The discussions following this subheading are based on the toxicity of nanoparticle components of nanocomposites. However, before that, let's look at some modern paradigms of toxicity.

TABLE 7.1

Some Products with Nanomaterials (Both Nanoparticles and Nanocomposites) and Their Fields of Application [10]

Fields of Application	Applications
Automotive	Surface coatings, tires, sensors, and windshields
Construction	Insulation; flame retardants; surface coatings; mortar
Electronics	Displays; data memory; laser diodes; fiber optics; optical switches; filters; conductive coatings; antistatic coatings; transistors
Engineering	Protective coatings for tools, machines; lubricant-free bearings
Food and Drink	Packaging; storage life sensors; additives; juice clarifiers
Medicine	Drug delivery systems; contrast medium; rapid testing systems; implants; antimicrobial agents; in-body diagnostic systems
Textiles	Coated fabrics (anti-wrinkle, stain resistant, temperature controlled)
Chemical	Fillers for paints; composite materials; impregnation of papers; adhesives
Cosmetics	Sunscreen; lipsticks; skin creams; toothpaste
Energy	Lighting; fuel cells; solar cells; batteries; capacitors
Household	Ceramic coatings for irons; odor removers; cleaners for glass, ceramics, metals
Sports	Ski wax; tennis rackets; golf clubs; tennis balls; coated glasses, goggles
Military	Neutralization materials for chemical weapons; bullet-proof protection

Source: Adapted with permission from Reference [10]. Copyright The Authors, some rights reserved; exclusive licensee [MDPI]. Distributed under a Creative Commons Attribution License 3.0 (CC BY) https://creativecommons.org/licenses/by/3.0/ (CC BY-NC-SA 3.0).

7.4 Modern Models and Paradigms of Toxicity

With an increase in the overall awareness of the health implications of nanotechnology and its products, more regulations are needed to justify the inclusion of specific products in the marketplace. Table 7.1 shows commercial products with single-nature nanoparticles and multicomponent nanocomposites as well as their fields of usage [10]. Indeed, a lot of products with nanoparticles and nanocomposites on the shelves are still unregulated. From water-proof fabrics, sunscreens, water filters, inks, tennis racquets/balls to nanosilver ointments, more harm has been done by not placing the health and safety of the public above profits. These cases raise more alarms on public safety than highlight the advantages of nanotechnology [11]. It is only fair that the enhancement of human knowledge through nanotechnology is accompanied by the provision of hazard-based data for each product constituent following careful regulatory research. From use to disposal of products with nanoparticles and nanocomposites, the law should enforce the role of science directly in the exposure, unintentional or intentional, of the public to harmful nanomaterials. The opportunities of nanotechnology are indeed enormous when we see them through the regulatory eyes of safety. So, let's briefly look at some paradigms of toxicity relating to nanoparticles and nanocomposites.

7.4.1 The Size-Shape Roles of Biological Membranes

By the old definition (and surprisingly recent misguided information), nanoparticles are characterized by their sizes being <100 nm. This definition is for practical purposes, they say, and it negates the fact that some nonstructured materials (e.g., tubes and fibers) may be measured in microscale in both cross-section and length. The thin and fibrous, flexible asbestos fibers may match up to this definition but not the other causative agents of fibrogenic and carcinogenicity. Otherwise known as the *fiber paradigm*, it applies to an illustrative example about inhaled fibers like those of asbestos. The fact that the chemical make-up of the fibers is not even considered, except when it contributes to their solubility is a problem. It appears that specialized cells of macrophages can take in as many fibers with no likelihood of harm. However, there is evidence-based research that shows that this paradigm also has relevance to erionite (a fibrous mineral). When exposed to erionite, some Turks suffered from a disease similar to that of asbestosis [11]. So, what happened here? When the phagocytic cells were exposed to these fibers, they spread in an attempt to engulf them and got injured in the process. This led to excessive releases of cytokines and oxidants while fibrosis and carcinogenesis were initiated. This means that humans may not be inhaling a sufficient dosage of certain nanofibers to cause disease; it is not a complete lack of nanomaterial toxicity.

7.4.2 Evidence of Numerous Toxicity Mechanisms

An increase in leukocyte and C-reactive protein counts is biochemically connected with exposure to air nanopollutants. Mills et al. [12] conducted epidemiological studies to this effect, albeit from a small population. Another study showed that these particles cause mild inflammations and increased white cell counts in alveolar fluid in response. Human trials have revealed the production of vasoconstrictor substances (e.g., epinephrine and norepinephrine) when exposed to exhaust gases from diesel engines, although without inflammatory reactions. This is suggestive of other possible toxicity mechanisms relating to air pollutant particulates. According to Seaton et al. [11], "(i) these are indirect effects of pulmonary inflammation and oxidative stress on atherothrombosis; and (ii) translocation of particles to blood vessels and a direct effect on their lining endothelial cells are pressing issues". There may also be a connection between exposure risk factors and heart attacks from atheromatous plaque on the artery walls in animal trials. The literature is filled with clinical examples of this nature. Apart from the aforementioned paradigms, there is also an *ultrafine hypothesis* that assumes that there is no significant medical correlation between toxicity of particulate air resides and their sizes (in nanometers).

7.5 Toxicity of 1D Nanocomposites and Their Components

So, what are nanocomposites? These are multiphase, solid materials with nanosized particles firmly or loosely incorporated within their bulk matrices to improve their gross physicomechanical and chemical properties. The properties altered by the association of these particles include, but are not limited to, optics, electrical and thermal conductivities, fluid permeability, and surface morphology. These improvements in properties are observed due to the alteration in the binding modes of these nanoparticles bonded within the bulk material. New nanostructured composites are still being produced for emerging applications in software engineering, battery technology, motor parts, etc. Their applicability may depend on their chemistry and size effects [13]. A particle size of less than 5 nm is required for nanocomposites with catalytic activities and less than 20 and 50 nm for those slated for magnetic and refractive index changes, respectively. To attain restrictive dislocation movements within bulk matrices for mechanical strengthening, particles of sizes less than 100 nm may be required. Structurally, nanocomposite materials have more than one phase with their nanocomponents in repeated distances apart. During their production, most of them are designed and built from nano-dimensional block phases with dissimilar chemistries that provide improved physical properties. Due to the high surface area of added nanoparticles, the gross reinforced properties of the final composite matrices are distinct in the macroscale. For instance, electrical and thermal conductivity are some of the properties conferred by carbon nanotubes (CNTs) during processing, compared with other nanoparticles that may only have to alter mechanical properties. In this case, during nano reinforcements, the confirmations of asymmetric nanoconstituents alter structures of the bulk matrices, in turn providing a mismatch in physical properties at nanointerfaces, including density per unit volume.

7.5.1 Ceramic and Metal Matrix Nanocomposites

When ceramic fibers (even carbon and carbon fibers) are incorporated within matrices of similar or dissimilar chemistry, they are classified as ceramic matrix composites (CMCs). In most CMCs, the primary components occupying most matrix volume could be nitrides, borides, or silicides incorporated within metals. The final composite product now possesses unique properties (optics, electricity, and magnetism) for a wide range of applications (e.g., corrosion resistance). From design to production, much has been done in making sure that most ceramic-metal nanocomposites come with constituent components that do not react. This is to keep the composites from losing their metallic character. With the high temperature required in their production, it can still prove prohibitive. One of the only ways to avoid this constraint is to ensure that the chosen metal and ceramic phases have completely immiscible phases, such as the Cu-O-Ti nanocomposites formed from immiscible TiO_2 and Cu constituent phases over large areas in the Gibbs' triangle [14]. Ceramic-matrix nanocomposites have been utilized on solid, layered, thin films deposited on underlying substrates.

In the same vein, metal matrix nanocomposites are also classified as reinforced materials. A typical example are the emerging new CNT-metal matrix composites. They are unique for their high tensile strength and electrical conductivity due to the presence of CNT within their matrices, and with strong interfacial adhesion between their constituent components. Boron and carbon nitrides have also been utilized to reinforce this class of composites in ongoing research [15]. Others include polymer nanocomposite matrices reinforced with tungsten disulfide nanotubes and silica-based materials reinforced with metal oxides. Like other nanocomposites, their constituent components collectively impact human health as well as that of the environment. In the next subsection, the toxicity and risk assessments of a few nanophase components of nanocomposites, with metallic ceramic properties, will be reviewed.

7.5.1.1 Toxicity Impacts of Metallic Nanoparticles from Nanocomposites

Metallic nanoparticles are a class of engineered metal matrix nanoparticles with significant toxic effects on our environments. As with other nanomaterials, there are causes for concern due to negative impacts in the light of environmental toxicity. Consider, for instance, electroplated Au (bulk Au), which is safe on coated wristwatches. However, the uptake of nano-Au is significantly toxic to cells [16]. In living systems, the migration and kinetics of metallic nanoparticles (NPs) are dependent on size-specific factors like physical dimensions. Relatively small nanoparticles (10–20 nm) are more easily taken up by living cells than larger ones (e.g., 100–200 nm). In a recent study by Connor et al. [17], authors studied the changes in human leukemia cells when exposed to nano-Au particles of varying morphologies and sizes. The authors observed very limited effects of nano-Au particles with spherical shapes as well as negligible effects when they were coated with surface modifiers. There are also similar studies featuring effects on immortal human keratinocytes. A similar study has been reported for an immune system cell model. However, there were no observed cytotoxic effects within the concentration range under study due to the reduced presence of reactive oxygen species (ROS) in cells [18].

Another group of researchers [19] who investigated the size effects of nano-Au on cellular uptake and toxicity found significant interactions within 24 h of exposure. In this study, nano-Ag of reduced size (15 nm) at 10–75 μg/mL showed greater oxidative stress levels compared with the same element with larger particles or doses. There was also a three-fold increase in ROS levels within cells treated with 50 μg/mL Ag particles with a 15-nm average size. Benn et al. [20] also reported an interesting study featuring the bulk release of nano-Ag from a textured fabric into laundry effluents. This study was designed by authors to mimic contaminated factory wastewater in manufacturing plants producing nanomaterials. The authors observed more release of Au in its ionic and colloidal forms. However, the release rates were not the same for all fabrics. In the end, the authors utilized their experimental results to designing a model that controlled and predicted nano-Ag leaching into wastewater lines. With the antibacterial ability of AgNP activity for a wide range of bacterial pathogens, there is also the call to monitor and reduce their toxicity against human cell lines, especially during *in vivo* studies.

In a study by Li et al. [21], the authors investigated the toxicity effects of Au/Ag nanoalloys on the growth process of cultured *Escherichia coli* cells. The morphology and elemental distribution of the spherical Au/Ag nanoalloys utilized in this study are shown in Figure 7.2a. Here, the authors observed that significantly lower toxicity could be achieved

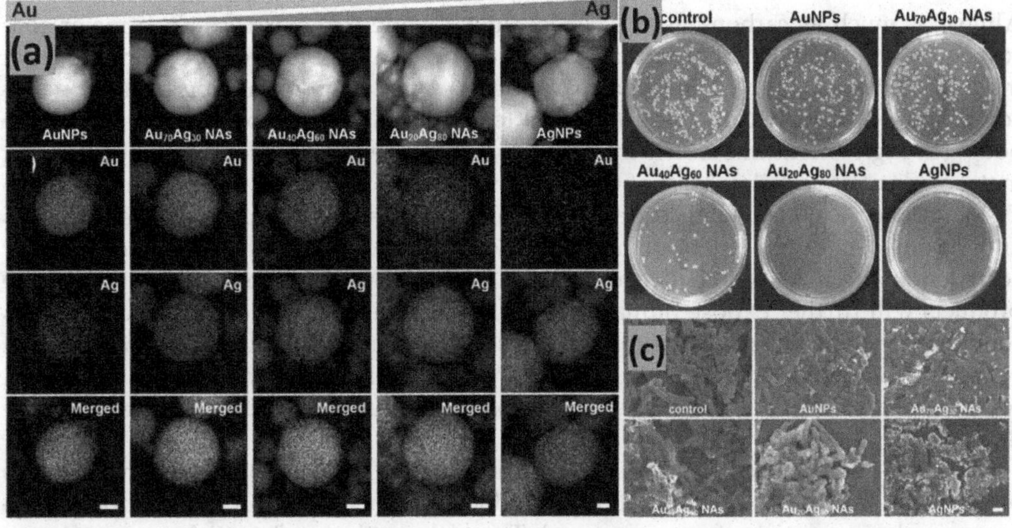

FIGURE 7.2
(a) Morphology and elemental distribution of spherical Au/Ag nanoalloys; images on the first and fifth columns are those of Au and AgNPs, respectively (all in 10-nm scale bars). (b) Pictorial representations and (c) SEM micrographs showing toxicity effects of nanoalloys on *E. coli* growth (Control: Sterile water; 1-μm scale bar). Adapted with permission from Reference [21] Copyright (2021), Elsevier.

by tuning the concentration of Au within the nanoalloy. The recorded significant bactericidal effect with no cytotoxicity for nanoalloys with 40–60% Au-Ag. Figure 7.2b further shows representations depicting toxicity effects of nanoalloys on *E. coli* growth at equal concentrations of 40–60%. These bacterial cells were incubated within a tryptone-soy broth agar for 37 °C for 12-hour nights. As revealed, the colony-forming units (CFUs) of bacteria in response to pure AuNPs- or Au70Ag30 nanoalloys were not different from those of the control. This effect was attributed to the inability of these nanoparticles to alter bacterial growth. Higher CFUs were observed at 60% Ag with no visible colonies at 85% Ag. Scanning electron microscopic (SEM) images further elucidated the antibacterial mechanism. The authors reported reduced negative surface charge in response to an increase in Ag concentration within nanoalloys and this must have led to lower repulsive forces against the negatively charged bacterial membrane. Hence, greater adherence of the nanoalloys to the surfaces of bacterial cells was observed in Figure 7.2c. A reduced number of particles could be observed for Au70Ag30 while more macroaggregates could be observed in Au40Ag60 and Au20Ag80 nanoalloys.

In another study, Ibrahim et al. [22] also investigated the toxicity effects of polypyrrole-gold nanocomposites (PPy-Au NCs) on Ehrlich Ascites carcinoma cells (Figure 7.3a–f). The toxicity tests were conducted in the dark. A greater proportion of viable cells (up to 100%) remained with 10 and 50% PPy NPs and PPy-Au nanocomposites whereas 100% shrinkage was observed in response to 90% PPy NPs and PPy-Au nanocomposites (see winding membranes in c and f). Concerning nanocomposites, other authors [23] have reported toxicity assays of Ni nanotubes and Ni@Au hybrid nanotubes (even after annealing) against Molt-4 cells. Figure 7.3g shows the variation in cell viability for a range of concentrations of nanocomposites within the study. In all, a reduced degree of cytotoxicity was observed for Ni@Au hybrid nanotubes even at lower concentrations.

7.5.1.2 Toxicity Impacts of Metal Oxide Nanoparticles from Nanocomposites

Almost as much as carbon-based particles, metal oxide nanoparticles are a major constituent of some nanocomposite materials in many commercial products, ranging from modern cosmetics to emerging pharmaceuticals. Apart from inhalation of these nanocomposites, exposure to the skin happens to be the most probable entryway into the body. Through lotions, the skin is intentionally introduced to nanocomposites containing metal oxide nanoparticles. Most of these creams contain varying concentrations of nano-TiO_2 and nano-ZnO as metal oxide. These nanoparticles are also present in sunscreens and in composites with other polymer nanomaterials added to confer resistance to water adhesion and to increase repellence to stains and dirt. An increase in demand for these products will further expose humans to the inherent risks of nanocomposite materials containing metal oxide nanoparticles. In this section, you will be introduced to published work from some researchers in this field. The focus will be on the toxicity levels and reported uptake capacities for named metal oxide nanoparticles within the nanocomposite system.

Earlier on, some authors [24] investigated the effects of TiO_2 nanoparticles in lung cells of lab mice. In this study, the authors introduced defined concentrations of only 2- to 5-nm-sized nanoparticles from an aerosol. The authors observed particle aggregation at significant degrees as well as recording inflammatory responses for models examined before 14 days from the last exposure and after a 3-week recovery period. In another study [25], the effects of TiO_2 nanoparticles on fibroblast cells of mice were also examined after a few hours of exposure. Here, the authors recorded cytotoxicity levels at higher TiO_2

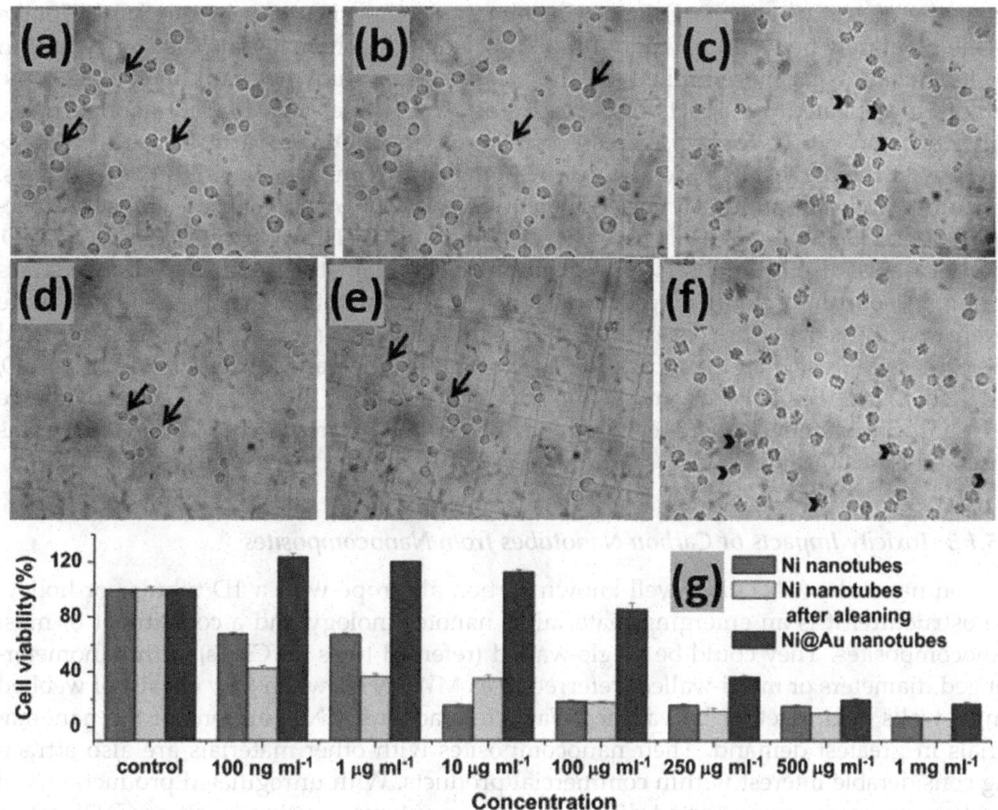

FIGURE 7.3
Pictorial representation of the toxicity effects of polypyrrole (PPy) nanoparticles (NPs) (first row) and PPy-Au nanocomposites (second row) on carcinoma cell line; cells were incubated with (a and d) 10% PPy NPs and PPy-Au nanocomposites, respectively, (b and e) 50% PPy NPs and PPy-Au nanocomposites, respectively, (c and f) 90% PPy NPs and PPy-Au nanocomposites, respectively (100% viable cells observed in a/d and b/e and 100% shrinkage in c/f). Adapted with permission from Reference [22] Copyright (2022), Elsevier. (g) cell viability per concentration of Ni and Ni@Au hybrid nanotubes. Adapted with permission from Reference [23] Copyright (2013), Elsevier.

concentrations. Between aggregated and homogeneous particles, this study demonstrated that inherent toxicity increased oxidative stress responses above 60 µg/mL TiO_2 within the nanocomposite. This was also accompanied by events leading to the release of lactate dehydrogenase. A similar observation has also been recorded for human non-tumorigenic lung epithelial cell lines. There was significant toxicity in this cell line accompanied by ROS-induced programmed cell death [26]. In this case, cytotoxicity of TiO_2 nanoparticles was linked with particulate penetration within cellular membranes *via* direct interaction with the cells. This is also accompanied by increased expression of the IL-8 gene responsible for inflammation.

Some metal oxide nanoparticles have also been reported to have adverse effects on plant growth. Al_2O_3 nanoparticles have been reported to significantly reduce root elongation in crops (e.g., carrots and cucumber). However, exposure to this nanomaterial did not affect the growth of some legumes (e.g., kidney beans) [27]. Exposure to TiO_2 nanoparticles improved the growth of spinach by supporting photosynthesis and nitrogen metabolism. However, the pathway for phytotoxicity is not widely reported, so very little is known about the effects of nanoparticles and their nanocomposites on phytotoxicity. Uptake by the plant may be well known, but the knowledge of nanomaterial persistency within the food chain is far more limited [28]. In a comparative study by Franklin et al. [29], the authors investigated the effects of exposure of a freshwater alga to ZnO nanoparticles, bulk ZnO, and $ZnCl_2$. Toxicity increased in the order: ZnO nanoparticles > bulk ZnO > $ZnCl_2$, with the highest recorded IC_{50} value being approximately 60 μg Zn/L after 3 days. In another study, the authors reported acute toxicity of ZnO nanoparticles against two bacterial species (*Staphylococcus aureus* and *E. coli*) and human T lymphocytes [30]. TiO_2, ZnO, and CeO_2 have also been observed to exhibit toxicity against and induce oxidative stress in human non-tumorigenic lung epithelial cell lines. ZnO nanoparticles exhibited more severe toxicity as they were observed to induce inflammation and cell death. CeO_2 induced less oxidative stress than the others [30]. Some metal oxide nanoparticles (including CuO, TiO_2, ZnO, $CuZnFe_2O_4$, Fe_3O_4, and Fe_2O_3) have also exhibited DNA damage and oxidative stress [31]. Similar effects were recorded for ZnO and TiO_2 particles, but lower toxicity was observed for Fe_3O_4 and Fe_2O_3.

7.5.1.3 Toxicity Impacts of Carbon Nanotubes from Nanocomposites

Carbon nanotube (CNT) is a well-known carbon allotrope with a 1D tubular or hollow nanostructure. It is an emerging material in nanotechnology and a constituent of most nanocomposites. They could be single-walled (referred to as SWCNTs) with nanometer-ranged diameters or multi-walled (referred to as MWCNTs), when they consist of webbed single walls held together by van der Waals interactions. CNTs are one of the nanomaterials in greatest demand. Their nanocomposites with other materials are also attracting considerable interest within commercial products. With unregulated production and careless use comes unrestricted disposal into ecosystems, causing harm as CNTs act as potentially risky toxins. For MWCNTs, their production on an industrial scale is growing faster than that of SWCNTs, and hence public exposure to MWCNTs will increase soon. Environmental uptake of CNTs has been the subject of many investigations in recent times while the quest to probe their potential hazards with related biological systems is also growing. CNT-based composites are known to induce inflammatory responses, chemical signaling cascades, cell death through apoptosis in T lymphocytes, and the induction of oxidative stress. MWCNTs are capable of initiating gene upregulation involved in response to stress in dermal fibroblasts [32].

This group of authors [32] demonstrated that C-based nanocomposites are cytotoxic at different capacities. Even when results from MWCNTs were indistinct, there were increased toxicity levels for nanoparticles and nanofibers in this study. Other authors [33] recorded no toxic effects on cells of the immune system (lymphocytes and neutrophils) were exposed to modified CNTs; this must have been due to the contribution of an added surface-modifier, polyethylene glycol (PEG). In this study, CNTs were modified with PEG, making it more soluble in water. Another set of researchers [34] studied the uptake of carbonaceous ultrafine particles within the human body. The lungs were the uptake organs in this study. The authors observed that adsorption was related to the size of the particles, with ultrafine particles being adsorbed faster than larger particles, migrating across the blood-brain barrier toward the nervous system. There was also more impact on the nervous system. The authors had also expected the release of clotting agents in response to exposure to CNTs, a phenomenon which was not observed. More often than not, the negative effects of the inhalation of asbestos fibers have been said to be synonymous with the effects of CNT nanocomposites [35]. This assumption could be due to the similarity of

CNTs with respect to their fibrous nature or dispersion patterns. However, more data are needed to support these claims for attributing asbestosis and lung cancer to nanofibers of both asbestos and CNTs.

A group of researchers [36] investigated the adsorption and health implications of SWCNTs on mice. The lungs were targeted as the primary impact organ in this study. The authors observed more SWCNT aggregates on the linings of the lungs, leading to pulmonary inflammation. In another study [35], *in vitro* cytotoxicity tests for some carbon-based materials (SWCNTs, MWCNTs, quartz, and C60) were investigated. The authors in this study observed high toxicity (up to 35%) at 11.30 $\mu g/cm^2$ SWCNTs. Cytotoxicity increased in the order: SWCNTs > MWCNTs > quartz > C60. Response in terms of DNA damage was observed in a study [37] involving mouse embryonic stem cells when exposed MWNTs for 2 h. Authors reported that these carbon-based materials could have accumulated within the cells, in turn, inducing programmed cell death while also leading to the release of the protein responsible for activation of the tumor suppressor gene. They also observed increased expression of protein for oxidative stress-induced DNA demethylation (8-oxoguanine-DNA glycosylase 1) after introducing defined levels of MWCNTs.

In a comparative study between unmodified MWCNTs and N-doped MWCNT materials by Carrero-Sanchez et al. [38], the authors recorded significant toxicological effects on mouse tracheal cells. The authors recorded deaths caused by shortness of breath at higher doses of the pure MWCNTs, but this was not observed for the N-doped material, where only inflammation of the granuloma occurred at higher concentrations. The doping process must have contributed to resistance to dyspnea compared with pure MWCNTs. Intratracheal injection of CNTs has also been reported to cause acute pulmonary edema and genotoxicity in outbred Wistar IGS rats [39] as well as alveolar macrophage activation in male ICR mice [40]. In both these latter studies, there were recorded cases of acute inflammatory responses and also activated transcription factor expression within days of CNT administration.

7.5.2 Stimulus-Responsive Nanocomposites

There is growing interest in nanocomposites capable of stimulus responses (e.g., magnetic, electrical, or mechanical fields) for various applications. With magnetism, successes have been recorded with several nanocomposites that are useful tools in applications involving responses to the magnetic nature of their constituent elements. These nanocomposites possess components that interact with one another between various interfaces, leading to these responses. In response to variation in magnetic field strengths, there are limitations to the affected areas within the composites where responses are felt. Most materials of this nature are incorporated with certain nanoparticles (of different types, sizes, and morphologies) which induce a response to magnetic fields. This alters the final nanostructures of most magnetic nanocomposites to core-shell, colloidal crystals, and even nanospheres due to the constituents of their bulk matrices [41]. These unique materials have extensive applications in catalysis and biomedicine. Most palladium-infused complexes within some nanocomposites alter the kinetics of certain reactions while some polymer matrices with magnetic nanorods aid in specialized protocols of therapeutic drug delivery. Some nanocomposites are also known for their heat-resistant properties. With careful choice of nanocomponents, polymer matrices can be made to adsorb and resist heat energies. The use of carbon quantum dots has been a success where resistance to high thermal conditions is an important benchmark, and they may also be toxic at specific concentrations; depending on the nanocomposites involved [42].

7.5.2.1 Toxicity Impacts of Carbon Quantum Dots from Nanocomposites

The constituent components of nanocomposites collectively impact human health as well as that of the environment. In this section, the toxicity and risk assessments of carbon quantum dots as a nanophase component of 1D nanocomposites with stimulus-responsive properties will be reviewed. Carbon quantum dots (CQDs) are small nanoparticles and nanocrystals of carbon with several thousands of atoms capable of characteristic quantum effects. Key among these unique quantum effects is lengthened fluorescence. CQDs also possess diverse properties based on their varying nanostructures. They have unique optical and electrical properties, while the carboxyl moieties on their surfaces make them soluble in water and highly biocompatible in living systems. When introduced into living tissues, CQDs can help in the detection or diagnosis purposes of cellular structures for biomedical imaging due to their unique fluorescence emissions as fluorophores [43].

In addition to biomedicine and drug delivery, CQDs are also widely utilized in catalysis, fingerprinting, and biological sensing. Since CQDs possess these distinct properties, this largely determines their toxicity and, in turn, results in differences in their inherent toxicity. Apart from the varying properties (size, charge, stability, etc.) for individual CQDs, the reason for reported differences in toxicity reports could be linked with factors related to the limited toxicology data and varying dosage responses during exposure trials [44]. A group of authors [44] investigated CQD exposure by leaching through skin cells in an earlier study, where the authors observed cytotoxicity and inflammatory potential of varying CQD concentrations within rat skins and in human epidermal keratinocytes (HEK). They also observed reduced porcine skin penetration for some CQDs within this study, but this was restricted only to the uppermost stratum corneum. The presence of skin defects (e.g., surface damage) permitted unrestricted exposure to CQDs and was the principal source of inflammatory responses.

Lovric et al. [45] studied the cellular accumulation of cadmium telluride (CdTe) QDs as a means of alleviating cellular toxicity due to QD exposure. CdTe QD concentration within cells was based on size-specific factors; these QDs were also detrimental to rat pheochromocytoma cells (PC12 and N9) at 10 μg/mL. Cytotoxicity was also directly assessed in terms of chromatin condensation and apoptosis. More of this activity was observed at $2r=2.2–5.2 \pm 0.1$ nm for both green- and red-emitting CQDs. CdTe QDs modified with antioxidants (e.g., *N*-acetylcysteine) reduced cytotoxicity. However, this was not found for Trolox-pretreated cells. Mercaptopropionic acid and cysteamine also had negative impacts on PC12 cells at the same concentration as CdTe QDs. Another group of authors [46] also studied the cytotoxicity effects of QD-capped mercaptoundecanoic acid and cysteamine after 12 h exposure. They observed significant toxicity against a murine T cell lymphoma cell line at 100 μg/mL concentration. The authors attributed this observation to the negative impact of the QD pretreatment agents, i.e., the QD capping materials. With cysteamine, the authors recorded reduced genotoxic activity at the same concentration and duration of exposure.

In another cytotoxicity study [47], the authors modified cadmium selenide zinc sulfide (CdSe/ZnS) QDs with mercaptoundecanoic acid. Some acid-coated CdSe/ZnS QDs were obtained and utilized in this study. The authors observed that CdSe/ZnS QDs were mostly toxic to the Hela immortal cell line and human hepatocytes at 100 μg/mL. This assessment was conducted using the MTT colorimetric assay for all cellular activities. With hepatocyte cells, another research group [48] demonstrated significant toxicity and they attributed this to surface parameters within the QDs. The authors mainly cited the cadmium ions released from degraded CdSe QDs as the principal reason for cytotoxicity. Gopee et al. [49] studied QD migration in skins of lab mice. They observed that for capped CdSe and CdS as well as PEG-coated QDs, the liver and kidney were among the secondary receiving organs of this exposure. They also observed that injected QDs migrated within the skin as far as the subcutis while also penetrating to lymph nodes *via* subcutaneous lymphatics.

In a study of the impact of QDs on HEKs (this time, cryopreserved neonatal HEKs were used) [50], the authors observed significant adverse cellular viability effects at 3 nM, with increased cytokine activity after 4 h. This was a cyclic tensile strain test also observed to increase cellular interactions between CQs and HEKs. The addition of strains led to cytokine activities as well as QD uptake, following added cyclic strains (4 h). This also impacted cell viability. In another study [51], the effect of UV radiation (UVR) was utilized to study the penetration of carboxylated QDs under the skin, using a murine model (SKH-1 mice). Results were collected after exposure durations of between 8 and 24 h. The authors observed greater penetration depths for all irradiated models over prolonged periods.

Chen et al. [52] investigated the toxicity toward Chinese rare minnow (*Gobiocypris rarus*) embryos after being exposed to InP/ZnS QDs. The authors observed a LC_{50} value of 1,678.007 nmol/L 72 h after fertilization. After exposure, these rare minnows exhibited reduced mobility and later died. There was also a reduced rate of hatching of cells that survived. A close observation of their morphology revealed prominent spinal curvature and significant bending of dorsal tails, including growths of cysts from QD-induced deformation (see Figure 7.4a). Whereas only normal, healthy features could be observed in the

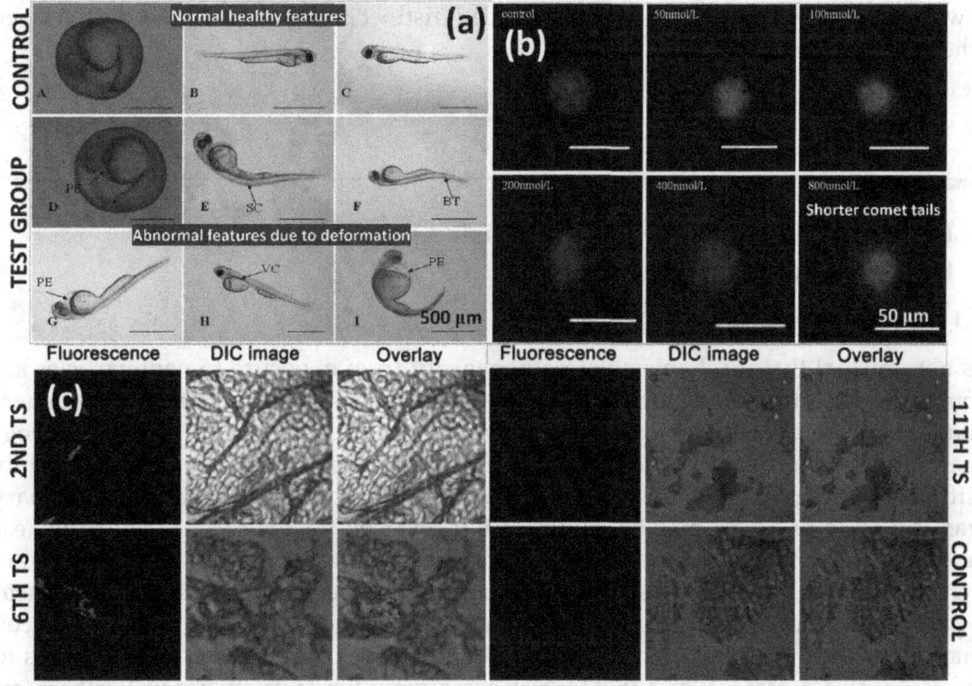

FIGURE 7.4
(a) Optical images showing a trend of morphological deformation on Chinese rare minnow embryos in response to concentrations of quantum dots (QDs): Normal features on the control group at 36 hpf (A), 72 hpf (B) and 96 hpf (C) whereas there are inherent abnormalities within the test group after exposure to InP/ZnS QDs: pericardial edema (PE, on D, G, I), spinal curvature (SC, on E), bent dorsal tails (BT, on E), and vitelline cyst (VC, on H). (b) Features from DNA damage showing the detrimental effects of InP/ZnS QDs on embryonic cells by comet assay (showing DNA strand breaks) at 96 hpf. Adapted with permission from Reference [52] Copyright (2018), Elsevier. (c) Confocal images of tape strips (TS) of skin layer after placing for 4 h following exposure to QDs: fluorescence, differential interference contrast (DIC), and fluorescence/DIC overlayed micrographs. Adapted with permission from Reference [53] Copyright (2010), Elsevier.

control population, there were inherent abnormalities within the test group upon exposure to InP/ZnS QDs. As observed, there was pericardial edema (on D, G, I in the figure), spinal curvature (on E), bent dorsal tails (on E), as well as a vitelline cyst (on H). Compared with InP/ZnS QDs, the effects of CuInS/ZnS QDs on the embryos were weaker. However, there was no change in the ROS-quenching enzyme superoxide dismutase (SOD) activity and levels of malondialdehyde (MDA) levels caused by ROS-induced membrane lipid peroxidation for both QDs. Genetic toxicity was not pronounced after exposure to < 800 nmol/L InP/ZnS QDs. Apart from changes in physical abnormalities, the authors also studied the damaging effects of QDs on DNA. As depicted in Figure 7.4b, organisms exposed to only 800 nmol/L InP/ZnS QDs developed shorter tails. The authors attributed this effect to DNA damage at 96 hours post-fertilization (hpf). There were no other observed changes in other treatments, according to the authors. There was no significant growth of larval tail and no significant changes ($p > 0.05$) after exposure to InP/ZnS QDs.

Jeong et al. [53] investigated the effects of QD penetration through human skin. The authors observed QD migration through the stratum corneum with significant under-skin accumulation observed after dermal exposure over an extended duration. Using the tape stripping method, the authors noted changes in the fluorescence intensities of QDs in the first 15 tapes as presented in Figure 7.4c. Skin furrows could be seen from the fluorescence intensities of the second tape strip. A fluorescence/differential interference contrast overlap was also imaged; the data presented revealed distinct penetration of QDs within layers of the skin.

7.6 Revisiting Theoretical Tools for Risk Assessments for Renewed Perspectives

7.6.1 Recognizing Risks

It is not unusual that mechanisms of pathogenicity from exposure to nanoparticles and nanocomposites are more complex than they seem. However, these effects also depend strongly on factors related to exposure routes, dosage, chemistry, and the host's biology. Regardless of the primary organ of exposure, the migration to other organs may exhibit a different toxicity mechanism. For instance, the conditions in the receiving (primary) organ (e.g., skin or lungs) may be oxidizing but reversed for the secondary organs (e.g., intestines). This change becomes the difference between reductive-led mechanisms and mechanisms based on oxidative stress. This makes the identification of hazards a factor for proper assessments of manufactured nanomaterials, no matter the target organ concerned. The fact that there are no clearly stated no-effect concentration (NEC) values for dose-response assessments as a tool for risk assessment limits its efficiency. Furthermore, this tool is also limited by the absence of appropriate dose-descriptors for so many engineered nanoparticles and nanocomposites [10, 54].

7.6.2 Tools for Risk Assessments

Recently, the Organization for Economic Cooperation and Development (OECD) provided the requisite devices and tools needed for risk assessments in their regulatory guidelines. This also includes those specified within the Registration, Evaluation, and Authorisation of Chemicals (REACH) documents from the European Union regulatory framework. In both cases, three concept models have been highlighted. (a) Regulatory toxicology tests cover standards for testing substances regarded as at-risk during toxicology. However, there are a few disadvantages to these tests, with the OECD still working on alternatives for both inhaled and non-inhaled particulate toxins in animal trials. (b) Another risk assessment device or tool is the quantitative structure-activity relationship (QSAR). This *in-silico* model seeks to provide insights into the properties of particulate toxins that affect biological activities and also simulates those of untested materials. It is a predictive tool for the remote study of the toxicity of chemicals and materials. (c) The final model is the physiologically based pharmacokinetic (PBPK) model, although there may not be a working PBPK model for body-bound nanoparticles and nanocomposites because of inherent inconsistencies with model transport equations needed to input the validity of nanomaterials [10, 54].

7.6.3 Assessment of Exposure to Toxins

This part simply addresses or approximates to the levels or dosages of nanoparticles and nanocomposites to which human populations and the environment may be exposed. The possible pathways of exposure are presented in Figure 7.5. Exposure assessment is an important element for studying the risks associated with nanomaterials, and this can be further divided into (i) occupational exposure assessment (OEA), (ii) environmental exposure assessment (EEA), and (iii) consumer exposure assessment (CEA); their discussions are beyond the scope of this chapter. In some cases, companies have revealed the high costs associated with monitoring nanomaterial-bound emissions within the workplace. This often surpasses inherent benefits except in cases where the nanoparticles and nanocomposites are assessed to be highly toxic. More attention could be focused on probing small masses during emission and surface areas within the background of ambient particles. However, there may be inherent issues with devices utilized in toxicology and

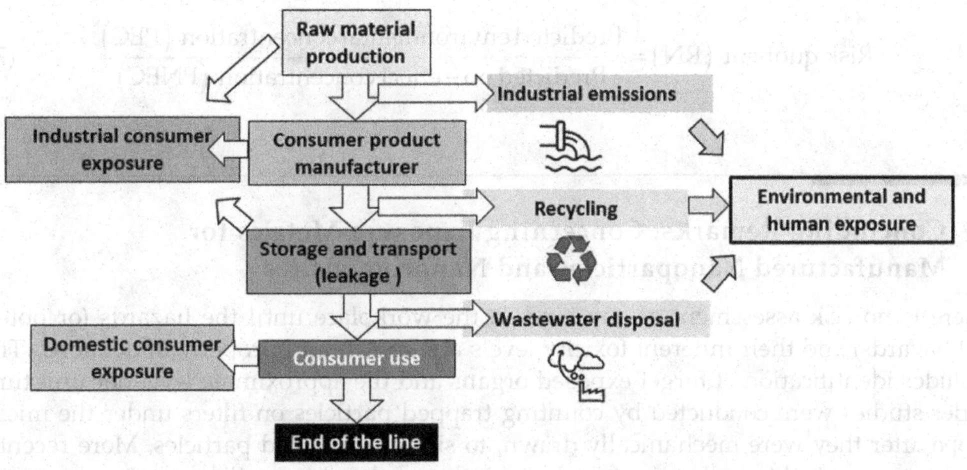

FIGURE 7.5
Pathways of exposure of toxic engineered nanoparticles and nanocomposites to humans and their environments. Adapted with permission from Reference [10]. Copyright The Authors, some rights reserved; exclusive licensee [MDPI]. Distributed under a Creative Commons Attribution License 3.0 (CC BY) https://creativecommons.org/licenses/by/3.0/ (CC BY-NC-SA 3.0).

environmental monitoring. This also includes when the prevalence of nanomaterials is estimated. This may hugely affect the exact dose measurements in media (e.g., air) of exposure of defined target organs. Exposure assessments also have their limitations as an element of risk assessment. This tool is hindered by drawbacks associated with measuring the exposure to nanomaterial within the environment, and it provides uncertainties on the effectiveness of the tool [10].

7.6.4 Risk Characterization

In addition to health risk strategies (e.g., assessment of hazard level and exposure control tools) for monitoring nanomaterial emissions, the quantitative estimation of risks may still be underdeveloped as a tool. This is the last procedural stage of risk assessment. It estimates inherent situations with adverse effects as well as their severities as they are projected to occur upon exposure of humans and their environment. Here, there is an actual risk quantification based on the data collected from the tools which were considered earlier. Risk characterization features a comparison between two distinct risk parameters, namely their quotient, otherwise defined as the risk quotient, as presented in Equation 7.1. This holds the key to understanding the need for risk reduction measures. Whereas predicted environment concentration (PEC) is the predicted level of toxin within a target environment, predicted no-effect concentation (PNEC) represents the level below which the toxic agent will not pose any risk to the environment. When RN < 1, there is no point initiating risk reduction measures; when RN values are greater than 1, the tendency toward risk is alarming and measures must be put in place to limit its effects, resulting in further testing [10, 54]. A typical example was reported by Müller and Nowack [55] for quantitative environmental risk assessment of some nanomaterials (e.g., Ag, TiO_2, and CNTs) in water. As an element of risk assessment, risk characterization is also challenged in some ways. Being that it uses the data trend from other tools of risk assessment, it also inherits their demerits [10]. Before a risk is identified, more should be done to further investigate the threat it poses to both humans and their environment.

$$\text{Risk quotient}\,(\text{RN}) = \frac{\text{Predicted environmental concentration}\,(\text{PEC})}{\text{Predicted no}-\text{effect concentration}\,(\text{PNEC})} \tag{7.1}$$

7.7 Concluding Remarks: Concerning Exposure Metrics for Manufactured Nanoparticles and Nanocomposites

There is no risk assessment of any kind in the workplace until the hazards (or potential hazards) and their inherent toxicity levels are known and properly understood. This includes identification of target exposed organs and the approximate level per unit time. Older studies were conducted by counting trapped particles on filters under the microscope after they were mechanically drawn, to simulate inhaled particles. More recently, parameters related to pollution were derived; those related to particle number concentration in air were further defined by counting. In most industries, researchers now derive a connection between inhaled masses and pneumoconiosis; this is a more accurate estimate than physical counting. This method has evolved over the years and now, workplace exposure limits (WELs) are the required standard for probing the exposure to nanoparticles and nanocomposites. Whether tracking the health effects of particles by mass concentration is better than by surface area is debatable. Particle counting is a worthy metric parameter since it directly links levels of particulate pollution with cardiovascular disease in elderly subjects. However, there is no literature or experimental evidence that one metric system should be preferred to the other. Maynard and Aitken [56] recommend that every metric related to particle counting, including mass concentration, and even counts per unit surface area should be collectively considered.

This has been an interesting discussion on the toxicity and risk assessments of 1D nanocomposites; a lot has been covered. In the end, progress in nanotechnology has also created single-nature nanoparticles and multi-component nanocomposites in routine household and industrial products. These products contain potentially hazardous substances with potential risks to humans and their environments. This calls for a better understanding of the implications of the presence of different factors to health effects, when a person is exposed to them, and when they are subsequently released into the environment. This chapter has highlighted current developments on paradigms of toxicity and theoretical tools to risk assessments of 1D nanocomposites. In addition to the listed examples, this chapter has also outlined recent advances and named constituents of 1D nanocomposites with adverse health implications identified in model cells, microorganisms, laboratory animals, humans, and plants.

Acknowledgments

The author acknowledges the University of Saskatchewan for providing the platform for this work.

References

1. P.C. Ray, H. Yu, P.P. Fu, Toxicity and environmental risks of nanomaterials: Challenges and future needs. *J. Environ. Sci. Health C Environ. Carcinog. Ecotoxicol. Rev.* 27(1) (2009) 1–35.
2. J.A. Cassell, DoD grants $3M to Study Nanoshells for Early Detection, Treatment of Breast Cancer. *NanoBiotech News* 1 (2003) 1–12.
3. http://www.luxresearchinc.com (Retrieved October 07 2021, 9 am).
4. http://www.nano.gov (Retrieved October 07 2021, 9.32 am).
5. http://www.wilsoncenter.org. 105. Report of the US chemical industry working group. 2003 (Retrieved October 07 2021, 9.30 am).
6. K.J. Lee, P.D. Nallathamby, L.M. Browning, C.J. Osgood, X. Xu, In vivo imaging of transport and biocompatibility of single silver nanoparticles in early development of zebrafish embryos. *ACS Nano* 1(2) (2007) 133–143.
7. L. Ding, J. Stilwell, T. Zhang, O. Elboundwarej, H. Jiang, J.P. Selegue, P.A. Cooke, J.W. Gray, F.F. Chen, Molecular characterization of the cytotoxic mechanism of multiwall carbon nanotubes and nano-onions on human skin fibroblast. *Nano Lett.* 5(12) (2005) 2448–2464.
8. C.W. Lam, J.T. James, R. McCluskey, R.L. Hunter, Pulmonary toxicity of single-wall carbon nanotubes in mice 7 and 90 days after intratracheal instillation. *Toxicol. Sci.* 77(1) (2004) 126–134.
9. W.G. Kreyling, M. Semmler-Behnke, W. Möller, Ultrafine particle-lung interactions: Does size matter? *J. Aerosol Med.* 19(1) (2006) 74–83.

10. D. Hristozov, I. Malsch, Hazards and risks of engineered nanoparticles for the environment and human health. *Sustainability* 1(4) (2009) 1161–1194.
11. A. Seaton, L. Tran, R. Aitken, K. Donaldson, Nanoparticles, human health hazard and regulation. *J. R. Soc. Interface* 7 (2010) S119–S129.
12. N.L. Mills, K. Donaldson, P.W. Hadoke, N.A. Boon, W. MacNee, F.R. Cassee, T. Sandstrom, A. Blomberg, D.E. Newby, Adverse cardiovascular effects of air pollution. *Nat. Clin. Pract. Cardiovasc. Med.* 6(1) (2009) 36–44.
13. O. Kamigaito, What can be improved by nanometer composites? *J. Jpn. Soc. Powder Metall.* 38 (1991) 315–321.
14. G. Effenberg, F. Aldinger, P. Rogl, Ternary alloys—a comprehensive compendium of evaluated constitutional data and phase diagrams: Volumes 1 and 2, ternary silver alloys. Edited by G. Petzow and G. Effenberg. Principal reviewer: H. L. Lukas. *VCH Verlagsgesellschaft*. Germany/ New York: Weinheim, Fed. rep. /VCH Publishers, 1988, Vol. 1, pp. 612, Vol. 2, 624.
15. S.R. Bakshi, D. Lahiri, A. Argawal, Carbon nanotube reinforced metal matrix composites - A review. *Int. Mater. Rev.* 55(1) (2010) 1–10.
16. R. Shukla, V. Bansal, M. Chaudhary, A. Basu, R. Bhonde, M. Sastry, Biocompatibility of gold nanoparticles and their endocytotic fate inside the cellular compartment: A microscopic overview. *Langmuir* 21(23) (2005) 10644–10654.
17. E.E. Connor, J. Mwamuka, A. Gole, C.J. Murphy, M.D. Wyatt, Gold nanoparticles are taken up by human cells but do not cause acute cytotoxicity. *Small* 1(3) (2005) 325–327.
18. S. Wang, W. Lu, O. Tovmachenko, U.S. Rai, H. Yu, P.C. Ray, Challenge in understanding size and shape dependent toxicity of gold nanomaterials in human skin keratinocytes. *Chem. Phys. Lett.* 463(1–3) (2008) 145–152.
19. C. Carlson, S.M. Hussain, A.M. Schrand, L.K. Braydich-Stolle, K.L. Hess, R.L. Jones, J.J. Schlager, Unique cellular interaction of silver nanoparticles: Size-dependent generation of reactive oxygen species. *J. Phys. Chem. B* 112 (2008) 13608–13619.
20. T.M. Benn, P. Westerhoff, Nanoparticle silver released into water from commercially available sock fabrics. *Environ. Sci. Technol.* 42(11) (2008) 4133–4139.
21. Z. Lin, Y. Luo, P. Liu, Y. Li, J. Yue, L. Jiang, Atomic-engineering Au-Ag nanoalloys for screening antimicrobial agents with low toxicity towards mammalian cells. *Colloids Surf. B Biointerfaces* 204 (2021) 111831.
22. J.S. Ibrahim, N. Hanafi, T.A. El-Tayeb, M.A. Sliem, Polypyrrole-gold nanocomposites as a promising photothermal agent: Preparation, characterization and cytotoxicity study. *Spectrochim. Acta A Mol. Biomol. Spectrosc.* 264 (2022) 120221.
23. X. Li, M. Wang, Y. Ye, K. Wu, Boron-doping Ni@Au nanotubes: Facile synthesis, magnetic property, and in vitro cytotoxicity on Molt-4 cells. *Mater. Lett.* 108 (2013) 222–224.
24. V.H. Grassian, P.T. O'Shaughnessy, A. Adamcakova-Dodd, J.M. Pettibone, P.S. Thorne, Inhalation exposure study of titanium dioxide nanoparticles with a primary particle size of 2 to 5 nm. *Environ. Health Perspect.* 115(3) (2007) 397–402.
25. C.Y. Jin, B.S. Zhu, X.F. Wang, Q.H. Lu, Cytotoxicity of titanium dioxide nanoparticles in mouse fibroblast cells. *Chem. Res. Toxicol.* 21(9) (2008) 1871–1877.
26. E.J. Park, J. Yi, K.H. Chung, D.Y. Rye, J. Choi, K. Park, Oxidative stress and apoptosis induced by titanium dioxide nanoparticles in cultured BEAS-2B cells. *Toxicol. Lett.* 180(3) (2008) 222–229.
27. R. Doshi, W. Braida, C. Christodoulatos, M. Wazne, G. O'Connor, Nano-aluminum: Transport through sand columns and environmental effects on plants and soil communities. *Environ. Res.* 106(3) (2008) 296–303.
28. D. Lin, B. Xing, Root uptake and phytotoxicity of ZnO nanoparticles. *Environ. Sci. Technol.* 42(15) (2008) 5580–5586.
29. N.M. Franklin, N.J. Rogers, S.C. Apte, G.E. Batley, G.E. Gadd, P.S. Casey, Comparative toxicity of nanoparticulate ZnO, Bulk ZnO, and $ZnCl_2$ to a freshwater microalga (*Pseudokirchneriella subcapitata*): The importance of particle solubility. *Environ. Sci. Technol.* 41(24) (2007) 8484–8490.
30. T. Xia, M. Kovochich, M. Liong, L. Madler, B. Gilbert, H. Shi, J.I. Yeh, J.I. Zink, A.E. Nel, Comparison of the mechanism of toxicity of zinc oxide and cerium oxide nanoparticles based on dissolution and oxidative stress properties. *ACS Nano* 2(10) (2008) 2121–2134.

31. H.L. Karlsson, P. Cronholm, J. Gustafsson, L. Moller, Copper oxide nanoparticles are highly toxic: A comparison between metal oxide nanoparticles and carbon nanotubes. *Chem. Res. Toxicol.* 21(9) (2008) 1726–1732.
32. A. Magrez, S. Kasas, V. Salicio, N. Pasquier, J.W. Seo, M. Celio, S. Catsicas, B. Schwaller, L. Forro, Cellular toxicity of carbon-based nanomaterials. *Nano Lett.* 6(6) (2006) 1121–1125.
33. H. Dumortier, S. Lacotte, G. Pastorin, R. Marega, W. Wu, D. Bonifazi, J.P. Briand, M. Prato, S. Muller, A. Bianco, Functionalized carbon nanotubes are non-cytotoxic and preserve the functionality of primary immune cells. *Nano Lett.* 6(7) (2006) 1522–1528.
34. V.M. Silva, N. Corson, A. Elder, G. Oberdorster, The rat ear vein model for investigating in vivo thrombogenicity of ultrafine particles (UFP). *Toxicol. Sci.* 85(2) (2005) 983–991.
35. G. Jia, H. Wang, L. Yan, X. Wang, R. Pei, T. Yan, Y. Zhao, X. Guo, Cytotoxicity of carbon nanomaterials: Single-wall nanotube, multi-wall nanotube, and fullerene. *Environ. Sci. Technol.* 39(5) (2005) 1378–1383.
36. C.W. Lam, J.T. James, R. McCluskey, R.L. Hunter, Pulmonary toxicity of single-wall carbon nanotubes in mice 7 and 90 days after intractracheal instillation. *Toxicol. Sci.* 77(1) (2004) 126–134.
37. L. Zhu, D.W. Chang, L. Dai, Y. Hong, DNA damage induced by multiwalled carbon nanotubes in mouse embryonic stem cells. *Nano Lett.* 7(12) (2007) 3592–3597.
38. J.C. Carrero-Sanchez, A.L. Elias, R. Mancilla, G. Arrellin, H. Terrones, J.P. Laclette, M. Terrones, Biocompatibility and toxicological studies of carbon nanotubes doped with nitrogen. *Nano Lett.* 6(8) (2006) 1609–1616.
39. J. Muller, F. Huaux, A. Fonseca, J.B. Nagy, N. Moreau, M. Delos, E. Raymundo-Pinero, F. Beguin, M. Kirsch-Volders, I. Fenoglio, B. Fubini, D. Lison, Structural defects play a major role in the acute lung toxicity of multiwall carbon nanotubes: Toxicological aspects. *Chem. Res. Toxicol.* 21(9) (2008) 1698–1705.
40. C.C. Chou, H.Y. Hsiao, Q.S. Hong, C.H. Chen, Y.W. Peng, H.W. Chen, P.C. Yang, Single-walled carbon nanotubes can induce pulmonary injury in mouse model. *Nano Lett.* 8(2) (2008) 437–445.
41. S. Behrens, Preparation of functional magnetic nanocomposites and hybrid materials: Recent progress and future directions. *Nanoscale* 3(3) (2011) 877–892.
42. V. Rimal, S. Shishodia, P.K. Srivastava, Novel synthesis of high-thermal stability carbon dots and nanocomposites from oleic acid as an organic substrate. *Appl. Nanosci.* 10(2) (2020) 455–464.
43. X. Gao, Y. Cui, R.M. Levenson, L.W. Chung, S. Nie, In vivo cancer targeting and imaging with semiconductor quantum dots. *Nat. Biotechnol.* 22(8) (2004) 969–976.
44. L.W. Zhang, W.W. Yu, V.L. Colvin, N.A. Monteiro-Riviere, Biological interactions of quantum dot nanoparticles in skin and in human epidermal keratinocytes. *Toxicol. Appl. Pharmacol.* 228(2) (2008) 200–211.
45. J. Lovric, H.S. Bazzi, Y. Cuie, G.R.A. Fortin, F.M. Winnik, D. Maysinger, Differences in subcellular distribution and toxicity of green and red emitting CdTe quantum dots. *J. Mol. Med. (Berl)* 83(5) (2005) 377–385.
46. A. Hoshino, K. Hanaki, K. Suzuki, K. Yamamoto, Applications of T-lymphoma labeled with fluorescent quantum dots to cell tracing markers in mouse body. *Biochem. Biophys. Res. Commun.* 314(1) (2004) 46–53.
47. A. Shiohara, A. Hoshino, K. Hanaki, K. Suzuki, K. Yamamoto, On the cytotoxicity of quantum dots. *Microbiol. Immunol.* 48(9) (2004) 669–675.
48. A.M. Derfus, W.C.W. Chan, S.N. Bhatia, Probing the cytotoxicity of semiconductor quantum dots. *Nano Lett.* 4(1) (2004) 11–18.
49. N.V. Gopee, D.W. Roberts, P. Webb, C.R. Cozart, P.H. Siitonen, A.R. Warbritton, W.W. Yu, V.L. Colvin, N.J. Walker, P.C. Howard, Migration of intradermally injected quantum dots to sentinel organs in mice. *Toxicol. Sci.* 98(1) (2007) 249–257.
50. J.G. Rouse, C.M. Haslauer, E.G. Loboa, N.A. Monteiro-Riviere, Cyclic tensile strain increases interactions between human epidermal keratinocytes and quantum dot nanoparticles. *Toxicol. In Vitro* 22(2) (2008) 491–497.
51. L.J. Mortensen, G. Oberdorster, A.P. Pentland, L.A. DeLouise, In vivo skin penetration of quantum dot nanoparticles in the murine model: The effect of UVR. *Nano Lett.* 8(9) (2008) 2779–2787.

52. Y. Chen, Y. Yang, F. Ou, L. Liu, X.H. Liu, Z.J. Wang, L. Jin, InP/ZnS QDs exposure induces developmental toxicity in rare minnow (*Gobiocypris rarus*) embryos. *Environ. Toxicol. Pharmacol.* 60 (2018) 28–36.
53. S.H. Jeong, J.H. Kim, S.M. Yi, J.P. Lee, J.H. Kim, K.H. Sohn, K.L. Park, M.K. Kim, S.W. Son, Assessment of penetration of quantum dots through in vitro and in vivo human skin using the human skin equivalent model and the tape stripping method. *Biochem. Biophys. Res. Commun.* 394(3) (2010) 612–615.
54. E. Nielsen, G. Ostergaard, J. Larsen, *Toxicological Risk Assessment of Chemicals: A Practical Guide.* New York: Informa Healthcare, 2007, pp. 2–3.
55. N. Mueller, B. Nowack, Exposure modeling of engineered nanoparticles in the environment. *Environ. Sci. Technol.* 42(12) (2008) 4447–4453.
56. A.D. Maynard, R.J. Aitken, Assessing exposure to airborne nanomaterials: Current abilities and future requirements. *Nanotoxicology* 1(1) (2009) 26–41.

8

One-Dimensional Polymeric Materials for Advanced Energy Applications

A.A.P.R. Perera, B.T. Punchihewa, Felipe de Souza, and Ram K. Gupta

CONTENTS

8.1 Introduction

From the many forms of energy available, electricity is one of the necessities of modern life. Often, the chemical energy stored in chemical compounds, such as biomass, petroleum, natural gas, and coal, is conventionally used to generate electricity. Thus, world power generation is primarily dependent on these conventional energy sources. However, the availability of conventional energy sources is finite. With the rising need for electricity and gas production, the demand for conventional energy sources such as coal and petroleum continues to soar. Environmental pollution and global warming due to excessive use of fossil fuels are some of the concerns. Therefore, the development of cost-effective, environmentally friendly electrical energy generation and storage technologies has become a global effort. Sunlight is one of the most abundant and daily available energy sources

DOI: 10.1201/9781003223764-8

which can be used to generate renewable and eco-friendly electricity that can be readily used in homes and businesses, for instance. Based on that, in 1954, researchers at Bell Labs, in the United States, developed the first photovoltaic cell using semiconductor materials to convert solar energy into electrical energy [1, 2]. The past 30 years have seen tremendous progress in research to improve the efficiency of solar cells. Research and development work is continuing with the aim of reducing costs and increasing efficiency to make solar energy more competitive to offer an alternative to the excessive use of non-renewable sources. Fuel cells are another technology that can provide a reliable green energy alternative to polluting processes based on combustion or incineration, from diesel and petrol engines to coal power plants. Instead of burning conventional fuel to generate electricity, fuel cells rely on an electrochemical reaction that does not produce greenhouse gases. Therefore, leading car companies like Honda, Toyota, and Ford are in the process of developing fuel cells to power their new generation of cars [3, 4].

Renewable energy is intermittent; for example, the sun can produce energy only at certain times, so, as a result, it is important to address the problem of energy storage to ensure a supply of sustainable electricity. Therefore, equipment such as batteries, capacitors, and supercapacitors are used to store this generated energy. Still, these devices have their limitations. Batteries that contain rare elements can be expensive. Supercapacitors store significantly less energy per unit load than electrochemical batteries [5]. As a result, there are constant innovations to find cheaper, more efficient, and more sustainable alternatives to energy storage devices. 1D structured materials, such as nanowires, nanobelts, nanofibers, nanorods, and nanorings, have received much attention due to their unique properties compared with their bulk counterparts. In that sense, these materials are within the nanoscale in two dimensions as their length is larger than their width. Thus, energy transport is confined within two dimensions, indicating that efficient energy transport is possible along the one controllable direction [6, 7]. The chemical, electrical, thermal, and optical energy transport properties inherent in 1D nanostructures allow these materials to be the building blocks for future generations of electronics and optoelectronic devices [8–11].

8.2 Synthesis and Characterization of 1D Polymeric Materials

A large number of synthetic methods have been established for the preparation of 1D polymeric materials in the form of fibers, wires, rods, belts, tubes, spirals, and rings. In this section, electrospinning, interfacial polymerization, sol-gel, and chemical vapor deposition (CVD) methods, and various characterization techniques are discussed.

8.2.1 Electrospinning

This technique is used to create fibers that can be used in a wide range of energy harvesting and energy storage applications. Electrospinning is a simple, continuous way of producing ultrathin fibers through an applied voltage that functions as a driving force. The basic electrospinning set-up for the synthesis of nanofibers consists of a high-voltage power supply, a needle spinneret, and a grounded conductive collector, as shown in Figure 8.1. The polymer solution, or the melt, is used as the starting material, which is applied to a syringe. During electrospinning, a polymer solution droplet is electrified to generate a jet, followed by stretching and elongation to generate the nanofibers.

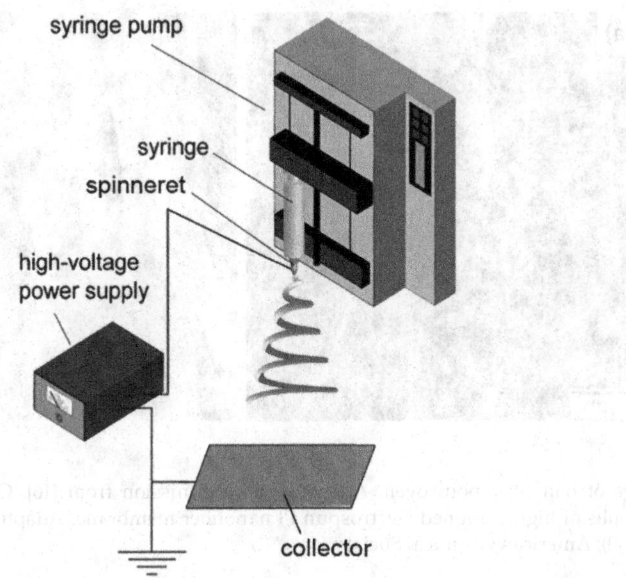

FIGURE 8.1
Schematic diagram of the electrospinning setup. Adapted with permission from [12]. Copyright (2017), American Chemical Society.

In general, the structure and morphology of the resulting nanofibers are determined by a set of parameters related to the polymeric solution/melt, processing parameters, and ambient conditions. Also, organic, inorganic, and composite polymers with attractive chemical, mechanical, and electrical properties, like high conductivity, high chemical resistance, and high tensile strength, have been spun into ultrafine fibers by the electrospinning process. Therefore, this technique has been extensively used to manufacture advanced materials for modified electrical devices like solar cells, fuel cells, batteries, and supercapacitors [13–15]. Polyaniline (PANI), polythiophene, polypyrrole (PPy), and polyimide (PI) are some of the conducting polymers that have been attempted to be used so far for electrospinning for energy applications.

Among them, PI nanofibers are widely used to prepare polymeric separators for batteries. In 2013, Miao et al. [16] fabricated PI nanofiber-based nonwoven polymeric separators for lithium-ion batteries (LiBs) *via* electrospinning (Figure 8.2a). They showed that PI nanofiber-based nonwovens were thermally stable at a high temperature of 500 °C, whereas the commercial Celgard membrane exhibited great shrinkage at 150 °C. In addition, this PI-based separator was evaluated to have greater capacity, lower resistance, and higher rate capability compared with the Celgard membrane separator [16, 17]. However, membranes made of randomly distributed nanofiber mats were not able to achieve the desired mechanical properties. Therefore, in 2013, another research team developed highly aligned PI-PANI composite nanofiber membranes *via* electrospinning to enhance the mechanical strength of the separator (Figure 8.2b) [18].

In 2021, Nanxi Dong and co-workers [19] fabricated a high-performance polymeric separator for LiBs by using the electrospun core-shell composite nanofiber of PI-ZrO_2, which exhibited a satisfactory battery capability (128.6 mAh/g@5C) that was able to surpass the capacitive performance of a membrane composed of a microporous polyolefin which displayed 95.2 mAh/g@5C. The shell of ZrO_2 played a major role in not only improving

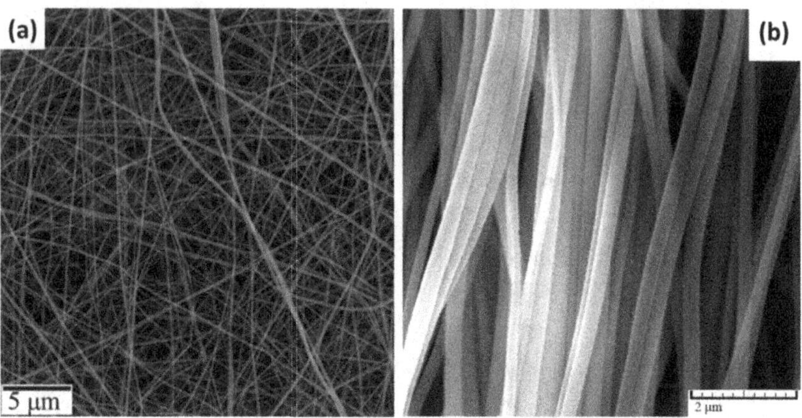

FIGURE 8.2
(a) SEM image of nanofiber nonwoven. Adapted with permission from [16]. Copyright (2013), Elsevier. (b) SEM micrographs of highly aligned electrospun PI nanofiber membrane. Adapted with permission from [18]. Copyright (2013), American Chemical Society.

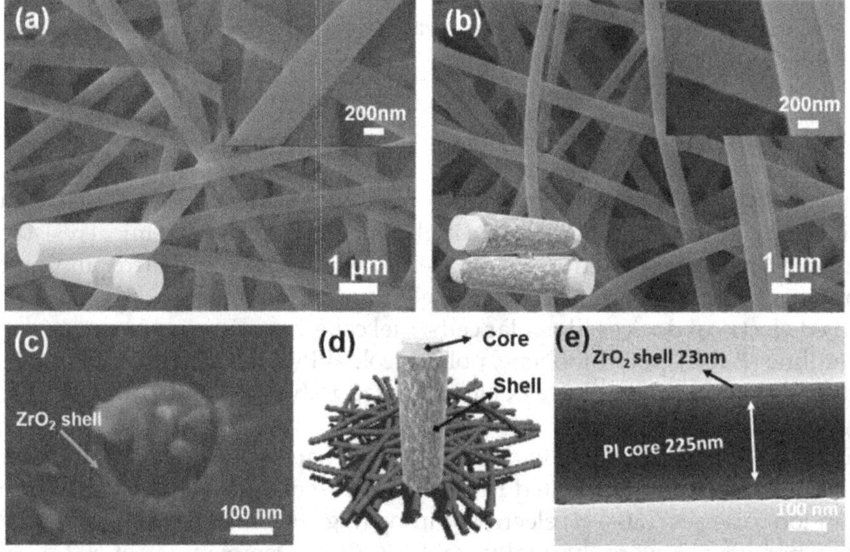

FIGURE 8.3
SEM images of the (a) pure PI core and (b) core-shell nanofiber membranes of ZrO_2@PI. (c) Images for the cross-section of the ZrO_2@PI composite nanofibers. (d) Scheme for the core-shell structure of Pi@ZrO_2 nanofibers. (e) TEM image of the nanofibers ZrO_2@PI displaying the thickness of both the ZrO_2 shell and the PI core. Reproduced with permission from [19]. Copyright (2021), American Chemical Society.

electrochemical performance but also in the mechanical properties due to proper interaction between the PI and ZrO_2, promoting an increase in hardness and strength. On top of that, flame retardancy was also improved, which served as a safety criterion for the use of the composite nanofiber LIBs. Figure 8.3 shows the scanning electron micrograph (SEM) and transmission electron micrograph (TEM) images of the nanofiber composite membranes.

8.2.2 Interfacial Polymerization

In interfacial polymerization (IP), two monomers are dissolved in two different immiscible solvents and react with each other at or near the interface. The polymerization can also occur at a liquid–gas or liquid–solid interface. The IP method is suitable for the synthesis of a large variety of polymers including polyamides, polyesters, polyurethanes, polyureas, polyimides, polysulfonamides, and polycarbonates, using a wide range of highly reactive monomers. Due to the use of highly reactive monomers, the reactions can take place rapidly even under mild reactive conditions and at high-end molecular weights [20]. IP is a technique widely used to create thin films and fibers of conducting polymers for electronic devices as it can produce high-quality nanofilms/fibers free of defects of uniform size and diameter along with an aligned structure, which generally leads to an improvement in conductivity. In addition, the morphology of the nanofiber can be controlled by adjusting the monomer concentration in the two adjacent phases, and by changing the solvent types.

PANI has gained more attention from scientists as a material for energy harvesting and storage devices. Due to its ease of synthesis, large-scale production, and environmentally friendly nature, IP is one of the most extensively used methods for the synthesis of PANI nanofibers [21]. In the IP method, polymerization of PANI occurs only at the interface of the organic/aqueous oxidant interface. The PANI generated leaves the interface at a certain stage of polymerization, thereby preventing the occurrence of "secondary growth" due to the controlled amount of reagents. Thus, interface polymerization can produce pure nanofibers without side chain growth.

8.2.3 Sol-Gel Method

The sol-gel method is one of the earliest and most commonly used wet chemical synthesis methods for metal oxide nanomaterials. In sol-gel chemistry, the precursor solution ('sol' – liquid phase colloidal particle dispersion) is converted into a gel (cross-linked, rigid, and porous inorganic network) through hydrolysis and polycondensation reactions. The general aqueous sol-gel process involves 1) precursor hydrolysis; 2) condensation followed by polycondensation to form particles; 3) gelation; and 4) drying under appropriate heating conditions. The sol-gel method is well suited to the synthesis of inorganic and inorganic/organic hybrid 1D polymers due to its low processing temperature (<100 °C) and its ability to adapt precursors to obtain different morphologies such as fibers, films, and monodispersed powders. In general, linear structures are obtained from the sol-gel reactions under acid catalysis conditions, whereas colloidal particles are obtained under the basic conditions [22–25].

Sol-gel methods can be easily combined with other techniques, such as electrospinning and template synthesis. Thus, sol-gel methods offer great flexibility in the production of various organic/inorganic polymer-metal oxide composite 1D structures, and related compounds. In recent years, there have been efforts to synthesize nanorods and nanofibers using the template-assisted sol-gel method in which the template can be placed directly onto a collector. The most commonly used templates are metal oxides (e.g., anodic aluminum oxide) and polymers [e.g., polyaniline, polycarbonate, poly(ethylene terephthalate), and polyimide] which have cylindrical nanopores of uniform diameter. These nanoscale cylindrical pores are filled with the desired material *via* the sol-gel process. In addition, the template can be dissolved to collect them after the formation of nanofibers or nanorods, as described in Figure 8.4. This approach is considered to be a negative template-based sol-gel synthesis [22, 24].

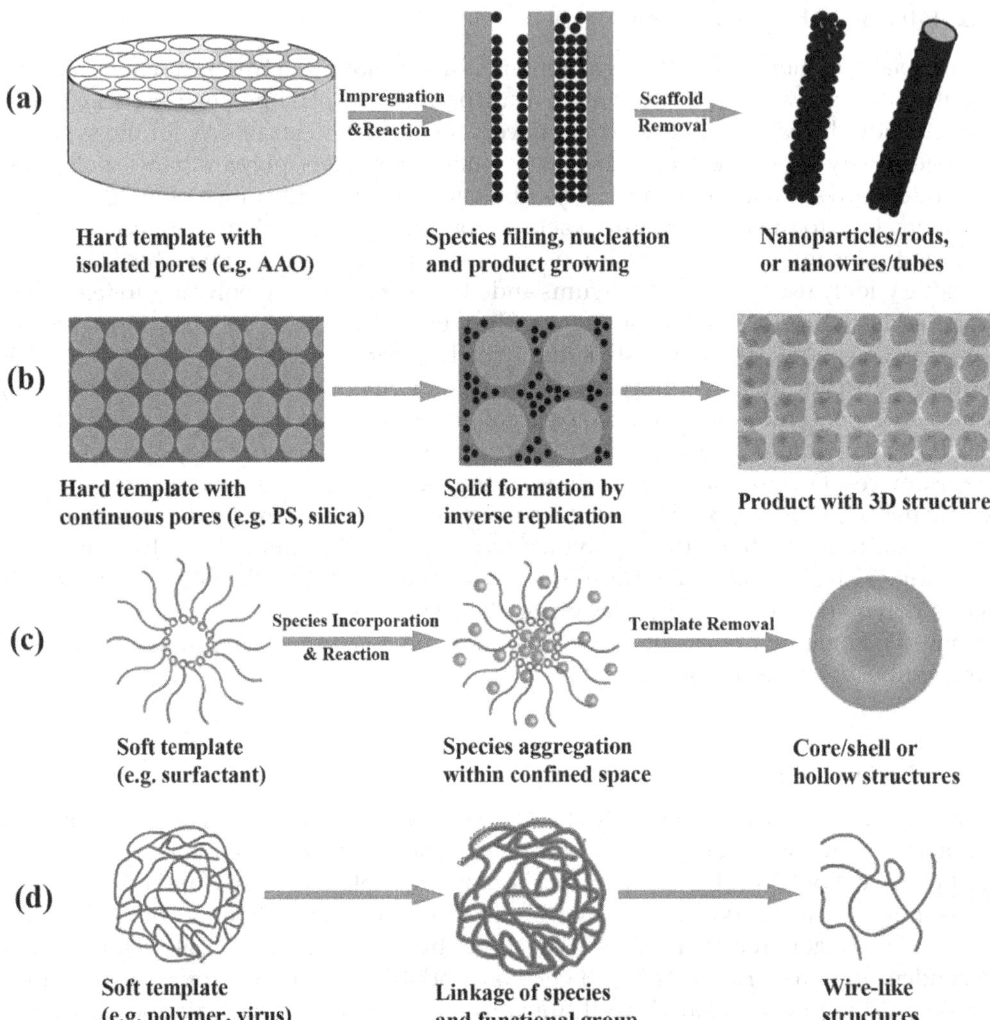

FIGURE 8.4
Schematic depiction of the (a) hard and (b) soft template synthesis. Reproduced with permission from [24]. Copyright (2008), American Chemical Society.

Another important variation of the sol-gel method is the sol-gel electrospinning technique. That approach was explored by Sung Seen Choi and co-workers, who showed the formation of silica nanofibers *via* the electrospinning technique and sol-gel method, using tetraethyl orthosilicate (TEOS)/ethanol/water/HCl solution with the molar ratios of 1/2/2/0.01, respectively. The synthetical approach of SiO nanofibers was performed by hydrolysis followed by a condensation reaction, which led to a highly branched chemical structure. The X-ray diffraction (XRD) patterns demonstrated an amorphous structure which could be related to the unpacked arrangement of Si–O chains. Yet, the fibrous morphology was confirmed through SEM [26]. In another approach, Tahira Pirzada synthesized polyacrylonitrile–silica composite nanofibers through controlled sol-gel electrospinning by using TEOS as a silica precursor and polyacrylonitrile [27]. These findings are encouraging for the generation of flexible electric devices in the future. Therefore, sol-gel methods play a key role in the development and manufacture of electrical devices like fuel cells, solar cells, and Li-ion batteries.

8.2.4 Chemical Vapor Deposition

Chemical vapor deposition (CVD) consists of the vaporization of a monomer under low pressure and high temperature to allow it to react with itself and adhere to a substrate. CVD has grown rapidly over the years and is a key technology used in the manufacture of many industrial products, such as semiconductors, electronic components, and corrosion-resistant coatings. Unlike other coating technologies, CVD is a solventless method that results in reduced costs and greater purity of the product. In addition, it is relatively versatile as a wide number of precursors can be used. Finally, thin and uniform films can be obtained. These factors allow CVD to be a competitive approach for industry. There are several categories of CVD, depending on the processes that initiate the chemical reaction. In a typical CVD polymerization process, a vapor of reactive monomer is produced in a specific manner and then subsequently or simultaneously introduced into the deposition phase. In the deposition phase, gas monomers are adsorbed onto the surface of the substrate to be polymerized. Among the various CVD polymerization techniques are two solvent-free CVD polymerization processes, namely "initiated CVD" (iCVD) and "oxidative CVD" (oCVD), which stand out. A variety of semiconducting and conducting conjugated polymeric thin-film materials, which are very difficult to process by typical solution-based methods, have been synthesized using oCVD and iCVD techniques.

The nature of a CVD structure can be controlled by properly manipulating the deposition parameters, such as temperature, pressure, supersaturation, and CVD reaction selectivity. Based on the operating parameters, CVD contains variable processes like atmospheric-pressure CVD (APCVD), low-pressure CVD (LPCVD), plasma-enhanced CVD (PECVD), and laser-enhanced CVD (LECVD). APCVD is a widely used method for the synthesis of carbon nanofibers and silicon nanowires at atmospheric pressure. Nan Liu successfully synthesized highly crystalline 1D graphene nanoribbons (GNRs) from palladium-incorporated poly-4-vinylphenol (Pd@PVP) electrospun polymeric nanofibers through a low-pressure, thermal CVD synthesis. First, the Pd@PVP nanofibrous wafer was fabricated by electrospinning (random and aligned fiber orientation) and then subjected to high-temperature CVD treatment in the presence of methane (carbon source) and hydrogen (carrier gas) in a furnace (Figure 8.5). This approach led to fibers with properties of high crystallinity (up to an adjustable width ~10 nm) and low resistivity (~ 4.8×10^{-6} m), making them suitable for electrical applications [28].

8.3 1D Polymers for Energy Generation

8.3.1 1D Polymeric Materials in Solar Cells

Si has been the most commonly used semiconductor material (SM) material, dominating the solar cell market for decades because it has an indirect band gap of 1.1 eV, a long lifetime, and high efficiency [29]. Due to the high initial cost of solar-grade Si wafers, however, a large number of alternative semiconductor materials and device fabrication methods

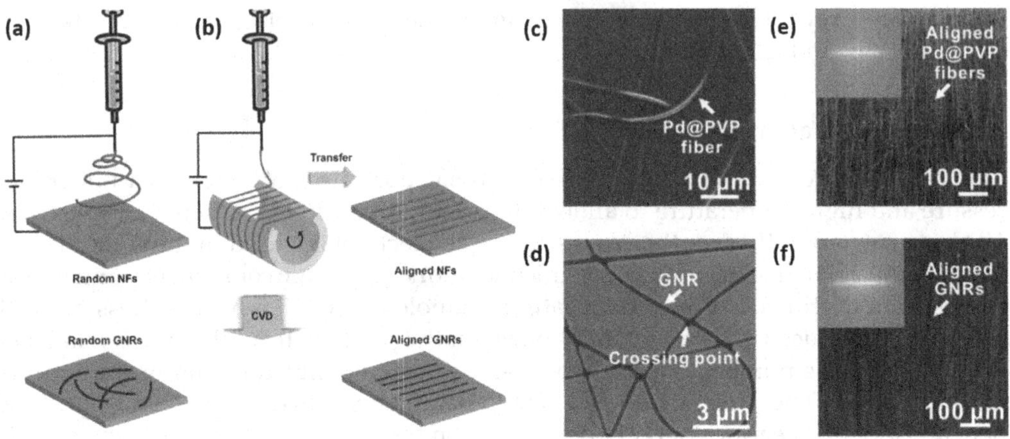

FIGURE 8.5
(a) GNR was obtained through the CVD process from electrospun polymer templates with a random distribution of nanofibers over the substrate. (b) Aligned nanofibers over a metallic rotor present a gap, followed by its application over the substrate. SEM images of (c) aligned PVP@Pd fibers over SiO_2/Si, and (d) GNR formed after the CVD process. Adapted with permission from [28]. Copyright (2014), American Chemical Society.

have recently been developed. The most common types of solar cells are crystalline silicon-based, thin-film, along with other high-efficiency ones based on gallium-arsenide, hetero-junction, etc. Furthermore, emerging cell technologies like perovskite, organic, dye-sensitized, and nanomaterial-based have the potential to be cheaper and more efficient than common types of solar cells. Since the nineteenth century, low-dimensional structures have dominated semiconductor technology, directing engineers to fabricate highly efficient solar cells by a "top-down" approach. Semiconductor nanowires/fibers are widely used to make such solar cells as they provide a natural freeway for the transport of electrons [2]. In this section, an overview of the development of semiconductor materials and their production technologies to achieve high power conversion efficiency (PCE) of silicon-based solar cells and bulk heterojunction polymers is provided, emphasizing the use of polymeric 1D materials.

8.3.1.1 1D Material in Silicon-Based Solar Cells

Among most functional semiconductor nanostructures, silicon nanowires (SiNWs) are the most studied and representative because of their versatile synthesis and excellent optical and electrical properties. Linyou Cao and his team demonstrated a strategy that utilized the powerful broadband, optical antenna effects of 1D semiconductor nanostructures to dramatically enhance the absorption of sunlight [8]. They demonstrated a commendable approach to grow SiNWs directly onto the Si substrate using the metal-assisted chemical etching method (with the desired, crystalline orientation and doping). Si-nanowire arrays ensure large active surface area, ultra-long axial, optical path length for efficient light-harvesting, ranging into the near-infrared (NIR) spectral region, superior surface anti-reflectivity, and surface defect-induced, higher electrical conductivity compared to Si wafers. Thus, this experiment confirms that there is a significant improvement in the optical current response of SiNWs compared with the bulk Si material. However, this technique has not yet been widely used for mass production since it has not been able to offset results with costs. Recently, many attempts have been made to develop SiNWs through a number of approaches, such as laser ablation, thermal evaporation/decomposition, and supercritical vapor-liquid-solid (VLS) synthesis.

8.3.1.2 1D Material in Bulk-Heterojunction Polymer Solar Cells (PSC)

A polymer solar cell is a multilayer structure consisting of at least a transparent front electrode, an active layer (semiconducting polymer material), and a back electrode printed onto a substrate (e.g., glass, plastic). The interfaces between these layers play a major role in PSC operation and performance. Each functional layer has a thickness of around 10–100 nm. Such a nano-thin-film structure makes PSCs more sensitive to various stress factors. For instance, PSCs undergo structural changes and degradation during operation [30].

It has recently been discovered that a highly efficient active layer can be made using polymeric fiber membranes, which can result in better device performance compared with other thin membranes. In 2018, a research group developed a new bulk-heterojunction (BHJ) PSC based on the use of organic donor nanofibers to optimize the morphology of the active layer [31]. They synthesized nanofibers using the electrospinning technique at ambient atmosphere, and fibers were collected on square polymer/TiO_2/indium tin oxide (ITO)/glass substrates. The nanofibers are composed of a mixture of poly[2-methoxy-5-(2′-ethylhexyloxy)-1,4-phenylene vinylene] (MEH-PPV), polyvinylpyrrolidone (PVP), and phenyl C61 butyric acid methyl ester-based polymers (PCBM). The use of a polymer blend is advantageous for maintaining the structure of layers better than a single polymer, even if the second polymer (e.g., PVP) is not photoactive. To maximize the adhesion of the fibers, a MEH-PPV/PVP/PCBM film is first deposited onto the TiO_2 layer. A layer of poly (3-hexylthiophene) P3HT and PCBM were applied as a backfill layer to interconnect the fibers and increase their acceptance range of incoming light frequencies. At the final step, MoO_3 and Ag were added as an electron-blocking layer and anode, respectively, by thermal evaporation. It was observed that the nanofiber active layer could enhance photon absorption and increase the PCE of solar cells relative to the thin film [30, 31].

8.3.2 Fuel Cells

A fuel cell is an electrochemical cell that generates electricity *via* a chemical reaction in the presence of an electrocatalyst. The electrochemical breakage of H_2O into H_2 and O_2, i.e., water splitting, is one of the most researched processes for the generation of renewable fuel. Researchers have also used hydrocarbons (methane, propane), alcohols (methanol), and solid carbon (municipal waste) as fuel to generate electricity. Similar to a general electrochemical cell, a fuel cell also contains three main compartments: anode, cathode, and electrolyte. Based on the type of electrolyte, electrode, fuel compatibility, and operating procedure, fuel cell systems can be categorized into five major types. Table 8.1 describes the major fuel cells, namely phosphoric acid fuel cell (PAFC), polymer electrolyte membrane fuel cell (PEMFC), alkaline fuel cell (AFC), molten carbonate fuel cell (MCFC), and solid-oxide fuel cell (SOFC). Each type of cell has its own strengths and weaknesses, making it more suitable for specific markets and applications.

The PEMFCs have a high focus over other fuel cell types with respect to portable power, stationary, and transport applications due to their operating temperature, lightweight cell components, and durability [32]. SOFC is another popular type of fuel cell, mainly used in power stations, military, and aerospace applications because it converts chemical energy into clean and efficient electricity. SOFC uses solid oxide material as an electrolyte and does not require expensive Pt catalysts because SOFC uses high operating temperatures.

TABLE 8.1
Description of Major Fuel Cell Types

	PEMFC	PAFC	AFC	MCFC	SOFC
Electrolyte	Polymer membrane	Liquid H_3PO_4 (immobilized)	Liquid KOH (immobilized)	Molten carbonate	Ceramic
Charge carrier	H^+	H^+	OH^-	CO_3^{2-}	O^{2-}
Operating temperature	80 °C	200 °C	60–220 °C	650 °C	600–1000 °C
Catalyst	Platinum	Platinum	Platinum	Nickel	Perovskites (ceramic)
Cell components	Carbon-based	Carbon-based	Carbon-based	Stainless steel-based	Ceramic- based
Fuel compatibility	H_2 and methanol	H_2	H_2	H_2 and CH_4	H_2, CO, CH_4

Higher operating temperatures create disadvantages and advantages. Decreases in cell interconnection and sealing, mechanical deterioration, and thermal expansion matching tasks are the main disadvantages of SOFCs. However, their advantages include higher electrical efficiency (about 50–60%) and fuel flexibility, which creates interest in SOFCs during the past few decades. Therefore, PEMFC and SOFC are discussed in this section.

8.3.2.1 Polymer Electrolyte Membrane Fuel Cells (PEMFCs)

The most significant feature of PEMFCs is the special, proton-conducting polymer electrolyte membrane. This membrane serves as the supporting component for the electrodes, separates the reactant gases, and conducts protons from anode to cathode. The polymer membrane used in PEMFCs is a thin (20–200 μm), flexible, and transparent material commonly made out of a fluorocarbon-based polymer. Either side of this membrane is coated with a thin layer of catalyst and porous carbon electrode material. This electrode–catalyst–membrane–catalyst–electrode sandwich structure is referred to as a membrane-electrode assembly (MEA). The electrochemical reactions (e.g., oxidation of hydrogen, methanol, ethanol, and reduction of oxygen) take place on the surface of catalysts at each three-phase boundary. Most PEMFC types run using pure H_2, thus the main electrochemical reactions that occur at the PEMFC can be briefly summarized in Equation (8.1):

$$\text{Overall cell reaction}: 2H_2 + O_2 \rightarrow 2H_2O + \text{heat} + \text{electrical energy} \quad (8.1)$$

Typical PEMFC electrolyte membranes are made out of perfluorinated copolymers with pendant sulfonic acid groups, such as Nafion®, Flemion®, 3M™, and Aquivion®. These polymer membranes have indispensable properties, such as high hydrolytic and oxidative stability and excellent proton conductivity. Although perfluorinated polymers offer some advantages, they also have some disadvantages, such as low proton conductivity and low humidity at high temperatures (>80 °C), which create barriers for commercialization [33]. As a result, extensive research has been conducted on the development of highly functional alternative membranes. Various types of electrospun-able polymer/Nafion composite fiber membranes have been developed as a way to overcome the shortcomings. Examples are composite Nafion® fibers made from polymers such as polyvinylpyrrolidone, polyethylene oxide, polybenzimidazole, poly(acetylene ether), and polyvinyl alcohol.

Silas Simotwo and Vibha Kalra tried to obtain nanostructured non-woven fiber mats of Nafion, PANI, and Nafion/PANI composite by single-step electrospinning technique for PEMFC applications. Poly(ethylene oxide) (PEO) was used to enhance chain entanglements of fibers [34]. Their results revealed that a Nafion/PANI composite nanofiber mat weight composition of 61/39 showed a proton conductivity of 0.032 S/cm and an electron conductivity of 0.041 S/cm. Also, a proton conductivity of 0.056 S/cm was obtained for a Nafion-PEO 98/2 fiber mat and an electron conductivity of 0.087 S/cm was recorded for a PANI-PEO (75/25 w/w) fiber mat. Furthermore, they reported that the Nafion/PANI composite nanofiber mat could potentially serve as excellent catalyst support for future PEMFCs.

Typically, the oxidation-reduction reactions in fuel cells in the absence of a catalyst are very slow. Pt is mainly used as an electrode catalyst for PEMFs because it has better electrochemical stability under strongly acidic conditions, and it has moderate binding energy with oxygen and hydrogen. However, scientists are developing carbon and other Pt-supporting materials for electrolyte catalysts in PEMFCs to minimize Pt usage by increasing the electrochemically active surface area for uniform dispersion of small Pt nanoparticles. When the Pt catalyst is supported by 1D carbon-based supports (carbon nanofibers, carbon nanotubes, GNRs, etc.), their physicochemical properties, electric conductivity, and catalytic activity are enhanced. On the other hand, the carbon-based support material can undergo corrosion. That leads to Pt agglomeration, and a loss of hydrophobic nature takes place. Therefore, carbon-based support materials doped with heteroatoms, for instance, F, N, B, S, and P, using wet chemical methods to improve electrochemical durability and catalytic activity, have been developed. A research group recently reported that fluorine-decorated graphene nanoribbon oxide-supported Pt (Pt/F-GNRO) (Figure 8.6) exhibited higher catalytic activity than Pt/C and showed the lowest carbon degradation rate and corrosion and excellent water management [35].

Polyacrylonitrile- (PAN-) based electrospun carbon nanofibers (e-CNF) formulated through electrospinning and carbonization were used for catalyst support to improve the performance of PEMFCs. PAN nanofibers were decorated with *in-situ*-generated Pt catalyst nanoparticles. Then, a PAN-based e-CNF membrane electrode was deposited onto a

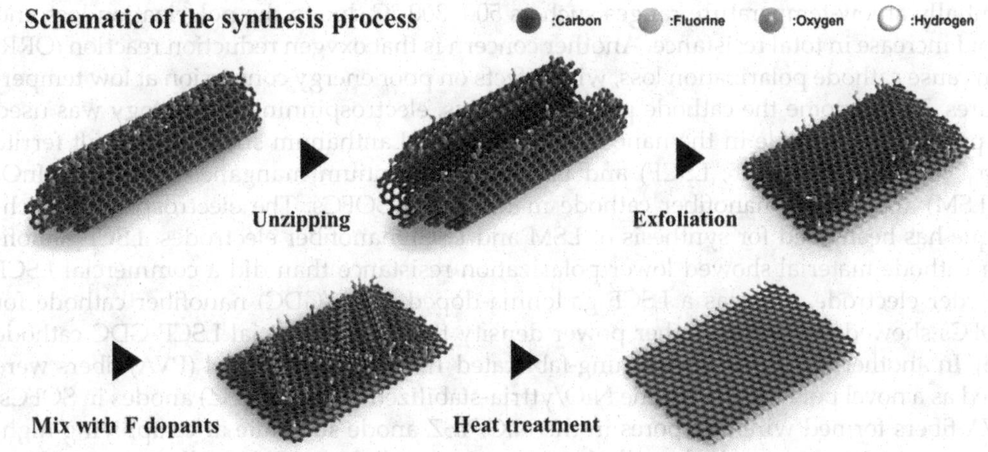

FIGURE 8.6
(a) Schematic diagram of F-GNROs synthesis method. Adapted with permission from [35]. Copyright (2021), American Chemical Society.

gas diffusion membrane after adding an ionomer binder. The improvement in the catalytic activity after introducing the conducting polymers to the Pt catalyst is contributed by several factors, i.e., (i) improved access to catalyst sites; (ii) high power at ultra-low Pt loadings; and (iii) improved durability of electrode material in PEMFCs compared with commercial Pt/XC-72R [36, 37].

8.3.2.2 Solid Oxide Fuel Cell

One of the main components of a solid oxide fuel cell (SOFC) is the solid, ceramic electrolyte such as yttria-stabilized zirconia (YSZ), which is the most common electrolyte used for the water-splitting process. The anodic and cathodic reactions of SOFCs are summarized below in Equations (8.2) and (8.3). Water generation happens at the anode and O^{2-} acts as a mobile conductor in SOFCs.

$$Anode:\ H_2 + O^{2-} \rightarrow H_2O\ + 2\,e- \tag{8.2}$$

$$Cathode:\ ½\,O_2 +\ 2e- \rightarrow O^{2-} \tag{8.3}$$

Ni-YSZ cermet (mixture of ceramic and metal) is also commonly used as an anode material in the SOFCs as Ni provides catalytic activity and conductivity whereas YSZ provides high porosity and high surface area for reaction, as well as high thermal expansion compatibility. The cathodic material in the SOFC is commonly a mixed ion-conducting and electronically conducting ceramic material. Lanthanum strontium ferrite (LSF), strontium-doped lanthanum manganite (LSM), and lanthanum strontium cobaltite ferrite (LSCF) have also shown good oxidation resistance and high catalytic activity in the cathodic environment of an SOFC. An SOFC has a high operation temperature (600–1000 °C). Due to this, light hydrocarbon fuel can be reformed internally inside the anode, while SOFCs can be refueled by reforming heavier hydrocarbons externally. However, the higher operating temperature can limit the applicability of SOFCs to portable power and transportation markets.

The electrochemical reaction kinetics and the efficiency of an SOFC can decrease exponentially at low-temperature ranges such as 500–800 °C due to the polarization loss and rapid increase in total resistance. Another concern is that oxygen reduction reaction (ORR) can cause cathode polarization loss, with effects on poor energy conversion at low temperatures. To overcome the cathode polarization loss, electrospinning technology was used to prepare the cathode in the nanoscale (nanofiber). Lanthanum strontium cobalt ferrite ($La_{(1-x)}Sr_xCo_{(1-y)}Fe_yFe_yO_3$ δ; LSCF) and lanthanum strontium manganese ($La_{(1-x)}Sr_xMnO_3$ δ; LSM) are used as nanofiber cathode materials for SOFCs. The electrospinning technique has been used for synthesis of LSM and LSCF nanofiber electrodes. LSCF nanofiber cathode material showed lower polarization resistance than did a commercial LSCF powder electrode, whereas a LSCF gadolinia-doped ceria (GDC) nanofiber cathode for SOFCs showed two times higher power density than a commercial LSCF-GDC cathode [38]. In another study, electrospinning-fabricated 1D polyvinyl alcohol (PVA) fibers were used as a novel pore-former for the NiO/yttria-stabilized zirconia (YSZ) anodes in SOFCs. PVA fibers formed wire-like pores in the NiO/YSZ anode substrate after applying high-temperature treatment, which easily forms continuous holes with a small amount of fiber. These wire-like pores in the anode improved the performance of SOFCs by facilitating the transport of fuel and by-products [39].

8.4 1D Polymers for Energy Storage

8.4.1 1D Polymeric Materials in Batteries

Batteries store and release electrical energy through a chemical reaction and contain three main active components, namely the electrolyte and two conductor electrodes. Batteries can be divided into two groups: primary batteries (non-rechargeable) and secondary batteries (rechargeable). Primary batteries are defined by the chemical reaction that happens at the irreversible electrode, whereas, in secondary batteries, the chemical reaction that happens at the electrodes is reversible. The main drawbacks associated with conventional batteries are low energy density, a limited number of full discharge cycles, and their environmentally unfriendly nature. To address that, researchers are mainly focusing on the development of metal-ion batteries, metal-air batteries, and metal-sulfide batteries.

8.4.1.1 Metal-Ion Batteries

The lithium-ion battery (LIB) is the most popular metal-ion type battery that plays an important role in the energy and transportation sectors due to the characteristics of high power and energy densities. The primary functional components of the lithium-ion cell are the carbon anode, metal oxide cathode, and a lithium salt dissolved in an organic solvent as an electrolyte. Graphite is the most popular anode material used in a lithium-ion cell. Layered oxide (lithium cobalt oxide) or spinel (lithium manganese oxide), or polyanion (lithium iron phosphate) are mainly used as the cathode material. The anode and cathode half-reactions are presented in Figure 8.7. During lithium-ion cell discharging, lithium ions are extracted from the anode and deposited onto the cathode, and the reverse process occurs during cell charging [40].

Electrospun polymers and their derivatives, for instance, PAN and polyvinylidenedifluoride (PVDF), provide a good nanoporous structure which improves the ionic conductivity of a membrane when soaked with liquid electrolyte. The tensile strength and modulus of the lithium-ion cell membrane can also be improved by the use of inter-fiber composite electrospun PVDF fibers under hot pressure. Uniform fiber diameter and uniform electrospun PVDF membrane thickness improve mechanical strength and provide higher charge and discharge capacities. Such factors allow the electrospun polymers to potentially surpass the electrochemical properties of conventional electrodes of powders and films due

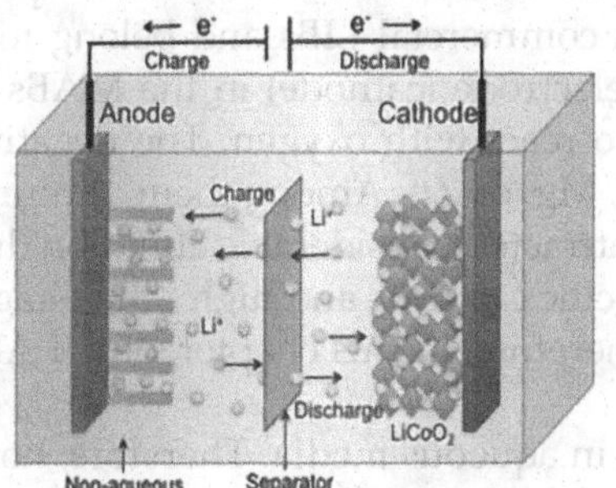

The cathode half-reaction(with charging being forward) is:
$LiCoO_2 \rightleftharpoons Li_{1-x}CoO_2 + xLi^+ + xe^-$

The anode half-reaction is:
$6C + xLi^+ + xe^- \rightleftharpoons Li_xC_6$

The overall reaction is:
$6C + LiCoO2 \rightleftharpoons Li_{1-x}CoO + xLi^+ + xe^-$

FIGURE 8.7
The working principle of LIBs. Adapted with permission from [41]. Copyright (2014), Royal Society of Chemistry.

to their high surface areas and electron transfer rate. In addition, an entangled network of nanofiber electrodes in lithium-ion cells facilitates access of the ions to the electrode and the electrolyte interface. Currently, researchers plan to use nanostructured materials as an alternative anode material for LIBs due to some disadvantages of graphite, such as safety and rate performance. Porous carbon nanofibers (CNFs) have been widely used to construct anodes for LIBs because CNFs have a large specific surface area, energy conversion density, extremely attractive surface activity, high rate capability, and better cycle performance.

However, some of the disadvantages of LIBs include the limited availability of lithium and some other transition metal used in LIBs, and frequent fires and explosions, which have directed scientists to think about alternative secondary batteries. Researchers have identified magnesium-ion batteries (MIBs), sodium-ion batteries (SIBs), potassium-ion batteries (PIBs), and aluminum-ion batteries (AIBs) as strong alternatives to LIBs due to their superior safety credentials, environmental friendliness, and low cost. PIBs and SIBs have a similar working principle to LIBs, and electrochemical energy is reversibly stored and released due to the transport of Na and K ions between anode and cathode. The use of polymeric nanofibers, such as hollow carbon nanofibers (HCNFs) fabricated by pyrolyzation of a PANI hollow nanofiber, as anode materials for Na storage has demonstrated high reversible charge capacity, rate capability, and superior cycling stability [42].

MIBs can be charged faster than LIBs as Mg does not form dendrites. Also, it is the second most abundant element on the earth's surface while exhibiting low toxicity. Therefore, it can be used to produce cheap metal-ion batteries. However, unstable electrolyte/electrode interfaces, the absence of suitable electrodes, and slow diffusion kinetics of Mg^{2+} in host materials are the main fundamental challenges for MIBs, which hinder the practical exploitation of MIBs. Over the past decade, scientists have used electrospun nanofiber materials to prepare advanced electrodes for MIBs and have obtained outstanding performance [43]. Al metal works as the negative electrode in AIBs and Al interchanges three electrons during the electrochemical process. The electron exchange at AIBs is, therefore, three times as high as the electron exchange at the Li electrode [44]. To test those values, scientists have used polymeric nanotubes of Co_9S_8@CNT–CNF to prepare binder-free new cathode electrodes (Figure 8.8) for AIBs and obtained excellent cycling stability and higher capacity (315 mAh/g at 100 mA/g) [45].

8.4.1.2 Metal-Air Batteries

Metal-air batteries (MABs) are powered by oxygen reduction and metal oxidation processes. They can exhibit higher energy densities than commercial LIBs and belong to a family of primary and secondary cells. The positive electrode (cathode) in the MABs is covered with a layer of carbon and precious metals to react with oxygen. The negative electrode (anode) is made of a metal such as Li, Na, Al, Mg, or Zn. Among them, lithium-air batteries (LABs) and zinc-air batteries (ZABs) have attracted considerable attention due to characteristics such as high energy density, high specific capacity, and high cell voltage. Moreover, ZABs have a lower cost rather than LIBs. Therefore, in this chapter, only LABs and ZABs are discussed.

Highly active metals such as Li and Na are unstable in aqueous media. Therefore, non-aqueous aprotic electrolyte media are required for such higher-activity metals, whereas relatively inactive metals like Zn, Al, Mg require alkaline, aqueous electrolyte media. MABs mainly contain four parts: a metal electrode, an electrolyte, a separator, and an air electrode. In most metal-gas batteries, the reactant O_2 used in the gas electrode is extracted

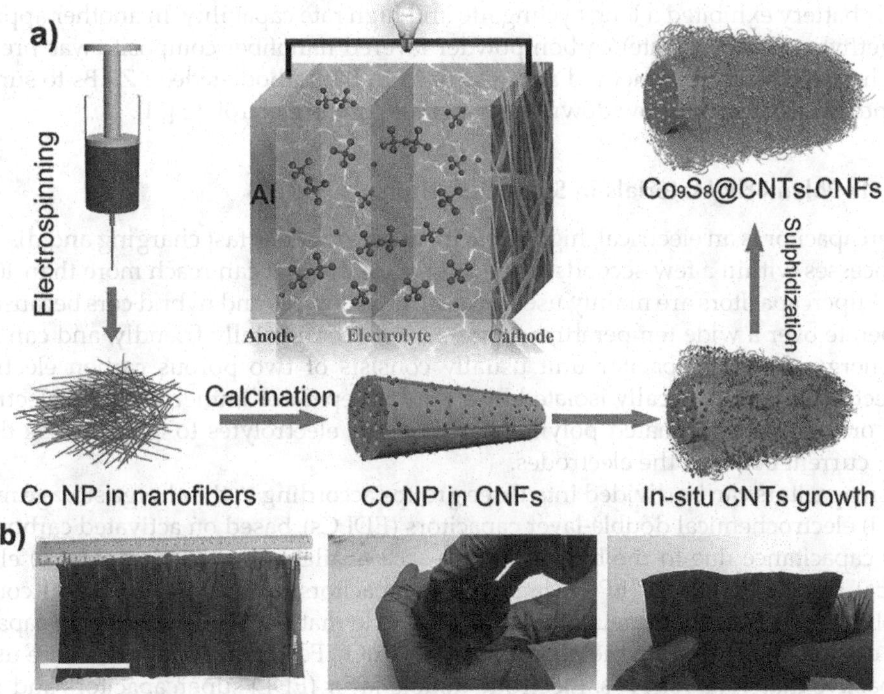

FIGURE 8.8
(a) Schematic illustration of the preparation process of the binder-free and free-standing Co_9S_8@CNT–CNF mats. (b) Photographs of highly flexible Co_9S_8@CNT–CNF before and after bending; scale bar is 10 cm. Reproduced with permission from [46]. Copyright (2017), John Wiley and Sons.

from the ambient air. Also, the air electrode usually consists of only the electrocatalyst and gas diffusion layers. Most MABs have very limited practical applications due to the challenges associated with discharge properties and rechargeability. Oxygen reduction reaction (ORR) and oxygen evolution reaction (OER) overpotentials and mass transport are the main issues associated with air electrodes. Other potential issues are corrosion, passivation, dendrite growth, and shape change of electrode materials. The stability, ionic conductivity, and oxygen solubility-related issues associated with electrolytes in MABs are other problematic characteristics that are being tackled by researchers. Scientists are increasingly focusing on nanofiber-based catalysts used for MABs due to their superior ion and electron transport properties and the formation of a continuous conductive network for the active material. Moreover, electrospun nanofiber-based air cathodes have exhibited good catalytic performance and higher cycling stability [47].

The electrochemical reactions at the gas cathodes of metal-gas electrodes involve the ORR and the OER. Those reactions have slow kinetics due to the complex mass transport and slow dynamics, which create a high overpotential that can limit the performance of MABs significantly. A binder-free, porous, N-doped Co carbon nanofiber cathode, synthesized by the electrospinning/plasma-etching method, exhibited excellent ORR performance in ZABs. Moreover, manganese sulfide and cobalt nanoparticles encapsulated by N-doped carbon nanofibers achieved high catalytic performance for ZABs [48, 49]. In another case, poly(1,4-anthraquinone) (P14AQ) polymerized *in situ* on carbon nanotubes (CNTs) was used as the anode and spinel cobalt manganese oxide supported on CNTs(CMO@CNT) was used as the catalyst cathode for an aqueous polymer-air battery [50]. The P14AQ/CNT air battery exhibited a long cycling life and high rate capability. In another approach, a poly(ethylene vinyl acetate)/carbon powder layered nanofiber composite was prepared using the electrospinning method and was used in the cathode-side of ZABs to suppress the dendrite growth and slow down the water loss of the electrolyte [51].

8.4.2 1D Polymeric Materials in Supercapacitors

A supercapacitor is an electrical, high-capacity device that has fast charging and discharging processes within a few seconds and a long cycle life that can reach more than 100,000 cycles. Supercapacitors are mainly used in electronic devices and hybrid cars because they can operate over a wide temperature range, are environmentally friendly, and can safely store energy. A supercapacitor unit usually consists of two porous carbon electrodes. The electrodes are electrically isolated by a porous separator immersed in an electrolyte. Metal- or carbon-impregnated polymers are used as electrolytes to facilitate the flow of electric current between the electrodes.

Supercapacitors can be divided into three groups according to the charge storage mechanisms: i) electrochemical double-layer capacitors (EDLCs), based on activated carbon with higher capacitance due to the higher surface area available for the electrode; ii) electrochemical pseudocapacitors (EPC) (redox supercapacitors), which use electrical conducting polymers or transition metal oxides as electrode materials; and iii) hybrid capacitors (HCs), carbon and transition metal, for instance, RuO_2, Fe_3O_4, and MnO_2, that are used as electroactive materials for HCs. Electrode double-layer (EDL) supercapacitors and physical dielectric capacitors (DCs) have some similarities and differences; for instance, both have similar electrostatic processes. However, the main difference is an insulating barrier in DCs that is replaced by electrolytes in EDL supercapacitors. Aqueous electrolytes and organic electrolytes are the main electrolytes used in supercapacitors. Salt, KOH, or H_2SO_4 are used as aqueous electrolytes, and about 1 V of the limit of operating voltage is required due to the electrochemical decomposition of water. On the other hand, the organic electrolyte has a higher operating voltage (2.5–3.5 V) than the aqueous electrolyte.

EDLC charging depends on the ion dissociation of the electrolyte salts, which develop a double-layer structure inside the pores in the carbon material with a high surface area close to the surface of the electrode. The thickness of the double-layer structure is inversely proportional to the double-layer capacity. Therefore, the focus on carbon nanotubes and conductive polymeric nanofibers has increased, because high surface area and small charge separation are essential to obtain high capacitance.

The charge storage in EPCs depends on the rapid Faradaic redox reaction that takes place between the electrode material and the electrolyte ions. Metal oxides and conductive polymers are mainly used as electroactive materials in EPCs. RuO_2 shows superior capacity and longer cycling life, but the high price of RuO_2 limits its commercial use. PANI, polypyrrole (PPy), and poly(3,4-ethylenedioxythiophene) (PEDOT) are used as the main electroactive conductive polymers for EPCs. However, limitations and challenges, including low energy density, lower capacitance, low rated voltage, and the need for flexible power storage devices, led scientists to conduct further research to improve the electrolytic material for supercapacitors. The focus on electrospun nanofibers for supercapacitors has grown over the past decade. Electrospun nanofiber formulation methods are excellent for fabricating supercapacitors because nanofibers have a small diameter that can be controlled by changing the procedure's parameters, leading to excellent flexibility and a higher surface area per unit volume. PANI is widely used as a conductive polymer to make electrospun nanofibers for electrode materials in supercapacitors due to its high conductivity and theoretical specific capacitance. However, the charge and discharge cycles of PANI remain low and the nature of the polymer electrode change during repeated fast redox reactions due to the ion doping and de-doping. As a result, researchers have proposed a combination of carbon material, such as CNTs or carbon fibers with PANI to improve capacitance performance and cycling stability. PANI-grafted CNTs with strong interfacial forces exhibit fast charge transfer, enhanced capacitance, and higher cyclability [52].

The solid-state electrolyte is another component that plays a major role in supercapacitive properties as their pore size must be compatible with the size of the ion that flows through it to create a better double-layer charge separation. However, preparing a carbon material with a higher surface area and a suitable pore size is challenging. Scientists have proposed to use polymer blend techniques to prepare materials with a high surface area containing carbon materials with controlled pore sizes. In this polymer blend, the different thermal stabilities of two polymers are vital to the functioning of the polymer blend. The polymer with higher thermal stability is used to prepare nanofibers, while the polymer with low thermal stability is used as a sacrificial material, which decomposes at high temperature and creates controlled pores. Nanofibers were prepared with 75 wt% of polyacrylonitrile (PAN) as the carbonizing polymer and 25 wt% of different sacrificial polymers including polyacrylic acid (PAA), polymethyl methacrylate, and polyethylene glycol, which exhibited the best cyclic voltammetry measurements [53].

PEDOT poly(styrene sulfonate) (PEDOT: PSS) nanofibers were prepared using the electrospinning technique and deposited onto flexible polyethylene terephthalate substrate to produce flexible electrodes for all-solid-state supercapacitors. Synthesized nanofibers exhibited higher conductivity than doped PEDOT: PSS films with PVA/phosphoric acid (PVA/H_3PO_4) gel polyelectrolyte. Furthermore, the advantages of solid electrolytes, such as lower internal corrosion, flexible packaging, and ease of handling without liquid leakage, led to an interest in the solid electrolyte from the scientific community. A flexible and foldable all-solid-state supercapacitor was synthesized using two main steps, consisting of electrospinning. A flexible PANI/carbonized PI (CPI) nanocomposite was used as a membrane electrode for the above-mentioned all-solid-state supercapacitor and a PVA/PAA nanofiber membrane reinforced with PVA/H_3PO_4 was used as the gel separator. PANI/CPI membrane electrodes exhibited faster electrolyte ion movement and electron transfer, higher specific capacitance, and longer cycle life, while the PVA/PAA nanofiber membrane reinforced with PVA/H_3PO_4 exhibited excellent mechanical properties [54].

8.5 1D Polymers for Wearable Devices

Wearable devices (WD) are electronic devices that are worn by an individual, usually close to the skin. Advances in WD are reshaping the way we live, work, and play. There are many attractive new wearable electronics on the market. Smartwatches, smart rings, core body temperature sensors, breathing trainers, head-mounted displays, smart earphones, and smart clothing are some of the most attractive wearable devices. The main function of each device is to constantly collect data to provide rapid, continuous, and long-term monitoring decisions. Also, each device needs its power supply and needs to be recharged regularly. As the number of WDs used in the future increases, so does the need to create an efficient charging device that can store a large amount of electricity at once. Therefore, one of the major issues facing manufacturers of these WDs is the proper and convenient powering of these devices. To date, batteries, capacitors, and supercapacitors are the most effective wireless energy conversion devices for WDs. Thus, the capacity of batteries and supercapacitors, as well as their lifespan, are directly affected to provide an undisrupted power supply to the WD. In addition, they must be small in size and resistant to external pressure and force [55, 56].

In the past few years, many research teams have been trying to develop high-performance, flexible, shock-resistant, electrically conductive, and inexpensive materials for energy conversion devices (ECDs). Conducting polymers, carbon nanomaterials, transition metal oxides, transition metal dichalcogenides, MXene, covalent organic frameworks, metal-organic frameworks, black phosphorus, and composite materials are the existing electrode materials in ECDs. Among the various modified advanced materials, fiber-shaped nanomaterials have significantly improved the performance of these ECDs [56–58]. This is because 1D nanofiber networks provide strong mechanical properties such as low-stress operation, arrest the rapid propagation of cracks or damage, exhibit outstanding fatigue resistance and damage tolerance, and promote charge transfer during the charge-discharge process. The large surface area of such a network with a porous structure increases the storage capacity for migrating of ions [57, 58]. Figure 8.9 shows a schematic illustration of a fiber-shaped 1D supercapacitor for LIBs.

Recently, there has been a growing interest in developing new nano-structured fiber parts based on CNT fibers for ECDs used in wearable technology since CNT fibers not only provide the desired mechanical properties but also have high thermal and electrical conductivities. According to data in the literature, it is known that the mechanical properties of CNT fiber electrodes depend on the fiber dimensions and the fiber-processing method [60]. Therefore, recent advances in the fabrication, structure, and properties of a CNT fiber-based battery, capacitor, and supercapacitor are summarized below, with the focus on the electrode material.

In 2013, a research team led by Prof. H. Peng fabricated a lightweight coaxial electro double-layer capacitor (EDLC) to power wearable devices [61]. It is designed by attaching a polymer gel between two lined CNT fibers and a sheet of electrodes. Both CNT fibers and the sheets were dry-spun from CNT arrays that had been synthesized by a CVD method.

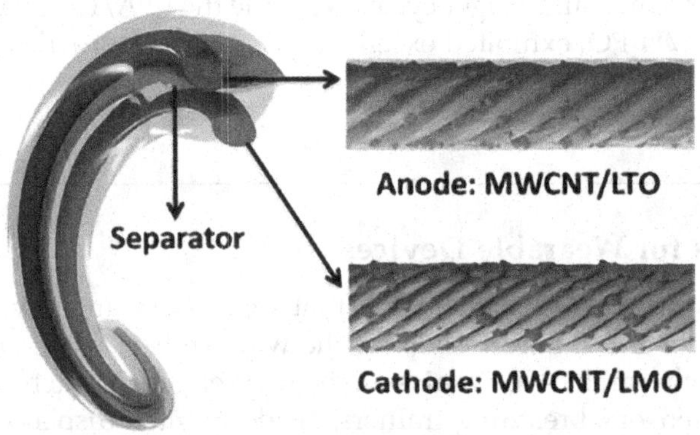

FIGURE 8.9
(a) Schematic illustration of a 1D lithium-ion battery. Multi-walled CNT/$Li_4Ti_5O_{12}$ (MWCNT/LTO) and MWCNT/$LiMn_2O_4$ (MWCNT/LMO) function as anode and cathode, respectively. (b) Schematic illustration of a self-healing supercapacitor. Aligned MWCNT sheets are wrapped onto the surface of the self-healing polymer fiber as the electrode. Adapted with permission from [59]. Copyright (2014), John Wiley and Sons Wiley Online Library.

These EDLC fibers exhibited energy densities up to 1.88 Wh/kg and power densities up to 755.9 W/kg, with the maximum discharge capacitance of 59 F/g. In 2014, the same group developed a new electrochromic fiber-shaped supercapacitor by electro-depositing PANI onto sheets of aligned CNTs as two electrodes. The resulting electrochromic fiber-shaped supercapacitor has achieved the specific capacitance of 255.5 F/g (0.1890 mF/cm 70% w/w of PANI) with an energy density of 12.75 Wh/kg and a power density of 1494 W/kg [62].

Compared with other battery types, lithium-ion polymer batteries (LIPBs) are used as a power supply type in WDs owing to their advanced features, such as high energy density, compactness, extremely thin structure, low weight, and high safety. The first CNT fiber-based LiPB appeared in 2012 [63]. It consists of a CNT/MnO_2 hybrid fiber electrode. This electrode was synthesized by depositing MnO_2 nanoparticles on the surface of CNT fibers through the electrochemical deposition method. The fiber-shaped battery was produced by twisting an aligned CNT/MnO_2 hybrid fiber (positive electrode) and a Li wire (negative electrode). The flower-shaped MnO_2 nanoparticles are uniformly spotted on the surface of the CNT fiber. The aligned CNT/MnO_2 hybrid fiber was further investigated with a Li wire as the negative electrode in 1 M $LiPF_6$ electrolyte solution. A specific capacity of 109.62 mAh/cm^3 or 218.32 mAh/g could be achieved based on the whole electrode. This battery presented 92.84 and 35.74 mWh/cm^3 charge and discharge energy densities, respectively, whereas the charge and discharge power densities were 3.87 and 2.43 W/cm^3, respectively, at 2×10^{-3} mA. To date, a team of researchers at MIT has developed the world's longest rechargeable LIB for wearable applications [64]. This Li-ion fiber battery has an ultra-long fiber form and is made of electroactive gels, particles, and polymers using the thermal drawing method. This Li-ion fiber battery demonstrated a discharge capacity of ~123 mAh and a discharge energy of ~217 mWh. All nanofiber devices have a high specific capacity and good flexibility of fiber electrodes, making them more appropriate for use in future wearable electronics applications.

8.6 Conclusion

Nanofibers, nanowires, and nanoribbons have emerged as exciting 1D polymeric nanomaterials for a broad spectrum of energy-related research and commercial applications. To date, various polymeric nanofibers, nanowires, and nanoribbons have been fabricated by electrospinning and non-electrospinning techniques (interfacial polymerization, sol-gel, CVD, metal-assisted chemical etching, etc.). Among them, electrospinning technology is widely used in the manufacture of commercial power equipment due to its production of very thin fibers with large surface areas, ease of processing for various purposes, and fiber production with superior mechanical properties.

The various 1D polymeric materials, such as Si-based 1D structures (inorganic), conductive polymer-based 1D structures (organic), and composite 1D structures (organic-metal/organic-inorganic) have been extensively studied for i) energy generation (solar cells, fuel cells) and ii) energy storage (batteries, supercapacitors) and wearable devices. PANI and polypyrrole-like conducting polymer nanofibers have made it possible to develop flexible, lightweight, and cost-effective alternatives for the electronic components of solar cells, fuel cells, supercapacitors, and batteries. It has been proved that 1D structures can enhance photon absorption and increase the PCE of solar cells relative to the film/bulk material. 1D structures such as nanofibers and nanowires enhance the proton-electron conductivity of the electrolytic membrane and act as excellent catalyst support for PEMFCs. In SOFCs, polymeric 1D structures serve as excellent heat-resistant electrode materials. Polymeric 1D structures consisting of batteries and supercapacitors have a higher energy density and have faster charging and discharging processes than conventional batteries. Moreover, these batteries are less expensive. The widespread usage of CNT fibers as an electrode in wearable devices is promoted by their high mechanical properties, long-term stability, and ease of charge transfer during the charge-discharge process. However, research on the functionality and deployment of integrated 1D fiber-shaped power harvesting and storage devices has reported less-than-expected. However, the devices containing the 1D polymer material discussed in this chapter will be a great foundation for providing a sustainable energy supply in the future.

References

1. H. Spanggaard, F.C. Krebs, A brief history of the development of organic and polymeric photovoltaics, *Sol. Energy Mater. Sol. Cells* 83(2–3) (2004) 125–146.
2. A. Goetzberger, C. Hebling, H.W. Schock, Photovoltaic materials, history, status and outlook, *Mater. Sci. Eng. R Rep.* 40(1) (2003) 1–46.
3. A.H. Tullo, A fuel cell in every car, *Chem. Eng. News Arch.* 79 (2001) 19–22.
4. Javaid Zaidi, T. Matsuura, *Polymer Membranes for Fuel Cells*. Spinger (2009).
5. J.M.D. Arcy, M.F. El-Kady, P.P. Khine, L. Zhang, S.H. Lee, N.R. Davis, D.S. Liu, M.T. Yeung, S.Y. Kim, C.L. Turner, A.T. Lech, P.T. Hammond, R.B. Kaner, Vapor-phase polymerization supercapacitors, *ACS Nano* 8 (2014) 1500–1510.
6. A. Nawaz, S. Goudarzi, M.A. Asghari, S. Pichiah, G.S. Selopal, F. Rosei, Z.M. Wang, H. Zarrin, Review of hybrid 1D/2D photocatalysts for light-harvesting applications, *ACS Appl. Nano Mater.* 4(11) (2021) 11323–11352.
7. Kenry, C.T. Lim, Nanofiber technology: Current status and emerging developments, *Prog. Polym. Sci.* 70 (2017) 1–17.
8. L. Cao, P. Fan, A.P. Vasudev, J.S. White, Z. Yu, W. Cai, J.A. Schuller, S. Fan, M.L. Brongersma, Semiconductor nanowire optical antenna solar absorbers, *Nano Lett.* 10(2) (2010) 439–445.
9. Q. Shu, J. Wei, K. Wang, H. Zhu, Z. Li, Y. Jia, X. Gui, N. Guo, X. Li, C. Ma, D. Wu, Hybrid heterojunction and photoelectrochemistry solar cell based on silicon nanowires and double-walled carbon nanotubes, *Nano Lett.* 9(12) (2009) 4338–4342.
10. X. Shen, B. Sun, D. Liu, S.T. Lee, Hybrid heterojunction solar cell based on organic-inorganic silicon nanowire array architecture, *J. Am. Chem. Soc.* 133(48) (2011) 19408–19415.
11. R.A. Puglisi, C. Bongiorno, S. Caccamo, E. Fazio, G. Mannino, F. Neri, S. Scalese, D. Spucches, A. La Magna, Chemical vapor deposition growth of silicon nanowires with diameter smaller than 5 nm, *ACS Omega* 4(19) (2019) 17967–17971.
12. J. Xue, J. Xie, W. Liu, Y. Xia, Electrospun nanofibers: New concepts, materials, and applications, *Acc. Chem. Res.* 50(8) (2017) 1976–1987.
13. T. Subbiah, G.S. Bhat, R.W. Tock, S. Parameswaran, S.S. Ramkumar, Electrospinning of nanofibers, *J. Appl. Polym. Sci.* 96(2) (2005) 557–569.
14. C. Wang, J. Wang, L. Zeng, Z. Qiao, X. Liu, H. Liu, J. Zhang, J. Ding, Fabrication of electrospun polymer nanofibers with diverse morphologies, *Molecules* 24(5) (2019) 834.
15. J. Xue, T. Wu, Y. Dai, Y. Xia, Electrospinning and electrospun nanofibers: Methods, materials, and applications, *Chem. Rev.* 119(8) (2019) 5298–5415.
16. Y.E. Miao, G.N. Zhu, H. Hou, Y.Y. Xia, T. Liu, Electrospun polyimide nanofiber-based nonwoven separators for lithium-ion batteries, *J. Power Sources* 226 (2013) 82–86.
17. Y. Ding, H. Hou, Y. Zhao, Z. Zhu, H. Fong, Electrospun polyimide nanofibers and their applications, *Prog. Polym. Sci.* 61 (2016) 67–103.

18. D. Chen, Y.E. Miao, T. Liu, Electrically conductive polyaniline/polyimide nanofiber membranes prepared via a combination of electrospinning and subsequent in situ polymerization growth, *ACS Appl. Mater. Interfaces* 5(4) (2013) 1206–1212.
19. N. Dong, J. Wang, N. Chen, B. Liu, G. Tian, S. Qi, G. Sun, D. Wu, In situ reinforcing: ZrO_2-armored hybrid polyimide separators for advanced and safe lithium-ion batteries, *ACS Sustain. Chem. Eng.* 9(18) (2021) 6250–6257.
20. C. Perignon, G. Ongmayeb, R. Neufeld, Y. Frere, D. Poncelet, Microencapsulation by interfacial polymerisation: Membrane formation and structure, *J. Microencapsul.* 32(1) (2015) 1–15.
21. Y. Wu, J. Wang, B. Ou, S. Zhao, Z. Wang, Some important issues of the commercial production of 1-D Nano-PANI, *Polymers (Basel)* 11(4) (2019) 681.
22. D. Gugulothu, A. Barhoum, R. Nerella, R. Ajmer, M. Bechelany, Fabrication of nanofibers: Electrospinning and non-electrospinning techniques. In: *Handb. Nanofibers*, ed. A. Barhoum, M. Bechelany, and A. S. H. Makhlouf, Springer Nature, Switzerland (2019) 45–77.
23. P. Nyamukamba, O. Okoh, H. Mungondori, R. Taziwa, S. Zinya, Synthetic methods for titanium dioxide nanoparticles: A review. In: *Titanium Dioxide - Material for a Sustainable Environment*, ed. P. Nyamukamba, O. Okoh, H. Mungondori, R. Taziwa, and S. Zinya, IntechOpen, London (2018).
24. F. Cheng, Z. Tao, J. Liang, J. Chen, Template-directed materials for rechargeable lithium-ion batteries, *Chem. Mater.* (2008) 667–681.
25. S.G. Ullattil, P. Periyat, *Sol-Gel Synthesis of Titanium Dioxide*, Springer (2017) 271–283.
26. S.-S. Choi, S.G. Lee, S.S. Im, S.H. Kim, Y.L. Joo, Silica nanofibers from electrospinning/sol-gel process, *J. Mater. Sci. Lett.* 22(12) (2003) 891–893.
27. T. Pirzada, S.A. Arvidson, C.D. Saquing, S.S. Shah, S.A. Khan, Hybrid carbon silica nanofibers through sol–gel electrospinning, *Langmuir* 30(51) (2014) 15504–15513.
28. N. Liu, K. Kim, P.C. Hsu, A.N. Sokolov, F.L. Yap, H. Yuan, Y. Xie, H. Yan, Y. Cui, H.Y. Hwang, Z. Bao, Large-scale production of graphene nanoribbons from electrospun polymers, *J. Am. Chem. Soc.* 136(49) (2014) 17284–17291.
29. S.E. Habas, H.A.S. Platt, M.F.A.M. van Hest, D.S. Ginley, Low-cost inorganic solar cells: From ink to printed device, *Chem. Rev.* 110(11) (2010) 6571–6594.
30. A. Gusain, R.M. Faria, P.B. Miranda, Polymer solar cells-interfacial processes related to performance issues, *Front. Chem.* 7 (2019) 1.
31. Z. Yang, M. Moffa, Y. Liu, H. Li, L. Persano, A. Camposeo, R. Saija, M.A. Iatì, O.M. Maragò, D. Pisignano, C.Y. Nam, E. Zussman, M. Rafailovich, Electrospun conjugated polymer/fullerene hybrid fibers: Photoactive blends, conductivity through tunneling-AFM, light scattering, and perspective for their use in bulk-heterojunction organic solar cells, *J. Phys. Chem. C* 122(5) (2018) 3058–3067.
32. Y. Wang, K.S. Chen, J. Mishler, S. Chan, X. Cordobes, A review of polymer electrolyte membrane fuel cells: Technology, applications, and needs on fundamental research, *Appl. Energy* 88(4) (2011) 981–1007.
33. S.-H. Shin, P.J. Nur, A. Kodir, D.-H. Kwak, H. Lee, D. Shin, B. Bae, Improving the mechanical durability of short-side-chain perfluorinated polymer electrolyte membranes by annealing and physical reinforcement, *ACS Omega* 4(21) (2019) 19153–19163.
34. S. Simotwo, V. Kalra, Dual nafion/polyaniline nanofibers architecture for applications in fuel cells electrodes, *ECS Trans.* 69(17) (2015) 943–953.
35. S. Jin, S.Y. Yang, J.M. Lee, M.S. Kang, S.M. Choi, W. Ahn, X. Fuku, R.M. Modibedi, B. Han, M.H. Seo, Fluorine-decorated graphene nanoribbons for an anticorrosive polymer electrolyte membrane fuel cell, *ACS Appl. Mater. Interfaces* 13(23) (2021) 26936–26947.
36. J.H. Park, Y.W. Ju, S.H. Park, H.R. Jung, K.S. Yang, W.J. Lee, Effects of electrospun polyacrylonitrile-based carbon nanofibers as catalyst support in PEMFC, *J. Appl. Electrochem.* 39(8) (2009) 1229–1236.
37. K. Waldrop, R. Wycisk, P.N. Pintauro, Application of electrospinning for the fabrication of proton-exchange membrane fuel cell electrodes, *Curr. Opin. Electrochem.* 21 (2020) 257–264.
38. J. Gong, P. Wu, Z. Bai, J. Ma, T. Li, Y. Yao, C. Jiang, Insight into the electrospinning process for SOFC cathode nanofibers, *J. Phys. Chem. C* 125(13) (2021) 7044–7053.

39. W. Pan, Z. Lü, K. Chen, X. Huang, B. Wei, W. Li, Z. Wang, W. Su, Novel polymer fibers prepared by electrospinning for use as the pore-former for the anode of solid oxide fuel cell, *Electrochim. Acta* 55(20) (2010) 5538–5544.
40. M. Endo, C. Kim, K. Nishimura, T. Fujino, K. Miyashita, Recent development of carbon materials for Li ion batteries, *Carbon N. Y.* 38 (2000) 183–197.
41. P. Roy, S.K. Srivastava, Nanostructured anode materials for lithium ion batteries, *J. Mater. Chem. A* 3(6) (2015) 2454–2484.
42. H. Han, X. Chen, J. Qian, F. Zhong, X. Feng, W. Chen, X. Ai, H. Yang, Y. Cao, Hollow carbon nanofibers as high-performance anode materials for sodium-ion batteries, *Nanoscale* 11(45) (2019) 21999–22005.
43. F. Shi, C. Chen, Z.-L. Xu, Recent advances on electrospun nanofiber materials for post-lithium ion batteries, *Adv. Fiber Mater.* 3(5) (2021) 275–301.
44. J. Verma, Metal-ion batteries for electric vehicles: current state of the technology, issues and future perspectives, *Nanoscale Adv.* 3 (2021) 3384–3394.
45. Y. Hu, D. Ye, B. Luo, H. Hu, X. Zhu, S. Wang, L. Li, S. Peng, L. Wang, A binder-free and free-standing cobalt Sulfide @ Carbon nanotube cathode material for aluminum-ion batteries, *Adv. Mater.* 1703824 (2017) 1–6.
46. Y. Hu, D. Ye, B. Luo, H. Hu, X. Zhu, S. Wang, L. Li, S. Peng, L. Wang, A binder-free and free-standing cobalt Sulfide@Carbon nanotube cathode material for aluminum-ion batteries, *Adv. Mater.* 30(2) (2018) 1703824.
47. H. Wang, Q. Xu, Materials design for rechargeable metal-air batteries, *Matter* 1 (2019) 565–595.
48. P. Rao, P. Cui, Z. Wei, M. Wang, J. Ma, Y. Wang, X. Zhao, Integrated N-Co/carbon nanofiber cathode for highly efficient zinc–air batteries, *ACS Appl. Mater. Interfaces* 11(33) (2019) 29708–29717.
49. Y. Wang, J. Fu, Y. Zhang, M. Li, F.M. Hassan, G. Li, Z. Chen, Continuous fabrication of a MnS/Co nanofibrous air electrode for wide integration of rechargeable zinc–air batteries, *Nanoscale* 9(41) (2017) 15865–15872.
50. Y. Li, L. Liu, C. Liu, Y. Lu, R. Shi, F. Li, J. Chen, Rechargeable aqueous polymer-air batteries based on polyanthraquinone anode, *Chem* 5(8) (2019) 2159–2170.
51. Z. Chen, X. Yang, W. Li, X. Liang, J. Guo, H. Li, Y. He, Y. Kim, Nanofiber composite for improved water retention and dendrites suppression in flexible zinc-air batteries, *Small* 17(39) (2021) e2103048.
52. W.-M. Chang, C.-C. Wang, C.-Y. Chen, Plasma-induced polyaniline grafted on carbon nanotube-embedded carbon nanofibers for high-performance supercapacitors, *Electrochim. Acta* 212 (2016) 130–140.
53. N.C. Abeykoon, J.S. Bonso, J.P. Ferraris, Supercapacitor performance of carbon nanofiber electrodes derived from immiscible PAN/PMMA polymer blends, *RSC Adv.* 5(26) (2015) 19865–19873.
54. Y.-E. Miao, J. Yan, Y. Huang, W. Fan, T. Liu, Electrospun polymer nanofiber membrane electrodes and an electrolyte for highly flexible and foldable all-solid-state supercapacitors, *RSC Adv.* 5(33) (2015) 26189–26196.
55. Y. Zhang, Y. Zhao, J. Ren, W. Weng, H. Peng, Advances in wearable fiber-shaped lithium-ion batteries, *Adv. Mater.* 28 (2016) 4524–4531.
56. Y. Zhou, C. Wang, W. Lu, L. Dai, Recent advances in fiber-shaped supercapacitors and lithium-ion batteries, *Adv. Mater.* 1902779 (2019) 1–24.
57. X. Tao, Study of fiber-based wearable, *Acc. Chem. Res.* 52(2) (2019) 307–315.
58. C. Lu, X. Chen, Latest advances in flexible symmetric supercapacitors: From material engineering to wearable applications, *Acc. Chem. Res.* 53(8) (2020) 1468–1477.
59. J. Ren, Y. Zhang, W. Bai, X. Chen, Z. Zhang, X. Fang, W. Weng, Y. Wang, H. Peng, Elastic and wearable wire-shaped lithium-ion battery with high electrochemical performance, *Angew. Chem.* 126 (2014) 7998–8003.
60. H. Sun, Y. Zhang, J. Zhang, X. Sun, H. Peng, Energy harvesting and storage in 1D devices, *Nat. Rev. Mater.* 2(6) (2017) 17023.

61. X. Chen, L. Qiu, J. Ren, G. Guan, H. Lin, Z. Zhang, P. Chen, Y. Wang, H. Peng, Novel electric double-layer capacitor with a coaxial fiber structure, *Adv. Mater.* 25 (2013) 6436–6441.
62. X. Chen, H. Lin, J. Deng, Y. Zhang, X. Sun, P. Chen, X. Fang, Z. Zhang, G. Guan, H. Peng, Electrochromic fiber-shaped supercapacitors, *Adv. Mater.* 26 (2014) 1–7.
63. J. Ren, L. Li, C. Chen, X. Chen, Z. Cai, L. Qiu, Y. Wang, Twisting carbon nanotube fibers for both wire-shaped, *Adv. Mater.* 25 (2013) 1155–1159.
64. T. Khudiyev, B. Grena, G. Loke, C. Hou, H. Jang, J. Lee, G.H. Noel, J. Alain, J. Joannopoulos, K. Xu, J. Li, Y. Fink, J.T. Lee, Thermally drawn rechargeable battery fiber enables pervasive power, *Mater. Today* 52 (2022) 80–89.

9

High-Performance Supercapacitors Based on One-Dimensional Polymeric Nanocomposites

Masoud Kasraie, Ting Zheng, Xiaodong Wang, Feiyan Su, and Parisa Pour Shahid Saeed Abadi

CONTENTS

9.1 Introduction

Supercapacitors, also known as electrochemical capacitors, are devices that store energy at the interface of an electrode and electrolyte. They have received much attention because of their high power storage capacity, rapid charging and discharging rates, long cycle life, flexible operating temperature, and environment-friendly nature [1]. The most common materials used within the electrode are conductive polymers, transition metal oxides, and carbon materials. Composites made of two or three of these categories of materials benefit from the unique characteristics of each material and result in more extraordinary properties.

When the size of the materials decreases to the nanoscale, their physical and chemical properties usually change. Reducing the size of materials increases the surface area, which improves the contact between electrodes and electrolytes. Hence, scientists use this concept to enhance the performance of energy storage and energy conversion devices using nanometer materials. 1D nanocomposites, such as nanowires, nanotubes, and nanofibers, in which one of the constituents is longer in one direction than in the other two directions, are particularly attractive for supercapacitors because of the large surface area and directionality in charge transfer and electric and electronic properties [2]. Among all 1D nanomaterials, 1D polymer electrodes have received significant attention in recent years due to their large capacitance, good electrical conductivity, ease of synthesis, and low cost.

9.2 Classifications of Supercapacitors

Supercapacitors are divided into two groups based on the storage mechanism: electric double-layer capacitors (EDLCs) and pseudocapacitors. EDLCs are mainly composed of

DOI: 10.1201/9781003223764-9

carbon materials such as carbon nanotubes (CNTs), graphene, and activated carbon, and operate with the accumulation of charges at the electrolyte/electrode interface without electron transfer. Due to the high surface area, excellent electrical conductivity, exceptional chemical stability, and reasonable cost, carbon materials have become widely used as electrode materials for supercapacitors [3]. CNTs, graphene, activated carbon, and carbon aerogel are the most favored materials for this purpose. Remarkable mechanical and electrical properties and high specific surface area of CNTs recommend them to be excellent candidates for the 1D component of a nanocomposite for an electrode in a supercapacitor. On the other hand, pseudocapacitors, mainly composed of metal oxides or conducting polymers [4], rely on electron transfer and the changes in the oxidation state of the electroactive materials [5]. High capacitance is the essential characteristic feature of materials that are commonly used in this type of supercapacitors. The drawback of these materials is the low conductivity and low cyclic stability. Therefore, hybrids of materials that operate within both storage mechanisms have been developed. The hybrids provide both high energy density and high power density. 1D polymer and metal oxide nanomaterials have been used within the structure of nanocomposites for pseudocapacitors and hybrids between pseudocapacitors and EDLCs.

9.3 1D Nanocomposites of Metal Oxides and Conducting Polymers

Conductive polymers have some advantages over other materials in terms of use as a base for capacitors, such as low self-discharge, relatively low cost, and high energy density because of redox reactions. Polythiophene (PT), polyaniline (PANI), polypyrrole (PPy), and their derivatives are the most commonly used conductive polymers for this purpose. Among conductive polymers, PPy, as a classical conductive polymer, has been widely used in microelectronics, electrochemistry, and biotechnology due to its good stability in air, high electrical conductivity, environmental non-toxicity, and reversible oxidation and reduction properties. The energy storage of PPy is achieved *via* the Faraday pseudocapacitance principle. The electrochemical reactions occur both on the surface in 2D and in the bulk of the material in quasi-2D; thus, PPy could store a high energy density of electric charge and produce higher Faraday pseudocapacitance. The specific capacitance of PPy nanofibers prepared by Arcila-Velez et al. [6] reached 236 F/g when the current density was 0.1 A/g in a 1 M H_2SO_4 electrolyte. Liu et al. [7] compared PPy nanotubes and PANI nanofibers used as electrodes in supercapacitors. They found that the specific capacitance of PPy nanotubes is as high as 463 F/g at a current density of 0.3 A/g, significantly higher than that of PANI nanofibers (243 F/g). PPy nanotubes also showed a higher long-term charge-discharge capacity.

The preparation of polymers with 1D nanostructures can be divided into two methods: template technique (e.g., using a porous film as a template, such as anodic alumina (AAO) or a porous copolymer film) and non-template technique (e.g., electrostatic spinning method, hydrothermal method, chemical vapor deposition method, or electrodeposition method) [8]. The template method was explored and studied by Martin and Parthasarathy [9] and has been widely applied in the synthesis of conductive polymers with nanofiber or nanowire arrays. The template method presents the advantages of simple operation, diversified functions, and controllable processes. During the preparation process, the voids of the template are filled with materials through pressure perfusion, gas-phase precipitation, chemical deposition, or electrochemical deposition, etc. Finally, the template is removed, and the material with the target morphology is formed. He et al. [10] prepared PPy nanowires using a template method and then synthesized MnO_2 nanosheets@PPy nanowire (PPy nanowire-supported MnO_2 nanosheets) composite using a hydrothermal method. As shown in Figure 9.1, MnO_2 nanosheets were uniformly grown on PPy nanowires. The specific capacity of MnO_2@PPy nanowire and pure PPy nanowire electrodes was 271 F/g and 23.3 F/, respectively, under a current density of 3 A/g. After 1,500 cycles of testing, the capacity of the pure PPy nanowire electrode decreased to 65.4% of the original capacity, whereas the MnO_2/PPy nanowire electrode still maintained 93.4% capacity, indicating that the specific capacity, rate performance, and cycle stability of the MnO_2/PPy nanowire composite electrode were markedly superior. The MnO_2 nanosheet structure was reported to buffer the collapse of the structure as the number of cycles increased and to prevent PPy nanowires from agglomeration and stacking. Thus, the MnO_2/PPy nanowire composite showed excellent electrochemical performance. Although the template method can produce nanotubes and nanowires of conductive polymers with various diameters (20 nm to a few microns), the template is expensive, and removing the template often damages the nanostructure of the polymer. Therefore, it is common to prepare conductive polymers with a 1D structure using non-template methods. PANI is an electroactive polymer and pseudocapacitor and has several advantages over other conductive polymers such as high conductivity, ease of synthesis, low cost, and relatively high capacitance. Chen et al. [11] used chemical polymerization methods to synthesize PANI on MnO_2 nanotubes and then mixed it with acetylene black and polytetrafluoroethylene (PTFE) to form the electrode materials. They used the PANI nanotubes as the electrode material and an ionic liquid as the electrolyte to fabricate supercapacitor devices with a remarkable energy density of 84 $Wh/kg_{(cell)}$ and a power density of 182 $kW/kg_{(cell)}$. Chiou et al. [12] used different doped acids ($HClO_4$, H_2SO_4, and camphorsulfonic acid [CSA]) and ammonium persulfate as oxidants to prepare polyaniline with one-dimensional nanostructure by dilute solution polymerization. The diameter of the polyaniline nanofibers obtained was in the range of 27–250 nm, 42–70 nm, and 17–50 nm after using $HClO_4$, H_2SO_4, and CSA, respectively. The results show that a low concentration of reactants is beneficial to the preparation of PANI nanofibers, whereas irregular aggregates and large particles of polymerization products are obtained if the concentration is high.

Ordered array refers to the directional arrangement of 1D nanostructures along a specific direction, which gives the macroscopic material anisotropic characteristics. The

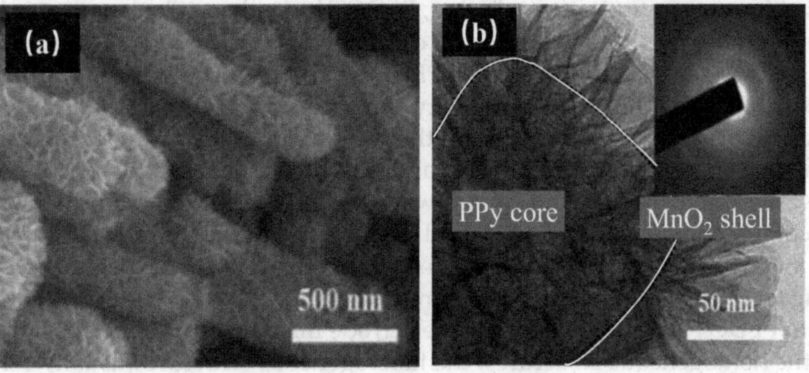

FIGURE 9.1
MnO_2 nanosheet@PPy nanowire composite electrode. Adapted with permission from [10] Copyright (2017) Nature Research.

directional arrangement of 1D nanostructures creates gaps between adjacent structures, promoting rapid ion transfer between the electrolyte and active materials [13]. Meanwhile, the directional movement of electrons shortens the transport distance and time, thus accelerating the electrochemical reaction kinetics. In recent years, the development of conductive polymers with an ordered array structure has become a research hotspot. Materials with this kind of structure are considered excellent for the next generation of supercapacitors. When conductive polymers have ordered 1D nanostructure, the charge and discharge rates of capacitors and the cycle stability of supercapacitors are significantly enhanced. The specific capacitance of ordered PANI nanomaterials was reported to be higher than that of disordered PANI nanomaterials [8]. The reasons are as follows: the small diameter of the ordered PANI nanoarray makes it ionically conductive and thus reduces the resistance to ion diffusion, while the gaps between ordered PANI nanoarrays can adapt to the volume change of the material, enhancing the electrochemical properties. The disordered PANI network structure is intertwined, which not only obstructs ion transport but also cannot adapt to the mechanical stress caused by volume changes. Therefore, 1D ordered PANI nanostructures serve as electrode materials with excellent electrochemical properties. The preparation of ordered 1D nanostructures could also be divided mainly into template and non-template methods. Li et al. [14] fabricated vertically oriented PANI nanowires on anodic alumina templates by electrochemical deposition as the electrode material for supercapacitors. The electrochemical deposition was carried out in aniline and sulfuric acid electrolyte at a voltage of 0.7 V, and the deposition time was 5,000 seconds. Subsequently, PANI nanowire was obtained by removing the AAO template in 0.01 M NaOH solution. The PANI nanowires obtained exhibited a specific capacity of 1142 F/g at a current density of 5000 mA/g, with only 5% capacity decay after 500 cycles. Despite the excellent electrochemical properties of these electrode materials, the use of AAO templates has limited their application in large-scale production. Therefore, Wang et al. [15] developed a template-free method to prepare PANI nanowires. PANI arrays were directly synthesized using a simple constant-current electropolymerization technique at 0.01 mA/cm current density for 1 hour. The PANI synthesized by this method has a specific capacity of 950 F/g at a current density of 1000 mA/g and decays by 16% after 500 cycles.

Compared with PPy and PANI, polythiophenes have a higher oxygenated doping potential, which provides them with the ability to be oxygenated into p-type and reduced into n-type doped states. Among all polythiophene derivatives, the 3, 4-ethylenedioxyethiophene polymer (PEDOT) meets the required properties of the electrode materials for supercapacitors. After n-type and p-type doping, PEDOT has enhanced properties, which have attracted much attention in the research for electrode materials for supercapacitors. D'Arcy et al. [16] synthesized vertically oriented PEDOT nanofibers with a large aspect ratio by vapor deposition. The electrical conductivity was measured to be as high as 130 S/cm. The PEDOT nanofibers synthesized by this method can easily be deposited into films to form supercapacitors without adding adhesives and conductive agents. After 10,000 charge and discharge cycles, the specific capacitance was at 92% of the initial value, showing good cycle stability. Although PEDOT is a good electrode material, its specific capacitance is relatively low. This may be caused by the high molecular weight of the monomer and the low doping level. Hence, it is necessary to improve the specific capacitance by adding other materials to PEDOT for the preparation of electrode materials for supercapacitors.

Unlike the homogeneous structures, the heterogeneous structures usually contain two or more components, and the biggest advantage is the synergistic interactions between the different components. For energy storage materials, this kind of synergistic effect can effectively reduce the surface energy, reduce the agglomeration of active materials, and alleviate the side reactions at the electrode/electrolyte interface. Thus, materials with heterogeneous structures always show better electrochemical performance. During the recharge-discharge process, the electrochemical performance of conducting polymers is weakened due to the expansion and contraction of their volume, which greatly limits their application as an electrode material for supercapacitors. Many researchers have proved that conductive polymer and metal oxide composites have improved cycle stability and specific capacitance compared with the individual constituents. These metal oxide particles, nanorods, or nanotubes have been used as the carrier materials of conducting polymers, increasing the volume of conducting polymers and providing a larger surface area and a shorter ion channel for the electrode. 1D heterogeneous nanomaterials have wide versatility and applicability as the key components in the fields of nanoelectronics, catalysis, chemical sensing, energy conversion, and storage. As a consequence, they have attracted extensive scientific and technical attention [17]. In particular, much effort is being made to synthesize core-shell heterostructures with excellent synergistic properties or multiple functions. The ordered metal oxide arrays (TiO_2, SnO_2, CoO, ZnO, etc.) are widely used as supporting materials for the polymerization of conducting polymers.

For example, Zhou et al. [18] grew a CoO array on nickel foam using hydrothermal methods combined with calcination, and then coated a layer of PPy on the surface by *in-situ* polymerization. The PPy was uniformly coated on a CoO nanowire array. The specific capacity of this composite was 647 F/g at a high current density of 50 mA/cm^2, a capacity value which is much higher than that of a pure CoO nanowire electrode (170 F/g). In addition, the capacity retention rate of the CoO/PPy nanowire electrode was 99.8% after 2,000 cycles. The excellent electrochemical performance of the CoO@PPy nanowire electrode is due to the enhanced electrical conductivity and mechanical stability of CoO from the PPy coating, which enables the electrode to maintain good structural integrity during the cycle testing. Yang et al. [19] reported that ZnO nanowires with a diameter of 80–100 nm and a length of 4–5 μm were prepared using a hydrothermal method. The prepared ZnO nanowires had a uniform structure and no impurities. Using an atom-transfer free radical polymerization method, the block copolymer PTFEMA-*b*- PSSNa was successfully grafted onto the ZnO surface, forming the core-shell structure of ZnO@PTFEMA-*b*-PSSNa. A PTFEMA polymer layer grafted onto ZnO surface was reported to alleviate the dissolution of ZnO nanowires effectively. Then, using PSSNa chains, the counterion dopant, ZnO@PEDOT/PSSNa, was successfully prepared, and the composite electrode obtained reached 101.34 F/g at a scan rate of 20 mV/s. Pruna et al. [20] deposited a ZnO nanorod array on an ITO template at constant pressure and then placed the ITO template covered with ZnO nanorods into a solution containing sodium *p*-toluenesulfonate, sodium dodecyl sulfonate, and pyrrole monomer for electrochemical deposition of PPy to obtain a ZnO/PPy core-shell structure. The thickness of the PPy on the ZnO was around 15–60 nm. They also prepared a core-shell ZnO/PPy- oxidized graphene (GO) structure in this study by adding GO to the solution. The ZnO/PPy-GO core-shell material presented a larger areal specific capacitance than ZnO/PPy at scan rates lower than 100 mV/s in the three-electrode cyclic voltammetry test in a 0.5 M $NaNO_3$ solution, but, when the scan rate increased from 10 mV/S to 100 mV /S, the ZnO/PPy-GO core-shell material showed only a 66% retention ratio of capacitance, whereas the ZnO/PPy core-shell material retained 71%.

Asymmetric supercapacitors (ASCs) use different materials for anode and cathode to achieve superior properties. Lu et al. [21] used Fe nanowire arrays on carbon cloth as supporting material to obtain α-Fe_2O_3 nanowire arrays and PANI film by electrodeposition, and the α-Fe_2O_3@PANI core-shell nanowire arrays prepared exhibited a specific capacitance of 103 mF/cm^2 under a current density of 0.86 mA/cm^2, which was significantly higher than the 33.93 mF/cm^2 obtained for pure α-Fe_2O_3 electrodes. After 2,500 cycles of charge and discharge, the specific capacitance experienced almost no loss. Then, an ASC device was made using the α-Fe_2O_3@PANI core-shell nanowire arrays operating as the anode and PANI nanowire arrays working as the cathode. Maximum energy density was measured to be 0.35 mW/cm^3 while maximum power density was 301.19 mW/cm^3. After 10,000 cycles, the cycling stability of the specific capacitance was 95.77%.

In recent years, TiO_2 nanowires have been widely used as electrode materials for supercapacitors due to their advantages of large specific surface area, good chemical and thermal stability, low synthesis cost, and fast charge transfer. They can be used as a support material to prepare ordered conducting polymer nanowires. The preparation methods of TiO_2 nanowires include the template method, hydrothermal method, anodic oxidation method, etc. [22, 23]. Among these synthesis methods, the anodic oxidation method shows great advantages in obtaining ordered TiO_2 nanowires of controllable size. Xie et al. [24] synthesized TiO_2 ordered nanowires by anodic oxidation with a Ti sheet first as a substrate and then used as a template to synthesize ternary coaxial PPy/TiO_2/PANI nanowires by electrochemical deposition. The specific capacitance of the PPy/TiO_2/PANI composite (497.1 F/g at 0.5 A/g) was higher than that of the PPy/TiO_2 composite (301.6 F/g at 0.5 A/g). In addition, the ternary coaxial PPy/TiO_2/PANI nanowires also presented good cycling performance; the capacitance remained at 72.3% of the initial value after 500 charge-discharge cycles, which shows promise for an electrode material for supercapacitors. Li et al. [25] synthesized an S-doped TiO_2 nanotube array using the interface engineering method. S can improve the conductivity of TiO_2 nanotubes and enhance the bonding ability between PANI and the TiO_2 surface, thus improving the capacitance and cycling stability of TiO_2/PANI composite materials. Then, PANI/S-TiO_2/Ti composites were prepared by electrochemical deposition, which resulted in large specific capacitance (91.9 mF/cm^2 at 0.01 mA/cm^2) compared with S-TiO_2/Ti fiber electrodes and other reported PANI/CNT fiber electrodes (38 mF/cm^2). Moreover, this 1D composite had good cycle stability (the specific capacitance was 93.78% of the initial value after 12,000 charge and discharge cycles). TiO_2/PANI nanoarray composites can be synthesized by electrochemical deposition and chemical oxidation, and their electrochemical properties are different. Still, their specific capacitance and cycle life are superior to those exhibited by pure PANI, as shown in Table 9.1.

TABLE 9.1

Electrochemical Properties of Metal Oxide/PANI Composites

Materials	Synthetic Methods	Current Density	Specific Capacitance	Testing Methods	Capacitance Retention	Reference
TiO_2 nanotubes/ PANI	Electrochemical deposition	3 A g^{-1}	740 F g^{-1}	Three electrode test system	87% (1,100 cycles)	[26]
PANI-APTES-TiO_2 NTA	Chemical oxidation	0.5 A g^{-1}	380 F g^{-1}	Three electrode test system	90.5% (1,100 cycles)	[24]
PPy/TiO_2/PANI coaxial nanotube arrays	Electrochemical deposition	0.5 A g^{-1}	497 F g^{-1}	Three electrode test system	72.3% (500 cycles)	[21]
PANI/H-TiO_2 nanotube arrays	Electrochemical deposition	1.2 A g^{-1}	872 F g^{-1}	Three electrode test system	70.2% (2,000 cycles)	[27]
TiO_2@PANI core-shell nanowire arrays	Electrochemical deposition	0.5 mA cm^{-2}	32.7 F g^{-1}	Three electrode test system	97% (10,000 cycles)	[28]

9.4 1D Nanocomposites of CNTs and Conducting Polymers

PANI has repeatedly been used with CNTs in supercapacitors. Several studies have been performed on fabricating PANI/CNT supercapacitors based on electrodeposition of PANI on aligned or wavy CNT films. Bavio et al. [29] synthesized a PANI-CNT nanostructure using an oxidative polymerization process in monomer acid solution, surfactant, and MWCNTs. A specific capacitance value of 1566 F/g at 0.3 mA/cm^2 was obtained. Additionally, high stability over 1,000 cycles showed capacitance retention of more than 85%. Most studies were focused on randomly oriented CNTs and PANI. However, Lin et al. [30] fabricated aligned CNT/PANI electrode-based supercapacitors and reported a large specific capacitance of 233 F/g. Ge et al. and Xie's team [31, 32] reported unique strategies to control morphology and orientation of SWCNTs in the PANI matrix, such as skeleton/kin strategy or using cellulose fiber network. Chen et al. [33] described a high-performance supercapacitor based on PANI/vertically aligned CNT nanocomposite electrodes. Vertically aligned CNTs were fabricated by the electrochemical induction of approximately 0.75 V. A large capacitance of 403.3 F/g was achieved, which is six-fold higher than that of disordered CNTs in $HCLO_4$ electrolyte. A 90.2% retention after 3,000 cycles at 4 A/g and a high energy density of 98.1 Wh/kg in $EMIBF_4$ organic electrolyte were the other two significant achievements of this study.

As shown in Figure 9.2, the straight conductive pathway inside vertically aligned CNTs helps charges move fast and enhances the electrochemical performance of the capacitor.

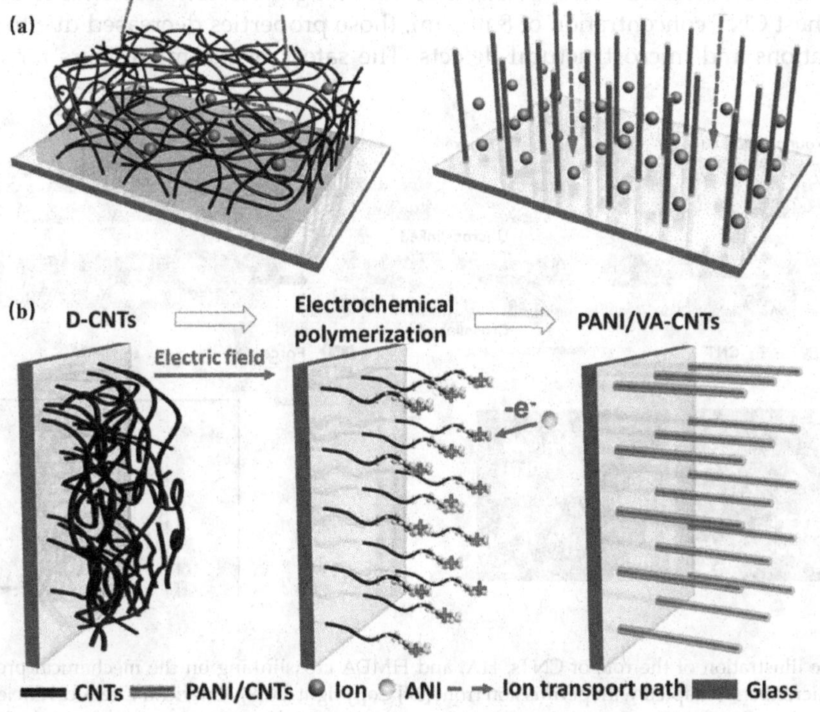

FIGURE 9.2
Schematic illustration of the preparation of PANI/VA-CNT films. Adapted with permission from ref [33].

The randomly oriented CNT film was used as the working electrode, with Af/AgCl as the reference electrode, platinum plate as the counter electrode in 0.1 M aniline, and 1 M $HClO_4$ as the electrolyte. The disordered CNTs were induced by the electric field and oriented along the electric field. In the meantime, aniline monomers were electrodeposited on the CNTs (Figure 9.2). Uniform coating of PANI on CNTs was reported. Chen et al. [34] electrodeposited PANI on the activated CNT hydrogel film and measured a high area-based capacitance of 680 mF/cm^2 in 1 M H_2SO_4 aqueous solution, with almost 100% capacitance retention after 500 cycles. Additionally, Liu et al. [35] reported high specific capacitance of 350 F/g, 91.9% capacitance retention after 1,000 cycles, and 17.2 μA loss of current by a uniform coating of PANI on CNTs.

Zheng et al. [36] used a combination of wet spinning and electrochemical deposition to fabricate hyaluronic acid (HA)/CNT as a highly conductive, stiff, and flexible support core for PANI particles, and reported improved cycling stability and flexibility of the nanocomposites under bending and twisting strains. The core was previously proven to be a biocompatible microfiber that actuates in a biological environment as a weak electrolyte with the application of a low potential of $\pm$ 1 V [37]. The effect of CNTs, HA, and crosslinking of the HA chains, using hexamethylenediamine (HMDA), on mechanical properties are depicted in Figure 9.3a; both HA crosslinking and an increase in CNT concentration increased the failure strain, which was explained by the role of alignment of CNTs in the strain and the effect of crosslinking on delaying failure. Elastic modulus is also increased by HA crosslinking due to the increased resistance to deformation in the case of crosslinked chains. Furthermore, 7 mg/ml was shown to be the optimum CNT concentration, achieving the best combination of elastic modulus, toughness, and electrical conductivity; at the highest CNT concentration of 8 mg/ml, those properties decreased due to possible agglomerations and microstructural defects. The safety of the microfibers for implants

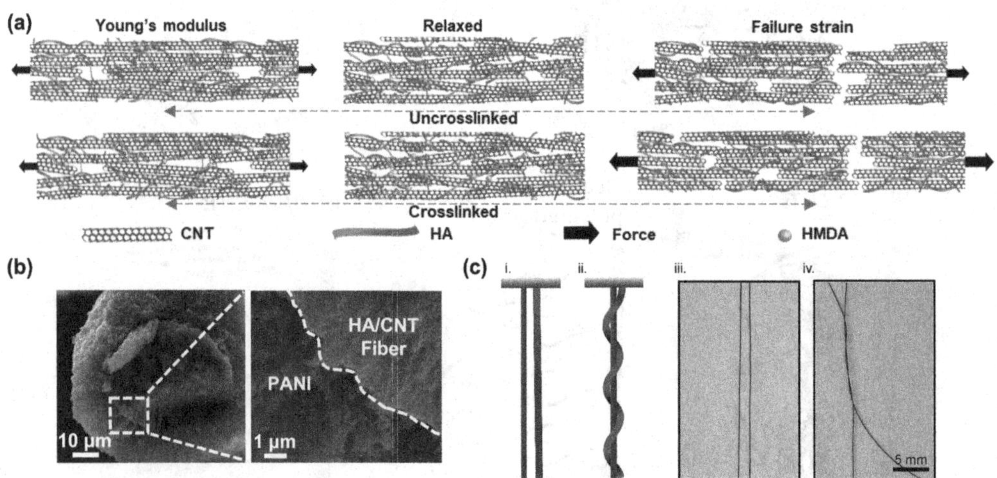

FIGURE 9.3
a) Schematic illustration of the role of CNTs, HA, and HMDA crosslinking on the mechanical properties of CNT/HA microfibers. Adapted with permission from [37] Copyright (2019) American Chemical Society, b) SEM micrographs of the HA/CNT core coated with PANI at two different magnifications, c) Schematic illustrations and optical photos of HA/CNT/PANI microfibers twisted around a wire to show flexibility; i) schematic illustration of the undeformed state, ii) schematic illustration of the twisted state, iii) optical photo of the undeformed state, and iv) optical photo of the twisted state. Images b and c are adapted with permission from [36] Copyright (2020) Elsevier.

and cell scaffolds was shown by testing *in-vivo* biocompatibility and *in-vitro* cell viability, respectively. The fibers were used as implants in the subcutaneous tissue in mice and the early inflammatory response to the fibers was resolved to show biocompatibility of both uncrosslinked and crosslinked fibers. Fibroblasts cultured on the fibers also showed viability scores greater than 80% in both uncrosslinked and crosslinked fibers. In the next work, PANI was deposited electrochemically on the wet-spun HA/CNT fibers and formed a porous layer on the outer surface of the HA/CNT fiber (Figure 9.3b) [36]. The specific capacitance of HA/CNT/PANI fibers was approximately 28000 mF/cm^2, which is approximately seven-fold greater than that of HA/CNT fibers in 0.5 M H_2SO_4, while 88.27% capacitance retention after 3,000 cycles at 100 mV/s was attained. The fibers showed flexibility during bending and twisting (Figure 9.3c) and 86% retention of capacitance over 100 bending cycles.

Stretchable supercapacitors can be functional under multiple modes of mechanical deformations, including bending, twisting, tension, and compression. The application of these devices is versatile and should be significant for powering wearable electronics and biomedical devices. Conductive polymers, carbon-based materials, and metal-based materials are the three essential components that have been used for fabricating stretchable supercapacitors. Each of these compounds has some advantages and disadvantages. For instance, carbon-based materials are the most favorable due to their high surface area, resulting in high capacitance. Still, they may have some issues, such as material processing and conductivity. Other limitations, like the low electrical conductivity of conductive polymers and low capacitance of metal-based materials, led to researchers combining all of these three materials to make better stretchable supercapacitors. Next, we discuss some of the stretchable polymer supercapacitors based on 1D carbon-based materials.

In addition to PANI, among all conductive polymers, PPy is one of the most widely used polymers for improving the capacitance of CNT electrodes. PPy has some advantages such as flexibility, stable chemical properties, easy preparation, and environmentally friendly composition. Zu's group [38] fabricated fiber-shaped supercapacitors (FSSs) by electrodepositing PPy on the surface of highly aligned discontinuous carbon fiber (CF/PPy) and using that as a core electrode, with LiCl/PVA as the electrolyte. MnO_2 was deposited on a CNT film with a thickness of 2 μm. CF/PPy core fiber was wrapped with the CNT/ MnO_2, with the electrolyte separator sandwich layer between them (Figure 9.4). An areal capacitance of 66.27 mF/cm^2 and area energy density of 23.56 $\mu Wh/cm^2$ was achieved for this particular FSS. An 83% capacitance retention after 5,000 cycles was reported. It is worth mentioning that Wang's group [39] employed a similar methodology without using CFs; they electrodeposited PPy and MnO_2 on a CNT film instead and obtained a stretchable, asymmetric supercapacitor. Wei et al. [40] synthesized CNT-PPy composite fibers using an electrochemical deposition method. The composite showed an exceptional specific capacitance of 302 F/g in a liquid electrolyte and 69 F/g in a gel electrolyte. The fiber-shaped supercapacitor showed a wide working temperature range from −27 to 61 °C, a high power density of 3.8 kW/kg, and a high energy density of 3.6 Wh/kg.

Li et al. [41] fabricated stretchable fiber supercapacitors based on a MnO_2-oxidized CNT fiber electrode, using a simple pre-straining-then-buckling method. CNT fibers were synthesized with an oxidation procedure that helps ion accessibility and increases the hydrophilicity of CNTs. As shown schematically in Figure 9.4b, two CNT fibers were coated with MnO_2 and placed on a pre-strained PDMS film covered with a gel electrolyte (LiCl-PVA). A specific volumetric capacitance of 409.4 F/cm^3 at 0.75 A/cm^3 was reported, which is an astonishing number and 33 times higher than that of pure CNT fiber-based supercapacitors. The fibers showed 40% stretchability.

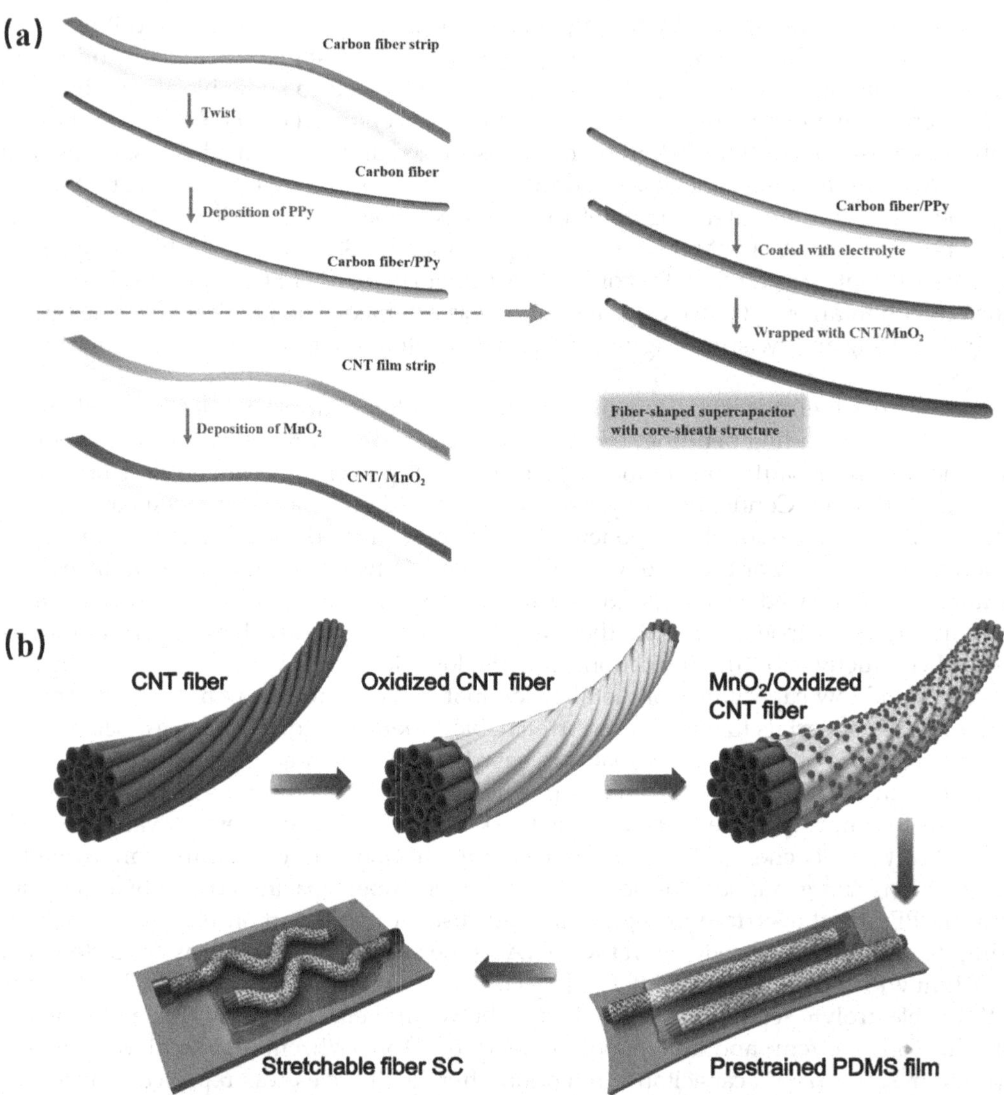

FIGURE 9.4
a) Schematic demonstration of CNT/MnO2, CF/PPy, and electrolyte separator structures. Adapted with permission from [38] Copyright (2021) American Chemical Society, b) Schematic demonstration of the fabrication process of the stretchable fiber supercapacitor. Adapted with permission from [41] Copyright (2017) Wiley.

Due to its large electroactive potential window, high conductivity, and high capacitance, Poly(3,4-ethylenedioxythiophene) (PEDOT) has attracted considerable attention for use as a pseudocapacitive polymeric conducting material. PEDOT, by itself, has several disadvantages such as requiring a complex fabrication process that takes hours or even days to complete and which needs a high temperature, exhibiting low solubility in solvents, and imperfect mechanical properties. To tackle these issues, PEDOT: polystyrene sulfonate (PSS) was fabricated. This complex is composed of two elements, the positively charged PEDOT and the water-soluble PSS. PSS has two significant roles: first, it helps the stabilization of dispersion of PEDOT, and second, it provides a base solution for forming an aqueous dispersion [42]. Yu et al. [43] fabricated a flexible, highly conductive MWCNT cellulose/PEDOT: PSS (MCPP) film. The MCPP film showed a high specific capacitance of 485 F/g at 1 A/g and excellent retention of 95% after 2,000 cycles at 2 A/g. They utilized the fabricated MCPP with two different solid electrolytes (Ni foams and PVA/KOH gel). The results showed a high volumetric capacitance of 50.4 F/cm^3 at 0.05 A/cm^3. Additionally, both electrodes and supercapacitors are environmentally stable and able to stay functional under large deformations.

9.5 Conclusions

Wherever charge transfer, conductivity, mechanical properties, large surface area, or a combination of these characteristics is in demand, nanomaterials, especially 1D geometries, exhibit great potential. In nanocomposites for electrodes in supercapacitors, they have sometimes been used as a supporting layer. In most cases, however, the unique properties of the nanomaterial, such as high values of capacitance, electrical conductivity, flexibility, or mechanical strength, have been the actual targets in the material selection process. The future directions of using 1D nanocomposite materials in high-performance supercapacitors are dictated by the applications in renewable energy devices and biotechnology. Improved energy density, cycle stability, and mechanical performance are some of the desired properties. Some of the future directions in this field are the development of more advanced devices that operate through a combination of EDLC and pseudocapacitance mechanisms, advanced manufacturing of the materials for electrodes, and new polymer and nanomaterial chemistries.

References

1. J.R. Miller, P. Simon, Materials science: Electrochemical capacitors for energy management, *Science.* 321(5889) (2008). https://doi.org/10.1126/science.1158736.
2. N. Phattharasupakun, J. Wutthiprom, S. Kaenket, T. Maihom, J. Limtrakul, M. Probst, S.S. Nagarkar, S. Horike, M. Sawangphruk, A proton-hopping charge storage mechanism of ionic one-dimensional coordination polymers for high-performance supercapacitors, *Chemical Communications.* 53(86) (2017). https://doi.org/10.1039/c7cc07490e.
3. J.R. Miller, P. Simon, Fundamentals of electrochemical capacitor design and operation, *Electrochemical Society Interface.* 17(1) (2008). https://doi.org/10.1149/2.F02081IF.
4. H. Ji, X. Zhao, Z. Qiao, J. Jung, Y. Zhu, Y. Lu, L.L. Zhang, A.H. MacDonald, R.S. Ruoff, Capacitance of carbon-based electrical double-layer capacitors, *Nature Communications.* 5 (2014). https://doi.org/10.1038/ncomms4317.
5. B.E. Conway, *Electrochemical Supercapacitors: Scientific Fundamentals and Technological.* New York: Springer, 1999.
6. M.R. Arcila-Velez, M.E. Roberts, Redox solute doped polypyrrole for high-charge capacity polymer electrodes, *Chemistry of Materials.* 26(4) (2014) 1601–1607. https://doi.org/10.1021/cm403630h.
7. J.H. Liu, J.W. An, Y.X. Ma, M.L. Li, R.B. Ma, M. Yu, S.M. Li, Electrochemical properties comparison of the polypyrrole nanotube and polyaniline nanofiber applied in supercapacitor, *EPJ Applied Physics.* 57(3) (2012). https://doi.org/10.1051/epjap/2012110368.

8. K. Wang, H. Wu, Y. Meng, Z. Wei, Conducting polymer nanowire arrays for high performance supercapacitors, *Small.* 10(1) (2014). https://doi.org/10.1002/smll.201301991.
9. R. V. Parthasarathy, C.R. Martin, Synthesis of polymeric microcapsule arrays and their use for enzyme immobilization, *Nature.* 369(6478) (1994). https://doi.org/10.1038/369298a0.
10. W. He, C. Wang, F. Zhuge, X. Deng, X. Xu, T. Zhai, Flexible and high energy density asymmetrical supercapacitors based on core/shell conducting polymer nanowires/manganese dioxide nanoflakes, *Nano Energy.* 35 (2017) 242–250. https://doi.org/10.1016/j.nanoen.2017.03.045.
11. W. Chen, R.B. Rakhi, H.N. Alshareef, Facile synthesis of polyaniline nanotubes using reactive oxide templates for high energy density pseudocapacitors, *Journal of Materials Chemistry A.* 1(10) (2013). https://doi.org/10.1039/c3ta00499f.
12. N.R. Chiou, A.J. Epstein, Polyaniline nanofibers prepared by dilute polymerization, *Advanced Materials.* 17(13) (2005) 1679–1683. https://doi.org/10.1002/adma.200401000.
13. N.R. Chiou, C. Lu, J. Guan, L.J. Lee, A.J. Epstein, Growth and alignment of polyaniline nanofibres with superhydrophobic, superhydrophilic and other properties, *Nature Nanotechnology.* 2(6) (2007) 354–357. https://doi.org/10.1038/nnano.2007.147.
14. G.Y. Zhao, H.L. Li, Preparation of polyaniline nanowire arrayed electrodes for electrochemical supercapacitors, *Microporous and Mesoporous Materials.* 110(2–3) (2008) 590–594. https://doi.org/10.1016/j.micromeso.2007.06.023.
15. K. Wang, J. Huang, Z. Wei, Conducting polyaniline nanowire arrays for high performance supercapacitors, *Journal of Physical Chemistry C.* 114(17) (2010) 8062–8067. https://doi.org/10.1021/jp9113255.
16. J.M. D'Arcy, M.F. El-Kady, P.P. Khine, L. Zhang, S.H. Lee, N.R. Davis, D.S. Liu, M.T. Yeung, S.Y. Kim, C.L. Turner, A.T. Lech, P.T. Hammond, R.B. Kaner, Vapor-phase polymerization of nanofibrillar poly(3,4- ethylenedioxythiophene) for supercapacitors, *ACS Nano.* 8(2) (2014) 1500–1510. https://doi.org/10.1021/nn405595r.
17. Q. Zhao, M. Zhao, J. Qiu, W.Y. Lai, H. Pang, W. Huang, One dimensional silver-based nanomaterials: Preparations and electrochemical applications, *Small.* 13(38) (2017). https://doi.org/10.1002/smll.201701091.
18. C. Zhou, Y. Zhang, Y. Li, J. Liu, Construction of high-capacitance 3D CoO@Polypyrrole nanowire array electrode for aqueous asymmetric supercapacitor, *Nano Letters.* 13(5) (2013). https://doi.org/10.1021/nl400378j.
19. F. Wang, X. Zhang, L. Yang, D. Xu, Y. Ma, D. Chen, L. Wang, C. Zhao, W. Yang, A scalable route to prepare core-shell structured ZnO@PEDOT nanowires and PEDOT nanotubes and their properties as electrode materials, *Applied Surface Science.* 370 (2016). https://doi.org/10.1016/j.apsusc.2016.02.159.
20. A. Pruna, Q. Shao, M. Kamruzzaman, J.A. Zapien, A. Ruotolo, Enhanced electrochemical performance of ZnO nanorod core/polypyrrole shell arrays by graphene oxide, *Electrochimica Acta.* 187 (2016) 517–524. https://doi.org/10.1016/j.electacta.2015.11.087.
21. X.F. Lu, X.Y. Chen, W. Zhou, Y.X. Tong, G.R. Li, α-Fe_2O_3@PANI core-shell nanowire arrays as negative electrodes for asymmetric supercapacitors, *ACS Applied Materials and Interfaces.* 7(27) (2015). https://doi.org/10.1021/acsami.5b03126.
22. C. Pan, Y. Hu, Z. Gong, Y. Yang, S. Liu, L. Quan, Z. Yang, Y. Wei, W. Ye, Improved blood compatibility and endothelialization of titanium oxide nanotube arrays on titanium surface by zinc doping, *ACS Biomaterials Science and Engineering.* 6(4) (2020). https://doi.org/10.1021/acsbiomaterials.0c00187.
23. H. Yang, Z. Tan, Y. Liu, Z. Ma, L. Zhang, Hierarchical wall formation of titanium oxide nanotube arrays using anodic oxidation, *IEEE Transactions on Nanotechnology.* 12(6) (2013). https://doi.org/10.1109/TNANO.2013.2273458.
24. S. Xie, M. Gan, L. Ma, Z. Li, J. Yan, H. Yin, X. Shen, F. Xu, J. Zheng, J. Zhang, J. Hu, Synthesis of polyaniline-titania nanotube arrays hybrid composite via self-assembling and graft polymerization for supercapacitor application, *Electrochimica Acta.* 120 (2014) 408–415. https://doi.org/10.1016/j.electacta.2013.12.067.
25. C. Li, Z. Wang, S. Li, J. Cheng, Y. Zhang, J. Zhou, D. Yang, D.G. Tong, B. Wang, Interfacial engineered polyaniline/sulfur-doped TiO_2 nanotube arrays for ultralong cycle lifetime

fiber-shaped, solid-state supercapacitors, *ACS Applied Materials and Interfaces*. 10(21) (2018). https://doi.org/10.1021/acsami.8b01160.
26. S.H. Mujawar, S.B. Ambade, T. Battumur, R.B. Ambade, S.H. Lee, Electropolymerization of polyaniline on titanium oxide nanotubes for supercapacitor application, *Electrochimica Acta*. 56(12) (2011). https://doi.org/10.1016/j.electacta.2011.02.043.
27. J. Chen, Z. Xia, H. Li, Q. Li, Y. Zhang, Preparation of highly capacitive polyaniline/black TiO2 nanotubes as supercapacitor electrode by hydrogenation and electrochemical deposition, *Electrochimica Acta*. 166 (2015). https://doi.org/10.1016/j.electacta.2015.03.058.
28. P. Zhang, Z. Liu, Y. Liu, H. Fan, Y. Jiao, B. Chen, Titanium Dioxide@Polyaniline core-shell nanowires as high-performance and stable electrodes for flexible solid-state supercapacitors, *Electrochimica Acta*. 184 (2015). https://doi.org/10.1016/j.electacta.2015.10.040.
29. M.A. Bavio, G.G. Acosta, T. Kessler, A. Visintin, Flexible symmetric and asymmetric supercapacitors based in nanocomposites of carbon cloth/polyaniline - Carbon nanotubes, *Energy*. 130 (2017). https://doi.org/10.1016/j.energy.2017.04.135.
30. H. Lin, L. Li, J. Ren, Z. Cai, L. Qiu, Z. Yang, H. Peng, Conducting polymer composite film incorporated with aligned carbon nanotubes for transparent, flexible and efficient supercapacitor, *Scientific Reports*. 3 (2013). https://doi.org/10.1038/srep01353.
31. Z. Niu, P. Luan, Q. Shao, H. Dong, J. Li, J. Chen, D. Zhao, L. Cai, W. Zhou, X. Chen, S. Xie, A "skeleton/skin" strategy for preparing ultrathin free-standing single-walled carbon nanotube/polyaniline films for high performance supercapacitor electrodes, *Energy and Environmental Science*. 5(9) (2012). https://doi.org/10.1039/c2ee22042c.
32. D. Ge, L. Yang, L. Fan, C. Zhang, X. Xiao, Y. Gogotsi, S. Yang, Foldable supercapacitors from triple networks of macroporous cellulose fibers, single-walled carbon nanotubes and polyaniline nanoribbons, *Nano Energy*. 11 (2015). https://doi.org/10.1016/j.nanoen.2014.11.023.
33. G. Wu, P. Tan, D. Wang, Z. Li, L. Peng, Y. Hu, C. Wang, W. Zhu, S. Chen, W. Chen, High-performance supercapacitors based on electrochemical-induced vertical-aligned carbon nanotubes and polyaniline nanocomposite electrodes, *Scientific Reports*. 7 (2017). https://doi.org/10.1038/srep43676.
34. S. Zeng, H. Chen, F. Cai, Y. Kang, M. Chen, Q. Li, Electrochemical fabrication of carbon nanotube/polyaniline hydrogel film for all-solid-state flexible supercapacitor with high areal capacitance, *Journal of Materials Chemistry A*. 3(47) (2015). https://doi.org/10.1039/c5ta05937b.
35. C. Meng, C. Liu, L. Chen, C. Hu, S. Fan, Highly flexible and all-solid-state paperlike polymer supercapacitors, *Nano Letters*. 10(10) (2010). https://doi.org/10.1021/nl1019672.
36. T. Zheng, X. Wang, Y. Liu, R. Bayaniahangar, H. Li, C. Lu, N. Xu, Z. Yao, Y. Qiao, D. Zhang, P. Pour Shahid Saeed Abadi, Polyaniline-decorated hyaluronic acid-carbon nanotube hybrid microfiber as a flexible supercapacitor electrode material, *Carbon*. 159 (2020). https://doi.org/10.1016/j.carbon.2019.11.074.
37. T. Zheng, P. Pour Shahid Saeed Abadi, J. Seo, B.H. Cha, B. Miccoli, Y.C. Li, K. Park, S. Park, S.J. Choi, R. Bayaniahangar, D. Zhang, S.H. Lee, C.K. Lee, A. Khademhosseini, S.R. Shin, Biocompatible carbon nanotube-based hybrid microfiber for implantable electrochemical actuator and flexible electronic applications, *ACS Applied Materials and Interfaces*. 11(23) (2019) 20615–20627. https://doi.org/10.1021/acsami.9b02927.
38. Y. Xu, Y. Yan, W. Lu, S. Yarlagadda, G. Xu, High-performance flexible asymmetric fiber-shaped supercapacitor based on CF/PPy and CNT/MnO_2 composite electrodes, *ACS Applied Energy Materials* (2021). https://doi.org/10.1021/acsaem.1c01648.
39. Q. Tang, M. Chen, C. Yang, W. Wang, H. Bao, G. Wang, Enhancing the energy density of asymmetric stretchable supercapacitor based on wrinkled CNT@MnO_2 cathode and CNT@polypyrrole anode, *ACS Applied Materials and Interfaces*. 7(28) (2015). https://doi.org/10.1021/acsami.5b03148.
40. F.M. Guo, R.Q. Xu, X. Cui, L. Zhang, K.L. Wang, Y.W. Yao, J.Q. Wei, High performance of stretchable carbon nanotube-polypyrrole fiber supercapacitors under dynamic deformation and temperature variation, *Journal of Materials Chemistry A*. 4(23) (2016). https://doi.org/10.1039/c6ta02437h.

41. M. Li, M. Zu, J. Yu, H. Cheng, Q. Li, Stretchable fiber supercapacitors with high volumetric performance based on buckled MnO_2/oxidized carbon nanotube fiber electrodes, *Small.* 13(12) (2017). https://doi.org/10.1002/smll.201602994.
42. L. V. Kayser, D.J. Lipomi, Stretchable conductive polymers and composites based on PEDOT and PEDOT:PSS, *Advanced Materials.* 31(10) (2019). https://doi.org/10.1002/adma.201806133.
43. D. Zhao, Q. Zhang, W. Chen, X. Yi, S. Liu, Q. Wang, Y. Liu, J. Li, X. Li, H. Yu, Highly flexible and conductive cellulose-mediated PEDOT:PSS/MWCNT composite films for supercapacitor electrodes, *ACS Applied Materials and Interfaces.* 9(15) (2017). https://doi.org/10.1021/acsami.7b01852.

10

One-Dimensional Polymeric Nanocomposites for Flexible Supercapacitors

Jose Garcia-Torres

CONTENTS

10.1 Introduction

Flexible and wearable electronics have experienced rapid growth during the past year due to their high impact in a modern society, like health monitoring, diagnosis, and treatment of diseases or human–machine interface [1, 2]. Ideally, wearable devices must show characteristics like flexibility, conformability to the human body, light weight, and maintenance of performance when subjected to mechanical deformations. However, those characteristics can't be achieved with traditional electronic materials (e.g., silicon, gallium arsenide) since they lead to rigid, bulky, and planar devices [3, 4]. For this reason, great efforts have been devoted to the development of flexible materials to avoid the mechanical mismatch between traditional rigid electronics and the soft human body [5]. These days, very interesting flexible and wearable electronic devices have been developed and even commercialized. But the rapid development of such wearable devices has also given rise to increasing demand for suitably thin, lightweight, and flexible energy storage devices (e.g., batteries, supercapacitors (SCs)). Unlike conventional planar structures, fiber-shaped energy storage devices have attracted increasing attention since they fit the former requirements. Fibers, with diameters ranging from tens to hundreds of micrometers, are inherently more flexible and lightweight than their bulk counterparts, which is the reason why they can be successfully used as the electrodes for flexible energy devices. Their fiber shape also allows easy integration into fabrics and clothes and other miniaturized devices, boosting their use in portable and wearable electronics [6, 7]. Moreover, the fibers are required to not

DOI: 10.1201/9781003223764-10

only be robust enough for proper incorporation within the former systems but also to show ideally no degradation after many mechanical deformations while maintaining the electrochemical performance.

Among the different energy devices, SCs are promising energy storage devices as they show high power densities, high charge-discharge rates, or long life cycles, making them very useful for applications where high power demands are needed, like electric grids, transportation, renewable energy production, aerospace systems, and portable and consumer electronics [8]. Moreover, SCs have been shown to be successfully fabricated into fiber shapes which expand their applicability even further. Since the development of the first 1D SC by Prof. Baughman's group at Texas [9] by assembling two composite fibers containing carbon nanotubes (CNT) dispersed in polyvinyl alcohol (PVA) polymer, many efforts have been dedicated to improving the electrochemical performance of this 1D SC, mainly in terms of capacitance and energy and power densities, but also rate capability and cycle life duration. Since one of the key components of SCs are electrodes, much effort has been focused to design and synthesize materials able to provide fibers with high electrical and electrochemical properties as well as flexibility to endure deformation without compromising those properties [7, 10]. In this regard, multiple materials have been explored as electrodes: (i) carbon nanomaterials (CNM), which show interesting properties like high electrical conductivity and mechanical stability but have low energy storage capabilities; (ii) transition metal oxides/hydroxides, which show high charge storage capacity but are highly resistive; and (iii) conducting polymers (CP), which show good electrical conductivity and energy storage but low cycling performance [10]. Therefore, a single material does not have the required properties, which is the reason why composite materials are preferred. A second strategy is based on the design of the SC and can be classified as symmetric and asymmetric SCs [11]. Depending on the charge storage mechanism, SCs can be classified as (i) electrochemical double-layer capacitors (EDLC) if the charge is stored through the adsorption/desorption of the electrolyte ions onto the electrode surface, or (ii) pseudocapacitors when the charge is stored through Faradaic reactions onto the electrode material [11]. Thus, symmetric SCs refer to devices where both electrodes are the same (either EDLC electrodes or pseudocapacitive electrodes); meanwhile, asymmetric SCs are built using electrodes with different energy storage mechanisms. Finally, a third strategy has been the configuration of the electrodes within the SC: parallel, twisted, or coaxial electrodes. These configurations result in different electrochemical performances (e.g., different capacitance or energy density) since they are limited by the electrode–electrolyte interface areas, which are different in each configuration [7].

In this chapter, the author first introduces basic concepts of the different energy storage devices (e.g., mechanisms, electrochemical parameters) to provide a foundation to understand the research approaches in the literature. Later, the main methodologies to prepare fibers and SCs are discussed for a better understanding of this chapter. Finally, the state-of-the-art progress in 1D SCs in terms of materials employed is discussed.

10.2 Energy Storage Systems: Basic Concepts and Comparison

Currently. the most common energy storage technologies available to power portable electronics are batteries and SCs, each of which shows advantages and disadvantages based on multiple properties (e.g., energy stored, power requirements, lifetime, temperature, size, or flexibility of the device). Thus, it is important to specify the different ranges of applications of each energy storage system in terms of performance parameters like energy and power densities. While the energy density is the total amount of energy stored in a given material volume or mass (Wh/cm^3 or Wh/kg), power density is defined as the energy transferable rate per unit volume or mass of the material (W/cm^3 or W/kg) [11]. Figure 10.1A shows the Ragone plot representing the energy and power densities for the different energy storage devices [12]. As can be seen, capacitors have the highest power values due to their rapid charge-discharge ability but show the lowest energy density. On the opposite side, batteries and fuel cells offer high specific energy densities and low specific powers. Meanwhile, the SCs bridge the gap between conventional capacitors and batteries since they have higher powers and longer lifetimes than batteries and higher energy densities than capacitors. Moreover, batteries and SCs also present different performances in terms of cyclability, charge and discharge times, or lifetimes, among others (Table 10.1). SCs can

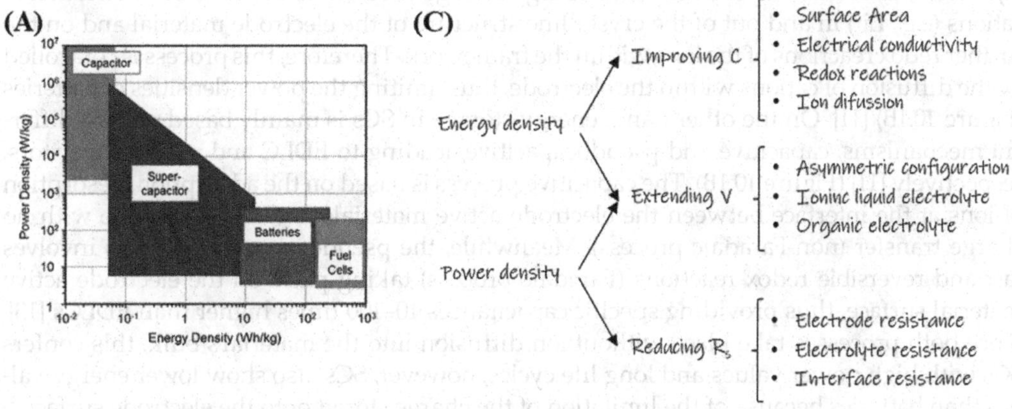

(B)

Types of energy storage device	Supercapacitor		Battery
	EDLC	Pseudocapacitor	
Natures	Non-faradaic	Faradaic	
Mechanisms	➢ High rate ➢ Low capacitance		Slow Intercalation Slow De-intercalation ➢ Low rate ➢ High capacitance

FIGURE 10.1

(A) Graph showing power density *vs* energy density relationship for the different energy storage systems. Adapted with permission from reference [12], Copyright (2013), Springer. (B) Schemes showing the mechanism of energy storage for electrochemical double layer SC, pseudocapacitors, and batteries. Adapted with permission from reference [11], Copyright (2016), Royal Society of Chemistry. (C) Scheme showing the main parameters influencing capacitance, energy density, and power density.

TABLE 10.1

Comparison of Charge-Discharge Times, Cycle Life, Life Time, and Efficiency for Batteries and Supercapacitors

Energy Storage System	Charge (s)	Discharge (s)	Cycle Life (Cycles)	Life Time (years)	Efficiency (%)
Battery	>1000	>1000	10^3	5	70–85
Supercapacitor	0.1–30	0.1–30	10^6	30	85–98

be charged and discharged at high rates and with high currents over short periods for several hundred thousand cycles, whereas batteries can deliver current over a longer time (up to several hours).

Such differences in the performance between batteries and SC arise from the energy storage mechanisms. On the one hand, the energy storage of batteries depends on the motion of cations (e.g., Li^+) in and out of the crystalline structure of the electrode material and on the further redox reactions of the ions within the framework. Therefore, this process is controlled by the diffusion of cations within the electrode, thus limiting the power densities of batteries (Figure 10.1B) [11]. On the other hand, energy storage in SCs is mainly based on two different mechanisms, capacitive and pseudocapacitive, leading to EDLC and pseudocapacitors, respectively [11] (Figure 10.1B). The capacitive process is based on the adsorption/desorption of ions at the interface between the electrode active materials and the electrolyte with no charge transfer (non-Faradaic process). Meanwhile, the pseudocapacitive process involves fast and reversible redox reactions (Faradaic process) taking place on the electrode active material surface, thus providing specific capacitances 10–100 times higher than EDLCs [13]. Since both processes take place without ion diffusion into the material's bulk, this confers SCs with high power values and long life cycles; however, SCs also show lower energy values than batteries because of the limitation of the charge stored onto the electrode surface.

Improving the SC electrochemical performance in terms of capacitance, energy, and power density is essential for translating such devices into real applications. 1D SCs are typically composed of two electrodes (cathode and anode), an electrolyte, and/or a current collector assembled in cylindrical geometry. Typically, a 1D SC uses gel electrolytes, acting simultaneously as both electrolyte and separator to avoid short circuits and chemical leakage. Although all of those components have been researched to improve their performance individually and therefore the overall performance of the SC, the electrode materials have been the most investigated. Other strategies to enhance the electrochemical performance involve SC configuration (e.g., parallel, two-ply, coaxial) and design (e.g., symmetric, asymmetric). Before explaining how materials can impact SC performance, the main electrochemical parameters (capacitance, energy density, and power density) will first be described. Capacitance (C), according to Eq (10.1), is the amount of stored charge (Q) per unit voltage (V), and Q is the current (I) multiplied by time (t):

$$C = Q/V = (I \cdot t)/V \tag{10.1}$$

Thus, the higher the stored charge, the higher the capacitance. As explained above, the two main mechanisms for charge storage are based on either the adsorption/desorption of ions or the redox reactions at the electrode/electrolyte interface. Meanwhile, the energy density (E) depends on the capacitance and the voltage range as expressed in Eq (10.2):

$$E = \tfrac{1}{2}\, C \cdot V^2 \tag{10.2}$$

Therefore, enhancing the capacitance or extending the voltage window has a positive impact on energy density. Meanwhile, power density (P) depends on the voltage and the internal resistance (R_s) as:

$$P = V^2 / 4R_s \tag{10.3}$$

Increasing the voltage window or reducing the internal resistance of the electrode material, the electrolyte, and the electrode–electrolyte interface will improve power density. Figure 10.1C shows the most studied approaches to improving capacitance, energy density, and power density.

10.3 Fabrication Methods for Fibers and Supercapacitors

Different techniques, like wet spinning, hydrothermal synthesis, or dry spinning (or film scrolling), have been used to fabricate polymeric nanocomposite fibers as electrodes for SCs [14–18]. These techniques allow the fabrication of fibers that will subsequently serve as a scaffold for deposition or coating with other (nano)materials to improve their properties. Of course, these coating techniques are many and varied, like dip coating, electrochemical deposition, or atomic layer deposition, among others. In this chapter, the author will discuss only the main techniques employed for fabricating the fiber itself.

10.3.1 Wet Spinning

Wet spinning is the main technique employed for the fabrication of fiber electrodes for SC applications (Figure 10.2A). This technique consists of the solidification of a liquid solution or dispersion, or spinning solution into a coagulation bath to induce fiber formation [14]. The spinning solution may contain either a dissolved polymer that can solidify or a dispersion of nanomaterials that can experience aggregation and precipitation into a 1D form. The fiber shape is conferred when the solution or dispersion is injected through a spinneret (e.g., a needle) into the coagulation bath. Solidification of the fiber can take place under different conditions, like changes in solubility, pH, temperature, or salt concentration, or the presence of a polymer in the coagulation bath. Since solidification occurs at the interface between the spinning and the coagulation solutions, continuous fiber removal is needed to ensure the availability of fresh solution at the interface to continuously form the fiber and to avoid clogging the spinneret. This can be performed either by continuously rotating the coagulation bath or attaching the fiber to a bobbin outside the bath to collect it. Wet spinning is a very versatile and easy technique since it allows control of the properties of the fibers by changing different parameters, such as the composition of the spinning and coagulation solutions and the spinning parameters (e.g., injection rate, rotation speed of the coagulation bath, or spinneret diameter). Those parameters are also critical in developing a successful fiber production method. Thus, wet spinning is the preferred technique to fabricate fibers from multiple materials (e.g., insulating polymers, CP, or nanomaterials) and compositions with tunable properties (Figure 10.2B,C). Wet spinning has also been proven to be successful in fabricating coaxial fibers (Figure 10.2D-F).

Wet spinning

(A) (B) (C) (D) (E) (F)

Hydrothermal synthesis

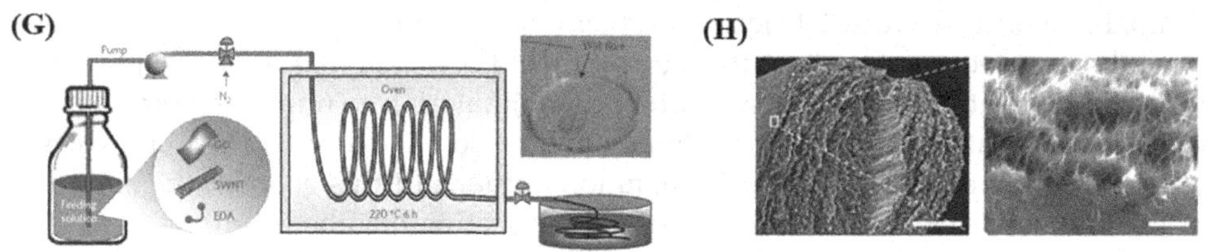

Dry spinning/Film scrolling

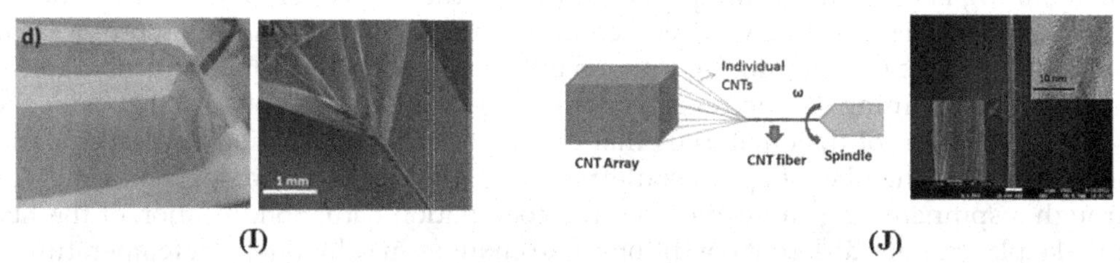

FIGURE 10.2
(A) Schematic of a wet-spinning process to obtain carbon-based fibers. (B) Optical microscope image. (C) Photograph of the as-spun fibers with different diameters. (A), (B) and (C) are adapted with permission from reference [14], Copyright (2015), Elsevier. (D) Schematic of a coaxial wet-spinning process using a two-opening needle. (E) Polarized-light optical microscopy image showing the core-shell structure. (F) Photograph of the as-spun coaxial fiber. (D), (E), and (F) are adapted with permission from reference [15], Copyright (2015), Nature Publishing Group. (G) Scheme showing the synthesis of carbon hybrid fibers by hydrothermal synthesis. (H) (Left) Scanning electron microscopy (SEM) image of the cross-section of fiber (scale bar: 15 μm) and (right) higher magnification SEM image showing the alignment of the CNTs along the fiber long axis (scale bar: 0.5 μm). (G) and (H) are adapted with permission from reference [16], Copyright (2015), Nature Publishing Group. (I) (Left) Free-standing, thin GO film and (Right) SEM image of the thin GO film being scrolled into a fiber. Adapted with permission from reference [17], Copyright (2015), American Chemical Society. (J) (Left) Scheme of the fiber-spinning process and (right) SEM image of an as-spun CNT fiber (insets are a magnified SEM image of an HRTEM (high-resolution transmission electron microscopy) image of an individual CNT). Adapted with permission from reference [18], Copyright (2011), Springer.

10.3.2 Confined Hydrothermal Process

Hydrothermal synthesis produces fibers by autoclaving aqueous solutions or dispersions at high vapor pressures [16]. First, a solution or homogeneous dispersion of the materials of interest is prepared and introduced into a capillary column acting as a hydrothermal microreactor. After the treatment, three-dimensional architectures are produced *via* self-assembly of the components (Figure 10.2G,H). This method has been less commonly used than wet spinning for the fabrication of fiber electrodes with applications in SCs.

10.3.3 Dry Spinning or Film Scrolling

This method is based on turning films into fibers (Figure 10.2I). First, a film of the desired material or materials is obtained and then twisted into fibers. To do so, one end of the film is fixed while a rotational force is applied to the other end. Moreover, the films must be flexible to avoid breaking during twisting [17]. An alternative method involves twisting fibers from CNT forests previously grown by chemical vapor deposition and further pulled and twisted to achieve CNT yarns (Figure 10.2J) [18]. During pulling and twisting, there is an entanglement of the CNTs to form a stable and continuous fiber. This method has been less commonly used than wet spinning for the fabrication of fiber electrodes with applications in SCs.

10.4 Materials in Supercapacitors

Electrode materials need to have the following properties to affect the important electrochemical parameters: (1) large specific surface areas (SSA) for ease of access to the adsorption/desorption processes of the electrolyte ions; (2) high electrical conductivity to achieve fast electron transfer and therefore a low charge transfer resistance; (3) high ionic conductivity for efficient ionic transport toward the electrode surface; and (4) structural stability to avoid dissolution of the materials, side reactions that hinder energy storage, or volume expansion/contraction leading to performance deterioration. In addition to these properties, a 1D SC is required to show mechanical flexibility to withstand repetitive mechanical deformations without compromising the electrochemical/energy storage performance. Thus, selection of appropriate materials is key to fabricating flexible fibers and SCs with the required mechanical, electrical, and electrochemical properties. Multiple materials have been employed (e.g., Carnon nanomaterials [CNM], conducting polymer [CP], metal nanomaterials, transition metal oxides) but none of them possess all the required properties, which explains why composite materials have been the materials most extensively used to get achieve flexible fibers [19, 20]. Many researchers have focused on composite 1D SCs containing current collectors (e.g., metallic wires, plastic wires coated with metals) other than fibers. Several reviews have focused on that issue [21, 22]. However, the presence of current collectors has some disadvantages as they contribute to the SC total weight and volume while not contributing practically to the capacitance, thus compromising the SC's electrochemical performance in terms of energy and power densities. For this reason, this chapter will focus on 1D SCs without current collectors.

10.4.1 Carbon-Based Fibers

CNMs are among the most widely employed materials with which to develop fibre-shape EDLC SCs due to properties like excellent electrical conductivity, great strength, high

SSA, or relatively low density [23]. Among them, CNTs and graphene have been widely employed as conductive materials for SC electrodes. Since the CNM energy storage mechanism occurs through the adsorption/desorption of ions at the electrode–electrolyte interface forming an electric double layer, fibers with high electrical conductivity, high porosity, and large, accessible SSA are needed [11].

CNTs are widely used in the development of flexible fibers since they can be assembled into fibers inheriting the shape and properties of the CNTs at the macroscale. One of the difficulties of working with CNTs in solution is their tendency to aggregate as a result of the maximized van der Waals interactions among the individual nanotubes due to the high SSA. Therefore, surfactants or polymer-stabilized dispersions are required to properly disperse and process CNTs. For example, the preparation of CNT-based composites improves the electrical conductivity and the specific capacitance [24], whereas the CNT orientation along the fiber confers good mechanical and electrical properties [25].

Dalton *et al.* synthesized the first 1D SC by assembling two CNT-PVA composite fibers by wet spinning [9]. The fibers were obtained by injecting the spinning solution (single-walled CNTs (SWCNTs) and lithium dodecyl sulfate as surfactant) into a cylindrical pipe where the PVA solution flowed, causing the collapse of the CNTs. Two composite fibers, each with 50 mm diameter and around 60 wt% SWCNTs, were separately coated with the gel electrolyte (19 wt% H_3PO_4 and 4 wt% PVA), twisted to form a SC, and recoated with the electrolyte (Figure 10.3A). The SC showed low gravimetric capacitance (5 F/g) and energy density (0.6 Wh/kg (at 1V)) but good cyclability as capacitance was constant over 1,200 charge-discharge cycles. Ren and co-workers reported the fabrication of aligned multi-walled CNT (MWCNT) fibers with controlled diameters in the range 2–30 μm and lengths up to 100 mm by scrolling MWCNT films previously obtained by chemical vapor deposition [26]. Figure 10.3B shows a scanning electron micrograph (SEM) image of the MWCNT fibers where uniform diameters and alignment of the MWCNT along the fiber long axis can be observed. Such alignment allowed high tensile strengths (1.3 GPa) and high electrical conductivities (10^3 S/cm) but low capacitances (15.61 F/g, 3.53 mF/cm^2) (Figure 10.3C). They reported good capacitance retention over 1,000 cycles (Figure 10.3D). The capacitance was even lower after assembling the fibers into an SC by twisting them with the PVA/H_3PO_4 gel electrolyte, showing a gravimetric capacitance of 13.31 F/g and an area-specific capacitance of 3.01 mF/cm^2. Bundling of CNTs, which leads to compact fibers, has been reported as the main origin of the low performance of CNT-based fibers [26].

Graphene is also a key material for SCs due to its high electrical conductivity, SSA, and charge storage capacity [23]. However, most reported graphene-based fibers have shown low capacitance values and energy densities due to the strong inter-sheet π-π stacking forces resulting in compact and low-porosity fibers [27, 28]. Since energy storage capacity is limited in electrochemical double layer supercapacitors (EDLSs) by the lower SSA, two different strategies have been followed: (i) the deposition of mesoporous carbon materials onto CNT or graphene fibers to increase the SSA; and (ii) the preparation of carbon-carbon composites to avoid bundling or stacking of the CNM.

With regard to the first strategy, Ren and co-authors developed fibers containing MWCNTs and ordered mesoporous carbon (OMC) with different concentrations of the OMC from 46 to 90% [29]. First, OMC was deposited by dip-coating an MWCNT film into an OMC/*N,N*-dimethylformamide solution for different periods of time and subsequently scrolling the film into a composite fiber (Figure 10.3E). They observed that the OMC particles were infiltrated inside the fiber and between the aligned MWCNTs, the nanotubes acting as a skeleton to support and connect the OMC particles through van der Waals interactions. Two composite fibers were coated with a PVA-H_3PO_4 gel electrolyte

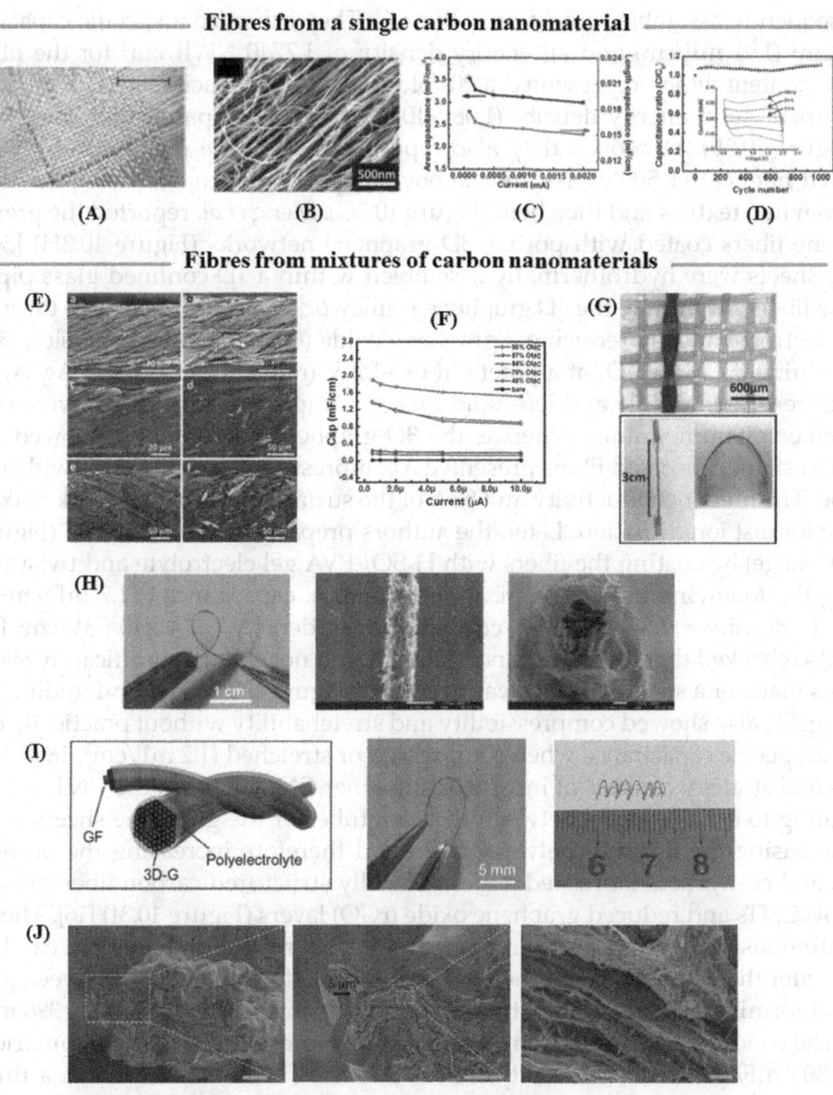

FIGURE 10.3
(A) Photograph showing two supercapacitors embedded into a textile (scale bar: 1 cm). Adapted with permission from reference [9], Copyright (2003), Nature Publishing Group. (B) SEM image showing aligned CNT along the fiber long axis. (C) Graph showing the dependence of the areal and length capacitances *vs* the applied current. (D) Dependence of the capacitance retention (C/C_0 ratio) with the number of charge/discharge cycles. Inset shows the cyclic voltammetric at different scan rates (1, 5, and 10 V s^{-1}). (B), (C) and (D) are adapted with permission from reference [23], Copyright (2013), Wiley-VCH Verlag GmbH). (E) SEM images of MWCNT-based fibers with increasing amounts of OMC on the fiber surface (0, 46, 70, 84, 87, and 90%). (F) Graph showing the specific capacitance *vs* current for the previous fibers. (G) (Top) Photograph of a 1D SC woven into a textile and (bottom) photographs of the 1D SC before and after bending. (E), (F), and (G) are adapted with permission from reference [29], Copyright (2013), Wiley-VCH Verlag GmbH. (H) (Left) Photograph of a bent graphene fiber. (Centre) SEM image of the side view, and (left) cross-section of the graphene fiber/3D-graphene composite fiber showing the core-shell structure. (I) (Left) Scheme illustrating a 1D SC fabricated from two twisted graphene fiber/3D-graphene composite fibers with polyelectrolyte. (Centre) Photograph of the bent SC and (right) photograph of the SC with a spring shape. (H) and (I) are adapted with permission from reference [30], Copyright (2013), Wiley-VCH Verlag GmbH. (J) SEM images at different magnifications of fibers fabricated from dispersions containing a mixture of GO and CNT. Scale bars 20 (left), 10 (centre) and 0.2 μm (right). Adapted with permission from reference [15], Copyright (2014), Nature Publishing Group.

and subsequently assembled into a wire-like SC. They achieved a specific capacitance of 39.7 mF/cm^2 (1.91 mF/cm) and an energy density of 1.77·10^{-6} Wh/cm^2 for the fiber with 87% OMC content which represented a 20-fold increase in capacitance (1.97 mF/cm^2) and a 16-fold increase in energy density (1.08 ×10^{-7} Wh/cm^2) compared with bare MWCNT fibers (Figure 10.3F). Moreover, they also reported capacitance retention of 87% for the composited fibers after 500 cycles and no obvious performance decrease after the fibers were woven into textiles and then bent (Figure 10.3G). Meng *et al.* reported the preparation of graphene fibers coated with porous 3D-graphene networks (Figure 10.3H) [30]. First, graphene sheets were hydrothermally assembled within a 1D confined glass pipeline to obtain the fibers. After that, the 3D graphene framework was deposited onto the graphene fiber by electrochemically reducing a graphene oxide (GO) aqueous dispersion (3 mg/ml GO) containing 0.1 M $LiClO_4$ at a potential of −1.2 V (reference electrode: Ag/AgCl (3 M KCl)) (Figure 10.3H, middle and left-hand images). Important differences were observed in the cyclic voltammograms. Whereas the 3D-graphene coated fibers showed a nearly rectangular shape, uncoated fibers presented a compressed voltammogram with a shuttle-like shape. The higher conductivity and SSA of the surface graphene network make it very accessible for fast ion diffusion. Later, the authors prepared a solid-state SC (Figure 10.3I, left-hand image) by coating the fibers with H_2SO_4-PVA gel electrolyte and twisting them, obtaining the following electrochemical performance: capacitance = 1.7 mF/cm^2 (19 μF/cm), energy density = 2.7 × 10^{-8} Wh/cm^2 and power density = 1.4 ×10^{-5} W/cm^2 [30]. The authors also checked that the capacitance of the SC did not change significantly even when a knot was made or a spring-like SC was prepared (Figure 10.3I, right and middle images). This spring SC also showed compressibility and stretchability without practically compromising the specific capacitance when compressed or stretched (1.2 mF/cm^2, 18 μF/cm) [30].

The second strategy consists of incorporating other CNMs within the CNT or graphene fibers aiming to act as spacers between the nanotubes or the graphene sheets, contributing to increasing the distance between them and therefore increasing the porosity and SSA. Dai and co-workers fabricated a hierarchically structured carbon fiber consisting of aligned SWCNTs and reduced graphene oxide (rGO) layers (Figure 10.3I) [16]. These fibers were continuously prepared following a hydrothermal method into a silica capillary column to confer the 1D shape. With the CNTs acting as effective spacers between graphene sheets and forming a conducting network, the authors measured an SSA of 396 m^2/g and an electrical conductivity of 102 S/cm. The fibers also presented a high volumetric capacitance of 305 mF/cm^3 (current density = 73.5 mA cm^{-3}) when measured in a three-electrode cell configuration and using H_2SO_4 as electrolyte, or 300 mF/cm^3 (current density = 26.7 mA/cm^3) in a two-electrode cell configuration, using PVA-H_3PO_4 as electrolyte [16]. The authors also developed a solid-state SC by placing two fibers in parallel onto a flexible poly(ethylene terephthalate) (PET) substrate and coating them with PVA-H_3PO_4. This micro-SC showed a volumetric energy density as high as 6.3 mWh/cm^3, ten times higher than commercial SCs and similar to thin-film lithium batteries, and a power density of 1085 mW/cm^3, a value similar to commercial SCs and two times higher than thin-film lithium batteries. The SC had excellent capacitance retention (93%) over 10,000 cycles [16].

Similarly, Liu's group also developed CNT/rGO fibers by wet spinning [31]. They directly spun the CNT-GO dispersion into a chitosan/acetic acid coagulation bath to obtain the fibers, which were subsequently washed and dried. After that, the fibers were immersed in a hydroiodic acid solution at 80 °C for 5 h to reduce the GO. The CNT/rGO fibers obtained showed better electrical conductivity (210 S/cm) and tensile strength (385.7 MPa) than the rGO fibers (53.3 S/cm, 193.3 MPa). The assembled SC showed high specific capacitance (351 μF/cm, 38.8 F/cm^3 at 50 mA/cm^3), energy density (1 mWh/cm^3), and power density

(0.5 W/cm^3), which the authors attributed to the greatly improved electrical conductivity. Garcia-Torres *et al.* developed composite fibers containing carbon nanoparticles (carbon black (CB)), CNTs, and chitosan by wet-spinning a CB/CNT dispersion into a chitosan coagulation bath [32]. In this study, the authors prepared fibers with different CB/CNT ratios, which affected the electrical conductivity and the electrochemical properties. The electrical conductivity was higher for the 10 % CB/90% CNT (14.7 S/cm) fiber than for the 100 % CNT fiber (10.4 S/cm) suggesting that the carbon nanoparticles acted as an effective dispersant for CNTs. The capacitance was improved 2.5 times for the10 % CB/90 % CNT fibers compared with the 100% CNT fibers. However, the gravimetric capacitance was still low (12.1 F/g). Gao *et al.* reported one of the highest energy densities for carbon composite fibers [15]. They reported the synthesis of a coaxial fiber by wet spinning where the core was formed from a GO and CNT dispersion and the shell from a carboxymethyl cellulose aqueous solution (Figure 10.3J). The assembled two-ply intertwined SC separated by PVA/H_3PO_4 gel electrolyte presented high capacitance (158 F/cm^3, 177 mF/cm^2) as well as high energy (3.5 mWh/cm^3) and power densities (18 mW/cm^3). The authors also showed that the capacitance of the fibers was constant even though the 1D SC was intertwined, knotted, or woven.

Despite the good electrochemical performance of all carbon composite fibers, capacitance and energy density are still limited by the electrical double-layer mechanism of CNMs. Thus, the incorporation of pseudocapacitive materials like transition metal oxides or CPs into carbon-based fibers has been revealed as an effective approach to generating high-performance electrodes and SCs as they can store 10–100 times more energy than CNMs [8]. In the next paragraphs, the author will discuss the two main groups of pseudocapacitive materials: (i) transition metal oxides/hydroxides and (ii) CPs.

10.4.2 Transition Metal Oxides-Carbon Composite Fibers

Some transition metal oxides (e.g., RuO_2, MnO_2) have been shown to be good candidates for improving energy storage capabilities based on the reversible redox reactions at the electrode surfaces [33, 34]. Among them, manganese dioxide (MnO_2) is under intense research as an active electrode material because of its high theoretical specific capacitance of –1380 F/g, low cost, environmentally benign nature, and abundance [33]. Moreover, it has multiple crystalline structures (e.g., α-, β-, δ-, γ-MnO_2), most of them showing tunnel structures with relatively large tunnels (e.g., α-MnO_2), which could be of value to facilitate ion diffusion inside the oxide and therefore mass transfer at the electrode surface. Two different strategies have been followed to incorporate MnO_2: (i) coating an existing fiber by using it as a template and (ii) incorporating the manganese oxide during fiber synthesis. For example, Zhou's group reported the deposition of MnO_2 onto the surface of carbon fiber by simply immersing it in a 0.1M $KMnO_4$/0.1 M Na_2SO_4 solution for 2 h where the following reaction took place (Figure 10.4A) [34]:

$$4MnO_4^- + 3C + H_2O \rightarrow 4MnO_2 + CO_3^{2-} + 2HCO_3^-$$

Then, a SC was fabricated by placing two parallel MnO_2/carbon fibers onto a PET film and coating them with PVA-H_2SO_4 gel electrolyte. The SC displayed an excellent volumetric capacitance of 2.5 F/cm^3 and a moderate energy density of 0.22 mWh/cm^3 (Figure 10.4B). García-Torres and Crean modified the CB/CNT/chitosan fiber surface with MnO_2 by chemically reducing $KMnO_4$ with ethanol by a simple electrodeless process [32]. They immersed the fibers in ethanol followed by the addition of 0.1 M $KMnO_4$ solution at

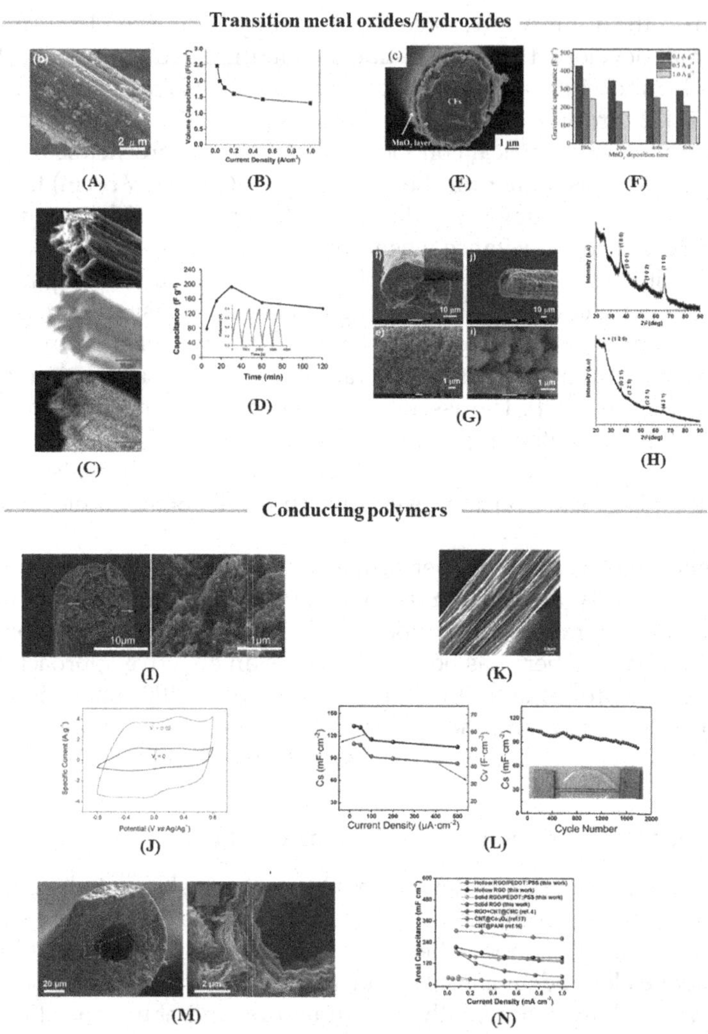

FIGURE 10.4
(A) SEM image of a carbon-based fiber after the deposition of MnO_2 due to the reduction of $KMnO_4$. (B) Dependence of volume capacitance with the applied current density. (A) and (B) are adapted with permission from reference [34], Copyright (2012), American Chemical Society. (C) (Top) SEM image of a CB/CNT/chitosan after the deposition of MnO_2. Energy-dispersive X-ray spectroscopy (EDX) map of carbon (center) and manganese (bottom). (D) Graph showing gravimetric capacitance *vs* $KMnO_4$ reduction time. (C) and (D) are adapted with permission from reference [32], Copyright (2018), Elsevier. (E) SEM image of the cross-section of a coaxial fiber. (F) Bar graph showing the dependence of the gravimetric capacitance with the MnO_2 deposition time at various current densities. Adapted with permission from reference [35], Copyright (2016), Elsevier. (G) SEM images of CB/CNT/chitosan fibers after the electrodeposition of MnO_2 from two different sources $Mn(NO_3)_2$ (left images) and $Mn(CH_3COOH)_2$ (right images). (H) X-ray diffractograms for the MnO_2-coated fibers using acetate (top) and nitrate salts (bottom). Adapted with permission from reference [36], Copyright (2018), Wiley-VCH Verlag. (I) SEM images of the cross-section of PEDOT-CNT composite fibers. (J) Comparison of the cyclic voltammograms of PEDOT and PEDOT-CNT fibers. Adapted with permission from reference [24], Copyright (2012), Royal Society of Chemistry. (K) SEM image of the GO-PEDOT fibers showing a wrinkled surface. (L) (Left) Dependence of the area and volumetric capacitance with the applied current density. (Right) Graph showing the cyclic performance with the number of charge/discharge cycles. Inset shows a photograph of the 1D SC. Adapted with permission from reference [44], Copyright (2017), Elsevier. (M) SEM image showing the hollow GO-PEDOT fibers. (N) Graph showing the variation of the areal capacitance with the current density for the different GO-PEDOT fibers. Adapted with permission from reference [45], Copyright (2016), Wiley-VCH Verlag.

different times (from 5 to 120 min), to obtain γ-MnO_2 (Figure 10.4C). The authors reported a 16-fold increase in capacitance compared with the uncoated CB/CNT/chitosan fibers, achieving a gravimetric capacitance of 195F/ g (Figure 10.4D). This increase was attributed to the reversible surface adsorption/desorption of alkali metal ions (M^+) on the manganese oxide [$MnO_{2(surface)} + M^+ + e^- \rightleftarrows (MnO_2^- M^+)_{surface}$] as well as a redox reaction involving Mn(III) and Mn(IV) oxidation states. An alternative method was the electrochemical deposition of MnO_2 to coat either nanoparticles or continuous layers onto the carbon-based fibers [35, 36]. For example, Zhang and co-workers developed MnO_2-carbon fiber yarn by a simple electrochemical method (Figure 10.4E) [35]. They controlled the deposition time to have the optimum, ultrathin MnO_2 nanosheets onto the fibers to have capacitance values as high as 428 F/g (based on the MnO_2 mass) (Figure 10.4F). They assembled two parallel fibers with polyvinyl pyrrolidone (PVP)/Na_2SO_4 gel electrolyte to form an SC displaying a high volumetric energy density of 3.8 mWh/cm^3 at a power density of 89 mW/cm^3. Garcia-Torres *et al.* also reported the electrochemical deposition of MnO_2 onto CB/CNT/chitosan fibers, endowing it with the ability to tune its crystalline structure (ε-MnO_2, γ-MnO_2) and morphology, depending on the manganese salt employed ($Mn(NO_3)_2$, $Mn(CH_3COO)_2$) (Figure 10.4G,H) [36]. Whereas the MnO_2 layer showed a columnar growth with nanoscale channels along it when the acetate salt was used, a globular morphology with flower-like nodules was seen in $Mn(NO_3)_2$ electrolyte. The two different crystalline structures resulted in different electrochemical performances, with the ε-MnO_2 coated fibers being superior as the presence of the channels through the MnO_2 layer led to an increased SSA and favored the diffusion and accessibility of the ions into this layer. The electrochemical deposition times were varied to achieve an optimal MnO_2 thickness. However, the low electrical conductivity of the oxides resulted in lower capacitance values than expected, because these nanostructures were wrapped in CP [36]. Thus, either PEDOT: PSS (poly(3,4-ethylenedioxythiophene): polystyrene sulfonate), PPy (polypyrrole), or a combination of both polymers resulted in the improvement in fiber electrochemical performance. PEDOT wrapping increased the electrical conductivity of the MnO_2, allowing a more effective ionic transport through the oxide layer and leading to a gravimetric capacitance of 291 F/g, whereas PPy improved fiber stability by enhancing capacitance retention up to 89%.

The second strategy has been the incorporation of manganese oxide within the fiber to avoid not only the usual limited active material loading in core-shell fibers (e.g., surface area is reduced as shell thickness increases) but also the lack of energy storage contribution of the inner core. In this sense, Crean's group developed a one-step wet-spinning process to fabricate fibers containing different CNMs to improve electrical conductivity and SSA and birnessite MnO_2 nanotubes to enhance specific capacitance [37]. The authors reported the successful incorporation of the functional nanomaterials within the fiber, achieving a huge increase in capacitance from 10 F/g for carbon-based fibers to 246 F/g for fibers containing 0.2 wt.% birnessite, which was attributed to the high specific capacitance of the birnessite phase and the high SSA of the tubular nanostructures.

Another transition metal oxide widely investigated has been RuO_2 because of its high theoretical capacitance (1400 F/g), high chemical stability, and fast charge/discharge rates [38]. In a two-step process, Zhai *et al.* synthesized multicomponent fibers, using graphene as the fiber backbone, CNTs to suppress the stacking of graphene sheets, ethylenediamine as a crosslinker, and RuO_2 as pseudocapacitive material [39]. They first synthesized the carbon-based fiber using a hydrothermal process, followed by the incorporation of RuO_2. The fibers were immersed into a $RuCl_3$ aqueous solution to incorporate Ru^{3+}, dried, and further treated with NaOH. The resulting $Ru(OH)_3$ was annealed at 150 °C in the air to obtain the RuO_2. After optimizing the RuO_2 content, they obtained composite fibers with gravimetric and volumetric capacitance values of 670 F/g and 912 F/cm^3, respectively. The

SC, which was built by placing two parallel fibers onto a PET substrate and coating with the gel electrolyte (PVA/H_3PO_4), delivered an energy density of 27.3 mWh/cm^3 at a power density of 147.7 mW/cm^3 and it retained more than half of the energy density (15.9 mWh/cm^3) when the power density was doubled (295.1 mW/cm^3). The main drawback of RuO_2 is its high cost, which limits its practical use.

Transition metal (hydro)oxides other than MnO_2 or RuO_2 have also been studied to increase the energy density, like nickel or cobalt oxides and hydroxides (e.g. NiO, $Ni(OH)_2$, $Co(OH)_2$) [40–42]. These oxides/hydroxides also show excellent theoretical capacitances, due to their layered structures with large interlayer spacings, and are cost effective for practical applications. For example, Liu and co-authors coated carbon fibers with $ZnCo_2O_4$ nanowires arrays by a hydrothermal process [40]. Then, several fibers were placed onto a PET substrate in an interdigitated design and subsequently coated with PVA/H_3PO_4 electrolyte resulting in an SC with a gravimetric capacitance of 610 mF/g. Gao *et al.* fabricated a multi-layered fiber consisting, from the inside to the outside, carbon fiber as the core, a metallic, thin Ni layer, a porous graphite nanoparticle-based layer, and an ultrathin Ni-Co layered double hydroxide as the outer layer [41]. Each of those layers played a key role in the whole performance of the fiber. The thin nickel layer was obtained by electrodeposition and served to improve the electrical conductivity without affecting the flexibility of the carbon fiber. On the other hand, the graphite nanoparticles were dip-coated onto the Ni layer, producing a highly porous carbon structure with large SSA and acted as a support to grow the Ni and Co hydroxides, also by electrodeposition. The multi-layered fibers showed very high capacitance values (1395.8 F/g at 2.5 A/g, 1.39 F/cm^2 at 2.5 mA/cm^2), very high capacitance retention (95.6%) for 5,000 cycles, and a fast charge transfer behavior. The fibers were assembled to prepare a solid-state SC using PVA/KOH as electrolyte to develop a high capacitance (28.67 mF/cm^2), energy density (9.57 μWh/cm^2), and power density (492.17 μW/cm^2)

10.4.3 Conducting Polymer-Based Fibers

Most of the composite fibers reported above consisted of non-conductive polymers like PVA, chitosan, or CMC, limiting their electrical conductivity due to the presence of insulating materials. Thus, the use of CPs (e.g., PPy, PANI (polyaniline), polyethylene dioxithiophene) instead is a very smart solution to further increase the levels of electrical conductivity but also the electrochemical properties due to their energy storage capability through Faradaic reactions while maintaining their flexibility [24, 43]. For example, Prof. Wallace's group prepared SWCNT and PEDOT: PSS fibers by the wet-spinning technique [24]. They first prepared dispersions containing PEDOT: PSS, SWCNTs, and sodium dodecyl sulfate surfactant to avoid nanotube aggregation at various nanotube volume fractions (V_f). Then, the formulations were wet spun into isopropanol as a fiber coagulation medium (Figure 10.4I). It was found that the addition of 0.02 V_f improved the electrical conductivity, gravimetric capacitance (three-electrode cell) as well as the Young's modulus, and the tensile strength from 200 S/cm, 15 F/g, 2.0 GPa and 130 MPa for the pure PEDOT: PSS fiber to 450 S/cm, 67 F/g, 5.2 GPa and 200 MPa for the PEDOT: PSS-SWCNT composite fiber, respectively (Figure 10.4J). Ding *et al.* fabricated rGO/PPy fibers by wet spinning where the spinning solution was a mixture of pyrrole monomer and GO and the coagulation bath was an aqueous $FeCl_3$ solution where the monomer was polymerized [43]. The fiber was further treated with hydroiodic acid to chemically reduce the GO. These fibers, in addition to having high conductivity and mechanical flexibility, showed an areal capacitance of 115 mF/cm^2 (current density: 0.2 mA/cm^2). Li and co-workers reported the synthesis of an rGO/PEDOT: PSS fiber by wet spinning [44]. The authors claimed that rGO restacking was avoided due to the presence of the PEDOT polymer and that the fibers showed a highly wrinkled surface with many pores inside, enhancing the SSA and therefore the electrochemical performance (Figure 10.4K). A parallel SC, consisting of two fibers placed onto a flexible substrate and coated with PVA/H_3PO_4 as gel electrolyte, showed an areal capacitance of 131 mF/cm^2 at 20 μA/cm^2 and delivered a specific energy density of 4.55 μWh/cm^2 (Figure 10.4L). Those values did not significantly change when the SC was bent or rolled up. To improve the electrolyte accessibility to the inner core, hollow composite fibers have been developed aiming to harness the energy storage capacity from the inner surface as well. Thus, Qu and co-workers synthesized hollow rGO/PEDOT: PSS fibers by injecting a mixture of PEDOT: PSS, GO, and vitamin C into a glass pipe at 90 °C to reduce the GO (Figure 10.4M) [45]. According to the authors, the release of gases (e.g., CO_2) during GO reduction led to the formation of the hollow structure. Two parallel fibers were assembled into an SC showing high areal capacitance (304.5 mF/cm^2 at 0.08 mA/cm^2), ultra-high energy density (27.1 μWh/cm^2) at a power density of 66.5 μW/cm^2, and excellent capacitance retention of 96% after 10,000 cycles (Figure 10.4N). These fibers were also mechanically robust and could be woven into clothes. CPs other than PEDOT: PSS, such as PANI, PPy, or polyacetylenes, have also been employed to obtain fiber electrodes due to their good intrinsic conductivity and flexibility [46, 47].

10.5 Conclusions

1D SCs have recently attracted significant interest to power flexible and wearable electronic devices due to properties like flexibility, light weight, high power density, high charge-discharge rates, and long capacitance retention. In this chapter, basic concepts about energy storage systems have been introduced to properly understand the research approach in the literature. Then, the main techniques used to fabricate fibers and 1D SCs have been described. And finally, the state of the art of the main materials employed to fabricate the fiber electrodes has been reviewed. Thus, CNMs are very useful because of their excellent electrical conductivity, high strength, or low density. However, fibers prepared from a single CNM are relatively compact leading to low SSA and therefore low capacitance values. Although coating the fibers with high-porosity CNM (e.g. graphene) or preparing fibers from two or more CNM mixtures provided higher energy storage, the energy storage performance was still low. Thus, the incorporation of pseudocapacitive materials, like transition metal oxides or CPs, into carbon-based electrodes has improved their charge storage. Transition metal oxides (e.g., MnO_2) increase the capacitance but, due to their low electrical conductivity, theoretical capacitance values cannot be achieved. On the other hand, CPs enhance the electrical conductivity and capacitance without compromising the flexibility required to develop 1D SCs.

References

1. S. Park, S.W. Heo, W. Lee, D. Inoue, Z. Jiang, K. Yu, H. Jinno, D. Hashizume, M. Sekino, T. Yokota, K. Fukuda, K. Tajima, T. Someya, Self-powered ultra-flexible electronics via nano-grating-patterned organic photovoltaics, *Nature* 561(7724) (2018) 516.

2. T.Q. Trung, N.-E. Lee, Flexible and stretchable physical sensor integrated platforms for wearable human-activity monitoring and personal healthcare, *Adv. Mater.* 28(22) (2016) 4338.
3. Z. Fan, J.C. Ho, Z.A. Jacobson, R. Yerushalmi, R.L. Alley, H. Razavi, A. Javey, Wafer-scale assembly of highly ordered semiconductor nanowire arrays by contact printing, *Nano Lett.* 8(1) (2008) 20.
4. D.-H. Kim, Y.-S. Kim, J. Amsden, B. Panilaitis, D.L. Kaplan, F.G. Omenetto, M.R. Zakin, J.A. Rogers, Silicon electronics on silk as a path to bioresorbable, implantable devices, *Appl. Phys. Lett.* 95(13) (2009) 133701.
5. W. Zeng, L. Shu, Q. Li, S. Chen, F. Wang, X.M. Tao, Fiber-based wearable electronics: A review of materials, fabrication, devices, and applications, *Adv. Mater.* 26(31) (2014) 5310.
6. S. Seyedin, T. Carey, A. Arbab, L. Eskandarian, S. Bohm, J.M. Kim, F. Torrisi, Fibre electronics: Towards scaled-up manufacturing of integrated e-textile systems, *Nanoscale* 13(30) (2021) 12818.
7. Y. Zhang, H. Wang, H. Lu, S. Li, Y. Zhang, Electronic fibers and textiles: Recent progress and perspective, iScience 24(7) (2021) 102716.
8. P. Simon, Y. Gogotsi, Materials for electrochemical capacitors, *Nat. Mater.* 7(11) (2008) 845.
9. A.B. Dalton, S. Collins, E. Muñoz, J.M. Razal, V.H. Ebron, J.P. Ferraris, J.N. Coleman, B.G. Kim, R.H. Baughman, Super-tough carbon-nanotube fibres, *Nature* 423(6941) (2003) 703.
10. W. Ma, Y. Zhang, S. Pan, Y. Cheng, Z. Shao, H. Xiang, G. Chen, L. Zhu, W. Weng, H. Bai, M. Zhu, Smart fibers for energy conversion and storage, *Chem. Soc. Rev.* 50(12) (2021) 7009.
11. Y. Wang, Y. Song, Y. Xia, Electrochemical capacitors: Mechanism, materials, systems, characterization and applications, *Chem. Soc. Rev.* 45(21) (2016) 5925.
12. C. Meng, O.Z. Gall, P.P. Irazoqui, A flexible super-capacitive solid-state power supply for miniature implantable medical devices, *Biomed. Microdevices* 15(6) (2013) 973.
13. A.G. Pandolfo, A.F. Hollenkamp, Carbon properties and their role in supercapacitors, *J. Power Sources* 157(1) (2006) 11.
14. S. Chen, W. Ma, Y. Cheng, Z. Weng, B. Sun, L. Wang, W. Chen, F. Li, M. Zhu, H.-M. Cheng, Scalable non-liquid-crystal spinning of locally aligned graphene fibers for high-performance wearable supercapacitors, *Nano Energy* 15 (2015) 642.
15. L. Kou, T. Huang, B. Zheng, Y. Han, X. Zhao, K. Gopalsamy, H. Sun, C. Gao, Coaxial wet-spun yarn supercapacitors for high-energy density and safe wearable electronics, *Nat. Commun.* 5 (2014) 3754.
16. D. Yu, K. Goh, H. Wang, L. Wei, W. Jiang, Q. Zhang, L. Dai, Y. Chen, Scalable synthesis of hierarchically structured carbon nanotube–graphene fibres for capacitive energy storage, *Nat. Nanotechnol.* 9(7) (2014) 555.
17. R. Cruz-Silva, A. Morelos-Gomez, H.-I. Kim, H.-K. Jang, F. Tristan, S. Vega-Diaz, L.P. Rajukumar, A. Laura Elías, N. Perea-Lopez, J. Suhr, M. Endo, M. Terrones, Super-stretchable graphene oxide macroscopic fibers with outstanding knotability fabricated by dry film scrolling, *ACS Nano* 8(6) (2014) 5959.
18. G. Sun, J. Zhou, F. Yu, Y. Zhang, J.H.L. Pang, L. Zheng, Electrochemical capacitive properties of CNT fibers spun from vertically aligned CNT arrays, *J. Solid State Electrochem.* 16(5) (2012) 1775.
19. J. Ren, Q. Xu, Y.-G. Li, Flexible fiber-shaped energy storage devices: Principles, progress, applications and challenges, *Flex. Print. Electron.* 3 (2018) 013001.
20. C. Meng, A review of flexible and weaveable fiber-like supercapacitors, *J. Postd Res.* 1(12) (2013) 16.
21. D. Yu, Q. Qian, L. Wei, W. Jiang, K. Goh, J. Wei, J. Zhang, Y. Chen, Emergence of fiber supercapacitors, *Chem. Soc. Rev.* 44(3) (2015) 647.
22. L. Dong, C. Xu, Y. Li, Z.-H. Huang, F. Kang, Q.-H. Yang, X. Zhao, Flexible electrodes and supercapacitors for wearable energy storage: A review by category, *J. Mater. Chem. A* 4(13) (2016) 4659.
23. A. Hirsch, The era of carbon allotropes, *Nat. Mater.* 9(11) (2010) 868.
24. R. Jalili, J.M. Razal, G.G. Wallace, Exploiting high quality PEDOT:PSS–SWNT composite formulations for wet-spinning multifunctional fibers, *J. Mater. Chem.* 22(48) (2012) 25174.

25. Y. Jang, S.M. Kim, G.M. Spinks, S.J. Kim, Carbon nanotube yarn for fiber-shaped electrical sensors, actuators, and energy storage for smart systems, *Adv. Mater.* 32(5) (2020) 1902670.
26. J. Ren, L. Li, C. Chen, X. Chen, Z. Cai, L. Qiu, Y. Wang, X. Zhu, H. Peng, Twisting carbon nanotube fibers for both wire-shaped micro-supercapacitor and micro-battery, *Adv. Mater.* 25(8) (2013) 1155.
27. Q. Li, N. Mahmood, J. Zhu, Y. Hou, S. Sun, Graphene and its composites with nanoparticles for electrochemical energy applications, *Nano Today* 9(5) (2014) 668.
28. X. Ji, Y. Xu, W. Zhang, L. Cui, J. Liu, Review of functionalization, structure and properties of graphene/polymer composite fibers, *Compos. A* 87 (2016) 29.
29. J. Ren, W. Bai, G. Guan, Y. Zhang, H. Peng, Flexible and weaveable capacitor wire based on a carbon nanocomposite fiber, *Adv. Mater.* 25(41) (2013) 5965.
30. Y.N. Meng, Y. Zhao, C.G. Hu, H.H. Cheng, Y. Hu, Z.P. Zhang, G.Q. Shi, L.T. Qu, All-graphene core-sheath, Microfibers for all-solid-state, stretchable fibriform supercapacitors and wearable electronic textiles, *Adv. Mater.* 25(16) (2013) 2326.
31. Y. Ma, P. Li, J.W. Sedloff, X. Zhang, H. Zhang, J. Liu, Conductive graphene fibers for wire-shaped supercapacitors strengthened by unfunctionalized few-walled carbon nanotubes, *ACS Nano* 9(2) (2015) 1352.
32. J. Garcia-Torres, C. Crean, Ternary composite solid-state flexible supercapacitor based on nanocarbons/manganese dioxide/PEDOT:PSS fibres, *Mater. Des.* 155 (2018) 194.
33. Q. Lv, S. Wang, H. Sun, J. Luo, J. Xiao, J. Xiao, F. Xiao, S. Wang, Solid-state thin-film supercapacitors with ultrafast charge/discharge based on N-doped-Carbon-Tubes/Au-nanoparticles-doped-MnO_2 nanocomposites, *Nano Lett.* 16(1) (2016) 40.
34. X. Xiao, T. Li, P. Yang, Y. Gao, H. Jin, W. Ni, W. Zhan, X. Zhang, Y. Cao, J. Zhong, L. Gong, W.-C. Yen, W. Mai, J. Chen, K. Huo, Y.-L. Chueh, Z.L. Wang, J. Zhou, Fiber-based all-solid-state flexible supercapacitors for self-powered systems, *ACS Nano* 6(10) (2012) 9200.
35. J. Zhang, X. Zhao, Z. Huang, T. Xu, Q. Zhang, High-performance all-solid-state flexible supercapacitors based on manganese dioxide/carbon fibers, *Carbon* 107 (2016) 844.
36. J. Garcia-Torres, C. Crean, Multilayered flexible fibers with high performance for wearable supercapacitor applications, *Adv. Sust. Syst.* 2(2) (2018) 1700143.
37. J. Garcia-Torres, A.J. Roberts, R.C.T. Slade, C. Crean, One-step wet-spinning process of CB/CNT/MnO_2 nanotubes hybrid flexible fibres as electrodes for wearable supercapacitors, *Electrochim. Acta* 296 (2019) 481.
38. L. Chen, Y. Hou, J. Kang, A. Hirata, T. Fujita, M. Chen, Toward the theoretical capacitance of RuO_2 reinforced by highly conductive nanoporous gold, *Adv. Energy Mater.* 3(7) (2013) 851.
39. S. Zhai, C. Wang, H.E. Karahan, Y. Wang, X. Chen, X. Sui, Q. Huang, X. Liao, X. Wang, Y. Chen, Nano-RuO_2-decorated holey graphene composite fibers for micro-supercapacitors with ultrahigh energy density, *Small* 14 (2018) 1800582.
40. B. Liu, D. Tan, X. Wang, D. Chen, G. Shen, Supercapacitors: Flexible, planar-integrated, all-solid-state fiber supercapacitors with an enhanced distributed-capacitance effect, *Small* 9(11) (2013) 1998.
41. L. Gao, J.U. Surjadi, K. Cao, H. Zhang, P. Li, S. Xu, C. Jiang, J. Song, D. Sun, Y. Lu, Flexible fiber-shaped supercapacitor based on nickel-cobalt double hydroxide and pen ink electrodes on metallized carbon fiber, *ACS Appl. Mater. Interfaces* 9(6) (2017) 5409.
42. L. Hua, Z. Ma, P. Shi, L. Li, K. Rui, J. Zhou, X. Huang, X. Liu, J. Zhu, G. Sun, W. Huang, Ultrathin and large-sized vanadium oxide nanosheets mildly prepared at room temperature for high performance fiber-based supercapacitors, *J. Mater. Chem. A* 5(6) (2017) 2483.
43. X. Ding, Y. Zhao, C. Hu, Y. Hu, Z. Dong, N. Chen, Z. Zhang, L. Qu, Spinning fabrication of graphene/polypyrrole composite fibers for all-solid-state, flexible fibriform supercapacitors, *J. Mater. Chem. A* 2(31) (2014) 12355.
44. B. Li, J. Cheng, Z. Wang, Y. Li, W. Ni, B. Wang, Highly-wrinkled reduced graphene oxide-conductive polymer fibers for flexible fiber-shaped and interdigital-designed supercapacitors, *J. Power Sources* 376 (2018) 117.

45. G. Qu, J. Cheng, X. Li, D. Yuan, P. Chen, X. Chen, B. Wang, H. Peng, A fiber supercapacitor with high energy density based on hollow graphene/conducting polymer fiber electrode, *Adv. Mater.* 28(19) (2016) 3646.
46. C. Li, H. Bai, G. Shi, Conducting polymer nanomaterials: Electrosynthesis and applications, *Chem. Soc. Rev.* 38(8) (2009) 2397.
47. Y. Zhao, Y. Ding, Y. Li, L. Peng, H.R. Byon, J.B. Goodenough, G. Yu, A chemistry and material perspective on lithium redox flow batteries towards high-density electrical energy storage, *Chem. Soc. Rev.* 44(22) (2015) 7968.

11

One-Dimensional Polymeric Nanocomposites for Rechargeable Batteries

Erlantz Lizundia and Manuel Salado

CONTENTS

11.1 Introduction

Since lithium-ion batteries (LIBs) were first commercialized by Sony, electrochemical energy storage has undergone unprecedented, rapid progress. In particular, several efforts to reduce global warming have driven a rapid evolution in the development and implementation of LIBs to power electric vehicles (EVs). Even though this technology offers strong robustness and reversibility, the depletion of the cobalt, lithium, and nickel reserves and the use of liquid electrolytes, which are composed of flammable organic solvents, seriously jeopardize battery safety and mid-term reliability.

Bearing this in mind, the different components of the battery (e.g., high voltage cathodes or chemically stable electrolytes) need to be modified to obtain higher energy densities. For that, sodium (Na), zinc (Zn), magnesium (Mg), and aluminum (Al) battery systems are being explored as alternative candidates due to these elements exhibiting lower cost, greater abundance, and low toxicity [1]. Lithium-sulfur (Li–S) and lithium-oxygen (Li–O_2) batteries have also received considerable attention because of their high energy density of 2600 Wh/kg and 3500 Wh/kg [2], respectively. In addition, the environmentally friendly nature and the use of non-toxic and low-cost sulfur make this element theoretically suitable for commercialization. However, "All that glitters is not gold". Several major challenges hinder their industrial development, including the formation of polysulphides, which cause severe capacity fading, low Coulombic efficiencies (CE), and large electrode volume expansion (80%) [3]. Regarding Li–O_2 batteries, sluggish kinetics of the oxygen reduction reaction (ORR) and oxygen evolution reaction (OER), resulting from the formation of insulating Li_2O_2, leads to inferior round-trip efficiency and cycling performance [4].

Monovalent metal ions (Na^+ and K^+) have the advantage of being quite similar in terms of the electrochemical processes to the mature and well-known lithium in LIBs. In addition, their high natural abundance, and consequent low cost make them promising

DOI: 10.1201/9781003223764-11

candidates (especially Na-based batteries) to replace LIBs in the future. Unfortunately, the higher redox potential of Na/Na^+ (–2.71 V), compared with Li/Li^+ (–3.04 V), entails a lower discharge voltage as well as low energy density [5]. On the other hand, multivalent metal ions (Mg^{2+} and Al^{3+}) possess high energy density as they have lower redox potential (–2.38 V and –1.66 V, respectively); however, they suffer from slow diffusion kinetics and poor cycling stability during the insertion/deinsertion process [6].

Researchers and industry are seeking novel alternatives to avoid the use of volatile and flammable solvents. In this scenario, one of the best candidates to fulfill these requirements are aqueous zinc (Zn) batteries due to their high theoretical capacity (820 mAh/g), low toxicity, high safety, and low cost [7]. As a technology in its infancy, material compositions need to be optimized to avoid irregular charge distribution at the electrodes and side reactions during plating/stripping. Until then, large overpotential and poor cycling stability due to the formation of dendrite growth need to be addressed.

As a multilayer stack system, degradation in LIBs is the result of a complex interaction among the different counterparts with various physical and chemical mechanisms. Birkl et al. [8] presented a seminal work where the different degradation mechanisms can be estimated by the open-circuit voltage (V_{OC}) of Li-ion cells and electrodes. Figure 11.1 represents the possible degradation mechanism depending on the interlayer where this phenomenon occurs. The understanding of the different causes, rates, and interdependencies of these degradation mechanisms could pave the way to developing novel chemistry and interface engineering into a new generation of energy storage systems.

The lifetime of the batteries is strongly influenced by the nature of the battery material components as well as the operating conditions, such as charge/discharge rates, voltage operation, and temperature. According to the scheme (Figure 11.1), capacity fading at the cathode is usually the result of the partial dissolution of the active material during cycling/storage or electrolyte oxidation. Therefore, a proper electron energy level alignment at the interface of the cathode will avoid changes in the electronic structure and defect formation induced by oxygen loss and phase transformation [9].

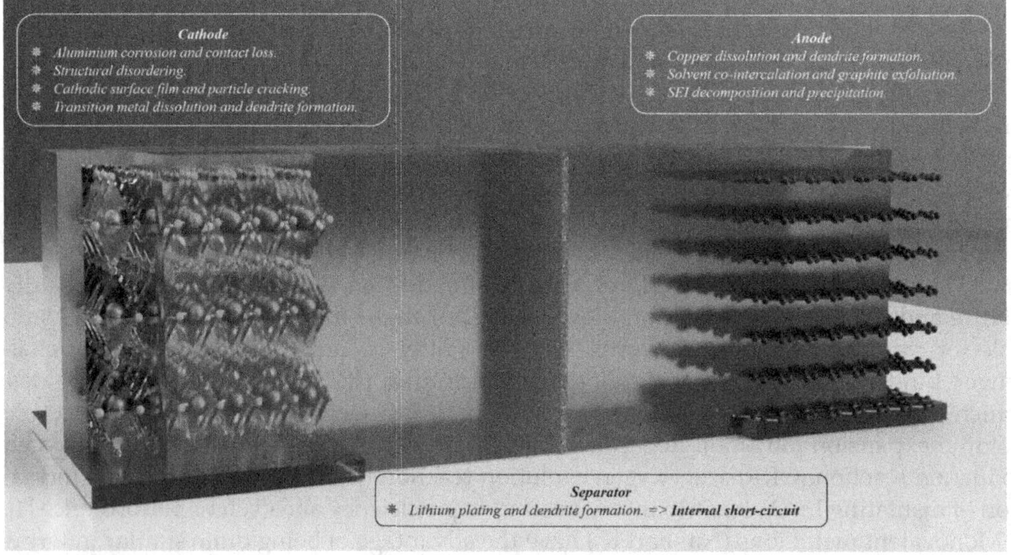

FIGURE 11.1
General degradation mechanism in LIBs.

On the other hand, the two major factors contributing to anode degradation in LIBs are the solid electrode/electrolyte interface (SEI) formed (boosted at higher temperatures) and lithium metal plating (intensified at low temperatures). In addition, nanostructured anodes for LIBs are greatly affected by volume expansion and consequently by mechanical stresses generated during the lithiation-delithiation process [10]. One of the best examples of volume expansion in anode material is silicon (about 400%) when a large number of Li atoms are inserted into the silicon structure during lithiation [11]. Taking into account that silicon has an excellent lithium storage capacity of 4200 mAh/g (approximately 10 times that of conventional graphite negative electrodes), Cui et al. [12] studied the fracture mechanism of Si nanostructures such as nanowires, nanotubes, and nanoparticles. The critical diameter for fracture was found to be around 360 nm during lithiation and even higher for delithiation. Nevertheless, the same research group recently used this strategy to embed Si nanoparticles between reduced graphene oxide layers. After lithiation, Si nanoparticles favored the storage of Li, improving the cycling performances, and reducing the overpotential [13].

Regarding electrolytes, both the solvent and the salt components of electrolytes are susceptible to decomposition during battery operation. Consider a general electrolyte composition of $LiPF_6$ as the conducting salt, and ethylene carbonate (EC), dimethyl carbonate (DMC), or ethyl methyl carbonate (EMC) as solvents. Generally, conducting salts are decomposed by hydrolysis, whereas the degradation mechanism of solvents will predominantly occur *via* polymerization of EC and transesterification of EMC. Moreover, a considerable part of the electrolyte is also consumed during the formation of the SEI [14].

As 1D-based materials provide large surface-to-volume ratios, the unidimensionality and high surface reactivity can be applied in a wide range of applications such as electro-catalysis, photo-catalysis, sensing, or energy storage. More specifically, anodes/cathodes can be designed with different morphologies to improve the rate capability, stability, and volumetric energy density of these electrodes [15]. Nevertheless, obtaining an uniform and homogeneous large area as well as controlled pore size also remains a tricky aspect. For the deposition of heavy metals and high-melting-point materials, such as graphite [16], molybdenum [17], or silicon carbide [18], magnetron sputtering is generally considered to have nanometer controllability and reproducibility of the deposited film thickness. Recently [19], this technique has been used to fabricate sulfur/carbon composite cathodes for Li-S batteries. The reduction of sulfur particles allowed a homogeneous distribution into the carbon matrix. As a result, magnetron-sputtered samples maintained a capacity above 83% at a current of 0.1 C. Other techniques, such as electrostatic spray deposition (EDS), atomic layer deposition, laser processing, physical vapor deposition, plasma-assisted synthesis, etc., can also be used. Sun et al. synthesized needle-like thin films of olivine-type $LiFePO_4$ using an oblique angled pulsed laser deposition (OAPLD) technique. The authors highlighted the importance of cathode material crystallinity to obtain optimum charging and discharging curves [20].

Although in the past few decades polymers have had a secondary function as battery components (binder or separator membranes), while inorganic intercalation compounds and carbon-based composites were used as electrode materials, the development of organic materials, such as conjugated carbonyls, bio-based nanoparticles, or conductive polymers represents a promising strategy to develop the next-generation sustainable LIBs [21]. In the case of soft materials, such as polymers, template methods and wet chemistry approaches can produce 1D structures, although large area uniformity and batch-to-batch reproducibility of 1D structures remain challenging. Specific techniques such as electrospinning allow the synthesis of nano-sized fibers by applying kilovolt voltages to viscous liquids at very low flow rates [22]. By tuning the electrospinning parameters, this technique can overcome some of the issues that diverse battery components present by controlling the diameter, porosity, composition, and structure. In addition, these techniques can be also applied to develop post-Li ion technologies. For instance, the electrospun 1D nanostructure can stabilize Na_xMO_2 for long-cycle sodium ion battery (SIB) cathodes. As a result, remarkably high race capacity (166.7 mAh/g at 0.1 C) and durable cycling life (80.8% capacity retention after 500 cycles) were achieved for the fibrous cathode [23].

In this chapter, novel strategies using 1D polymeric nanocomposites in the three main components (electrodes, electrolyte, and separator) of a battery are explained (Sections 11.2–11.4), paying special attention to the development of new interfaces between the components to favor ionic transport as well as to improve battery safety. To conclude, future perspectives are offered so that these issues can be overcome (Section 11.5).

11.2 1D Polymer Nanocomposite for Battery Electrodes

The use of 1D nanostructured electroactive materials has the potential to improve the charge transfer rate by enhancing both the ionic (Li^+, Na^+, Zn^{2+}) intercalation/deintercalation and electronic transference processes, while the issues regarding electrode volume changes are reduced. Chan et al. provided an early example of how nanowires can be applied to accommodate large strains with no pulverization thanks to the short Li intercalation distance [24]. Despite the well-known volume changes of silicon (up to 400%), a silicon nanowire anode enabled an efficient 1D electron transport, good electrode/electrolyte interfacial area, and facile strain relaxation. With this in mind, bottom-up electrospinning and gas-phase polymerization strategy were followed to coat 1D spinel $ZnFe_2O_4$ with a polypyrrole shell [25]. The 1D structure, together with the 2–4 nm polypyrrole shell, solved the poor electrical conductivity and severe volume change characteristic of $ZnFe_2O_4$, shortening Li^+ transport length and thus offering a LIB anode capable of providing large, reversible and stable capacities (881 mAh/g after 220 cycles at 1 A/g). The polypyrrole coating enhanced the specific surface area from 42.6 m^2/g for bare $ZnFe_2O_4$ to 80.8 m^2/g, which also ensured an improved electrode-electrolyte contact.

1D polymeric nanoparticles can be simultaneously applied as templates and precursors to transfer their structural order into solidified materials at the macroscopic scale. Benefiting from the spindle-shaped characteristics of cellulose nanocrystals (CNCs), bio-based nanoparticles can be extracted following the controlled hydrolysis of cellulose, Nguyen et al. prepared free-standing mesoporous titanium carbide anodes having a chiral nematic order [10]. Thanks to the mesoporous architecture, enabling a stable physical structure to accommodate the volume changes upon cycling, the titanium carbide free-standing anode displays an extremely weak capacity fade upon galvanostatic charge-discharge (0.019% per cycle). A similar chiral templating was pursued later to obtain chiral nematic mesoporous black titanium dioxide (TiO_{2-x})/carbon composites [26]. As schematically summarized in Figure 11.2a, gelatin-functionalized CNCs are self-assembled into helices and after titania (titanium dioxide) incorporation, their carbonization results in chiral TiO_2/C replicas. An additional, hydrothermal surface reduction of white titania by ascorbic acid affords free-standing chiral nematic black TiO_{2-x}. As shown by SEM images, these materials keep the original chiral structure characteristic of CNCs. After five cycles at a current density of 100 mA/g, the discharge capacities were 351 mAh/g for the TiO_{2-x}/C and ≈131 mAh/g for the white (conventional) TiO_2/C. Oxygen-deficient black titania was

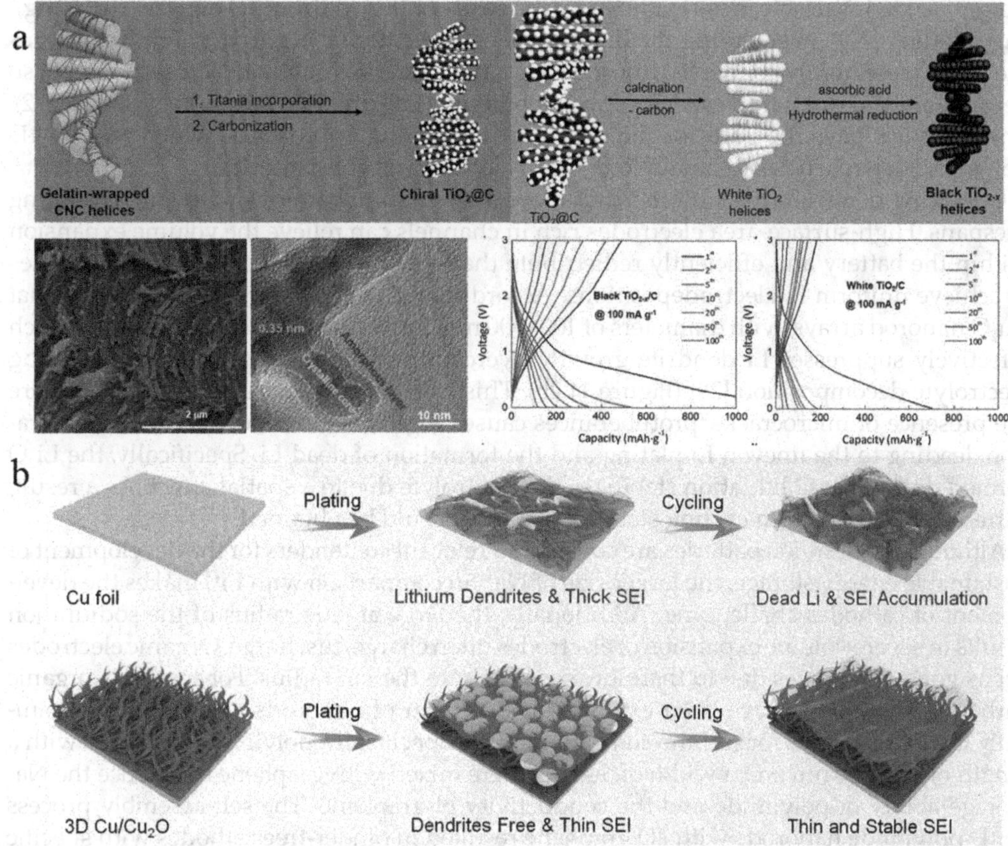

FIGURE 11.2
a) Fabrication approach to achieve chiral nematic mesoporous black TiO_{2-x}/C freestanding materials. SEM image shows the chiral nematic structure, while the TEM image reveals the presence of an external amorphous layer onto the crystalline core. Galvanostatic charge and discharge curves for black and white titania at a current density of 100 mA·g^{-1} are shown. Adapted with permission from [26]. Copyright (2019) Wiley. b) Schematic comparison showing the Li plating and stripping onto a flat Cu foil (with uneven lithium nucleation, dendrite formation, and thick SEI) and onto a Cu/Cu_2O nanorod array-based 3D electrode (uniform and dendrite-free Li deposition, with a thin and stable SEI). Adapted with permission from [28]. Copyright (2021) American Chemical Society.

found to enable faster Li^+ diffusion. Interestingly, the achieved TiO_{2-x} nanorods presented a crystallized TiO_2 core and an amorphous TiO_{2-x} outer layer resulting in an increased storage capacity due to the pseudocapacitive lithium storage of the disordered shell. The chiral mesoporous matrix provides conducting support to facilitate the electron/ion transfer, accelerates electrolyte penetration and Li^+ diffusion through enlarged pores, and minimizes volume changes during Li^+ insertion/extraction. Not only CNCs but also cellulose nanofibers (CNFs) are useful in the development of 1D polymeric battery electrodes. For instance, flexible, free-standing, electrically conductive electrodes based on Li_2S were synthesized [27] using a porous and mechanically flexible CNF structure which, upon carbonization at elevated temperatures and active material infiltration (Li_2S-ethanol solution), a free-standing electrode for Li-S batteries is obtained. This approach offers a cheap alternative to conventional procedures, and Li_2S is also an interesting material for next-generation sustainable batteries with high specific energy as no toxic transition metals, such as Ni or

Co, are needed. Variable mass loadings can be easily obtained (from 1.3 mg/cm^2 to 3.2 mg/cm^2 in study [27]), overcoming the limitations of conventional electrodes. Importantly, as the pores were highly connected, most of the active material remained accessible to Li^+, so near-theoretical discharge capacities were obtained (1200 mAh/g after 100 cycles at C/2). These studies pave the path for the design of templating approaches, which rely on 1D polymeric nanoparticles to obtain long life cycle, rechargeable batteries.

Obtaining dendrite-free batteries can allow the design of batteries with long operating lifespans. High-surface-area electrodes rich in channels can relieve the volume expansion within the battery and efficiently redistribute the local current density on the electrodes to achieve uniform Li electrodepositions. Accordingly, Li et al. very recently showed that Cu_2O nanorod arrays, with diameters of 100–300 nm, form a jungle-like porous layer which effectively suppresses Li dendrite growth, overcoming volume expansion and limiting electrolyte decomposition [28] (Figure 11.2b). This is in contrast to the bare Cu foil, where the presence of microcracks/protuberances causes high local Li^+ and electron concentration, leading to the uneven Li plating and the formation of dead Li. Specifically, the Li_2O formed during Cu_2O lithiation stabilizes the electrolyte due to a spatial effect. As a result, extraordinary long-term cycling stability of 1,700 h could be obtained.

Although sodium-ion batteries are considered relevant contenders for the development of sustainable energy storage, the larger size of Na^+ (in comparison with Li^+) makes the development of cathodes challenging. Additionally, the large atomic radius of the sodium ion results in severe volume expansion of electrodes upon charge/discharge. Organic electrodes act as good candidates due to their low sensitivity to the ion radius. For instance, organic carbonyl compounds have been exploited in the form of nanorods to develop mechanically flexible cathodes for sodium-ion batteries [29]. Specifically, polyimide nanorods with a length of 100–200 nm and a width of 6–8 nm were mixed with graphene to balance the Na^+ storage ability of polyimide and the conductivity of graphene. The self-assembly process of 1D polyimide nanorods with 2D graphene resulted in binder-free cathodes with specific surface areas of 152.2 m^2/g and having improved transport efficiency and Na^+ diffusion, achieving capacities of 101 mAh/g at 500 mA/g after 1,000 cycles.

One of the main bottlenecks toward practical implementation of aqueous zinc-ion batteries is their poor rate performance arising from slow diffusion kinetics of the high charge/radius ratio of Zn^{2+}. To achieve efficient pathways for the insertion/extraction of divalent Zn^{2+}, polyaniline nanorod array cathodes were prepared by chemical oxidation polymerization [30]. The large contact area between the liquid electrolyte and the 170 nm long nanorods developed promotes the diffusion of electrolyte ions, delivering a discharge capacity of 178 mAh/g at 100 mA/g. Active material dissolution is also considered a pressing issue to be solved shortly to develop reliable ZIBs. In this sense, Glatz et al. used 1D CNCs to construct an electrostatic filter capable of suppressing the organic cathode diffusion into the liquid electrolyte [31]. As a result, the organic 1,4 bis(diphenylamino)benzene (BDB) cathode delivers nearly a 2-electron redox capacity of 125 mAh/g at an average voltage of 1.25 V, showing a long-term cyclability with 75% retention after 1,000 cycles.

11.3 1D Polymer Nanocomposite for Battery Separators

One of the key components in rechargeable batteries is the separator, as its physicochemical properties have a strong impact on battery capacity, cycle life, and safety [32]. The ideal battery separator is characterized by good electrochemical stability, high mechanical strength, good thermal resistance, and high electrolyte uptake. The most common materials used as a microporous separator at low temperatures (<100 °C) comprise nonwoven fibers (e.g. nylon, cotton, polyesters, glass), polymer membranes (e.g. polyethylene (PE), polypropylene (PP), poly(tetrafluoroethylene) (PTFE), poly(vinyl chloride) (PVC)), and naturally occurring substances (e.g. cellulose, rubber, chitin) [32]. To overcome the degradation mechanisms caused by dendrite puncture, two strategies can be implemented: 1) the functional modification of the aforementioned traditional separators, and (2) the development of novel solid-state electrolytes (explained in Section 11.4).

Regarding the functional modification of the separators, one possible solution is to introduce ceramic nanoparticles, such as Al_2O_3 or SiO_2, so that the electrolyte wettability and thermal stability are improved. As an example, a two-step process where the separator was first treated with plasma and then immersed into a dispersion of SiO_2 nanoparticles results in nanoparticles electrostatically adsorbed not only onto the exterior of the membrane but also within its pores. The thickness and depth of the coatings can be easily controlled by tuning the Z-potential of the nanoparticles [33]. Although this strategy extends battery cycle life (nearly 135 h *vs.* 25 h (bare separator) at 90 mV) [34], the nanoparticle layer has an excessive number of voids so Li dendrites can go through them, increasing significantly the risk of short-circuiting due to dendrite penetration across the separator. On the other hand, ultralong-SiO_2 NWs were synthesized for both high-safety and high-power lithium-metal batteries [35]. As a result, the high porosity allowed increases in electrolyte uptake and retention of up to 325% and 63%, respectively. Furthermore, suppression of lithium dendrites and improved ionic conductivity (2.71 S/cm), as compared with commercial PP separators (0.56 S/cm), was achieved.

Sustainable batteries are attracting increasing levels of research attention, where the use of non-degradable and critical raw materials is ideally avoided. For instance, 1D natural minerals (including halloysite nanotubes, attapulgite, sepiolite) have unique micro- and nano-pore structures and are richly abundant, low cost, and environmentally friendly. In this sense, halloysite (HNT) ($Al_2(OH)_4Si_2O_5{\cdot}2H_2O$) with a hollow, tubular structure was incorporated into PVDF-CA (3% HNT) to improve the lithium ion transportation efficiency [36]. The modified separator presents higher tensile strength (from 5.3 to 7.6 MPa) and a 9% lower thermal shrinkage at 225 °C. Furthermore, the Lewis acid site on the surface of HNT can inhibit decomposition by the electrolyte. Recently, Jia et al. [37] synthesized eco-friendly xonotlite nanowires ($Ca_6Si_6O_{17}(OH)_2$)/wood pulp fibers through a simple paper-making process. Remarkably, negligible thermal shrinkage was observed at 600 °C and the ionic conductivity was higher than that of Celgard separators. It also allowed maintenance of 91% of the initial capacity after 100 charge-discharge cycles at 0.2 C thanks to its high electrolyte uptake (198%) with 20% of wood pulp fiber content.

The excellent chemical resistance and the good adhesive properties make polymers among the best candidates to coat materials for separator modification. In addition, porous polymer nanofibers synthesized by electrospinning present superior ionic conductivity and electrolyte wettability properties. Aramid nanofibers (ANF) offer good mechanical properties, excellent thermal and chemical stability, and extraordinary interfacial bonding when compared with other kinds of nanofillers. For those reasons, ANF attracted great interest for use in various applications; in particular, ANF was utilized as a pore-size-conditioning agent, to reduce the large pore size of the polyphenylene sulfide- (PPS-) based separator in rechargeable LIBs [38]. ANF was also used to improve the mechanical properties of bacterial cellulose (BC) [39]. The best result was achieved with a filler loading of 2 wt% of ANF, exhibiting a discharge capacity of about 146 mAh/g (capacity retention of

93% *vs* 86.5 % of the BC separator) after 100 cycles. The high interfacial bonding of the polymer with natural cellulose as a separator component, PVDF/triphenyl phosphate (TPP)/cellulose acetate (CA) nanofiber membrane was fabricated by one-step electrospinning [40]. In addition to the excellent mechanical properties, high porosity, and enhanced electrolyte uptake, each component of the membrane contributes to better safety. In particular, CA improved the thermal stability, whereas TPP induced flame resistance.

11.4 1D Polymer Nanocomposite for Battery Electrolytes

Because of the intrinsic limitations and safety risks (Figure 11.1) of liquid-phase electrolytes (organic solvents mainly), these systems do not currently fulfill the fundamental requirements for future applications of energy storage devices in applications such as EVs. On the other hand, solid electrolytes, and in particular polymeric ones offering the ability to form thin films, and due to the existence of amorphous regions, will favor the ionic transport and solvation ions [41].

To date, great efforts have been made to discover new research paths to design solid-state electrolytes (SSEs), to identify an integrated strategy that allows: (i) modification of the electrolyte chemistry to tune the metal-ion (Li or Na) mobility through it, and at the same time (ii) also achieve control over the mechanical and interfacial resistance properties when integrating the SSEs into a LIB. The inorganic or organic nature of current SSEs pose some intrinsic limitations in terms of solid-to-solid interfacial resistance for (Na^+, Li^+) transport (inorganic SSEs) or lowering the ionic conductivity due to crystalline phases (polymer-based SSEs).

Depending on the nature of the filler (Figure 11.3), it is possible to classify the solid electrolyte into liquid-containing electrolytes (e.g. gel polymers) and dry electrolytes (e.g. nanocomposites with inactive fillers, nanocomposites with active fillers, such as metal organic frameworks (MOFs), or inorganic electrolytes) [42].

The filler shape and size can influence the ionic conductivity of the polymer electrolyte. It has been shown that small fillers (10–20 nm) increase the conductivity by up to one order of magnitude compared with fillers of 10 μm [44]. This improvement is ascribed to the higher surface area of the smaller filler, emphasizing the importance of "nano-scaled" particles in dictating the ionic transport behavior of polymer electrolytes.

Contrary to spherical nanoparticles, 1D fillers show unique superiorities, such as cross-linked structures for external stress-buffering and large draw ratios for internal stress dispersion. Therefore, nanowires (NWs) have been extensively used as nanocomposite fillers to improve the mechanical properties of the polymer matrix to counteract the addition of small organic molecules or ionic liquids, used as plasticizers, to increase their ionic conductivity. Recently, Gan et al. [45] used silica-coated silver nanowires (AgNWs@SiO_2) to incorporate these hybrids into a PVDF-HFP matrix. Although the electrolyte uptake decreases from 270% at 0 wt% to 160% at 10 wt% because of the addition of high loadings of AgNWs@SiO_2, a greater update than the 87% of the Celgard separator was achieved. Among the different compositions studied, the 5 wt% loading of AgNWs@SiO_2 presented the best balance between mechanical properties (120% strain) and thermal conductivity. The high thermal conductivity results obtained are also beneficial for improving the ionic conductivity (4 mS/cm) and the cycling performance of lithium metal batteries (LMBs), especially at higher temperatures.

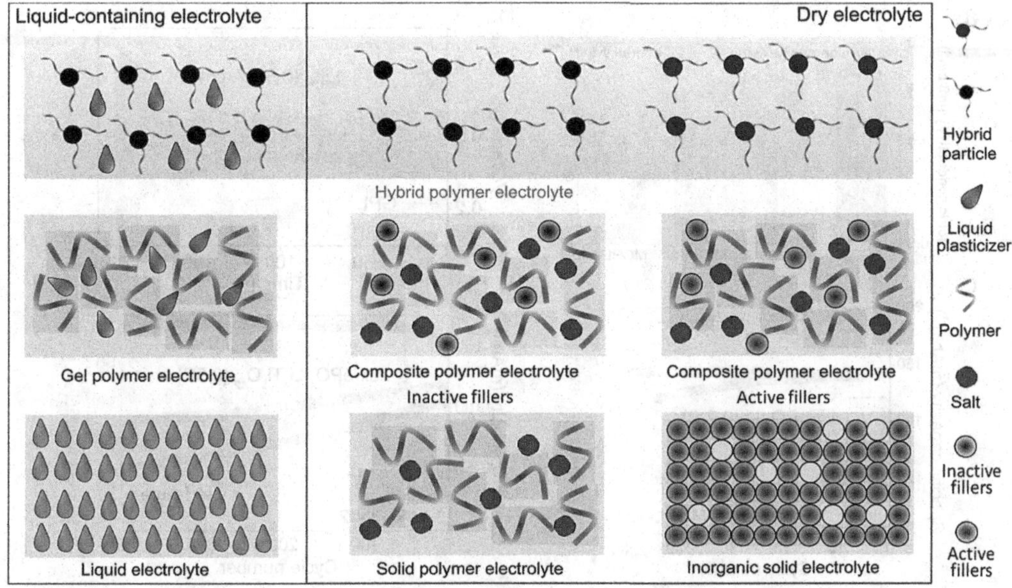

FIGURE 11.3
Classification of electrolytes for rechargeable batteries. Adapted with permission from [43]. Copyright (2020); exclusive licensee IOP Publishing Limited. Distributed under a Creative Commons Attribution License 4.0 (CC BY) https://creativecommons.org/licenses/by/4.0/.

The variation in the chemical composition of the filler and the diversity of the matrix allowed the formation of multiple systems, suggesting the potential to improve the electrochemical properties of these solid electrolytes. For instance, the incorporation of palygorskite ($(Mg_2Al_2)Si_8O_{20}(OH)_2{\cdot}4H_2O$) NWs into a PVDF matrix not only improved the elastic modulus by 200% but also enhanced the transference number of Li^+ from 0.21 to 0.54 due to interaction between palygorskite and ClO_4^- ions [46]. In addition, Yan et al. concluded that, with the incorporation of 5 wt% of the ceramic $Li_7La_3Zr_2O_{12}$ (LLZO) garnet-type NWs into polyacrylonitrile (PAN), Li^+ ionic conductivity increased by three orders of magnitude (1.31 $\times10^{-4}$ S/cm) when compared with the polymer and Li^+ salt alone. According to solid-state NMR results, LLZO NWs partially transformed the surroundings of the PAN polymer filler, favoring the Li^+ diffusion in the modified regions [47]. This diffusion mechanism can also be applied to next-generation sodium-ion batteries with embedded, engineered, ultrafast 1D percolation networks. For that, Devi et al. [48] developed a novel composite PEO-PVP-$NaPF_6$ SPEs with 1.0 wt% indium arsenide (InAs) NWs, achieving an ionic conductivity of 1.5 $\times10^{-4}$ S/cm and an electrochemical stability window of 4.10 V at 40 °C.

Recently, the versatility of metal-organic frameworks (MOFs) allowed the possibility of designing 1D crystal structures favoring the transport of metal ions between the electrodes [49]. Increased ion-transport efficiency was achieved when a Zr-based MOF, UiO-66, was incorporated into a gel polymer electrolyte (composed of PVDF-HFP and $LiClO_4$ in propylene carbonate). More specifically, high ionic conductivity (>10^{-3} S/cm), high t_{Li}^+ value (0.64), and low activation energy (<0.1 eV) were achieved (Figure 11.4). Moreover, the use of the MOF-stabilized electrolyte-electrode interfaces reduced the polarization and thus improved the cycle lifespan for the batteries.

In the case of dry, polymer electrolytes, Zhang et al. [51] synthesized PEO/Zn-based MOF-5 nanorods solid polymeric electrolyte (SPE). The geometrical modification resulted

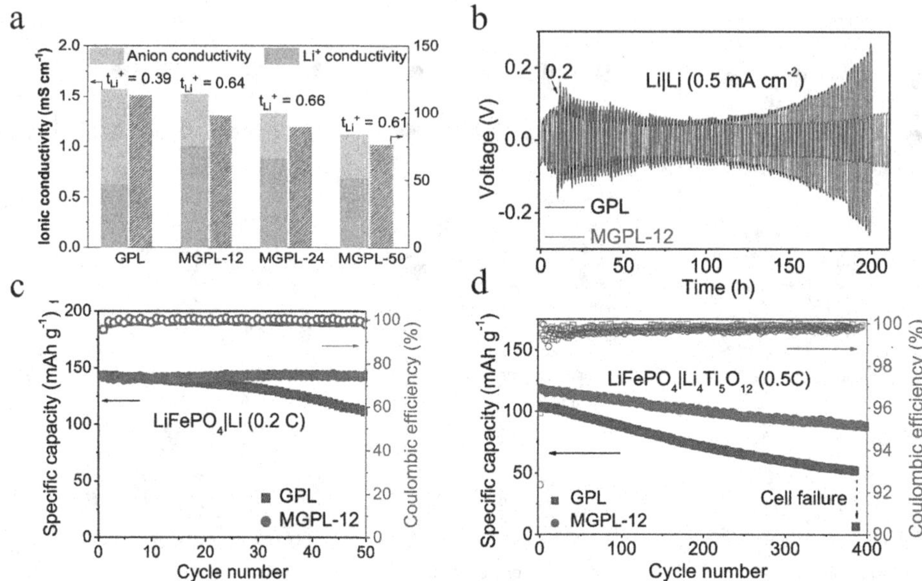

FIGURE 11.4
a) Ionic conductivity and thermal activation at room temperature of electrolyte membranes. b) Polarization tests of symmetric cells for a period over 200 h at 0.5 mA/cm². c) and d) Galvanostatic cycling of $LiFePO_4$|Li cell under 0.2 C and $LiFePO_4|Li_4Ti_5O_{12}$ cells under 0.5 C, respectively. Adapted with permission from [50]. Copyright (2020) American Chemical Society.

in a major improvement of the capacity as well as the cycle life, with a special discharge capacity of 137.9 mAh/g at 0.1 C and a coulombic efficiency above 99.5%.

11.5 Future Needs and Prospects

1D polymeric nanostructures have been proven to be efficient at enabling improved electrochemical performances relative to conventional electrodes, separators, and electrolytes. However, future energy storage requires the implementation of additional functions as batteries are playing a pivotal role in the decarbonization and electrification of the worldwide economies. In this sense, energy density, operation lifespan, safety, and environmental sustainability are regarded as some of the most urgent topics for research and development.

The energy density of the batteries is considered one of the prime requisites for effective commercialization. Advances in the field may arise from the development of cathode materials able to work at high voltages so that the overall battery volume can be reduced for a given delivered energy (being especially interesting for electric vehicles or portable electronics). This can be accomplished using 1D polymeric nanostructures, as they enable the formation of battery cathodes with tailored geometries. Increasing cathode mass loading to levels > 5 mg/cm² (as opposed to conventional values which often remain below 1 mg/cm²) is also of paramount relevance. In this sense, 1D structures are key enablers as they facilitate ionic diffusion and electrical conductivity by developing highly porous and high specific surface area materials, overcoming the classical drawbacks faced when achieving large mass loadings. Solving the rapid capacity fade suffered by certain battery chemistries represents a future research challenge that needs to be addressed. This undesired capacity fade results in batteries with short operation life spans, making them unsuitable for many applications. Interwoven 1D nanofiber structures with hierarchical mesoporous morphologies have been proven to be efficient for prolonged cycling, where a continuous framework, presenting opened pores for lithiation and subsequent delithiation processes, outperforms conventional lithium-ion batteries anodes [52].

With regard to battery safety, replacing the conventional microporous membrane-liquid electrolyte pair with gel-like electrolytes is considered to be the next step. In this framework, 3D percolative networks composed of 1D polymeric nanofibers have the potential of upholding large amounts of liquid electrolytes without collapse to render ionically conducting gel polymer electrolytes. These structures not only avoid undesired liquid electrolyte leakage but also effectively block dendritic growth. Although polyethylene oxide has traditionally been the most investigated polymer to that end, alternative materials need to be investigated in the near future. Moreover, most of these 3D percolative networks can be synthesized so that they are mechanically flexible, opening up the pathway to the fabrication of wearable batteries.

It should be also considered that conventional electrolytes, composed of a salt dissolved in a solvent, typically display severe ion concentration gradients, which are exacerbated at high charge/discharge rates. This is because the transference number of the anion is higher than that of the cation (PF_6^- and Li^+, respectively, for lithium-ion batteries), which favors metal dendritic growth and represents serious safety limitations. To avoid these issues, single-ion conducting polymer electrolytes are being developed, where 1D polymers show promising performances [53]. 1D polymers enable the regulation of the hierarchical structure and interface of the electrolyte, resulting in batteries that display improved cyclability and enhanced rate capability in comparison with commercial polyolefin-liquid electrolyte systems. In addition, the incorporation of nanofillers, such as hydroxyapatite nanowires, serves to further boost the electrochemical performance and safety delivered by these single-ion conducting electrolytes [54]. As shown in Figure 11.5a–c, a markedly improved metal deposition is achieved by prevention of ion concentration gradients, polarization reduction, and avoidance of ramified metal deposition, which are then translated into a long-term, stable stripping/plating of ions, which can enable the path for the implementation of alkali metal batteries, a promising technology for next-generation high-energy batteries.

Insofar as 1D polymeric nanocomposites enable the fabrication of battery anodes and cathodes, their use can reduce the pressure over toxic, expensive, and scarce materials. In this sense, on 3 September 2020, the European Commission (EC) provided a list of materials considered to be critical raw materials (CRMs), given the environmental unsustainability of their extraction and supply chain insecurity. Lithium, manganese, nickel, natural graphite, and cobalt can be found within this list (see Figure 11.5d for their supply risk and economic importance assessment). Therefore, developing 1D polymeric nanocomposites with competitive electrochemical performance to replace the $LiCoO_2$, $LiMn_2O_4$, or $Li(Ni,Mn,Co)O_2$ widely used in LIBs [1], $NaCo_2O_4$ or $Na_3V_2(PO_4)_3$ applied in NIBs [1], or the XV_2O_5 (X = Zn, Mg, Ca, Na) commonly found in ZIBs [7] should be a priority. The issues related to CRMs are becoming increasingly alarming as, if no viable alternatives are developed, their scarcity may increase even further, given the global demand trends for batteries.

As the battery field is being developed and the different battery chemistries are understood and exploited, further efforts are now focusing on the end-of-life of the batteries.

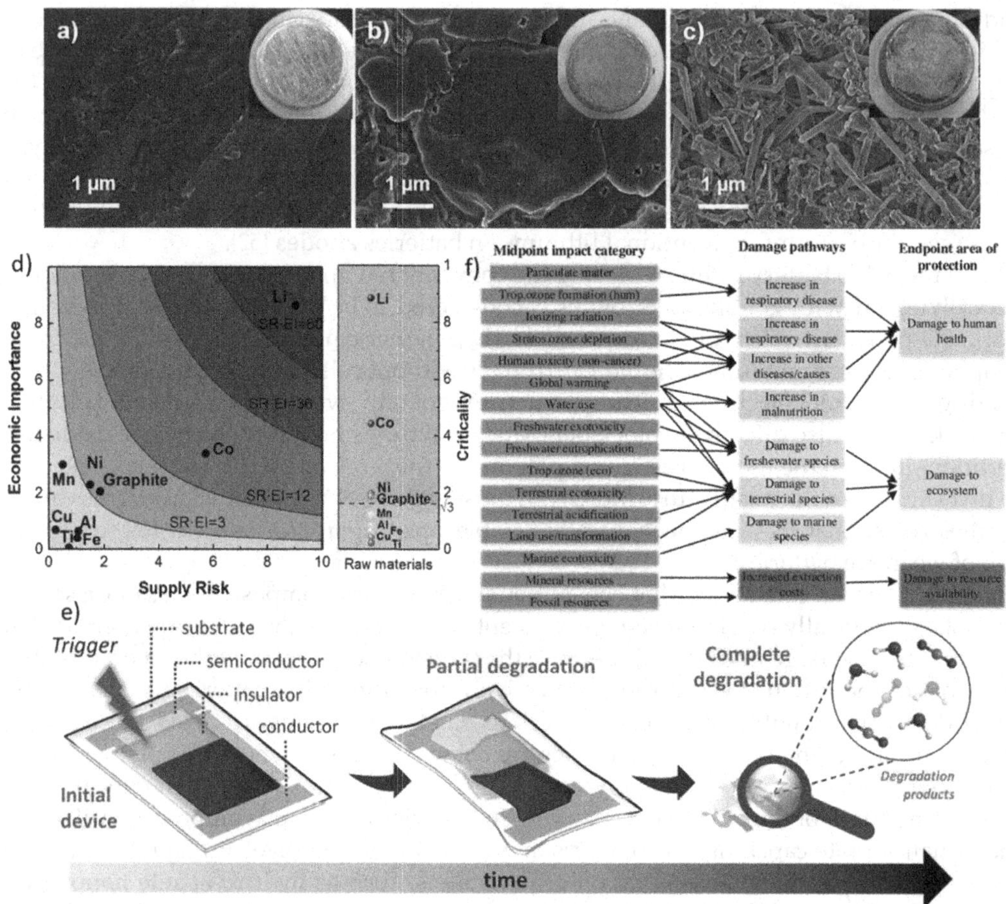

FIGURE 11.5
From a to c: morphological observations of Li metal surfaces. a) Before galvanostatic cycling; b) after galvanostatic cycling using a single-ion conducting cellulosic electrolyte; and c) after galvanostatic cycling using the conventional microporous Celgard membrane soaked in 1 M $LiPF_6$ EC-DEC (1:1 v/v) as electrolyte. Reproduced with permission from [55]. Copyright (2019) American Chemical Society. d) Supply risk and economic importance of nine materials used in the battery field. A criticality evaluation from 0 (non-critical) to 10 (very critical) is shown. Reproduced with permission from [56]. Copyright (2019) Elsevier. e) Schematic process showing the concept behind transient batteries, which are degraded into non-toxic by-products once a trigger is applied. Reproduced with permission from [57]. Copyright (2019) Wiley-VCH. f) Environmental impact categories according to the Global ReCiPe2016 method. Reproduced with permission from [58]. Copyright (2021) Elsevier.

Although the traceability of large batteries aimed for electric vehicles or stationary energy storage can be accomplished relatively easily given their large size, batteries powering small electronics are usually discarded under uncontrolled conditions. Accordingly, these batteries, containing harmful organic electrolytes and toxic compounds, may end up and accumulate in marine or land environments. To avoid this undesirable scenario, scientists are working on the development of transient batteries, which essentially are batteries capable of undergoing controlled (bio)degradation processes after a stable operation period, leaving behind harmless residues (Figure 11.5e) [57]. To that end, battery components need to be rethought and redesigned, in which polymers play a pivotal role. Given the extensive choice of biodegradable polymers which, blended with different additives and/or nanofillers, result in tailored physicochemical and electrochemical properties, 1D polymeric nanostructures present an exciting future to develop new materials for battery applications. Not only biodegradable but also renewable materials are required to develop high-performance sustainable batteries. Polymers are a relevant class of materials that can be renewable in origin. Especially interesting are polysaccharides, long-chain polymeric carbohydrates made of monosaccharide units that remain linked through glycosidic linkages. Polysaccharides such as cellulose (mainly derived from agroforestry sources), chitin (mostly originating from the exoskeletons of arthropods), or agar and carrageenan (derived from algae) present the advantages that they could be processed into 1D materials. For example, cellulosic or chitin nanoparticles can be readily extracted using chemically or mechanically induced deconstruction strategies. Additionally, upon a proper chemical modification, many polysaccharides can be dissolved in water or other liquids, enabling the fabrication of 1D electrospun nanofibers. Importantly, the functional groups of polymers of natural origin, such as –OH, $-SO_3H$, $-CONH_2$, $-NH_2$, or $-CONH_{,}^{-}$ allow a high affinity with the liquid electrolyte and ensure a preferential interaction with the anion of the salt to improve the solubility of the salt and the cationic transport through the electrolyte. Thanks to these qualities, it is possible to obtain higher ionic conductivities that exceed by an order of magnitude those obtained in conventional systems.

Environmental sustainability should not only be limited to the development of battery components relying on non-toxic materials but also the impact of different synthetic procedures should be analyzed. In this sense, life cycle assessment (LCA) methodology results are especially useful as they enable the quantification of the environmental impacts of a given product or service along its whole life cycle, from the extraction of raw materials to final battery disposal. Interestingly, as summarized in Figure 11.5f, the environmental impacts can be quantified, using not only CO_2-equivalents as the main parameter but also additional categories such as mineral resource scarcity, terrestrial acidification, water consumption, or freshwater eutrophication [59]. This allows a holistic understanding of the impacts of battery, as not only global warming potential will be relevant to the new, sustainability landscape, but also other, associated impacts. Additionally, if the impacts of the fabrication procedures are fully disclosed, it may be easier to adopt resource-efficient processes.

To translate the benefits provided by 1D polymeric nanostructures into commercial devices and make their advantages a tangible reality, laboratory-scale research should be moved into industrial-scale fabrication. In this sense, much of the research lies in Technology Readiness Levels (TRL) between 1 and 3, or 4 at best. Therefore, work at technological demonstrations of TRL 6 should be desirable so that 1D polymeric nanostructures can be truly implemented.

11.6 Conclusions

In this chapter, we summarize the recent developments in one-dimensional nanocomposites as applied to battery electrodes, electrolytes, and membranes. The inherent advantages of nanoscale materials are combined with the versatility of polymers (regarding processing and property tailorability) to render innovative materials for electrochemical energy storage. Specifically, the large surface-to-volume ratio offers many advantages over conventional battery materials, which are translated into a reduced volume expansion upon cycling, an easier diffusion of electrolyte ions, an extended contact area between the liquid electrolyte and the active materials, or the reduction of ion concentration gradients by homogeneous and nanoporous morphologies. One-dimensional polymeric nanocomposites also enable the development of gel-like or solid-like electrolytes with competitive electrochemical properties (ionic conductivity) compared with conventional polyolefin-liquid electrolyte systems, but with the advantage of providing safer batteries. Additionally, nature provides a wide range of polymers of renewable origin that can be implemented into 1D structures, although their potential in the battery field is yet to be explored.

All these features make one-dimensional nanocomposites attractive candidates for the next-generation batteries which provide safer, higher-performance, and more sustainable energy storage solutions. Future research should be extended from lithium-ion batteries technology to more earth-abundant chemistries such as sodium-ion batteries or zinc-ion batteries. Metal-sulfur and metal-air batteries also offer lots of innovations regarding 1D polymeric materials, especially as air cathodes, where highly porous structures are required.

References

1. J.W. Choi, D. Aurbach, Promise and Reality of Post-lithium-Ion Batteries with High Energy Densities, *Nat. Rev. Mater.* 1(4) (2016) 16013.
2. P.G. Bruce, S.A. Freunberger, L.J. Hardwick, J.-M. Tarascon, Li–O_2 and Li–S Batteries with High Energy Storage, *Nat. Mater.* 11(1) (2012) 19–29.
3. Y. Hu, W. Chen, T. Lei, Y. Jiao, J. Huang, A. Hu, C. Gong, C. Yan, X. Wang, J. Xiong, Strategies Toward High-Loading Lithium–Sulfur Battery, *Adv. Energy Mater.* 10(17) (2020) 2000082.
4. W.-J. Kwak, Rosy, D. Sharon, C. Xia, H. Kim, L.R. Johnson, P.G. Bruce, L.F. Nazar, Y.-K. Sun, A.A. Frimer, M. Noked, S.A. Freunberger, D. Aurbach, Lithium–Oxygen Batteries and Related Systems: Potential, Status, and Future, *Chem. Rev.* 120(14) (2020) 6626–6683.
5. M. Li, J. Lu, X. Ji, Y. Li, Y. Shao, Z. Chen, C. Zhong, K. Amine, Design Strategies for Nonaqueous Multivalent-Ion and Monovalent-Ion Battery Anodes, *Nat. Rev. Mater.* 5(4) (2020) 276–294.
6. Y. Liang, H. Dong, D. Aurbach, Y. Yao, Current Status and Future Directions of Multivalent Metal-Ion Batteries, *Nat. Energy* 5(9) (2020) 646–656.
7. L.E. Blanc, D. Kundu, L.F. Nazar, Scientific Challenges for the Implementation of Zn-Ion Batteries, *Joule* 4(4) (2020) 771–799.
8. C.R. Birkl, M.R. Roberts, E. McTurk, P.G. Bruce, D.A. Howey, Degradation Diagnostics for Lithium Ion Cells, *J. Power Sources* 341 (2017) 373–386.
9. M.R. Palacín, Understanding Ageing in Li-Ion Batteries: A Chemical Issue, *Chem. Soc. Rev.* 47(13) (2018) 4924–4933.
10. T.-D. Nguyen, E. Lizundia, M. Niederberger, W.Y. Hamad, M.J. MacLachlan, Self-Assembly Route to TiO_2 and TiC with a Liquid Crystalline Order, *Chem. Mater.* 31(6) (2019) 2174–2181.
11. H. Schmidt, B. Jerliu, E. Hüger, J. Stahn, Volume Expansion of Amorphous Silicon Electrodes during Potentiostatic Lithiation of Li-Ion Batteries, *Electrochem. Commun.* 115 (2020) 106738.
12. S.W. Lee, M.T. McDowell, L.A. Berla, W.D. Nix, Y. Cui, Fracture of Crystalline Silicon Nanopillars during Electrochemical Lithium Insertion, *Proc. Natl. Acad. Sci.* 109(11) (2012) 4080–4085.
13. H. Wang, X. Cao, H. Gu, Y. Liu, Y. Li, Z. Zhang, W. Huang, H. Wang, J. Wang, W. Xu, J.-G. Zhang, Y. Cui, Improving Lithium Metal Composite Anodes with Seeding and Pillaring Effects of Silicon Nanoparticles, *ACS Nano* 14(4) (2020) 4601–4608.
14. J. Henschel, F. Horsthemke, Y.P. Stenzel, M. Evertz, S. Girod, C. Lürenbaum, K. Kösters, S. Wiemers-Meyer, M. Winter, S. Nowak, Lithium Ion Battery Electrolyte Degradation of

Field-Tested Electric Vehicle Battery Cells – A Comprehensive Analytical Study, *J. Power Sources* 447 (2020) 227370.

15. T.-T. Nguyen, A. Demortière, B. Fleutot, B. Delobel, C. Delacourt, S.J. Cooper, The Electrode Tortuosity Factor: Why the Conventional Tortuosity Factor is Not Well Suited for Quantifying Transport in Porous Li-Ion Battery Electrodes and What to Use Instead, *npj Comput. Mater.* 6(1) (2020) 123.
16. S.A. Grudinkin, A.Y. Vinogradov, Fabrication of Carbon Nanowalls by Radio Frequency Magnetron Sputtering of Graphite Target in Argon Plasma, *J. Phys. Conf. Ser.* 1697(1) (2020) 12108.
17. J. Wang, S. Liu, Y. Wang, T. Wang, S. Shang, W. Ren, Magnetron-Sputtering Deposited Molybdenum Carbide MXene Thin Films as a Saturable Absorber for Passively Q-Switched Lasers, *J. Mater. Chem. C* 8(5) (2020) 1608–1613.
18. L. Lai, X.J. Wang, J.J. Hao, The Study of SiC Thin Films Produced by Magnetron Sputtering, *Key Eng. Mater.* 609–610 (2014) 82–87.
19. S. Yan, X. Yan, Y. Tong, M. Zhang, J. Liu, Application of Magnetron Sputtering Technique to Fabricate Sulfur/Carbon Composites Cathode Plates of Lithium-Sulfur Battery, *Mater. Lett.* 273 (2020) 127888.
20. J. Sun, K. Tang, X. Yu, H. Li, X. Huang, Needle-Like $LiFePO_4$ Thin Films Prepared by an Off-Axis Pulsed Laser Deposition Technique, *Thin Solid Films* 517(8) (2009) 2618–2622.
21. Y. Lu, J. Chen, Prospects of Organic Electrode Materials for Practical Lithium Batteries, *Nat. Rev. Chem.* 4(3) (2020) 127–142.
22. F. Shi, C. Chen, Z.-L. Xu, Recent Advances on Electrospun Nanofiber Materials for Post-lithium Ion Batteries, *Adv. Fiber Mater.* 3(5) (2021) 275–301.
23. Y. Liu, Q. Shen, X. Zhao, J. Zhang, X. Liu, T. Wang, N. Zhang, L. Jiao, J. Chen, L.-Z. Fan, Hierarchical Engineering of Porous P_2-$Na_{2/3}Ni_{1/3}Mn_{2/3}O_2$ Nanofibers Assembled by Nanoparticles Enables Superior Sodium-Ion Storage Cathodes, *Adv. Funct. Mater.* 30(6) (2020) 1907837.
24. C.K. Chan, H. Peng, G. Liu, K. McIlwrath, X.F. Zhang, R.A. Huggins, Y. Cui, High-Performance Lithium Battery Anodes Using Silicon Nanowires, *Nat. Nanotechnol.* 3(1) (2008) 31–35.
25. L. Hou, R. Bao, D. Kionga Denis, X. Sun, J. Zhang, F. Uz Zaman, C. Yuan, Synthesis of Ultralong $ZnFe_2O_4$@polypyrrole Nanowires with Enhanced Electrochemical Li-Storage Behaviors for Lithium-Ion Batteries, *Electrochim. Acta* 306 (2019) 198–208.
26. T.-D. Nguyen, J. Li, E. Lizundia, M. Niederberger, W.Y. Hamad, M.J. MacLachlan, Black Titania with Nanoscale Helicity, *Adv. Funct. Mater.* 29(40) (2019) 1904639.
27. F. Wu, E. Zhao, D. Gordon, Y. Xiao, C. Hu, G. Yushin, Infiltrated Porous Polymer Sheets as Free-Standing Flexible Lithium-Sulfur Battery Electrodes, *Adv. Mater.* 28(30) (2016) 6365–6371.
28. Q. Li, Z. Cao, G. Liu, H. Cheng, Y. Wu, H. Ming, G.-T. Park, D. Yin, L. Wang, L. Cavallo, Y.-K. Sun, J. Ming, Electrolyte Chemistry in 3D Metal Oxide Nanorod Arrays Deciphers Lithium Dendrite-Free Plating/Stripping Behaviors for High-Performance Lithium Batteries, *J. Phys. Chem. Lett.* 12(20) (2021) 4857–4866.
29. J. Ma, Y. Kong, Y. Luo, Y. Huang, S. Han, Flexible Polyimide Nanorod/Graphene Framework as an Organic Cathode for Rechargeable Sodium-Ion Batteries, *J. Phys. Chem. C* 125(12) (2021) 6564–6569.
30. X. Fu, W. Zhang, B. Lan, J. Wen, S. Zhang, P. Luo, R. Zhang, S. Hu, Q. Liu, Polyaniline Nanorod Arrays as a Cathode Material for High-Rate Zinc-Ion Batteries, *ACS Appl. Energy Mater.* 3(12) (2020) 12360–12367.
31. H. Glatz, E. Lizundia, F. Pacifico, D. Kundu, An Organic Cathode Based Dual-Ion Aqueous Zinc Battery Enabled by a Cellulose Membrane, *ACS Appl. Energy Mater.* 2(2) (2019) 1288–1294.
32. P. Arora, Z. (John) Zhang, Battery Separators, *Chem. Rev.* 104(10) (2004) 4419–4462.
33. J. Fang, A. Kelarakis, Y.-W. Lin, C.-Y. Kang, M.-H. Yang, C.-L. Cheng, Y. Wang, E.P. Giannelis, L.-D. Tsai, Nanoparticle-Coated Separators for Lithium-Ion Batteries with Advanced Electrochemical Performance, *Phys. Chem. Chem. Phys.* 13(32) (2011) 14457–14461.
34. K. Liu, D. Zhuo, H.-W. Lee, W. Liu, D. Lin, Y. Lu, Y. Cui, Extending the Life of Lithium-Based Rechargeable Batteries by Reaction of Lithium Dendrites with a Novel Silica Nanoparticle Sandwiched Separator, *Adv. Mater.* 29(4) (2017) 1603987.

35. Q.-C. Du, M.-T. Yang, J.-K. Yang, P. Zhang, J.-Q. Qi, L. Bai, Z. Li, J.-Y. Chen, R.-Q. Liu, X.-M. Feng, Z.-D. Huang, T. Masese, Y.-W. Ma, W. Huang, Bendable Network Built with Ultralong Silica Nanowires as a Stable Separator for High-Safety and High-Power Lithium-Metal Batteries, *ACS Appl. Mater. Interfaces* 11(38) (2019) 34895–34903.
36. H. Xu, D. Li, Y. Liu, Y. Jiang, F. Li, B. Xue, Preparation of Halloysite/Polyvinylidene Fluoride Composite Membrane by Phase Inversion Method for Lithium Ion Battery, *J. Alloys Compd.* 790 (2019) 305–315.
37. S. Jia, S. Yang, M. Zhang, K. Huang, J. Long, J. Xiao, Eco-Friendly Xonotlite Nanowires/Wood Pulp Fibers Ceramic Hybrid Separators through a Simple Papermaking Process for Lithium Ion Battery, *J. Membr. Sci.* 597 (2020) 117725.
38. B. Yang, L. Wang, M. Zhang, J. Luo, Z. Lu, X. Ding, Fabrication, Applications, and Prospects of Aramid Nanofiber, *Adv. Funct. Mater.* 30(22) (2020) 2000186.
39. Y. Yang, C. Huang, G. Gao, C. Hu, L. Luo, J. Xu, Aramid Nanofiber/Bacterial Cellulose Composite Separators for Lithium-Ion Batteries, *Carbohydr. Polym.* 247 (2020) 116702.
40. Y. Chen, L. Qiu, X. Ma, L. Dong, Z. Jin, G. Xia, P. Du, J. Xiong, Electrospun Cellulose Polymer Nanofiber Membrane with Flame Resistance Properties for Lithium-Ion Batteries, *Carbohydr. Polym.* 234 (2020) 115907.
41. M. Forsyth, L. Porcarelli, X. Wang, N. Goujon, D. Mecerreyes, Innovative Electrolytes Based on Ionic Liquids and Polymers for Next-Generation Solid-State Batteries, *Acc. Chem. Res.* 52(3) (2019) 686–694.
42. P. Yao, H. Yu, Z. Ding, Y. Liu, J. Lu, M. Lavorgna, J. Wu, X. Liu, Review on Polymer-Based Composite Electrolytes for Lithium Batteries, *Front. Chem.* 7 (2019) 522.
43. N. Boaretto, L. Meabe, M. Martinez-Ibañez, M. Armand, H. Zhang, Review—Polymer Electrolytes for Rechargeable Batteries: From Nanocomposite to Nanohybrid, *J. Electrochem. Soc.* 167(7) (2020) 70524.
44. M.A.K.L. Dissanayake, P.A.R.D. Jayathilaka, R.S.P. Bokalawala, I. Albinsson, B.-E. Mellander, Effect of Concentration and Grain Size of Alumina Filler on the Ionic Conductivity Enhancement of the $(PEO)_9LiCF_3SO_3$:Al_2O_3 Composite Polymer Electrolyte, *J. Power Sources* 119–121 (2003) 409–414.
45. H. Gan, J. Yuan, Y. Zhang, S. Li, L. Yu, J. Wang, J. Hu, N. Yang, Z. Xue, Electrospun Composite Gel Polymer Electrolytes with High Thermal Conductivity Toward Wide Temperature Lithium Metal Batteries, *ACS Appl. Energy Mater.* 4(8) (2021) 8130–8141.
46. P. Yao, B. Zhu, H. Zhai, X. Liao, Y. Zhu, W. Xu, Q. Cheng, C. Jayyosi, Z. Li, J. Zhu, K.M. Myers, X. Chen, Y. Yang, PVDF/Palygorskite Nanowire Composite Electrolyte for 4 V Rechargeable Lithium Batteries with High Energy Density, *Nano Lett.* 18(10) (2018) 6113–6120.
47. T. Yang, J. Zheng, Q. Cheng, Y.-Y. Hu, C.K. Chan, Composite Polymer Electrolytes with $Li_7La_3Zr_2O_{12}$ Garnet-Type Nanowires as Ceramic Fillers: Mechanism of Conductivity Enhancement and Role of Doping and Morphology, *ACS Appl. Mater. Interfaces* 9(26) (2017) 21773–21780.
48. C. Devi, J. Gellanki, H. Pettersson, S. Kumar, High Sodium Ionic Conductivity in PEO/PVP Solid Polymer Electrolytes with InAs Nanowire Fillers, *Sci. Rep.* 11 (2021) 20180.
49. Z. Ye, Y. Jiang, L. Li, F. Wu, R. Chen, Rational Design of MOF-Based Materials for Next-Generation Rechargeable Batteries, *Nano Micro Lett.* 13(1) (2021) 203.
50. X. Lu, H. Wu, D. Kong, X. Li, L. Shen, Y. Lu, Facilitating Lithium-Ion Conduction in Gel Polymer Electrolyte by Metal-Organic Frameworks, *ACS Mater. Lett.* 2(11) (2020) 1435–1441.
51. Z. Zhang, J.-H. You, S.-J. Zhang, C.-W. Wang, Y. Zhou, J.-T. Li, L. Huang, S.-G. Sun, Metal Organic Framework Nanorod Doped Solid Polymer Electrolyte with Decreased Crystallinity for High-Performance All-Solid-State Lithium Batteries, *ChemElectroChem* 7(5) (2020) 1125–1134.
52. P.F. Teh, Y. Sharma, S.S. Pramana, M. Srinivasan, Nanoweb Anodes Composed of One-Dimensional, High Aspect Ratio, Size Tunable Electrospun $ZnFe_2O_4$ Nanofibers for Lithium Ion Batteries, *J. Mater. Chem.* 21(38) (2011) 14999–15008.
53. C. Li, B. Qin, Y. Zhang, A. Varzi, S. Passerini, J. Wang, J. Dong, D. Zeng, Z. Liu, H. Cheng, Single-Ion Conducting Electrolyte Based on Electrospun Nanofibers for High-Performance Lithium Batteries, *Adv. Energy Mater.* 9(10) (2019) 1803422.

54. H. Li, D. Wu, J. Wu, L.-Y. Dong, Y.-J. Zhu, X. Hu, Flexible, High-Wettability and Fire-Resistant Separators Based on Hydroxyapatite Nanowires for Advanced Lithium-Ion Batteries, *Adv. Mater.* 29(44) (2017) 1703548.
55. C. Hänsel, E. Lizundia, D. Kundu, A Single Li-Ion, Conductor Based on Cellulose, *ACS Appl. Energy Mater.* 2(8) (2019) 5686–5691.
56. J. Song, W. Yan, H. Cao, Q. Song, H. Ding, Z. Lv, Y. Zhang, Z. Sun, Material Flow Analysis on Critical Raw Materials of Lithium-Ion Batteries in China, *J. Clean. Prod.* 215 (2019) 570–581.
57. N. Mittal, A. Ojanguren, M. Niederberger, E. Lizundia, Degradation Behavior, Biocompatibility, Electrochemical Performance, and Circularity Potential of Transient Batteries, *Adv. Sci. (Weinh)* 8(12) (2021) 2004814.
58. X. Shu, Y. Guo, W. Yang, K. Wei, G. Zhu, Life-Cycle Assessment of the Environmental Impact of the Batteries Used in Pure Electric Passenger Cars, *Energy Rep.* 7 (2021) 2302–2315.
59. I. Rey, C. Vallejo, G. Santiago, M. Iturrondobeitia, E. Lizundia, Environmental Impacts of Graphite Recycling from Spent Lithium-Ion Batteries Based on Life Cycle Assessment, *ACS Sustain. Chem. Eng.* 9(43) (2021) 14488–14501.

12

One-Dimensional Polymeric Nanocomposites and Other Low-Dimensional Materials for Flexible Batteries

Anna Basa and Krzysztof Winkler

CONTENTS

12.1 Introduction

Li-ion systems are considered to be the best storage and power systems available for the near future. Li-ion batteries are more compact and are lighter than other conventional batteries. They offer a high operating voltage of ~4 V and deliver high specific energy and capacity. Furthermore, during charging, Li-ion batteries consume up to 30% less energy than lead-acid batteries. Additionally, their long lifetimes enable their versatile use. They are widely used in consumer electronics (smartphones, tablets, etc.), as well as in transport devices (cars, scooters, electric bicycles). The Li-ion battery was first commercialized in 1991 by Sony, initiating significant developments in this field of technology. Trends to improve these batteries have progressed in different directions. An important area of research includes the search for new electroactive cathode and anode materials, such as metal oxides, metal sulfides, MXenes, and conducting polymers, with the goals of ensuring effective ion and electron transport and storage of Li ions in the electrode material. The energy storage capability is influenced by the architecture of the electroactive materials, conductivity, and specific surface area. Therefore, significant attention has been focused on designing electroactive material structures to ensure a well-developed surface, achieving good thermal and electric conductivity, reducing ionic diffusion path length, and promoting electron transfer. This role is performed by nanoparticle materials, both nanoparticles

DOI: 10.1201/9781003223764-12

of electroactive materials and nanoparticle additives, by ensuring the appropriate morphology and physicochemical properties of the electrode material.

Carbon nanoparticles, including graphene and carbon nanotubes (CNTs), are particularly useful in the formation of high-performance alkali metal-ion batteries [1–3]. Both graphene and CNTs provide an ideal conducting matrix that enhances the utilization of electroactive materials [4]. They also significantly improve the stability of the electrode material during charging and discharging processes [5]. Carbon nanomaterials can also be functionalized to more effectively accommodate electroactive materials and enhance the conductivity of carbon nanoparticle scaffolds [6].

A more recent research direction for charge storage devices is focused on the formation and properties of flexible systems [7, 8]. Such devices can be used in touch screens, wearable communicators, roll-up displays, smart electronics, and smart textiles. Moreover, flexible batteries are important not only for portable devices but also for the future of electric and hybrid electric vehicles. Flexible batteries must undergo frequent mechanical deformations, such as bending, folding, twisting, and stretching. Such deformations can be classified as elastic or plastic deformations. In the former case, the shape of the material is recoverable after the removal of external stress. Conversely, plastic deformations are irreversible. A variety of carbon nanomaterials, including graphene sheets, CNTs, carbon paper, and buckypaper, are commonly used in flexible electronic devices. Owing to its 1D structure, electrical and mechanical properties, CNTs have been extensively employed for the preparation of different electrode materials for flexible alkali metal-ion batteries [9]. Vertically aligned CNT structures not only serve as a flexible environment for the electroactive materials but also provide electrically conductive networks and can accommodate the volumetric expansion of electroactive materials because of their porous structures.

In this chapter, we systematically summarize the progress in the application of 1D composite materials to flexible alkali metal-ion batteries. The predominant focus is on 1D polymeric composites. However, the major aspects of the use of these materials in alkali metal-ion batteries are discussed in a broader context, encompassing other 1D electroactive material structures. As most of the 1D flexible composite materials are based on CNTs, the structure and properties of these carbon nanomaterials are described in this chapter, and a variety of flexible anodic and cathodic materials based on CNTs are described. The prototypes of flexible and bendable batteries are also highlighted. Finally, we conclude by describing the progress and main challenges of flexible alkali metal-ion batteries.

12.2 Critical Parameters for Flexible Batteries

12.2.1 Geometric Parameters

When an external force is applied to the batteries, it causes the deformation presented in Figure 12.1a. The flexibility (f) of the batteries can be characterized by four geometric parameters (r, l, θ, and h) [10]. The first parameter (r), which has a significant impact on flexibility, is the bending radius of curvature (measured on the inside of the curvature). A smaller radius leads to a more elastic battery. The size of the deformation area depends on the bend angle (θ). Another essential parameter is the length of the device (l). The shorter the length, the greater the percentage of the device's area that will be deformed. Thin materials are more easily deformable, so the battery thickness (h) also strongly affects the

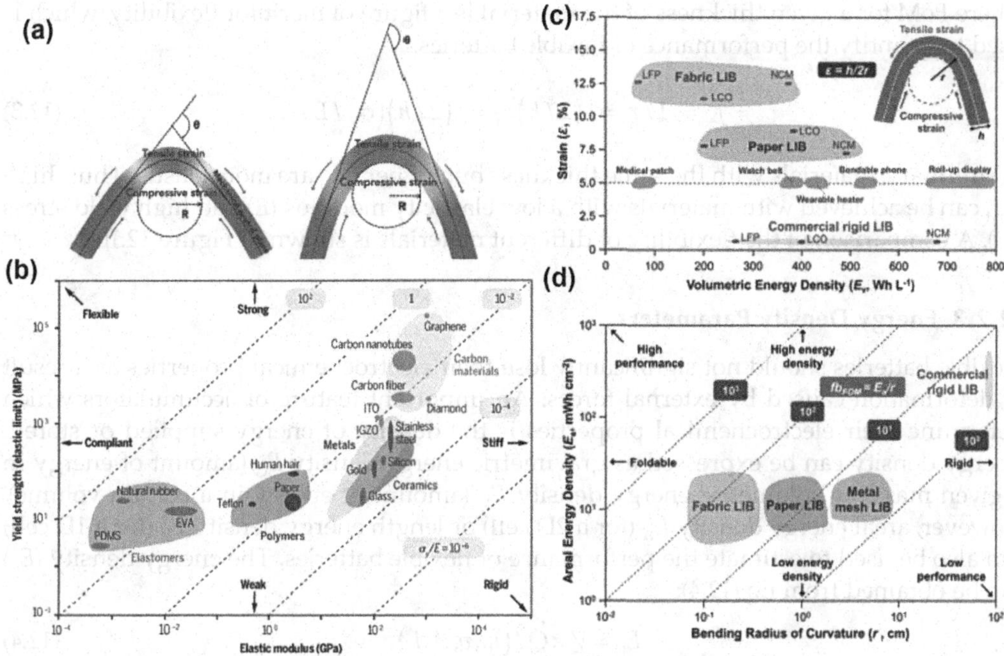

FIGURE 12.1
(a) Schematics of l, θ, and r parameters and strain distribution to describe the flexibility of working devices. Adapted with permission from ref. [10]. Copyright Elsevier (2019). (b) Flexibility figure of merit (f_{FoM}) for selected materials. Adapted with permission from ref. [11]. Copyright Science (2019). (c) Dependence of volumetric energy density on the strain of various electrode materials and batteries. (d) A figure of merit for the flexible batteries. Adapted with permission from ref. [12]. Copyright Elsevier (2020).

flexibility. The effect of these geometric parameters on the flexibility of 1D energy storage devices is shown in Figure 12.1a.

12.2.2 Mechanical Parameters

The four geometric parameters described above can be used to describe the mechanical properties of the batteries [11]. The tensile/compressive strain (ε) can be determined using the radius (r) and thickness (h), according to Eq (12.1):

$$\varepsilon = \left[2\pi(r + h/2) - 2\pi r\right]/2\pi r = h/2r \tag{12.1}$$

For each material, there is the yield strain (ε_y) above which the material cannot return to its original shape after deformation. Therefore, to ensure the correct operation of flexible accumulators the tensile stress (σ) should not be higher than the yield stress (σ_y) which characterized the elastic limit. Both parameters, stress, and strain, are used to calculate the (Young's) modulus of elasticity (E), according to Eq (12.2):

$$E = \sigma/\varepsilon \tag{12.2}$$

Flexibility is inversely proportional to the bending radius of curvature in Eq (12.3). The radius can be calculated from Eq (12.1); subsequently, Eq (12.2) can be transformed to obtain the yield strain, which can be a substitute in Eq (12.3). Ultimately, the flexibility directly depends on the reciprocal of the thickness ($1/h$) and ratio σ_y/E (defined as a f_{FoM}), where FoM for a given thickness of the material is a figure of merit for flexibility, which is used to quantify the performance of flexible batteries.

$$f = 1/r = (2/h)\varepsilon_y = (2/h)(\sigma_y/E) \tag{12.3}$$

In summary, materials with the same thickness but higher f_{FoM} are more elastic. Thus, high f_{FoM} can be achieved with materials with a low elasticity modulus (E) and high yield stress (σ_y). A comparison of the flexibility of different materials is shown in Figure 12.1b.

12.2.3 Energy Density Parameters

Flexible batteries should not significantly lose their electrochemical properties as a result of deformation caused by external forces. An important feature of accumulators which determine their electrochemical properties is the density of energy supplied or stored. Energy density can be expressed as gravimetric energy density E_G (amount of energy in a given mass) or volumetric energy density E_V (amount of energy in a certain volume). However, areal energy density E_A (for a 2D cell) or length energy density E_L (for a 1D cell) can also be used to estimate the performance of flexible batteries. The energy density (E_n) can be obtained from Eq (12.4):

$$E_n = V \times Q/(m, v, A, L), \tag{12.4}$$

where V is the voltage, Q is the capacity, m is the mass, v is the volume, A is the area, and L is the length.

The rating and selection of materials for flexible batteries should be guided by both energy density and tensile stress parameters. Therefore, taking these parameters into account, Eq (12.5) can be proposed [12] to describe the f_b parameter, which is useful for evaluating the performance of the flexible batteries (FBs):

$$f_b = \varepsilon E_V = (h/2r)(VQ/Ah) = (VQ/A)/2r = 1/2(E_A/r), \tag{12.5}$$

where f_b is the parameter describing the capacity performance of the flexible battery and A is the projective area of the battery cell. Therefore, the FoM of flexible batteries could be determined using the ratio E_A/r according to Eq (12.6):

$$f_{bFoM} = E_A/r \tag{12.6}$$

In Figure 12.1c, volumetric energy density *versus* strain of various electrode materials and batteries is presented. Additionally, a figure of merit for flexible batteries is shown in Figure 12.1d.

12.3 Flexible 1D Materials for Batteries

12.3.1 Carbon Nanotubes in Flexible Batteries

12.3.1.1 General Characterization of Carbon Nanotubes

CNTs have attracted significant attention in flexible battery technology owing to their distinctive structure, and mechanical, chemical, and physical properties. They can be divided into single-walled CNTs (SWCNTs) or multi-walled CNTs (MWCNTs). SWCNTs can be considered as carbon cylinders that are formed by the rolling of graphene sheets. The atomic arrangement in such a structure and chirality of the nanotube depends on the way the sheet of graphene is rolled. The chirality of the nanotube is determined by the chiral vector, C_h, and the chiral angle θ (Figure 12.2) [13]. The chiral vector is defined as follows in Eq (12.7):

$$C_h = na_1 + ma_2 \tag{12.7}$$

For a chiral angle of 0°, a "zigzag"-structured CNT is formed. In this case, the *m* integer is equal to 0°. If θ = 30°, the "armchair" geometry of the carbon atom is observed. In this case, both integers (*n* and *m*) are the same. The chirality vector also defines the nanotube diameter. Concentric SWCNTs form MWCNTs (Figure 12.2). The interlayer distance of such a structure is close to 0.34 nm.

The unique 1D structure of CNTs offers several significant advantages, particularly from the perspective of their application in flexible battery production, as follows:

(i) They exhibit a relatively good electrical conductivity. A specific conductivity of 10^6 S/m was reported for SWCNTs and 10^5 S/m for MWCNTs. Therefore, CNTs can effectively participate in electron transfer during the battery charging/discharging processes. Depending on the *m* and *n* indexes, CNTs can exhibit either a semiconducting or a metallic nature. If the difference between *n* and *m* is divisible by three, the nanotube exhibits metallic properties. In all other cases, SWCNTs exhibit semiconducting properties. A significantly high current density of 10^9 A/cm^2 was observed for metallic nanotubes [14]. In the case of semiconducting nanotubes, the electron mobility in the range of 10^5 cm^2/V was reported [15]. During a typical nonselective synthesis, two-thirds of the SWCNTs exhibited semiconducting properties and one-third exhibited metallic behavior. The conducting property of MWCNTs depends on the number of shells. MWCNTs of smaller sizes behave like the corresponding SWCNTs. In the case of MWCNTs, each shell exhibits metallic

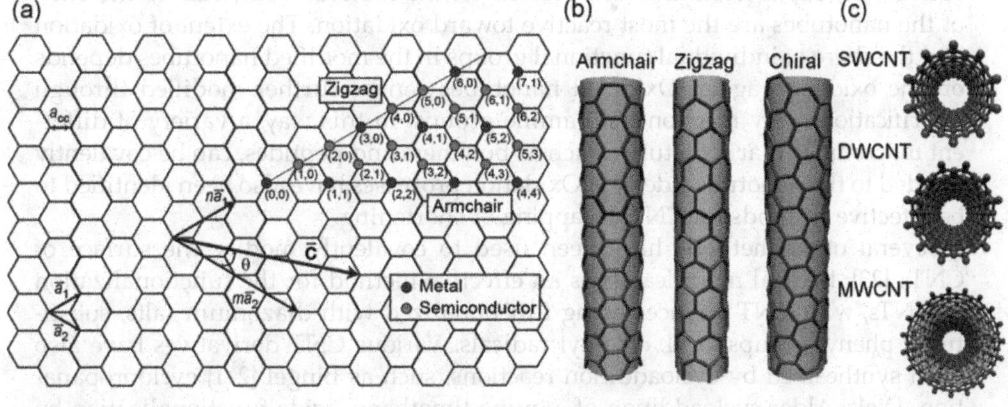

FIGURE 12.2
(a) Graphene plane with chiral vector C indicated how different values of the integers *n* and *m* affect the conducting properties. (b) Examples of different structures of SWCNTs (the (4,4) armchair shape, (6,0) zigzag shape, and (5,3) chiral shape). (c) "Ball and Stick" representation of CNT. Adapted with permission from ref. [13]. Copyright American Chemical Society (2016).

behavior. All tubes forming an MWCNT are involved separately in the charge transfer process.

(ii) CNTs are considered to be strong materials, particularly in the axial direction. The tensile strength depends on the number of defects and type of nanotube. The typical values are in the range of 50–500 GPa and 10–50 GPa for SWCNTs and MWCNTs, respectively [16]. The Young's modulus ranges from 0.6–0.7 TPa for SWCNTs [17]. It changes with the diameter of the nanotube, and, for MWCNTs, it reaches a value in the range 1–1.2 TPa [18]. For comparison, it can be noted that Young's modulus for typical polymers, such as polypropylene, polyethylene, polystyrene, and polytetrafluoroethylene, is in the range 0.5–3 GPa [19]. Only graphene has a Young's modulus that is comparable to that of nanotubes. In the radial direction, CNTs are rather soft, resulting in a high radial elasticity of these structures. Therefore, CNTs are flexible and are not destroyed during bending or other deformations. Such properties are particularly important for their applications to flexible batteries. Additionally, nanotubes are characterized by high thermal conductivity, greater than 200 W/mK for SWCNTs and 3000 W/mK for MWCNTs, so that incorporating nanotubes into various materials can double the thermal conductivity of the resulting composites [20].

(iii) CNTs can be covalently or noncovalently functionalized. In a noncovalent modification, van der Waal forces and π-π interactions are responsible for the interaction between CNTs and the modifying molecules of particles. During this modification, the surface of the CNTs remains unchanged. CNTs are noncovalently functionalized with surfactants, polymers, proteins, and other biologically active compounds.

In the case of covalent functionalization, the desired functional groups are attached to the sidewall of the nanotubes or their tips. In contrast to the noncovalent modification, the covalent functionalization of CNTs results in defect formation at their surface. For the oxidation of CNTs in concentrated sulfuric acid, a mixture of concentrated sulfuric and nitric acids or hydrogen peroxide is commonly used for CNT surface functionalization. Oxidation of CNTs results in the production of different functional groups (–COOH, –OH, –CHO) at the nanotube surface [21, 22]. Defects that are present on the nanotube sidewall or the ends of the nanotubes are the most reactive toward oxidation. The extent of oxidation and the share of individual functional groups in the modified nanotubes depends on the oxidation agent. Oxidized nanotubes can be further modified through esterification or by reaction with amine groups. In this way, a variety of different macromolecular structures, such as polymers and peptides, can be covalently bonded to the nanotube sidewall. Oxidation processes have also been identified to be effective methods for CNT decapping or shortening.

Several other methods have been used to covalently modify the surface of CNTs [23]. Radical modification is an effective method for the functionalization of CNTs, with CNT surfaces being functionalized with diazonium salts, substituted phenyl groups, aryl, or alkyl radicals. Various CNT derivatives have also been synthesized by cycloaddition reactions, such as Bingel [2+1] cyclopropanation, Diels–Alder cycloaddition of *o*-quinodimethane, azide functionalization by 1,3-dipolar cycloaddition, or functionalization of the CNT sidewall with nitrile imides. Electrophilic addition of chloroform in the presence of Lewis acid results in the modification of the CNT sidewall with hydroxyl groups. Polyacylete nanotubes can be formed under the Friedel–Crafts acylation reaction.

Systems containing CNTs and polymers have been intensively studied [24]. Such structures are important from the perspective of their application to flexible electrode production. Polymers can be attached noncovalently and covalently to the surface of CNTs. In the case of noncovalent modification, van der Waals and π-π forces are responsible for the interaction of polymeric chains with nanotube sidewalls. Polymeric chains enable the formation of stable dispersions of nanotubes in solutions and facilitate the subsequent processing of carbon structures. A variety of conducting polymers, such as polypyrrole, polyaniline, polythiophene, and CNT composites, have been developed for use in charge storage devices. Covalent bonding of the polymer to the surface of the nanotubes can be accomplished in two ways, which are classified as "grafting to" or "grafting from." In the former case, the polymeric chain with reactive functional groups reacts with the double bonds of the CNT sidewalls to form a hybrid system. The "grafting to" procedure was applied to form CNTs functionalized with poly(methylmethacrylate), poly(vinylcarbazole), and poly(butadiene). In the "grafting from" method, the monomer of the polymerization process is covalently attached to the nanotube sidewall in the first step. Next, polymerization is initialized on the surface of the nanotubes. The atom transfer radical polymerization method has also been used to form hybrids of CNTs and polyacrylates. Conducting polymers, such as polyaniline [25] and polythiophene [26], were also covalently bonded to the carbon nanotubes.

(iv) CNTs can also form a variety of assembled macroscopic structures. A spaghetti-like nanotube network (Figure 12.3a) [27] has a less organized structure. Such a morphology is usually formed during the arch-discharge method of nanotube formation. Chemical vapor deposition and pyrolysis techniques enable the formation of a significantly more regular and aligned structure of the nanotube network composed of carbon nanotubes joined end-to-end through van der Waals forces. The well-aligned, straight carbon nanotube network formed by chemical vapor deposition is shown in Figure 12.3b [28]. This method allows the formation of carbon nanotubes over a large area of a variety of substrates with good uniformity in diameter, length, straightness, and side density. 3D films of aligned carbon nanotubes are another important and technologically useful structure. In this case, carbon tubes are deposited in parallel to the substrate surface. These films, formed by pyrolysis of iron (II) phthalocyanine, are shown in Figure 12.3c [29]. Materials composed of multilayers of such parallel, ordered CNTs exhibit flexibility and elastic behavior upon bending. They also exhibit high conductivity in the range of 2000 S/cm. CNTs can also form 3D sponges, aerogels, and foams. CNT sponges, with high porosity and controlled nanotube orientation, can endure compressive stress of 32 MPa at maximum strain.

The structure of CNT films and the density of the carbon material package influence their electrical conductivity. The total resistance through the charge transfer path, R_{total}, of the CNT film depends on the sum of the resistance of the contact between the nanotubes, $R_{CNT\text{-}CNT}$, and the resistance of individual CNTs, R_{CNT}:

$$R_{total} = R_{CNT-CNT} + R_{CNT} \quad (12.8)$$

The resistance of the junction between CNTs is significantly higher than that of individual nanotubes. Therefore, the electrical conductivity of a CNT film is generally lower than that of the individual particles. In the case of SWCNTs, the network is composed of metallic

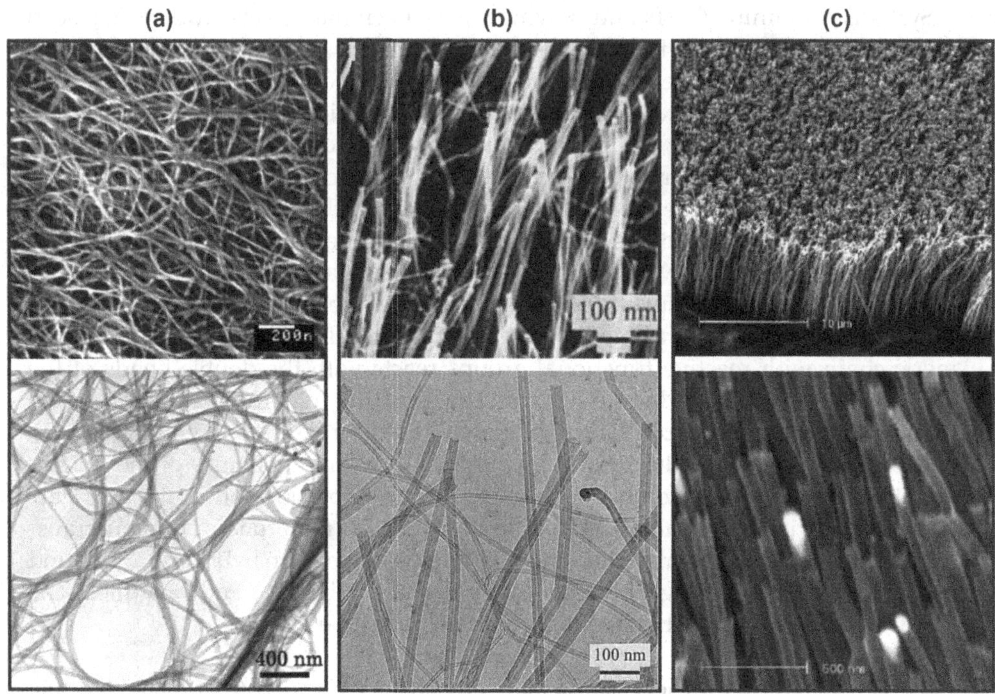

FIGURE 12.3
TEM and SEM images of (a) the electrophoretically deposited spaghetti-like SWCNT film. Adapted with permission from ref. [27]. Copyright Elsevier (2006); (b) aligned carbon nanotube arrays grown by chemical vapor deposition. Adapted with permission from ref. [28]. Copyright Elsevier (1999). (c) SEM images of the uniformity-aligned carbon nanotubes film grown by pyrolysis. Adapted with permission from ref. [29]. Copyright Elsevier (2000).

and semiconducting nanotubes, and the conductivity of such a layer depends on the ratio of the amounts of the two forms in the material. The density of the film also influences the conducting properties of the layer composed of nanotubes. It was also reported that films formed from longer CNTs exhibit lower resistance than films created from shorter nanotubes because of the longer bundle that conducts charge carriers in the first case.

CNTs are also electrochemically active. Unmodified CNTs exhibit double-layer capacitance properties. Films composed of these structures exhibit specific capacitance in the range of 4–135 F/g in the case of MWCNT. Maximum specific capacitances of 113 F/g or 180 F/g were obtained for SWCNTs film in H_2SO_4 or KOH solutions, respectively [30]. Covalent modification of CNTs with electrochemically active functional groups, such as –COOH and –CHO, results in Faradaic pseudocapacitive behavior and enhancement of specific capacitance of films composed of these oxidized carbon nanotubes. Faradaic reactions are accompanied by counter-ion transport and can significantly improve the charge storage properties of electrode materials.

12.3.1.2 Brief Overview of Electrode Materials for Flexible Batteries

12.3.1.2.1 Carbon Nanotube-Based Flexible Anodes for Alkali-Metal Batteries

In the case of alkali metal-ion batteries, CNTs are primarily used as a component in the formation of composites, providing high conductivity, large real surface area, and superior mechanical properties, particularly for flexible batteries. There are two general approaches for the formation of flexible CNT-based composite electrodes: (i) the electrochemically active material can be deposited on top of the CNT layer to form a two-layer structure, or (ii) it can be incorporated into a network of CNTs. The second approach is used more frequently. It offers a high electroactive area for alkali metal-ion doping, uniform distribution of electroactive material in the CNT matrix, and good mechanical properties of the composite. The electronic interaction between the CNT and the electroactive material can also positively contribute to the electrochemical properties of the composite. The CNT matrix can also suppress the volume expansion of the incorporated material during the alkali metal-ion doping process. They can also contribute to the capacity of electroactive materials.

The different materials shown in Figure 12.4 can be incorporated into a CNT polymeric matrix to form flexible electrochemically active materials. These composites can serve as both anodic and cathodic materials. Composites of CNTs and transition metal oxides are frequently used as anodic materials for flexible alkali metal-ion batteries. Composites containing SnO_2 [31], Fe_3O_4 [32, 33], MnO_2 [34], Mn_3O_4, [35], TiO_2 [36], or super-aligned carbon nanotube (SACNT) films are examples of such flexible anodic materials. The CNT network serves as both a conducting material and a structural scaffold. Therefore, in such systems, binders, extra-conductive additives, and current collectors are not required. Mn_2O_3 was deposited inside the SACNT film by the decomposition of $Mn(NO_3)_2$. The *in-situ* sol-gel method was applied to form a TiO_2/SACNT composite [36]. Both composites exhibited superior capacitance properties. This film demonstrates a high capacity of 190 mAh/g at 1 C. It also shows a significantly stable cyclic performance during charging/discharging, with an unchanged structure of the electroactive material after 1,000 cycles.

The composite of Fe_3O_4 and carbon nanotubes exhibits a capacity of 656 mAh/g and stable cyclic performance upon charging and discharging [32]. Fe_3O_4/MWCNT composites containing over 90 wt% of Fe_3O_4 formed by a simple co-precipitation method show a lithium-ion storage capacity of 1093 mAh/g at 100 mA/g and high coulombic efficiency of ca. 98% [33]. Composites of MnO_2 and CNTs have also been extensively studied as anode materials for lithium batteries. A capacity of 903 mAh/g at 0.24 A/g was reported for MnO_2/CNT hybrid material [34]. In electrochemical synthesis, the structure and properties of transition metal oxide within the nanotube network can be controlled. For example, the electrochemical properties of MnO_2/CNT composite formed under galvanostatic deposition of MnO_2 can be tuned by changing the current density applied.

The sulfides and selenides of transition metals at higher valence states can also serve as anodic materials in flexible alkali metal-ion batteries. Metallic octahedral $CoSe_2$ threaded by N-doped CNTs can be used as flexible anodes for potassium-ion batteries [37]. In this composite, octahedral $CoSe_2$ particles are arranged among the CNTs, leaving a zigzag void space among the particles. Such a structure allows volume expansion after the incorporation of alkali metal ions. The octahedral cobalt particles in the size range of 150–300 nm can accommodate large, alkali metal cations. Additionally, the $CoSe_2$/N-doped CNT composite possesses elastic properties that are necessary for application to flexible batteries.

Considerable attention has also been focused on $Li_4Ti_5O_{12}$ and MWCNTs. A composite composed of nano-sized $Li_4Ti_5O_{12}$ that is distributed in the MWCNT network synthesized by the sol-gel method exhibited a high capacity of 145 mAh/g at 5 C and good capacity retention of 142 mAh/g after 500 cycles [38]. A similar capacity performance was reported for a $Li_4Ti_5O_{12}$/MWCNT composite formed by the controlled hydrolysis of tetrabutyl titanate anchored to MWCNT sidewalls [39]. A considerably stable capacity of 152 mAh/g at 5 C was reported for this material.

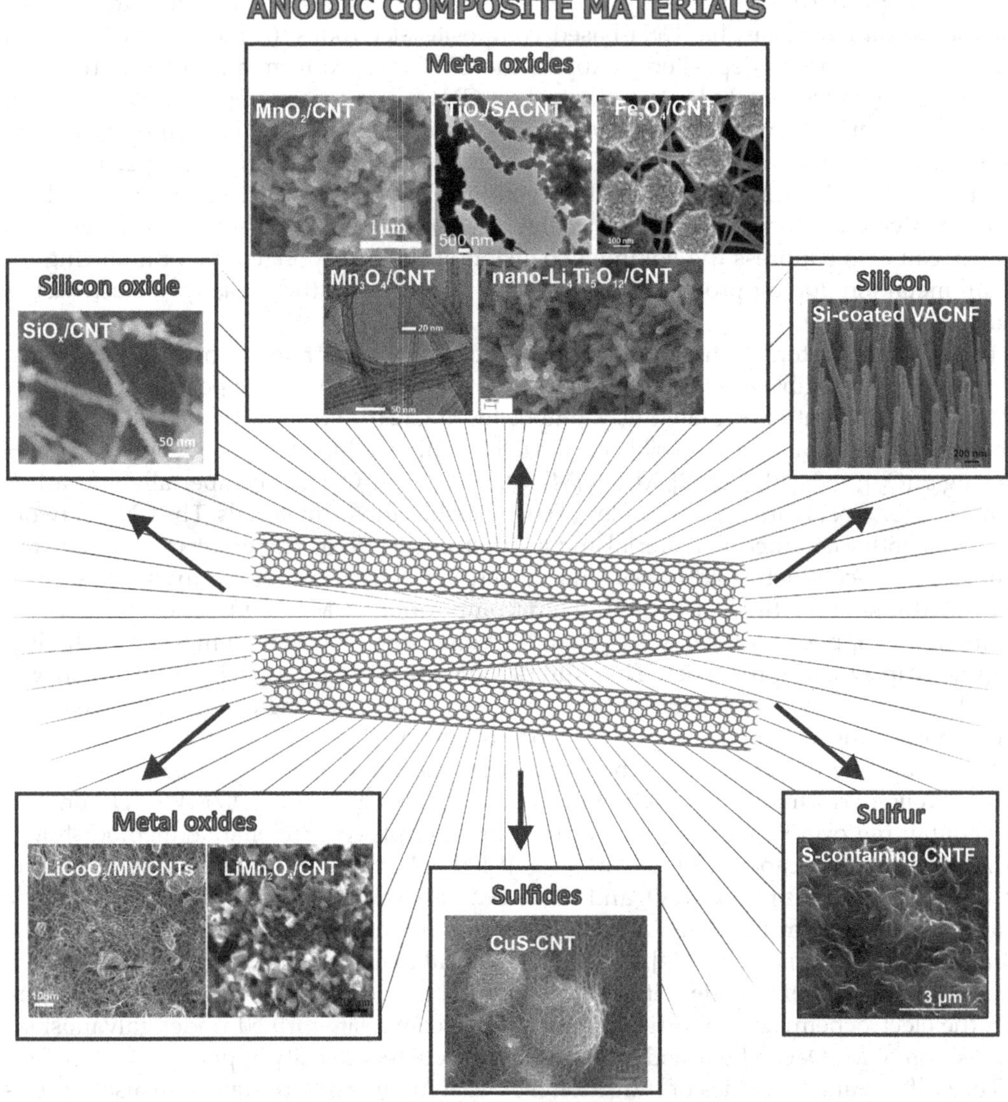

FIGURE 12.4
Selected carbon nanotube-based 1D structures used in flexible alkali metal-ion batteries.

Considerable attention has been focused on silicon-based anodes. A very high theoretical capacity of 4200 mAh/g was reported for silicon, making it a promising material for high-capacity alkali metal-ion batteries. However, the insertion of Li and other alkali metal ions into the silicon structure results in significant volumetric changes and, consequently, the pulverization of electroactive materials. Silicon-based anodes suffer from low cyclic stability upon repetitive charging and discharging. To overcome this limitation, the size of the silicon particles used for anode construction was decreased. Additionally, these particles were incorporated into a network of conducting materials to improve their cyclic performance and capacity. A composite of SiO_x and CNTs was also tested as anodes for flexible electrodes. Superior mechanical flexibility and structural integrity have been reported for paper-like and free-standing SiO_x/MWCNT composite films [40]. This material also shows superior properties for Li-ion storage with a high and reversible capacity of 1240 mAh/g at 100 mA/g, with significantly better rate performance of 441 mAh/g at 2 A/g, and superior cyclic stability compared to noncomposite film. CNTs were also used to form a multiphase Si@C@CNT@C composite as an anodic material for Li-ion batteries [41, 42]. This composite exhibited an initial discharge capacity of 620 mAh/g, which dropped to 563 mAh/g after 60 cycles at 100 mA/g.

Silicon was also deposited on the vertically aligned carbon nanotube film [43]. Deposited silicon creates a designed architecture of pores between vertically oriented carbon nanotubes. The size of these pores gradually increases from the top to the bottom (Figure 12.5). Such morphology allows for effective ion transport and provides void space to accommodate volume changes during charging and discharging, resulting in a very high capacity in the range of 3500 mAh/g_{Si} for 1C and good capacity retention. The typical performance of CNT/Si anode is shown in Figure 12.5.

12.3.1.2.2 Carbon Nanotube-Based Flexible Cathodes for Alkali-Metal Batteries

Similar to CNT-based flexible anodes, CNT networks offer several advantages for cathodic materials, such as improved electrical conductivity for electrode materials with lesser conductivity, an environment for electrochemically active materials deposited in the form of nanoparticles, suppression of the volume expansion of the incorporated material during the alkali metal-ion doping process, and improvement of the capacity of the electrode. However, most importantly, the CNT network provides mechanical flexibility owing to

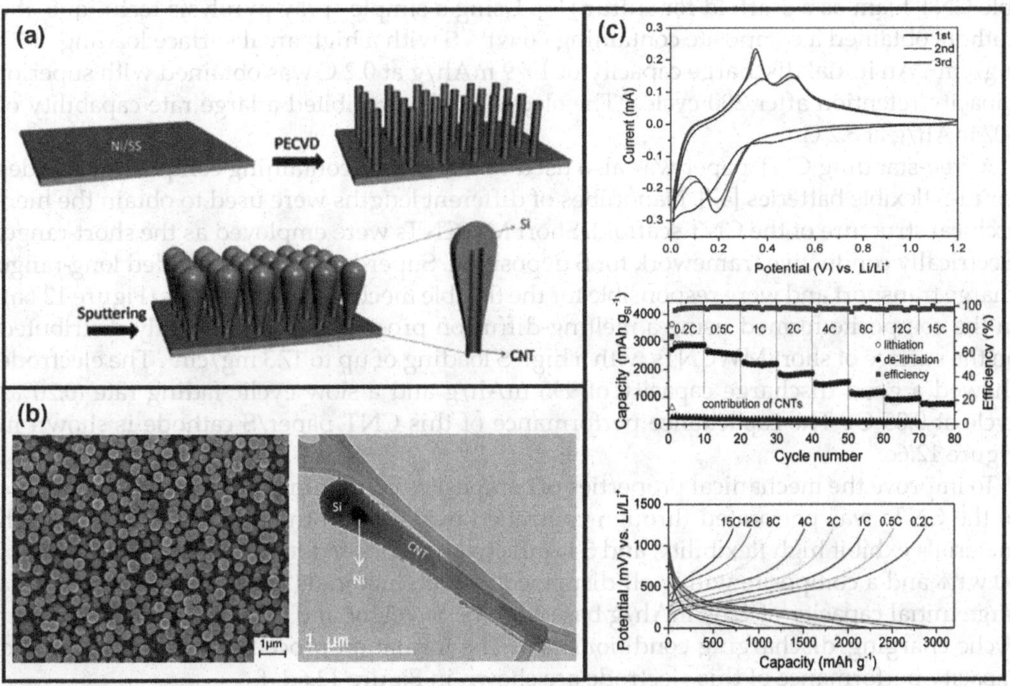

FIGURE 12.5
(a) Schematic of fabrication, (b) morphology (SEM and TEM images) and (c) electrochemical performance of a CNT–silicon anode Adapted with permission from ref. [43]. Copyright Elsevier (2013).

its robust properties. Some of the most important systems that can be used as cathodes in flexible alkali metal-ion batteries are shown in Figure 12.4.

A high-voltage, free-standing flexible cathode was fabricated from $LiCoO_2$ and MWCNT paper. This cathode shows a high rate capacity of 441 mAh/g at 4 C and 4.3 V, and high capacity retention [44]. A similar binder-free and flexible cathode for application in Li-ion batteries was formed by $LiMn_2O_4$ hydrothermal deposition in the MWCNT network. Initial charge and discharge capacities of 126 and 109 mAh/g, respectively, were reported for a current density of 22 mA/g. This cathode also exhibited a good cycling performance.

A free-standing MWCNT network was used to encapsulate CuS for flexible and foldable electrodes in Li-ion batteries [45]. This electrode, containing up to 70 wt% of electroactive material, exhibited a high initial specific capacity of 479 mAh/g and relatively good cycling stability. The good capacity performance of this composite can be attributed to the flexible and highly porous structure of the material, which aids in achieving fast Li^+ ion diffusion and stabilization of the CuS phase within the CNT network.

Considerable attention has been focused on flexible cathodes containing Li-S as an electroactive material. Such a cathode offers the unique advantage of an extremely high specific capacity of 1672 mAh/g. However, this material loses its electrochemical activity during repeated discharging and charging cycles because of the formation of high-order polysulfides (Li_2Sx, where x= 4–8) during the discharge process. These species tend to dissolve in the electrolyte and reduce the amount of electroactive material. Polysulfides can also migrate to the anode of the battery and deposit on its surface. Consequently, the effective surface area of the electrode diminishes. To overcome these problems, carbon nanostructured materials have been used as a component in composites containing Li-S cathodic materials. For example, Ummenthala et al. used free-standing, mechanically flexible CNT foam as a scaffold for sulfur [46]. Using a simple spray pyrolysis technique, the authors obtained a composite containing 66 wt% S with a high areal surface loading of 7.1 mg/cm^2. An initial discharge capacity of 1379 mAh/g at 0.2 C was obtained with superior capacity retention after 200 cycles. The electrode also exhibited a large rate capability of 507 mAh/g at 3.2 C.

A free-standing CNT paper was also used to fabricate S-containing composite cathodes for Li-S flexible batteries [47]. Nanotubes of different lengths were used to obtain the hierarchical structure of the CNT scaffold. Short MWCNTs were employed as the short-range, electrically conductive framework for S deposition. Super-long CNTs provided long-range charge transport and were responsible for the flexible mechanical properties (Figure 12.6a). In the composite formed using a melting-diffusion process, S was uniformly distributed on the surface of short MWCNTs with a high S loading of up to 17.3 mg/cm^2. The electrode showed a good discharge capacity of 995 mAh/g and a slow cyclic fading rate (0.20%/cycle at 0.05 C). The capacitance performance of this CNT paper/S cathode is shown in Figure 12.6c.

To improve the mechanical properties of composites containing CNTs and S, the surface of the CNTs was perforated through controlled oxidation in the air to form pores. Such materials exhibit high flexibility, and S is effectively deposited to form a high S loading of 60 wt% and a composite with well-dispersed S. Such materials in Li-S batteries deliver a large initial capacity of 1264 mAh/g based on the S weight and good performance under cyclic charging/discharging conditions [48]. The formation procedure, morphology, and capacity performance of this electrode are shown in Figure 12.6d–f.

Improvements in the mechanical and capacitance properties of Li-S cathodes can be achieved by incorporating another material that exhibits high flexibility and a high surface area into a composite of CNTs and S. In this way, a three-component composite of

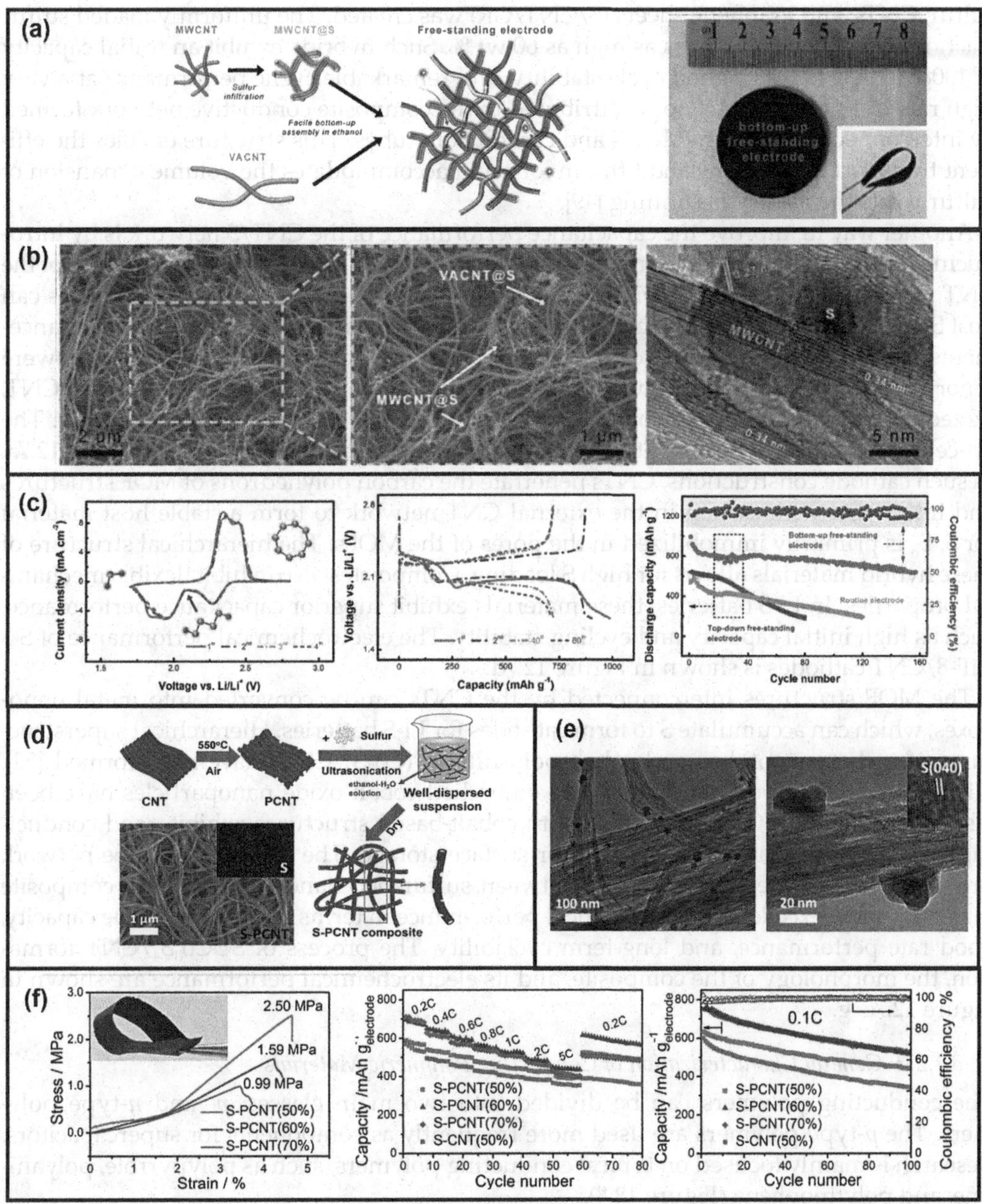

FIGURE 12.6
(a) Schematic illustration of the bottom-up approach of the hierarchical, free-standing electrode with ultrahigh sulfur-loading capability. Grey spheres represent lithium ions and electrons. (b) SEM and TEM images of the bottom-up free-standing electrode. (c) The electrochemical performance of the bottom-up free-standing electrodes with ultrahigh sulfur-loading including cyclic voltammetry (CV) profiles, cycling performance at a current density of 0.05 C compared with the performance of the top-down free-standing electrode, and of a routinely prepared electrode blade on aluminum foil, and corresponding galvanostatic charge-discharge at different cycles. Adapted with permission from ref. [47]. Copyright WILEY-VCH Verlag GmbH & Co (2014) (d) Synthesis and characterization of the S-PCNT composites. (e) TEM images of S-PCNT composite containing 60 wt% of sulfur. (f) mechanical properties, rate performance, and cycle performance at 0.1 C of the S-PCNT composite. Adapted with permission from ref. [48]. Copyright American Chemical Society (2014).

sulfur, CNTs, and graphene sheets (S/CNT/GR) was created. The uniformly loaded sulfur fraction in this composite was as high as 60 wt %. Such hybrids exhibit an initial capacity of 1008 mAh/g at 0.3 C, good cycle stability, and remarkable cyclic performance at a very high rate of 1 C. Such behavior is attributed to the composite conductive network formed by interconnected graphene sheets and carbon nanotubes. This structure enables the efficient transport of electrons and lithium ions and accommodates the volume expansion of sulfur while the battery is running [49].

Another way to improve the capacitance performance of the CNT/S network is by introducing nanoparticles of metal-organic frameworks (MOFs) into the porous structure of the CNT network. MOFs that exhibit a high specific surface area and well-defined pores can host S and immobilize it in the composite network. In these systems, significant enhancements in the conductivity of S cathodes and prevention of polysulfide dissolution were reported [49–51]. The formation of the three-component composites, S@HKUST-1/CNT, S@zeolitic imidazolate framework-8/CNT, and S@MOF-5/CNT, have been reported. The procedures for the formation of these composites are schematically shown in Figure 12.7a. In such cathode constructions, CNTs penetrate the carbon polyhedrons of MOF structures and intertwine with carbon in the external CNT network to form a stable host material for S. S_8 is primarily immobilized in the pores of the MOFs. The hierarchical structure of these hybrid materials allows for high S loading. Composites also exhibit flexible mechanical properties. In Li-S batteries, these materials exhibit superior capacitance performance, such as high initial capacity and cycling stability. The electrochemical performance of S@ZIF-8/CNT cathodes is shown in Figure 12.7d.

The MOF structures interconnected by the CNTs can be converted into metal nanoboxes, which can accumulate S to form cathodes for Li-S batteries. Hierarchical superstructures of carbon nanotubes and cobalt polysulfides (Co_3S_4/CNT) have been formed [52]. Nitrogen-doped carbon nanotubes with embodied cobalt oxide nanoparticles have been also synthesized (CoO/N-CNT) [53]. Both cobalt-based structures exhibit good conducting properties and large, inner space for surface storage. The carbon nanotube network provides effective electron transport between sulfur-rich nanoparticles. Both composite materials show excellent electrochemical performance in terms of high reversible capacity, good rate performance, and long-term cyclability. The process of S@Co_3S_4/CNT formation, the morphology of the composite, and its electrochemical performance are shown in Figure 12.7e–g.

12.3.1.2.3 General Characterization of Conducting Polymeric Materials

The conducting polymers can be divided into two main classes: *p*- and *n*-type polymers. The *p*-type polymers are used more frequently as components for supercapacitors. Research is mainly focused on typical conducting polymers, such as polypyrrole, polyaniline, and polythiophene (Figure 12.8).

The *n*-type conducting polymers remain less explored in comparison with *p*-type polymers. Studies of *n*-doped polymers are more difficult than those of *p*-doped polymeric materials. In the case of organic, conducting polymers, very negative potentials are required for *n*-doping, and reduced materials are very reactive and sensitive to traces of oxygen and water. Recently, fullerene-based, conducting polymers have attracted a lot of attention. Because of the relatively good electron-accepting properties of fullerene moieties, these polymers belong to the *n*-doping class of materials.

The most promising polymer configuration in terms of the charge storage device is the *n-p* type configuration. Poly(3,4-(ethylenedioxy)thiophene), and their derivatives exhibit

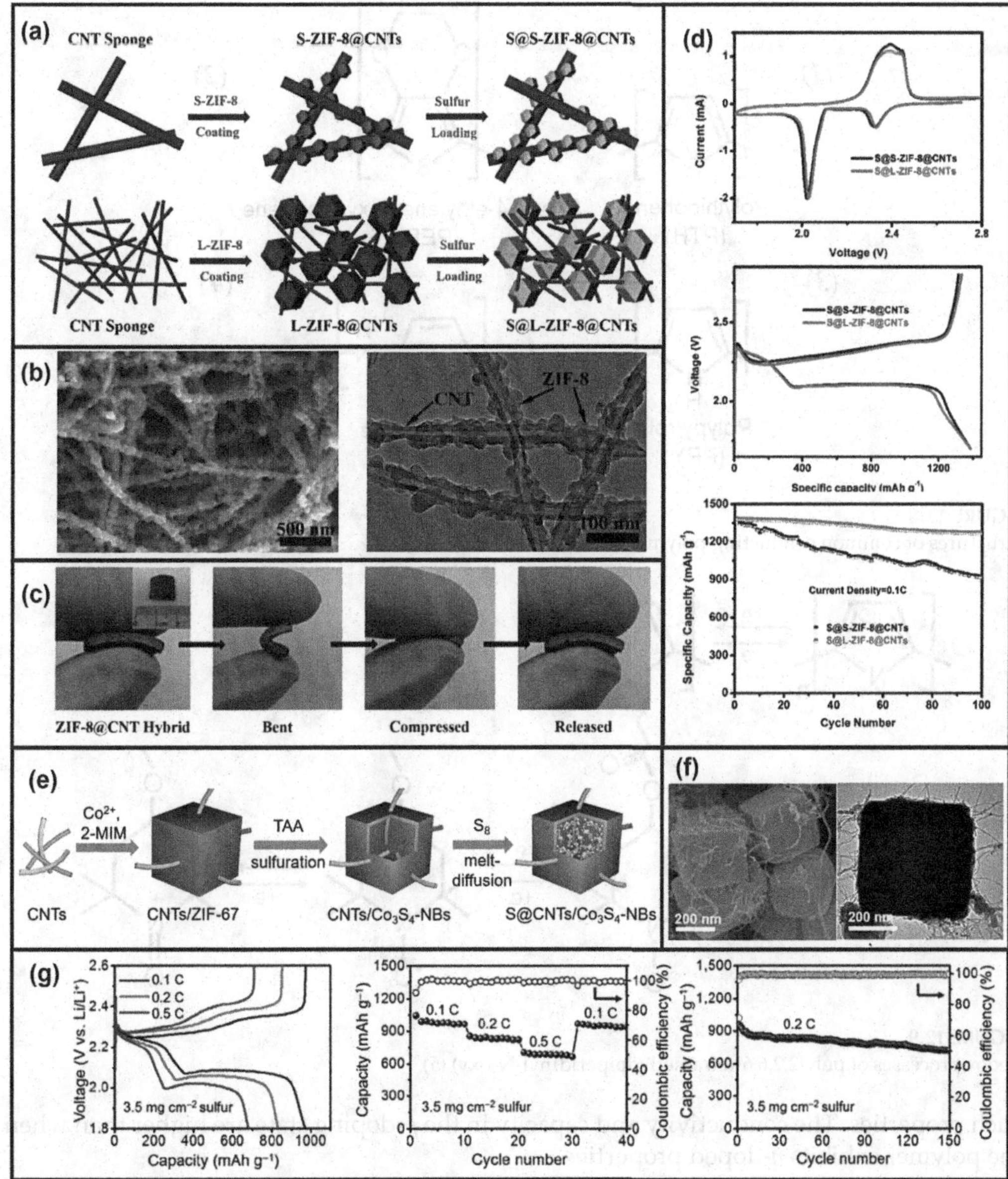

FIGURE 12.7
(a) Typical hybrid structures formed by growing smaller (S-ZIF-8) or larger (L-ZIF-8) particles on the 3D CNT network after sulfur loading. (b) Morphology characterization of the porous hybrid structures with smaller ZIF-8 particle sizes. (c) Photographic illustration of the flexible and compressible properties of the ZIF-8@CNT hybrid. (d) The electrochemical properties of the 3D porous hybrid electrodes including CV profiles, galvanostatic charge/discharge curves, and cycling performance at 0.1 C. Adapted with permission from ref. [50]. Copyright WILEY-VCH Verlag GmbH & Co (2018). (e) Schematic illustration of the S@CNTs/Co_3S_4–NBs hybrid material formation. (f) SEM and TEM images of as-obtained S@CNTs/Co_3S_4–NBs after the filling of sulfur. (g) The electrochemical performance of S@Co_3S_4/CNTs electrode: voltage profiles, rate capacities at various current densities and cycling performance at 0.2 C. Adapted with permission from ref. [52]. Copyright American Chemical Society (2017).

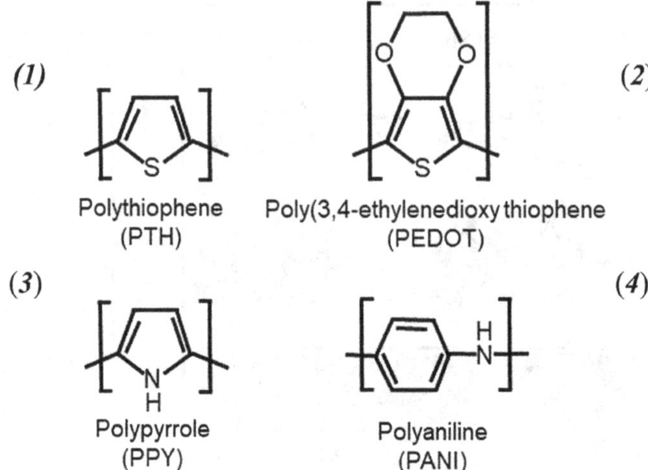

FIGURE 12.8
Structures of common conducting polymers.

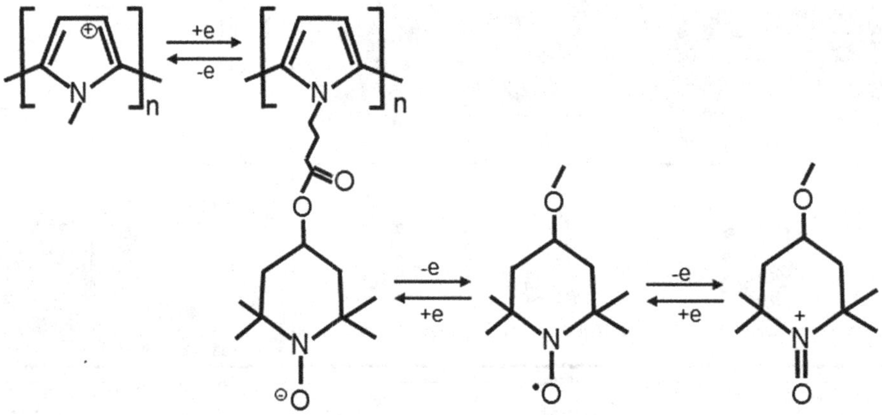

FIGURE 12.9
Redox processes of poly(2,2,6,6-tetramethylpiperidinyl-N-oxy) (5).

such properties. The conductivity and capacity in the *p*-doping state are higher than when the polymer exhibits *n*-doped properties.

The modification of pyrrole and polyaniline results in the formation of a large group of new monomers for the polymerization process. However, the electrochemical properties of polymers formed from these monomers have rarely been investigated. Radical polymers formed from 2,2,6,6-tetramethylpiperidine-1-oxyl (TEMPO)-pyrrole derivatives (5) were used as cathode material for organic radical batteries [54]. Because of the presence of two redox centers, more charge can be accumulated in the polymeric material during oxidation processes that involve the pyrrole and nitroxyl anion according to the reaction presented in Figure 12.9.

Electropolymerization of nitrobenzoyl pyrrole derivatives (*6*) and (*7*) results in the formation of polymeric films that exhibit stable and reversible electrochemical behavior over a wide range of sweep rates [55]. Similar electrochemical behavior was also reported for poly(thienylpyrrole) polymers [56] (Figure 12.10).

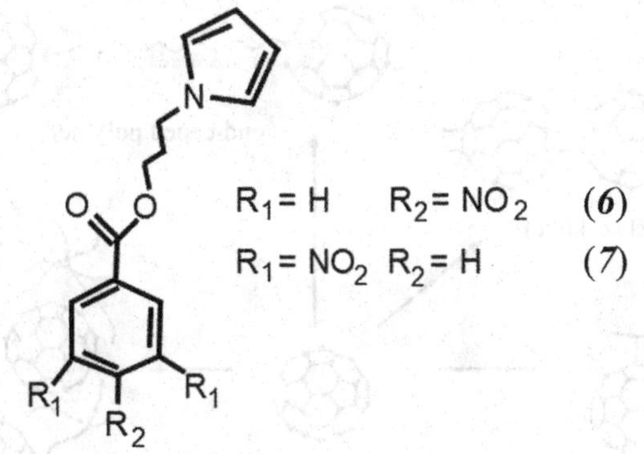

R_1= H R_2= NO_2 (*6*)

R_1= NO_2 R_2= H (*7*)

FIGURE 12.10
Nitrobenzoyl pyrrole derivatives used as monomers of polymerization.

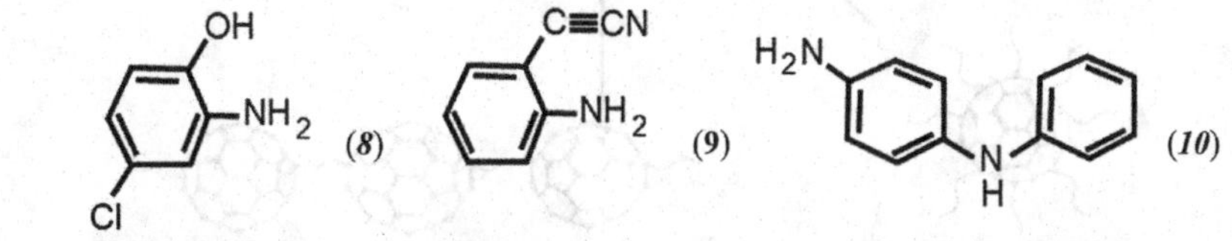

(*8*) 2-Hydroksyl-5-chloroaniline (HCA)

(*9*) 2-And-3-ethynyloaniline

(*10*) 2-Phenyl-1,4-phenylenediamine

FIGURE 12.11
Aniline-based monomers.

The structure and physicochemical properties of aniline-based polymeric materials can be also controlled by the covalent modification of the phenyl ring. Substituted polyanilines overcome the polyaniline insolubility problems and enhance the processability of this material in industrial procedures. They can also change the potential oxidation of monomers and, therefore, the rate of polymerization. Some recently investigated aniline derivatives are presented in Figure 12.11.

Polythiophene, polydioxythiophene, and their derivatives exhibit electrochemical activity both in positive and negative potential ranges. Poly(dithieno[3,4-b:3′,4′-d]thiophene) is active at both positive and negative potentials. However, the relatively negative potential range of this material reduction results in the low stability of *n*-doped materials and sensitivity to oxygen and traces of water. Consequently, the material has high self-discharge and low cycle life in charge storage devices. The application of these polymers in their reduced state also requires the use of special solvents and supporting electrolytes with more negative breakdown potentials. Moreover, the conductivity of the *n*-doped form of polythiophene derivatives was found to be lower than that of *p*-doped polythiophene-based polymers.

Fullerenes are chemically similar to unsaturated hydrocarbons. They can form different polymeric structures [57, 58]. The most common structures among the polymers containing fullerene moieties are shown in Figure 12.12. Because of the relatively good electron-accepting properties, fullerene-based polymers exhibit *n*-doped properties. They also can be electrochemically reduced at less negative potentials in comparison to the π-conjugated organic polymers, such as thiophene and polydioxythiophene, and, therefore, exhibit better stability upon reduction.

The electrochemical properties of other fullerene-based, polymeric structures, such as polymers containing fullerene moieties covalently linked to the conductive polymeric

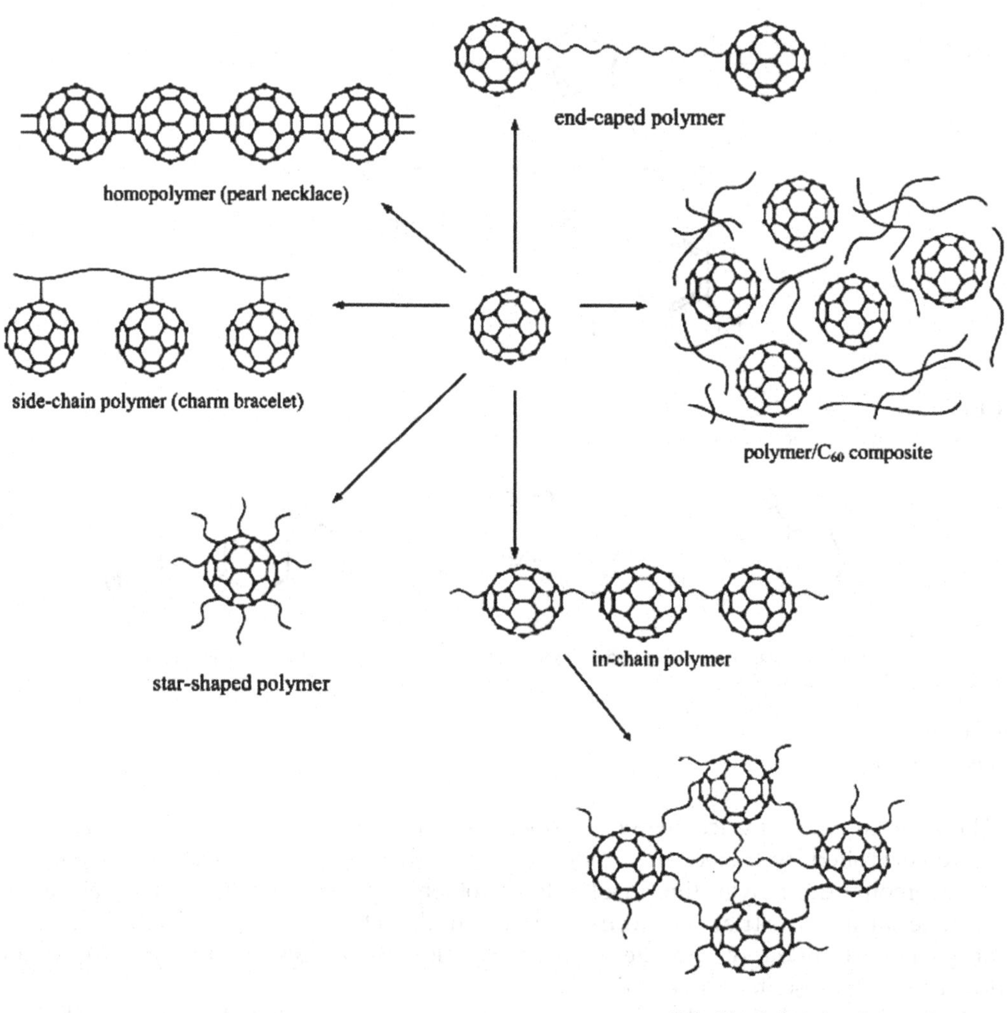

FIGURE 12.12
The most common structures of the polymers contain fullerene moieties. Adapted with permission from ref. [58]. Copyright American Chemical Society (2016).

backbone and main-chain fullerene polymers also show that these materials can be used as electroactive components in charge storage devices. In these systems, fullerene moieties are very often incorporated into the polymeric chain that exhibits electron-donating properties, and these materials show electrochemical activity both in positive and negative potential ranges. Examples of monomers used for polymers containing fullerene moieties covalently linked to the conductive polymeric backbone formation are presented in Figure 12.13.

12.3.1.2.4 Composites of Carbon Nanotubes and Conducting Polymers for Flexible Batteries

Organic conducting polymers, such as polyaniline (PA), polypyrrole (PPY), polythiophene (PTH), and poly(3,4-ethylenedioxythiophene) (PEDOT), and their derivatives which exhibit *p*-doping properties were also used as cathodes in alkali-metal ion batteries. They

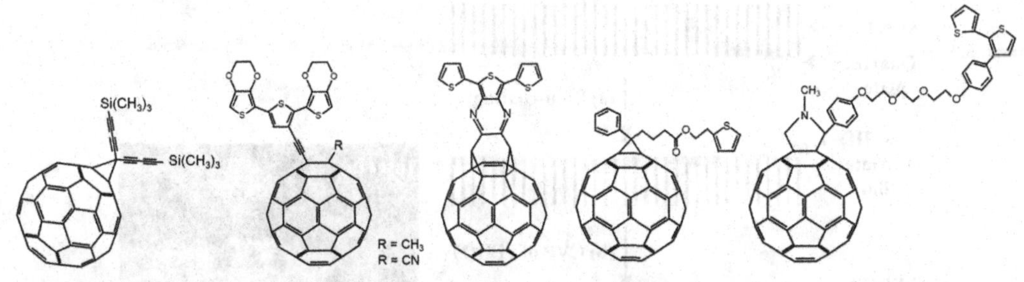

(*11*) (*12*) (*13*) (*14*) (*15*)

FIGURE 12.13
Fullerene-based monomers of the formation of polymeric chains containingfullerene moieties covalently linked to the conductive polymeric backbone.

can conduct both electrons through π-conjugated polymeric chains and ions through a porous network. They also have many other properties, such as high surface area, that allow them to be used in batteries. The mechanical properties of these materials depend on their morphology and structure. However, in most cases, they show limited flexibility. To use them successfully in flexible batteries, composites of polymeric materials with carbon nanotubes are produced. Conducting polymers can be deposited on the top of carbon nanotube network to form two-layer structures, in the empty places of the carbon nanotube network in the form of nanoparticles or core-shell structures in which the carbon nanotube is completely wrapped with the polymer. Composites of SACNTs and conducting polymers exhibit flexible properties.

Composites of conducting polymers with CNTs exhibit excellent electrical conductivity and good stability. For example, fabrication of polythiophene/multi-walled CNT composites (polythiophene/MWCNTs) through a novel electropolymerization process in an ionic liquid micro-emulsion results in highly conducting material formation, in which the MWCNTs were uniformly coated with a thin layer of polythiophene with a thickness of 2–3 nm [59]. The capacitance behavior of this composite material showed favorable capacitance behavior with a high electron transfer rate and low resistance.

The flexible, lightweight, highly conductive, and mechanically robust electrode formed from an aligned carbon nanotube, PEDOT, and poly(vinylidene fluoride) PVDF was used as a flexible cathode for a Li-ion battery [60]. The procedure of this nanostructured electrode formation is shown in Figure 12.14. The CNT array was deposited on a quartz plate by the spin-coating method. PEDOT was synthesized on the CNT array by chemical vapor-phase polymerization. This membrane electrode shows electronic conductivity of 200 S/cm. It also exhibits a high initial discharge capacity of 1150 mAh/g at a current density of 0.1 mA/cm^2. This capacity decreases in the first few charge/discharge cycles, finally reaching a constant value of 265 mAh/g.

Three-component materials were also used for flexible electrodes for alkali metal-ion batteries. The electrochemically deposited graphene oxide (GO)/PEDOT composite exhibits a maximum specific capacitance of 115 F/g at a current density of 0.3 A/g, and an energy density and power density of 13.60 Wh/kg and 139 W/kg, respectively [61]. However, if CNTs are added to the composite, then capacitive properties are significantly increased because the CNTs improve the electric conductivity of the composite and contribute to its low resistance. Chen et al. [62] prepared the PEDOT/MWCNTs/rGO nano-composite by reducing the mixture of graphene oxide with hydrazine in the presence of multi-walled

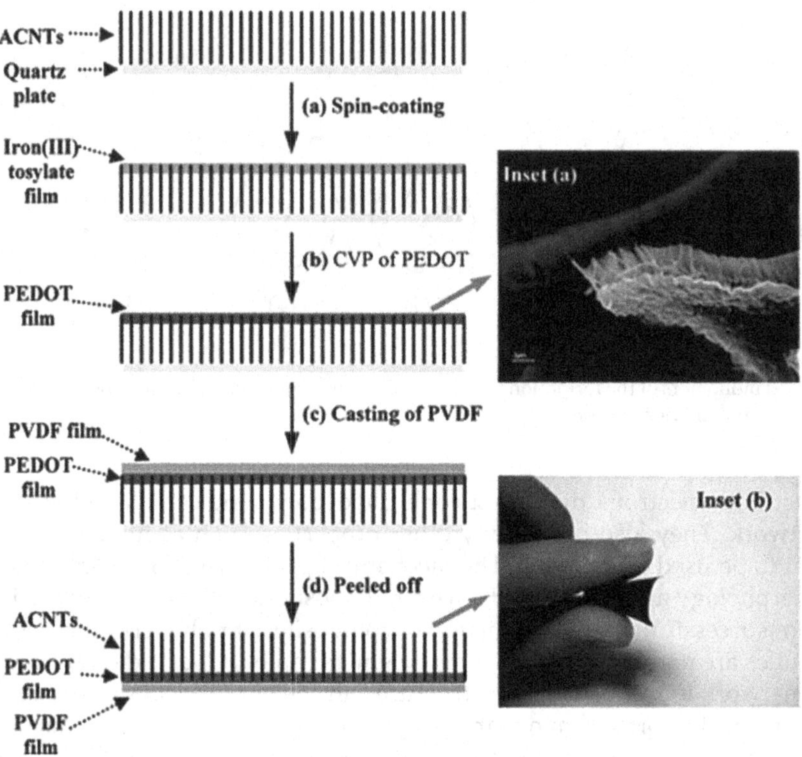

FIGURE 12.14
Schematic description of the procedures for the preparation of a free-standing and highly conductive ACNT/PEDOT/PVDF membrane electrode. Adapted with permission from ref. [60]. Copyright American Chemical Society (2007).

CNTs and the anchored poly(3.4-ethylenedioxythiophene) hollow spheres. This composite was characterized electrochemically. The electrode made with the nano-composite exhibited a specific capacitance of 225 F/g at a current density of 0.1 A/g and exhibited a significant pseudo-capacitance energy storage mechanism during the charge/discharge process. Abdah et al. [63] reported that the specific capacitance of a nano-composite of polypyrrole/GO/MWCNT prepared by a single-step potentiostatic electrodeposition was 360 F/g, whereas, for the PPY/MWCNT and polypyrrole/GO, the specific capacitance was lower, at 207 and 139 F/g, respectively.

A flexible, integrated, high-performance electrode for the energy storage devices based on a novel PANI/SWCNT/cloth composite was also constructed [64]. An ordinary wiper cloth was chosen as a flexible substrate because of its soft and robust mechanical properties, and its conductivity was achieved by being dip-coated in SWCNT ink. Subsequently, polyaniline nanowires were deposited onto the surface of the SWCNT/cloth composite through chemical polymerization. The composite obtained showed much better capacitance performance than the SWCNT/cloth and polyaniline/cloth and had an improved specific capacitance of 410 F/g, a high rate capability, and high stability. The assembled supercapacitor device possessed considerable flexibility under various mechanical bending angles from 0° to 180°.

Newly discovered, sponge-like, and compressible nanotube-based materials for supercapacitor electrodes have recently been produced [65]. These electrodes were fabricated from a

3D CNT network and two materials exhibiting pseudo-capacitance properties, polypyrrole and MnO_2, arranged in a hierarchical structure, CNT@PPY@MnO_2 (Figure 12.15). The core/double-shell skeleton of this material is responsible for its distinctive mechanical properties, such as high flexibility and the possibility of severe compression without collapse. On the other hand, the shell sequence is responsible for its high capacitance. The highest capacitance of approximately 325 F/g was reported for the CNT@PPY@MnO_2 sponge in a slow charging/discharging process. Prolonged compression of the sponge does not significantly influence the specific capacitance. The sponge exhibits similar voltammetric characteristics at a compressive strain of 50% compared with the original state (Figure 12.15). Even if its volume has been greatly reduced, the sponge still retains more than 90% of the initial specific capacitance [65]. The proposed core/double-shell structure can be used in the formation of other sponge materials for the fabrication of compressible supercapacitors with enhanced volume-normalized capacitance.

An interesting approach to the formation of flexible anodic materials for alkali-metal ion batteries was presented by Wang and co-workers. They produced a three-component composite of chemically bonding phosphorus, carbon nanotubes, and cross-linked polymeric

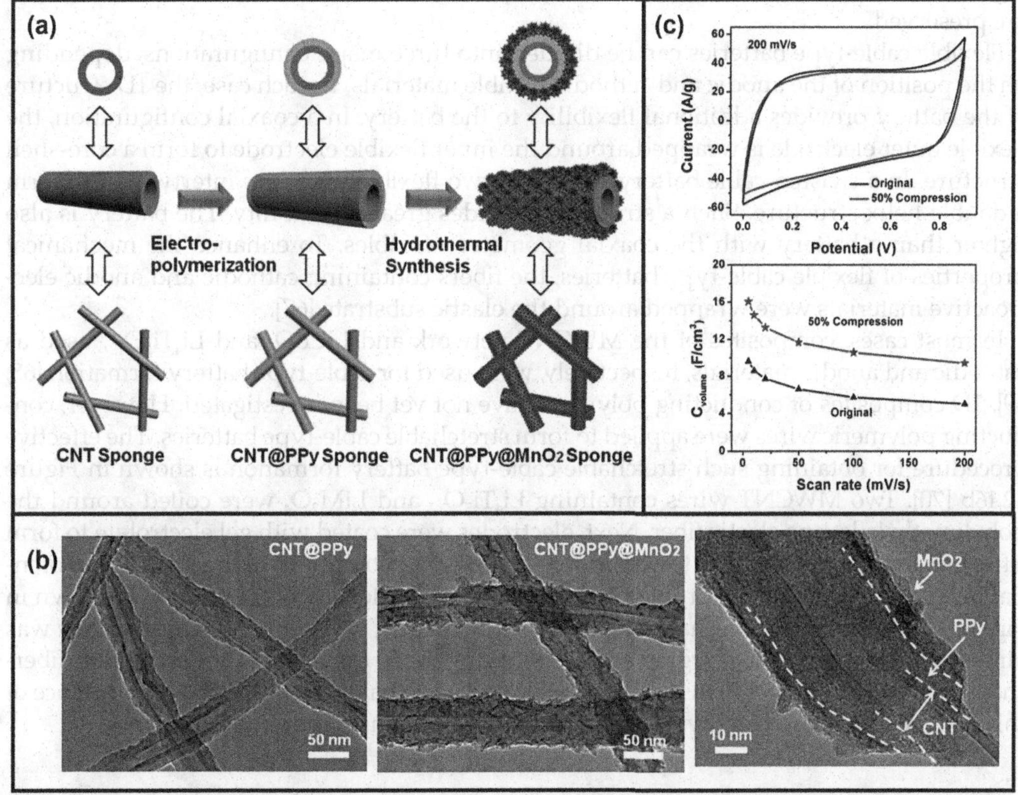

FIGURE 12.15
(a) Schematic illustration for the CNT@PPY@MnO_2 core/double-shell sponge formation. (b) TEM images of the: CNT@PPY sponge, CNT@PPY@MnO_2 sponge, and high-magnification TEM image showing the core/double-shell structure. (c) CV responses of a CNT@PPY@MnO_2 sponge in original and compressed states and volume-normalized capacitances of the sponge in original and compressed states at different scan rates. Reproduced with permission from ref. [65]. Copyright American Chemical Society (2014).

blender [66]. The P-O-C bond formed during mechanical synthesis ensures effective contact between phosphorus and carbon nanotubes. The composite was tested as an anodic material for the Na-ion battery. The unique electrode framework permitted a high reversible capacity of 2134 mAh/g and good cycling stability. Amorphous red phosphorus and carbon nanotube composite was formed by the planetary ball-milling technique. In this composite, phosphorus is uniformly distributed in a conducting carbon nanotube scaffold. It was shown that the presence of the CNT network dramatically improved the capacity performance of this anodic material for Li-ion batteries. A high initial charge capacity of 2133 mAh/g at a current density of 0.05 C and good cycling stability was reported for this material.

12.4 Flexible Batteries Based on CNT-Containing Composites

Composites of aligned carbon nanotube networks with incorporated electroactive materials can be used to form flexible batteries. Generally, depending on the geometry of the battery, they can be divided into plane structures and cable-type batteries shown in Figure 12.16a [67]. In all cases, the flexible properties of the cathodic and anodic materials are preserved.

Flexible cable-type batteries can be divided into three major configurations, depending on the position of the anodic and cathodic flexible materials. In each case, the 1D structure of the battery provides additional flexibility to the battery. In a coaxial configuration, the flexible outer electrode is wrapped around the inner flexible electrode to form a core-shell structure. In a twisted-cable battery structure, two flexible wires are intertwined to form a double-helix structure. Such a structure provides greater flexibility. The battery is also lighter than a battery with the coaxial geometry of cables. To enhance the mechanical properties of flexible cable-type batteries, the fibers containing cathodic and anodic electroactive materials were wrapped around the elastic substrate [67].

In most cases, composites of the MWCNT network and $LiCoO_2$ and $Li_4Ti_5O_{12}$, used as cathodic and anodic materials, respectively, were used for cable-type battery formation [68, 69]. 1D composites of conducting polymers have not yet been investigated. However, conducting polymeric wires were applied to form stretchable cable-type batteries. The effective procedure for obtaining such stretchable cable-type battery formation is shown in Figure 12.16b [70]. Two MWCNT wires containing $Li_4Ti_5O_{12}$ and LiM_2O_4 were coiled around the polydimethylsiloxane elastic fiber. Next, electrodes were coated with gel electrolyte to form a full battery. This battery was tested under galvanostatic conditions in the 1.5 to 3.2 V potential range. A typical voltage profile obtained at a current density of 0.1 mA/cm is shown in Figure 12.16c. The initial discharge capacity of 91.3 mAh/g was obtained. This capacity was almost stable upon cyclic charging and discharging for 50 cycles (Figure 12.16c). This fiber-shaped lithium-ion battery was highly flexible and stretchable. The capacity performance of this battery was unaffected by its mechanical deformations (Figure 12.16d).

12.5 Conclusion and Outlooks

This chapter presented the recent developments in the synthesis, characterization, and application of novel conducting polymer materials for flexible batteries. Research in

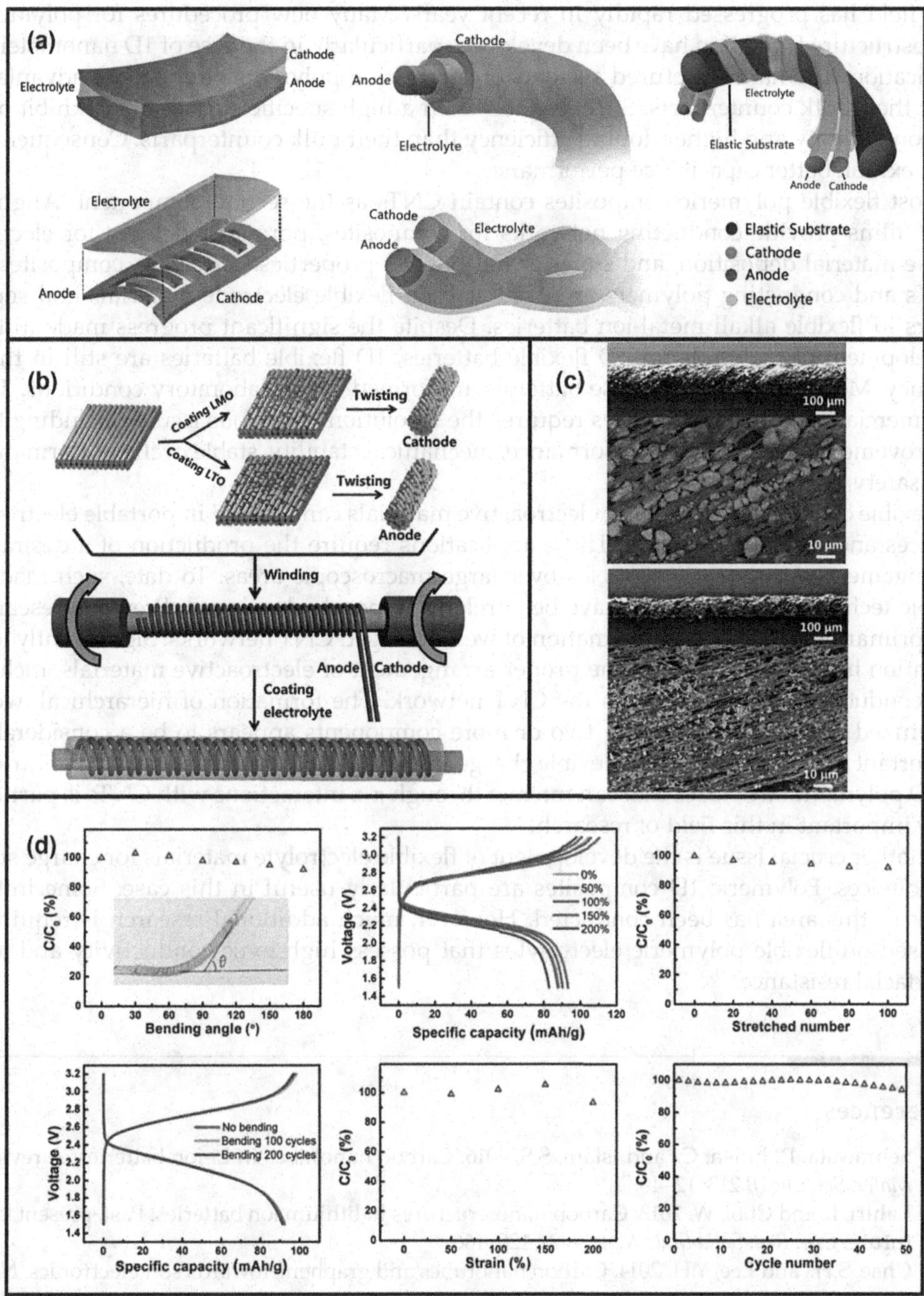

FIGURE 12.16
(a) Design approaches of plane and cable-flexible electrochemical energy storage devices. Reproduced with permission from ref. [67]. Copyright The Royal Society of Chemistry (2018) (b) Schematic illustration of the fabrication of the super-stretchy lithium-ion battery based on the MWCNT/$LiMn_2O_4$ composite fiber as a positive electrode and MWCNT/$Li_4Ti_5O_{12}$ composite fiber as a negative electrode. (c) Scanning electron microscopy (SEM) images of MWCNT/$LiMn_2O_4$ and MWCNT/$Li_4Ti_5O_{12}$ composite fibers. (d) Influence of mechanical deformation on the capacity properties of lithium-ion batteries. Reproduced with permission from ref. [70]. Copyright The Royal Society of Chemistry (2014).

this field has progressed rapidly in recent years. Many new procedures for polymeric nanostructure formation have been developed, particularly in the case of 1D nanomaterial fabrication. The nanostructured systems of conducting polymers offer a clear advantage over their bulk counterparts. Such systems with a high specific surface area exhibit better conductivity and higher doping efficiency than their bulk counterparts. Consequently, they exhibit better capacitance performance.

Most flexible polymeric composites contain CNTs as the second component. Aligned CNT films provide conducting networks for composites, porous structures for electroactive material deposition, and superior mechanical properties. Therefore, composites of CNTs and conducting polymers are used as both flexible electrode materials and separators in flexible alkali metal-ion batteries. Despite the significant progress made in the development of materials for 1D flexible batteries, 1D flexible batteries are still in their infancy. Most prototypes of these batteries are formed under laboratory conditions. The commercialization of 1D batteries requires the resolution of various issues, including the improvement of capacitance performance, mechanical stability, stable cyclic performance, and safety issues.

Flexible composites containing electroactive materials can be used in portable electronic devices and electronic devices. These applications require the production of a desirable arrangement of polymeric particles over large macroscopic areas. To date, such macroscopic technological processes have been relatively poorly developed. Previous research has primarily focused on the formation of well-arranged CNT networks. Significantly less attention has been focused on the proper arrangement of electroactive materials, including conducting polymers within the CNT network. The formation of hierarchical, well-organized structures containing two or more components appears to be a considerably important issue for research on flexible charge storage devices. The application of a variety of 1D polymeric structures that can interact through π-π interactions with CNTs is particularly important in this field of research.

Another crucial issue is the development of flexible electrolyte materials for charge storage devices. Polymeric 1D composites are particularly useful in this case. Some initial work in this area has been conducted. However, much additional research is required, focused on flexible polymeric electrolytes that possess high ionic conductivity and low interfacial resistance.

References

1. Sehrawata, P., Julien, C., and Islam, S.S. 2016. Carbon nanotubes in Li-ion batteries: A review. *Mater. Sci. Eng. B* 213: 12–40.
2. Lahiri, I., and Choi, W. 2013. Carbon nanostructures in lithium ion batteries: Past, present, and future. *Crit. Rev. Solid State Mat.* 38(2): 128–166.
3. Chae, S.H., and Lee, Y.H. 2014. Carbon nanotubes and graphene towards soft electronics. *Nano Converg.* 1(1): 15.
4. Xiang, X.F., Li, Z.D., Xie, K., Jiang, J.Z., Chen, J.J., Lian, P.C., Wu, J.S., Yu, Y., and Wang, H.H. 2012. Graphene sheets as anode materials for Li-ion batteries: Preparation, structure, electrochemical properties and mechanism for lithium storage. *RSC Adv.* 2(17): 6792–6799.
5. Chen, T., Cheng, B., Zhu, G., Chen, R., Hu, Y., Ma, L., Lv, H., Wang, Y., Liang, J., Tie, Z., Jin, Z., and Liu, J. 2016. Highly efficient retention of polysulfides in "sea urchin"-like carbon nanotube/nanopolyhedra superstructures as cathode material for ultralong-life lithium–sulfur batteries. *Nano Lett.* 17: 437–444.

6. Punetha, V.D., Rana, S., Yoo, H.J., Chaurasia, A., McLeskey Jr., J.T., Ramasamy, M.S., Sahoo, N.G., and Cho, J.W. 2017. Functionalization of carbon nanomaterials for advanced polymer nanocomposites: A comparison study between CNT and graphene. *Prog. Polym. Sci.* 67: 1–47.
7. Fang, Z., Wang, J., Wu, H., Li, Q., Fan, S., and Wang, J. 2020. Progress and challenges of flexible lithium ion batteries. *J. Power Sources* 454: 227932.
8. Qian, G., Liao, X., Zhu, Y., Pan, F., Chen, X., and Yang, Y. 2019. Designing flexible lithium-ion batteries by structural engineering. *ACS Energy Lett.* 4(3): 690–701.
9. Ahmad, S., Copic, D., George, C., and de Volder, M. 2016. Hierarchical assemblies of carbon nanotubes for ultraflexible Li-ion batteries. *Adv. Mater.* 28(31): 6705–6710.
10. Li, H., Tang, Z., Liu, Z., and Zhi, C. 2019. Evaluating flexibility and wearability of flexible energy storage devices. *Joule* 3(3): 613–619.
11. Peng, J., and Snyder, G.J. 2019. A figure of merit for flexibility. *Science* 366(6466): 690–691.
12. Chang, J., Huang, Q., and Zheng, Z. 2020. A figure of merit for flexible batteries. *Joule* 4: 1337–1349.
13. Yu, L.P., Shearer, C., and Shapter, J. 2016. Recent development of carbon nanotube transparent conductive films. *Chem. Rev.* 116(22): 13413–13453.
14. Yao, Z., Kane, C.L., and Dekker, C. 2000. High-field electrical transport in single-wall carbon nanotubes. *Phys. Rev. Lett.* 84(13): 2941.
15. Roch, A., Greifzu, M., Talens, E.R., Stepien, L., Roch, T., Hege, J., Nong, N.V., Schmiel, T., Dani, I., Leyens, C., Jost, O., and Leson, A. 2015. Ambient effects on the electrical conductivity of carbon nanotubes. *Carbon* 95: 347–353.
16. Jeon, I.Y., Chang, D.W., Kumar, N.A., et al. Functionalization of carbon nanotubes. Chapter 5. In Yellampalli S., Ed., *Carbon Nanotubes - Polymer Nanocomposites.* Rijeka, Croatia: InTech, 2011, pp. 91–110.
17. Guanghua, G., Tahir, C., and William, A.G. 1998. III. Energetics, structure, mechanical and vibrational properties of single-walled carbon nanotubes. *Nanotechnology* 9(3): 184–191.
18. Hernandez, E., Goze, C., Bernier, P., and Rubio, A. 1998. Elastic properties of C and BxCyNz composite nanotubes. *Phys. Rev. Lett.* 80(20): 4502–4505.
19. Ashby, M.F. *Materials and the Environment 2013,* Elsevier 2013, Chapter 9 Material Selection Strategies, pp. 227–273.
20. Kim, P., Shi, L., Majumdar, A., and McEuen, P.L. 2001. Thermal transport measurements of individual multiwalled nanotubes. *Phys. Rev. Lett.* 87(21): 215502.
21. Likodimos, V., Steriotis, T.A., Papageorgiou, S.K., Romanos, G.E., Marques, R.R.N., Rocha, R.P., Faria, J.L., Pereira, M.F.R., Figueiredo, J.L., Silva, A.M.T., and Falaras, P. 2014. Controlled surface functionalization of multiwall carbon nanotubes by HNO_3 hydrothermal oxidation. *Carbon* 69: 311–326.
22. Sezer, N., and Koc, M. 2019. Oxidative acid treatment of carbon nanotubes. *Surf. Interfaces* 14: 1–8.
23. Karousis, N., Tagmatarchis, N., and Tasis, D. 2010. Current progress on the chemical modification of carbon nanotubes. *Chem. Rev.* 110(9): 5366–5397.
24. Liu, P. 2005. Modifications of carbon nanotubes with polymers. *Eur. Polym. J.* 1(11): 2693–2703.
25. Xiong, S., Zhang, X., Wang, R., Lu, Y., Li, H., Liu, J., Li, S., Qiu, Z., Wu, B., Chu, J., Wang, X., Zhang, R., Gong, M., and Chen, Z. 2019. Preparation of covalently bonded polyaniline nanofibers/carbon nanotubes supercapacitor electrode materials using interfacial polymerization approach. *J. Polym. Res.* 26(4): 90.
26. Maubane, M.S., Mamo, M.A., Nxumalo, E.N., van Otterlo, W.A.L., and Coville, N.J. 2012. Tubular shaped composites made from polythiophene covalently linked to Prato functionalized N-doped carbon nanotubes. *Synth. Met.* 162(24): 2307–2315.
27. Zhao, X., Ohkohchi, M., Inoue, S., Suzuki, T., Kadoya, T., and Ando, Y. 2006. Large-scale purification of single-wall carbon nanotubes prepared by electric arc discharge. *Diam. Relat. Mater.* 15(4–8): 1098–1102.
28. Pan, Z.W., Xie, S.S., Chang, B.H., Sun, L.F., Zhou, W.Y., and Wang, G. 1999. Direct growth of aligned open carbon nanotubes by chemical vapor deposition. *Chem. Phys. Lett.* 299(1): 97–102.

29. Li, D.C., Dai, L., Huang, S., Mau, A.W.H., and Wang, Z.L. 2000. Structure and growth of aligned carbon nanotube films by pyrolysis. *Chem. Phys. Lett.* 316(5–6): 349–355.
30. An, K.Y., Kim, W.S., Park, Y.S., Choi, Y.C., Lee, S.M., Chung, D.C., Bae, D.J., Lim, S.C., and Lee, Y.H. 2001. Supercapacitors using single-walled carbon nanotube electrodes. *Adv. Mater.* 13(7): 497–500.
31. Fu,Y., Ma, R., Shu, Y., Cao, Z., and Ma, X. 2009. Preparation and characterization of SnO_2/carbon nanotube composite for lithium ionbattery applications. *Mater. Lett.* 63(22): 1946–1948.
32. He, Y., Huang, L., Cai, J.S., Zheng, X.M., and Sun, S.G. 2010. Structure and electrochemical performance of nanostructured Fe_3O_4/carbon nanotube composites as anodes for lithium ion batteries. *Electrochim. Acta* 55(3): 1140–1144.
33. Abbas, S.M., Ali, S., Niaz, N.A., Ali, N., Ahmed, R., and Ahmad, N. 2014. Superior electrochemical performance of mesoporous Fe_3O_4/CNT nanocomposites as anode material for lithium ion batteries. *J. Alloys Compd.* 611: 260–266.
34. Zhang, X., Wang, T., Jiang, C., Zhang, F., Li, W., and Tang, Y. 2016. Manganese dioxide/carbon nanotubes composite with optimized microstructure via room temperature solution approach for high performance lithium-ion battery anodes. *Electrochim. Acta* 187: 465–472.
35. Luo, S., Wu, H., Wu, Y., Jiang, K., Wang, J., and Fan, S. 2014. Mn_3O_4 nanoparticles anchored on continuous carbon nanotube network as superior anodes for lithium ion batteries. *J. Power Sources* 249: 463–469.
36. Zhu, K., Luo, Y., Zhao, F., Hou, J., Wang, X., Ma, H., Wu, H., Zhang, Y., Jiang, K., Fan, S., Wang, J., and Liu, K. 2018. Free-standing, binder-free titania/super-aligned carbon nanotube anodes for flexible and fast-charging Li-ion batteries. *ACS Sustain. Chem. Eng.* 6(3): 3426–3433.
37. Yu, Q., Hu, J.H., Lao, C.Y., Lao, C.Y., Gao,Y., Li, P., Liu, Z., Suo, G., He, D., Wang, W., and Yin, G. 2018. Metallic octahedral $CoSe_2$ threaded by N-doped carbon nanotubes: A flexible framework for high-performance potassium-ion batteries. *Adv. Sci.* 5(10): 1800782.
38. Huang, J., and Jiang, Z. 2008. The preparation and characterization of $Li_4Ti_5O_{12}$/carbon nanotubes for lithium ion battery. *Electrochim. Acta* 53(26): 7756–7759.
39. Ni, H., and Fan, L.Z. 2012. Nano-$Li_4Ti_5O_{12}$ anchored on carbon nanotubes by liquid phase deposition as anode material for high rate lithium-ion batteries. *J. Power Sources* 214: 195–199.
40. Guo, W., Yan, X., Hou, F., Wen, L.,Dai, Y., Yang, D., Jiang, X., Liu, J., Liang, J., and Dou, S.X. 2019. Flexible and free standing SiO_x/CNT composite films for high capacity and durable lithium ion batteries. *Carbon* 152: 888–897.
41. Li, L., Liu, P., Zhu, K., Wang, J., Tai, G., and Liu, J. 2017. Flexible and robust N-doped carbon nanofiber film encapsulating uniformly silica nanoparticles: Free-standing long-life and low-cost electrodes for Li-and Na-ion batteries. *Electrochim. Acta* 235: 79–87.
42. Shao, L., Shu, J., Wu, K., Lin, X., Li, P., Shui, M., Wang, D., Long, N., and Ren, Y. 2014. Low pressure preparation of spherical Si@C@CNT@C anode material for lithium-ion batteries. *J. Electroanal. Chem.* 727: 8–12.
43. Fan, Y., Zhang, Q., Xiao, Q., Wanga, X., and Huang, K. 2013. High performance lithium ion anodes on carbon nanotube-silicon core shell nanowires with controlled morphology. *Carbon* 59: 264–269.
44. Xie, M., Li, B., and Zhou, Y. 2015. Free-standing high-voltage $LiCoO_2$/multi-wall carbon nanotube paper electrodes with extremely high areal mass loading for lithium ion batteries. *J. Mater. Chem. A* 3(46): 23180–23184.
45. Zhang, Y., Li, Y., Wang, Y., Guo, F., Liu, W., Pei, H., Yin, G., Ye, D., Yu, S., and Xie, J. 2019. A flexible copper sulfide @ multi-walled carbon nanotubes cathode for advanced magnesium-lithium-ion batteries. *J. Colloid Interface Sci.* 553: 239–246.
46. Ummethala, R., Fritzsche, M., Balach, J.T., Oswald, S., Nowak, R., Sobczak, N., Kaban, I., Rümmeli, M.H., and Giebele, L. 2018. Lightweight, free-standing 3D interconnected carbon nanotube foam as a flexible sulfur host for high performance lithium-sulfur battery cathodes. *Energy Storage Mater.* 10: 206–215.
47. Yuan, Z., Peng, H.J., Huang, J.Q., Liu, X.Y., ang, D.W., Cheng, X.B., and Zhang, Q. 2014. Hierarchical free-standing carbon-nanotube paper electrodes with ultrahigh sulfur-loading for lithium–sulfur batteries. *Adv. Funct. Mater.* 24(39): 6105–6112.

48. Sun, L., Wang, D., Luo, Y., Wang, K., Kong, W., Wu, Y., Zhang, L., Jiang, K., Li, Q., Zhang, Y., JWang, J., and Fan, S. 2015. Sulfur embedded in a mesoporous carbon nanotube network as a binder-free electrode for high-performance lithium–sulfur batteries. *ACS Nano* 10(1): 1300–1308.
49. Mao, Y., Li, G., Guo, Y., Li, Z., Liang, C., Peng, X., and Lin, Z. 2017. Foldable interpenetrated metal-organic frameworks/carbon nanotubes thin film for lithium-sulfur batteries. *Nat. Commun.* 8: 14628.
50. Zhang, H., Zhao, W., Zou, M., Wang, Y., Chen, Y., Xu, L., Wu, H., and Cao, A. 2018. 3D, mutually embedded MOF@ carbon nanotube hybrid networks for high-performance lithium-sulfur batteries. *Adv. Energy Mater.* 8(19): 1800013.
51. Liu, Y., Li, G., Chen, Z., Chen,Z., and Pengl, X. 2017. Strings of porous carbon polyhedrons as self-standing cathode host for high-energy-density lithium–sulfur batteries. *Angew. Chem.* 129: 6272–6276.
52. Chen, T., Zhang, Z., Cheng, B., Chen, R., Hu, Y., Ma, L., Zhu, G., Liu, J., and Jin, Z. 2017. Self-templated formation of interlaced carbon nanotubes threaded hollow Co_3S_4 nanoboxes for high-rate and heat-resistant lithium–sulfur batteries. *J. Am. Chem. Soc.* 139(36): 12710–12715.
53. Khan, I.A., Nasim, F., Choucair, M., Ullah, S., Badshah, A., and Nadeem, M.A. 2016. Cobalt oxide nanoparticle embedded N-CNTs: Lithium ion battery applications. *RSC Adv.* 6(2): 1129–1135.
54. Zhang, S., Lv, G., Wang, G., Zhu, K., Yu, D., Sha, J., Wang, Y., and Liu, Y. 2015. Facile preparation and fluorescence properties of a soluble oligopyrrole derivative. *J. Photochem. Photobiol. A: Chem.* 309: 30–36.
55. Su, P.G., Chiu, S.R., and Lin, Y.T. 2016. Flexibility and electrical and humidity-sensing properties of N-substituted pyrrole derivatives and composite films of Au nanoparticles/N-substituted pyrrole derivatives. *Sens. Actuat. B: Chem.* 224: 833–840.
56. Yigit, D., Hacloglu, S.O., Gullu, M., and Toppare, L. 2015. Novel poly(2,5-dithienylpyrrole) (PSNS) derivatives functionalized with azobenzene, coumarin and fluorescein chromophore units: Spectroelectrochemical properties and electrochromic device applications. *New J. Chem.* 39(5): 3371–3379.
57. Giacalone, F., and Martin, N. 2006. Fullerene polymers: Synthesis and properties. *Chem. Rev.* 106(12): 5136–5190.
58. Balch, A.L., and Winkler, K. 2016. Two-component polymeric materials of fullerenes and the transition metal complexes: A bridge between metal–organic frameworks and conducting polymers. *Chem. Rev.* 116(6): 3812–3882.
59. Zhang, H., Hu, Z., Li, M., Hu, L., and Jiao, S. 2014. A high-performance supercapacitor based on a polythiophene/multiwalled carbon nanotube composite by electropolymerization in an ionic liquid microemulsion. *J. Mater. Chem. A* 2(40): 17024–17030.
60. Chen, J., Liu, Y., Minett, A.I., Lynam, C., Wang, J., and Wallace, G.G. 2007. Flexible, aligned carbon nanotube/conducting polymer electrodes for a lithium-ion battery. *Chem. Mater.* 19(15): 3595–3597.
61. Alabadi, A., Razzaque, S., Dong, Z., Wang, W., and Tan, B. 2016. Graphene oxide-polythiophene derivative hybrid nanosheet for enhancing performance of supercapacitor. *J. Power Sources* 306: 241–247.
62. Chen, J., Jia, C., and Wan, Z. 2014. Novel hybrid nanocomposite based on poly(3,4-ethylenedio xythiophene)/multiwalled carbon nanotubes/graphene as electrode material for supercapacitor. *Synth. Met.* 189: 69–76.
63. Abdah, M.A.A.M., Razali, N.S.M., Lim, P.T., Kulandaivalu, S., and Sulaiman, Y. 2018. One-step potentiostatic electro-deposition of polypyrrole/graphene oxide/multi walled carbon nanotubes ternary nanocomposite for supercapacitor. *Mater. Chem. Phys.* 219: 120–128.
64. Wang, K., Zhao, P., Zhou, X., Wu, H., and Wei, Z. 2011. Flexible supercapacitors based on cloth-supported electrodes of conducting polymer nanowire array/SWCNT composites. *J. Mater. Chem.* 21(41): 16373–16378.
65. Li, P., Yang, Y., Shi, E., Shen, Q., Shang, Y., Wu, S., Wei, J., Wang, K., Zhu, H., Quan Yuan, Q., Cao, A., and Wu, D. 2014. Core-double-shell, carbon nanotube@polypyrrole@MnO_2 sponge

as freestanding, compressible supercapacitor electrode. *ACS Appl. Mater. Interfaces* 6(7): 5228–5234.

66. Song, J., Yu, Z., Gordin, M.L., Li, X., Peng, H., and Wang, D. 2015. Advanced sodium ion battery anode constructed *via* chemical bonding between phosphorus, carbon nanotube, and cross-linked polymer binder. *ACS Nano* 9(12): 11933–11941.
67. Sumboja, A., Liu, J., Zheng, W.G., Zong, Y., Zhang, H., and Liu, Z. 2018. Electrochemical energy storage devices for wearable technology: A rationale for materials selection and cell design. *Chem. Soc. Rev.* 47(15): 5919–5945.
68. Ren, J., Zhang, Y., Bai, W., Chen, X., Zhang, Z., Fang, X., Weng, W., Wang, Y., and Peng, H. 2014. Elastic and wearable wire-shaped lithium-ion battery with high electrochemical performance. *Angew. Chem.* 126: 7998–8003.
69. Zhang, Y., Zhao, Y., Cheng, X., Weng, W., Ren, J., Fang, X., Jiang,Y., Chen, P., Zhang, Z., Wang, Y., and Peng, H. 2015. Realizing both high energy and high power densities by twisting three carbon-nanotube-based hybrid fibers. *Angew. Chem. Int. Ed.* 54(38): 11177–11182.
70. Zhang, Y., Bai, W., Ren, J., Weng, W., Lin, H., Zhitao Zhang, Z., and Peng, H. 2014. Super-stretchy lithium-ion battery based on carbon nanotube fiber. *J. Mater. Chem. A* 2(29): 11054–11059.

13

One-Dimensional Polymeric Nanocomposites for Overall Water-Splitting Applications

Soheila Sharafinia, and Alimorad Rashidi

CONTENTS

13.1 Introduction

Water electrolysis *via* renewable electricity is a method to convert water to H_2 and O_2 gases. The normal electrolysis process is applied to generate O_2 in different applications such as aerospace and naval [1]. Hydrogen can be utilized as the alternative energy carrier with an energy content higher than other fuels (e.g., 118 (hydrogen), 44 (gasoline) MJ/

DOI: 10.1201/9781003223764-13

kg at 298 K) [2]. Hydrogen is one of the most accessible elements on earth, mainly in the form of water and hydrocarbons [2]. Hydrogen can be created *via* the kind of techniques such as photolytic, electrolytic, or thermal methods from water, biomass, or fossil fuels. Steam reforming of methane is one of the industrial hydrogen manufacturing methods. But this method generates hazardous gases and leads to the greenhouse effect. So, researchers are looking for a green and environmentally friendly way to overcome these problems.

Among different sources, water is one of the main and long-term sources for hydrogen production due to its abundance and accessibility. Hydrogen production from water has many advantages such as reducing the use of fossil fuels, and also preventing CO_2 emissions. Mainly, hydrogen production from water is done by various methods such as steam reforming of fossil fuels and water splitting [2, 3]. In the steam reforming method of fossil fuels, most of the pollutants enter into the atmosphere [4]. Therefore, the electrolysis of water *via* water splitting as a method of hydrogen manufacture has attracted much attention. Water splitting is carried out by photocatalytic water splitting, chemical conversion of biomass, or thermochemical processes. According to the operating temperature, there are several types of electrolyzers as follows [5, 6]:

i) The temperature < 50–80 °C for alkaline water electrolyzers (AWE);
ii) The temperature in the range of 50–80 °C for the proton exchange membrane electrolyzers (PEMWE);
iii) The temperature > 600–800 °C for solid oxide vapor electrolyzer cells (SOEC).

The use of nanotechnology is known to be a new way to produce nanomaterials that greatly contribute to the development of hydrogen as an alternative energy carrier with a higher energy content than other fuels [7]. Therefore, we review the latest advances in the application of nanotechnology to the development of one-dimensional nanomaterials for the process of water splitting and energy generation. The development of one-dimensional nanomaterials, such as polymer dots, carbon quantum dots, and nanofibers, has attracted a lot of attention in the field of water-splitting mechanisms. Researchers are particularly interested in using these one-dimensional polymers.

13.2 One-Dimensional Polymers

Polymers with nanoscale size and macro-scale lengths are called one-dimensional (1D) polymers. These materials include nanorods, nanotubes, nanorods, and nanowires. In recent years, the energy crisis due to dependence on traditional fossil fuels has become a global worry, so that numerous researchers are looking for a way to develop new technologies for the production of emerging materials with potential applications as sustainable, green, and renewable energy sources. 1D polymer nanomaterials have attracted much attention due to their economic efficiency, dimensional-controlled capability, and industrial-scaled manufacturing capability [8]. These nanomaterials have achieved significant progress in recent years due to their large surface area and excellent mechanical, electrical, and structural properties. There are various methods for the fabrication of 1D polymers, which are listed in Figure 13.1.

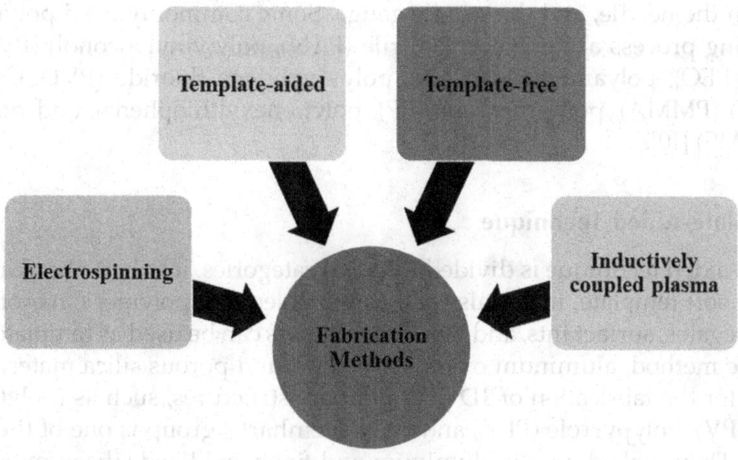

FIGURE 13.1
The methods of fabrication of 1D polymers.

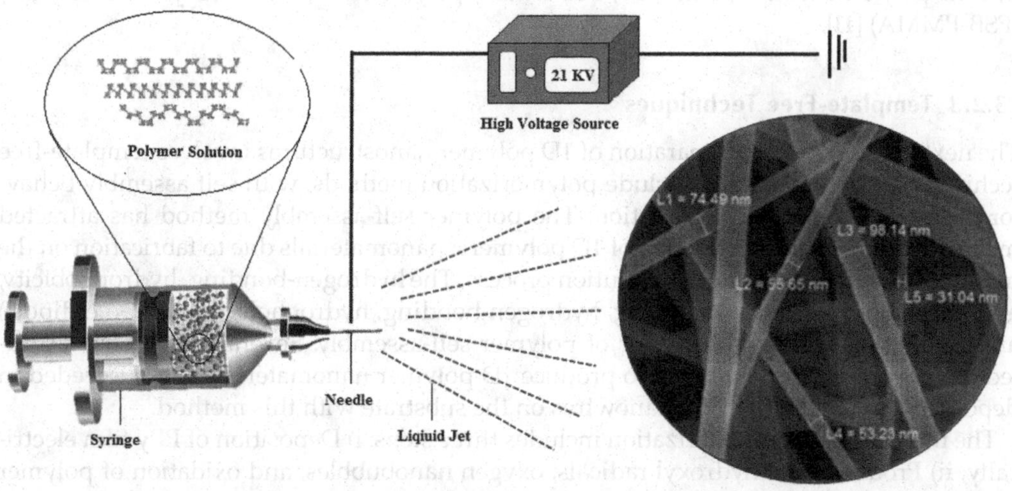

FIGURE 13.2
The steps of the production of 1D polymer nanomaterials through the electrospinning method. Adapted with permission from reference [10]. Copyright (2021) Elsevier.

13.2.1 Electrospinning Technique

Electrospinning is a straightforward, conventional, versatile, and easy method for the fabrication of nanofibers that uses electrical forces. The advantage of this method is that fibers with controllable diameters are synthesized. Also, electrospinning does not require high temperatures or the use of coagulation chemistry to produce fibers. Figure 13.2 illustrates three steps of the synthesis of polymer nanomaterials by the electrospinning method: i) Increase the length of the charged jet along a straight line; ii) Increase the electrical flexural instability and further increase the length of the jet. iii) Solidify the jet to 1D polymer nanomaterials on the collector [9].

Various parameters influence the electrospinning process, such as molecular weight of the polymer, solution viscosity, flow concentration, electric potential, the collector distance from the needle, and the needle gauge. Some commonly used polymers for the electrospinning process are polyacrylonitrile (PAN), poly(vinyl alcohol) (PVA), polyethylene oxide (PEO), polyamic acid (PAA), polyvinylidene fluoride (PVDF), poly(methyl methacrylate) (PMMA), polypropylene (PP), poly(3-hexylthiophene), and polydimethylsiloxane (PDMS) [10].

13.2.2 Template-Aided Technique

The template-aided technique is divided into two categories, involving hard and soft templates. In the soft template, materials such as biomolecules, polymeric nanofibers, liquid crystals, aggregates, surfactants, and block copolymers can be used as templates. But in the hard-template method, aluminum oxide, zinc oxide, and porous silica materials are used as templates for the fabrication of 1D polymer nanostructures, such as (poly(*p*-phenylenevinylene) (PPV), polypyrrole (PPy), and PAN. Steinhart's group is one of the pioneers of this method. They soaked anodic aluminum and finely oxidized silicon molds in a polymer solution, and succeeded in making polymeric materials in nanoscale. Currently, the polymeric materials used to manufacture 1D polymers using template-assisted methods include poly(3-hexyl thiophene), PP, PDMS, and poly(styrene-block-methyl methacrylate) (PSB-PMMA) [11].

13.2.3 Template-Free Techniques

The next method for the preparation of 1D polymer nanostructures involves template-free techniques, which generally include polymerization methods, with self-assembly behavior of the polymer chain in solution. The polymer self-assembly method has attracted much attention for the production of 1D polymeric nanomaterials due to fabrication on the industrial scale, low cost, and dissolution process. The hydrogen-bonding, hydrophobicity, and intermolecular interactions (e.g. hydrogen-bonding, hydrophobic, and dipole-dipole) are factors that affect the behavior of polymer self-assembly. Several groups have studied the polymerization methods to produce 1D polymer nanomaterials and succeeded in depositing 1D conductive PPy nanowires on the substrate with this method.

The mechanism of polymerization includes three steps: i) Deposition of PPy film electrically; ii) Production of hydroxyl radicals, oxygen nanobubbles, and oxidation of polymer monomers; and iii) Growth of 1D PPy nanowires [12].

Likewise, PAN nanoparticles form heterogeneous nucleation on the surface of materials by electrodeposition under high current density conditions. As a result, they have reduced the surface energy barriers to growth under PAN by creating active nucleation centers.

13.2.4 Inductively Coupled Plasma Technique

In recent years, the inductively coupled plasma technique has been recognized as a clean, one-step method to produce 1D polymer nanostructures on an industrial scale [13]. Afterward, a set of 1D polymer nanomaterials based on the variety of polymers, such as poly(2-methoxy-5-2′-ethylhexyloxy)-1,4-phenylenevinylene) (MEH-PPV), polystyrene (PS), PMMA, PVDF, poly(1-methyoxy-2-propyl acetate) (SU8), PPy, poly(styrenesulfonate) (PEDOT: PSS), and poly(3,4 ethylenedioxythiophene) are manufactured on the surfaces of the polymer film. Figure 13.3 shows SEM images of PVDF nanofibers. The mechanism used in this method is through the etching reaction between the plasma and the polymer surface. This reaction involves wet chemical etching and physical milling of ion beams.

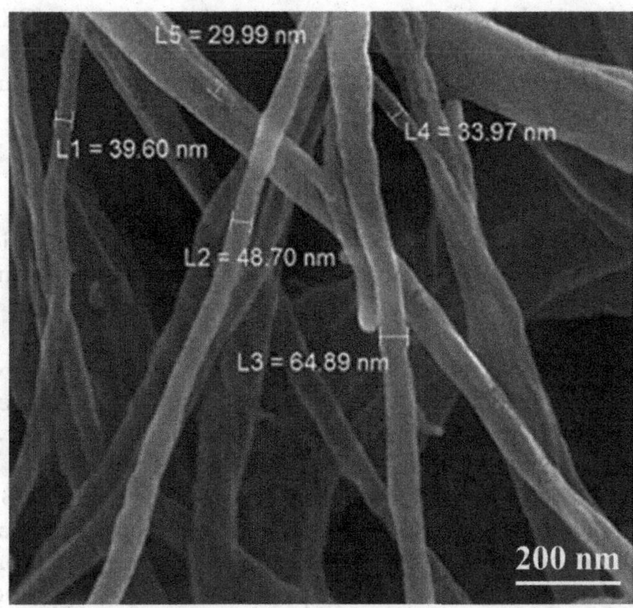

FIGURE 13.3
SEM image for conductive PVDF polymers prepared by the electrospinning method. Adapted with permission from reference [10]. Copyright (2021) Elsevier.

Generally, the Ar, O_2, and CF_4 gases are used to create radicals and plasma. Also, by adjusting the vacuum strength, gas velocity, and pressure, the production of 1D polymer nanomaterials in the polymer surface can be controlled.

13.3 Water Splitting

Energy demand and environmental pollution have been affected by the rapidly growing population, the continuous progression of human standards, and the social economy [14]. Therefore, this issue has become a challenge for researchers to discover alternative technologies to produce clean and sustainable electricity. With increasing environmental pollution and the seriousness of the energy crisis, the use of new renewable energy sources has been urgently brought up. One of the oxidation products that has been considered as a pure, excellent fuel, without the involvement of any carbon, is hydrogen. Also, the enthalpy of the combustion of hydrogen fuel is higher than for other chemical fuels. Despite the great attention paid to the production of hydrogen fuel, the sustainable production of hydrogen is still a major challenge. Electrolysis of water is one of the best processes to produce hydrogen by splitting water.

Hydrogen gas produced by the electrolysis method has different applications as follows:

i) It can be utilized as hydrogen fuel;
ii) Combined with oxygen, it produces oxyhydrogenated gas, which is used in welding [14].

The electron transfers include anodic oxygen evolution reactions (OER) and cathodic hydrogen evolution reactions (HER). Many noble metals, such as Ru, Ir, and Pt, are used as electrocatalysts. These noble metals offer excellent catalytic performance, but they are very expensive.

13.3.1 Fundamentals for Neutral Water Splitting

The water-splitting reaction consists of two simultaneously half-cell reactions, with HER at the negative (cathode) and OER at the positive electrodes (anode). According to Figure 13.4, for complete water splitting, a bifunctional catalyst is used, which plays the roles of both cathode and anode.

The chemical reaction of water splitting to produce hydrogen and oxygen molecules at standard temperature (room temperature) and pressure (atm) proceeds in different ways and pHs [15]. Eqs. (1)–(3) show two HER, OER reactions, and the overall reaction in sodium phosphate buffer (neutral medium). Theoretically, the amount of standard oxidization potential required for a water-splitting reaction is equal to 1.23 V [14]. Also, the standard reduction potential of HER is 0 V. But, in practice, an additional potential must be used to compensate for the potential damage caused by the kinetic constraints created due to the complex ion and electron transfer processes. Generally, in the water-splitting process, many events, such as heat release, production of bubble, electrolyte diffusion blockage, electrode wire and resistances, and concentration of solution must be considered [16]. These disadvantageous agents create an additional potential compared with the standard potential, which is named overpotential (η). Electrocatalysts can be used for this purpose [14]:

$$\text{HER}: 4\text{H}(\text{aq})^{+} + 4\text{e}^{-} \rightarrow 2\text{H}(\text{g})_{2} \quad \text{E}^{\cdot} = 0.00\,\text{V}(\text{Reduction}) \tag{13.1}$$

$$\text{OER}: 2\text{H}_2\text{O}(\text{l}) \rightarrow \text{O}_2(\text{g}) + 4\text{H}^{+}(\text{aq}) + 4\text{e}^{-} \quad \text{E}^{\cdot} = 1.23\ \text{V}(\text{Oxidation}) \tag{13.2}$$

$$\text{Overall}: 2\text{H}_2\text{O}(\text{l}) \rightarrow \text{O}_2(\text{g}) + 2\text{H}_2(\text{g}) \quad \text{E}^{\cdot} = 1.23\,\text{V} \tag{13.3}$$

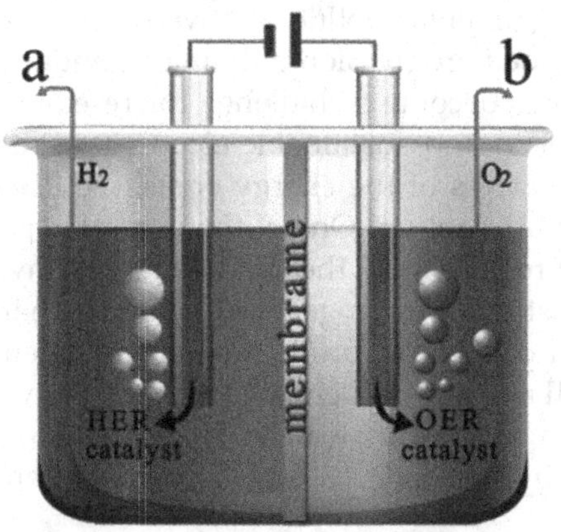

FIGURE 13.4
Schematic representation of two electrodes for overall water splitting. A) HER and B) OER. Adapted with permission from reference [14]. Copyright (2020) Elsevier.

13.3.2 Mechanism of HER

One of the key factors affecting many processes, such as fuel production from CO_2 reduction, corrosion, metal deposition, water, and alkaline chlorine electrolysis, is HER [17]. As has been mentioned, the water-splitting reaction occurs in different ways and under different pHs. Under acidic conditions, many protons are available and, as a result, HER is greatly facilitated, whereas achieving HER in a neutral electrolyte with low potential is relatively difficult. Based on various experimental and theoretical studies performed on HER, it has been determined that the mechanism of HER occurs in three-electron transfer steps: the Volmer step, water decomposition, and formation of a reaction hydrogen intermediates [18]. HER corresponds to the Volmer–Tafel or Volmer–Heyrovsky steps, which are described by Eqs. (13.4)–(13.6) [18].

$$\text{Volmer step}: H_2O + O^- \rightarrow H^* + OH^- \tag{13.4}$$

$$\text{Tafel step}: 2H^* \rightarrow H_2 + 2^* \tag{13.5}$$

$$\text{Heyrovsky step}: H^* + H_2O \rightarrow e^- + OH^- \tag{13.6}$$

The mechanism of the HER on the surface of an electrode is shown in Figure 13.5. HER is carried out by the formation of hydrogen intermediates, associated with electron transfer by proton discharge. This step is named Volmer and is followed by the Tafel or Heyrovsky (second electron transfer) steps [16, 19]. In the Tafel step, a hydrogen atom combines with a proton to form a hydrogen molecule [16]. The final step is the Heyrovsky

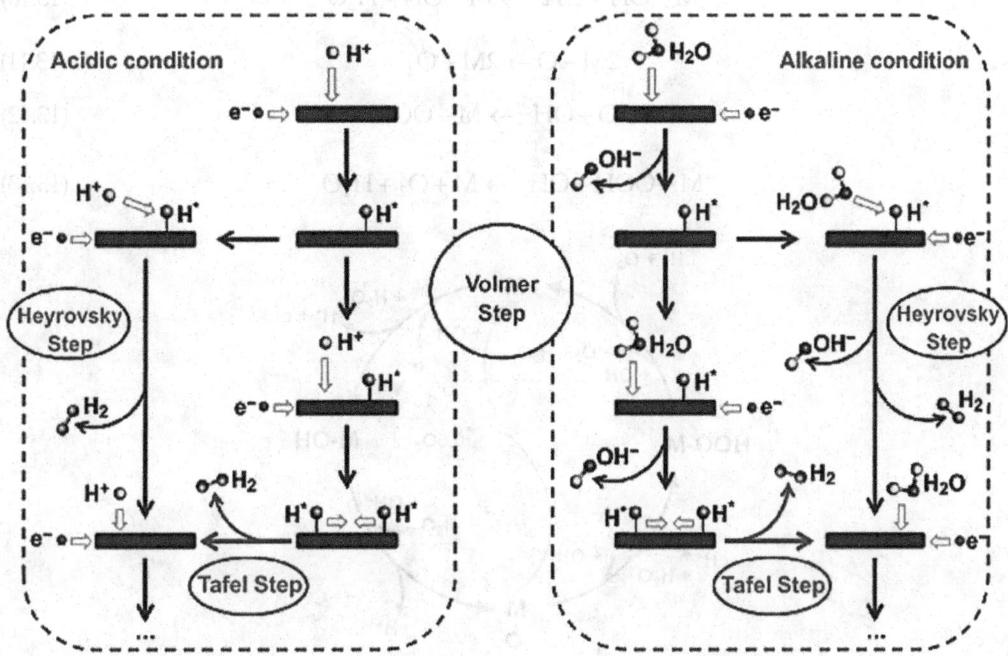

FIGURE 13.5
The mechanism of the HER on the surface of an electrode. Adapted with permission from reference [20]. Copyright (2018) Springer.

step, in which a hydrogen molecule is formed by the bonding together of two adsorbed hydrogen atoms [16].

As shown in the above equations, the H_2O molecules are involved in HER, so that the kinetics of HER in acidic electrolytes are much slower than in neutral electrolytes. Therefore, it can be said that the degradability of water molecules and the free energy of hydrogen adsorption (ΔG_H) affect the performance of electrocatalysts in neutral media. One of the main parameters in controlling the evolution of hydrogen under neutral conditions is the formation and desorption of OH species. With the adsorption of water molecules by the electrochemical method, hydroxide atoms are adsorbed to the surface, which must be desorbed from the surface. Thus, surface sites produce H_2 by formation of H* [14].

13.3.3 Mechanism of OER

Performance OER in acidic or alkaline media is as shown in Figure 13.6 (OH*, OOH*, and O* species act as intermediates, and M is the electrocatalyst surface) and the following equations:

$$\text{Acidic media}: \ 2H_2O \rightarrow O_2 + 4H^+ + 4e^- \tag{13.7}$$

$$\text{Alkaline media}: 4OH^- \rightarrow O_2 + 2H_2O + 4e^- \tag{13.8}$$

The proposed mechanisms for OER under alkaline conditions are as follows:

$$M + OH^- \rightarrow M - OH \tag{13.9}$$

$$M - OH + OH^- \rightarrow M - OH + H_2O \tag{13.10}$$

$$2M - O \rightarrow 2M + O_2 \tag{13.11}$$

$$\text{or}\, M - O + OH^- \rightarrow M - OOH + e^- \tag{13.12}$$

$$M - OOH + OH^- \rightarrow M + O_2 + H_2O \tag{13.13}$$

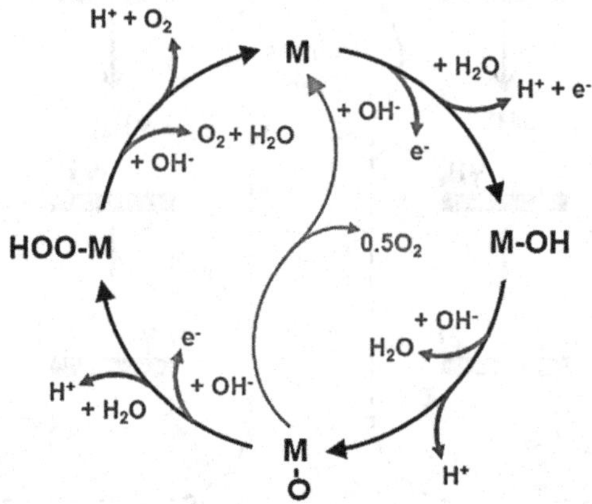

FIGURE 13.6
The mechanism of OER in the acidic and alkaline media. Adapted with permission from reference [21]. Copyright (2018) American Chemical Society.

According to Figure 13.6, OER can proceed in the following two ways:

1) Production of O_2 by direct combination of two O*;
2) Formation of OOH* intermediate [21].

Although OER includes both pathways, the interaction of M with intermediates (OH*, OOH*, and O*) is an important factor for the overall catalytic ability of OER. OER is a multi-electron reaction that consists of several, successive, reaction steps, including electron transfer and chemical steps (e.g., decomposition or bonding reactions) [22]. The Tafel gradient is used to obtain the OER mechanism on an electrocatalyst. If this slope is small, the rate-determining step occurs in the OER path, which exhibits efficient OER kinetics.

13.4 Nafions

In recent years, membrane materials, including cation-exchange membranes (CEMs), such as Nafion, and anion-exchange membranes (AEMs), have been extensively used in water-splitting electrolyzers. Fluoropolymer-copolymers, such as PFS polymer [23], are known as Nafion (brand name). These polymers have ionic properties and are called ionomers, having high thermal and chemical stability, excellent mechanical strength, and proton conductivity. However, a drawback of PFS membranes is that they are known to lose water.

These membranes maintain efficient ionic conduction of the products by the physical separation between the anolyte and the catholyte. But the function of a membrane is limited, although it operates under acidic (CEM) or alkaline (AEM) media. As a result, it limits the choice of materials and photoelectrode catalysts. In addition, unipolar membrane cells exhibit a change in voltage and hence instability over time by creating a pH gradient [24]. Therefore, for these reasons, we need to look for a suitable alternative for these membranes. One of the best alternatives to Nafions is 1D polymers. The remarkable thing about these polymers is that they are not limited to acidic or alkaline media.

13.5 Polymer Dots

These mineral and nanocarbon particles are known as quantum dots (QDs). These particles are usually thinner than 5 nm and are classified as carbon-based quantum dots, such as inorganic semiconductor nanoparticles, carbon dots (CDs), graphene quantum dots (GQDs), and polymer dots (PD) [25]. PD is a carbon and quantum dot material that combines the properties of a polymer and the fluorescence properties of a quantum dot [25].

Polymer dots are obtained using conjugated and non-conjugated polymers. Conjugated polymers have excellent bonding, photoelectric properties to form polymer dots. Polymer dots have advantages such as advanced semiconducting property, processability, solvent miscibility, water-solubility, easy surface modification, good optical properties, and eco-friendly nature. Polymer dots are synthesized by various methods, such as hydrothermal methods, nano-precipitation method (dispersion of a monomer or polymer in solvents), and hydrothermal treatment. Conjugated polymers for the formation of PD

TABLE 13.1

Comparison of QD and PD Characteristics

Properties	QD	PD
Diameter	5–10 nm	5–50 nm
Conformation	Inorganic materials	Non-conjugated/*p*-conjugated polymers
Quantum yield	0.1–0.9	0.1–0.6
Optical properties	Valance-conduction bands	π–π^*
Doping	Ions	Molecules
Polydispersity	Low	High

include polyaniline, polypyrrole, polythiophene, and their derivatives. Non-conjugated polymers for the formation of PD nanostructures include polyvinyl alcohol, polyacrylamide, polyethylene glycol (PEG), and polysaccharides [25, 26]. The PD and QD characteristics are compared in Table 13.1 [25]. Figure 13.7 also illustrates several advantages of this PD. Therefore, it can be said that PD-based materials have several valuable properties compared to carbon dots and mineral quantum dots. Polymer dot-based nanocomposites are used in fluorescence imaging and biosensing, biological imaging, drug delivery, and water splitting.

13.6 Bifunctional Carbon Quantum Dots (CQDs)

CQDs are electroactive species extensively used in electrocatalysis processes because of their provision of active sites for electrocatalytic reactions. The electrocatalysis activity of these species depends on circumambient edges and functional groups available on their surfaces. In 2017, Lv et al. [27] introduced bifunctional precious-metal-free electrocatalyst, nitrogen-doped graphene quantum dots (NGQDs) and Ni_3S_2 on nickel foam (NF), which can be applied as a catalyst for overall water splitting. Their results revealed that such a system had an overpotential of 218 mV for HER and 216 mV for OER which is capable of driving a current density of 10 mA/cm^2 in alkaline solutions, separately. For this purpose, to achieve a 10 mA/cm^2 current density for overall water splitting, a cell voltage of 1.58 V is required. It was observed that, between Ni_3S_2 and NGQDs, there was a synergetic effect that played an important role in promoting the improvement of the overall water splitting performance, HER, and OER of this catalyst. In 2018, Tian et al. [28] prepared a nanosheet array of $NiCo_2P_2$ and GQDs. Their results showed that $NiCo_2P_2$/GQDs are more active than $NiCo_2O_4$/GQDs and free-GQDs $NiCo_2P_2$ as well as, more importantly, Pt/C/RuO_2. The excellent performance of $NiCo_2P_2$/GQDs is attributed to the presence of GQDs in the final morphology because GQDs increase electron current transfers, and ameliorate catalytic properties.

13.6.1 1D Carbon Nanocomposite

Carbon is a promising element which is capable of generating various shapes, such as rings, sheets, bulky balls, or long chains. Important allotropes of carbon can be

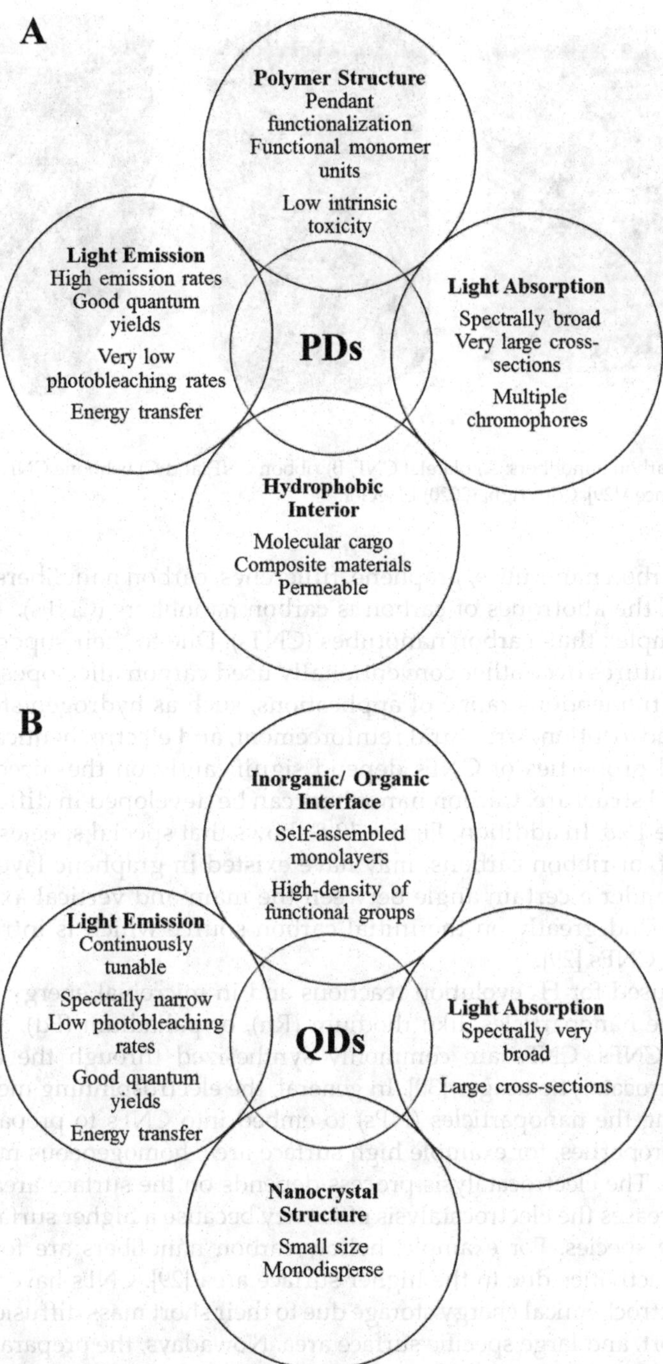

FIGURE 13.7
Advantages of using A) polymer dots or B) quantum dots.

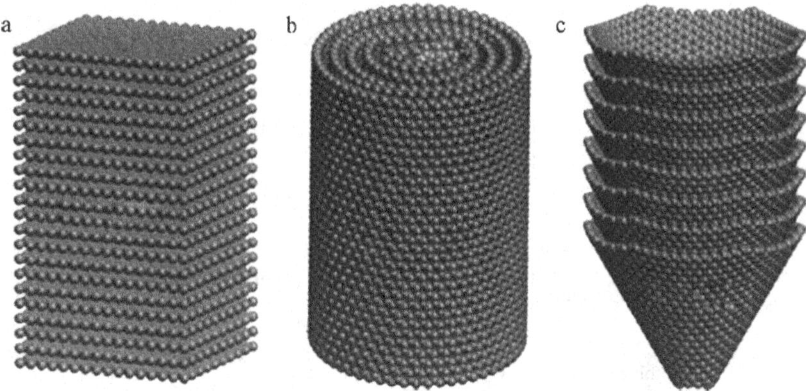

FIGURE 13.8
Different forms of carbon nanofibers: A) platelet CNF, B) ribbon CNF, and C) fishbone CNF. Adapted with permission from referenceA[29]. Copyright (2020) Elsevier.

referred to as carbon nanotubes, graphene, fullerenes, carbon nanofibers, and diamond carbons. One of the allotropes of carbon is carbon nanofibers (CNFs). These are more structurally complex than carbon nanotubes (CNTs). Due to their superior mechanical and chemical features over other conventionally used carbon allotropes, CNFs are able to be used in a tremendous range of applications, such as hydrogen storage, nanodevices, selective adsorption, structural reinforcement, and electrochemical catalysis [29]. The mechanical properties of CNFs depend significantly on the direction of carbon fibers in the final structure. Carbon nanofibers can be developed in different shapes, as shown in Figure 13.8. In addition, Figure 13.8 shows that special species of carbon, like tubular, platelet, or ribbon carbons, may have existed in graphene layers. These compounds orient under a certain angle between the main and vertical axes, while these orientations depend greatly on the initial carbon source which is introduced during the synthesis of CNFs [29].

CNFs can be used for H_2 evolution reactions and in microbial energy harvesting. For these aims, some nanoparticles, like rhodium (Rh), or palladium (Pd), are doped in or anchored onto CNFs. CNFs are commonly synthesized through the electrospinning method for electrocatalytic usages [30]. In general, the electrospinning method is an easy strategy, allowing the nanoparticles (NPs) to embed into CNFs to prepare metal/CNFs with desirable properties, for example high surface area, homogeneous morphology, and small grain size. The electrocatalysis process depends on the surface area. High specific surface area increases the electrocatalysis efficiency because a higher surface area is available for reactive species. For example, hollow carbon nanofibers are found to achieve higher catalytic activities due to the higher surface area [29]. CNFs have attracted broad attention for electrochemical energy storage due to their short mass diffusion distance, fast electron transport, and large specific surface area. Nowadays, the preparation of low-cost and robust bifunctional electrocatalysts is a challenge. It has been found that a synergistic effect may be a successful alternative route to generating effective electro-catalysts. Synergistic effects are combined effects of two active ingredients with similar, different, or related outcomes. Li et al. [31] found that Ni and Ru nanoparticles doped in or embedded within nitrogen-doped carbon nanofibers had synergistic effects which can result in considerable electrocatalytic activity [29, 31].

13.7 Nanofibers Used in Water Splitting

Nowadays, nitrides, sulfides, carbides, phosphides, metal alloys, and more composite compounds have been used to prepare various low-cost electrocatalysts [32]. In the present subsection, we briefly focus on the nanofiber-based electrocatalysts and offer a summary of their catalytic activates and structural effects toward HER and OER (Table 13.2) [32].

13.7.1 Application of Nanofiber-Based Electro-Catalysts in HER

In water splitting, HER takes place at the cathode. This process includes hydrogen adsorption and production. For a favorable HER electrocatalyst, ΔGH^*, the free energy of hydrogen intermediates, should be near to zero, while the energy of the hydrogen binding should not be extremely small or high [33]. Based on this, numerous nanofiber-based electrocatalysts for HER have been proposed and designed. CNF base electrocatalysts could rapidly transfer mass or electrons due to their unique overlapping identity. In addition, CNFs as support can avoid the agglomeration and coagulation of active agents, facilitate their use, and then promote the stability and activity of the electrocatalysts. Therefore, CNF electrocatalysts with optimized electronic properties synergistically regulate the ΔGH^*, leading to an increase in the kinetics of the catalytic reaction [32].

13.7.1.1 Noble Metals

Noble metals like Pt are still the best electrocatalysts for HER, because of their low Tafel slope (~28 mV/dec) and almost zero over-potential. Due to the high cost of Pt, numerous non-precious-metal-based electrocatalysts have been prepared [32]. There are a few

TABLE 13.2
Summary of Nanofiber-Based Electrocatalysts Used in HER and OER

HER		OER	
Electrocatalysts	**Tafel Slope (mV/dec)**	**Electrocatalysts**	**Tafel Slope (mV/dec)**
AuCu/CNFs	70	$CoFe_2O_4$/CNFs	82.15
Au/MoS_2/CNFs	126	$CuFe_2O_4$/CNFs	93.97
$Ca_3Co_4O_9$/CNFs	76	$NiFe_2O_4$/CNFs	98.22
CNT-Ni/NiO NFs	67	$MnFe_2O_4$/CNFs	113.62
CNTs/P/Co_3O_4 NRs	102	$CoFe_2O_4$/NCNFs	80
Co_6W_6C/NCNFs	85 101	$CoFe_2$/$CoFe_2O_4$ NFs	73.34
Co/CeO_2/CNFs	54	$CuFe_2O_4$/CNFs	89.43
Co/CoP/CNFs	50	NCNFs	–
Cu/CuP/CNFs	80	Cu/PVdF/HFP/CNFs	66.3
Co/CoP/NCNFs	51.3 60.3	Fe/CoO/CNFs	60
CoS_2/CNFs	73	$FeNi(PO_x)_n$/CNFs	30
Cu/CNFs	152	NF-PVP/$CoFe_{1.3}$ NFs	47.43
Fe_3C/Mo_2C/HNCNFs	43	FeCoNi/CNFs	57
Ir/IrO_2 NFs	30.1	FeNiP/NCNFs	47
$Ir_{0.80}Ru_{0.20}O_y$ NFs	31.5	Co_3O_4/CeO_2/NCNFs	85

common upgrade techniques to reduce the noble metal content in electrocatalysts. The first method is to combine noble metals with other non-precious active metals or carbon. For example, Du et al. [34] prepared CNFs (Pt NPs/CNFs) by step-by-step method (electrospinning, reduction, and carbonization processes). No aggregation of Pt NPs and low Pt mass loading was observed in the prepared electrocatalyst, with high catalytic efficiency for HER. The second technique is pore engineering, in which the pores are made on the surface of support nanofibers and noble metals are placed in the pores of the supports which greatly enhances electrocatalytic performance. HPCNFs/CNFs is a typical example that was used as appropriate support to graft the octahedral Pt-Cu nanocrystals (Figure 13.9A–D). An excellent electrocatalytic activity with low overpotential was obtained from the $Pt_1Cu_{1.1}$/HPCNF system and this system was stable after 10,000 cycles in H_2SO_4. Thus, these systems are better than the Pt/C as well as most of the supported catalysts of HER (Figure 13.9E–G) [32].

13.7.1.2 The Alloys of Transition Metals

Nickle, cobalt, ferrous iron, copper, etc. are transition metals that are abundant in the Earth's crust. Because of their cost effectiveness and promising catalytic activity, they

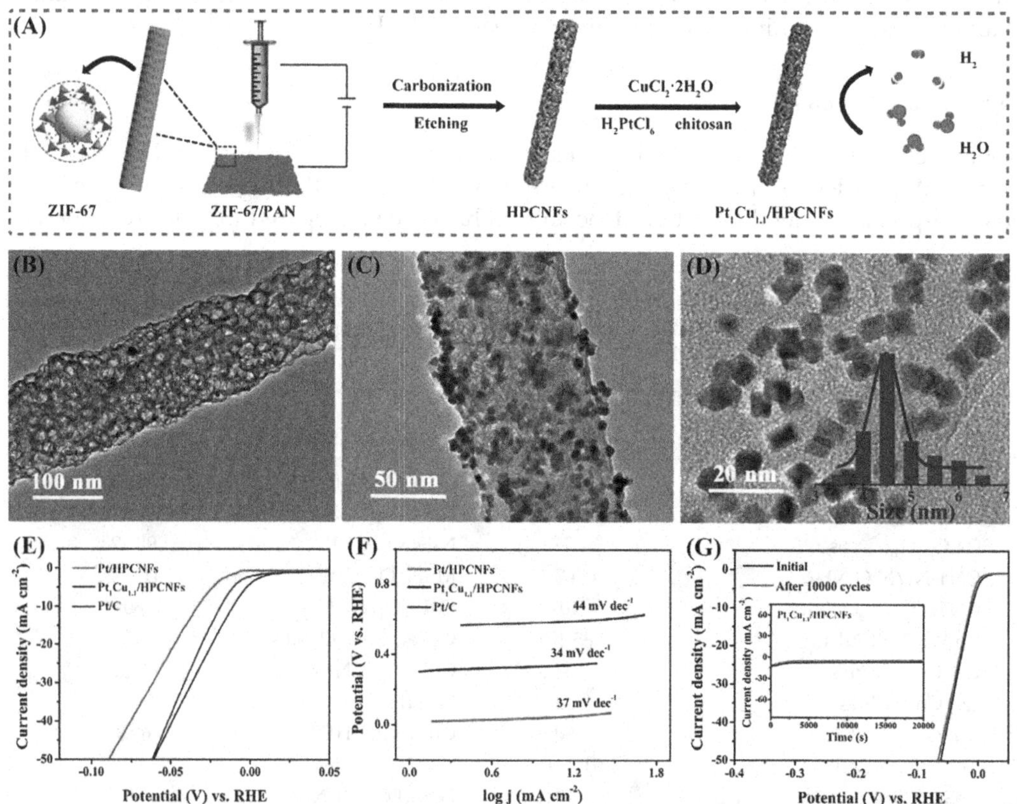

FIGURE 13.9
A) The creation procedure of $Pt_1Cu_{1.1}$/HPCNFs; TEM images of B) HPCNFs and C, D) $Pt_1Cu_{1.1}$/HPCNFs, inset: size distribution histogram of Pt-Cu nanocrystals; E-G) HER performance tests in 0.5 M H_2SO_4; E) linear sweep voltammetry (LSV) curves, F) Tafel plots, and G) stability, inset: chronoamperometric response curve. Adapted with permission from reference [32]. Copyright (2022) Elsevier.

are alternative options to HER electrocatalysts for noble metal-based electrocatalysts [35, 36]. Most of the transition metals show poor electrocatalytic activities in acidic media because they get corroded and deactivated. During long operations, these transition metals exhibit aggregation and decomposition, thus reducing their electro-catalytic efficiency [35]. Interestingly, when the transition metal species are anchored or doped with CNFs, they provide better electrocatalytic activities and avoid all of the above-mentioned issues [35]. For instance, the nanocomposites of CNFs with nickel, cobalt, or copper have been synthesized by electrospinning followed by pyrolysis, and showed high electrocatalytic performance toward HER [35, 36]. Also, it was found that nanocarbon nanotube/nanofiber (NCNT/NF) hybrids and the formation of metal alloys like uniform FeCo Ni_3Co, and Ni_3Fe showed enhanced catalytic reaction kinetics and durability for water splitting [35]. For example, $Ni_{1.5}Co_{0.5}$/NCNT/NFs composites, the logical combination of Ni_3Co with NCNT/NFs, resulted in hierarchical nanostructures. This composite has several advantages, like increased active sites, lowered electron transition direction, accelerated ion diffusion, and amplified mechanical strength, lower overpotential (114 mV at 10 mA/cm^2), and a lesser Tafel slope (117 mV/dec) (Figure 13.10A–D) [35]. The systems consisting of metal alloys and CNFs are more favorable for the HER process (Figure 13.10E–F), and exhibit prominent structural resistance during the electrolysis of 45,000 s (Figure 13.10G). In this item, the synergistic interaction between nickel and cobalt regulates the electronic structure of the Ni_3Co alloy and therefore changes the adsorption energy of the intermediates, as a result leading to the favorable catalytic activity in HER [35]. In addition, the alloys of PtNi, RuNi, PtCu, and AuCu NPs supported by CNFs are known to be highly active HER electrocatalysts [32].

13.7.1.3 Transition Metal Composites

Transition metal oxides (TMO) and hydroxides have become popular as catalysts for HER and OER [32]. TMOs and hydroxides have weak electrocatalytic performance in HER because of their low electronic conductivity, thus must be combined with conductive carbon and/or metals. For instance, NFs such as NiOOH/$Ni(OH)_2$/CNF, WO_{3-x}/CNF, and NiO/Ni/CNF show excellent performance for HER [37]. The effect of metal/metal oxide (M/MO) composition has been widely investigated As electrocatalysts for HER. MO species lead to hydrogen-mediated H_2O adsorption cleavage and OH^- (at the M/MO interface). MO is positively charged and absorbs OH^- and H_2-mediator transfers to the location near M, thereby modulating ΔGH^* catalysts [32]. Some transition metal composites, for example, phosphides, carbides, and chalcogenides have attracted a lot of attention as electrocatalysts in the field of HER. Among these electrocatalysts are the compounds based on W and Mo paradigms [32]. In recent years, transition metal chalcogenides (MX_2 = MS_2 and MSe_2) with interlayer chemical bonds and graphene-like sheet structures have attracted considerable attention in electrocatalysis. However, based on computational experiments, they show weak catalytic activity, which depends on the number of exposed edge locations [32]. Therefore, it is essential to fabricate MX_2 catalysts equipped with edge sites accessible through electronic engineering. MoS_2 nanocrystals have a structure similar to rose petals that are placed on electron CNFs. These catalysts showed excellent catalytic activity for HER. Also, experiments on WS_2 in the field of the production of layered nanostructures with the ability to increase the edges were performed to achieve adjustable electronic properties and accessible active centers [32]. Single-layer WS_2 nano-sheets were uniformly bonded to the hollow NCNFs by electrospinning to form WS_2/ HNCNFs (Figure 13.11A–D). The WS_2/ HNCNF produced showed a larger magnified surface area,

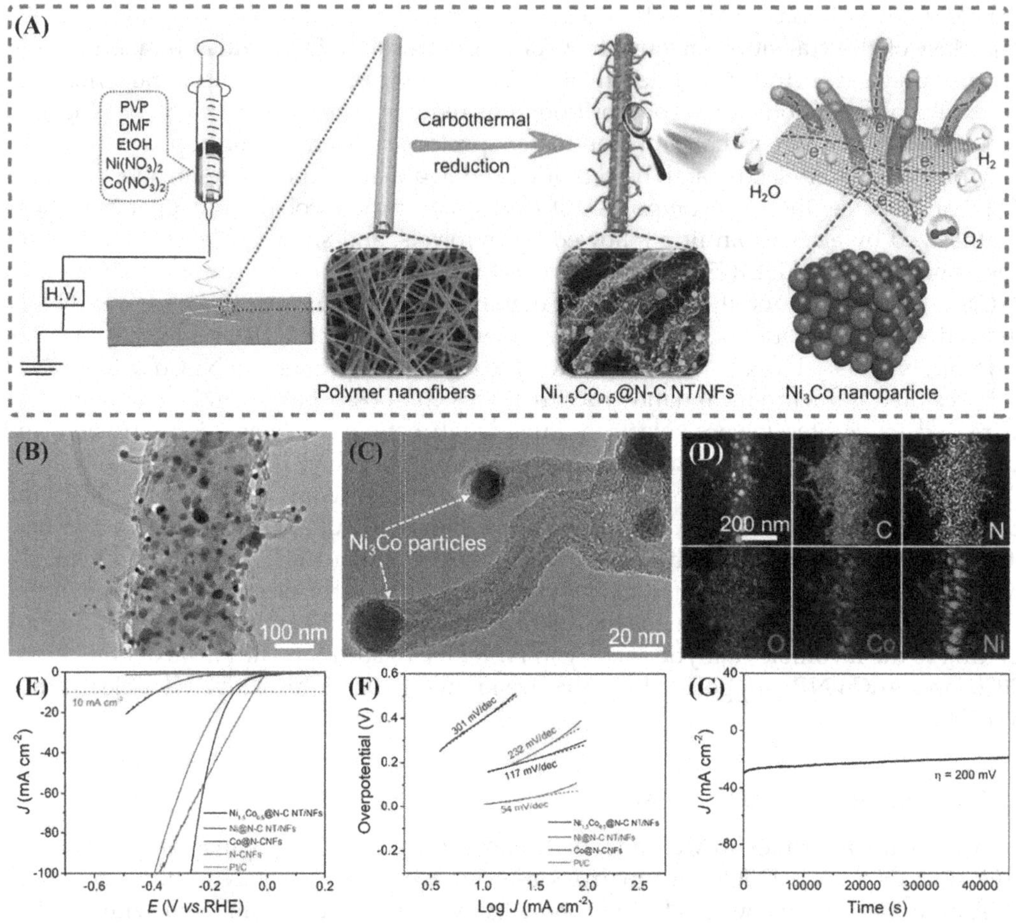

FIGURE 13.10
A) The production steps of $Ni_{1.5}Co_{0.5}$/NCNT/NFs; B, C) TEM images at 100 and 20× magnifications; D) HADDF-STEM and elemental mapping images; E-G) HER measurements in 0.1 M KOH; E) LSV polarization curves; F) Tafel plots and G) time-dependent current density curve at 200 mV. Adapted with permission from reference [32]. Copyright (2022) Elsevier.

a lesser Tafel slope (60 mV/dec), a lower potential (280 mV at 10 mV/cm^2), a reduced load resistance, and increased HER performance compared to the WS_2 and WS_2/ NCNF species (Figure 13.11E–G).

Transfer metal selenides have stronger metal properties than metal sulfides due to their exclusive electronic properties. Therefore, electron transfer and oxidation reactions are easily performed by these metals [32]. Liu et al. [38] reported a useful and stable electrocatalyst based on porous carbon fibers hybridized with $MoSe_2$ plates (i.e. $MoSe_2$/PCFs) for HER. In this combination, PCFs are obtained from polystyrene fibers with a coating of polydopamine (PDA) [38]. As a result, the connection of PDA/PS-derived PCFs with $MoSe_2$ has resulted in a synergy with the carbon substrate. The results also show that transition metal carbides have excellent conductivity and long-term constancy in the extensive range

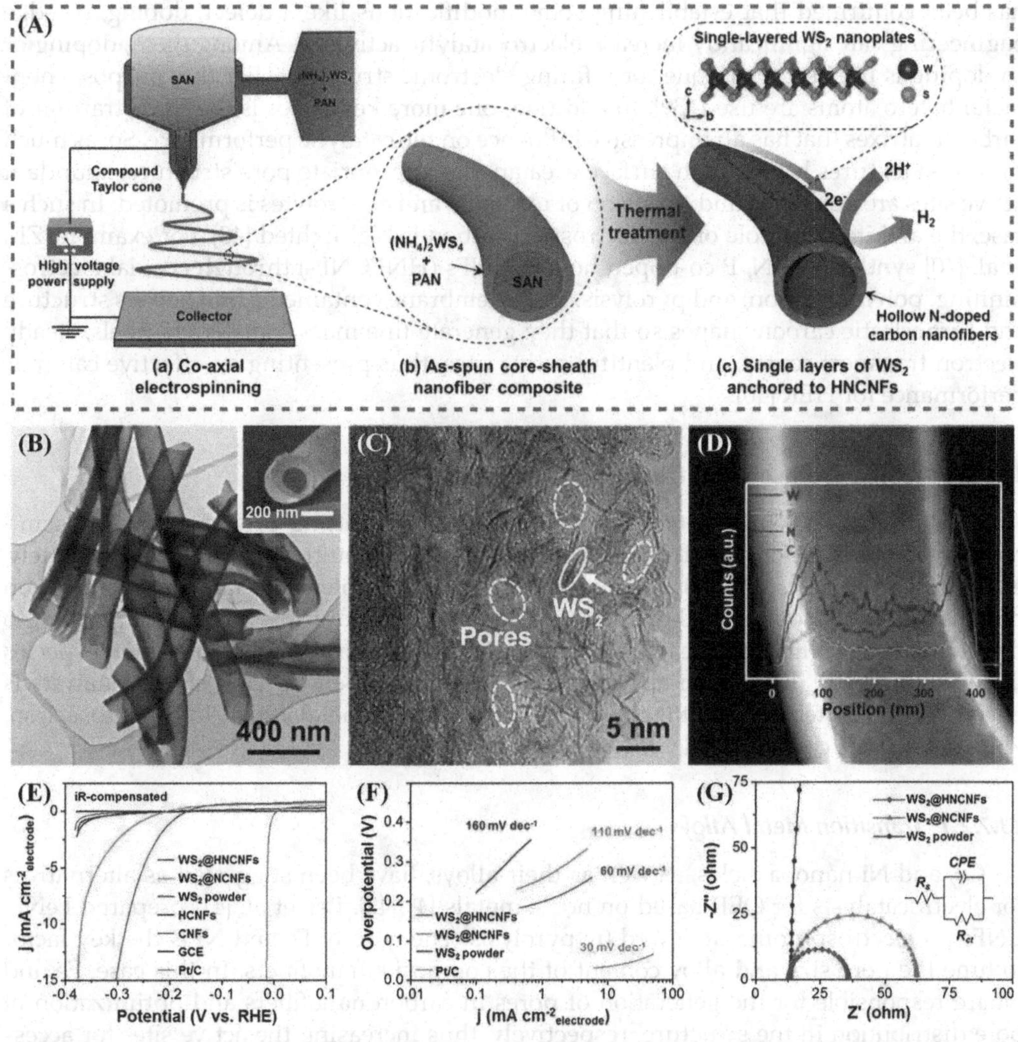

FIGURE 13.11
A) Schematic illustration of the synthetic route of WS2/HNCNFs; B) TEM; C) HRTEM, and D) STEM images of WS2/HNCNFs, insets: the corresponding SEM image and EDS line-scan profile; E-G) Comparison of HER performance of various electrocatalysts in 0.5 M H2SO4: E) LSV curves, F) Tafel plots, and G) electrochemical impedance spectroscopy (EIS) curves. Adapted with permission from reference [32]. Copyright (2022) Elsevier.

of pH. So they have Pt-like electronic behavior, which makes them a good alternative to Pt-based electrocatalysts [38].

13.7.1.4 Metal-Free Carbons

Among the promising substances that have been extensively used as non-metal electrocatalysts in HER are carbon species because of their favorable properties, such as chemical stability, low cost, mechanical robustness conductivity, and tailorable architectures, although some carbon species, like pristine carbon, usually show an inert nature in the adsorption of hydrogen, thus resulting in poor catalytic activity toward HER [32]. It has been confirmed that establishing some modifications like a defect, doping, or edge engineering can significantly increase electrocatalytic activities. Among them, doping or co-doping is the best technique for refining electronic structures. For this purpose, non-metal hetero atoms are used [39]. In addition, one more key factor is the mass transfer of carbon matrixes that has an impressive influence on the catalytic performance. So, as much as carbon textures have a high surface area and an appropriate pore structure, abundant active sites are accessible and diffusion of reactants and electrolytes is promoted. In such a case, the architectural role of 1D electrospun carbon is highlighted [40]. For example, Zhi et al. [40] synthesized N, P co-doped hollow CNFs (HNPCNFs) through coaxial electrospinning, polymerization, and pyrolysis. This membrane contained a fine porous structure and turbostratic carbon shapes so that they generate fine mass transfer channels, steady electron transport routes, and plentiful active sites, thus presenting an effective catalytic performance for HER [40].

13.7.2 Nanofiber-Based Electrocatalysts for OER

Another important semi-reaction is OER in the water-splitting process [40]. This semi-reaction includes a four-electron transfer process. The multi-step OER, unfortunately, exhibits slow kinetics and usually needs a greater overpotential than the two-electron process in HER to cope with the kinetic barrier [32]. Therefore, it is far more important to prepare effective OER electrocatalysts, although, for this purpose, the noble metal species like RuO_2 and IrO_2 have been applied. The use of nanofibers as OER electrocatalysts is effective and is often applied together with the transition metal alloys. In this subsection, we discuss some conventional OER catalysts.

13.7.2.1 Transition Metal Alloys

Fe, Co, and Ni nanoparticles, as well as their alloys, have been suggested as alternatives for electrocatalysts for OER based on noble metals [41, 42]. Lee et al. [41] prepared FeNi/CNFs *via* electrospinning, followed by pyrolysis. The ratio of Fe and Ni is the key factor to tune the pore size and alloy content of the composite nanofibers. In this case, Fe and Ni are responsible for the generation of pores in carbon nanofibers and optimization of pore distribution in the structure, respectively, thus increasing the active sites for access to electrolyte and reactants. The physicochemical analyses revealed that the ratio 50 wt% FeNi/CNFs expressed excellent catalytic operation and improved long-time stability of the electrocatalysts for OER, ascribed to the homogenous distribution effect of Fe and Ni in the carbon nanofibers [41]. On other hand, alloys exhibited splendid electrocatalytic performance toward OER. For instance, FeCoNi/CNFs show a far lower overpotential, better durability, and lower Tafel slope than FeCo/CNFs systems and IrO_2.

13.7.2.2 Transition Metal Oxides

Nowadays, TMOs are another alternative for OER electrocatalysts based on non-precious metals in alkaline media. Perovskite-type oxides (ABO_3) are good candidates in this respect, due to their cost effectiveness and high stability [43]. Xiong et al. [43], through adjusting the amount of metal salt mass in the solution for electrospinning, could synthesize various $La_{0.8}Sr_{0.2}Co_{0.2}Fe_{0.8}O_{3-\delta}$ (LSCF) catalysts with versatile shapes like a particle, rod, or fiber. They found that, with increasing salt concentration from 10 wt% to 20 wt%, to 30 wt%, the morphology of LSCF was changed from porous nanofibers to nanorods to nanoparticles, respectively [43]. Nanofiber-shaped LSCF has promising features like high specific surface area, and high electron-mass transfer, thus resulting in better performance of OER catalytic activity than nanorods and nanoparticles, the difference possibly being related to the superlative synergy of Fe with Co. Also, spinel oxides, especially MFe_2O_4 (M = Co, Ni, Cu, and Mn, etc.) ferrites, due to their attractive advantages, such as environmental friendliness, low cost, ease of production, favorable stability, and rich redox chemistry, like perovskite oxides, have broadly been used as OER electrocatalysts [44].

13.7.2.3 Other Transition Metal Composites

Among the transition metal composites with activity in the field of HER and water oxidation, metal carbides and phosphides are attracting considerable attention [45]. Transition metal phosphides are better qualified than metal oxides due to their higher electronic conductivity and faster charge transport [45]. In addition, bimetallic phosphides commonly outperform the metal species of the respective unit, due to the design of electronic structures and precise adjustment of the kinetic barriers of the intermediate [46]. For instance, FeNiP NPs are fixed in NCNF where FeNiP crystals are held tightly by multilayer graphene (Figure 13.12A–C). As shown in Figure 13.12D–E, these NFs showed high electrocatalytic activity for OER with a low Tafel slope (47 mV/dec) and very low potential (300 mV at 10 mA/cm²). Like $Ni_{1.5}Co_{0.5}$/NCNT/NFs, Fe_3C nanoparticles are embedded in the NCNT/NFs structure through electrospinning and carbonation in the presence of melamine. These NFs showed excellent electrocatalytic performance in OER due to their powerful synergistic effects, fast electron transfer, and the presence of sufficient active centers [32].

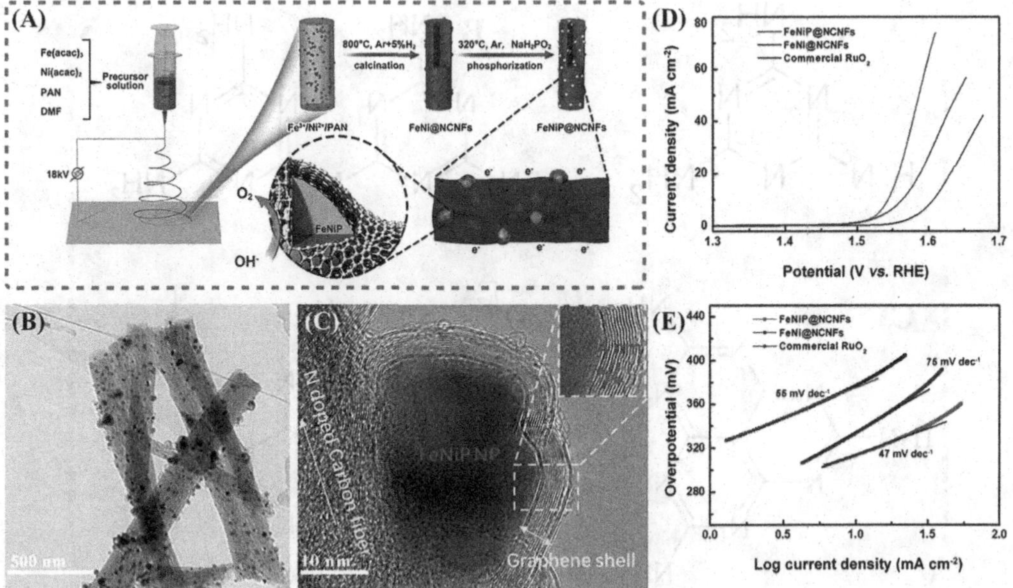

FIGURE 13.12
A) Image of the production route of FeNiP/NCNFs; B) TEM and C) HRTEM images of FeNiP/NCNFs; D) Polarization plots of various OER catalysts; E) Tafel plots. Adapted with permission from reference [32]. Copyright (2022) Elsevier.

13.8 Polymeric Carbon Nitride (PCN)

PCNs are known as non-metallic catalysts, which have attracted the attention of many scientists in the field of photocatalysis due to their high abundance on the Earth, very high corrosion resistance, excellent electronic structure, and good physical and chemical properties [47]. With the discovery of melamine, malam, melem, and melon by Libig and Berzlius in 1834, PCNs were introduced for the first time (Figure 13.13A–D).

Notably, the history of using PCNs in water splitting dates back to 2009, when Wang and his colleagues first used it for HER [48]. The photocatalytic activity of 1D materials can be enhanced by adjusting their size and shape. Scientists use PCN nanotubes to improve light efficiency, as the interior of these nanomaterials leads to multiple scatter or reflection of light. Li et al. have produced PCN nanotubes *via* the hydrothermal-thermal compaction process of the urea-melamine precursors [49]. The synthetic PCN nanotube with vacant spaces improves the adsorption of electrons produced by light. Also, the tubular structure leads to shortening of the propagation distance of the carriers produced from the bulk phase to the surface and finally separating the carriers. In further research, Pi et al. [50] synthesized 1D PCN nanorods using steps as follows: i) Dispersing bulk melamine in water; ii) Electrospinning of melamine-water mixture; and iii) Calcination of the melamine-water NFs [50]. These 1D nanorods increase the light and absorption capacity. On the other hand, oxygen atoms doped on these 1D nanorods separate photogenic carriers, thus improving the photocatalytic capacity. Furthermore, Zhao et al. [51] and Liu et al. [52] have prepared 1D PCN nanoribbons through microwave methods and hydrothermal methods (from seaweed), respectively. Figure 13.14A–B shows a uniform, seaweed-like structure. The TEM

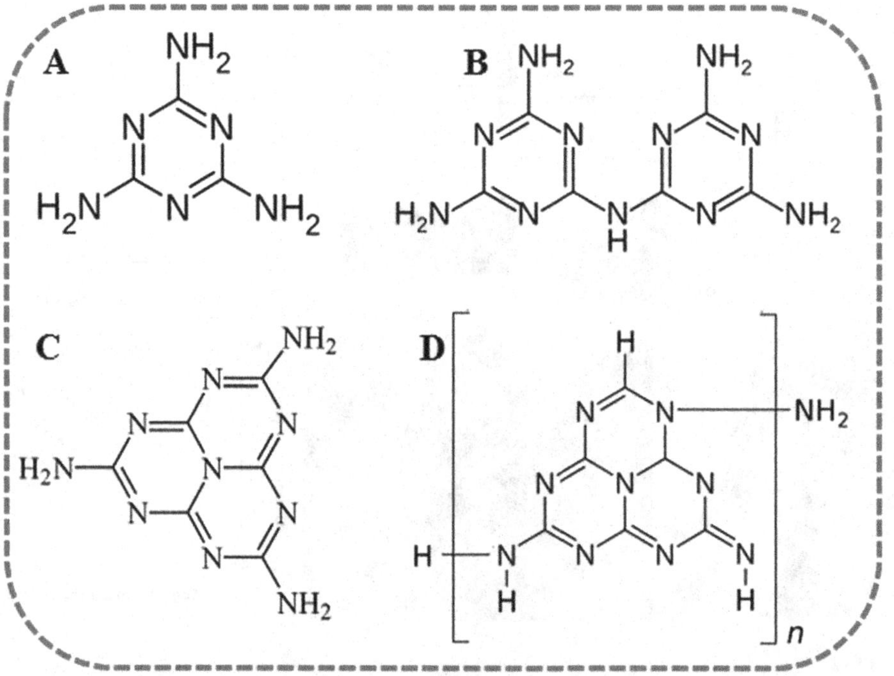

FIGURE 13.13
Molecular structures of the A) melamine, B) melam, C) melem, and D) melon.

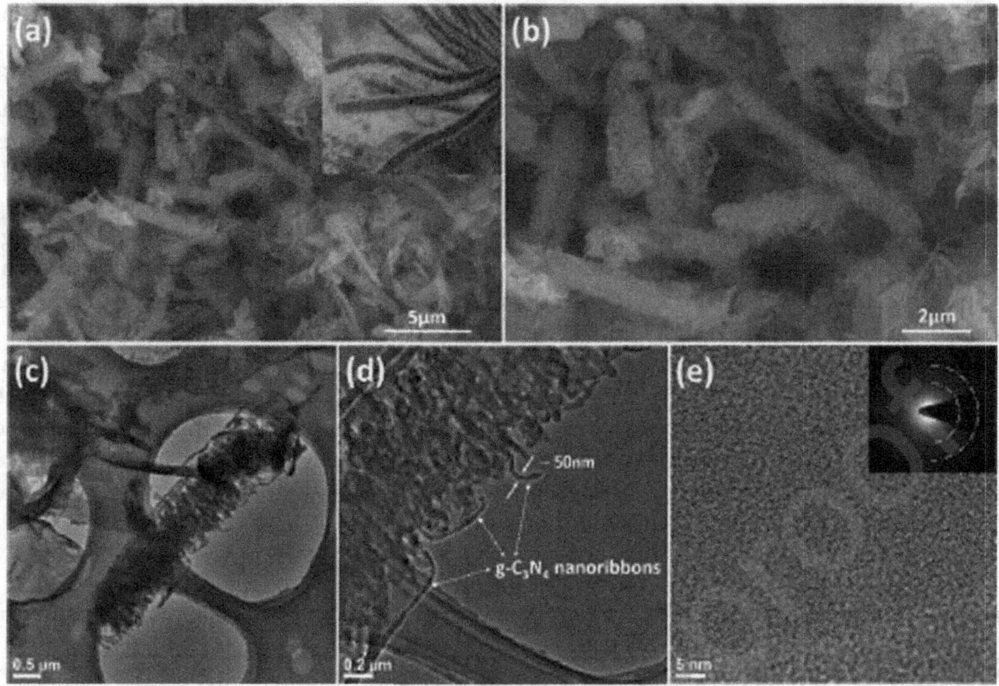

FIGURE 13.14
A and B) SEM images (inset is a seaweed); C- E) TEM images of PCN nanoribbons with multiple magnifications. Adapted with permission from reference [52]. Copyright (2020) Elsevier.

results indicate that the seaweed-like structure prepared is assembled by PCN nanoribbons with a diameter of ~50 nanometers (Figure 13.14C–D). In Figure 13.14E, the high-resolution TEM image shows that the PCN nanoribbons possesses two weak diffraction rings related to the (002) and (100) planes of PCN.

13.9 Conclusion

Water splitting is one of the best methods to create H_2 from water and thus produce renewable fuels. Therefore, water splitting has attracted much attention as a source of H_2 production. Among the substances that are widely used in water-splitting, electrolyzers of various membrane materials, such as Nafions and AEMs, are very popular. Despite their high thermal and chemical stability, excellent mechanical strength, and high proton conductivity, Nafions reduce efficient ionic conductivity by the physical separation between anolyte and catholyte, thereby limiting membrane performance. Recently, impressive studies have been performed on the design and fabrication of 1D polymer nanomaterials as an active and resistant electrocatalyst in the process of water-splitting. Due to their inherent properties, these materials can solve many problems related to other counterparts. As a result, they have been considered promising candidates in the water-splitting process. 1D materials include CQDs, for example, mineral semiconductor NPs, CDs, GQDs, and PDs. PDs are widely used in water splitting due to properties such as advanced semiconductivity, increased processing capability, solvent solubility, water solubility, easy surface modification, good optical properties, and eco-friendly nature.

In addition, carbon allotropes, including carbon nanotubes, graphene, fullerenes, CNF, diamond carbon, and CNT, due to their superior mechanical and chemical properties compared with other carbon allotropes, have excellent applications such as hydrogen storage, nanostructures, selective adsorption, structural reinforcement, and electrochemical catalysis. CNFs could rapidly carry out mass/electron transport in electrodes due to the unique overlapping identity. In addition, CNFs as support can avoid the agglomeration and coagulation of active agents, facilitate their usage, and then promote the stability and activity of the electrocatalysts. Therefore, CNFs electrocatalysts with optimized electronic properties synergistically regulate the ΔGH^{*}, thus accelerating the catalytic reaction kinetics.

Although, changes in the length, diameter, and aspect ratio of 1D polymer nanomaterials can achieve very high performance, so far the uniformity of existing materials is still not good enough for sustainable applications. Furthermore, crystallinity and surface defects of these materials are critical properties which affect charge transfer and surface reaction kinetics. Accordingly, considerable research efforts are required to control the uniform surface properties of these nanomaterials.

References

1. Lindquist GA, Xu Q, Oener SZ, Boettcher SW, Membrane electrolyzers for impure-water splitting, *Joule*, 4(12), 2020, 2549–2561.
2. Vincent I, Bessarabov D, Low cost hydrogen production by anion exchange membrane electrolysis: A review, *Renewable and Sustainable Energy Reviews*, 81, 2018, 1690–1704.
3. Thomas C, Kuhn Jr I, James B, Lomax Jr F, Baum G, Affordable hydrogen supply pathways for fuel cell vehicles, *International Journal of Hydrogen Energy*, 23(6), 1998, 507–516.
4. Cao Y-C, Wu X, Scott K, A quaternary ammonium grafted poly vinyl benzyl chloride membrane for alkaline anion exchange membrane water electrolysers with no-noble-metal catalysts, *International Journal of Hydrogen Energy*, 37(12), 2012, 9524–9528.
5. López-Fernández E, Gil-Rostra J, Espinós JP, González-Elipe AR, Yubero F, de Lucas-Consuegra A, $Cu_xCO_{3-x}O_4$ ultra-thin film as efficient anodic catalysts for anion exchange membrane water electrolysers, *Journal of Power Sources*, 415, 2019, 136–144.
6. Xiao L, Wu S-Y, Li Y-R, Advances in solar hydrogen production via two-step water-splitting thermochemical cycles based on metal redox reactions, *Renewable Energy*, 41, 2012, 1–12.
7. Daulbayev C, Sultanov F, Bakbolat B, Daulbayev O, 0D, 1D and 2D nanomaterials for visible photoelectrochemical water splitting: A review, *International Journal of Hydrogen Energy*, 45(58), 2020, 33325–33342.
8. Lin D, Liu Y, Cui Y, Reviving the lithium metal anode for high-energy batteries, *Nature Nanotechnology*, 12(3), 2017, 194–206.
9. Reneker DH, Yarin AL, Electrospinning jets and polymer nanofibers, *Polymer*, 49(10), 2008, 2387–2425.
10. Sharafinia S, Farrokhnia A, Lemraski EG, Optimized safranin adsorption onto poly (vinylidene fluoride)-based nanofiber via response surface methodology, *Materials Chemistry and Physics*, 276, 2021, 125407.
11. Alexandris S, Papadopoulos P, Sakellariou G, Steinhart M, Butt H-Jr, Floudas G, Interfacial energy and glass temperature of polymers confined to nanoporous alumina, *Macromolecules*, 49(19), 2016, 7400–7414.
12. Debiemme-Chouvy C, Template-free one-step electrochemical formation of polypyrrole nanowire array, *Electrochemistry Communications*, 11(2), 2009, 298–301.

13. Morber JR, Wang X, Liu J, Snyder RL, Wang ZL, Wafer-level patterned and aligned polymer nanowire/micro-and nanotube arrays on any substrate, *Advanced Materials*, 21(20), 2009, 2072–2076.
14. Peng J, Dong W, Wang Z, Meng Y, Liu W, Song P, Liu Z, Recent advances in 2D transition metal compounds for electrocatalytic full water splitting in neutral media, *Materials Today Advances*, 8, 2020, 100081.
15. Schalenbach M, Zeradjanin AR, Kasian O, Cherevko S, Mayrhofer KJ, A perspective on low-temperature water electrolysis–challenges in alkaline and acidic technology, International, *Journal of Electrochemical Science*, 13(2), 2018, 1173–1226.
16. Li X, Hao X, Abudula A, Guan G, Nanostructured catalysts for electrochemical water splitting: Current state and prospects, *Journal of Materials Chemistry A*, 4(31), 2016, 11973–12000.
17. Danilovic N, Subbaraman R, Strmcnik D, Chang KC, Paulikas A, Stamenkovic V, Markovic NM, Enhancing the alkaline hydrogen evolution reaction activity through the bifunctionality of Ni $(OH)_2$/metal catalysts, *Angewandte Chemie*, 124(50), 2012, 12663–12666.
18. Vilekar SA, Fishtik I, Datta R, Kinetics of the hydrogen electrode reaction, *Journal of the Electrochemical Society*, 157(7), 2010, B1040.
19. Huang Q, Ye Z, Xiao X, Recent progress in photocathodes for hydrogen evolution, *Journal of Materials Chemistry A*, 3(31), 2015, 15824–15837.
20. Wei J, Zhou M, Long Y, Xue Y, Liao H, Wei C, Xu, ZJ, Heterostructured electrocatalysts for hydrogen evolution reaction under alkaline conditions, *Nano-Micro Letters*, 40(10), 2018, 1–15.
21. Jin H, Guo C, Liu X, Liu J, Vasileff A, Jiao Y, Zheng Y, Qiao S, Emerging two-dimensional nanomaterials for electrocatalysis, *Chemical Reviews*, 118(13), 2018, 6337–6408.
22. Trotochaud L, Young SL, Ranney JK, Boettcher SW, Nickel–iron oxyhydroxide oxygen-evolution electrocatalysts: The role of intentional and incidental iron incorporation, *Journal of the American Chemical Society*, 136(18), 2014, 6744–6753.
23. Siracusano S, Baglio V, Nicotera I, Mazzapioda L, Aricò A, Panero S, Navarra M, Sulfated titania as additive in nafion membranes for water electrolysis applications, *International Journal of Hydrogen Energy*, 42(46), 2017, 27851–27858.
24. Chabi S, Wright AG, Holdcroft S, Freund MS, Transparent bipolar membrane for water splitting applications, *ACS Applied Materials and Interfaces*, 9(32), 2017, 26749–26755.
25. Kausar A, Polymer dots and derived hybrid nanomaterials: A review, *Journal of Plastic Film and Sheeting*, 37(4), 2021, 510–528.
26. Wang C, Xu Z, Cheng H, Lin H, Humphrey MG, Zhang C, A hydrothermal route to water-stable luminescent carbon dots as nanosensors for pH and temperature, *Carbon*, 82, 2015, 87–95.
27. Lv J-J, Zhao J, Fang H, Jiang L-P, Li L-L, Ma J, Zhu J-J, Incorporating nitrogen-doped graphene quantum dots and Ni3S2 nanosheets: A synergistic electrocatalyst with highly enhanced activity for overall water splitting, *Small*, 1, 2017, 1700264.
28. Tian J, Chen J, Liu J, Tian Q, Chen P, Graphene quantum dot engineered nickel-cobalt phosphide as highly efficient bifunctional catalyst for overall water splitting, *Nano Energy*, 48, 2018, 284.
29. Yadav D, Amini F, Ehrmann A, Recent advances in carbon nanofibers and their applications–A review, *European Polymer Journal*, 138, 2020, 109963.
30. Lu H, Fan W, Huang Y, Liu T, Lotus root-like porous carbon nanofiber anchored with CoP nanoparticles as all-pH hydrogen evolution electrocatalysts, *Nano Research*, 11(3), 2018, 1274–1284.
31. Li M, Wang H, Zhu W, Li W, Wang C, Lu X, RuNi nanoparticles embedded in N-doped carbon nanofibers as a robust bifunctional catalyst for efficient overall water splitting, *Advanced Science*, 7(2), 2020, 1901833.
32. Zhang Z, Wu X, Kou Z, Song N, Nie G, Wang C, Verpoort F, Mu S, Rational design of electrospun nanofiber-typed electrocatalysts for water splitting: A review, *Chemical Engineering Journal*, 428, 2021, 131133.
33. Wei Y, Soomro RA, Xie X, Xu B, Design of efficient electrocatalysts for hydrogen evolution reaction based on 2D MXenes, *Journal of Energy Chemistry*, 55, 2021, 244–255.

34. Yang T, Du M, Zhu H, Zhang M, Zou M, Immobilization of Pt nanoparticles in carbon nanofibers: Bifunctional catalyst for hydrogen evolution and electrochemical sensor, *Electrochimica Acta*, 167, 2015, 48–54.
35. Yang T, Du M, Zhu H, Zhang M, Zou M, Immobilization of Ni_3Co nanoparticles into N-doped carbon nanotube/nanofiber integrated hierarchically branched architectures toward efficient overall water splitting, *Advanced Science*, 7(1), 2020, 1902371.
36. Wang J, Zhu H, Chen J, Zhang B, Zhang M, Wang L, Du M, Small and well-dispersed Cu nanoparticles on carbon nanofibers: Self-supported electrode materials for efficient hydrogen evolution reaction, *International Journal of Hydrogen Energy*, 41(40), 2016, 18044–18049.
37. Patil B, Satılmış B, Khalily MA, Uyar T, Atomic layer deposition of NiOOH/Ni $(OH)_2$ on PIM-1-based N-doped carbon nanofibers for electrochemical water splitting in alkaline medium, *ChemSusChem*, 12(7), 2019, 1469–1477.
38. Yan J, Zhang Y, Huang Y, Miao YE, Liu T, $MoSe_2$ nanosheets grown on polydopamine-derived porous fibers: A high-performance catalyst for hydrogen evolution reaction, *Advanced Materials Interfaces*, 4(6), 2017, 1600825.
39. Wang X, Vasileff A, Jiao Y, Zheng Y, Qiao SZ, Electronic and structural engineering of carbon-based metal-free electrocatalysts for water splitting, *Advanced Materials*, 31(13), 2019, 1803625.
40. Gao Y, Xiao Z, Kong D, Iqbal R, Yang Q-H, Zhi L, N,P Co-doped hollow carbon nanofiber membranes with superior mass transfer property for trifunctional metal-free electrocatalysis, *Nano Energy*, 64, 2019, 103879.
41. An X, Shin D, Jeong J, Lee J, Metal-derived mesoporous structure of a carbon nanofiber electrocatalyst for improved oxygen evolution reaction in alkaline water electrolysis, *ChemElectroChem*, 3(10), 2016, 1720–1724.
42. An X, Shin D, Ocon JD, Lee JK, Son Y-I, Lee J, Electrocatalytic oxygen evolution reaction at a FeNi composite on a carbon nanofiber matrix in alkaline media, *Chinese Journal of Catalysis*, 35(6), 2014, 891–895.
43. Wang Z, Li M, Liang C, Fan L, Han J, Xiong Y, Effect of morphology on the oxygen evolution reaction for $La_{0.8}Sr_{0.2}Co_{0.2}Fe_{0.8}O_{3-\delta}$ electrochemical catalyst in alkaline media, *RSC Advances*, 6(73), 2016, 69251–69256.
44. Wu L, Shi L, Zhou S, Zhao J, Miao X, Guo J, Direct growth of $CoFe_2$ alloy strongly coupling and oxygen-vacancy-rich $CoFe_2O_4$ porous hollow nanofibers: An efficient electrocatalyst for oxygen evolution reaction, *Energy Technology*, 6(12), 2018, 2350–2357.
45. Qian M, Cui S, Jiang D, Zhang L, Du P, Highly efficient and stable water-oxidation electrocatalysis with a very low overpotential using FeNiP substitutional-solid-solution nanoplate arrays, *Advanced Materials*, 29(46), 2017, 1704075.
46. Li Y, Dong Z, Jiao L, Multifunctional transition metal-based phosphides in energy-related electrocatalysis, *Advanced Energy Materials*, 10(11), 2020, 1902104.
47. Zhang D, He W, Ye J, Gao X, Wang D, Song J, Polymeric carbon nitride-derived photocatalysts for water splitting and nitrogen fixation, *Small*, 17(13), 2021, 2005149.
48. Wang X, Maeda K, Thomas A, Takanabe K, Xin G, Carlsson JM, Domen K, Antonietti M, A metal-free polymeric photocatalyst for hydrogen production from water under visible light, *Nature Materials*, 8(1), 2009, 76–80.
49. Li Y, Gu M, Shi T, Cui W, Zhang X, Dong F, Cheng J, Fan J, Lv K, Carbon vacancy in C_3N_4 nanotube: Electronic structure, photocatalysis mechanism and highly enhanced activity, *Applied Catalysis B: Environmental*, 262, 2020, 118281.
50. Zeng Y, Liu X, Liu C, Wang L, Xia Y, Zhang S, Luo S, Pei Y, Scalable one-step production of porous oxygen-doped g- C_3N_4 nanorods with effective electron separation for excellent visible-light photocatalytic activity, *Applied Catalysis B: Environmental*, 224, 2018, 1–9.
51. Zhao Y, Zhao F, Wang X, Xu C, Zhang Z, Shi G, Qu L, Graphitic carbon nitride nanoribbons: Graphene-assisted formation and synergic function for highly efficient hydrogen evolution, *Angewandte Chemie International Edition*, 53(50), 2014, 13934–13939.
52. Liu Y, Guo X, Chen Z, Zhang W, Wang Y, Zheng Y, Tang X, Zhang M, Peng Z, Li R, Huang Y, Microwave-synthesis of g- C_3N_4 nanoribbons assembled seaweed-like architecture with enhanced photocatalytic property, *Applied Catalysis B: Environmental*, 266, 2020, 118624.

14

One-Dimensional Polymeric Nanocomposites for Fuel Cells

Ozge Surucu and Filiz Kuralay

CONTENTS

14.1 Introduction

There is an urgent need for research into the development of clean and renewable energy sources for a rapidly growing economy and for sustainable life quality. Among these advances, fuel cells are taking the lead as good alternatives to commonly used petroleum energy technologies, with many advantages such as environmental friendliness and high energy efficiency [1–3]. Fuel cells are one of the fascinating applications of electrochemical systems which can also serve as effective practices such as batteries and capacitors [4, 5]. In these systems, chemical energy is directly converted to electrical energy *via* electrode reactions. Fuel cells consist of an anode, a cathode, and an electrolyte, similar to other types

DOI: 10.1201/9781003223764-14

of electrochemical systems. In these systems, positively charged hydrogen ions (H^+) and electrons are most commonly generated during a reaction of the fuel in the presence of a catalyst. The ions formed move from the anode part to the cathode part *via* the electrolyte with the flow of electrons moving in the same direction through an external circuit. At this time, direct current electricity is produced. As a result of the electrochemical reaction and in the presence of another catalyst, water and possibly other products [such as nitrogen dioxide (NO_2)] are formed. Heat is also produced as a result of the electrochemical reaction that takes place [6, 7].

Fuel cells can be divided into different categories based on their operating conditions, such as moisture content, pressure, temperature, and the structure of fuel cells [8, 9]. The main classification depends on the electrolyte. For example, proton-exchange membrane fuel cells (PEMFCs) have multiple functions including inhibition of the fuel flow between the cathode and anode, blockade of the electrons, and transfer of the protons [10–12]. They are also known as polymer electrolyte membrane fuel cells or solid polymer fuel cells. They are known to be environmentally friendly because of their zero emissions of greenhouse gases. In this type of fuel cell, the composition of membranes are the major route to improving the performance of the fuel cell. Polymer-based materials have been commonly used for this aim. Alkaline fuel cells (AFCs) use alkaline media, such as potassium hydroxide (KOH) [13]. Direct methanol fuel cells (DMFCs) are portable electronic devices characterized by their compact design, high energy density, and start-up at room temperature [14]. Phosphoric acid fuel cells (PAFCs) in which the primary composition is phosphoric acid (H_3PO_4), have been one of the most commonly used and commercialized fuel cell types [15]. Molten carbonate fuel cells (MCFCs) depend on the separation of carbon dioxide (CO_2) through electricity generation. These fuel cells use lithium carbonate (Li_2CO_3) and potassium carbonate (K_2CO_3) as electrolytes and they are operated at higher temperatures (around 650 °C) than the other fuel cell types [16]. In solid oxide fuel cells (SOFCs), the electrolyte is a solid ceramic inorganic oxide [17]. This type of fuel cell is also applied at high temperatures in the range 750–1000 °C. Of all these fuel cells, PEMFCs and SOFCs have been the most convenient ones for successful commercialization.

Fuel cells can also be classified according to the nature of the polymer used in the fuel cells. Therefore, different polymer nanocomposites are designed for a large family of fuel cells. A polymer nanocomposite is a combination of polymer and nanomaterials having at least one nanometer dimension (0–100 nm) [18, 19]. The presence of a nanomaterial in a polymer composite enhances the performance of the resulting material in terms of mechanical and thermal properties, degradability, and its ability to act as a gas barrier or flame retardant. As well as polymeric structures, such as their nanofiber and nanotube forms used in the fuel cell technology known as one-dimensional materials, nanomaterial classification can also be made according to their dimensions, namely zero-dimensional, one-dimensional, and two-dimensional. All these materials have unique properties compared to their three-dimensional (bulk) forms [20]. Therefore, their combination with polymers and integration of these nanocomposites into the one-dimensional polymeric nanocomposite-based fuel cells offer remarkable advantages and superiorities. In the mention of one-dimensional nanomaterials, the thickness and width are in the nanometer (nm) range while the length is in micrometer (μm) scale or greater. Nanotubes, nanorods, nanowires, and nanofibers are the forms chiefly used as one-dimensional nanomaterials. The most important examples of one-dimensional nanomaterials inclu single-walled carbon nanotubes (SWCNTs), multi-walled carbon nanotubes (MWCNTs), polyaniline nanofibers (PANI), cellulose nanofibers, carbon nanofibers (CNFs), nanorods, and metal nanotubes [21–23].

This chapter summarizes the one-dimensional polymeric nanocomposites used in the field of electricity production, with the help of fuel cell technology. Various nanocomposites are systematically under debate based on the different polymeric structures and the one-dimensional nanomaterials combined with the polymers. Recent contributions to the energy field are given and the outputs of these developments are discussed by mentioning the properties such as efficiency, reproducibility, robustness, stability, cost-effectiveness, as well as describing the modification-preparation steps.

14.2 Polymeric Nanocomposites

The development of polymeric nanocomposites is critical for the advancement of fuel cells. The reliability and large-scale commercialization of fuel cells depend on high proton conductivity, thermal, mechanical, and chemical stability, as well as low cost and resistance to deformation [24]. To increase the conductivity and mechanical strength of fuel cells, polymeric nanocomposites are coming into prominence, and one-dimensional polymeric nanocomposites are focusing most on the CNTs combined with various polymeric structures, such as perfluorosulfonic acid (PFSA) ionomers [25], functionalized polyaromatic systems [26], conducting polymers [27], synthetic polymers [28], and biopolymers [29]. Detailed research on polymeric nanocomposites has shown that the CNTs are promising and advanced nanomaterials for modification. Therefore, in terms of one-dimensional polymeric nanocomposites, studies on polymer-CNTs will be outlined in this part of the chapter.

14.2.1 PFSA-Based Polymeric Nanocomposites

14.2.1.1 Nafion-CNT Nanocomposites

Membranes are very important targets of PEMFCs to achieve improvement of the performance of the cell. Therefore, one of the main tasks in fuel cell technology is to enforce the mechanical, chemical, and thermal properties of the polymer membrane materials that conduct protons through the membrane. Dependency on water content is also considered to be an indicator of PEM performance. In this context, Nafion, which is a kind of PFSA-based polymer, has been widely used in fuel cell applications. Its potential in the fuel cell field originated from its high mechanical strength, its high chemical and thermal stability, and its proton conductivity. The weak aspects of Nafion membranes, such as high fuel permeability and high production cost, can be overcome using additive materials, especially with the addition of CNTs. Nafion-based fuel cell membrane technology has been one of the pioneering research topics of all time. In one of the studies, Liu and his co-workers fabricated a Nafion-functionalized MWCNT membrane for PEMFCs [25]. The ozone-mediated process, which relied on the radical addition reaction between the structures, was used to functionalize MWCNTs. The composite membrane prepared enhanced the mechanical strength (1% CNTs by weight) and proton conductivity (5-fold increment at 60 °C) in a significant way. The nanocomposite membrane yielded a current density of 1556 mA/cm^2 at 0.6 V and a maximum power density of 650 mW/cm^2. In the experiments, the average diameter of the MWCNTs were 10–50 nm with a length of 1–25 μm. A Nafion-CNT dispersion (1 kDA by weight) was prepared in isopropanol (C_3H_7OH) in an ultrasonic bath (1 h). Then, the dispersion was cast on a glass plate to dry at 100 °C for 1 day. After this step, the annealing step was carried out under vacuum for 4 h. The resulting membranes were finally protonated by boiling them in sulfuric acid (H_2SO_4). Characterizations of the nanocomposite-based membranes were realized with X-ray photoelectron spectroscopy (XPS), high-resolution transmission electron microscopy (HR-TEM), thermogravimetric analysis (TGA), scanning electron microscopy (SEM), and scanning transmission electron micrographs (STEM). In another study, Sigwadi et al. studied the addition of zirconium oxide (ZrO_2)-CNT filler combined with Nafion membrane in terms of crystallinity and stability [30]. The incorporation of ZrO_2-CNTs improved the mechanical properties and water retention of the Nafion membrane. According to this paper [30], one-dimensional polymer and nanotubes were combined with ZrO_2 in one-dimensional or three-dimensional structures. ZrO_2 nanoparticles were synthesized by the precipitation of zirconium hydroxide [$Zr(OH)_4$] by the addition of sodium hydroxide (NaOH) to zirconium oxychloride hydrate ($ZrOCl_2.8H_2O$) under continuous stirring. CNTs used in the study were prepared by aerosol-assisted chemical vapor deposition (CVD). The nanotubes obtained were dispersed in ZrO_2 nanoparticles at 80 °C for 2 h. After centrifugation and calcination at 600 °C, the Nafion membrane was treated with ZrO_2-CNT in methanol (CH_3OH). Characterization of the membranes formed in this way was performed with Fourier transform infrared (FTIR) spectroscopy, X-ray diffraction (XRD) analysis, atomic force microscopy (AFM), SEM, and TGA. Tensile mechanical testing to determine the improved mechanical properties was also carried out. This study [30] was then followed by the contact angle measurements to measure water retention of the prepared membranes. The proton conductivity of the nanocomposite membrane was found to be 0.113 S/cm. After these studies, CH_3OH permeation of the membranes was examined at various temperatures. There was no CH_3OH crossover at lower temperatures (30 °C and 60 °C), although it wash 1.02×10^{-6} cm^2/s at a temperature of 80 °C, using 5 M CH_3OH [30].

Recently, multilayer sulfonated carbon nanotubes (Su-CNTs) were incorporated with Nafion membranes by using layer-by-layer assembly (1, 10, 20, 35, 50, and 80 layers) for their possible use as PEMs [31]. The key point in the study has relied on the variation of layer thicknesses, ranging from 40 μm to 0.5 μm. The results have pointed out that the thin layers of 1 μm have boosted the nanotubes' alignment and proton-conducting pathways in the lateral direction by improving the proton conductance. In addition, lateral alignment has yielded advanced mechanical and electrochemical properties such as long-term performance stability. The preparation of the Su-CNTs was performed by chemical functionalization. The MWCNTs used in the experiments had outer diameters of 10–20 nm and an average length of 0.5–2.0 μm. The functionalized nanotubes were dispersed in tetrahydrofuran (THF) and then, the dispersion was combined with pure Nafion solution with sonication for 3 h. Layer-by-layer assembly of the Su-CNTs and Nafion nanocomposite membranes was achieved by depositing them on a flat-bottomed quartz dish. Figure 14.1 presents the layer-by-layer assembly preparation of the nanocomposite membranes (Figure 14.1A) and distribution of Su-CNTs in a Nafion matrix (Figure 14.1B). It was observed that the mass of the functionalized CNTs remained stable at high temperatures such as 300 °C. These stable, material-based Nafion membranes reached a proton conductivity value of 0.33 S/cm at 152 °C as a result of their high water-binding capacity (with the 80 layers of the assembly). The break stress of the thicker-layered membranes (50-layer and 80-layer) increased to 40 MPa in the lateral direction.

In addition to CNT incorporation into the Nafion membrane system, Vinothkannan and his colleagues have used Nafion nanocomposite-based system embedded with graphite

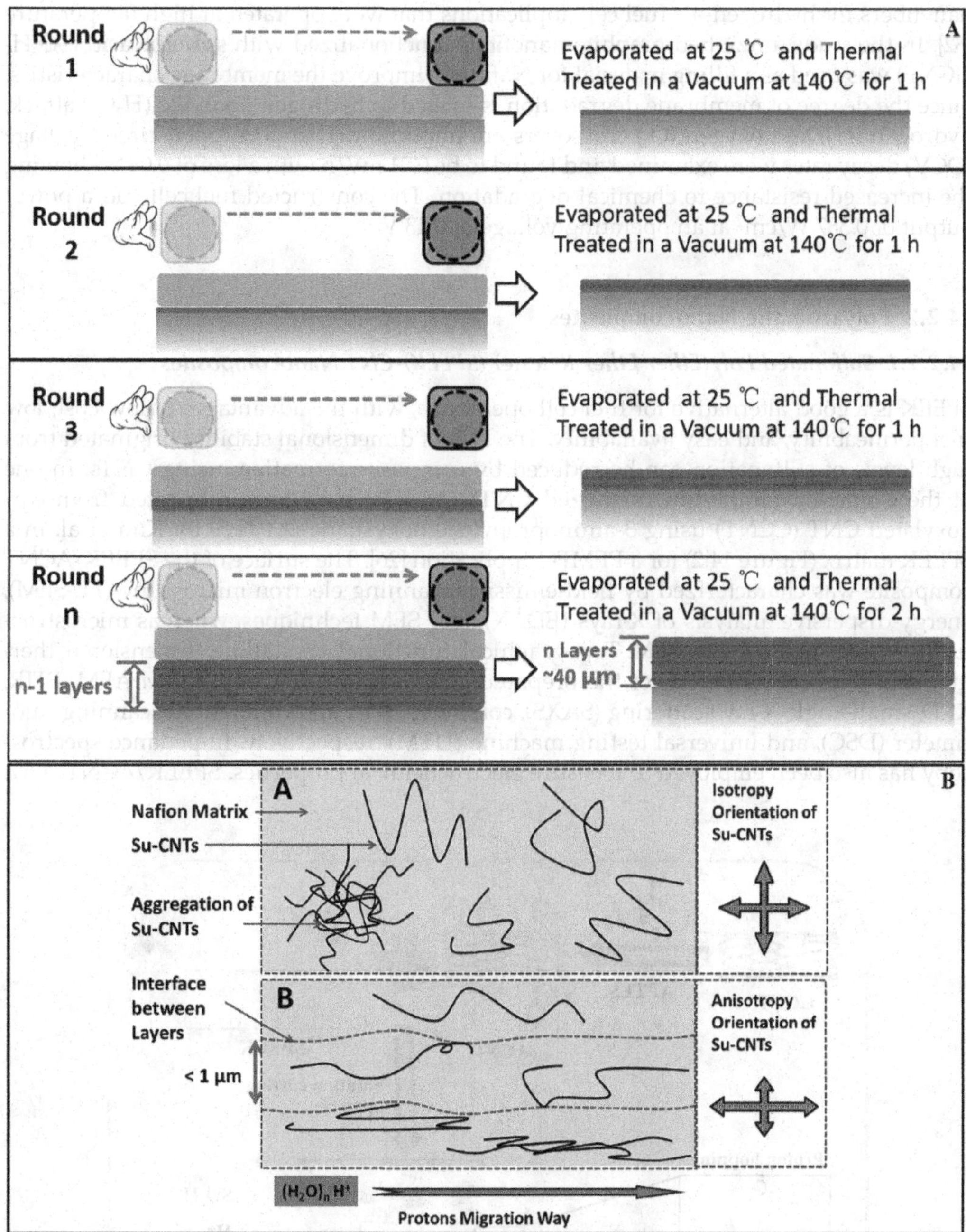

FIGURE 14.1
A) Schematic illustration of the preparation of multilayer Su-CNT/Nafion composite membranes by using a layer-by-layer assembly. B) Schematic illustration of distributions of Su-CNTs (A) in the Su-CNT/Nafion composite membranes, and (B) the multi-ultrathin-layer Su-CNT/Nafion composite membranes. Adapted with permission from reference [31]. Copyright (2019) Elsevier.

nanofibers for hydrogen-air fuel cell applications that were operated at high temperature [32]. In the study, unzipped graphite nanofiber functionalized with sulfonic acid (SO_3H-UGNF) was used as a filling material for Nafion to improve the membrane characteristics. Since the degree of membrane degradation is related to hydrogen peroxide (H_2O_2) attack, hydrogen (H_2) and oxygen (O_2) crossovers are important. Therefore, open-circuit voltage (OCV) decay rates were examined and found to be 0.31 mV/h with a loss of 0.08 V, showing the increased resistance to chemical degradation. The constructed fuel cell had a power output of 0.387 W/cm^2 at an operating voltage of 0.43 V.

14.2.2 Polyaromatic Nanocomposites

14.2.2.1 Sulfonated Poly(Ether Ether Ketone) (SPEEK)-CNT Nanocomposites

SPEEK is a good alternative for fuel cell operations, with the advantages of low cost, low fuel permeability, and easy availability. The lack of dimensional stability originated from high levels of sulfonation can be reduced by composite formation, using CNTs. In one of the examples, amine-functionalized CNTs (ACNTs) have been fabricated from carboxylated CNT (CCNT) using 3-aminopropyltriethoxysilane (APTES) by Kim et al. in a SPEEK matrix (Figure 14.2) for a PEMFC application [26]. The surface of the SPEEK/ACNT composite was characterized by field-emission scanning electron microscopy (FE-SEM), energy-dispersive analysis of X-rays (EDAX), and SEM techniques, whereas microstructure was investigated with TEM. Topographical, functional, crystalline, dimensional, thermal, and physical properties of the prepared composites were captured *via* AFM, FTIR, XRD, small-angle X-ray scattering (SAXS), combined TGA and differential scanning calorimeter (DSC), and universal testing machine (UTM), respectively. Impedance spectroscopy has also been employed to measure electrochemical properties. SPEEK/ACNT (1.5%

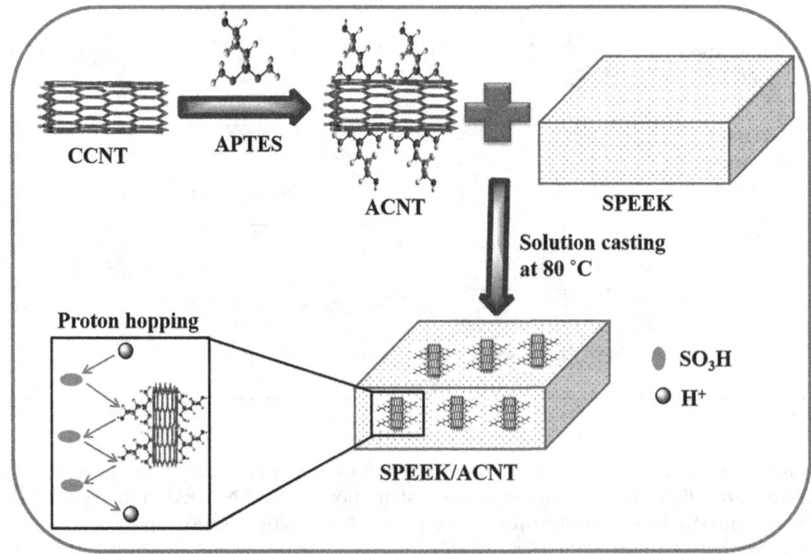

FIGURE 14.2
Preparation process of SPEEK/ACNT composite membrane and its proton hopping. Adapted with permission from reference [26]. Copyright (2020) Elsevier.

by weight) composite resulted in a peak power density of 51.1 mW/cm^2 at a load current density of 101.2 mA/cm^2 in a PEMFC including 20% relative humidity at 60 °C. At the same time, SPEEK/ACNT composite exhibited a slight deformation in morphology with improved durability. The proposed SPEEK/ACNT composite showed enhanced proton conductivity and water uptake, suggesting a good performance for PEMFCs.

14.2.2.2 Polybenzimidazole (PBI)-CNT Nanocomposites

PBI membranes are good candidates for proton conductors with the doping of acids. PBI-based membranes show excellent oxidative and thermal stability as well as low CH_3OH crossover. Eren et al. fabricated PBI-wrapped MWCNTs for platinum (Pt)-based fuel cell applications [33]. Four-step physicochemical analysis has been applied to define the active sites, and cyclic voltammetry (CV) technique was performed to detect H_2 oxidation and durability performance of Pt-PBI/MWCNT, including 43.0 m^2/g of electrochemically active surface area (ECSA). After 1,000 cycles, Pt-PBI/MWCNT had retained 80% of its ESCA, contributing increased durability of Pt-PBI/MWCN relative to its counterparts. The temperature dependence of the Pt-PBI/MWCNT electrocatalyst was investigated with high-temperature proton exchange membrane (HT-PEM) fuel cell performance tests at 150–180 °C under the moisture-free H_2/air condition. PBI/MWCNT improved the durability of noble catalysts and increased the conductivity at high operating temperatures of HT-PEM fuel cells.

14.2.2.3 Sulfonated Polysulfone (SPS)-CNT Nanocomposites

Polysulfone (PS) is a polyaromatic polymer showing good thermal, chemical, and mechanical stability, but low proton conductivity. To solve the conductivity problem of PS, the sulfonation process was applied, and SPSs served as good candidates for membranes for fuel cells. A novel SPS-based nanocomposite filled with polydopamine-modified CNT (PDCNT) was introduced by Altaf et al. for DMFC [34]. The PDCNT composite membranes have been characterized by XRD for structural properties, FTIR for functional groups, energy dispersive X-ray spectrometer (EDS) for elemental analysis, TEM for topographical analysis, and SEM for morphological analysis. The PDCNT (0.5 wt%) membrane increased the proton conductivity by 43%, from 0.0850 S cm^{-1} to 0.1216 S cm^{-1}, in comparison to bare SPS membrane and decreased the CH_3OH permeability.

14.2.2.4 Sulfonated Poly(Arylene Ether Sulfone) (SPAES)-CNT Nanocomposites

SPAES is a valuable, polyaromatic polymer with excellent oxidative, hydrolytic, thermal, and mechanical stability, high glass transition temperature, and high processability. Sung et al. fabricated a composite membrane incorporating hydrophilic oligomer-g-CNTs with SPAES for PEMFCs [35]. Different molecular weights of SPAES hydrophilic oligomers, containing 3 kDa, 7 kDa, and 15 kDa, were combined with CNTs to create hydrophilic oligomer-g-CNTs. The composite membrane displayed great mechanical properties and single-cell performance through reinforcing agents. Using the sulfonic acid groups on the hydrophilic oligomer-g-CNTs, powerful water transport channels were created, holding much more water in a low-humidity environment. The improved composite membrane enhanced proton conductivity by 40% in comparison to the bare membrane at an 80 °C and 50% relative humidity environment. The current density of hydrophilic oligomer (15 k)-g-CNT (1.0 wt%) was 86% higher than that of the bare membrane at 0.6 V.

14.2.3 Conducting Polymer-Based Nanocomposites

14.2.3.1 PANI-CNT Composites

PANI is a widely used conducting polymer with high stability, processability, and biocompatibility. However, PANI is a poor conductor and has a small surface area. Therefore, conductive materials such as CNTs have been successfully combined with PANI to increase the conductivity and performance of PANI. Wu et al. developed a novel CNT/PANI nanocomposite film by layer-by-layer self-assembling onto an indium tin oxide (ITO) anode, using the graft polymerization of PANI and the modification of APTES for microbial fuel cells (MFCs) [27]. A (CNTs/PANI)n/APTES/ITO composite with twelve layers showed better electrochemical behavior, such as high current density and high output power density. In another example of PANI-CNT composites, Yellappa et al. introduced PANI and PANI/CNT composites onto the surface of stainless steel mesh (SSM) for MFC application [36]. PANI and PANI/CNT composite materials were created by oxidative chemical polymerization and coated onto an SSM (SSM-PANI/CNT) anode. SSM-PANI/CNT showed maximum power and current density, as well as cell voltage, but decreased internal resistances, external losses, and activation energy when compared with SSM/PANI and bare SSM (Figure 14.3). At the same time, composite materials were characterized by analytical techniques such as FTIR, FE-SEM, XRD, ultraviolet-visible (UV-Vis) spectroscopy, and electrochemical techniques such as CV and electrochemical impedance spectroscopy (EIS). Thus, the highly conductive SSM-PANI/CNT composite material formed a higher surface area directing the microbe-electrode interactions toward greater electrocatalytic activity, enhanced power production, and wastewater treatment.

14.2.3.2 Polypyrrole (PPy)-CNT Composites

PPy is one of the most attractive conducting polymers, with excellent biocompatibility, conductivity, and environmental safety. The modification of PPy by CNTs is leading to the construction of new composite materials with improved performances. A vertical CNT (VCNT)/PPy composite was generated by Zhao et al. for MFCs [37]. The CVD method was used to accumulate CNTs on the CNFs vertically, and the resulting composites increased the efficiency of extracellular electron transfer. The performances of VCNT/PPy- modified CNFs were investigated using FTIR, XPS, and FE-SEM techniques. The proposed composite had enhanced the electricity generation performance as well as the positive charge and biocompatibility. The power density of the modified composite obtained was 2.63-fold higher (1876.62 mW/m^2) than that of the unmodified fiber.

14.2.4 Synthetic Polymer-Based Nanocomposites

14.2.4.1 Poly(Vinyl Alcohol) (PVA)-CNT Composites

PVA is a water-soluble synthetic polymer. PVA-based membranes possess good proton conductivity and fuel permeability. Despite its powerful properties, the weakness of its mechanical and thermal properties can be overcome by modification with materials such as CNTs. Chebil et al. reported a simple and low-cost synthesis method for PVA/MWCNT nanocomposites with different compositions of MWCNT (0, 1, 3, 4, or 5 wt%) [28]. Nanocomposite films of PVA/MWCNT with five mass ratios were prepared using a solution-casting method. The optical properties of composites were characterized by FTIR, UV-Vis spectroscopy, and photoluminescence (PL) spectroscopy whereas the electrical

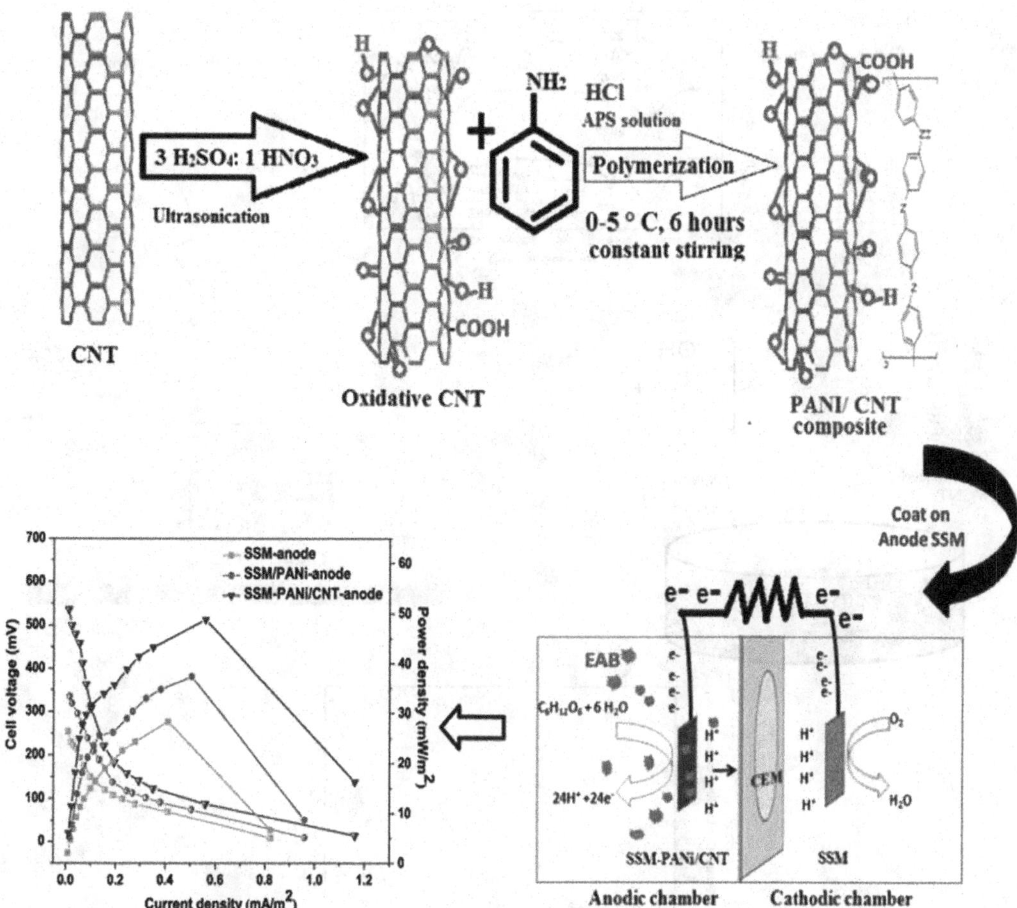

FIGURE 14.3
Fabrication of PANI/CNT and PANI composite materials, coating onto SSM as anodes, and polarization behaviors of SSM, SSM/PANI, and SSM-PANI/CNT. Adapted with permission from reference [36]. Copyright (2019) Elsevier.

properties were recorded *via* current-voltage characteristics. PVA and PVA/MWCNT nanocomposites showed similar absorption peaks in FTIR spectra, while PVA emission decreased with MWCNT concentration in PL spectra, and the electrical conductivity increased with a rise in MWCNT concentration (Figure 14.4). The proposed PVA/MWCNT (5 wt%) nanomaterial proved to be a promising alternative for energy applications, with its lowest energy bandgap and highest conductivity.

14.2.4.2 Polyester-CNT Composites

Polyester is a biodegradable synthetic polymer, holding functional groups in the main chain. Shape memory is the most important ability of polyesters and such an ability provides good shape-memory nanocomposite fabrication for fuel cell applications. Qi et al. have developed electroactive shape-memory nanocomposites of bi-component biodegradable polyesters composed of poly(propylene carbonate) (PPC) and poly(lactic acid) (PLA) combined with MWCNTs [38]. The composite blend obtained showed high values

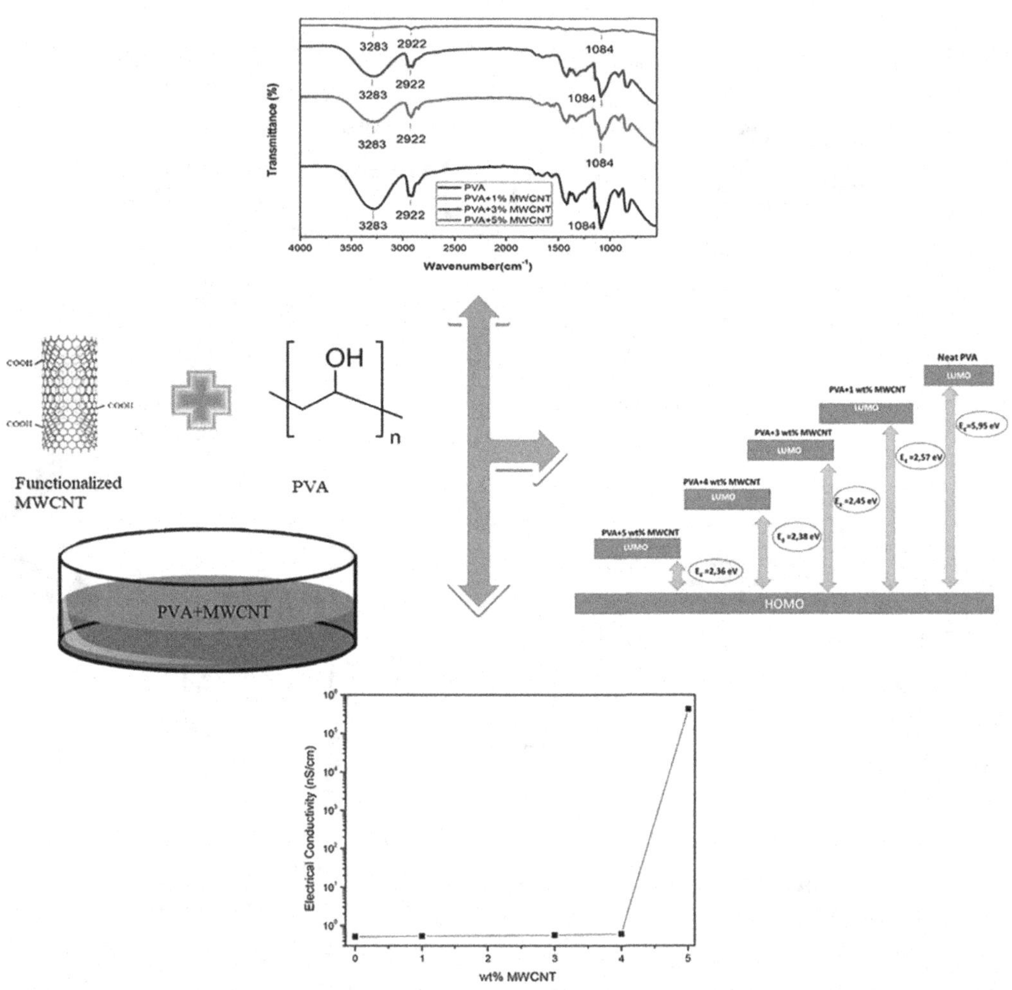

FIGURE 14.4
Preparation of PVA/MWCNT composite films using a solution-casting method, FTIR spectra, PL spectra of PVA and PVA/MWCNT nanocomposites, and semi-log plot of the electrical conductivity as a function of the MWCNT weight ratio. Adapted with permission from reference [28]. Copyright (2019) Elsevier.

for electrical conductivity, recovery force, and rapid electroactive shape-memory behavior. In another example of polyester composite, Wu et al. proposed CNF-reinforced unsaturated polyester (UP) composites, using amine (–NH_2)-functionalized CNT with ethylenediamine (CNT-EDA) [39]. Four types of polymers were used and arranged as epoxy, UP (MR13006), and vinyl esters (M7270 and R806). In the chemical reaction diagram of M7270-CNTs-EDA in the CNF/UP composites (Figure 14.5), the oxidation of the CNF surface with numerous oxygen-containing groups facilitated the reaction of M7270, MR13006, and CNT-EDA. The M7270 molecular chain connected the CNF *via* hydroxyl (–OH) groups to the UP *via* unsaturated bonds. In this way, the surface roughness, the number of polar functional groups, and the wettability of CNFs were increased using CNT-containing sizing agents. At the same time, interfacial adhesion was enhanced using advanced surface energy, as well as mechanical interlocking and chemical bonding.

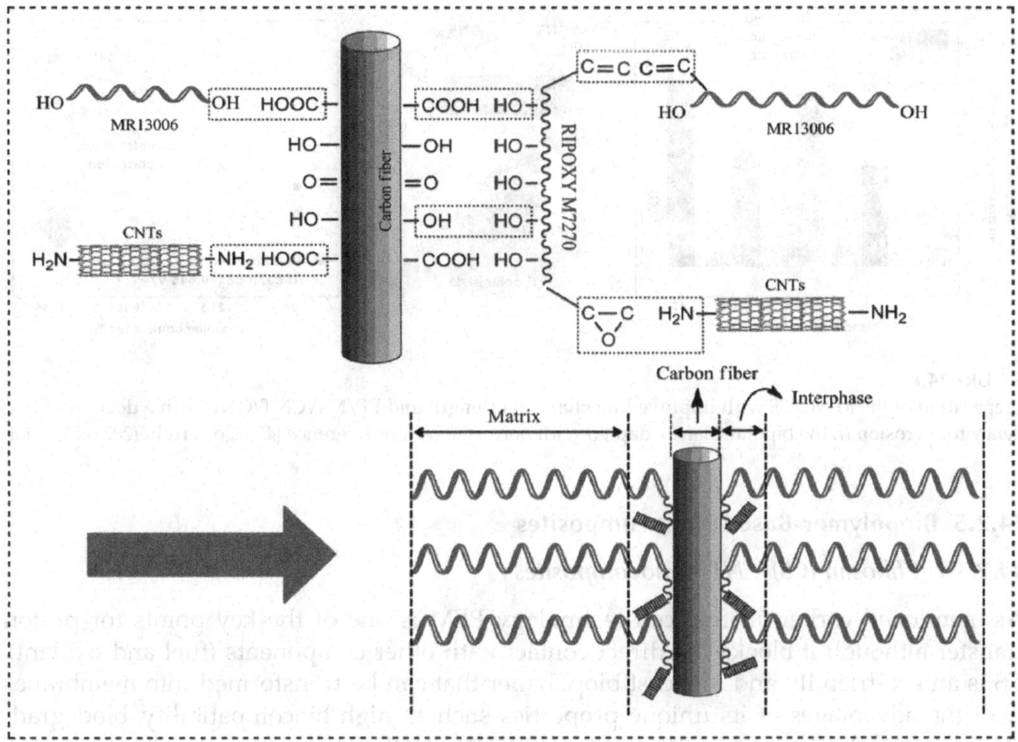

FIGURE 14.5
The chemical reaction diagram for M7270-CNTs-EDA in the CNF/UP composites. Adapted with permission from reference [39]. Copyright (2018) Elsevier.

14.2.4.3 Polypropylene (PP)-CNT Composites

PP is a nonpolar, synthetic polymer possessing a high chemical, mechanical, and thermal stability, low density and cost, and excellent processability. PP-based CNT nanocomposites have been widely studied to improve performance characteristics. In one of these studies, Herrera et al. developed a hybrid PP/MWCNT nanocomposite as well as a CNF for PEMFCs [40]. PP/MWCNT and PP/MWCNT/CNF nanocomposites were fabricated at different loadings in bipolar plates (Figure 14.6). The mechanical properties and corrosion behavior was enhanced using the hybrid nanocomposites as good candidates for bipolar plates. The PP/20MWCNT, PP/21.5MWCNT, and PP/15MWCNT/15CNF nanocomposites were selected as the most effective bipolar plates in PEMFCs for their superior mechanical and corrosion properties. MWCNT loading up to 20 wt% in the PP composite improved the flexural and tensile strength, elastic modulus, and microhardness by 30, 56, 47, and 71%, respectively. Radzuan et al. proposed bipolar plates consisting of PP and milled carbon fiber (MCF) in the presence of CNT or graphene nanoplatelets (xGNP) as secondary fillers for PEMFC applications [41]. This study showed the superiority of one-dimensional materials (CNTs) over two-dimensional materials (GNPs), exhibiting the higher conductivity of the CNT composites (14.8 S/cm) than the xGNP composites (4.9 S/cm). At the same time, flexural strengths were obtained as 99.6 MPa and 172.5 MPa for CNT and xGNP, respectively. In addition to electrochemical conductivity and mechanical properties, the effect of filler orientation was also evaluated and the highly orientated composite materials showed a better performance than randomly orientated fillers.

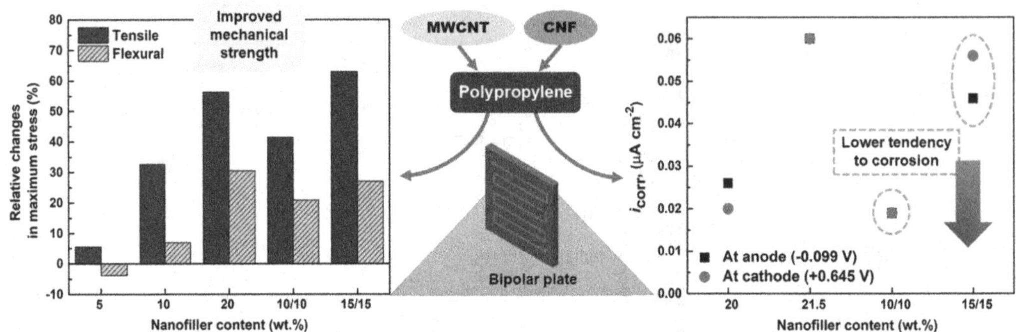

FIGURE 14.6
Preparation of PP/MWCNT with improved mechanical strength and PP/MWCNT/CNF with a decreased tendency to corrosion in the bipolar plate. Adapted with permission from reference [40]. Copyright (2021) Elsevier.

14.2.5 Biopolymer-Based Nanocomposites

14.2.5.1 Chitosan (CS)-CNT Nanocomposites

As mentioned earlier, in fuel cell technology PEM is one of the key points for proton transfer although it blocks the direct contact with other components (fuel and oxidant). CS is an eco-friendly and low-cost biopolymer that can be transformed into membranes with the advantages of its unique properties such as high biocompatibility, biodegradability, hydrophilicity, and antimicrobial activity. At the same time, the deficiencies of CS (such as its low proton conductivity) can be remedied using modifiers like CNTs to increase mechanical strength, conductivity, and robustness. Wang et al. discovered a CS/CNT matrix (Figure 14.7) using a solution-casting method, followed by cross-linking for fuel cell applications [29]. To prepare CNT fluids, pure CNTs were oxidized (o-CNT) using an acid mixture and dispersed in DC5700 $[(CH_3O)_3Si(CH_2)_3N^+(CH_3)_2(C_{18}H_{37})Cl^-]$ to obtain DC7500-grafted CNTs (DC5700-g-CNTs). CNT fluids were finally generated from a DC5700-g-CNTs and NPES $[C_9H_{19}\text{-}C_6H_4\text{-}O(CH_2CH_2O)_{10}SO_3^-Na^+]$ mixture. Sulfur trioxide ($-SO_3$) groups in the CNT fluids and $-NH_2$ groups in the CS matrix created a strong interaction. The CS/CNT fluid composite membrane has exhibited a high proton conductivity, power density, and a reduced CH_3OH crossover.

In another study by Liu et al., CNTs were coated with silica (SiO_2) using the sol-gel method and used for the construction of CS/SiO_2-coated CNT (CS/SCNT)-based membranes for PEMFC studies [42]. MWCNTs with 40–60 nm diameter and 5–15 μm length were treated with HNO_3 and H_2SO_4 at a temperature of 100 °C at first, and then suspended in hydrochloric acid (HCl) to form oxidized nanotubes. Oxidized CNTs were combined with tetraethyl orthosilicate (TEOS) to fabricate SiO_2-coated MWCNTs. Afterward, their nanocomposites with CS were developed. For this goal, an SCNT dispersion prepared in acetic acid (CH_3COOH) was added into CS/CH_3COOH solution (2 wt%) under ultrasonic treatment. The membranes were prepared by solution casting followed by soaking in H_2SO_4 (0.5 M). Characterization studies were carried out using SEM, EDAX, and TGA. The water uptake properties of the fabricated membranes were also examined. The studies revealed the morphological and mechanical properties of the membranes as well as their oxidative and thermal stability. Their proton conductivity was compared with the proton conductivity of the CS membrane. Improved proton conductivities were achieved in all forms of the nanocomposite membranes, ranging from 0.015 S/cm to 0.025 S/cm. The CNT

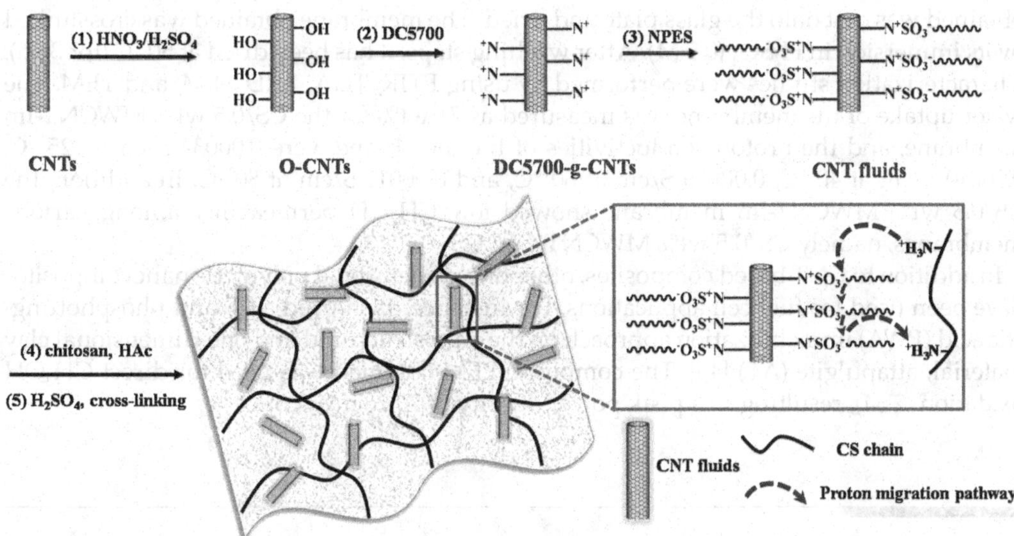

FIGURE 14.7
Preparation of CS/CNT fluid composite membranes. Adapted with permission from reference [29]. Copyright (2018) Elsevier.

loading (5 wt%) also increased the oxidation stability of the membranes at 80 °C from 125 min to 217 min. MWCNTs treated with 1,3-propane sultone were used for CS incorporation by Ahmed et al. to determine their use in fuel cells as PEMs [43]. For sulfonation studies, MWCNTs were dissolved in 1,3-propane in toluene. This step was coupled with the reflux step at a temperature of 110 °C (for 24 h). After that, centrifugation was carried out and the resulting product was dried. Subsequently, a mixture of CNTs, ethanol (C_2H_5OH), water (H_2O), and ammonium (NH_4^+) were prepared, and 3-methacryloxypropyltrimethoxysilane (MPS) was added to this mixture and stirred for 24 h. Modified CNTs have been treated with styrene and divinylbenzene (DVB) in the presence of azobisisobutyronitrile (AIBN) in acetonitrile (C_2H_3N), using a distillation procedure. After washing, the resulting CNTs were acid treated (H_2SO_4) and dried in a vacuum oven. For incorporation of these CNTs into the CS, CNTs sonicated in C_2H_5OH were added to CS/CH_3COOH solution and ultrasonicated for 30 min. The solution obtained was transferred onto a glass plate and dried. Prepared membranes were characterized by FTIR, SEM, EDX, XRD, and TGA analyses. When the CNT concentration increased from 1 wt% to 5 wt%, proton conductivities of the membranes increased from 0.016 S/cm to 0.025 S/cm. The prepared CNT-based membrane exhibited improved mechanical and thermal properties because of the electrostatic interactions between the $-NH_2$ groups of CS and the sulfonate groups ($-SO_3H$) of CNTs.

More recently, Tohidian and Ghaffarian proposed nanocomposite polyelectrolyte membranes consisting of MWCNTs modified by imidazole (MWCNT-Im) and CS for DMFC [44]. In their study, Im addition showed better dispersion of the nanofiller; in addition, it improved the proton conductivity owing to the H_2 bonding between Im- and CS-modified CNTs. Furthermore, Im addition reduced the permeability to CH_3OH. As the starting point, –OH-modified MWCNTs (MWCNTs-OH) were combined with 1H-imidazole-2-acetic acid (Im-COOH) in dichloromethane (CH_2Cl_2) and stirred. After the drying step of the MWCNT-Im, specified amounts of the MWCNT-Im and MWCNTs-OH were added into the CS solution (1% by weight per volume) prepared in CH_3COOH. The solution obtained was cast onto the glass plate and dried. The membrane obtained was crosslinked by its immersion in H_2SO_4 (0.5 M). After washing steps, it has been dried at 50 °C (for 36 h). Characterization studies were performed by using FTIR, TGA, XRD, SEM, and TEM. The water uptake of the membrane was measured as 71 wt% for the CS/0.5 wt% MWCNT-Im membrane, and the proton conductivities of the membrane were 0.00034 S/cm at 25 °C, 0.00010 S/cm at 40 °C, 0.00029 S/cm at 60 °C, and 0.00011 S/cm at 80 °C. In addition, the CS/0.5 wt% MWCNT-Im membrane showed low CH_3OH permeability among various membranes, namely CS/0.5 wt% MWCNT-OH, CS.

In addition to CNT-based composites, other one-dimensional polymeric nanocomposites have been used for fuel cell applications. For instance, Tsen used a CS and phosphotungstic acid (PWA) immobilization approach on the nanostructured and one-dimensional clay material, attapulgite (AT) [45]. The composite PEM obtained was used for direct CH_3OH oxidation (2 M), resulting in a peak power density of 70.26 mW/cm^2.

14.3 Conclusions

Energy sources and their production have been crucial matters from the beginning of time, with the discovery of fire. Fossil fuels and nuclear power are the main effective ways of energy production. However, their direct and indirect damage to the environment and health are inevitable. In addition to this, fossil fuel sources are limited. Therefore, renewable energy sources are in demand. Fuel cells, that have great potential to overcome this need for conventional energy sources, are attracting enormous interest. This technology offers a more environmentally friendly approach: it can be in stationary or portable versions and it has high efficiency for electrical energy production. Since the discovery of nanotechnology, studies on energy applications have questioned the integration of high-tech nanomaterials into this field. The studies in the field of fuel cells have focused on the development of anode or cathode materials or electrolytes to improve the performance of the fuel cell. In this context, the use of one-dimensional nanostructures, and in particular the remarkable one-dimensional polymeric structures, combined with one-dimensional nanostructures such as CNTs, inorganic semiconducting nanotubes, metallic nanotubes, or wires, have come to the fore. In addition to achieving superior performance, generating and storing the energy at a low cost are also important tasks in energy production and storage. This chapter outlines the recent advances in one-dimensional polymeric nanocomposites for fuel cells. The studies point out the favorable utilization of one-dimensional polymeric nanocomposites with enhanced mechanical, chemical, and thermal properties. It can be concluded that developments in the synthesis, preparation, and characterization of one-dimensional nanomaterials will provide unique benefits for practical and efficient fuel cell applications.

Acknowledgments

Filiz Kuralay expresses her gratitude to the Turkish Academy of Sciences (TÜBA) as an associate member.

References

1. Steele BCH, Heinzel A. Materials for fuel-cell technologies. *Nature* 2001;414(6861):345–352.
2. Lin B, Qiu L, Lu J, Yan F. Cross-linked alkaline ionic liquid-based polymer electrolytes for alkaline fuel cell applications. *Chem Mater* 2010;22(24):6718–6725.
3. Zhou M, Zhou N, Kuralay F, Windmiller JR, Parkhomovsky S, Valdés-Ramírez G, Katz E, Wang J. A self-powered "sense-act-treat" system that is based on a biofuel cell and controlled by Boolean logic. *Angew Chem Int Ed* 2012;124:2740–2743.
4. Wang S, Xu Z, Wu X, Zhao H, Zhao J, Liu J, Yan C, Fan X. Analyses and optimization of electrolyte concentration on the electrochemical performance of iron-chromium flow battery. *Appl Energy* 2020;271:115252.
5. BoopathiRaja R, Parthibavarman M. Desert rose like heterostructure of $NiCo_2O_4$/NF@PPy composite has high stability and excellent electrochemical performance for asymmetric super capacitor application. *Electrochim Acta* 2020;346:136270.
6. Pan Z, Huang B, An L. Performance of a hybrid direct ethylene glycol fuel cell. *Int J Energy Res* 2019;43(7):2583–2591.
7. Muthukumar M, Rengarajan N, Velliyangiri B, Omprakas MA, Rohit CB, Kartheek Raja U. The development of fuel cell electric vehicles – A review. *Mater Today Proc* 2021;45:1181–1187.
8. Zhou M, Kuralay F, Windmiller JR, Wang J. DNAzyme logic-controlled biofuel cells for self-powered biosensors. *Chem Commun* 2012;48(32):3815–3817.
9. Authayanun S, Imorb K, Arpornwichanop A. A review of the development of high temperature proton exchange membrane fuel cells. *Chin J Catal* 2015;36(4):473–483.
10. Neburchilov V, Martin J, Wang H, Zhang J. A review of polymer electrolyte membranes for direct methanol fuel cells. *J Power Sources* 2007;169(2):221–238.
11. Ijaodola OS, El-Hassan Z, Ogungbemi E, Khatib FN, Wilberforce T, Thompson J, Olabi AG. Energy efficiency improvements by investigating the water flooding management on proton exchange membrane fuel cell (PEMFC). *Energy* 2019;179:246–267.
12. Sarker M, Rahman A, Mojica F, Mehrazi S, Kort-Kamp WJM, Chuang PYA. Experimental and computational study of the microporous layer and hydrophobic treatment in the gas diffusion layer of a proton exchange membrane fuel cell. *J Power Sources* 2021;509:230350.
13. Gülzov E. Alkaline fuel cells. *Fuel Cells* 2004;4(4):251–255.
14. Lue SJ, Wang WT, Mahesh KPO, Yang CC. Enhanced performance of a direct methanol alkaline fuel cell (DMAFC) using a polyvinyl alcohol/fumed silica/KOH electrolyte. *J Power Sources* 2010;195(24):7991–7999.
15. Cheng S, Zhao G, Gao M, Shi Y, Huang M, Marefati M. A new hybrid solar photovoltaic/phosphoric acid fuel cell and energy storage system; energy and exergy performance. *Int J Hydrog Energy* 2021;46(11):8048–8066.
16. Spinelli M, Bona DD, Gatti M, Martelli E, Viganò F, Consonni S. Assessing the potential of molten carbonate fuel cell-based schemes for carbon capture in natural gas-fired combined cycle power plants. *J Power Sources* 2020;448:227223.
17. Ormerod RM. Solid oxide fuel cells. *Chem Soc Rev* 2003;32(1):17–28.
18. Jordan J, Jacob KI, Tannenbaum R, Sharaf MA, Jasiuk I. Experimental trends in polymer nanocomposites. *Mater Sci Eng A* 2005;393(1–2):1–11.
19. Kausar A. Polymer and nanobelt derived nanomaterials: Opening doors to revolutionary stadia. *Polym Plast Technol Mater* 2021;60(2):117–131.
20. Mu J, Wang J, Hao J, Cao P, Zhao S, Zeng W, Miao B, Xu S. Hydrothermal synthesis and electrochemical properties of V_2O_5 nanomaterials with different dimensions. *Ceram Int* 2015;41(10):12626–12632.
21. Long YZ, Li MM, Gu C, Wan M, Duvail JL, Liu Z, Fan Z. Recent advances in synthesis, physical properties and applications of conducting polymer nanotubes and nanofibers. *Prog Polym Sci* 2011;36(10):1415–1442.

22. Li P, Chen W. Recent advances in one-dimensional nanostructures for energy electrocatalysis. *Chin J Catal* 2019;40(1):4–22.
23. Zhu S, Nie L. Progress in fabrication of one-dimensional catalytic materials by electrospinning technology. *J Ind Eng Chem* 2021;93:28–56.
24. Sopian K, Wan Daud WR. Challenges and future developments in proton exchange membrane fuel cells. *Renew Energy* 2006;31(5):719–727.
25. Liu YL, Su YH, Chang CM, Suryani, Wang DM, Lai JY. Preparation and applications of nafion-functionalized multiwalled carbon nanotubes for proton exchange membrane fuel cells. *J Mater Chem* 2010;20(21):4409–4416.
26. Kim AR, Vinothkannan M, Song MH, Lee JY, Lee HK, Yoo DJ. Amine functionalized carbon nanotube (ACNT) filled in sulfonated poly(ether ether ketone) membrane: Effects of ACNT in improving polymer electrolyte fuel cell performance under reduced relative humidity. *Compos B Eng* 2020;188:107890.
27. Wu W, Niu H, Yang D, Wang S, Jiang N, Wang J, Lin J, Hu C. Polyaniline/carbon nanotubes composite modified anode via graft polymerization and self-assembling for microbial fuel cells. *Polymers* 2018;10(7):759.
28. Chebil A, Doudou BB, Dridi C, Dammak M. Synthesis characterization, optical and electrical properties of polyvinyl alcohol/multi-walled carbon nanotube nanocomposites: A composition dependence study. *Mater Sci Eng B* 2019;243:125–130.
29. Wang J, Gong C, Wen S, Liu H, Qin C, Xiong C, Dong L. Proton exchange membrane based on chitosan and solvent-free carbon nanotube fluids for fuel cells applications. *Carbohydr Polym* 2018;186:200–207.
30. Sigwadi R, Dhlamini MS, Mokrani T, Nemavhola F. Enhancing the mechanical properties of zirconia/Nafion® nanocomposite membrane through carbon nanotubes for fuel cell application. *Heliyon* 2019;5(7):e02112.
31. Yin C, Xiong B, Liu Q, Li J, Qian L, Zhou Y, He C. Lateral-aligned sulfonated carbon-nanotubes/nafion composite membranes with high proton conductivity and improved mechanical properties. *J Membr Sci* 2019;591:117356.
32. Vinothkannan M, Kim AR, Ramakrishnan S, Yu YT, Yoo DJ. Advanced nafion nanocomposite membrane embedded with unzipped and functionalized graphite nanofibers for high-temperature hydrogen-air fuel cell system: The impact of filler on power density, chemical durability and hydrogen permeability of membrane. *Compos B Eng* 2021;215:108828.
33. Eren EO, Özkan N, Devrim Y. Polybenzimidazole-modified carbon nanotubes as a support material for platinum-based high-temperature proton exchange membrane fuel cell electrocatalysts. *Int J Hydrog Energy* 2021;46(57):29556–29567.
34. Altaf F, Gill R, Batool R, Ur-Rehman Z, Majeed H, Abbas G, Jacob K. Synthesis and applicability study of novel poly(dopamine)-modified carbon nanotubes based polymer electrolyte membranes for direct methanol fuel cell. *J Environ Chem Eng* 2020;8(5):104118.
35. Sung IH, Yu DM, Yoon YJ, Kim TH, Lee JY, Hong SK, Hong YT. Preparation and properties of sulfonated poly(arylene ether sulfone)/hydrophilic oligomer-g-CNT composite membranes for PEMFC. *Macromol Res* 2013;21(10):1138–1144.
36. Yellappa M, Sravan JS, Sarkar O, Reddy YVR, Mohan SV. Modified conductive polyaniline-carbon nanotube composite electrodes for bioelectricity generation and waste remediation. *Bioresour Technol* 2019;284:148–154.
37. Zhao N, Ma Z, Song H, Xie Y, Zhang M. Enhancement of bioelectricity generation by synergistic modification of vertical carbon nanotubes/polypyrrole for the carbon fibers anode in microbial fuel cell. *Electrochim Acta* 2019;296:69–74.
38. Qi X, Dong P, Liu Z, Liu T, Fu Q. Selective localization of multi-walled carbon nanotubes in bi-component biodegradable polyester blend for rapid electroactive shape memory performance. *Compos Sci Technol* 2016;125:38–46.
39. Wu Z, Cui H, Chen L, Jiang D, Weng L, Ma Y, Li X, Zhang X, Liu H, Wang N, Zhang J, Ma Y, Zhang M, Huang Y, Guo Z. Interfacially reinforced unsaturated polyester carbon fiber composites with a vinyl ester-carbon nanotubes sizing agent. *Compos Sci Technol* 2018;164:195–203.

40. Ramírez-Herrera CA, Tellez-Cruz MM, Pérez-González J, Solorza-Feria O, Flores-Vela A, Cabañas-Moreno JG. Enhanced mechanical properties and corrosion behavior of polypropylene/multi-walled carbon nanotubes/carbon nanofibers nanocomposites for application in bipolar plates of proton exchange membrane fuel cells. *Int J Hydrog Energy* 2021;46(51):26110–26125.
41. Radzuan NAM, Sulong AB, Somalu MR, Abdullah AT, Husaini T, Rosli RE, Majlan EH, Rosli MI. Fibre orientation effect on polypropylene/milled carbon fiber composites in the presence of carbon nanotubes or graphene as a secondary filler: Application on PEM fuel cell bipolar plate. *Int J Hydrog Energy* 2019;44(58):30618–30626.
42. Liu H, Gong C, Wang J, Liu X, Liu H, Cheng F, Wang G, Zheng G, Qin C, Wen S. Chitosan/silica coated carbon nanotubes composite proton exchange membranes for fuel cell applications. *Carbohydr Polym* 2016;136:1379–1385.
43. Ahmed S, Ali M, Cai Y, Lu Y, Ahmad Z, Khannal S, Xu S. Novel sulfonated multi-walled carbon nanotubes filled chitosan composite membrane for fuel-cell applications. *J Appl Polym Sci* 2019;136(22):47603.
44. Tohidian M, Ghaffarian SR. Polyelectrolyte nanocomposite membranes based on chitosan and surface modified multi-walled carbon nanotubes for use in fuel cell applications. *J Macromol Sci Part A* 2021;58(11):778–791.
45. Tsen WC. Composite proton exchange membranes based on chitosan and phosphotungstic acid immobilized one-dimensional attapulgite for direct methanol fuel cells. *Nanomaterials* 2020;10(9):1641.

15

1D Polymers for High-Performance Photovoltaics

Roger Gonçalves and Kaique Afonso Tozzi

CONTENTS

15.1 Contextualization

The development of science and technology requires that there is also a development in energy supply. Currently, fossil fuels and hydroelectric power generation form the basis for this development. According to the International Energy Agency (IEA), it is estimated that, by 2030, oil demand will increase by 1.7% per year, reaching the level of 15.3 billion tons of oil equivalents per year. Oil is the main fuel used and will remain so until supply is restricted [1]. The theme of fuel scarcity has become more apparent and of greater concern. Therefore, developing technologies for renewable energy and the treatment of waste is a great challenge in different countries. There is, therefore, a pressure to increase development and an even greater pressure that all this demand does not destroy the environment.

For this situation not to worsen, it is necessary to implement alternative, low-cost, renewable energy sources. In this way, solar energy presents itself as an ecologically viable alternative with great potential for being a renewable source of energy which has little environmental impact. However, its use has some drawbacks: the high cost of existing technologies for energy conversion and a weak policy to encourage use. Thus, more and more research is being carried out to obtain economically viable solar cells. The construction of nanostructured electrochemical energy-converting devices has shown remarkable results due to their intrinsic properties of low dimensionality, quantum confinement, and increased electrical, electronic, and ionic conduction rates [2]. Among the wide range of possible materials available to construct these systems, it is important to highlight the conductive polymers that can be used in, among other devices, photovoltaic cells with low cost, high chemical stability, and good durability.

DOI: 10.1201/9781003223764-15

15.2 Introduction

Photovoltaic cells (PVs) convert solar energy into electrical energy through electronic phenomena that occur in the active material, a semiconductor with unique properties. The first observation of photoconductivity in an organic compound (anthracene) was reported in 1906 [3]. Photoconductivity in the poly(*N*-vinyl-carbazole) (PVK) polymers was first shown in 1957 [4]; from there, these conductive polymers and small organic molecules began to be used for this purpose. PVK polymers were chosen as a possible replacement for silicon solar cells already consolidated in the market, due to the low manufacturing cost, the high potential for miniaturization, as well as environmental advantages, a prerequisite that has become quite important in recent years. However, for organic photovoltaic (OPV) cells to truly take the lead in the market, some problems need to be overcome, such as lifetime and efficiency factors that must be increased. From a marketing point of view, three aspects need to be fulfilled simultaneously for this industrial niche to be reached: low cost, high efficiency, and high durability. Research is being carried out daily to increase these important factors. However, the techniques used to develop these new devices make this technology increasingly distant from the renewable energy market since costs have not been prioritized.

In this sense, research in the nanotechnology field has become a viable alternative, as they end up planning and optimizing the different components of an organic solar cell in a better way. The inherent versatility of conductive polymers certainly characterizes them as an important candidate in this area of research. Although conductive polymers are considered one-dimensional materials due to their very high length-to-width ratio, a polymeric matrix is a 3D structure in which other materials can be dispersed to increase energy conversion efficiency.

Currently, four generations of photovoltaic devices are recognized. The pioneering devices ("first-generation devices") used silicon plates as the photoactive material. The second-generation devices emerged to solve some problems arising from the first, such as the mass of the device and its lack of flexibility. The second-generation devices use inorganic semiconductor thin films and are currently commercially available. The third-generation devices allowed a big step in the evolution of this energy conversion technology. Films of organic molecules and conductive polymers are used, which are generally associated with two types of semiconductors to reduce the effect of recombination [5]. The fourth-generation devices arose from the need to make the third-generation devices cheaper, combining the qualities of organic materials, such as low cost and ease of availability, with the qualities of inorganic materials, such as durability and high conversion efficiency [6].

These last two generations of devices are based on the idea of using materials with different electronic affinities and ionization potentials, favoring the dissociation of the pairs connected at the interface between the two. Thus, electrons will have a greater affinity with more electronegative materials (n-type semiconductors) and holes with materials with higher ionization potential (p-type semiconductors). Due to the existence of this interface, recombination is less frequent in this type of device [7]. These devices can be divided into bilayer or monolayer with volume heterojunction (bulk heterojunction, BHJ). In first-generation devices, efficiency is limited by the thickness of the active layer. If it is too thick, recombination will be more frequent than dissociation, as the excitons will have to diffuse a long way to reach the collection interface of charge carriers.

Therefore, the BHJ type improves its efficiency through the mixture of materials with good values of electron affinity. In this way, the donor/receiver interfaces, where the dissociation of excitons occurs are distributed throughout the volume of the active layer, not only at its threshold. Therefore, an aggregate structure is formed, and in this case, the charge dissociation is facilitated by the interfacial increase of the p-n regions [6]. Furthermore, since the relative position of the donor LUMO (lowest occupied molecular orbital) and the acceptor LUMO is crucial for an efficient charge transfer between donor and acceptor, it is necessary to find the appropriate acceptor for each polymeric donor used [8].

Polythiophene derivatives, in general, are more studied as an active layer in OPVs, as these are chemically and electrochemically stable to air and moisture in both doped and undoped states [9–11]. However, unlike other conductive polymers, polythiophenes are very difficult to obtain in an aqueous medium and, when obtained in these solutions, do not exhibit adequate properties [12, 13]. Therefore, synthetically they are usually obtained in organic solvents such as propylene carbonate, acetonitrile, and chloroform [14].

Although a 4-carbon alkyl group is enough to achieve good solubility [15], poly(3-hexylthiophene) (P3HT) is the most exploited derivative thanks to its absorption range in the visible region and the native properties of P3HTs and its balance between solubility and crystallinity, because the larger the alkyl group, the more difficulties will occur in this packaging [16]. In addition, it has a high mobility of holes and an efficient absorption in the solar spectrum [10, 17, 18]. All these properties are highly dependent on the organization of the polymer chain, which is better in crystalline materials [19]. However, the synthesis of polymers with highly organized chains is not a trivial process. The presence of defects in the chains leads to high recombination rates of extinctions and slow transport of load [20]. Thus, to minimize the recombination rate and increase cargo transport, various e-transport materials have been proposed. Carbon materials are the most used for this purpose, particularly soluble fullerene derivatives [21–23]. It has been described that the photovoltaic characteristics of polymer composites/carbon materials present better optical absorption, rapid electron transfer, and adequate dielectric constant, resulting in good electronic mobility. Although the highest efficiency in OPV devices was obtained using polymer/fullerene composites [24, 25], Nardes *et al.* [26] reported an improvement in light absorption and photocurrent generation, using the self-organization of individual components in a straight-joint-like structure. It is important to note that fullerene derivatives are better dispersed in the amorphous parts of the P3HT [27].

Therefore, a better understanding of the interaction between the materials must be gained to optimize both properties and obtain a synergistic effect. This chapter aims to show the innovations in the field of OPVs based on conducting polymers using new techniques, so that these devices can become viable alternatives for the renewable energy industry.

15.2.1 The Morphological and Conformational Question

Polymers, especially the conductive ones, are materials with very peculiar characteristics. They have the mechanical properties of most conventional polymers, with electrical conduction properties inherent to their chemical structure. Concurrent to their conductive characteristic, polymers exhibit other properties such as electrochromism, supercapacitance, and photoconduction. As with other semiconductor materials, the bandgap governs its properties; however, when it comes to polymers, the conformation of their chains directly affects their electronic properties and consequently their bandgap. Some studies have demonstrated that a conformational rearrangement can change from its conductivity to its photoconductive properties [28]. Therefore, the structural design, conformation, and arrangement of composites based on conducting polymers is the way to obtain increasingly more efficient optoelectronic devices.

Like carbon nanotubes, the conductive polymer is considered to be a 1D material since the length of the polymer chain is infinitely greater than its width, even in polymers from more complex monomers. However, when preparing an electrode or a conductive polymer film, there is not only one polymeric chain but also an entanglement of polymeric chains that interact with each other to make up the properties of that material at the macroscopic level.

Thus, although the basic structure of the polymer is 1D, the whole material, especially its composites, are necessarily 3D materials, as the entire structure counts and contributes to the observed properties. Therefore, polymer modeling is a complicated task requiring 3-axis thinking. This chapter will address polymer-derived materials as 3D materials, including modeling and simulation methods to predict and explain their properties.

15.3 Transparent and Low-Bandgap Devices

In any semiconductor material, the most important parameter is the electronic bandgap (E_g) and the position of highest occupied and lowest unoccupied bands (HOMO and LUMO, respectively) versus vacuum, as these are the ones that govern the photoelectrochemical properties of these materials, as well as choosing compatible materials from which to build more efficient composites. So, one great advantage of semiconducting polymers is the ability to tune the bandgap and the position of the HOMO and LUMO levels by molecular chemical design. Such intrinsic abilities of this class of materials opens up the possibility of creating materials with a low bandgap, which can characterize an alternative to maximize the power conversion efficiency of photovoltaic devices.

Several strategies can be adopted to narrow the bandgap. In the case of polymers, some easily controllable factors can be mentioned: bond-length, planarity of the conjugated structure, substituents, intermolecular interactions, and donor-acceptor structure. Several of these parameters will be covered throughout the chapter, describing how they can help design new materials [29]. One of the first applications of low-bandgap polymers is in transparent photovoltaic devices and final conductive layers. The polymer will gain transparency in the visible range by changing the optical absorption to red or near-infrared. In addition, flexibility, processability, low cost, and the potential to replace transparent conductive oxides have been the main drivers for developing transparent conducting polymers. However, it is still important that durability is also an issue.

The idea of semi-transparent solar cells has been explored extensively in recent years. Applications can be as varied as possible, including electric windows for photovoltaic energy integrated into buildings, automobiles, or greenhouses, or integrated into many portable devices. The basic idea of this type of cell is to assemble it as a sandwich, absorbing radiation in the infrared range, maintaining transparency to the visible range between two semi-transparent electrodes. The transparency and color of the photovoltaic cell can be adjusted by varying the thickness and composition of the active layer. However, the main disadvantages of these OPVs are the semi-transparent electrodes, which can cause electrical problems and optical losses. There is a negative relationship between transparency and the device's power conversion efficiency.

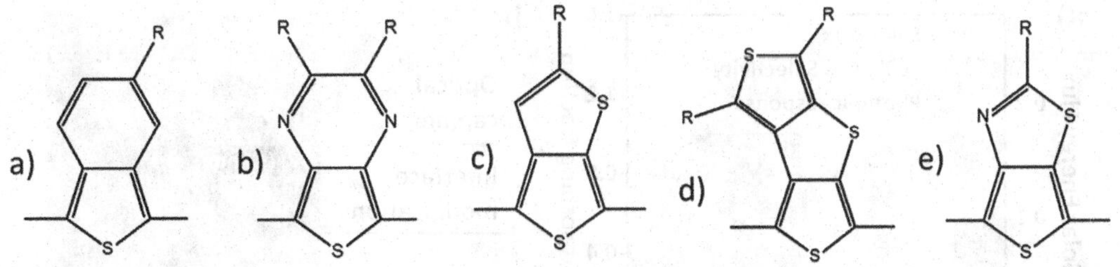

FIGURE 15.1
a) Benzothiophene, b) thienopyrazine, c) thienothiophene, d) dithienothiophene, e) thienothiazole.

Recently, semi-transparent solar cells with an energy conversion efficiency of up to 10% and visible transparencies up to 30% have been reported in the literature [30]. To obtain such results, Zhu and collaborators developed a ternary device of a low-bandgap polymer, a non-fullerene acceptor, and another near-infrared sensitive acceptor. The cell obtained has considerable absorbance only above 690 nm, peaking mainly at 815 nm. Different proportions of molecules were studied until the optimal combination culminated in this result. It is stated in the literature that the resulting quinoid bonds predominately induce the low gap. In addition, the quinoid form can be stabilized by fusing another aromatic ring to the polymer backbone, which is more effective in thiophene rings, as shown in Figure 15.1. However, many of these monomers have handling problems that make them a little unstable for the generation of good films. These problems were solved in BHJ-type cells that demonstrated excellent compatibility between polymers obtained by these monomers and fullerene-type acceptors. Thus, it is more common to find acceptors derived from C60 in the literature.

Another method to achieve high efficiencies in semi-transparent cells is to optimize the out-of-visible absorption bands through selective absorption. This is to modify the various components of a BHJ device to absorb different spectral regions while keeping the device as transparent as possible. For example, it is possible to cite the work of Li et al. [31], who planned the device previously, calculating the theoretical absorption of the materials that would compose it. After that, light absorption was optimized by controlling the morphology, while increased electrical properties were due to a thin layer of transparent Ag (Figure 15.2). Semi-transparent devices hold great promise and can already be found in cities worldwide, being incorporated into both vehicles and mobile phone charging units. The more versatile the applications of a given device, the greater the attention this device will attract.

15.4 Designed Devices for Indoor Applications

The growing demand for devices driven by low power consumption is closely associated with research into indoor solar cells, as these materials can use the light in indoor environments to generate electricity. The environment where these devices are used does not always have the usual source as output (sunlight). They have different requirements because they must be functional even with artificial light (which has different spectra), making them very attractive to Industry 4.0 and advanced materials research. Environmental sensors

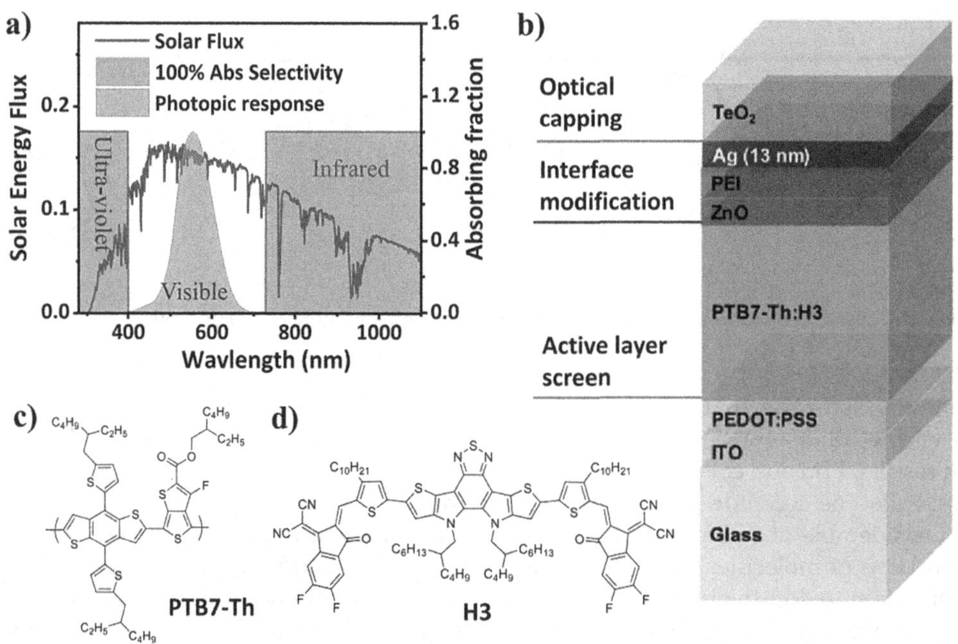

FIGURE 15.2
a) Absorbing profile of absolute absorbing selectivity as functions of the solar flux, b) device structure and molecular structures of c) PTB7-Th and d) H3 polyurethane. Reproduced with permission from reference [31]. Copyright (2021) Wiley Online Library.

and Internet of Things (IoT) technologies are the most promising applications regarding these photovoltaic materials.

Although silica and III-V semiconductors are probably the most used materials for outdoor solar cells, their incompatibility to produce suitable home devices in large-scale processes makes them a less researched option for IoT applications, as the first one has a bandgap that does not match the spectrum emitted by sources such as LEDs (light-emitting diodes) or fluorescent lamps. Properties like light weight, high power energy conversion, and ease of manufacture are highly desired in this case. For this reason, printable solar cells based on quantum dots (QDPV), organic (OPV), or perovskite (PPV) materials are at the core of the indoor photovoltaic trend. But the lack of understanding of the possible harm that the usage of these materials in environments that are prone to heat or humidity stresses, their toxicity after being degraded, and their disposal after use are still a challenge to overcome before these materials can be fully commercialized and be part of domestic life.

Based on these three kinds of materials, we will explain their advantages, disadvantages, and perspectives to be applied as future indoor solar cells. Organic photovoltaic systems made of polymer blends or polymer/fullerenes are already settled as promising materials for indoor solar cells. $PC_{71}BM$ fullerene is one of the most used fullerene acceptors, while PBDB-TF polymer is used as a donor. As they have tunable properties, it is possible to reach a power conversion efficiency (PCE) higher than 17% through physical-chemical modifications, such as intermolecular packing and phase separation of the active layer [32]. The Voltage at Open Circuit (VOC) plays a key role in enhancing some properties for all the OPV systems. This value can be controlled by tailoring the acceptor-donor energy dynamic (HOMO-LUMO bandgap value). In this sense, Cho *et al.* [33] used a Y-Th2 acceptor in PM6:Y6 non-fullerene system, reaching a PCE of up to 22.72% under LED light at 1000 lx. With this, the authors broadened the absorption in visible light and tailored the energy level alignment between donor and acceptor, influencing the charge transport in the active layer. Using a similar strategy, Ma *et al.* [34] reported a high-efficiency indoor OPV device based on a polymer/non-fullerene system PM6:Y6-O that has an alignment in the electron transporting interlayer. A high PCE of 30.1% was obtained with a fill factor of 76%. The main disadvantage of these systems is the difficulty of reaching transparency and achieving the same PCE as their inorganic counterparts.

Perovskite solar cells are well known and one of the most researched topics in the energy field. Their most attractive properties are their long-term stability and the possibility of achieving PCEs of 25%. As discussed in the OPV, one of the most controlled parameters for these materials is the crystallinity of the perovskite structure and charge carriers' dynamic. Perovskite materials have an ABX_3 structure, where A and B are cations, and X is usually a halide. So, engineering this kind of composition is the key to tailoring its optoelectronic properties. Nowadays, the most common perovskite materials for IPV are methylammonium lead iodide $CH_3NH_3PbI_3$ ($MAPbI_3$) and its triple-cation tailored structures, with a bandgap of 1.6 eV [35]. The main advantage is the possibility of reaching transparency in a wide range of combinations and the higher PCE values compared with OPVs. In contrast, the main disadvantages are that they are made from inorganic and non-eco-friendly substances as well as the high leak current associated with their usage.

Quantum dots (QDs) are now a popular topic in the research field due to their unique properties as they have all dimensions in quantum confinement. They are versatile and can be synthesized from inorganic and organic sources since they can be formed from carbonized particles. Compared with perovskites and organic semiconductors, QDs have a lower photon energy threshold for exciton generation, enabling them to go beyond the Shockley–Queisser limit [36]. Another great advantage is the possibility to control the bandgap of these materials as this value is directly related to their size. However, the PCE is still lower (maximum 13%) when compared with the other indoor PV materials, but they still have a competitive value when used in low light environments. Figure 15.3 gives a summary of the current state-of-the-art of these three kinds of indoor photovoltaic materials [37].

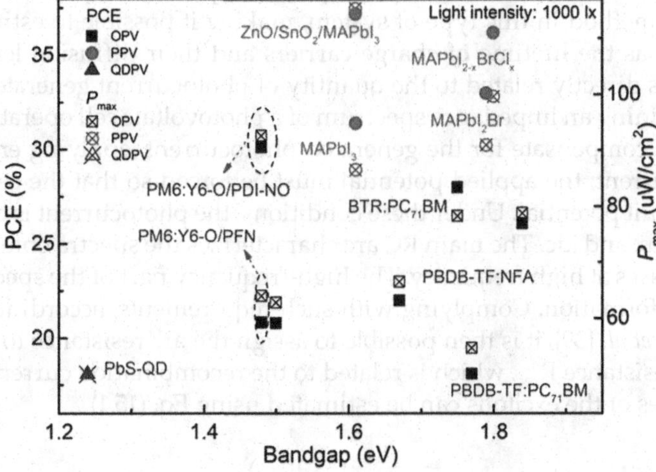

FIGURE 15.3
Chart of current state-of-the-art about indoor photovoltaic materials. Reproduced with permission from reference [37]. Copyright (2020) Royal Society of Chemistry.

It must be noted that the device composition, engineering, and architecture are all correlated to achieving high PCE values. As these materials are promising and some of them (mainly the QDPV) still need to be further studied, there is a great expectation that their performances will be enhanced in the future to be used in domestic life applications.

15.5 Impedance Spectroscopy Applied to Photovoltaics

Electrochemical impedance spectroscopy is a versatile technique for characterizing interfaces and modified electrodes. Its use for photoanode and photocathode characterization is not new and is widespread in the literature. Measuring with or without incident light can provide a realistic idea of how the material generates the photocurrent. However, it is possible to obtain much more information with a certain mathematical treatment. As an example, it is possible to cite the Mott–Schottky graph, which, from a potential scan at a given frequency corresponding to the capacitance response of the electric double-layer, it is possible to estimate the flat band potential, which provides an idea of the position of the HOMO or LUMO of the semiconductor, as well as the density of charge carriers. This charge carrier is directly related to the energy conversion capacity and conductivity of the semiconductor material. Thus, even as an approximation, it is still an important electronic parameter to be studied.

Thus, Aloui *et al.* [38], using Schottky's ratio and impedance measurements, were able to demonstrate in devices based on P3HT:fullerene derivatives that, depending on the added percentage and purity of the acceptor, P3HT:fullerene has a strong influence on the number of charge carriers and also on their lifetime. Furthermore, the electronic parameters proved to be quite consistent with the photoelectrochemical properties, highlighting the interest in performing impedance spectroscopy in organic solar cells. On the other hand, in the works by Boix *et al.* [20], it was demonstrated by measuring capacitance as a function of temperature that, depending on the number of defects and the region where they occur, improvements in the properties of the devices can be observed.

In addition, electrochemical impedance spectroscopy can be used as an electronic characterization method in this type of system, making it possible to estimate important parameters such as the lifetime of charge carriers and their diffusion length [39], these being parameters directly related to the quantity of photocurrent generated by the active layer. When obtaining an impedance spectrum of a photovoltaic cell operating under light, it is necessary to compensate for the generated photocurrent, applying enough potential to produce no current; the applied potential must be forced so that the cell works as if it were at open circuit potential. Under these conditions, the photocurrent is cancelled by the recombination flux and jdc. The main RC arc characterizes the spectra obtained with other secondary processes at high frequency. The high-frequency part of the spectrum may contain transport information. Complying with such requirements, according to the authors Garcia-Belmonte *et al.* [39], it is then possible to assign the arc resistance to the differential recombination resistance R_{rec}, which is related to the recombination current density j_{rec}, so that the properties of the excitons can be estimated using Eq. (15.1).

$$R_{rec} = \frac{1}{A}\left(\frac{dj_{rec}}{dV_{oc}}\right) \tag{15.1}$$

where *A* is the electrode area.

Furthermore, the capacitance obtained under these conditions can be associated with the capacitance of the excitons. Therefore, by calculating the time constant associated with $(RC)_{rec}$, it is possible to estimate the lifetime of the electron-hole pair.

Additionally, using the data obtained by impedance at each potential, it was also possible to calculate the density of states (DOS) spectrum by applying Eq. (15.2), proposed by Gmucová *et al.* [40] and Nádaždy *et al.* [41]. DOS describes the number of electronic states per power range. So, the high value of the spectrum at a given energy means that there are too many states occupied or available for occupation.

$$d(E_{NF} = eU) = \frac{1}{ek_{ct}[E]AR_{ct}} \tag{15.2}$$

where k_{tc} is the charge transfer constant, *[E]* is the concentration of the supporting electrolyte, *A* is the geometric area of the electrode, and R_{tc} is the load transfer resistance. The spectrum obtained gives an idea of where the band edges are and, therefore, using this information, it is possible to estimate the working region

Through these unusual applications of electrochemical impedance spectroscopy data, it is possible to see how this technique is powerful and yet unexplored; thus, it is necessary to reinvent the techniques already used so that they can provide as much information as possible and be able to characterize deep into the new photo materials.

15.6 Quantum Methods Applied to OPVs

A very interesting way to plan new materials and nanocomposites for application in photovoltaic electrodes is using quantum methods because, using these, it is possible to predict both the absorption spectrum and the compatibility between the materials that make up these composites to maximize the synergistic effect among them. Therefore, this section addresses some of the approaches used to study these new materials theoretically.

Due to their high charge-carrying conductivity, fullerene derivatives are the most sought-after candidates for applying electron acceptors in BHJ (bulk heterojunction) solar cells. However, inspired by nature, the use of porphyrins and derivatives has become popular over the years. In this sense, Cooling *et al.* [42] studied the steric effects of lateral groupings in applied porphyrins used in composites with polymer matrix. Three different tetraryl-porphyrins with side groups of *p*-tolyl, 3,5-dimethylphenyl, and 3,5-di-*t*-butylphenyl were obtained so that the different levels of steric hindrance could be studied in the organization of porphyrins and their dispersion in composites with poly[2-methoxy-5-(20 -ethylhexyloxy)-1,4-phenylenevinylene] (MEH-PPV) and PCBM. Although several parameters were determined experimentally, due to observations made regarding the behavior of the series of porphyrins obtained, the DFT (Density Functional Theory) method was used to confirm the hypothesis that there is a dominant contribution of porphyrin to the mechanism of generation of load, along with the presence of significant differences in the electronic structure of the porphyrins, which resulted in the changes observed in VOC. However, compared to experimental and theoretical data, it was concluded that porphyrin does not have dominant participation in the generation of charges. Its electronic structure is little affected by the different functional groups. Thus, it was possible to attribute the differences observed to morphological differences, as well as to the existence of pinholes. Finally, through the series of experiments performed, the authors concluded that it is possible to change both the recombination rate and the charge separation efficiency by changing the composite morphology induced by steric effects of one of the compounds used. Such a conclusion would not be possible if a quantum mechanical method had not been used to confirm or refute hypotheses.

To the same end, Schiros *et al.* [43] used the DFT technique to compare electronic coupling and charge transfer rates at the interface between fullerene-based acceptors and different morphological arrangements of hexabenzocoronene (HBC), which, although not polymeric materials, nor 2D semi-rigid structures that can provide valuable information about how the morphological arrangement of interaction between materials may or may not favor better charge separation. Once again, combining experimental data with observations from quantum-mechanical simulations, the authors correlated the changes observed in the photoelectric properties of the materials obtained with the different morphologies of each of the HBCs with fullerene (Figure 15.4). As shown in Figure 15.4a, the charge transfer rate was the most affected property, as confirmed by the theoretical data, and thus, there was a great improvement in the external quantum efficiency (EQE) and PCE of the device. Such results proved to be very promising in guiding the design of new nanocomposites to improve solar cell performance.

Other studies involving conducting polymers have also shown, using theoretical methods such as DFT and Hartree–Fock, that different morphological and conformational arrangements profoundly change the electronic characteristics of the conducting polymer, impacting both its electrochemical and photoelectrochemical properties [44–46].

P3HT is one of the materials most used in polymer-based organic solar cells due to its unique hole-transporting properties; naturally, it is an object of study even in the theoretical field. Thus, Khlaifia *et al.* [47] investigated the electronic coupling of P3HT with small electron transport molecules. The authors simulated only the 3HT monomer interacting

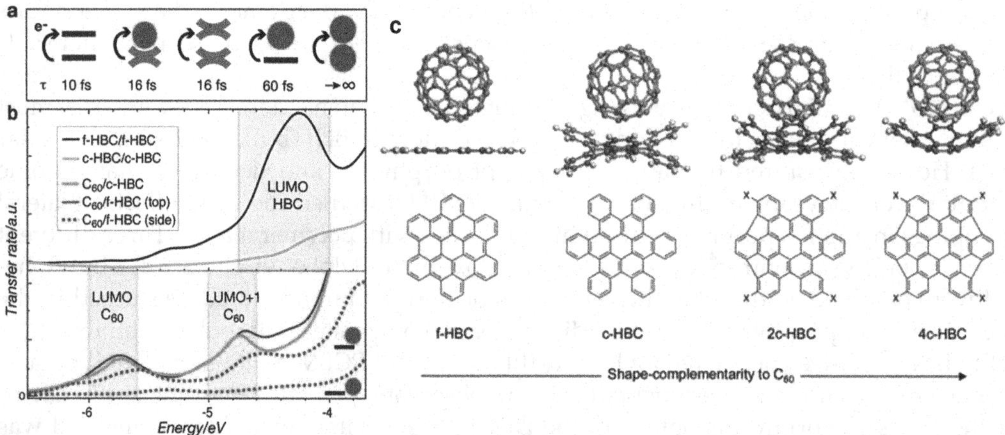

FIGURE 15.4
Experimental and computed charge transfer rates as a function of contortion: a) electron transfer rates for the HBC/fullerene series, b) electronic transfer rates, determined by the TD-DFT method, and c) schematic representation of HBC molecules with varying degrees of contortion and shape. Reproduced with permission from reference [43]. Copyright (2013) Wiley Online Library.

with such molecules to save on computational cost. However far from the practical reality, it is an interesting approach to predict the best donor-acceptor pair to give better results. Using a protocol based on more robust functions, pairs with optimized geometric arrangements were simulated to calculate the composites' bandgap (Figure 15.5). It was observed that, in those arrangements in which HOMO and LUMO are present in different molecules, the bandgap became wider, in addition to suggesting that the charge separation efficiency will be optimized. Such a conclusion agrees with the work of Gonçalves, Moraes, and Pereira [48], studying the electronic interaction between P3HT and reduced graphene oxide (rGO), using TD-DFT (Time-Dependent DFT), relating the electronic changes from pure materials to the composite with the increased observed photocurrent. This study demonstrates that the best separation of charges is the best factor for determining an improved photoelectrode performance, even allowing the prediction of electronic promotions other than HOMO-LUMO.

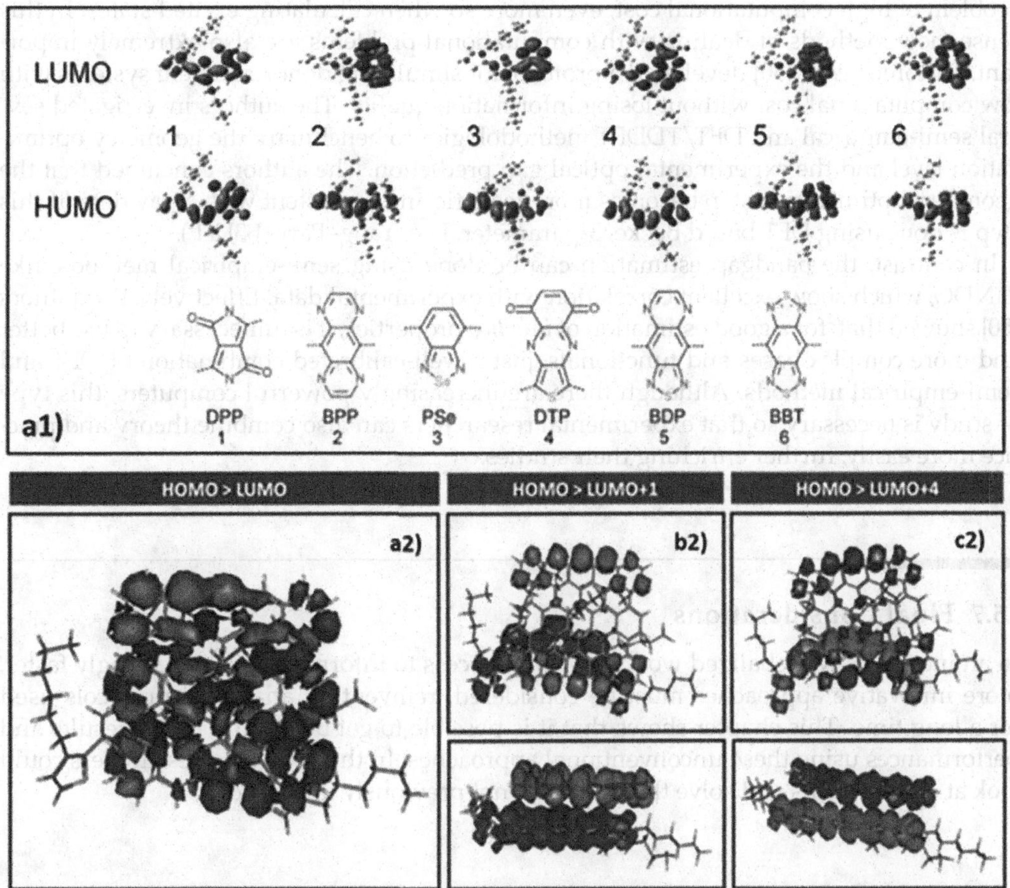

FIGURE 15.5
a1) Frontier molecular orbital spatial distribution for several donor molecules. Reproduced with permission from reference [47]. Copyright (2017) Wiley Online Library. The bottom figure illustrates the transition orbitals a2) HOMO and LUMO, b2) HOMO-LUMO+1, and c2) HOMO-LUMO+4, molecular orbitals involved in electronic promotions between P3HT and rGO. Reproduced with permission from reference [48]. Copyright (2016) Elsevier.

In addition to predicting new composites, theoretical methods are also very useful in modeling new polymers, predicting monomers that are promising for application in a particular device. This is the case of the work of Mestiri *et al.* [49], who modeled a co-polymer with donor and acceptor properties in the same molecule. It is important to emphasize that such amphoteric properties are only possible in macromolecules such as polymers. Their large size allows intramolecular interactions that give this type of property, such as the optoelectronic properties of 2,2′-(1,4-phenylene)bis(quinazolin-4(3H)-imine (PBQI) derivatives, the thiophene-alt-PBQI (TH-PBQI) and *N*-vinylcarbazole-alt-PBQI (VK-PBQI) copolymers, using DFT and TD-DFT. The new polymers obtained with this architecture showed less bandgap energy, with large optical absorption. However, they present charge donor and acceptor; the authors claim that better use of the material would be in photovoltaic devices of the BHJ type, using PCBM as the electron acceptor.

It is important to point out that quantum-mechanical methods are strong allies for experimental data; however, purely theoretical studies are also important for the progress of this research area. Nevertheless, theoretical studies of complex systems have the problem of high computational cost, even more so when calculating excited states. In this sense, new methods of dealing with computational problems are also extremely important. Tortorella *et al.* [50] developed a protocol for simulating donor-recipient systems with low computational cost without losing information quality. The authors investigated several semi-empirical and DFT/TDDFT methodologies to benchmark the geometry optimization level and the experimental optical gap prediction. The authors concluded that the geometry optimization step becomes more realistic and consistent with X-ray data, if this step is done using DFT-based Becke, 3-parameter, Lee–Yang–Parr (B3LYP).

In contrast, the bandgap estimation can be done using semi-empirical methods, like ZINDO, which show excellent correlation with experimental data. Effectively, the authors [50] showed that, for a good estimation of *in-silico* properties, it is unnecessary to use better and more complex bases and functionals, just a well-calibrated combination of DFT and semi-empirical methods. Although there are increasingly powerful computers, this type of study is necessary so that experimental researchers can also combine theory and practice more easily, further enriching their studies.

15.7 Final Considerations

In an increasingly globalized world, in which access to information is increasingly faster, more innovative approaches must be considered, reinventing analytical protocols used for a long time. This chapter shows that it is possible to get better and better results and performances using these unconventional approaches. In this way, new researchers could look at old problems and solve them in ways not previously imagined.

Acknowledgments

This study was financed in part by the Coordenação de Aperfeiçoamento de Pessoal de Nível Superior – Brasil (CAPES) – Finance Code 001 and CNPq. The authors would also like to thank FAPESP (grant no. 2018/09761-8) funding agency.

References

1. J. C. Bernède, "Organic photovoltaic cells: History, principle and techniques," *Journal of the Chilean Chemical Society*, vol. 53, no. 3, pp. 1549–1564, Sep. 2008, doi: 10.4067/S0717-97072008000300001.
2. A. J. Heeger, "Semiconducting polymers: The Third Generation," *Chemical Society Reviews*, vol. 39, no. 7, p. 2354, 2010, doi: 10.1039/b914956m.
3. A. Pochettino, "Sul comportamento foto-elettrico dell'antracene," *Accademia dei Lincei Rendiconti*, vol. 15, pp. 355–363, 1906.
4. H. Hoegl, "On photoelectric effects in polymers and their sensitization by dopants 1," *The Journal of Physical Chemistry*, vol. 69, no. 3, pp. 755–766, Mar. 1965, doi: 10.1021/j100887a008.
5. Y.-W. W. Su, S.-C. C. Lan, and K.-H. H. Wei, "Organic photovoltaics," *Materials Today*, vol. 15, no. 12, pp. 554–562, Dec. 2012, doi: 10.1016/S1369-7021(13)70013-0.
6. F. M. Nodari, *Propriedades opto-elétricas de dispositivos fotovoltaicos orgânicos em bicamadas*. Universidade Federal do Paraná, 2005.
7. D. J. Coutinho, *Estudo e caracterização de dispositivos fotovoltaicos orgânicos (OPV) baseados em heterojunção de volume*. Universidade de São Paulo, 2011.
8. F. B. Kooistra *et al.*, "Increasing the open circuit voltage of bulk-heterojunction solar cells by raising the LUMO level of the acceptor," *Organic Letters*, vol. 9, no. 4, pp. 551–554, Mar. 2007, doi: 10.1021/ol062666p.
9. A. Facchetti, "Polymer donor–polymer acceptor (all-polymer) solar cells," *Materials Today*, vol. 16, no. 4, pp. 123–132, Apr. 2013, doi: 10.1016/j.mattod.2013.04.005.
10. A. Marrocchi, D. Lanari, A. Facchetti, and L. Vaccaro, "Poly(3-hexylthiophene): Synthetic methodologies and properties in bulk heterojunction solar cells," *Energy and Environmental Science*, vol. 5, no. 9, p. 8457, 2012, doi: 10.1039/c2ee22129b.
11. T.-P. Nguyen, "Polymer-based nanocomposites for organic optoelectronic devices: A review," *Surface and Coatings Technology*, vol. 206, no. 4, pp. 742–752, Nov. 2011, doi: 10.1016/j.surfcoat.2011.07.010.
12. S. Das, D. P. Chatterjee, R. Ghosh, and A. K. Nandi, "Water soluble polythiophenes: Preparation and applications," *RSC Advances*, vol. 5, no. 26, pp. 20160–20177, 2015, doi: 10.1039/C4RA16496B.
13. F. Beck and U. Barsch, "The role of water in the electrodeposition and doping of polythiophene and two of its derivatives," *Die Makromolekulare Chemie*, vol. 194, no. 10, pp. 2725–2739, Oct. 1993, doi: 10.1002/macp.1993.021941008.
14. M. Jeffries-El, R. D. McCullough, S. Dong, and W. Zhang. *Conjugated Polymers: Theory, Synthesis, Properties, and Characterization*, Third Edition, T. A. Skotheim and J. R. Reynolds, Eds. New York: CRC Press, 2006, pp. 348–396.
15. Y. Pang, X. Li, H. Ding, G. Shi, and L. Jin, "Electropolymerization of high quality electrochromic poly(3-alkyl-thiophene)s via a room temperature ionic liquid," *Electrochimica Acta*, vol. 52, no. 20, pp. 6172–6177, Jun. 2007, doi: 10.1016/j.electacta.2007.04.015.
16. D. Kekuda, H.-S. Lin, M. Chyi Wu, J.-S. Huang, K.-C. Ho, and C.-W. Chu, "The effect of solvent induced crystallinity of polymer layer on poly(3-hexylthiophene)/C70 bilayer solar cells," *Solar Energy Materials and Solar Cells*, vol. 95, no. 2, pp. 419–422, Feb. 2011, doi: 10.1016/j.solmat.2010.05.055.
17. R. Bkakri, A. Sayari, E. Shalaan, S. Wageh, A. A. Al-Ghamdi, and A. Bouazizi, "Effects of the graphene doping level on the optical and electrical properties of ITO/P3HT:Graphene/Au organic solar cells," *Superlattices and Microstructures*, vol. 76, pp. 461–471, Dec. 2014, doi: 10.1016/j.spmi.2014.10.016.
18. C. Kang, S. Kim, Y. Hong, and C. Lee, "Frequency analysis on poly(3-hexylthiopene) rectifier using impedance spectroscopy," *Thin Solid Films*, vol. 518, no. 2, pp. 889–892, Nov. 2009, doi: 10.1016/j.tsf.2009.07.110.
19. Y. Harima, X. Jiang, R. Patil, K. Komaguchi, and H. Mizota, "Influence of film structure on mobilities of charge carriers in conducting polymers," *Electrochimica Acta*, vol. 52, no. 28, pp. 8088–8095, Nov. 2007, doi: 10.1016/j.electacta.2007.07.007.

20. P. P. Boix, J. Ajuria, I. Etxebarria, R. Pacios, and G. Garcia-Belmonte, "Kinetics of occupancy of defect states in poly(3-hexylthiophene): Fullerene solar cells," *Thin Solid Films*, vol. 520, no. 6, pp. 2265–2268, Jan. 2012, doi: 10.1016/j.tsf.2011.09.044.
21. J. Roncali, "Molecular bulk heterojunctions: An emerging approach to organic solar cells," *Accounts of Chemical Research*, vol. 42, no. 11, pp. 1719–1730, Nov. 2009, doi: 10.1021/ar900041b.
22. M. Lanzi, E. Salatelli, T. Benelli, D. Caretti, L. Giorgini, and F. P. Di-Nicola, "A regioregular polythiophene-fullerene for polymeric solar cells," *Journal of Applied Polymer Science*, vol. 132, no. 25, p. n/a-n/a, Jul. 2015, doi: 10.1002/app.42121.
23. G. Dennler, M. C. Scharber, and C. J. Brabec, "Polymer-fullerene bulk-heterojunction solar cells," *Advanced Materials*, vol. 21, no. 13, pp. 1323–1338, Apr. 2009, doi: 10.1002/adma.200801283.
24. M. T. Dang, L. Hirsch, and G. Wantz, "P3HT:PCBM, best seller in polymer photovoltaic research," *Advanced Materials*, vol. 23, no. 31, pp. 3597–3602, Aug. 2011, doi: 10.1002/adma.201100792.
25. M. A. Brady, G. M. Su, and M. L. Chabinyc, "Recent progress in the morphology of bulk heterojunction photovoltaics," *Soft Matter*, vol. 7, no. 23, p. 11065, Nov. 2011, doi: 10.1039/c1sm06147j.
26. A. M. Nardes *et al.*, "Free carrier generation in organic photovoltaic bulk heterojunctions of conjugated polymers with molecular acceptors: Planar versus spherical acceptors," *ChemPhysChem*, vol. 15, no. 8, pp. 1539–1549, 2014, doi: 10.1002/cphc.201301022.
27. J. M. Lobez, T. L. Andrew, V. Bulović, and T. M. Swager, "Improving the performance of P3HT–fullerene solar cells with side-chain-functionalized poly(thiophene) additives: A new paradigm for polymer design," *ACS Nano*, vol. 6, no. 4, pp. 3044–3056, Apr. 2012, doi: 10.1021/nn204589u.
28. J. Heinze, B. Frontana-Uribe, and S. Ludwigs, "Electrochemistry of conducting polymers–Persistent models and new concepts," *Chemical Reviews*, vol. 110, no. 8, pp. 4724–4771, Aug. 2010, doi: 10.1021/cr900226k.
29. G. P. Kini, S. J. Jeon, and D. K. Moon, "Latest progress on photoabsorbent materials for multifunctional semitransparent organic solar cells," *Advanced Functional Materials*, vol. 31, no. 15, pp. 1–32, 2021, doi: 10.1002/adfm.202007931.
30. C. Zhu *et al.*, "Spin-coated 10.46% and blade-coated 9.52% of ternary semitransparent organic solar cells with 26.56% average visible transmittance," *Solar Energy*, vol. 204, pp. 660–666, Jul. 2020, doi: 10.1016/j.solener.2020.05.027.
31. Y. Li *et al.*, "High-performance semi-transparent organic photovoltaic devices via improving absorbing selectivity," *Advanced Energy Materials*, vol. 11, no. 11, pp. 1–10, 2021, doi: 10.1002/aenm.202003408.
32. Y. Cui *et al.*, "Single-junction organic photovoltaic cells with approaching 18% efficiency," *Advanced Materials*, vol. 32, no. 19, pp. 1–7, 2020, doi: 10.1002/adma.201908205.
33. Y. Cho *et al.*, "Guest-oriented non-fullerene acceptors for ternary organic solar cells with over 16.0% and 22.7% efficiencies under one-sun and indoor light," *Nano Energy*, vol. 75, p. 104896, 2020, doi: 10.1016/j.nanoen.2020.104896.
34. L. K. Ma *et al.*, "High-efficiency indoor organic photovoltaics with a band-aligned interlayer," *Joule*, vol. 4, no. 7, pp. 1486–1500, 2020, doi: 10.1016/j.joule.2020.05.010.
35. Q. Fu *et al.*, "Recent progress on the long-term stability of perovskite solar cells," *Advanced Science*, vol. 5, no. 5, 2018, doi: 10.1002/advs.201700387.
36. M. C. Beard, J. M. Luther, O. E. Semonin, and A. J. Nozik, "Third generation photovoltaics based on multiple exciton generation in quantum confined semiconductors," *Accounts of Chemical Research*, vol. 46, no. 6, pp. 1252–1260, 2013, doi: 10.1021/ar3001958.
37. X. Hou *et al.*, "Indoor application of emerging photovoltaics - Progress, challenges and perspectives," *Journal of Materials Chemistry A*, vol. 8, no. 41, pp. 21503–21525, 2020, doi: 10.1039/d0ta06950g.
38. W. Aloui, A. Ltaief, and A. Bouazizi, "Electrical impedance studies of optimized standard P3HT:PC70BM organic bulk heterojunctions solar cells," *Superlattices and Microstructures*, vol. 75, pp. 416–423, Nov. 2014, doi: 10.1016/j.spmi.2014.07.051.
39. G. Garcia-Belmonte, P. P. Boix, J. Bisquert, M. Sessolo, and H. J. Bolink, "Simultaneous determination of carrier lifetime and electron density-of-states in P3HT:PCBM organic solar cells

under illumination by impedance spectroscopy," *Solar Energy Materials and Solar Cells*, vol. 94, no. 2, pp. 366–375, 2010, doi: 10.1016/j.solmat.2009.10.015.

40. K. Gmucová, V. Nádaždy, F. Schauer, M. Kaiser, and E. Majková, "Electrochemical spectroscopic methods for the fine band gap electronic structure mapping in organic semiconductors," *The Journal of Physical Chemistry C*, vol. 119, no. 28, pp. 15926–15934, Jul. 2015, doi: 10.1021/acs.jpcc.5b04378.
41. V. Nádaždy, F. Schauer, and K. Gmucová, "Energy resolved electrochemical impedance spectroscopy for electronic structure mapping in organic semiconductors," *Applied Physics Letters*, vol. 105, no. 14, p. 142109, Oct. 2014, doi: 10.1063/1.4898068.
42. N. Cooling *et al.*, "A study of the factors influencing the performance of ternary MEH-PPV:porphyrin:PCBM heterojunction devices: A steric approach to controlling charge recombination," *Solar Energy Materials and Solar Cells*, vol. 95, no. 7, pp. 1767–1774, 2011, doi: 10.1016/j.solmat.2011.01.046.
43. T. Schiros *et al.*, "Donor-acceptor shape matching drives performance in photovoltaics," *Advanced Energy Materials*, vol. 3, no. 7, pp. 894–902, 2013, doi: 10.1002/aenm.201201125.
44. R. Gonçalves, W. A. A. Christinelli, A. B. B. Trench, A. Cuesta, and E. C. C. Pereira, "Properties improvement of poly(o-methoxyaniline) based supercapacitors: Experimental and theoretical behaviour study of self-doping effect," *Electrochimica Acta*, vol. 228, pp. 57–65, Feb. 2017, doi: 10.1016/j.electacta.2016.12.188.
45. R. Gonçalves, R. S. Paiva, T. M. Lima, M. W. Paixão, and E. C. Pereira, "Carbon nitride/polypyrrole composite supercapacitor: Boosting performance and stability," *Electrochimica Acta*, vol. 368, p. 137570, Feb. 2021, doi: 10.1016/j.electacta.2020.137570.
46. R. S. Bhatta and M. Tsige, "Chain length and torsional dependence of exciton binding energies in P3HT and PTB7 conjugated polymers: A first-principles study," *Polymer*, vol. 55, no. 11, pp. 2667–2672, 2014, doi: 10.1016/j.polymer.2014.04.022.
47. D. Khlaifia, F. Massuyeau, C. P. Ewels, J. L. Duvail, E. Faulques, and K. Alimi, "DFT modeling of novel donor-acceptor (D-A) molecules incorporating 3-hexylthiophene (3HT) for bulk heterojunction solar cells," *ChemistrySelect*, vol. 2, no. 31, pp. 10082–10090, 2017, doi: 10.1002/slct.201701481.
48. R. Gonçalves, F. C. C. Moraes, and E. C. C. Pereira, "Reduced graphene oxide enhancing the photoelectrochemical properties of poly(3-hexylthiophene)," *Carbon*, vol. 109, pp. 57–64, Nov. 2016, doi: 10.1016/j.carbon.2016.07.041.
49. T. Mestiri, S. Ghomrasni, D. Khlaifia, and K. Alimi, "DFT modeling of new (2,2′-(1,4-phenylene)bis(quinazolin-4(3H)-imine) (PBQI) based copolymers for photovoltaic applications," *Physica B: Condensed Matter*, vol. 544, no. Mar., pp. 1–9, 2018, doi: 10.1016/j.physb.2018.05.009.
50. S. Tortorella, M. M. Talamo, A. Cardone, M. Pastore, and F. De Angelis, "Benchmarking DFT and semi-empirical methods for a reliable and cost-efficient computational screening of benzofulvene derivatives as donor materials for small-molecule organic solar cells," *Journal of Physics Condensed Matter*, vol. 28, no. 7, 2016, doi: 10.1088/0953-8984/28/7/074005.

16

One-Dimensional Polymeric Nanocomposites for Photovoltaic Devices

Seyyed Alireza Hashemi, Sonia Bahrani, Seyyed Mojtaba Mousavi, Mohammad Arjmand, and Seeram Ramakrishna

CONTENTS

16.1 Introduction

Polymeric solar cells (PSC) have been identified as alternative energy sources, instead of inorganic photovoltaic configurations, due to their low cost, ease of processing, reduced environmental hazard, light weight, and mechanical flexibility. Organic/polymeric solar cells have been extensively developed over the past few decades. These platforms enable roll-to-roll processing to improve the manufacturing speed and reduce the total upkeep cost of photovoltaic platforms for general use [1]. Despite the potential of PSCs for mass-scale production, however, they face serious challenges, among which short life service and low power conversion efficiency (PCE) should be mentioned. Hence, improving the durability, lifetime, and total PCE of such configurations has become a major challenge in the industrialization of PSCs.

Recent improvements in the structure and components of PSCs led to the formation of polymeric photovoltaic configurations with PCEs of about 16.5% [2] and 17.3% [3] through the single-junction and tandem configuration, respectively. As a result, the devices developed are three-fold stronger than early PSCs, highlighting the requirement for practical strategies to improve the capability of polymeric photovoltaic devices. In this regard, improvement of fused-ring electron acceptors [4] led to a rise in the PCE of devices to more than 17% *via* tandem bulk-heterojunction and single-junction platforms [3]. On the other hand, an increase in the stability and development of large-area roll-to-roll processing approaches became possible by improving non-fullerene and fullerene-based acceptors, allowing the commercialization of PSCs [5].

Increases in the PCE and durability of PSCs have become possible owing to the development of non-fullerene electron acceptors with broader adsorption rates in the spectrum of sunlight, reduced dissipation in the charge separation process, improved morphology of the

DOI: 10.1201/9781003223764-16

respective layers, device engineering, and improvement of interfacial layers. Accordingly, the integration of small perylene bisimide molecules into the zinc oxide interfacial layer increases the PCE of the optimum devices to about 16% [6]. This progress exhibited the potential of PSCs to reach a PCE of more than 20% in the next few years in response to practical strategies and material processing approaches. It is essential to optimize multiple factors of PSCs to obtain a balance among exciton splitting, light-harvesting, and carrier transportation/collection [7].

In addition, the main parts of polymeric photovoltaic devices, such as small electron donor molecules and non-fullerene or fullerene electron acceptor molecules, experienced massive advances toward improving the PCE and durability of PSCs. Correspondingly, the all-polymeric solar cell (APSC) is a type of photovoltaic platform that consumes both polymeric donor and acceptor. The advantages of APSCs over conventional organic solar cells are their tunable light-harvesting rate, mechanical/morphological robustness, long-term stability, and compatibility with devices with large, active surface areas [8].

In the current chapter, first, the mechanism of function of PSCs is explained in detail, and thence effective, one-dimensional, polymeric platforms for the development of polymeric, photovoltaic configurations are described, and their pros and cons are assessed. Next, recent progress in flexible PSCs is evaluated, and the platforms developed are examined through an in-depth assessment.

16.2 Light-Harvesting Mechanism of Polymeric Solar Cells

In this section, the light-harvesting mechanism of PSCs was assessed to gain an insight into the working principles of such photovoltaic configurations. A view of a typical polymeric solar cell is demonstrated in Figure 16.1a. As depicted, the active photovoltaic layer of PSCs consists of a polymer blend composed of polymeric donor and acceptor sandwiched between electron/hole transport layers. In bulk heterojunctions, donor and acceptor polymers generate a bi-continuous interpenetrating configuration that provides a large active surface area and proper phase separation at the nanoscale. Such a configuration allows effective exciton splitting and charge collection. Furthermore, interfacial layers positioned below and above the bulk heterojunction, e.g., small molecules, metal oxides (MoO_3 and ZnO), and transparent polymers (PEDOT: PSS), act as selective charge filters that lead to charge extraction and hence generation of current output [8].

The operational mechanism of PSCs could be divided into four main steps, including (Figure 16.1b): 1) generation of strong coulombic bounds of electron-hole pairs due to the photoexcitation of the acceptor and/or donor, viz., excitons; 2) infiltration of excitons into the interfaces of the donor-acceptor complex that leads to their dissociation to generate free electrons and holes; 3) separation of drifted charges through acceptor and donor polymeric domains; and 4) collection of separated electrons and holes at the cathode and anode of the photovoltaic configuration [8].

In Figure 16.2c, a common current density–voltage plot is illustrated that shows the performance of a PSC upon its exposure to sunlight. The short circuit current (J_{sc}) corresponds to the maximum amount of flowed current in the photovoltaic device that originated from the internal field. Domination of J_{sc} occurs *via* the photoexcitation density, charge separation of quantum efficiency, and charge carrier mobility of holes or electrons. To achieve a

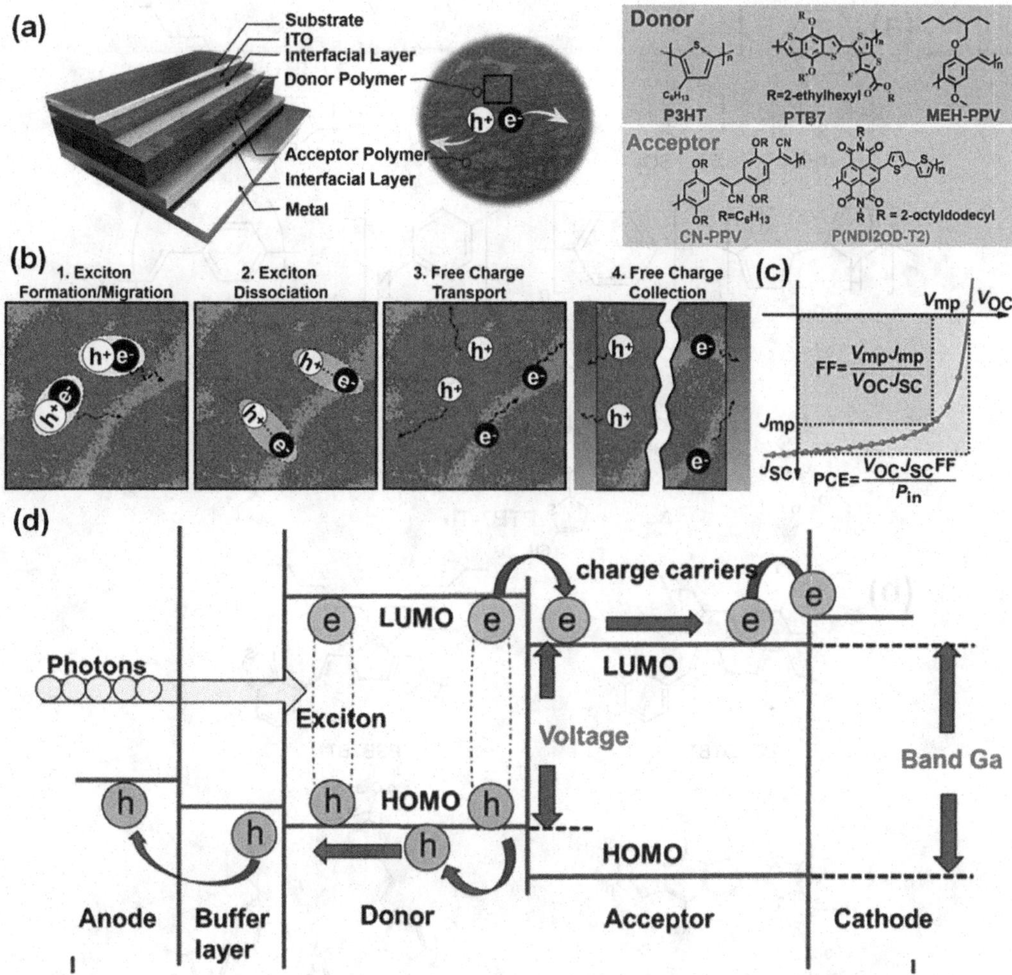

FIGURE 16.1
(a) Operational mechanism and structure of a typical PSC along with common donors and acceptors, and (b) charge generation mechanism in a bulk heterojunction PSC, grey areas show the acceptor and donor polymeric domains; (c) figures-of-merit of typical solar cells including open-circuit voltage (V_{oc}), short circuit current (J_{sc}), fill factor (FF), and power conversion efficiency (PCE). Adapted with permission from reference [8], copyright (2019), Wiley. (d) Working mechanism of polymeric photovoltaic devices. Adapted with permission from reference [9], copyright (2016), Elsevier.

high light-harvesting rate, the preparation of an active layer with a broader light absorption spectrum is essential. Moreover, the open-circuit voltage (V_{oc}) comes from splitting electron and hole quasi-Fermi levels [8]. Therefore, the V_{oc} could be measured *via* Eq. (16.1):

$$V_{oc} = \frac{1}{e}\left(E_{Fn} - E_{Fp}\right) \tag{16.1}$$

where E_{Fn} and E_{Fp} correspond to the quasi-Fermi levels of electrons and holes, respectively. Another major parameter is the fill factor (FF). This parameter could be measured *via* the competition between the recombination and extraction of free charges, known as the ratio

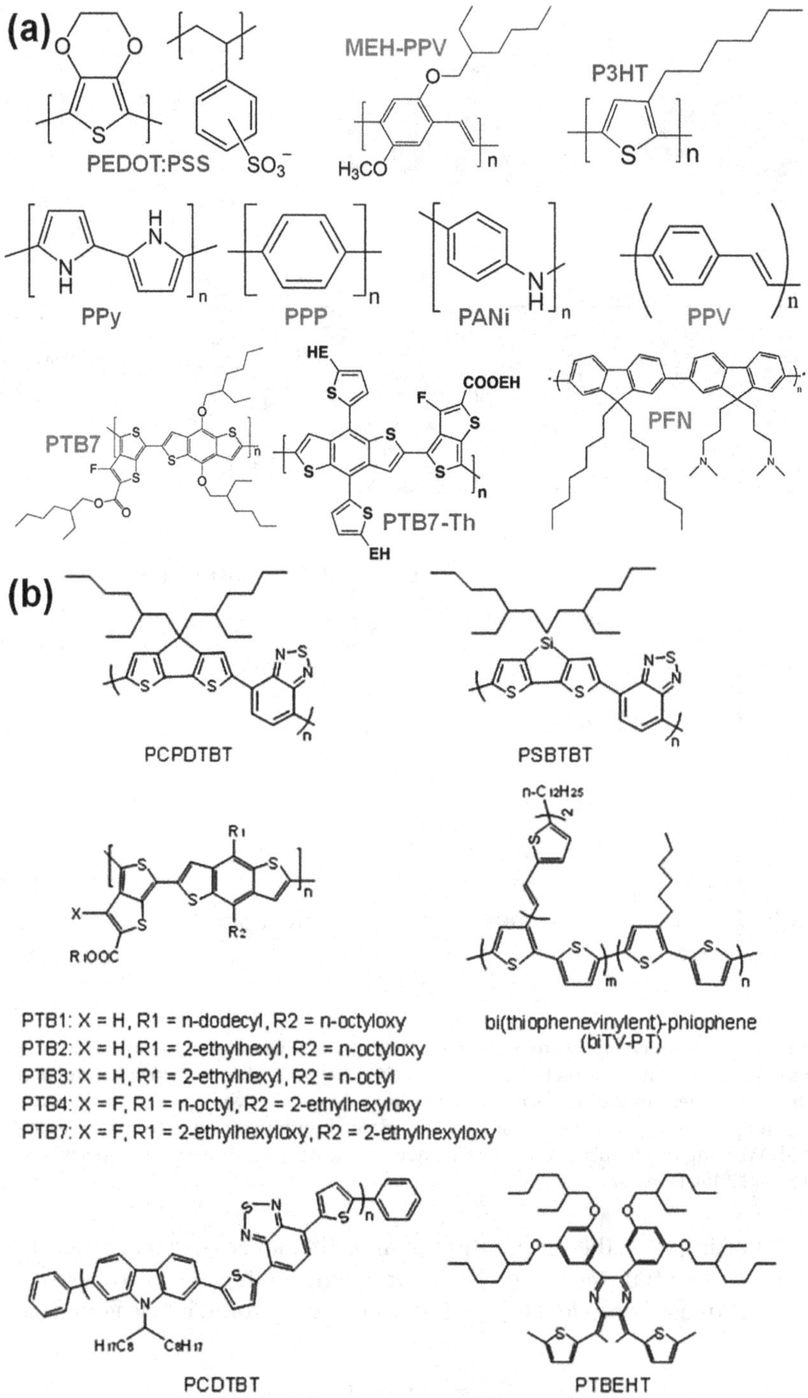

FIGURE 16.2
(a) Chemical structure of typical conjugated polymers for PSCs, and (b) the molecular structure of vital polymers with low bandgaps Adapted with permission from reference [27], copyright (2012), Wiley.

between a solar cell's maximum power output (P_m) and V_{oc} and J_{sc}. For example, the FF could be measured as described in Eq. (16.2) [8]:

$$FF = \frac{P_m}{V_{oc}J_{sc}} = \frac{V_{mp}J_{mp}}{V_{oc}J_{sc}} \tag{16.2}$$

where V_{mp} and J_{mp} are the voltage and current density at the maximum power point, respectively. The power conversion efficiency (PCE) as the main parameter of solar cells is known as the ratio between the P_m and the input power of the incident light (P_{in}). PCE can be measured as described in Eq. (16.3) [8]:

$$PCE = \frac{P_m}{P_{in}} \times 100 = \frac{V_{oc}J_{sc}FF}{P_{in}} \tag{16.3}$$

In Figure 16.1e, the operational mechanism of PSCs is demonstrated. As depicted, emitted photons were adsorbed to the surface of the photovoltaic device *via* the acceptor and/or donor polymeric substrates, promoting the adsorbed electrons to enter the lowest unoccupied molecular orbital (LUMO) and leaving the holes in the highest occupied molecular orbital (HOMO), leading to the generation of excitons, viz., electron-hole pairs, bounded via the coulomb attraction forces because of the low dielectric constant of polymeric compounds. Subsequently, the excitons generated diffuse into the interface between the acceptor and donor and thence separate into free charge carriers, leading to electricity generation by transporting and collecting these charge carriers at the anode and cathode interfaces. In PSCs, the active layer composed of donor and acceptor molecules is a semiconductor polymeric composite, in which the HOMO and LUMO are attributed to the valence and conduction bands within polymeric photovoltaic devices. Excitons have a short life span of about $<$ 1 ns that demands prompt dissociation to prevent recombination of charges [9]. Another vital issue for generating free charge carriers and their effective collection is that the distance between the nearest acceptor/donor interface and the excitons formed must be within their acceptable diffusion length (L_D) [10]. Mikhnenko et al. [11] demonstrated that, in various prototypes, the narrow bandgap of polymeric substrates, viz. L_D, is within the 10 nm range that highlights the criticality of controlling the morphology of the active layer within the nanoscale range.

The term "donor" is attributed to compounds that can become highly ionized and willing to donate their electrons to a compound with a high electron affinity, i.e., an acceptor, with strong electron-accepting specifications, which could thereby split the excitons into collectible free charge carriers. The variation between the LUMO energy level of the acceptor and donor leads to a driving force that could promptly transfer electrons. For an optimal PCE, the definition of the variation between the HOMO and LUMO, i.e., the optical bandgap, is a crucial matter. Selection of proper opposite electrodes with favorable functions guarantees the transportation and collection of free charge carriers at the electrode interfaces. In this matter, the V_{oc} corresponds to the variation between the LUMO of the acceptor layer and the HOMO of the donor layer. Furthermore, the J_{sc} is defined as the overall number of adsorbed photons and quantum efficiency [12].

Several loss sources are also defined in PSCs that can reduce the total PCE of the device, among which the exciton decay, non-absorbed photons, and biomolecular recombination can be mentioned. The recombination can be categorized into two main groups, namely non-geminate and geminate recombination, highlighting the crucial role of the recombination process in the dissipation of generated charge carriers [13].

16.3 Effective One-Dimensional Polymers for Solar Cells

APSCs are considered to be promising photovoltaic configurations because of their application in developing flexible configurations by dint of their potential advantages, including proper light-harvesting rate *via* a polymeric donor, a chemically tunable polymeric acceptor, and ideal mechanical robustness [14–16]. However, the type of polymers used and their specifications greatly affect PSCs' PCE and total performance. For instance, alternating the composition of bithiophene polymers and naphthalene diimide could promote the PCE of APSCs to about 11% [17, 18].

APSCs are a type of photovoltaic cell consisting of a bulk heterojunction active layer composed of a polymeric donor and acceptor. The concept of an all-polymer solar cell can be compared with the small molecule concept in three types of bulk heterojunction substrates: 1) a polymeric donor with a small molecule acceptor; 2) a small molecule acceptor with a polymeric acceptor; and 3) an all-small-molecule bulk heterojunction film. PSCs have a better potential for controlling the viscosity of the blend during the mass-scale production of polymeric photovoltaic cells [1].

Naphthalene diimide-based polymers have been widely utilized as potential polymeric acceptors in APSCs due to their robust electron affinity and boosted electron mobility (μ_e) [8, 19]. The superior electron mobility of naphthalene diimide-based acceptors originated from the self-assembled crystalline structure generated *via* the intermolecular interaction between the planar units of the naphthalene diimide polymer chains [20]. Furthermore, the high crystallinity of the polymeric acceptor can improve the electron mobility through an increase in the level of crystal domains over the common exciton diffusion length (~20 nm) [21]. However, an increase in the crystalline domain of polymeric acceptors could lead to an undesirable morphology because of the low intermixing with polymeric donors that lead to greater phase separation [22]. This process leads to ineffective charge transportation and exciton dissociation between the polymeric acceptor and donor, resulting in a decrease in the J_{sc} and FF of PSCs [23]. More importantly, polymeric active layers are rigid and brittle because of their high crystallinity, leading to the generation of cracks within the polymeric film after being exposed to morphological stress or mechanical load [24].

The n-type polymer acceptors can be categorized into various types, including benzothiadiazole, cyanated polyphenylenevinylenes, diketopyrrolopyrrole, indacenodithiophene, bithiophene imide, and rylene diimide-based polymers. In this regard, the electronic, morphological, and optical characteristics along with the chemical structure of these polymeric compounds can be modulated to produce various types of donors. From a mechanistic point of view, controlling the molecular weight of the acceptor and donor polymers is of great importance. In addition, the relative ratio of each one of acceptor/donor polymers could significantly affect the intermolecular assembly, which is a highly critical factor for the charge transfer capability and phase separation of the active layer. Large-scale phase separation of PSCs can be governed by engineering the molecular weight, solution processing protocol, and blends ratio. The morphology of PSCs can also be subtly tuned *via* experimental post-treatment modification. The electron transport rate can be suppressed in a bulk heterojunction polymer by the Poole–Frenkel effect and a decline in the energetic disorder [1].

PSCs can be better engineered to become well-suited for practical applications. In this matter, most of the PSCs are assessed under laboratory conditions *via* standard solar simulators under a nitrogen atmosphere. On the other hand, these devices should tolerate harsh conditions and be utilized under indoor low-light illumination, and atmospheric conditions. To sustain the optimum lifetime of PSCs, practical improvements have been considered to circumvent various degradative processes by optimizing the morphology of active layers and packing to prolong the lifetime of the resulting photovoltaic configurations [7]. Along with conventional approaches for optimizing the molecular packing within the active layers, removing the electron acceptor component near the top electrode leads to the generation of durable and robust PSCs that can even work underwater without the need for encapsulation [25].

The optimum conditions of various steps of PSCs, such as exciton splitting, charge collection, light adsorption, and migration of charge carriers, are often mismatched. For instance, the wide absorption spectrum of polymeric compounds originated from their intrinsic disorder within the solid film might boost the light-harvesting rate and increase the overall rate of nonradiative decays of excitons because of the distribution of the lower energy transition, which is known as a major cause of significant losses in the V_{oc} [26]. In addition, some donors with small molecules that exhibit ordered packing compared with polymers could lead to a decline in the level of energy disorder and thereby decrease the nonradiative decay related to excitons and minimize losses in the driving force of the resulting configuration. Hence, an in-depth understanding of loss management at the molecular level could guide material designers to improve the life service and performance of PSCs. In Figure 16.2a, a view of frequently used polymeric compounds in the PSCs can be seen, including PEDOT: PSS, MEH-PPV, P3HT, polypyrrole (PPy), poly(*p*-phenylene) (PPP), polyaniline (PANi), poly(p-phenylenevinylene) (PPV), PTB7, PTB7-Th, and PFN. In Figure 16.2b, the chemical structures of important polymers with low bandgaps is presented.

Figure 16.3a presents some frequently used donor and acceptor polymers. As the diffusion length (L_D) is limited to about 10 nm, obtaining a proper, phase-separated, donor-acceptor domain is crucial to developing APSCs with boosted PCE. Controlling the morphology of polymeric layers remains an important challenge in developing high-performance PSCs [28]. The domain's connectivity is also considered to be a critical parameter since it governs the charge flow, thereby charging carrier mobility toward the respective electrode, and avoiding recombination of charge carriers.

The charge flow could be strongly affected by the generation of crystalline domains, aggregates, and the orientation situation of a π-conjugated polymer's backbone with regard to the electrode surface [29]. Enhancing the PCE of PSCs passes through the optimization of the morphology of the bulk-heterojunction that depends on the following approaches: i) selecting multi-block/random co-polymers (especially for the acceptor) and engineering the side-chains of the π -conjugated polymeric core; ii) enhancing the molecular mass (M_n) of active layers; and iii) considering an effective blend-processing approach such as use of solvent additives. These approaches could govern crucial aspects of PSCs and enhance their morphologies toward boosting the optical adsorption characteristic of the resulting configuration [8]. In Figure 16.3b, images of recently developed polymeric donors according to molecular mass are presented.

Many approaches have been attempted to develop 1D polymeric configurations for solar cells. Ahmad et al. [30] developed a facile production approach toward more stable lead-free methyl ammonium bismuth chloride in the form of a 1D polymeric platform for perovskite solar cells. The lead-free polymeric layer developed showed a perfect open-circuit voltage of 430 mV. In another attempt, Gu et al. [31] developed a novel type of donor-acceptor polymer based on the alkylthienyl asymmetric benzodithiophene along with $DT_{ff}BT$ *via* the Pd-catalyzed Stille process. The configuration developed is blended with $PC_{71}BM$, while the resulting device showed values for FF, V_{oc}, J_{sc}, and PCE of 70.4 %, 0.87 V, 15.06 mA.cm^{-2}, and 9.22%, respectively. The outcome of this study revealed the prominent role of alkylthienyl units in boosting the photovoltaic performance of polymeric solar cells.

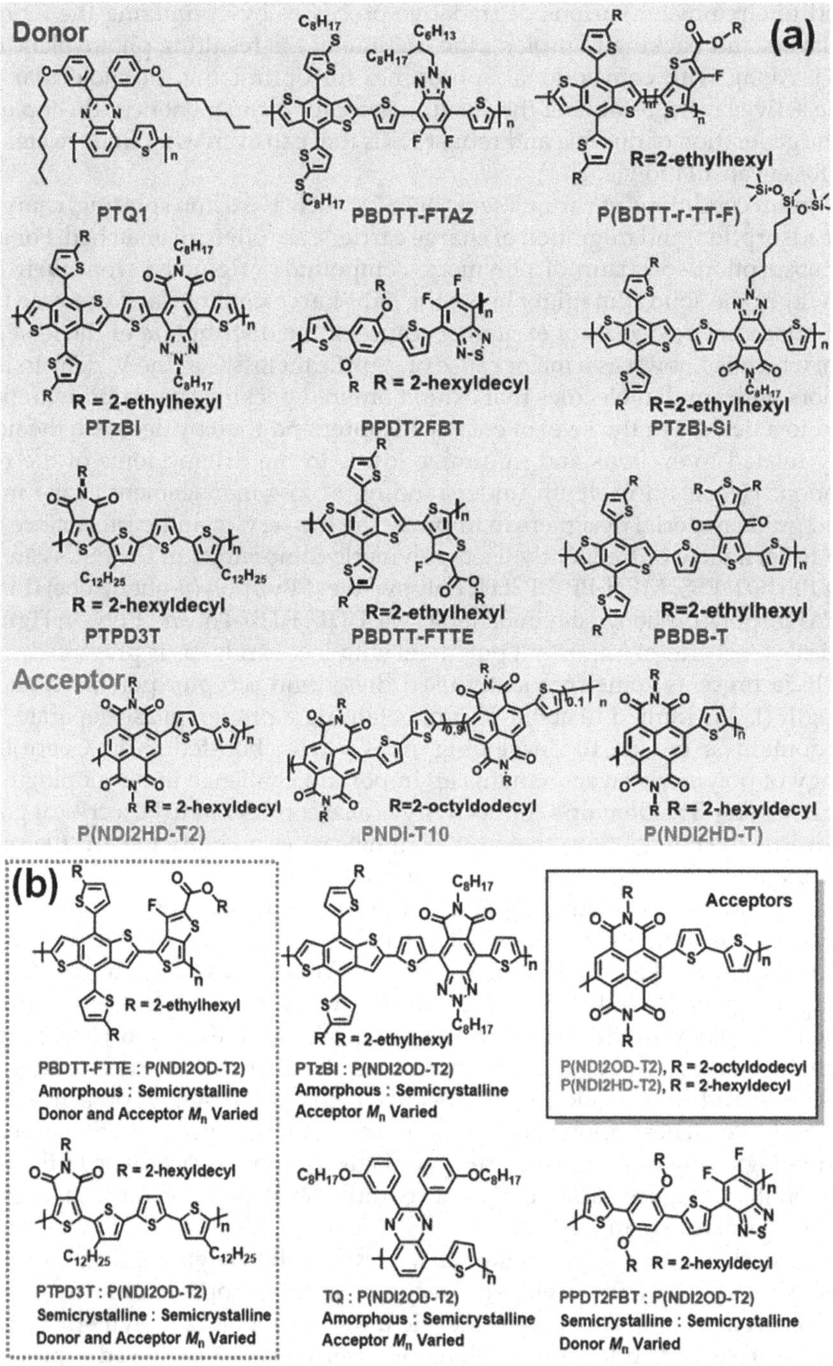

FIGURE 16.3
(a) Chemical structure of some frequently used polymeric donors and polymeric acceptors, and (b) recently developed polymeric donors assessed as a function of molecular mass. Adapted with permission from reference [8], copyright (2019), Wiley.

Han et al. [32] boosted the capability of PSCs by introducing 1D PEDOT nanorods to the buffer layer. Accordingly, they used 1D PEDOT nanorod along with the PEDOT: PSS configuration as the anode buffer substrate in the PSCs. The 1D PEDOT layer considerably improved the photovoltaic capability of the cell and boosted the polarity/conductivity of the buffer layer, which enhanced the V_{oc}, J_{sc}, and PCE of the resulting configuration. The development of nanowire-shaped polymers through donor-acceptor conjugated polymeric structures is considered to be an interesting approach for improving the PCE of bulk-heterojunction solar cells. To achieve this aim, Lee et al. [33] developed a single-junction cell composed of donor-acceptor conjugated nanowire-shaped polymer, viz., P4TNTz-2F, which presented a low bandgap of 1.59 eV and deep-lying HOMO of −5.46 eV within a rigid/planar backbone structure. The active layer developed, i.e., P4TNTz-2F:$PC_{71}BM$, completely adsorbed the incident light, leading to a boosted J_{sc} of 19.45 $mA.cm^{-2}$, as well as a superior PCE of about 10.62% under optimum conditions. The high PCE of the photovoltaic configuration developed is due to the integration of 1D narrow nanowire-shaped polymers with PC71BM, which facilitates the charge separation rate, viz., QPL of about 98% and charge transport rate of 8.31×10^{-3} cm^2 V^{-1} s^{-1} for the polymeric photovoltaic device. This nanowire-shaped polymeric platform within the active layer can be prepared *via* the film-forming protocol at a moderate blending temperature (about 70 °C), highlighting its excellent reproducibility, adaptable with mass-scale production approaches. One-dimensional polymeric configurations could modify the light-harvesting rate and morphological features of the resulting polymeric photovoltaic devices.

In another attempt, Liu et al. [34] developed a single-junction polymeric solar cell consisting of 1D and 2D conjugated polymers. By coupling non-conjugated alkoxy chains of a 1D (PTB7) polymeric configuration with conjugated 5-alkyl thiophene-2-yl chains, a low bandgap donor-acceptor conjugated polymeric configuration was developed (PTB7-DT). This configuration showed boosted charge carrier mobility, narrower bandgap, and tighter π-π stacking interaction. The photovoltaic device developed also exhibited a high PCE of 10.12% as a result of the modification steps employed. Similarly, Wang et al. [35] developed a one-dimensional 2Å-thick substrate composed of an active ternary layer based on PEDOT as donor and $PC_{71}BM$ along with non-fullerene ITIC as the polymeric acceptor. Whereas the PEDOT: ITIC revealed poor low-thickness tolerance, PEDOT: $PC_{71}BM$ exhibited better thick-film capability with PCEs of 5.14 % and 7.55% at thicknesses of 110 nm and 300 nm, respectively. The integrated configuration composed of PEDOT: $PC_{71}BM$:ITIC with a blend ratio of 1:1.2:0.3 exhibited a much higher PCE of 8.42% at 300 nm thickness compared with the other polymeric configurations developed. The hybrid polymeric platform showed ideal thickness tolerance ranging from 110 nm to 520 nm along with an ideal morphological arrangement. The ternary active layer developed improved the J_{sc} due to an increase in the carrier recombination and photo-responses.

Improving the π -conjugation rate in the benzodithiophene side chains could improve the effectiveness of the resulting polymeric solar cells. Zhu et al. [36] employed a protocol in which, with the symmetric breaking of benzodithiophene units, a novel 1D–2D monomer was developed, composed of alkoxy groups as the 1D part and alkoxybiphenyl as the 2D part. Additionally, a donor-acceptor conjugated polymeric configuration with moderate bandgap was developed with asy-BDTBP (donor) and DTffBT (acceptor). Integration of the developed blend with PC71BM improved the PCE to about 8.45 % and exhibited values for V_{oc}, J_{sc}, and FF of 0.838 V, 14.35 $mA.cm^{-2}$, and 70.27%, respectively. However, integrating the developed polymeric blend with non-fullerene acceptors, such as ITIC, could reduce the PCE to 6.35% and adversely affect the light-harvesting rate. On the other hand, upon matching the bandgap of ITIC layer with wide bandgap donor-acceptor P2 polymer, a boosted PCE of 10.04%, along with values for V_{oc}, J_{sc}, and FF of 0.873 V, 17.60 mA.cm^{-2}, and 65.37%, respectively, can be obtained. This work demonstrated the importance of rational designs of polymeric blends for boosting the PCE and overall light-harvesting rate of polymeric solar cells.

16.4 Flexible Polymeric Solar Cells

PSCs are considered to be ideal configurations for designing and developing flexible photovoltaic devices due to their superior flexibility, light weight, low processing temperature, and semi-transparency. Most of the polymeric-based solar cells contain a hetero-junction composed of a conjugated polymer or molecule as the electron donors and fullerene-based compounds as electron acceptors. However, PSCs that employ fullerene-based compounds suffer from a number of disadvantages, including fixed energy levels, weak light-harvesting rate, and chemical constitutions that adversely affect the J_{sc} and V_{oc} of the resulting configuration. Additionally, fullerene-based devices exhibit brittle structure, poor flexibility, and low stretchability compared with other solar cells, highlighting the requirement for tailoring their structure and composition to cover their drawbacks [37].

Various attempts were carried out to improve the life service and efficiency of PSCs. In a study by Kim et al. [38], they improved the PCE and mechanical robustness of PSCs by tailoring the intermolecular interactions between the main substrates. They controlled the blended mixture of the polymeric acceptor *via* the introduction of a fluorinated co-polymer to the (E)-1,2-bis(3-fluorothiophen-2-yl)ethene (FTVT) and naphthalene diimide. The resulting copolymer exhibited lower crystallinity and greater electron mobility than non-fluorinated co-polymers. Furthermore, the blend developed showed ideal morphology that could improve the interconnectivity of active chains with a polymeric donor, increasing the charge transport rate and enhancing the mechanical robustness. The flexible PSC developed *via* the enhanced polymeric acceptor exhibited PCE values of about 5.11–7.14%, with excellent mechanical stability capable of maintaining 81% of its primary performance even after 1,000 bending cycles with a bending radius of 8 mm. In Figures 16.4a–b, the chemical structure of the polymeric compounds used in the active layer and the 3D structure of the photovoltaic configuration can be seen, respectively. Additionally, in Figure 16.4c, the normalized PCE of the PSC as a function of the bending radius and bending cycle is demonstrated.

Chirilă et al. [39] developed a flexible polymeric photovoltaic configuration consisting of polycrystalline Cu(In, Ga)Se_2 adsorber to improve the PCE of PSCs. Such a platform could also significantly lower the production cost of PSCs by keeping their PCE at the optimum level. However, the flexibility of the substrates is also a crucial matter in the design and development of PSCs for real-life applications. For this aim, by identifying a strong gradient in the composition of the adsorber layer, the main reason behind the inferior performance of this layer is identified, which paves the way to boosting the PCE by addressing its weaknesses. In Figure 16.4d–e, a view of the flexible solar cell developed and its respective layers is presented, exhibiting a much higher PCE of about 18.7%. Additionally, in Figure 16.4f, a view of the flexible ITO-free PSCs can be seen that is composed of a highly conductive PEDOT: PSS layer and an Ag electrode; such electrodes provide a great asset for PSCs by improving their transparency, lowering the sheet resistance, and boosting the flexibility of the final device.

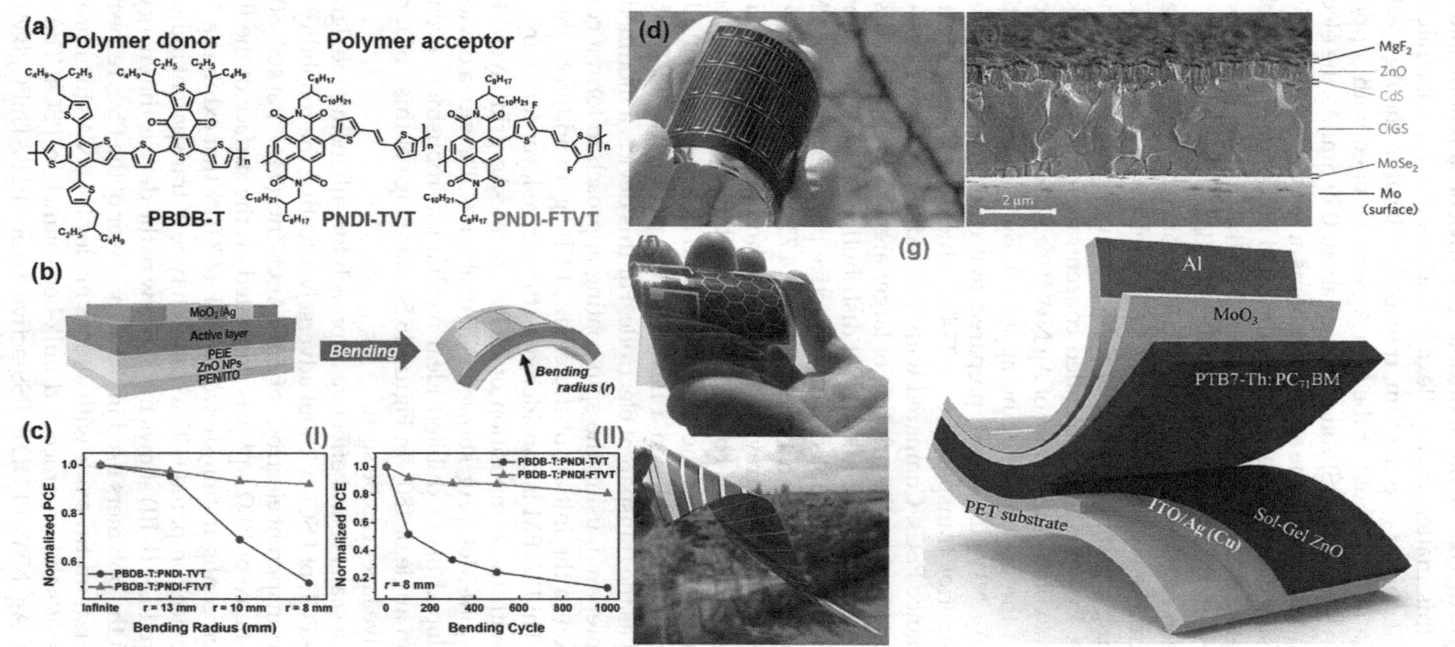

FIGURE 16.4
(a) Chemical structure of donor (PBDB-T) and acceptor (PNDI–TVT, and PNDI–FTVT) polymers, (b) the 3D structure of the PSCs developed under bending, (c) normalized PCE of the configuration developed (I) at diverse bending radius for 1,000 cycles and (II) at various bending cycles at the bending radius of 8 mm. Adapted with permission from reference [38], copyright (2019), American Chemical Society; (d) optical image of the flexible solar cell's configuration on a PI substrate, and (e) cross-sectional image of the PSC prepared on a PI substrate with efficiency more than 18%. Adapted with permission from reference [39], copyright (2011), Nature Publishing Group; (f) a view of an ITO-free flexible PSC with superiorly conductive PEDOT: PSS layer and Ag grid as the front electrode. Adapted with permission from reference [41], copyright (2013), Wiley; (g) the configuration of a flexible PSC consisted of Al-MoO_3-active layer (PTB7-Th: $PC_{71}BM$-sol-gel-ZnO-ITO/Ag(Cu)-PET. Adapted with permission from reference [40], copyright (2018), Elsevier.

Huang et al. [40] developed a flexible PSC configuration composed of Al-MoO_3-PTB7-Th: $PC_{71}BM$-sol-gel-ZnO-ITO/Ag(Cu) coated on a flexible polythylene terephthalate (PET) substrate (Figure 16.4g). In this matter, highly flexible and conductive substrates were developed by introducing Cu dopant into the thin, transparent layer of the Ag electrode. Correspondingly, the percolation threshold of the thin Ag layers doped with copper was set at 6 nm with a root mean square (RMS) roughness of about 0.19 nm. As the electrode which was developed transmitted 80% of the emitted photons at the wavelength of 550 nm and exhibited a fine sheet resistance of 14.1 $\Omega.sq^{-1}$, the fabricated films showed ideal chemical, mechanical, and thermal stability. The PSC developed by the Cu-doped Ag electrode reached a PCE of 7.53%, using the active layer of PTB7-Th: $PC_{71}BM$.

Various protocols have been employed to develop high-performance PSCs with appropriate flexibility and PCE. In this regard, Wang et al. [42] developed a flexible PSC with a PCE of 13.61% by developing a patternable and scalable, gravure-printed, Ag nanowire (NW) transparent electrode. By varying the blended concentration of Ag NW inks, the viscosity, surface tension, and volatilization rate of Ag NW-based inks were tuned to meet the essential requirement of the gravure-printing approach. By following the optimum process, uniformly coated Ag NW films were prepared that exhibited sheet resistance of 10.8 $\Omega.sq^{-1}$ along with superior transparency of 95.4%, which could be a great asset for producing high-performance PSCs. Compared with the traditional processes, such as spin-coating, gravure-printing enables the printing of larger areas with uniform distribution of additives, which leads to improved ink droplet distribution on the given substrate. The transparent electrode developed, coupled with the active polymeric layer of PM6:Y6, showed the optimum PCE of 15.28% with an average PCE of 14.75% $\pm$ 0.35%. Because of the ideal distribution rate of Ag NWs *via* the gravure-printing technique, a boosted PCE with sufficient flexibility and transparency was achieved for the resulting photovoltaic configuration [42]. Figure 16.5a shows an image of the layered PSC and its active compounds.

In another attempt, Lu et al. [43] developed a flexible PSC based on the PET substrate imprinted with a silver grid as the transparent electrode. For this device, a modified composition of PEDOT: PSS is prepared, using the spin-coating approach to form a compact layer with adequate adhesion to the initial substrate, viz., PET-Ag. In addition, the polymeric composition of PTB7-Th: $PC_{71}BM$ is considered as the active layer for charge separation. The resulting PSC, with the composition of PET substrate-Ag grid layer-PEDOT: PSS-ZnO-PFN-PTB7-Th: $PC_{71}BM$-MoO_3-Ag, showed PCE of about 6.51% with a device area of 2.25 cm^2. The PSC developed also exhibited ideal mechanical robustness even after 1,000 bending cycles with an angle of 180°. In Figure 16.5b, an image of this PSC and the energy level of each respective layer can be seen.

Shen et al. [44] developed a scalable and effective approach for enhancing the light-harvesting rate and thereby the PCE of PSCs without adversely affecting the rollability of the device. For this aim, a more transparent electrode of ZnO-AgO-ZnO (93.67% at 400–800 nm) with a negligible sheet resistance of 10 $\Omega.sq^{-1}$ was decorated on the surface of the flexible PET substrate. The optimized configuration showed a PCE of 8.04% that exhibits a 36.27% increase in the PCE of the device compared with a common type of transparent indium-tin oxide-based solar cell. Figure 16.5c (I–III) shows the 3D view of the device, the energy level of each respective layer, and the polymers used in the active layer of the PSC, respectively. Yang et al. [45] replaced the indium-tin-oxide with anodes made of Ag NWs to improve the flexibility and performance of solution-processed bulk-heterojunction PSCs. The device configuration composed of Ag NWs-PEDOT: PSS-active layer (PTB7-Th:$PC_{71}BM$)-CaAl exhibited PCE values of 2.8% and 2.5% upon its assembly on glass and PET substrates,

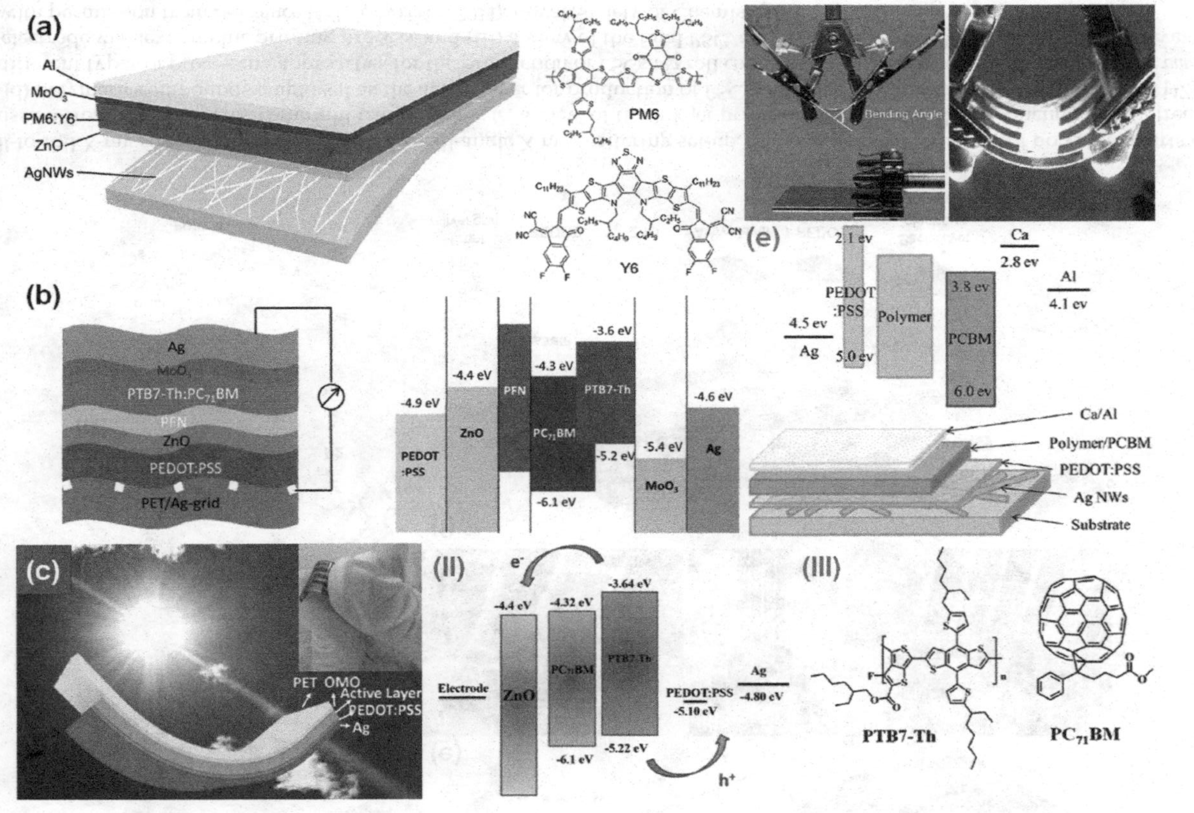

FIGURE 16.5
(a) 3D structure of a flexible PSC with PM6 and Y6 as donor and acceptor polymers, respectively. Adapted with permission from reference [42], copyright (2021), Wiley; (b) 2D structure of a flexible PSC composed of Ag-MoO_3-active layer (PTB7-Th: PC71BM)-PFN-ZnO-PEDOT:PSS-Ag-grid-PET along with the energy level of each respective layer. Adapted with permission from reference [43], copyright (2017), Elsevier; (c) (I) the 3D structure of a flexible organic photovoltaic configuration consisted of Ag-PEDOT: PSS-active layer-OMO-PET, (II) the energy level of each respective layer, and (III) chemical structure of polymeric structures used within the active layer. Adapted with permission from reference [44], copyright (2020), MDPI; (d) a view of the experimental approach for measuring the parameters of flexible PSCs, and (e) the structure and energy level of each component of the PSC consisting of Ca/Al-polymer/PCBM-PEDOT: PSS-Ag NWs-initial substrate. Adapted with permission from reference [45], copyright (2011), American Chemical Society.

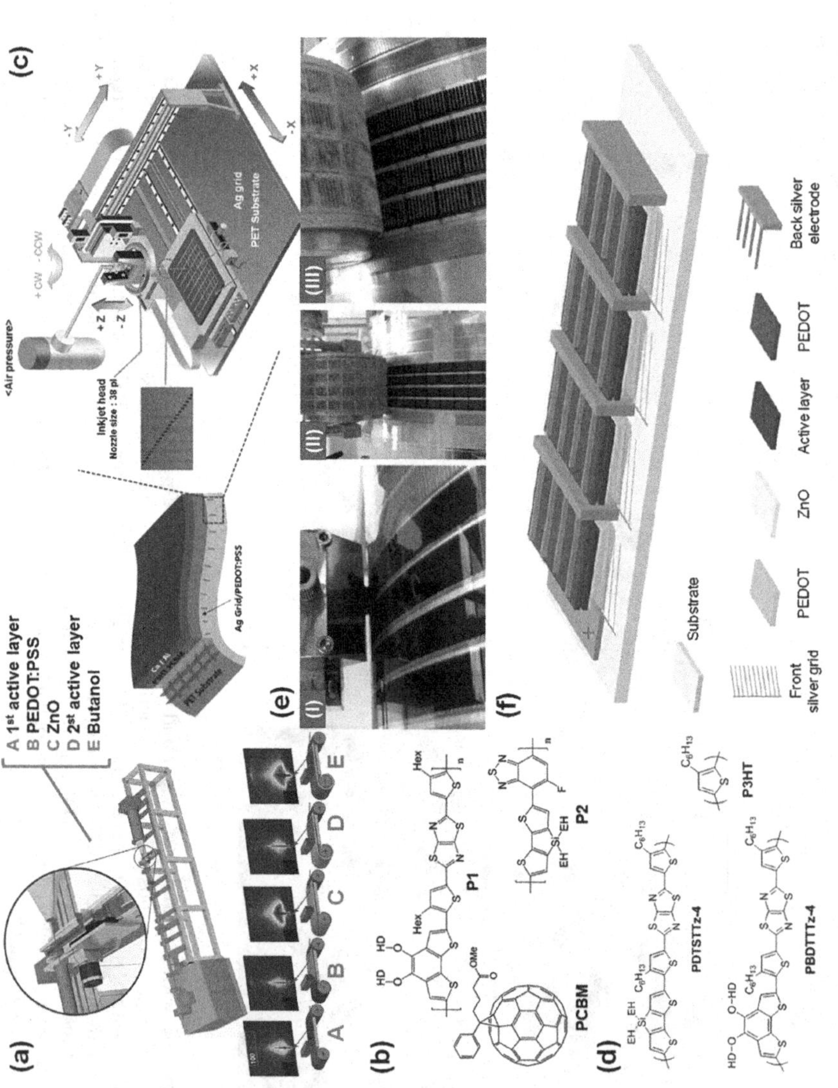

FIGURE 16.6
(a) A view of the roll-to-roll X-ray configuration along with a small-angle X-ray scattering set-up, and (b) chemical structure of polymeric parts of the active layer. Adapted with permission from reference [46], copyright (2015), Wiley; (c) a view of the ink-jet printing approach for the production of ITO-free polymeric photovoltaic devices, and (d) polymeric compounds employed as the active layer for production of PSCs. Adapted with permission from reference [47], copyright (2012), IOP Publishing; (e) different types of processing approaches for the production of PSCs *via* roll coatings including (I) coating the active layer *via* slot-die approach, (II) printing the Ag electrode via flexographic printing process, and (III) a view of the final PSC, and (f) a view of the 3D structure of PSC developed *via* a scalable approach. Adapted with permission from reference [48], copyright (2014), Royal Society of Chemistry.

respectively. In Figures 16.5d–e, a view of the optical image –and composition of this PSC along with the energy level of each respective layer can be seen, respectively.

Polymeric solar cells are adaptable to mass-scale production which could reduce their total manufacturing cost and greatly increase the speed of fabrication, which is a great asset for translating the PSCs to real-life applications. For this goal, various types of approaches were developed for mass-scale production of PSCs, among which the roll-to-roll production procedure (Figure 16.6a–b) [46], ink-jet printing method (Figure 16.6c–d) [47], and slot-die method (Figure 16.6e–f) [48] can be mentioned.

16.5 Conclusion

Harvesting electricity from sunlight's photons via photovoltaic devices is one of the most important techniques for addressing the ever-growing worldwide energy requirement and environmental concerns. PSCs are considered to be promising alternatives to the common photovoltaic configurations as a result of their low production cost, ease of fabrication *via* solution processing approaches, flexibility, versatility in design of polymeric materials, negligible ecological impacts, and mechanical robustness. However, PSCs suffer from a number of disadvantages, such as low PCE, short life service, and low stability under harsh, environmental conditions, which require practical approaches to overcome these serious problems. To address these issues, the design and development of effective polymeric donors/acceptors, improving the transparency and conductivity of electrode layers, minimizing charge carrier recombination/losses, and improving the stability of PSCs would significantly boost their capability and make them ideal candidates for light harvesting. In this chapter, the mechanism of light harvesting by PSCs is carefully assessed, and recently developed polymeric compounds for the active layer of this type of solar cells are evaluated in detail. Some practical approaches for mass-scale production of effective PSCs with boosted PCE were then examined and discussed.

References

1. H. Yin, C. Yan, H. Hu, J.K.W. Ho, X. Zhan, G. Li, S.K. So, Recent progress of all-polymer solar cells–From chemical structure and device physics to photovoltaic performance, *Materials Science & Engineering R: Reports* 140 (2020) 100542.
2. Y. Cui, H. Yao, J. Zhang, T. Zhang, Y. Wang, L. Hong, K. Xian, B. Xu, S. Zhang, J. Peng, Z. Wei, F. Gao, J. Hou, Over 16% efficiency organic photovoltaic cells enabled by a chlorinated acceptor with increased open-circuit voltages, *Nature Communications* 10(1) (2019) 1–8.
3. L. Meng, Y. Zhang, X. Wan, C. Li, X. Zhang, Y. Wang, X. Ke, Z. Xiao, L. Ding, R. Xia, H.L. Yip, Y. Cao, Y. Chen, Organic and solution-processed tandem solar cells with 17.3% efficiency, *Science* 361(6407) (2018) 1094–1098.
4. C. Yan, S. Barlow, Z. Wang, H. Yan, A. Jen, S. Marder, X. Zhan, Non-fullerene acceptors for organic solar cells, *Nature Reviews Materials* 3 (2018) 18003.
5. P. Cheng, X. Zhan, Stability of organic solar cells: Challenges and strategies, *Chemical Society Reviews* 45(9) (2016) 2544–2582.
6. X. Wen, A. Nowak-Król, O. Nagler, F. Kraus, N. Zhu, N. Zheng, M. Müller, D. Schmidt, Z. Xie, F. Würthner, Tetrahydroxy-perylene bisimide embedded in a zinc oxide thin film as an

electron-transporting layer for high-performance non-fullerene organic solar cells, *Angewandte Chemie* 131 (2019) 13185–13189.

7. L.X. Chen, *Organic Solar Cells: Recent Progress and Challenges*, ACS Publications, 2019.
8. G. Wang, F.S. Melkonyan, A. Facchetti, T.J. Marks, All-polymer solar cells: Recent progress, challenges, and prospects, *Angewandte Chemie International Edition* 58(13) (2019) 4129–4142.
9. Q. Wang, Y. Xie, F. Soltani-Kordshuli, M. Eslamian, Progress in emerging solution-processed thin film solar cells–Part I: Polymer solar cells, *Renewable & Sustainable Energy Reviews* 56 (2016) 347–361.
10. M.C. Scharber, N.S. Sariciftci, Efficiency of bulk-heterojunction organic solar cells, *Progress in Polymer Science* 38(12) (2013) 1929–1940.
11. O.V. Mikhnenko, H. Azimi, M. Scharber, M. Morana, P.W. Blom, M.A. Loi, Exciton diffusion length in narrow bandgap polymers, *Energy & Environmental Science* 5(5) (2012) 6960–6965.
12. C. Winder, N.S. Sariciftci, Low bandgap polymers for photon harvesting in bulk heterojunction solar cells, *Journal of Materials Chemistry* 14(7) (2004) 1077–1086.
13. C.M. Proctor, M. Kuik, T.Q. Nguyen, Charge carrier recombination in organic solar cells, *Progress in Polymer Science* 38(12) (2013) 1941–1960.
14. H. Kang, W. Lee, J. Oh, T. Kim, C. Lee, B.J. Kim, From fullerene–polymer to all-polymer solar cells: The importance of molecular packing, orientation, and morphology control, *Accounts of Chemical Research* 49(11) (2016) 2424–2434.
15. W. Lee, J.H. Kim, T. Kim, S. Kim, C. Lee, J.S. Kim, H. Ahn, T.S. Kim, B.J. Kim, Mechanically robust and high-performance ternary solar cells combining the merits of all-polymer and fullerene blends, *Journal of Materials Chemistry A* 6(10) (2018) 4494–4503.
16. T. Kim, J. Choi, H.J. Kim, W. Lee, B.J. Kim, Comparative study of thermal stability, morphology, and performance of all-polymer, fullerene–polymer, and ternary blend solar cells based on the same polymer donor, *Macromolecules* 50(17) (2017) 6861–6871.
17. J.R. Moore, S. Albert-Seifried, A. Rao, S. Massip, B. Watts, D.J. Morgan, R.H. Friend, C.R. McNeill, H. Sirringhaus, Polymer blend solar cells based on a high-mobility naphthalenediimide-based polymer acceptor: Device physics, photophysics and morphology, *Advanced Energy Materials* 1(2) (2011) 230–240.
18. Z. Li, L. Ying, P. Zhu, W. Zhong, N. Li, F. Liu, F. Huang, Y. Cao, A generic green solvent concept boosting the power conversion efficiency of all-polymer solar cells to 11%, *Energy & Environmental Science* 12(1) (2019) 157–163.
19. K.J. Baeg, M. Caironi, Y.Y. Noh, Toward printed integrated circuits based on unipolar or ambipolar polymer semiconductors, *Advanced Materials* 25(31) (2013) 4210–4244.
20. A. Das, S. Ghosh, H-bonding directed programmed supramolecular assembly of naphthalenediimide (NDI) derivatives, *Chemical Communications* 52(42) (2016) 6860–6872.
21. K.D. Deshmukh, T. Qin, J.K. Gallaher, A.C. Liu, E. Gann, K. O'Donnell, L. Thomsen, J.M. Hodgkiss, S.E. Watkins, C.R. McNeill, Performance, morphology and photophysics of high open-circuit voltage, low band gap all-polymer solar cells, *Energy & Environmental Science* 8(1) (2015) 332–342.
22. C.R. McNeill, Morphology of all-polymer solar cells, *Energy & Environmental Science* 5(2) (2012) 5653–5667.
23. C.R. McNeill, N.C. Greenham, Conjugated-polymer blends for optoelectronics, *Advanced Materials* 21 (2009) 3840–3850.
24. X. Gu, Y. Zhou, K. Gu, T. Kurosawa, Y. Guo, Y. Li, H. Lin, B.C. Schroeder, H. Yan, F. Molina-Lopez, C.J. Tassone, C. Wang, S.C.B. Mannsfeld, H. Yan, D. Zhao, M.F. Toney, Z. Bao, Roll-to-roll printed large-area all-polymer solar cells with 5% efficiency based on a low crystallinity conjugated polymer blend, *Advanced Energy Materials* 7(14) (2017) 1602742.
25. J. Kong, D. Nordlund, J.S. Jin, S.Y. Kim, S.-M. Jin, D. Huang, Y. Zheng, C. Karpovich, G. Sertic, H. Wang, J. Li, G. Weng, F. Antonio, M. Mariano, S. Maclean, T. Goh, J.Y. Kim, A.D. Taylor, Underwater organic solar cells via selective removal of electron acceptors near the top electrode, *ACS Energy Letters* 4(5) (2019) 1034–1041.
26. S.M. Tuladhar, M. Azzouzi, F. Delval, J. Yao, A.A. Guilbert, T. Kirchartz, N.F. Montcada, R. Dominguez, F. Langa, E. Palomares, J. Nelson, Low open-circuit voltage loss in solution-processed small-molecule organic solar cells, *ACS Energy Letters* 1(1) (2016) 302–308.

27. P. Kumar, S. Chand, Recent progress and future aspects of organic solar cells, *Progress in Photovoltaics: Research & Applications* 20(4) (2012) 377–415.
28. C. Lee, Y. Li, W. Lee, Y. Lee, J. Choi, T. Kim, C. Wang, E.D. Gomez, H.Y. Woo, B.J. Kim, Correlation between phase-separated domain sizes of active layer and photovoltaic performances in all-polymer solar cells, *Macromolecules* 49(14) (2016) 5051–5058.
29. N. Zhou, A. Facchetti, Naphthalenediimide (NDI) polymers for all-polymer photovoltaics, *Materials Today* 21(4) (2018) 377–390.
30. K. Ahmad, S.N. Ansari, K. Natarajan, S.M. Mobin, Design and synthesis of 1D-polymeric chain based $[(CH_3NH_3)_3Bi_2Cl_9]$ n perovskite: A new light absorber material for lead free perovskite solar cells, *ACS Applied Energy Materials* 1(6) (2018) 2405–2409.
31. C. Gu, D. Liu, J. Wang, Q. Niu, C. Gu, B. Shahid, B. Yu, H. Cong, R. Yang, Alkylthienyl substituted asymmetric 2D BDT and DTBT-based polymer solar cells with a power conversion efficiency of 9.2%, *Journal of Materials Chemistry A* 6(5) (2018) 2371–2378.
32. Y.K. Han, M.Y. Chang, W.Y. Huang, H.-Y. Pan, K.S. Ho, T.H. Hsieh, S.Y. Pan, Improved performance of polymer solar cells featuring one-dimensional PEDOT nanorods in a modified buffer layer, *Journal of the Electrochemical Society* 158(3) (2011) K88.
33. J. Lee, D.H. Sin, B. Moon, J. Shin, H.G. Kim, M. Kim, K. Cho, Highly crystalline low-bandgap polymer nanowires towards high-performance thick-film organic solar cells exceeding 10% power conversion efficiency, *Energy & Environmental Science* 10(1) (2017) 247–257.
34. C. Liu, C. Yi, K. Wang, Y. Yang, R.S. Bhatta, M. Tsige, S. Xiao, X. Gong, Single-junction polymer solar cells with over 10% efficiency by a novel two-dimensional donor–acceptor conjugated copolymer, *ACS Applied Materials & Interfaces* 7(8) (2015) 4928–4935.
35. Z. Wang, H. Jiang, L. Zhang, X. Liu, J. Sun, Y. Cao, J. Chen, 1D/2A ternary blend active layer enables as-cast polymer solar cells with higher efficiency, better thickness tolerance, and higher thermal stability, *Organic Electronics* 61 (2018) 359–365.
36. T. Zhu, D. Liu, K. Zhang, Y. Li, Z. Liu, X. Gao, X. Bao, M. Sun, R. Yang, Rational design of asymmetric benzodithiophene based photovoltaic polymers for efficient solar cells, *Journal of Materials Chemistry A* 6(3) (2018) 948–956.
37. S.A. Hashemi, S. Ramakrishna, A.G. Aberle, Recent progress in flexible–wearable solar cells for self-powered electronic devices, *Energy & Environmental Science* 13(3) (2020) 685–743.
38. M. Kim, H.I. Kim, S.U. Ryu, S.Y. Son, S.A. Park, N. Khan, W.S. Shin, C.E. Song, T. Park, Improving the photovoltaic performance and mechanical stability of flexible all-polymer solar cells via tailoring intermolecular interactions, *Chemistry of Materials* 31(14) (2019) 5047–5055.
39. A. Chirilă, S. Buecheler, F. Pianezzi, P. Bloesch, C. Gretener, A.R. Uhl, C. Fella, L. Kranz, J. Perrenoud, S. Seyrling, R. Verma, S. Nishiwaki, Y.E. Romanyuk, G. Bilger, A.N. Tiwari, Highly efficient Cu (In, Ga) Se 2 solar cells grown on flexible polymer films, *Nature Materials* 10(11) (2011) 857–861.
40. J. Huang, X. Liu, Y. Lu, Y. Zhou, J. Xu, J. Li, H. Wang, J. Fang, Y. Yang, W. Wang, R. Tan, W. Song, Seed-layer-free growth of ultra-thin Ag transparent conductive films imparts flexibility to polymer solar cells, *Solar Energy Materials & Solar Cells* 184 (2018) 73–81.
41. D. Angmo, F.C. Krebs, Flexible ITO-free polymer solar cells, *Journal of Applied Polymer Science* 129(1) (2013) 1–14.
42. Z. Wang, Y. Han, L. Yan, C. Gong, J. Kang, H. Zhang, X. Sun, L. Zhang, J. Lin, Q. Luo, C. Ma, High power conversion efficiency of 13.61% for 1 cm^2 flexible polymer solar cells based on patternable and mass-producible gravure-printed silver nanowire electrodes, *Advanced Functional Materials* 31(4) (2021) 2007276.
43. S. Lu, J. Lin, K. Liu, S. Yue, K. Ren, F. Tan, Z. Wang, P. Jin, S. Qu, Z. Wang, Large area flexible polymer solar cells with high efficiency enabled by imprinted Ag grid and modified buffer layer, *Acta Materialia* 130 (2017) 208–214.
44. W. Shen, G. Zhao, X. Zhang, F. Bu, J. Yun, J. Tang, Using dual microresonant cavity and plasmonic effects to enhance the photovoltaic efficiency of flexible polymer solar cells, *Nanomaterials* 10(5) (2020) 944.
45. L. Yang, T. Zhang, H. Zhou, S.C. Price, B.J. Wiley, W. You, Solution-processed flexible polymer solar cells with silver nanowire electrodes, *ACS Applied Materials & Interfaces* 3(10) (2011) 4075–4084.

46. H.F. Dam, T.R. Andersen, E.B. Pedersen, K.T. Thydén, M. Helgesen, J.E. Carlé, P.S. Jørgensen, J. Reinhardt, R.R. Søndergaard, M. Jørgensen, E. Bundgaard, F.C. Krebs, J.W. Andreasen, Enabling flexible polymer tandem solar cells by 3D ptychographic imaging, *Advanced Energy Materials* 5(1) (2015) 1400736.
47. S.I. Na, D.W. Park, S.S. Kim, S.Y. Yang, K. Lee, M.H. Lee, ITO-free flexible polymer solar cells with ink-jet-printed Ag grids, *Semiconductor Science & Technology* 27(12) (2012) 125002.
48. J.E. Carlé, M. Helgesen, M.V. Madsen, E. Bundgaard, F.C. Krebs, Upscaling from single cells to modules–fabrication of vacuum-and ITO-free polymer solar cells on flexible substrates with long lifetime, *Journal of Materials Chemistry C* 2 (2014) 1290–1297.

17

One-Dimensional Polymeric Nanocomposites for Flexible Solar Cells

Shveta Sharma, Richika Ganjoo, Abhinay Thakur, and Ashish Kumar

CONTENTS

17.1 Introduction

Polymers are frequently reinforced with fillers of varying sizes to overcome some limitations and to increase their range of applications. An entirely new type of polymer composite has emerged using nanoscale fillers to enhance polymer mechanical and physical properties [1]. The nanofillers come in a wide range of isotropic to highly anisotropic sheet-like or needle-like morphologies, with at least one distinctive length scale of the order of nanometers. Nanoscience and nanotechnology have plenty of potential for developing novel polymer nanocomposites with novel features by combining nanoscale fillers with polymer components [2]. Polymer nanocomposites can be divided into three groups based on the size of the nanoscale fillers scattered within them. In the first type, fillers in polymeric matrices can be two-dimensional (2D) nanoscale and can be found as thin

DOI: 10.1201/9781003223764-17

sheets a few nanometers thick and hundreds to thousands of nanometers long, such as layered silicate, graphene, or MXene [3]. In the second type, a very long-dimensional structural unit is present in which one dimension is substantial as compared to the other two (which are measured in nanometers); nanofibers or nanotubes come under this category of nanofillers, e.g., nanofibers and nanotubes of carbon or halloysite nanotubes as nanofillers are used by researchers to obtain desired material properties [4]. Nanocomposites with nanoscale fillers of three dimensions in nanometers are the third type. Quantum dots, nanoclusters of medium conductivity, silica of spherical shape are examples of nanoscale fillers of this kind and have iso-dimensions with a low aspect ratio.

Furthermore, the necessity for new things is because, to date, the world relies on fossil fuel as a significant power generation source for its energy needs. According to statistics obtained in 2019, coal burning still provides 38% of the world's total power supply. However, solar energy, wind energy, rain energy, tidal energy, and geothermal heat have enormous potential to replace traditional fossil fuels [5] ultimately. The most potential sustainable source of clean energy has long been identified as sunlight and the recent past decades were devoted to studying solar cells. According to estimates, total global photovoltaic (PV) capacity had reached 512 GW in 2014 and supplies electricity to the world's fast-increasing population. With the sincere efforts of research and development and government subsidies, solar panel prices decreased from approximately $100 in 1975 to $3 in 2014 [6]. Polysilicon-based panel installation has become a lucrative industry for most solar plant operations. For example, Seville, Spain's solar PV facility, produces enough electricity for 6,000 families each year. With a total area of 43 km^2 and a capacity to generate over 1500 MW of electricity yearly, China's Tengger desert solar farm is the world's sixth-largest solar farm.

Using thin-film cells as a power source, mechanically solid and flexible units are created that are ideal for a wide range of outdoor applications where weight and elasticity and retention of the original shape are critical [7]. Numerous vehicle manufacturers have begun including solar panels as an additional charging option on the roofs of their electric vehicles. Sunlight may be harnessed from an umbrella's flexible solar panels to illuminate the street at night while also powering portable electronic devices like cell phones and laptop computers. Methods like solution-phase spin-coating, vacuum-phase deposition, printing, etc., are utilized to form thin-layer solar cell with the help of substrate and other functional layers, with the substrate being generally flexile and providing mechanical support and environmental protection [8]. Two electrodes are used to carry the charge and electrically complete the external circuits. These electrodes are made up of oxides, thin metal wires or sheets of nanoscale size, and polymers, but the only condition is the conductive nature of all the chosen materials. To absorb sunlight *via* an active semiconductor layer, to convert light into photoelectric charge carriers, one of the electrodes must be transparent to visible light, along with these functional materials generally being classified into inorganic or organic, with semiconductors made up of both inorganic and organic material, which is also very important because efficiency is somehow related to them. Cadmium telluride and copper indium gallium diselenide are the most prevalent inorganic semiconductors [9]. Donor and acceptor organic chemicals are used in organic semiconductors. These can be low-molecular-weight organic molecules and polymers, and metal halide perovskite is commonly employed in hybrid semiconductors for flexible solar cells. All types of inorganic, organic or hybrid active materials have equal potential to be used in their flexible form and to achieve very high efficiencies. [8]. The following equation Eq. (17.1) provides efficacy η of a solar cell [10]:

$$P = FF \times |J_{SC}| \times V_{OC} \tag{17.1}$$

where FF = fill factor,
$|J_{sc}|$ = value of short-circuit current density,
V_{oc} = open-circuit photovoltage,
P = power density of incident light.

In principle, the maximum $|J_{sc}|$ is determined by the extent to which the absorption window of the dye overlaps the solar spectrum [11]. So, one can say that high-efficiency semiconductors should be made up of nanostructured material [12]. A thin-film solar cell is formed when many functional layers are attached over a flexible substrate and it can be achieved by employing varied techniques like a vacuum-phase deposition, solution-phase spin-coating, or printing processes. The complete cell is supported mechanically and protected from the environment by a flexible substrate.

17.2 One-Dimensional Nanostructured Materials for Flexible Solar Cells

Solar cells have recently attracted a lot of interest because of their potential to help solve global environmental problems. Organic photovoltaics (OPVs), like dye-sensitized OPVs, have made high power conversion efficiency (PCE) and endurance highly noteworthy. The maximum PCEs achieved by the dye-sensitized solar cells (DSSCs) is 11%, but the maximum efficiency of organic thin-film solar cells is only 7%, which therefore needs to be improved compared with silicon-based solar cells. On the other hand, because of their low weight, greater flexibility, color adjustability, and ease of processing, these organic materials outperform silicon and hold great promise for widespread applications [13]. Nanorod/nanowire grids of low-bandgap semiconductors have outstanding optoelectric properties, allowing them to give excellent results in solar cells [14].

17.2.1 Carbon Nanotubes (CNTs)

As the name suggests, carbon nanotubes (CNTs) are made up of cylindrical-shaped carbon tubes on a nano-scale, which are arranged hexagonally, and the length-to-diameter ratio of CNTs has been boosted to new heights (up to 132,000,000:1) [15]. Because of their length, they are well suited for usage in electrical, optical, and sensing applications. CNTs possess the highest tensile strength and elastic modulus of any material found so far [6]. Even a CNT's tensile strength exceeds that of stainless steel by a factor of 16. The sp^2 bonding present in the CNT structure helps to increase the power of a composite made up of nanotubes and polymers, which in turn increase the flexibility of OPVs. These tubular fibers can be as tiny as a nanometer or as long as thousands of micrometers, which undoubtedly affects the aspect ratio. It has been shown that CNTs are good thermal and electrical conductors because of their high aspect ratio [6]. Compared to copper, the thermal conductance power of CNTs is five times greater, and this high aspect ratio also helps increase its effective life.

17.2.1.1 Structure and Classification of CNTs

Based on their diameter, hexagon ring arrangement, and length, metallic or semiconducting possibilities exist for CNTs. The electrical behavior and efficiency of these tubes depend upon their chirality. The carbon atom arrangement and general defects can significantly impact the mechanical strength of CNTs. Based on graphene layers, CNTs are mainly classified into two categories, namely single-walled CNTs (SWCNTs) or multi-walled CNTs (MWCNTs), having a roll of a single layer or many rolled layers, respectively.

17.2.1.2 Single-Walled Carbon Nanotubes (SWCNTs)

For SWCNTs, graphene (layered graphite), with the width of a single C atom, is kept in cylindrical form, and this wrapped structural unit is explained with indices (n, m). Generally, vectors along the hexagonal directions are denoted by these indices. Values of these indices also help to provide an idea about the diameter and electronic properties of the nanotubes. Due to their unique features, SWCNTs have received more research attention than MWCNTs [16]. For example, MWCNTs have a zero gap, whereas, in SWCNTs, their band gap ranges from zero to 2 eV that can change their electrical properties, showing the unique characteristics of metallic or semiconducting materials.

17.2.1.3 Multi-Walled Carbon Nanotubes (MWCNTs)

Multi-walled nanotube structures are often described using the Russian doll model or the parchment model. However, the doll model is most popular and is widely employed. Nanotube sheets are placed in a concentric cylinder pattern in the doll model. This is like putting one tube inside the other. In the parchment model, graphene sheets are wrapped up like sheets, precisely like rolling paper. Researchers found the spacing between MWCNTs to be 3.52 Å, the same as the distance between the graphene layers. Thus, carbon nanotubes are a blessing to the present world because of their extraordinary thermal conductivity, good operational life, slow speed of degradation, high aspect ratio values, high stability and conductivity, and robust structure. CNTs come in SWCNTs, DWCNTs, and MWCNTs. In these, the different types of CNTs (SWCNTs, DWCNTs and MWCNTs) are prepared by the rolling together of a single sheet, two sheets, and many sheets, respectively.

Graphene sheets are used to make the CNTs, spatially arranged in SWCNTs (single sheet hexagonal structure), DWCNTs (two sheets are involved), and MWCNTs (multiple sheets are present) [16]. Graphene nanotubes (CNTs) are widely utilized in electronics, including sensors, solar cells, transistors, and other components [16]. Organic solar cells (OSCs) use CNTs because of their low resistance and high specular transmittance capacity in the UV-to-MIR spectrum. Furthermore, their extraordinary degree of flexibility and pretty straightforward synthesis qualify them for use in solar cells. The fabrication processes include chemical vapor deposition, electric arc discharge electrolysis, catalyst arrays, and low-temperature solid pyrolysis.

In contrast, the deposition techniques employed for deposition of CNTs include dip coating, spin coating, spray coating, vacuum filtration, sol-gel method, electrophoretic as well as sputter deposition [17]. The electrodes should have different work functions, one with high and the second with low work function. Numerous research groups are currently improving thin-film optoelectronic qualities for future solar/photovoltaic cells across the globe, and improving the free charge generation and percolation of excitons through materials to electrodes is a fascinating task. The bulk-heterojunction (BHJ) way is the most promising and advanced strategy in organic photovoltaic (OPV) [18]. The organic film's nanoscale structure is a crucial component. CNTs have the advantages of a wide-ranging bandgap because of different sizes and spatial arrangements over the structures like fullerenes and graphene, a situation which is fascinating for the researchers. Along with that, CNTs tendency to accept electrons and having a larger available surface to achieve exciton dissociation makes CNTs a good choice for the formation of nanocomposites with polymers (with some conductance); it offers an excellent chance for exciton dissociation [18]. Organic cells include the CNTs attached inside a BHJ photoactive film and nanocomposites made up of CNTs and polymeric material, as a hole extraction layer.

17.2.1.4 CNTs in Flexible Solar Cells

Techniques commonly employed to assimilate CNTs within photoactive BHJs include the addition of nanotubes in a solution of polymers (electron-donating in nature). After that, the substrate is spin-coated. On the top of the substrate, the electrode made of transparent material is present [19]. Khan et al. demonstrated that MWCNTs at 4% concentration resulted in a brief drop in circuit current from 8.64 to 6.99 mA/cm^2, attributed to increased MWCNT quantity [16]. Zhang et al. employed carbon nanotubes and graphene sheets as both anodes and cathodes in traditional OPVs (Figure 17.1). They produced and transferred large-area, continuous, and extremely transparent carbon nanotube and graphene sheets onto flexible substrates [19]. Multiple characterizations were performed on these thin films to determine their suitability as electrode materials. A study [19] demonstrated the flexible and semitransparent OPV devices formed with electrodes made up of carbon and in which the base electrode was made up of graphene films on polyethylene naphthalate (PEN), and the top layer was formed with many layers of graphene films or carbon nanotube thin films. Initially, CVD graphene and CNT films were utilized to create OPV

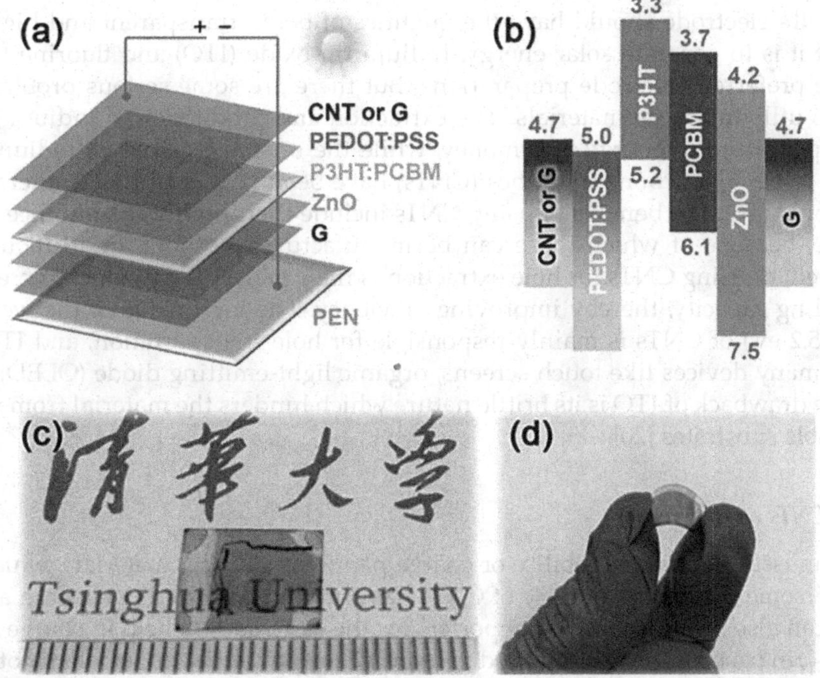

FIGURE 17.1
(a) Diagram of modified OPV and (b) diagrammatic presentation of an energy level of solar cells constructed using carbon nanotubes or graphene behaving as anodes and graphene as the cathode. (c) Image of a translucent solar cell with an electrode made up of carbon. (d) The solar cell device formed exhibits excellent flexibility. Adapted with permission from reference [19]. The article was published using CCBY License.

cells to test the viability of carbon-based film substrates as all-carbon electrodes. To absorb incoming radiation and produce excitons, a combination of C60-based derivatives (phenyl-C61-butyricacid methyl ester; PCBM) and P3HT (poly (3-hexyl) thiophene) was used as an active layer. Figure 17.1a illustrates the schematic design of these systems. The bottom electrodes were constructed using graphene sheets that were then spin coated with ZnO nanoparticles to serve as an electron transport layer on a flexible PEN wafer.

This technique was performed twice to get uniformly distributed nanoparticles. CNT or graphene films were made to adhere as the top transparent electrode and came after both the active layer and the hole transport layer of poly(3,4-ethylenedioxythiophene): poly(styrenesulfonate) (PEDOT: PSS). In Figure 17.1b, where the energy bands are shown, the value of 4.7 eV for work functions was fixed for carbon nanotube and graphene films, and it entirely depended upon their thickness and purity. According to the complete energy band diagram, it was theoretically feasible to employ all-carbon electrodes in this configuration. Figure 17.1c, d illustrates semitransparent and flexible solar cells made up of purely carbon electrodes.

The greatest performance achieved with all-carbon electrodes was 0.63%, which was equivalent to OPVs employing pure CVD graphene sheets as anodes. To achieve greater efficiencies, the parameters film hydrophilicity, smoothness, and conductivity may all be improved. This research shows that graphitic nanomaterials may be used as all-carbon electrodes in photovoltaics and thin-film solar cells, which will be produced in the future.

17.2.1.5 CNTs as a Hole Extraction Layer or the Transparent Conducting Electrode

A solar cell's electrode should have the qualities of being transparent and highly conductive if it is to generate solar energy. Indium tin oxide (ITO) and fluorine tin oxide (FTO) are preferred electrode preparations. But there are some serious problems with respect to utilizing these materials. The extraction and production of indium involves heavy expenditure of energy and money, while the unstable nature of indium is also a serious issue [6]. Carbon nanotubes (CNTs) have been proposed for an alternative to conventional ITO. The benefits of using CNTs include their high transmittance and low resistance, because of which these can be manufactured without any difficulty. One more benefit of using CNTs for hole extraction is their morphology, which increases the light-holding capacity, thereby improving efficiency. The high value of the work function (4.7–5.2 eV) of CNTs is mainly responsible for hole transportation, and ITOs have replaced many devices like touch screens, organic light-emitting diode (OLED), etc. [6]. One more drawback of ITO is its brittle nature which hinders the material from working with flexible substrates [20].

17.2.1.6 CNTs as Additives

There is an issue of charge mobility or charge pathways in other materials, which CNTs easily overcome. Therefore, the role of CNTs has not been restricted to electron acceptors but they can also be employed as supporters for the electrodes to help in charge transfer. C_{60} and C_{70}, in particular, are often used to improve the properties of carbon nanotubes. To circumvent fullerene's limited carrier dispersion and to minimize the rejoining of the electron-hole pair following exciton breakdown, nanotube imaging was used in P3HT-PCBM devices. Nanotube imaging increased hole extraction when P3HT wrapped the SWNTs [18]. Short circuits resulting from contact between the hole-transporting SWNTs and the electron-transporting PCBM should be avoided due to the wrapping polymer's shielding effect. Although this application method seems promising, only slight gains in PCE were obtained.

17.2.2 Nanowires for High Efficiency Flexible Solar Cells

Many innovative mechanical, optical, and electrical features present in semiconductor nanowires (NWs) are absent from their thin-film counterparts. They can give a genuinely exciting alternative to constructing solar cells on silicon with greater efficiencies and cost-effectiveness [21]. Silicon-based solar cells are prevalent because of their low cost and reliable working, but they give only 8–19% efficiency. To achieve high efficiencies, III–V multijunction cells with many Ge (Germanium) substrate layers are required, which becomes a costly affair.

New inventions in this field are desired and research is continually going on to find the ideal material that should be light in weight, cost effective, easily transportable, and flexible. So, in this field, Schubert investigated NWs of one-dimensional structure, which can be easily dispersed in the polymers [22].

17.2.2.1 Challenges in Making NW Solar Cells

Despite the numerous advantages of employing NWs in the production of solar cells, several obstacles require considerable work to overcome.

17.2.2.1.1 Characterization

Quantifying theoretical properties for nanowires by employing conventional electrical transport measurements, like the Hall effect, etc., is not very easy to achieve [23]. Field-effect measurements can be used for NWs, but the chances of error because of modeling assumptions are always there [24]. To optimize NW formation and configuration necessities for development and manufacture, knowledge of charge carrier durations, charge density, layering, doping concentrations, and interfacial velocities is needed.

17.2.2.1.2 Surface Passivation

NWs have a large surface-to-volume ratio with sufficient contact area, which is more significant than for other thin-film structures, and these advantages are because of their shape. The crystal surface alters the electrical structure by breaking the three-dimensional bulk periodicity [25, 26]. These surface states can trap carriers, which affects carrier capacity and involves the complete designing of high surface recombination velocity, so that NWs have very small carrier lifetimes (picoseconds), which also influences cell output [27]. As a result, a lot of work has been reported in the literature related to surface passivation to enhance the efficiencies. But the main challenge is to quantify the level of passivation accurately. Therefore, it is difficult to determine how much improvement is required in the passivation process based on comparative research.

17.2.2.1.3 P–i–n Junctions

The intrinsic/undoped layer is placed between a p- and an n-doped area of a p–i–n junction, and it is a critical component of most solar cells. The Fermi levels in each doped region must be set to the same value. The built-in charge is essential for emitting electrons and holes in their proper locations. When connected to other circuits, already-presented charge carriers could help drive the current [12, 21]. Doping is responsible for the better functioning of p–i–n junctions; NWs are different from thin-film structures in respect of their properties, so doping cannot be easily studied in the case of NWs, which can influence the doping, efficiency, etc. [21]. For example, it has been found that the ionization energy of dopants increases as the NW diameter decreases, leading to a lower carrier density than predicted. Dopant inclusion was also affected by the NW's "self-purification effect", in which impurities with high formation energy are ejected from the nanocrystal, lowering the system's Gibbs free energy [28,29]. The high surface-to-volume ratio of NWs, combined with their small size, can impact the performance of the junction. Carriers get trapped in the surface states, which lowers the adequate doping level [30]. It is more challenging to maintain a balance in the p and n areas in terms of effective dopant concentration caused by the more diminutive size of NWs. Also, quantity imbalance between the dopants results in difficulty forming the p-n junction. There is no direct application to NWs of past thin-film knowledge gained. There have been no studies of a reliable technique to assess the quality of a p–i–n junction. Even though atom-probe measurements are pretty helpful in studying dopant incorporation, they cannot reveal information on the density of ionized dopants [31]. As a result, the systematic optimization of p–i–n junctions will necessitate more investigation. Off-axis electron holography can make development of a high-quality junction accessible. However, thin-film solar cell manufacturing is a well-established technique. The processes (etching and annealing) employed for the formation of a thin-film device destroy the junction. These processes increase the cost, and a lot of time is wasted on NW devices.

The axial structure can be helpful to achieve the high fill factor of 80% [32], whereas radial p–i–n-type junction formation gives a fill factor lower than 70% [33]. Also, a fill factor of 80.5% by employing single NWs has been achieved by the radial p–i–n structure [34]. Compared with other planar materials of the same volume, NWs can significantly improve light absorption. Also, the extraction technique for photon-generated carriers can be substantially improved by using NWs, increasing efficiency. To date, the efficiency of single-NW solar cells was measured to be 10.2% and 40% for horizontal and vertical structures, respectively; an efficiency of 18.9% for the large-area design indicates the high probability of new solar cells using the NW structure with greater efficiencies. Researchers need to improve characterization, p-n junctions, manufacturing technologies, and surface passivation for nanowire solar cell devices [21].

17.2.3 Halloysite Nanotubes (HNTs) in Flexible Solar Cells

There are several advantages to using dye-sensitized solar cells over conventional crystalline silicon solar cells, like being easy to build with high cost effectiveness and high conversion efficiency [35]. The standard structure of DSSCs includes three components: mesoscopic metal-oxide layer deposited on a transparent conducting oxide substrate, dye in liquid electrolyte, and Pt on TCO to act as the counter electrode (CE) [36]. DSSC performance is greatly influenced by the characteristics of the CE layer, such as catalyst nature, conductivity, low charge carrier mobility, high aspect ratio, transparency, and mechanical and chemical stability [37]. But the high cost and corrosion of Pt restrict its usage on flexible polymeric substrates, which paves the way for new materials like conducting polymers [38] and carbon-based nanomaterials [39]. In a work by Gemeiner et al., screen-printing inks were formulated, in which process polymer poly(3,4-ethylenedioxythiophene): poly(styrene-sulfonate) was used along with other constituents, like halloysite nanotubes as a filler, and different organic additives, in which polymer has conductivity and nanotubes were an insulator. Generally, counter electrodes (CEs), made up of FTO substrates, are used in dye-sensitized solar cells (DSSCs), but in this study new formulated ink was coated on the counter electrodes (FTO substrates) and then it was attached to the dye-sensitized solar cells as counter electrodes. Halloysite nanotubes were responsible for the increased stability of ink at high temperature, which was verified by employing various techniques like thermogravimetric analysis (TGA) and x-ray photo spectroscopy (XPS) and results showed that, through the homogenization process, the nanocomposite structure of halloysite and polymer was there. Also, both were attached through physical interactions, and even HNTs with a very low weight percentage of the order of 1% resulted in a very high value of conductance (of the order of 381 S/cm) [35].

17.3 Application of One-Dimensional Metal Oxide Nanotubes, Nanowires, Nanoribbons, and Nanorods in Solar Cells

17.3.1 Titanium Oxide and Zinc Oxide

One-dimensional (1D) nanostructures, among many kinds of nanostructured materials, are potential components for DSSCs owing to their various performance and processing advantages [40]. Due to directed smooth electron mobility, such nanostructures can improve electron transport characteristics. As a result, they speed up electron transport and reduce the likelihood of recombination [41]. Furthermore, their shape provides a high surface area, which has a direct impact on the efficiency of dye-sensitized solar cells (DSSCs) [42]. Furthermore, because of their massive size, nanostructures add to the light-scattering phenomenon, increasing the light absorption capacity of the dye molecule [43]. A new study has discovered that DSSCs made of nanotubes have the potential to increase the photo-conversion efficiency of DSSCs. Changes in density and location of structural flaws and crystallinity among the constituents of the redox electrolyte all influence the transport or recombination process [44]. Nanotubes are another kind of 1D nanostructure. Using an electrochemical method, Xie et al. developed coaxial TiO_2/ZnO nanotube arrays. The fabrication of ZnO films within the existing TiO_2 pores was successful [44]. The inner layers of the ZnO layer are rough and nanoporous because they are made up of extremely small ZnO nanoparticles. The layer is suitable for dye loading as well as for transporting electrolytes. When compared with pure TiO_2 NTs as a photoanode, this photoanode had a 40% higher energy conversion rate of 2.8%. In another study, Liu et al. used ZnO to create TiO_2 nanotube arrays with an average pore diameter and length of 83 nm and 14 m, respectively. The material is supported on a Ti substrate and is a product of the electrodeposition method (photo-anode electrode). When ZnO was deposited on TiO_2 nanotubes with a 60-minute deposition period, it was discovered that the conversion efficiency of bare TiO_2 nanotubes had doubled [45].

Liu et al. measured DSSCs using rutile TiO_2 nanorod arrays on FTO substrate [46]. To build a DSSC, a photoanode of TiO_2 nanorods with a thickness of 4 m was utilized. When irradiated with 100 mW cm^{-2} AM1.5 light, the $TiCl_4$-treated cell had a voltage of 0.71 V. The EQE is roughly 50% at the peak of dye absorption, and 3.0% was discovered to be the total power conversion efficiency. Compared with pure TiO_2 nanoparticles, nanoribbon-based DSSCs showed a 60 % increase in energy conversion. Due to their enhanced charge transport and segregation properties, restrictions for back-charge transfer, and fewer intraband sites, nanoribbon-containing cells have a higher voltage, greater electron lifetime, and more extended electron carrier mobility [46]. Li et al. [47] investigated (Figure 17.2)

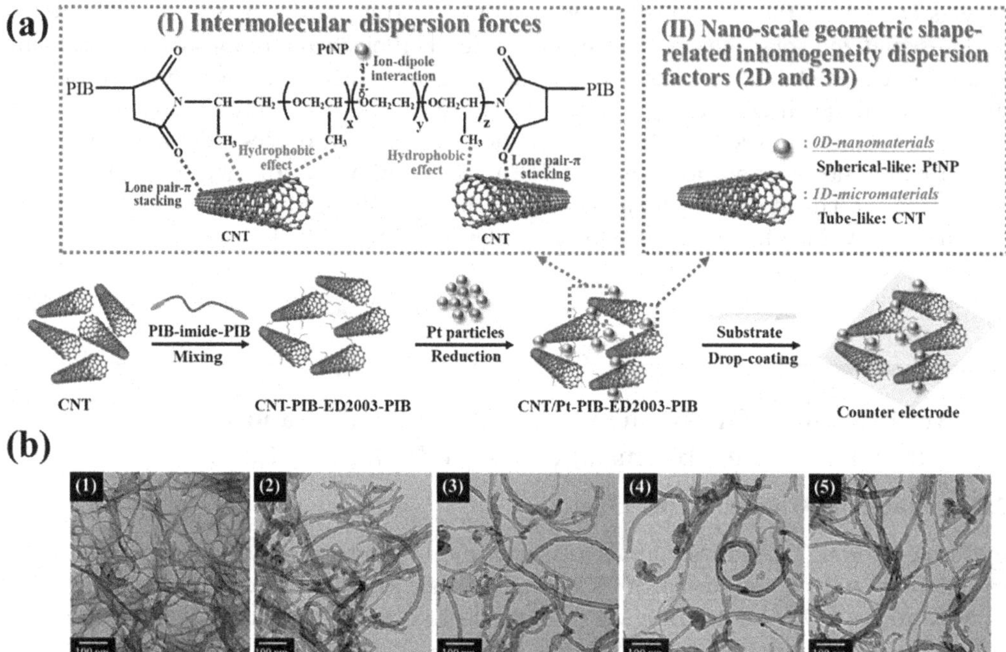

FIGURE 17.2
(a) Intermolecular forces between CNTs and PtNPs and the polymer-assisted dispersant PIB-ED2003-PIB. (b) TEM images of carbon nanotubes with (1) no dispersant, (2) PIB-M1000, (3) PIB-ED900-PIB, (4) PIB-ED2003-PIB, and (5) PIB-D2000-PIB added at 2:1, 1:1, 1:2, and 1:2 ratios, respectively. Adapted with permission from reference [50], The article was published under CCBY License.

polymer-assisted dispersants that were utilized to stabilize platinum nanoparticle (PtNP)/carbon nanotube (CNT) nanohybrids and the forces that existed between them were supposed to be non-covalent bond forces.

These dispersants were designed to replace the conventional fluorine-doped tin oxide (FTO) glass in classic dye-sensitized solar cells (DSSCs) and the main usage of these was to be utilized in the preparation of the counter electrode. PtNPs/CNT nanohybrids were chosen for the making of auxiliary electrodes for a dye-sensitized solar cell because of their versatile properties, like large surface area associated with these nanohybrids, appreciable conductance, and value of redox potential, which are very helpful in enhancing the efficiency of solar cells. According to the conductivity data, the resistance of these PtNP/CNT nanohybrid films decreased to 7.25 Ohm approximately and an acceptable efficiency of the range of 6.28 % was shown by the cell, when carbon nanotubes were allowed to merge with the Pt metallic nanoparticles with a 5:1 weight ratio.

The conversion efficiency of the hybrid material generated by PtNPs/CNTs on the conventional FTO substrate was likewise outstanding (8.45%) when employing the FTO- bearing auxiliary electrode of the substrate. A large-area dye-sensitive cell was also created, with a conversion efficiency of 7.95% in the cell (with measurement 8 × 8 cm). As a result, the standard Pt auxiliary electrode may be substituted with a PtNP/CNT nanohybrid sheet, which increases the photoelectric conversion efficiency of dye-sensitive cells while also lowering costs [48]. The effect of the dispersed polymer in it on the dispersion of CNTs was demonstrated by studying the intermolecular dispersion forces and transmission electron microscopic (TEM) images of carbon nanotubes as shown in Figures 17.2(a) and 2(b), respectively.

17.3.2 Copper Oxides

For dye-sensitized solar cells, CuO nanoneedle arrays were employed [49]. The cathode is a 1D CuO film (nanoscale, with a needle-like structure)/Cu substrate with an ITO/TiO_2/dye/E/CuO/Cu structure. Under one sun exposure (100 mW cm^{-2}), the variation of current (I)–voltage (V) was calculated in cells.

According to the results, high efficiency (about 1.12%) occurred along with a 0.37 fill factor when CuO nanoneedles were used as the cathode. This complete efficiency (η) was more significant than the past reported efficiencies of 0.12% to 0.29% and the 0.8% achieved with Cu_2O film. Under the same lighting conditions, with conventional photovoltaic cells employing a Pt cathode (ITO/TiO_2/dye/E/Pt/ITO), the maximum efficiency and fill factor came out to be 7.21% and 0.79, respectively. The one-dimensional CuO nanoneedles have a needle-like tip firmly connected to the Cu surface, which may be the primary reason for superior performance compared with prior findings [48]. Thus, the nanoneedles used improved the diffusion length of the electrons and the adsorption time of the dye. In reality, decreasing the resistance between the nanoneedles and substrates during heat oxidization and preventing the development of oxide layers may significantly increase the whole sun efficiency (η) [50].

17.3.3 Effective Use of Carbon Nanotube/Graphene Nanocomposite Counter Electrodes in Dye-Sensitized Solar Cells

Using carbon nanotube/graphene nanocomposites as efficient counter electrodes in dye-sensitive solar cells is a novel approach. Velten et al. successfully demonstrated the replacement of Pt catalyst used in the counter electrode of a dye-sensitized solar cell (DSSC) (Gr-F) with that of NPS made of dry-spun MWNT films and graphene flakes. In the iodide/triiodide redox (I^-/I_3^-) procedure, the efficiency of this counter electrode in limiting triiodide was examined in tandem by employing the dry-spun carbon MWNT strips alone and then graphene flakes used independently. This nanocomposite, when deposited on top of fluorinated tin-oxide-coated glass in the triiodide reduction reaction in DSSCs, showed exceptionally high and increased catalytic behavior and energy conversion efficacy (7.55%) when compared with MWNTs alone (6.62%) or graphene alone (6.62%). Slender atomic boundaries arising through graphene flake architecture and defective regions in carbon MWNTs allow for higher catalytic performance, which is measured in terms of improved electrical conductivity across the carbon MWNT bundle and the graphene flake [7].

17.4 Conclusion

In conclusion, 1D nanomaterials are particularly flexible in carrier separation/collection and photon absorption due to their large aspect ratio and anisotropic features, which may increase solar energy harvesting efficiency. Customizing 1D nanomaterials to hierarchies or hybrid systems with additional elements – such as zero-dimensional (0D) and 1D hybrid materials or core-shell frameworks – could improve light absorbance and carrier collection efficiency. The use of 1D nanomaterials to build cheap, potential PEC and PV solar cells as a result of bandgap engineering and nanostructure design is exciting. This chapter, on the other hand, provides an in-depth discussion of the important materials employed in the manufacture of flexible solar cells. (substrates, electrodes, and conductive compounds, to name but a few examples). Considering advances throughout flexible substrate technology, a new and superior substrate is still required to achieve the desired low-cost and elastic substance. Stainless steel is likely to become a prominent component in the not-too-distant future. It is critical to creating a low-cost technology for applying moisture barrier film to plastic PI substrates. Manufacturing large-scale, flexible ceramic/glass substrates cheaply is challenging for flexible ceramic/glass substrates. For materials which are very active conductors, we looked at hydrogenated amorphous silicon (a-Si:H), copper indium gallium selenide (CIGS), organometal halide perovskites, and conjugated polymer semiconductors. A-Si: H and CIGS are assumed to propel the industry forward in the future. Non-fullerene organic and perovskite solar cells will gain popularity due to their better performance and ease of manufacture. The most popular electrode materials will continue to be transparent, conducting oxides. The search for new technologies, like conducting polymers and nanocomposite materials, is anticipated to increase in the coming years.

References

1. Fu S, Sun Z, Huang P, Li Y, & Hu N (2019). Some basic aspects of polymer nanocomposites: A critical review. *Nano Materials Science, 1*(1), 2–30. https://doi.org/10.1016/j.nanoms.2019.02.006.
2. Yousif Q A, & Agbolaghi S (2020). A comparison between functions of carbon nanotube and reduced graphene oxide and respective ameliorated derivatives in perovskite solar cells. *Macromolecular Research, 28*(5), 425–432. https://doi.org/10.1007/s13233-020-8054-8.
3. Lee K M, Hu C W, Chen H W, & Ho K C (2008). Incorporating carbon nanotube in a low-temperature fabrication process for dye-sensitized TiO_2 solar cells. *Solar Energy Materials and Solar Cells, 92*(12), 1628–1633. https://doi.org/10.1016/j.solmat.2008.07.012.
4. Ramos J C, Flores J R, Turlakov G, Moggio I, Arias E, & Rodríguez G (2020). Self-assembly of a poly (phenyleneethynylene) on multiwall carbon nanotubes: Correlation of structural and optoelectronic properties towards solar cells application. *Journal of Molecular Structure, 1222*, 128845. https://doi.org/10.1016/j.molstruc.2020.128845.
5. Ates M, Eker A A, & Eker B (2017). Carbon nanotube-based nanocomposites and their applications. *Journal of Adhesion Science and Technology, 31*(18), 1977–1997. https://doi.org/10.1080/01694243.2017.1295625.
6. Alturaif H A, ALOthman Z A, Shapter J G, & Wabaidur S M (2014). Use of carbon nanotubes (CNTs) with polymers in solar cells. *Molecules, 19*(11), 17329–17344. https://doi.org/10.3390/molecules191117329.
7. Velten J, Mozer A J, Li D, Officer D, Wallace G, Baughman R, & Zakhidov A (2012). Carbon nanotube/graphene nanocomposite as efficient counter electrodes in dye-sensitized solar cells. *Nanotechnology, 23*(8), 085201. https://doi.org/10.1088/0957-4484/23/8/085201.
8. Li X, Li P, Wu Z, Luo D, Yu H Y, & Lu Z H (2020). Review and perspective of materials for flexible solar cells. *Materials Reports: Energy*. https://doi.org/10.1016/j.matre.2020.09.001.
9. Han N, Wang F, & Ho J C (2012). One-dimensional nanostructured materials for solar energy harvesting. *Nanomaterials and Energy, 1*(1), 4–17. https://doi.org/10.1680/nme.11.00005.
10. Abdulrazzaq O A, Saini V, Bourdo S, Dervishi E, & Biris A S (2013). Organic solar cells: A review of materials, limitations, and possibilities for improvement. *Particulate Science and Technology, 31*(5), 427–442. https://doi.org/10.1080/02726351.2013.769470.

11. Anandan S, Wen X, & Yang S (2005). Room temperature growth of CuO nanorod arrays on copper and their application as a cathode in dye-sensitized solar cells. *Materials Chemistry and Physics, 93*(1), 35–40.
12. Li Y, Yang X Y, Feng Y, Yuan Z Y, & Su B L (2012). One-dimensional metal oxide nanotubes, nanowires, nanoribbons, and nanorods: Synthesis, characterizations, properties and applications. *Critical Reviews in Solid State and Materials Sciences, 37*(1), 1–74. https://doi.org/10.1080/10408436.2011.606512.
13. Sagawa T, Yoshikawa S, & Imahori H (2010). One-dimensional nanostructured semiconducting materials for organic photovoltaics. *The Journal of Physical Chemistry Letters, 1*(7), 1020–1025. https://doi.org/10.1021/jz100065u.
14. Liu R, Shen Z, Wan Z, Zhu L, Chen J, Dong C, Chen W, Cao W, Chen B, Yuan X, Ding B, Yang S, Chen T, Tian X, Chen C, & Wang M (2021). Nanoarray heterojunction and its efficient solar cells without negative impact of photogenerated electric field. *Communications on Physics, 4*(1), 1–12. https://doi.org/10.1038/s42005-021-00678-1.
15. Wang X, Li Q, Xie J, Jin Z, Wang J, Li Y, & Fan S (2009). Fabrication of ultralong and electrically uniform single-walled carbon nanotubes on clean substrates. *Nano Letters, 9*(9), 3137–3141.
16. Khan D, Ali Z, Asif D, Panjwani M K, & Khan I (2021). Incorporation of carbon nanotubes in photoactive layer of organic solar cells. *Ain Shams Engineering Journal, 12*(1), 897–900. https://doi.org/10.1016/j.asej.2020.06.002.
17. Kislyuk V V, & Dimitriev O P (2008). Nanorods and nanotubes for solar cells. *Journal of Nanoscience and Nanotechnology, 8*(1), 131–148.
18. Berson S, de Bettignies R, Bailly S, Guillerez S, & Jousselme B (2007). Elaboration of P3HT/CNT/PCBM composites for organic photovoltaic cells. *Advanced Functional Materials, 17*(16), 3363–3370.
19. Zhang Z, Lv R, Jia Y, Gan X, Zhu H, & Kang F (2018). All-carbon electrodes for flexible solar cells. *Applied Sciences, 8*(2), 152.
20. Van De Lagemaat J, Barnes T M, Rumbles G, Shaheen S E, Coutts T J, Weeks C, Igor L, Peltola J, & Glatkowski P (2006). Organic solar cells with carbon nanotubes replacing in_2O_3:Sn as the transparent electrode. *Applied Physics Letters, 88*(23), 233503.
21. Zhang Y, & Liu H (2019). Nanowires for high-efficiency, low-cost solar photovoltaics. *Crystals, 9*(2), 87. https://doi.org/10.3390/cryst9020087.
22. Valente J, Godde T, Zhang Y, Mowbray D J, & Liu H (2018). Light-emitting GaAs nanowires on a flexible substrate. *Nano Letters, 18*(7), 4206–4213.
23. Blomers C, Grap T, Lepsa M I, Moers J, Trellenkamp S, Grutzmacher D, & Schapers T (2012). Hall effect measurements on InAs nanowires. *Applied Physics Letters, 101*(15), 152106.
24. Wunnicke O (2006). Gate capacitance of back-gated nanowire field-effect transistors. *Applied Physics Letters, 89*(8), 083102.
25. Ashkenasy G, Cahen D, Cohen R, Shanzer A, & Vilan A (2002). Molecular engineering of semiconductor surfaces and devices. *Accounts of Chemical Research, 35*(2), 121–128.
26. Dalmau Mallorquí A, Epple F M, Fan D, Demichel O, & Fontcuberta i Morral A (2012). Effect of the pn junction engineering on Si microwire-array solar cells. *Physica Status Solidi (A), 209*(8), 1588–1591.
27. Parkinson P, Joyce H J, Gao Q, Tan H H, Zhang X, Zou J, Jagadish C, Herz L M, & Johnston M B (2009). Carrier lifetime and mobility enhancement in nearly defect-free core–shell nanowires measured using time-resolved terahertz spectroscopy. *Nano Letters, 9*(9), 3349–3353.
28. Dalpian G M, & Chelikowsky J R (2006). Self-purification in semiconductor nanocrystals. *Physical Review Letters, 96*(22), 226802.
29. Norris D J, Efros A L, & Erwin S C (2008). Doped nanocrystals. *Science*, 319(5871), 1776.
30. Chia A C, & LaPierre R R (2012). Analytical model of surface depletion in GaAs nanowires. *Journal of Applied Physics, 112*(6), 063705.
31. Fiordaliso E M, Balogh Z I, Kasama T, LaPierre R, & Aagesen M (2016, November). Mapping electrostatic potentials across the p-n junction in GaAs nanowires by off-axis electron holography. In *European Microscopy Congress 2016: Proceedings* (pp. 743–744). Weinheim, Germany: Wiley-VCH Verlag GmbH & Co. KGaA, 743–744.

32. Van Dam D, Van Hoof N J, Cui Y, Van Veldhoven P J, Bakkers E P, Gomez Rivas J, & Haverkort J E (2016). High-efficiency nanowire solar cells with omnidirectionally enhanced absorption due to self-aligned indium–tin–oxide mie scatterers. *ACS Nano, 10*(12), 11414–11419.
33. Mariani G, Zhou Z, Scofield A, & Huffaker D L (2013). Direct-bandgap epitaxial core–multishell nanopillar photovoltaics featuring subwavelength optical concentrators. *Nano Letters, 13*(4), 1632–1637.
34. Zhang Y, Sanchez A M, Aagesen M, Huo S, Fonseka H A, Gott J A, & Liu H (2019). Growth and fabrication of high-quality single nanowire devices with radial p-i-n junctions. *Small, 15*(3), 1803684.
35. Gemeiner P, Kulicek J, Syrovy T, Haz A, Khunova V, Hatala M, Mikula M, Hvojnik M, Gál L, Jablonský M, & Omastová M (2019). Screen-printed PEDOT: PSS/halloysite counter electrodes for dye-sensitized solar cells. *Synthetic Metals, 256*, 116148.
36. Saranya K, Rameez M, & Subramania A (2015). Developments in conducting polymer based counter electrodes for dye-sensitized solar cells–An overview. *European Polymer Journal, 66*, 207–227.
37. Wu J, Lan Z, Lin J, Huang M, Huang Y, Fan L, Luo G, Lin Y, Xie Y, & Wei Y (2017). Counter electrodes in dye-sensitized solar cells. *Chemical Society Reviews, 46*(19), 5975–6023.
38. Wei W, Wang H, & Hu Y H (2014). A review on PEDOT-based counter electrodes for dye-sensitized solar cells. *International Journal of Energy Research, 38*(9), 1099–1111.
39. Brennan L J, Byrne M T, Bari M, & Gunko Y K (2011). Carbon nanomaterials for dye-sensitized solar cell applications: A bright future. *Advanced Energy Materials, 1*(4), 472–485.
40. Tiwari J N, Tiwari R N, & Kim K S (2012). Zero-dimensional, one-dimensional, two-dimensional and three-dimensional nanostructured materials for advanced electrochemical energy devices. *Progress in Materials Science, 57*(4), 724–803.
41. Roy P, Kim D, Lee K, Spiecker E, & Schmuki P (2010). TiO_2 nanotubes and their application in dye-sensitized solar cells. *Nanoscale, 2*(1), 45–59.
42. Yu K, & Chen J (2009). Enhancing solar cell efficiencies through 1-D nanostructures. *Nanoscale Research Letters, 4*(1), 1–10.
43. Chen H Y, Kuang D B, & Su C Y (2012). Hierarchically micro/nanostructured photoanode materials for dye-sensitized solar cells. *Journal of Materials Chemistry, 22*(31), 15475–15489.
44. Xie Y L, Li Z X, Xu Z G, & Zhang H L (2011). Preparation of coaxial TiO_2/ZnO nanotube arrays for high-efficiency photo-energy conversion applications. *Electrochemistry Communications, 13*(8), 788–791.
45. Liu R, Yang W D, Qaing L S, & Liu H Y (2012). Conveniently fabricated heterojunction ZnO/TiO_2 electrodes using TiO_2 nanotube arrays for dye-sensitized solar cells. *Journal of Power Sources, 220*, 153–159.
46. Liu B, & Aydil E S (2009). Growth of oriented single-crystalline rutile TiO_2 nanorods on transparent conducting substrates for dye-sensitized solar cells. *Journal of the American Chemical Society, 131*(11), 3985–3990.
47. Meng Q B, Takahashi K, Zhang X T, Sutanto I, Rao T N, Sato O, Fujishima A, Watanabe H, Nakamori T, & Uragami M (2003). Fabrication of an efficient solid-state dye-sensitized solar cell. *Langmuir, 19*(9), 3572–3574.
48. Dabera G D M, Jayawardena K I, Prabhath M R, Yahya I, Tan Y, Nismy N A, Shiozawa H, Sauer M, Ruiz-Soria G, Ayala P, Stolojan V, Adikaari A A, Jarowski P D, Pichler T, & Silva S R (2013). Hybrid carbon nanotube networks as efficient hole extraction layers for organic photovoltaics. *ACS Nano, 7*(1), 556–565.
49. Zhu K, Neale N R, Miedaner A, & Frank A J (2007). Enhanced charge-collection efficiencies and light scattering in dye-sensitized solar cells using oriented TiO_2 nanotubes arrays. *Nano Letters, 7*(1), 69–74.
50. Li J W, Chen Y S, Chen Y F, Chen J X, Kuo C F J, Chen L Y, & Chiu C W (2021). Enhanced efficiency of dye-sensitized solar cells based on polymer-assisted dispersion of platinum nanoparticles/carbon nanotubes nanohybrid films as FTO-free counter electrodes. *Polymers, 13*(18), 3103.

18

Recent Development in One-Dimensional Polymer-Based Nanomaterials for High-Performance Solar Cells

Karapagavinayagam Petchimuthu, Baby Suneetha Ragupathy, Joseph Sahaya Anand, Suguna Perumal, and Vedhi Chinnapiyan

CONTENTS

18.1 Introduction

The scale of measurements used to determine nanotechnology's usefulness is generally 100 nm. Nanoscience has probably been around since the early twentieth century, but the concept that working at such a small scale may provide any technical advantages arrived much later. Richard Feynman, a physicist, initially proposed it in a presentation to the American Physical Society in 1959, entitled 'There's plenty of room at the bottom—an invitation to enter a new area of physics' [1–3]. Materials in the nano dimension are expected to have the ability to improve energy efficiency, assist clean-up of the environment, and resolve significant health issues. It is also expected to considerably improve industrial output at much lower costs. This technology will help in the production of smaller, cheaper, lighter, and more useful products, and will take less energy and raw resources to create, according to proponents of the technology.

Polymer nanomaterials (NMs) are one of the most researched materials in various fields. The potential of polymeric NMs and nanoparticles (NPs) to improve current medicine

DOI: 10.1201/9781003223764-18

and energy has piqued people's curiosity. Particle size, shape, material selection, and processing techniques are all study topics of interest for identifying the optimal nanosystem for more effective energy harvesting and precisely targeted delivery of therapeutic applications.

Concerns about the protracted impacts of CO_2 and other greenhouse gases, as well as the diminishing supply of non-renewable resources, are driving massive research into renewable energy technology. Solar cells, which capture energy directly from the sun and convert it to electricity, are widely regarded as an important component of renewable energy transmission and distribution. Despite tremendous cost reductions in solar modules over the past several decades, major advances in device performance and manufacturing processes are still required for photovoltaic systems to be financially viable for gigawatt uses. The technique of absorbing and transforming energy to power generation is known as energy harvesting (also known as energy scavenging or ambient power). The energy would either be utilized immediately or saved for later. Regenerative braking is the process of capturing small amounts of ambient energy that would otherwise be dissipated or squandered as heat, vibration, light, or other forms of waste. Energy harvesting technology is still very much in its infancy, and it will not be able to address all of our energy concerns, although, when it comes to supplying energy to low-power gadgets, it has great promise. With the fast-growing Internet of Things (IoT) sector, this technology shows enormous promise. It may be important to develop an energy storage method that takes numerous transceivers in different environments such as photovoltaic (PV) cells [4]. Farinholt et al. describe a smart sensor system that uses a hybrid algorithm to extract energy from numerous sources by mixing various transducer types. Light, the energy of motion, heat energy, and radio wave energy are some of the types of energy that can be captured. Based on their properties, energy recovery sources can be grouped into two types [5].

In 1839, Alexandre-Edmund Becquerel discovered that 'electrical currents arose from certain light-driven chemical reactions', and comparable phenomena were later discovered in a solid (selenium) by other scientists. However, it was not until the late 1940s that the industry saw the creation of the foremost hard-condition strategy, which prepared the method for the silicon solar cell with a 6% effectiveness. The first silicon solar cell was crucial in the development of solar technology because it represented the power exchange unit of a PV system, but it also had valuable consequences. The silicon materials are not used individually; instead, they are put together into components.

This chapter will provide a swift outline of the concepts and history of PV energy systems, as well as a detailed examination of the numerous existing technologies, including their advantages and limitations, as well as photovoltaic applications. Deliberations of the current state, growth, and relevance of various PV and solar energy technologies are included. This chapter provides a comprehensive overview of current photovoltaic (PV) technology. It focuses on the four main forms of PV now in use: crystalline, thin-film, compound, and nanotechnology. The goal of ongoing PV technological innovation is not only to increase cell performance but also to lower module production costs, making it more viable for a wide range of applications.

Solar panels are currently commercially accessible; nevertheless, research and development efforts are ongoing to enhance this power harvester. Natural sources, like sunlight, wind, and geothermal heat, are freely available from the environment. Artificial sources are not a natural element of the environment but ones that are produced as a result of human or system activity. They include human motion, pressure on floors/shoe inserts while walking or running, and system vibration while operating.

18.2 High-Performance Solar Cells

A photovoltaic (PV) cell is an energy harvesting technique that enables the photovoltaic effect to transform renewable electricity into usable power. PV cells come in a variety of shapes and sizes, but they always rely on semiconductors to interact with photons from the sun to generate an electric current [6–8]. The performance is determined under ideal laboratory circumstances and represents the PV cell or module's maximum attainable efficiency. Temperature, irradiance, and spectrum all have an impact on actual efficiency. Solar cells with commonly produced photovoltaics have an efficiency of 14–22%. Solar cells made of amorphous silicon have a conversion efficiency of only 6%. Experimental multiple-junction concentrating photovoltaics have attained a 44.0% efficiency in experimental situations [9–11]. PV cell efficiency can be improved in a variety of ways, each of which comes at a greater cost. The number of semiconductor impurities and crystal structural deformations can be reduced, for example, and monocrystalline cells, or "single-crystal" cells, can be used to do this. A cell that is more pure and homogeneous has a better likelihood of interacting with photons coming in. A more efficient semiconducting semiconductor, such as gallium arsenide, is another option. Gallium arsenide has an ideal bandgap of 1.4 electron volts (eV), which allows a greater percentage of the sun's energy to be harvested although it is much more uncommon and costly than silicon. Many layers of semiconductor material, or p-n junctions, can also be employed to boost electrochemical performance. Because each junction has a distinct energy bandgap, these multi-junction cells can achieve greater efficiency with energy extraction from numerous parts of the solar spectrum. Concentrated photovoltaics can also help to boost efficiency. This method entails concentrating the sun's radiation using a variety of techniques to improve the amount of energy reaching the solar cell [12].

18.3 Photovoltaic Cells and Mechanism of Their Functioning

For thermal uses, substantial progress has recently been made with 1D structures (such as nano/microfibers, nanowires, nanotubes, nanorods, and so on) with polymer-based nanomaterials. The synthesized methodologies for diverse 1D nano-scaled polymer materials, as well as the thermoelectric improved performance that come from the distinct structures, will be discussed in detail.

A PV cell, commonly known as the solar cell, is an electronic instrument that produces energy whenever it is exposed to light rays or photons. The photovoltaic effect was discovered in 1839 by the French physicist Alexandre-Edmond Becquerel, but it was not until the 1960s that photovoltaic cells found their first direct implementation. Solar cells composed of PV cell modules began to appear on rooftops at the end of the 1980s. Since the advent of the twenty-first century, photovoltaic capacity has been steadily expanding, fueled by the construction of enormous solar farms. The solar cell, also called a photovoltaic cell, works on the principle of the conversion of sunlight (photons) directly into electricity (voltage and electric current) at the atomic level. A photovoltaic cell is made up of semiconductor materials that can absorb photons from the sun and produce an electrical current. Photons are atoms that travel at 3×10^5 km/s and convey solar radiation. In the 1920s, Albert Einstein called these photons 'grains of light'. When photons strike a semiconductor material such

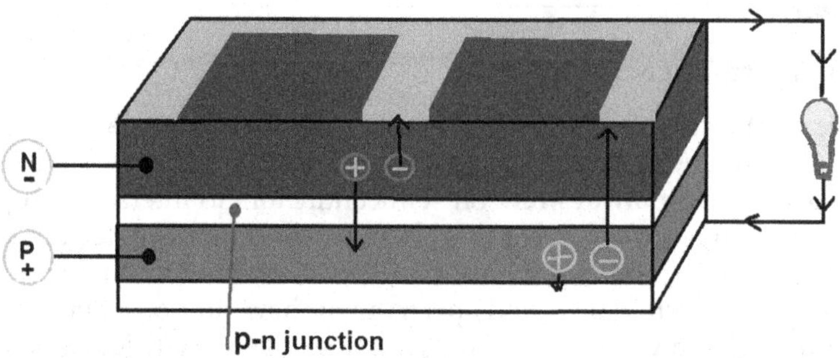

FIGURE 18.1
Schematic model of a photovoltaic cell.

as silicon, the electrons in the material's atoms are released, leaving a space called a 'hole'. The released electrons randomly move in search of a new 'hole' to fill. To generate an electric charge, the electrons must all flow in the same direction. So, the important factors required for solar energy conversion are the use of a suitable semiconductor material and electron flow in a single direction. One such semiconductor material which satisfies these conditions is silicon. On the surface that is exposed to the elements, silicon is blasted with phosphorus ions, which already have one more electron than silicon, while the other side is loaded with a central atom, which has one electron. The resultant layered silicon works like a battery, with the layer with more electrons acting as the negative terminal (n) and the layer with fewer electrons acting as the positive terminal (p) (p). An electric field is created at the point where the two layers meet. An electromagnet is used to sweep electrons to the n-side when they have been excited by photons, while holes drift to the p-side. While transferring electrical energy to the open circuit, the electron-hole pairs are guided to the electrical connections from both ends, resulting in a direct current. An anti-reflective layer is applied on top of the cell to prevent photon loss changing the surface reflection (Figure 18.1).

Deposition of 1D polymers on inorganic or carbon nanomaterials by *in-situ* polymerization is now the most widely utilized process for fabricating composite solar cells. The coated polymers should have an organized alignment and be elongated throughout to obtain superior thermoelectric efficiency, which decreases the energy needed for inter-chain and intra-chain bouncing and increases the polymer's carrier concentration. The *in-situ* polymerization reaction can induce monomer subunits to polymerize and extend systematically on the surface of nanofillers as a result of interface connections (conjugation, hydrogen bonding, van der Waals forces, etc.) between polymer and nanofillers. Furthermore, by altering the weight ratios of monomers to nanofillers, the density of the polymeric matrix can be easily changed.

18.4 Applications of Photovoltaic Cells

Thousands of acres of PV panels generate utility-scale power, ranging from tens of megawatts to more than a gigawatt. These big systems deliver power into municipal or regional grids utilizing fixed or sun-tracking panels. Extending power lines to all the regions where energy is required is always uneconomical, cumbersome, or even unpractical. PV could be the answer for isolated sites such as rural residences, villages in developing countries, lighthouses, offshore oil platforms, desalination facilities, and out-of-the-way health clinics. Vehicles, including cars and boats, can benefit from mobility PVs. Solar panels or trickle-charging batteries can be installed in automobile sunroofs to meet onboard power needs. To assist in the powering of high-altitude aeroplanes, lightweight PVs may fit the curvature of aeroplane wings. PV panels on roofs or the ground can provide electricity to structures. PV material can be used as windows, roof tiles, or cladding to serve a dual role in a building's structure. PV panels can also be used to shade and power awnings and parking structures.

Solar rooftop panels have been erected. As a sun radiator, such panels are being used to boil water, which is safe to swim in. They are also capable of generating electricity. In the case of a power outage, individuals can save this heat in a rechargeable battery pack and consume it. People can also store this energy and use it to generate electricity in their houses, saving money and cutting their electricity bills. Solar energy is used in automobiles [13–16]. This energy is stored in a battery or used to power a motor. Ed Passerini was the first person to build a solar car. The first electric vehicle was introduced in 1977. Solar-powered calculators employ photovoltaics. These calculators are powered by solar energy. Calculators require energy to function, which is provided by the sun's light. Because servicing is easy and affordable, the total purchase price is low. Solar power is among the most effective, non-renewable energy options.

18.5 Advantages of 1D Organic Photovoltaic (OPV) Materials for Energy Harvesting

The 1D organic photovoltaic technology is a relatively new type of thin-film solar cell in terms of performance, stability, and maturity in the PV sector. It has photoactive organic layers consisting of p-type and n-type semiconducting polymeric and oligomeric chains and bucky-balls. The energy levels of the highest occupied molecular orbitals (HOMOs) and lowest unoccupied molecular orbitals (LUMOs) in such polymeric OPV materials are identical to those of ordinary inorganic semiconductors' valence and conduction bands. The frequency optical bandgap across their LUMO and HOMO states are used to select, characterize, and create organic transistors (*via* different chemical syntheses). Poly(*p*-phenylenevinylene) (PPV), poly(2-methoxy-5-(2-ethylhexyloxy)-1-4-phenylenevinylene) (MEH-PPV), and poly[2-methoxy-5-(3′,7′-dimethyloctyloxy)]-*p*-1-4-phenylenevinylene) (MDMO-PPV) [17] are some examples of commonly utilized organic donors (p-type).

Pochettino was the first to describe the first organic solar cell (OSC) developed on anthracene in 1906 [18]. Following his finding, a significant amount of research attempts have been made to create higher semiconducting materials that have improved the performance of organic electronic devices. OSCs, organic light-emitting diodes (OLEDs), and stacked storage architectures are examples of such systems [18]. In terms of energy conversion efficiencies, Organic photovoltaic (OPV) technology has advanced tremendously recently, changing from the basic science to being a rather dependable application of solar energy.

Power conversion efficiencies (PCEs) of organic photovoltaic (PV) cells have steadily grown, with experimental solar panel effectiveness reaching 10% in 2012, and rising to

11.2% in OSCs with reversed circuit topologies and 12% in OPV systems, with an advantage being its ability as acceptor materials lately. Plans to create semiconductor polymers, OPV devices, and inorganic-organic hybrid architectures are still being researched around the globe. As a result, OSCs and other natural device systems that are linked with wearable electronics are being tested.

The following list consists of the most popular research types of OSC structures:

(i) Solar cells that are dye-sensitized (DSSCs)
(ii) Metal/polymer junction photovoltaic systems with PCEs less than 0.1% percent
(iii) Single-layer semiconducting solar cell systems with PCEs as low as 1%
(iv) Bulk-heterojunction (BHJ) OSC systems with a high degree of interactions between electron-donating components due to high-contacting surface area among donor-acceptor equipment

18.6 Operating Principle of Dye-Sensitized Solar Cells

To provide the infinite range required to absorb organic molecules, nanocrystalline TiO_2 is placed on the electrode material (conducting electrode) (sensitizers). Organic materials are stimulated from the highest occupied molecular orbitals (HOMO) to the lowest unoccupied molecular orbitals (LUMO) when photons are absorbed (LUMO). Figure 18.2 depicts a schematic illustration of this procedure. The dye molecules, which are the photocatalyst, are oxidized when an electron is inserted into the conduction band of the wide bandgap semiconductor nanostructured TiO_2 film. The injected electron is transferred between the TiO_2 nanoparticles before being removed and provided as electrical energy to a load. Between the TiO_2 photoelectrode and the carbon-covered counter electrode, electrolytes containing I^-/I_3^- redox ions are used as an electron mediator. The oxidized dye molecules (photosensitizer) are then regenerated by receiving electrons from the I^- ion redox mediator, which

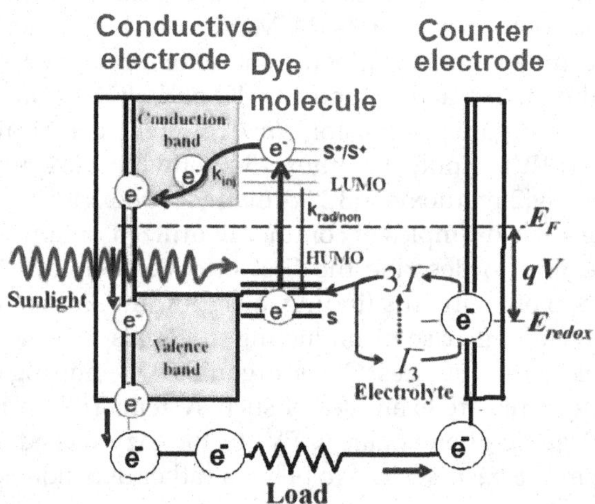

FIGURE 18.2
Illustration of a dye-sensitized solar cell.

is then oxidized to I_3^- (tri-iodide ions). The I_3^- ion replaces the internally given electron of the solar cells–dye-sensitized devices with one from the external load and returns to the I^- ion. The I_3 substitutes the internally donated electron of the solar cells–dye-sensitized devices with that from the external load and gets reduced back to the I ion. The electron movement in the conduction band of the wide-bandgap, nanostructured semiconductor is accompanied by the charge-compensating cations in the electrolyte layer that diffuses close to the nanoparticle surface. Thus, the electric power generated in DSSCs causes no permanent chemical transformation or change.

18.7 Types of Solar Cells

PV cells are divided into three categories: crystalline silicon cells, thin-film cells, and organic solar cells. Their conversion efficiencies are constantly improving. Many people are unaware that those solar cells are more sophisticated than they appear, and that there are several varieties of solar cells. When one looks at the various types of solar cells that are available (Figure 18.3), it becomes easier to understand them and choose the one that best meets one's needs.

18.7.1 Crystalline Silicon Cells

The very first era of PV technology was based on a crystalline structure, with solar cells composed of silicon (Si) and assembled into PV modules. This innovation, on the other hand, is not obsolete; it is always being improved to increase its productivity and quality. The next researchers explicitly used mono-crystalline, multi-crystalline, and emitter wrap-through (EWT) cells, all of which fall under the umbrella of silicon crystalline structures.

Around 90% of the world's photovoltaics are made of silicon, and crystalline silicon cells are used in around the same percentage of household solar panel systems. Monocrystalline and polycrystalline cells are made from crystalline silicon cells and are shown in Figure 18.4. Solar cells include silicon in a variety of forms. The purity of the silicon, however, is the most important factor. This is because it has a direct impact on the efficiency of the solar cell. In this situation, purity refers to the alignment of the silicon molecules. The purer the resultant silicon, the better the alignment. This, in turn, leads to higher rates of sunlight conversion into energy. As previously stated, the degrees of efficiency work in tandem with the purity of the silicon molecules—and improving purity

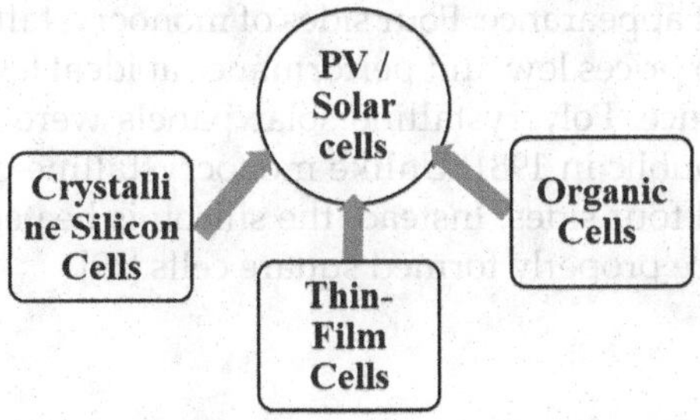

FIGURE 18.3
Different types of photovoltaic cell.

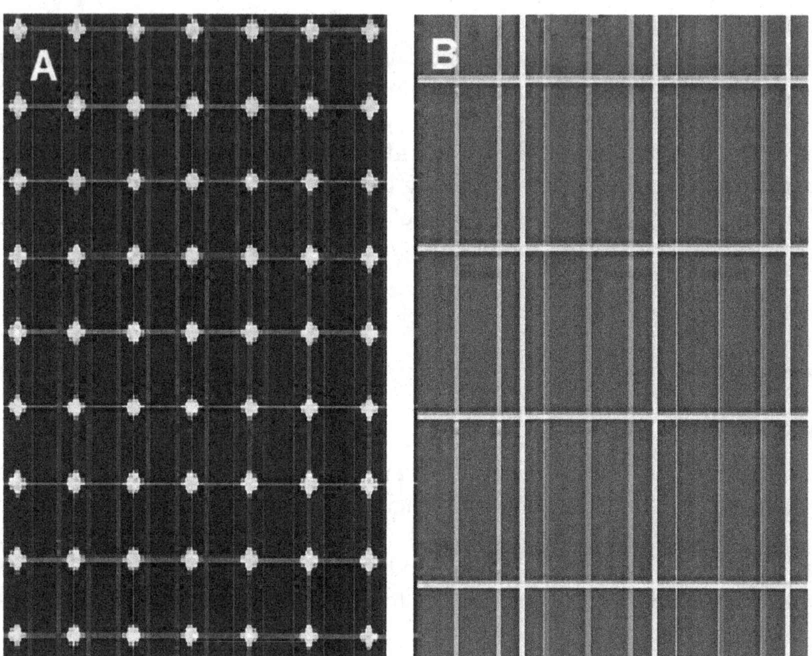

FIGURE 18.4
(a) mono and (b) polycrystalline form of a photovoltaic panel.

can be rather costly. However, it may come as a surprise to hear that efficiency is not the primary motivator for solar energy investors. For most potential customers, the price and the amount of time it would take are the most essential factors. The crystalline silicon cell is perhaps the most prevalent type of cell, accounting for around 80% of the market, and will remain the leading company until a more efficient and less expensive PV technology is produced. It mainly relies on crystalline Si p–n junctions. Because of the silicon material, existing efforts to improve efficiency are restricted by the power generated by photons, which reduces as frequencies increase. Furthermore, different frequency light produces a thermal loss, which causes the cell to heat up, lowering its efficiency. Under standard test conditions (STC), the maximum efficiency of a mono-crystalline silicon solar cell has been approximately 23%, with the highest measured being 24.7 percent (under STC). Self-losses are produced as a consequence of solar cell impedance, solar irradiance reflection, and metal contacts available on the top side. Monocrystalline solar cells are usually made using single-crystal silicon. They are frequently colored and have a cylindrical form and so the cells have a distinct appearance. Four sides of monocrystalline cells are chopped out by manufacturers to keep prices low and performance at ideal levels, which gives the cells their distinctive appearance. Polycrystalline solar panels were first made commercially available to the general public in 1981. Unlike monocrystalline cells, polycrystalline cells do not require cutting all four sides. Instead, the silicon is heated and placed into square molds which then produce properly formed square cells [19].

18.7.2 Thin-Film Cells

A thin-film solar cell is a device that makes use of layers of micron-thick photon-absorbing material placed over a flexible substrate to convert light energy into electrical energy (*via* the photovoltaic effect). Thin-film solar panels are made up of thin slices of semiconducting material that are adhered to a solid support. Compared with silicon wafers, thin-film cells decrease the number of transistor components needed for each cell, lowering the cost of solar cells. Materials used in thin-film PV cells include gallium arsenide (GaAs), copper, cadmium telluride (CdTe), copper indium diselenide ($CuInSe_2$), or titanium dioxide (TiO_2).

Thin-film solar cells were first developed in the United States in the 1970s by researchers at the University of Delaware's Institute of Energy Conversion. With the advances in technology in the early twenty-first century, the worldwide thin-film photovoltaic market grew at an unprecedented rate and is expected to continue to rise due to their low cost and high efficiency. Thin-film solar cells are produced in a variety of shapes and sizes. Although thin-film solar cells come in a variety of shapes and sizes, the most prevalent type contains cadmium telluride. They are less expensive than silicon thin-film cells.

In contrast with crystalline silicon cells, thin-film solar cells provide the possibility of lowering PV array costs by lowering material and production costs without affecting cell lifespan or posing any environmental risks. Unlike crystalline solar cells, which are made by sandwiching bits of semiconductors between windowpanes, coating panels are produced by depositing thin layers of specific materials on glass or stainless steel (SS) substrates using sputtering techniques. The benefit of this system is that the depth of the substratum is only a few microns (less than 10 μm) thick, as opposed to crystallized chips, which can be many as 100 microns thick; furthermore, films can be formed on SS plates, allowing the fabrication of elastic PV panels.

Cadmium telluride thin-films have maximum effectiveness of more than 22.1% (the percentage of photons hitting the surface of the cell that are transformed into an electric current). Cadmium telluride thin-film solar panel techniques had the least environmental impact and shortest payback period of any thin-film solar cell system on the market by 2014. (Payback time is the amount of time it takes for the electricity generated by the solar panel to cover the cost of buying and installing the panel [20, 21].)

18.7.3 Organic Solar Cells

Thin films (typically 100 nm) of semiconductor materials such as polymer and low-concentration substances such as pentacene, multiphenylene vinylene, copper phthalocyanine (a blue or green organic pigment), and carbon fullerenes are used to produce organic as well as polymeric photovoltaic energy. The maximum accuracy yet attained with organic semiconductor materials is 4–5%; nonetheless, the material's mechanical flexibility and disposability are what piqued the interest of researchers. Because they are mostly constructed of plastic rather than conventional silicon, the production is less expensive and has fewer technological obstacles (no need for elevated temperature or elevated-vacuum conditions are required). An organic solar cell, which is also known as a plastic solar cell, is a type of photovoltaic cell that makes use of organic electronics, a division of electronics concerned with conductive organic polymers or small organic molecules, for absorption of light and charge transport to generate electricity from sunlight *via* the photovoltaic effect. The majority of organic photovoltaic cells are made up of polymer solar cells. Organic solar cells use chemicals that are inexpensive and can be solution-processed at high speed, resulting in low production costs on a commercial scale. Organic solar cells, together with the flexibility of organic molecules, have the potential to be cost effective for photovoltaic applications. Electronic tunability can be achieved through engineering the molecule (e.g., modifying the size and substituents on the polymers). As a result, organic molecules have a high optical absorption coefficient, meaning that a considerable amount of light can be absorbed with a smaller amount of material, typically of the order of hundreds of nanometers. Current research is mostly focused on getting a sense of how these devices work to find a solution to increase the effectiveness of organic cells. At this moment, it is not possible to propose a self-contained theoretical model that can explain all, or even most, of the above-mentioned experiments. The energetical condition in organic and macromolecular materials is extremely complicated due to intramolecular and intermolecular interactions, numerous local disorders, and the presence of impurities and chemical dopants.

18.8 One-Dimensional Conducting Polymeric Nanomaterials for Dye-Sensitized Solar Cells (DSSCs)

Dye-sensitized solar cells (DSSCs) have emerged as an economically and technically viable alternative to p-n junction photovoltaics. Electrons transported by photoexcitation of dye molecules adsorbed onto the nanostructured wide-bandgap material photoelectrode generate power in dye-sensitized solar cells. The oxidized dye molecules are regenerated by electrons obtained from the electrolyte, which are then reduced by electrons reaching the cell's counter-electrode. Optical absorption and carrier transportation are not done concurrently in dye-sensitized solar cells. Because nanostructured DSSCs mimic photosynthesis in plants, they employ a multidisciplinary method that incorporates biological extraction, chemistry, physics, earth sciences, and electron transport concepts. Using light-sensitized organic dyes in electrochemical cells, a process which was discovered in the late 1960s, electricity can be generated. Chlorophyll (the main light-absorbing plant pigment involved in photosynthesis) was isolated from spinach at the University of California, Berkeley, and, in 1972, the first chlorophyll-sensitized zinc oxide (ZnO) electrode was synthesized. Photons were transformed into electricity for the first time by injecting electrons from excited dye molecules into a wide-bandgap semiconductor material. In recent years, efficiencies comparable with non-DSCCs have been reported for DSSCs; however, owing to some of the restrictions connected with these cells, they still require additional development. Stability failure can be divided into two categories in terms of limits: (i) external stability restrictions or (ii) inherent stability limitations. In addition, due to the energy mismatch between the oxidized dye and the electrolyte, a significant amount of energy is lost during the regeneration process. As a result, various electrolytes have been devised to improve the efficacy of these cells. In a 1D p-n junction diode model, an ideal PV cell illuminated current-voltage (I-V) characteristics and output dark can be easily solved. This can be achieved by necessarily adopting a simple 1D p-n junction diode model for a silicon solar cell which involves constant doping in the n- and p-type regions, and a blunt doping step transition, along with additional simplified assumptions.

A basic DSSC contains dye-sensitized semiconductor electrodes, oxidative electrolytes, and reference electrodes [22, 23]. The compound absorbs the light, which excites it and injects electrons into the semiconductors. At the very same time, the oxidized dye cation is reduced by the electrolyte solution, which contends for recombination with the electron transition. Charged particles gathered at the semiconductor electrodes travel across the external circuit before returning to the cell's counter electrode to reduce the oxidized electrolyte. Because of the competition between electron separation and recombination accompanying various electron transfer operations, the electrode of the dye-sensitized semiconductor determines the efficiency of DSSCs [24–28].

In brief, the oxidizes obtain positive ions from the excitons from the photo-catalyst dye, and all these charged particles may be locked by hydroxylated dye cations or anions in analyte or the Earth's crust state in TiO_2 (e.g., oxygen vacancy) before being transferred to an electrical current, a practice called as recombination. The type and shape of nanocrystalline semiconductor oxides control the type and shape of electron transport processes in semiconductor electrodes, which have a major impact on DSSC energy conversion. When nanoparticles and 1D hierarchical structures are compared, 1D hierarchical structures show better photovoltaic properties. As a result, a realistic solution for circumventing the limitations of nanoparticles and one-dimensional nanostructures as photoanode materials is now available. Nonetheless, the grain boundary effect rises as grains form between nanoparticles and nanorods, which is inescapable in a matrix of modular design. Hierarchic 1D nanomaterials were formed by blending nanoparticles with nanowires or nanostructured to solve the shortcomings of nanomaterials and 1D nanostructured materials. These substances were employed to take benefit of the efficient charge transfer and large surface area of nanoparticles and 1D nanostructures [29]. The interpenetration of the electrolyte in the 1D hierarchical structure is improved, and the interfacial area is expanded, potentially resulting in greater electron segregation and faster electron travel.

The use of one-dimensional (1D) nanostructures as photoanodes in effective DSSCs has seen a lot of early progress. DSSCs utilize a variety of randomly distributed and tightly coupled 1D nanostructures, as well as their composites with nanoparticles. Isotropic 1D nanotubes and nanotubes, uniformly dispersed 1D nanotubes and nanorods, randomly oriented 1D hierarchical TiO_2, and tightly coupled 1D hierarchical TiO_2 are shown to be the most efficient.

The photocurrent and photovoltage of a DSSC are significantly affected by the photoanode. The micromorphology and crystallographic structure of TiO_2 is a critical factor as the fast electron transfer kinetics of TiO_2 are required as a material in photoanodes to minimize photoelectron recombination. This results in the need for an efficient synthetic procedure that synthesizes titania nanomaterials with precise geometries and large surface areas that have optimum electron transfer kinetics [30].

1D TiO_2 nanostructures were shown to get a minimal number of grains and can behave as if crystalline, allowing the intergranular effect to be reduced and fast electron transport to be achieved [31–33]. At higher temperatures, the rod shape might be preserved. As a result, many studies concentrated on TiO_2 nanorods. Single-crystal anatase titanium dioxide nanorods were produced by a surfactant-assisted hydrothermal process, according to Jiu et al. [34, 35]. Their approach produced short rods with a diameter of 20–30 nm and a length of 100–150 nm. The existence of clean lattice strips suggested that high-crystalline-phase nanorods with fewer flaws had been achieved, which is advantageous for the transfer of electrons in the rod. The use of TiO_2 nanorods as a thin layer of dye-sensitized solar cells resulted in a light-to-electricity conversion yield of 7.06%. Nanorods have the same benefits as nanotubes, such as fast electron transport, low mixing, and durability.

On tin oxide- (SnO_2-) coated conductive glass slides, polycrystalline films are electrospun cathodically. The thickness of the MoS_2 films was estimated, and it was discovered that the film is an n-type semiconductor. The fill factor, open-circuit voltage, short circuit current, and effectiveness were estimated based on the photoelectrochemical cell investigations. the manufacture of semiconductor materials, particularly metal oxide chalcogenide thin films, and their characterization to assess their suitability for use as a photoelectrochemical (PEC) solar cell [36–39].

TiO_2 nanoparticles were the first documented 1D nanomaterials for use as a bifunctional photoanode, with excellent photon-generated electron production and positive light dispersion properties. Uchida et al. [40] devised a simple one-step hydrothermal process for producing TiO_2 nanotubes [41]. The photoelectrochemical characteristics of TiO_2 nanotubes with high distribution and crystallization might be improved. Such TiO_2 nanotubes (TNTs) with a high aspect ratio and a high specific surface area have been described by Adachi et al. [42] and Ohsaki et al. [43]. One-dimensional TiO_2 nanoparticles have a minimal number of intergranular and can behave as a single crystal, reducing intergranular effects and allowing for rapid electron transfer [44]. With high thermal calcination, the rod could maintain its shape. Several studies have been conducted on TiO_2 nanorods. By using TiO_2 nanorods as a thin layer of dye-sensitized solar panels, satisfaction and social conversion yield of 7.06% was attained. Nanorods have the same advantages as nanotubes, including fast electron transfer, low recombination, and durability.

The energy production efficiency of solar cells noted here already is higher than that of most reported PV cells based on nanostructured materials; however, more detailed studies are still underway to optimize the effectiveness by retaining the increasing application requirements [45–47]. The reported performance is better than that of a planar CdS/CdTe cell with comparable CdTe film thickness, but it is still lower than that of cells with ideal CdTe film thicknesses. As the simulation shows, the cell efficiency can be easily improved in the future by additional device and material optimization. Use top electrical connections with good optical clarity and lower parasite contact resistance, for illustration, efficiency can easily be boosted up to >10%.

18.9 Conclusion and Future Perspectives

Due to the obvious simplicity of efficient charge synthesis and movement inside the cell structure, dye-sensitized solar cells (DSSCs) made of nanocrystals exhibit a high efficiency. The DSSC is an eco-conscious, and high-efficiency solar cell. The nano-structured dye-sensitized solar cell (DSSC) is a low-cost alternative to p–n junction photovoltaic cells. In reality, dye-sensitized solar cells are green in color, resembling a leaf. The interaction of light with a variety of different materials can also create electricity. Organic lead compounds and halide elements like chlorine, bromine, and iodine can be used to make perovskite solar cells, which are known for their unique crystal structure. They are reasonably inexpensive to manufacture and have efficiency comparable to commercially available silicon cells, but their lifetime is currently restricted. Organic solar cells are made up of layers of polymers and can be mass-produced at a low cost. Although these cells may be made as a semi-transparent film, their efficiencies are poor. Semiconducting titanium dioxide with a one-molecule-thick coating of a suitable dye, which can function as a 'sensitizer', can be used to make DSSCs. These cells have modest efficiency, but they cannot tolerate direct sunshine without deteriorating. Nanotechnology is used to alter semiconducting materials at extremely small sizes with quantum dots. Nanoparticles, which are made up of 10,000 atoms and can be tuned to cover different sections of the sun's spectrum to improve solar energy absorption. Despite theoretically achieving great efficiency, laboratory test efficiencies are still quite low.

References

1. There's plenty of room at the bottom: An invitation to enter a new field of physics. Eng Sci Mag 22 February 1960; 23, USA: California Institute of Technology. Text reprinted in: Gilbert HD, ed. Miniaturization. New York: Reinhold Publishing Corp.
2. N. Taniguchi, On the basic concept of nanotechnology. In: *Proceedings of the International Conference on Production Engineering.* Tokyo: JSPE, 1974, 18–23.
3. N. Taniguchi, *Nanotechnology, Integrated Processing Systems for Ultra Precision and Ultra Fine Products.* Oxford University Press, 1996.
4. J. Zhu, Y. Zongfu, S. Fan, and Y. Cu, Nanostructured photon management for high performance solar cells, *Materials Science and Engineering: R: Reports*, vol. 70, pp. 330–340, 2010.
5. K. M. Farinholt, N. Miller, W. Sifuentes, J. MacDonald, G. Park, and C. R. Farrar, Energy harvesting and wireless energy transmission for embedded SHM sensor nodes, *Structural Health Monitoring*, vol. 9, no. 3, pp. 269–280, 2010.
6. J. Burschka, N. Pellet, S. J. Moon, R. Humphry-Baker, P. Gao, M. K. Nazeeruddin, M. Grätzel, Sequential deposition as a route to high-performance perovskite-sensitized solar cells, *Nature*, vol. 499, no. 7458, pp. 316–319, 2013.
7. H. Zhou Interface engineering of highly efficient perovskite solar cells, *Science*, vol. 345, no. 6196, pp. 542–546, 2014.
8. N. Pellet, P. Gao, G. Gregori, T. Yang, M. K. Nazeeruddin, J. Maier, M. Grätzel, Mixed-organic-cation perovskite photovoltaics for enhanced solarlight harvesting, *Angewandte Chemie International Edition*, vol. 53, no. 12, pp. 3151–3157, 2014.
9. O. Schultz, A. Mette, R. Preu, S. W. Glunz, Silicon solar cells with screen-printed front side metallization exceeding 19% efficiency, the compiled state-of-the-art of PV solar technology and deployment. 22nd European Photovoltaic Solar Energy Conference, EU PVSEC 2007. Proceedings of the International Conference, pp. 980–983, 2007.
10. Z. Shahan, Sun power panels awarded Guinness World Record, *Reuters.com*, 2013.
11. UD-led team sets solar cell record, joins DuPont on $100 million project, *UDaily*. University of Delaware, 2007.
12. S. Peake, *Renewable Energy: Power for a Sustainable Future*, 4th ed. Oxford: Oxford University Press, 2018.
13. A. G. Aberle, Surface passivation of crystalline silicon solar cells: A review, *Progress in Photovoltaics: Research and Applications*, vol. 8, no. 5, pp. 473–487, 2000.
14. A. R. Burgers, New metallisation patterns and analysis of light trapping for silicon solar cells. *Energieonderzoek Centrum Nederland*, 2005. PhD thesis.
15. A. Luque and S. Hegedus, *Handbook of Photovoltaic Science and Engineering*. Hoboken, NJ, Wiley, 2011.
16. M. Taguchi, A. Yano, S. Tohoda, K. Matsuyama, Y. Nakamura, T. Nishiwaki, K. Fujita, and E. Maruyama, 24.7% record efficiency hit solar cell on thin silicon wafer, *IEEE Journal of Photovoltaics*, vol. 4, no. 1, pp. 96–99, 2014.
17. M. A. Green, K. Emery, Y. Hishikawa, W. Warta, and E. D. Dunlop, Solar Cell Efficiency Tables (version 47), *Progress in Photovoltaics: Research and Applications*, vol. 24, pp. 3–11, 2016.
18. K. Yoshikawa, H. Kawasaki, W. Yoshida, T. Irie, K. Konishi, K. Nakano, T. Uto, D. Adachi, M. Kanematsu, H. Uzu, and K. Yamamoto, Silicon heterojunction solar cell with interdigitated back contacts for a photoconversion efficiency over 26%, *Nature Energy*, vol. 2, no. 5, p. 17032, 2017.
19. S.-Y. Lee, H. Choi, H. Li, K. Ji, S. Nam, J. Choi, S.-W. Ahn, H.-M. Lee, and B. Park, Analysis of a-Si:H/TCO contact resistance for the Si heterojunction back-contact solar cell, *Solar Energy Materials and Solar Cells*, vol. 120, pp. 412–416, 2014.
20. N. Mingirulli, J. Haschke, R. Gogolin, R. Ferre, T. F. Schulze, J. Dusterhoft, N. P. Harder, L. Korte, R. Brendel, and B. Rech, Efficient interdigitated back-contacted silicon heterojunction solar cells, *Physica Status Solidi – Rapid Research Letters*, vol. 5, pp. 159–161, 2011.

21. M. A. Green, K. Emery, Y. Hishikawa, W. Warta, and E. D. Dunlop, Solar cell efficiency tables, *Progress in Photovoltaics: Research and Applications*, vol. 30, pp. 3–12, 2022.
22. S. S. Hegedus, and W. N. Shafarman, *Progress in Photovoltaics: Research and Applications*, vol. 12, no. 23, pp. 155–176, 2004.
23. G. Dennler, K. Forberich, M. C. Scharber, C. J. Brabec, I. Tomis, K. Hingerl, and T. Fromherz, Angle dependence of external and internal quantum efficiencies in bulk-heterojunction organic solar cells, *Journal of Applied Physics*, vol. 102, no. 5, p. 054516, 2007.
24. A. Hagfeldt, and M. Grätzel, Molecular photovoltaics, *Accounts of Chemical Research*, vol. 3, no. 5, pp. 269–277, 2000.
25. J. Bisquert, D. Cahen, G. Hodes, S. Rühle, and A. Zaban, Physical chemical principles of photovoltaic conversion with nanoparticulate, mesoporous dye-sensitized solar cells, *Journal of Physical Chemistry: Part B*, vol. 108, no. 24, pp. 8106–8118, 2004.
26. M. Grätzel, Photoelectrochemical cells, *Nature*, vol. 414, no. 6861, pp. 338–344, 2001.
27. K. Pan, Y. Dong, C. Tian, W. Zhou, G. Tian, B. Zhao, H. Fu, TiO_2-B narrow nanobelt/TiO_2 nanoparticle composite photoelectrode for dye-sensitized solar cells, *Electrochimica Acta*, vol. 54, no. 28, pp. 7350–7356, 2009.
28. S. A. Haque, E. Palomares, B. M. Cho, A. N. Green, N. Hirata, D. R. Klug, J. R. Durrant, Charge separation versus recombination in dye-sensitized nanocrystalline solar cells: The minimization of kinetic redundancy, *Journal of the American Chemical Society*, vol. 127, no. 10, pp. 3456–3462, 2005.
29. T. Berger, T. Lana-Villarreal, D. Monllor-Satoca, and R. Gómez, An electrochemical study on the nature of trap states in nanocrystalline rutile thin films, *Journal of Physical Chemistry C*, vol. 111, no. 27, pp. 9936–9942, 2007.
30. M. J. Bierman, and S. Jin, Potential applications of hierarchical branching nanowires in solar energy conversion, *Energy and Environmental Science*, vol. 2, no. 10, pp. 1050–1059, 2009.
31. E. Enache-Pommer, J. E. Boercker, and E. S. Aydil, Electron transport and recombination in polycrystalline TiO_2 nanowire dye-sensitized solar cells, *Applied Physics Letters*, vol. 91, no. 12, 2007. Article ID 123116.
32. M. Kläui, H. Ehrke, U. Rüdiger et al., Direct observation of domain-wall pinning at nanoscale constrictions, *Applied Physics Letters*, vol. 87, no. 10, 2005. Article ID 102509.
33. S. H. Kang, S. H. Choi, M. S. Kang, J.-Y. Kim, H.-S. Kim, T. Hyeon, Y.-E. Sung, Nanorod-based dye-sensitized solar cells with improved charge collection efficiency, *Advanced Materials*, vol. 20, no. 1, pp. 54–58, 2008.
34. J. Jiu, F. Wang, S. Isoda, and M. Adachi, Highly efficient dye-sensitized solar cells based on single crystalline TiO_2 nanorod film, *Chemistry Letters*, vol. 34, no. 11, pp. 1506–1507, 2005.
35. J. Jiu, S. Isoda, M. Adachi, and F. Wang, Dye-sensitized solar cells based on a single-crystalline TiO_2 nanorod film, *Journal of Physical Chemistry: Part B*, vol. 110, no. 5, pp. 2087–2092, 2006.
36. T. J. Sahaya Anand, Synthesis and characterization of MoS2 films for photoelectrochemical cells, *Sains Malaysiana*, vol. 38, no. 1, pp. 85–89, 2009.
37. S. M. Ho, T. Anand, and Joseph Sahaya, A review of chalcogenide thin films for solar cell applications, *Indian Journal of Science and Technology*, vol. 8, no. 12, pp. 1–6, 2015.
38. S. M. Delphine, M. Jayachandran, C. Sanjeeviraja, A. W. Almusallam, and S. M. Delphine, Study on (mo/W) Se2 layered compound semi conductors useful for Photoeletrochemical solar cells, *International Journal of ChemTech Research*, vol. 3, no. 2, pp. 846–852, April–June 2011.
39. T. J. S. Anand, S. Shariza, S. I. Abd Razak, and C. Vedhi, Effect of deposition time and microstructure on transition metal molybdenum chalcogenides for photovoltaic applications, *Journal of Advanced Manufacturing Technology*, vol. 14, pp. 1–14, 2020.
40. S. Uchida, R. Chiba, M. Tomiha, N. Masaki, and M. Shirai, Application of titania nanotubes to a dye-sensitized solar cell, *Electrochemistry*, vol. 70, no. 6, pp. 418–420, 2002.
41. T. Kasuga, M. Hiramatsu, A. Hoson, T. Sekino, and K. Niihara, Formation of titanium oxide nanotube, *Langmuir*, vol. 14, no. 12, pp. 3160–3163, 1998.
42. M. Adachi, Y. Murata, I. Okada, and S. Yoshikawa, Formation of titania nanotubes and applications for dye-sensitized solar cells, *Journal of the Electrochemical Society*, vol. 150, no. 8, pp. G488–G493, 2003.

43. Y. Ohsaki, N. Masaki, S. Yanagida et al., Dye-sensitized TiO_2 nanotube solar cells: Fabrication and electronic characterization, *Physical Chemistry Chemical Physics*, vol. 7, no. 24, pp. 4157–4163, 2005.
44. E. Enache-Pommer, J. E. Boercker, and E. S. Aydil, Electron transport and recombination in polycrystalline TiO_2 nanowire dye-sensitized solar cells, *Applied Physics Letters*, vol. 91, no. 12, p. 3, 2007. Article ID 123116.
45. M. Kläui, H. Ehrke, U. Rüdiger et al., Direct observation of domain-wall pinning at nanoscale constrictions, *Applied Physics Letters*, vol. 87, no. 10, p. 3, 2005. Article ID 102509.
46. S. H. Kang, S. H. Choi, M. S. Kang, J. -Y. Kim, H. -S. Kim, T. Hyeon, Y. -E. Sung, Nanorod-based dye-sensitized solar cells with improved charge collection efficiency, *Advanced Materials*, vol. 20, no. 1, pp. 54–58, 2008.
47. Z. Fan, and J. C. Ho, Self-assembly of one-dimensional nanomaterials for cost-effective photovoltaics, *International Journal of Nanoparticles*, vol. 4, nos. 2/3, pp. 64–183, 2011.

19

One-Dimensional Polymeric Nanocomposites for Heavy Metal Detection

Golnaz Heidari, Zari Fallah, and Ehsan Nazarzadeh Zare

CONTENTS

19.1 Introduction

Heavy metals are among the most challenging toxins to deal with, owing to their toxic and non-biodegradable nature, as well as their bioaccumulation in biological environments [1]. They are released to the environment by both industrial activities and end-user materials and contaminate the environment when exposed to water through a variety of routes, including smelting, metallurgical mining, industrial chemical production, pulp industry, agriculture waste management, and so on. Heavy metal ions (HMIs) (like Cd^{2+}, As^{3+}, Pb^{2+}, Hg^{2+}, and Cu^{2+}) in polluted water can cause serious health problems in humans and have a negative impact on marine life. Nervous system problems, skin conditions, depression, headaches, and cancer are possible side effects of long-term exposure to heavy metal ions [2]. Detection of heavy metals, therefore, is of crucial importance in controlling environmental quality to improve public health. HMI detection continues to be a challenge due to the low concentrations of these ions, as well as the need for complicated sampling methods and the use of expensive equipment. Ion exchange, chemical precipitation, adsorption, reverse osmosis, electrochemical technology, and membrane filtration are just a few of the techniques available for removing heavy metals from water. At present, atomic absorption or emission spectroscopies, X-ray fluorescence, electrochemical sensor, optical way, and so on are the most used and respected conventional methods for HMI detection. Despite their high sensitivity and selectivity, these techniques require time, infrastructure in the form of expensive instrumentation, and manpower. So, electrochemistry-based and optical methods for analytical purposes were developed to address these limitations. The need

DOI: 10.1201/9781003223764-19

for complex apparatus and experimental methods was eliminated because optical sensors exhibit a specific color change when interacting with the target metal ion. These colorimetric sensors can conduct non-destructive detection without the need for any instruments, and they may be used by individuals with little technical knowledge [3].

A three-electrode system, comprising a reference electrode (RE), a working electrode (WE), and a count electrode (CE), is typically used for electrochemical detection. For particular detection of metal ions, the WE can be changed using various materials. The three 'S's, selectivity, sensitivity, and stability, are the essential features for investigating various sensors. Surface modification, data analytical methods, studies on unique sensing materials, measuring instruments, sensor fabrication techniques, control of sensor systems, and other efforts have been made to better resolve these problems. With the advancements in nanotechnology every year, much literature on one-dimensional (1D) nanostructured materials, like nanobelts, nanotubes, nanorods, and nanowires, has been published. As these materials have their unique structure, they show interesting physical and chemical features. A 1D nanostructure's initial feature is its high aspect ratio and very small dimensions, which allow it to successfully transfer electrical carriers in a single, controllable direction, making it excellent for electric currents in integrated nanoscale systems. The device function of 1D nanostructures, which can be used as device elements in a variety of nanodevices, is the second characteristic [4, 5]. Furthermore, with the advent of extremely selective nanomaterials, the number of selective electrode systems for various ionic species is growing. These materials can be simply put on electrode surfaces for the construction of sensing electrodes and the selective and sensitive detection of HMIs. The sensor's performance can be improved even more by combining different materials.

The synthesis, characterization, and application of 1D polymeric nanocomposites are an illustration of how nanotechnology and nanoscience have advanced in the past two decades. Because of the unique combination of electrical properties of polymers and the massive surface area of nanoparticles, the use of 1D nanostructures in the development of various sensors is gaining prominence. Incorporating a minimum of one additional component into polymers to produce a nanocomposite is another effective method for improving or extending the usefulness of polymers [6]. Different components are combined with polymers, including carbon nanotubes, metal nanomaterials, metal oxides, chalcogenides, biological materials, multi-components, etc. These nanocomposites are expected to have different characteristics resulting from each component's synergistic impact. The first part of this chapter discusses the fabrication of 1D polymeric nanocomposites and their properties. The second part discusses optical and electrochemical detection of HMIs using these nanocomposites.

19.2 Fabrication of One-Dimensional (1D) Polymeric Nanocomposites

There are two essential issues to consider when fabricating one-dimensional polymeric nanocomposites: how to create one-dimensional polymer nanostructures and how to integrate other components into these polymers. Several synthetic methods for fabricating different 1D polymer nanoparticles have been explored so far. Template-based synthesis and template-free synthesis are the two major methods for controlling the synthesis of 1D nanostructures which are discussed briefly here.

19.2.1 Template-Based Synthesis

Template-based synthesis is one of the most widely used methods for manufacturing many types of 1D nanomaterials, including functional composites, structural ceramics, metals, and inorganic semiconductors. As this is a potent and controlled technique for creating nanostructured materials, it is a crucial and successful approach for creating 1D polymer nanostructures with controllable forms and adjustable sizes and dimensions. The traditional hard-template approach, the soft-template method, and the wire-template method are all examples of template-based synthesis. The hard template method entails polymerizing nanostructures within the template membrane's pores or channels, then removing the template. A monomer solution is supplied into the pores of the templates, and the monomer is subsequently polymerized in the channels by chemical and/or electrochemical methods. As shown in Figure 19.1A, the anodic aluminum oxide (AAO) can be utilized as a hard template to make 1D robust polymer nanowires [7]. Although the hard template approach is the most effective way to create highly oriented conducting polymers, it is very difficult to remove the template. To eliminate the shortcomings of the hard template approach, the development of conducting polymer nanostructure synthesis

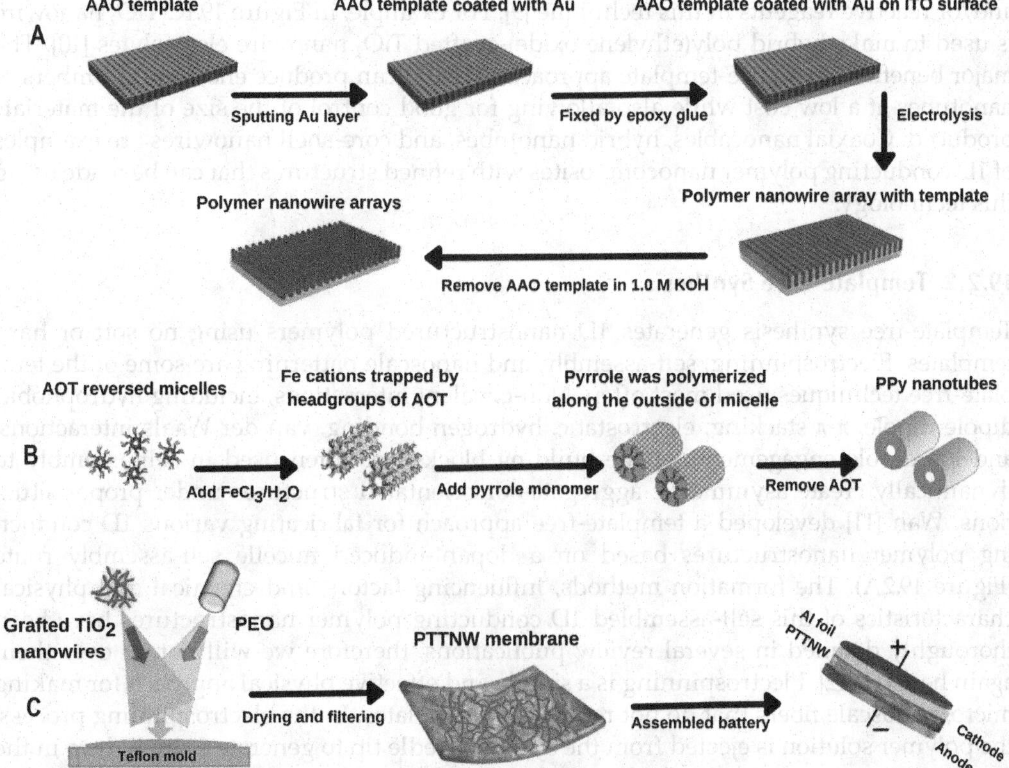

FIGURE 19.1
Template-based synthesis. (A) A sequence of procedures for creating polymer nanowire arrays using a hard template. Adapted with permission from reference [7]. Copyright (2020) American Chemical Society. (B) Reverse microemulsion polymerization of polypyrrole nanotubes using a soft template. Adapted with permission from reference [8]. Copyright (2003) Royal Society of Chemistry. (C) Electrolyte synthesis and use of a polymer electrolyte membrane modified with isocyanate-grafted Ti^{3+} doped TiO_2 nanowires. Adapted with permission from reference [10]. Copyright (2021) Elsevier.

may well be evaluated using a soft template technique. The ordered aggregates self-assembled from amphiphilic molecules *via* hydrogen bonds, van der Waals forces, and hydrophilic–hydrophobic interactions, among other things, are the soft templates utilized in this approach. 1D polymer nanostructures have been synthesized using colloidal particles, surfactants, oligomers, soap bubbles, colloids, and structure-directing molecules as soft templates. To make 1D polymeric nanostructures, soft templates guided by surfactants are commonly used. The shape and size of 1D nanostructured products are greatly influenced by the type of the surfactants as well as their concentrations. For example, reverse microemulsion polymerization in an apolar solvent was used to make polypyrrole (PPy) nanotubes and the generated micelles act as soft templates to facilitate pyrrole monomer polymerization catalyzed by Fe^{3+} ions along the templates' outer surface (Figure 19.1B) [8]. Nevertheless, the soft-template method's applicability is limited due to the soft templates' low stability and the difficulty in fine-tuning the shape and size of the resulting products. The novel wire-template approach has emerged as one of the most efficient options for the controlled synthesis of 1D polymeric nanostructures as it includes the benefits of both soft templates and hard templates. In theory, this wire-template technique uses templates made from pre-mature 1D nanomaterials for growing polymers on their surfaces to produce 1D nanocomposites. The utilized templates serve as physical scaffolds and/or reactive reagents in this technique [9]. For example, in Figure 19.1C TiO_2 nanowire is used to make hybrid poly(ethylene oxide)-grafted TiO_2 nanowire electrolytes [10]. The major benefit of this wire-template approach is that it can produce enormous numbers of nanotubes at a low cost while also allowing for good control of the size of the materials produced. Coaxial nanocables, hybrid nanotubes, and core-shell nanowires are examples of 1D conducting polymer nanocomposites with refined structures that can be made using this technology.

19.2.2 Template-Free Synthesis

Template-free synthesis generates 1D nanostructured polymers using no soft or hard templates. Electrospinning, self-assembly, and nanoscale patterning are some of the template-free techniques used most often. Non-covalent interactions, including hydrophobic, dipole-dipole, π-π stacking, electrostatic, hydrogen bonding, van der Waals interactions, and ion-dipole engagements of the building blocks, are often used in self-assembly to dynamically create asymmetric aggregates or orientated structures under proper situations. Wan [11] developed a template-free approach for fabricating various 1D conducting polymer nanostructures based on a dopant-induced micelle self-assembly route (Figure 19.2A). The formation methods, influencing factors, and chemical and physical characteristics of this self-assembled 1D conducting polymer nanostructures have been thoroughly detailed in several review publications, therefore we will not go over them again here [11, 12]. Electrospinning is a simple and effective physical approach for making micro/nanoscale fibers that do not require any templates. In the electrospinning process, the polymer solution is ejected from the syringe needle tip to generate a small drop in the presence of a strong electric field, and chargeable liquid jets are subsequently released from the cone. The jet is subjected to an elongation and instability process, allowing it to grow very thin and long. On a grounded collector, the solvent evaporates, leaving a collection of charged polymer fibers (Figure 19.2B). The comprehensive information affecting the electrospinning process is discussed in review papers and we will not go into detail here [13, 14].

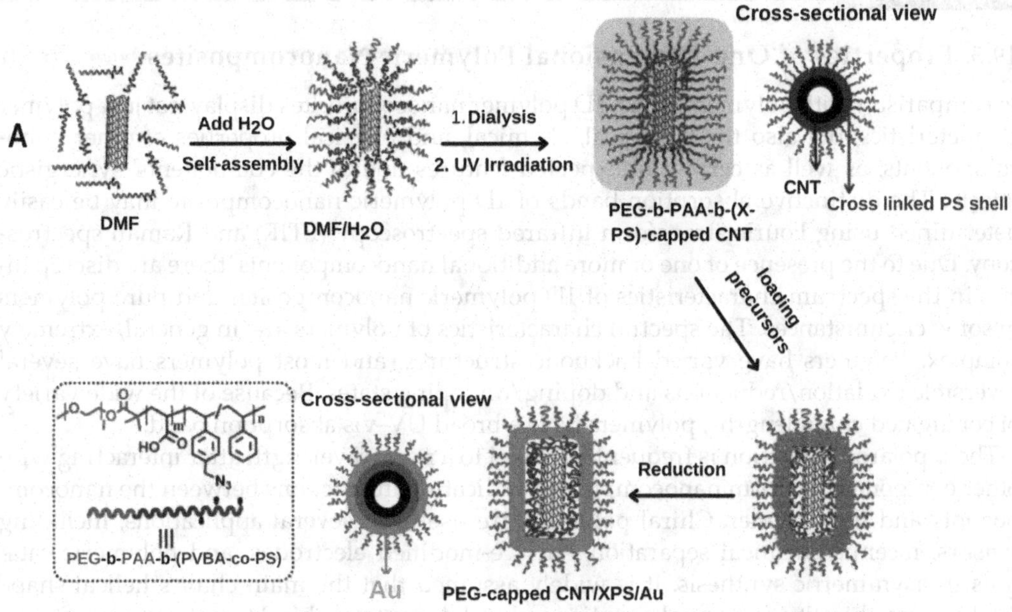

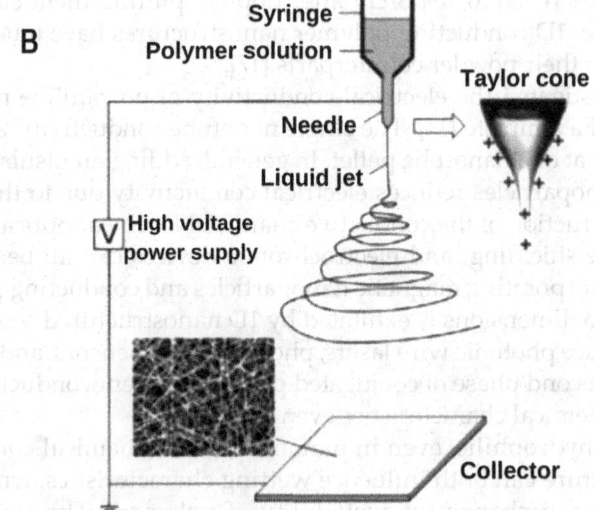

FIGURE 19.2
Template-free synthesis. (A) A step-by-step depiction of the production of a 1D carbon nanotube (CNT)/polymer/Au nanocomposite using a self-assembly technique. Adapted with permission from reference [15]. Copyright (2021) Elsevier. B) The basic electrospinning arrangement for direct fabrication of a hollow nanofiber. Adapted with permission from reference [16]. Copyright (2004) WILEY-VCH Verlag GmbH & Co. KGaA, Weinheim.

19.3 Properties of One-Dimensional Polymeric Nanocomposites

In comparison with polymers alone, 1D polymer nanocomposites display not just polymer characteristics, but also the biological, chemical, and physical properties of other nanocomponents, as well as certain unexpected features due to the components' synergistic effects. The distinctive absorption bands of 1D polymeric nanocomposite may be easily determined using Fourier-transform infrared spectroscopy (FTIR) and Raman spectroscopy. Due to the presence of one or more additional nanocomponents, there are discrepancies in the spectrum characteristics of 1D polymeric nanocomposites and pure polymers in some circumstances. The spectral characteristics of polymers are, in general, extremely complex. Polymers have varied backbone structures, and most polymers have several reversible oxidation/reductions and doping/undoping states. Because of the wide variety of conjugated chain lengths, polymers have a broad UV–vis absorption band.

The π-polaron transition is frequently moved to a new wavelength after interacting with other components to form nanocomposites, indicating interactions between the nanocomponents and the polymer. Chiral polymers are useful in several applications, including sensors, racemic chemical separation, surface-modified electrodes, and polymeric catalysts in asymmetric synthesis. It is widely assumed that the main chain's helical shape might cause chirality. In several situations, chiral dopants or chiral templates promote the helical shape of polymer chains. When another component is combined with polymers, the electrical conductivity of the polymers is increased or decreased. The electrical conductivity of polymers can reach 100 to 104 S/cm after doping, putting them close to the "metallic" conducting regime. 1D conducting polymer nanostructures have much greater electrical conductivities than their powder counterparts [17].

Chen and colleagues investigated the electrical conductivity of polyaniline nanotubes produced without the use of a template [18]. The single nanotube conductivity was found to be two-fold higher than that of a nanotube pellet. In general, adding an insulating component into 1D polymer nanoparticles reduces electrical conductivity due to the insulating component's partial obstruction of the conductive channel. Nonlinear optical systems, electromagnetic interference shielding, and electrochromic devices can all benefit from composite nanomaterials incorporating magnetic nanoparticles and conducting polymers. Photonic confinement in two dimensions is exhibited by 1D nanostructured semiconductors, which are ideal to produce photonic wire lasers, photochemical sensors, and photodetectors. The addition of the second phase of conjugated polymers to semiconductors might change their physical and chemical characteristics even further.

Polymers, in general, are hydrophilic, even in nanofiber films. Chemical composition and geometrical microstructure can both influence wetting characteristics, while incorporating other components can change the wettability of polymers. The specific surface area of nanoparticles is an essential characteristic that influences the catalytic and adsorptive behavior of metallic powders and ceramics, for example. The specific surface area of polymer nanofibers was determined by fiber diameters and doping state. The surface area of polyaniline nanofibers, for example, was 41.2 m^2/g before base dedoping, rising to 49.3 m^2/g following dopant removal, which was ascribed to an increase in the free volume of the polyaniline nanofibers [19]. Furthermore, polymer nanocomposites loaded with 1D core-shell structured $NaNbO_3$@SiO_2 nanowires (NN@SiO_2 NWs) improved their energy storage capabilities substantially. The laboratory findings and finite element modeling support the superiority of the 1D core-shell structured NN@SiO_2 NWs in increasing the energy density of the nanocomposite polymer [20]. Lu et al.

investigated the properties of one-dimensional polymeric nanocomposites in detail in their research [21].

19.4 Heavy Metal Detection

19.4.1 Optical Detection of Heavy Metals

Optical detection has the distinct benefits of high selectivity, high sensitivity, low cost, and a simple operating mode when compared with other detection systems. Optical sensors for detecting different heavy metal ions were quickly developed. The concepts of optical detection are typically focused on quantum dots, photoinduced electron transfer, fluorescence resonance energy transfer, intramolecular charge transfer, nanoparticles with optical characteristics, and the optical properties of rhodamine and its derivatives. Fluorescence and colorimetric detection are the two optical techniques most used for detecting heavy metal ions [22]. Fluorescence sensing is based on analyte-induced changes in fluorophores physicochemical properties, including anisotropy, lifetime, and fluorescence intensity, which are all connected to charge or energy transfer processes. Light-emitting diodes (LEDs), laser sources, and mercury and xenon arc lamps are some of the most used excitation sources for fluorescence-based detection [23]. When it comes to creating fluorescence sensors for heavy metal detection, scientists have focused on the Förster resonance energy transfer (FRET) mechanism. The dipole-dipole interaction between an acceptor and an energy donor causes FRET, which is determined by a 1/d6 separation distance dependency [24].

Metal ions are detected selectively using colorimetry or changes in the UV absorption spectrum in colorimetric detection. The collective oscillation of free conduction electrons in the noble metal in resonance with incoming electromagnetic radiation causes surface plasmon resonance (SPR) [24]. For example, in the absorption spectra of 20 nm monodispersed Au particles in an aqueous solution, there is a prominent SPR peak, and the aqueous solution has a "red wine" hue. The color of the solution changes due to a shift in the SPR peak once the Au NPs aggregate to some extent. Colorimetric sensors for the detection of different analytes are built on this concept. This sensor detects analytes directly, visually, and quickly, lowering expense. Several heavy metals, such as Hg^{2+}, Cr^{3+}, Cu^{2+}, Pb^{2+}, and As^{3+}, have been monitored using the colorimetric detection approach so far. Other optical techniques, such as surface-enhanced Raman scattering spectroscopy and circular dichroism spectroscopy, can all be utilized to identify metal ions selectively.

Given the increasing interest in optical sensors, various chemicals have been employed to detect metal ions, including inorganic nanoparticles, hydrogel film, organic dyes, composites, and polymers. Notably, sensing investigations must be carried out in a clean aquatic environment, since this is critical for detecting metal ions in groundwater, biological systems, and polluted soil [25, 26]. Furthermore, colorimetric sensing techniques based on chromo-ionophores have recently attracted a lot of attention since they provide visible color changes, which are excellent for real-time monitoring. As a result, numerous fluorescence-based and liquid-based colorimetric sensors based on multiple probe molecules have been created; however, these sensing systems are soluble in organic solvents, rendering them useless for sensor applications [27]. Because of these challenges, utilizing fluorescent small-molecule probes to identify metal ions selectively and sensitively is difficult.

Polymer-based sensors are the most sought-after materials for the next generation, offering aqueous phase solubility, signal amplification, high structural stability, enhanced selective binding efficiency, and improved biocompatibility by incorporating appropriate receptor units into a functional polymeric building block [28]. Taking use of the 1D nanoscale receptor layers and high surface area/volume ratio is one way to improve the characteristics of this type of sensor. For example, because of the characteristics linked to the geometry of this class of nanomaterials, nanofibers produced by electrospinning seem to be a very appealing alternative in this instance. Most nanofiber-based ion-selective sensors are optical, including the use of optical, spectral receptors that are influenced by analyte ion contact. These sensors are like colorimetric approaches that use tailor-made or custom-made complexing ligands. Polymeric nanofibers modified post-spinning with colorimetric dyes or doped nanofibers have been proposed as nanofiber sensors. These sensors are made utilizing a variety of receptors, not just ionophores, and benefit from the porous structure and high surface area of nanofiber mats [29]. Matysiak and colleagues [30] also looked at how conductive polymers affected the optical characteristics of electrospun polyacrylonitrile nanofibers packed with polyaniline, polythiophene, or polypyrrole. They showed that the inclusion of nanoparticles of conductive polymers in the polyacrylonitrile warp has a favorable impact on the absorption of the produced PAN/PT, PAN/PPy, and PAN/PANI nanofibers, according to Abs (λ) spectra. Furthermore, optical constant measurements demonstrated a rise in the dielectric constant and light refraction coefficient.

Sivaraman et al. [31] describe how to immobilize a chromo-ionophoric probe directly onto a porous polymer monolithic template to produce a low-cost, simple, portable, and safe solid-state optical sensor material (Figure 19.3). The sensor was solid-state, colorimetric sensing equipment capable of detecting trace Cu^{2+} concentrations in water samples visible to the human eye. The polymer monolith material's host template might be a better optical solid-state sensor material for detecting Cu^{2+} ions, with excellent sensitivity and selectivity.

Polypyrrole (PPy), polyaniline, polythiophene, and polypyrrole-chitosan are sensitive to Ni, Hg, Cu, Zn, Pb, and Au ions because they are conducting polymers with good environmental stability and the capability to detect the innate sensitivity to heavy metal ions. Polypyrrole with an multiwalled carbon nanotubes (MWCNT) composite has been included to enhance the selectivity and sensitivity of sensors primarily by interfacial interactions among conducting polymer and MWCNTs because delocalized π electrons in PPy and MWCNTs could get along with each other in a nanocomposite to decrease the system's energy and build a PPy/CNT nanocomposite. For example, to detect heavy metal ions through the surface plasmon resonance method, the gold layer was modified with polypyrrole multi-walled carbon nanotube composite layers (Figure 19.4). The MWCNTs increased the SPR sensor's sensitivity, enabling the concentration of the specified ions to be detected. The SPR sensor provides good sensitivity, stability, repeatability, and portability. SPR is affected by changes in the imaginary and real portions of the analyte refractive index. The selected studies using 1D polymeric nanocomposites in optical techniques for heavy metal detection are summarized in Table 19.1.

19.4.2 Electrochemical Detection of Heavy Metals

Electrochemical detection techniques for HMs have received a great deal of attention due to the benefits of the ability to perform simultaneous multi-metal detection, ease of operation, high sensitivity, and compact equipment size. Electrochemical detection is a type of instrumental analysis based on the electrochemical characteristics of electrolytes in the

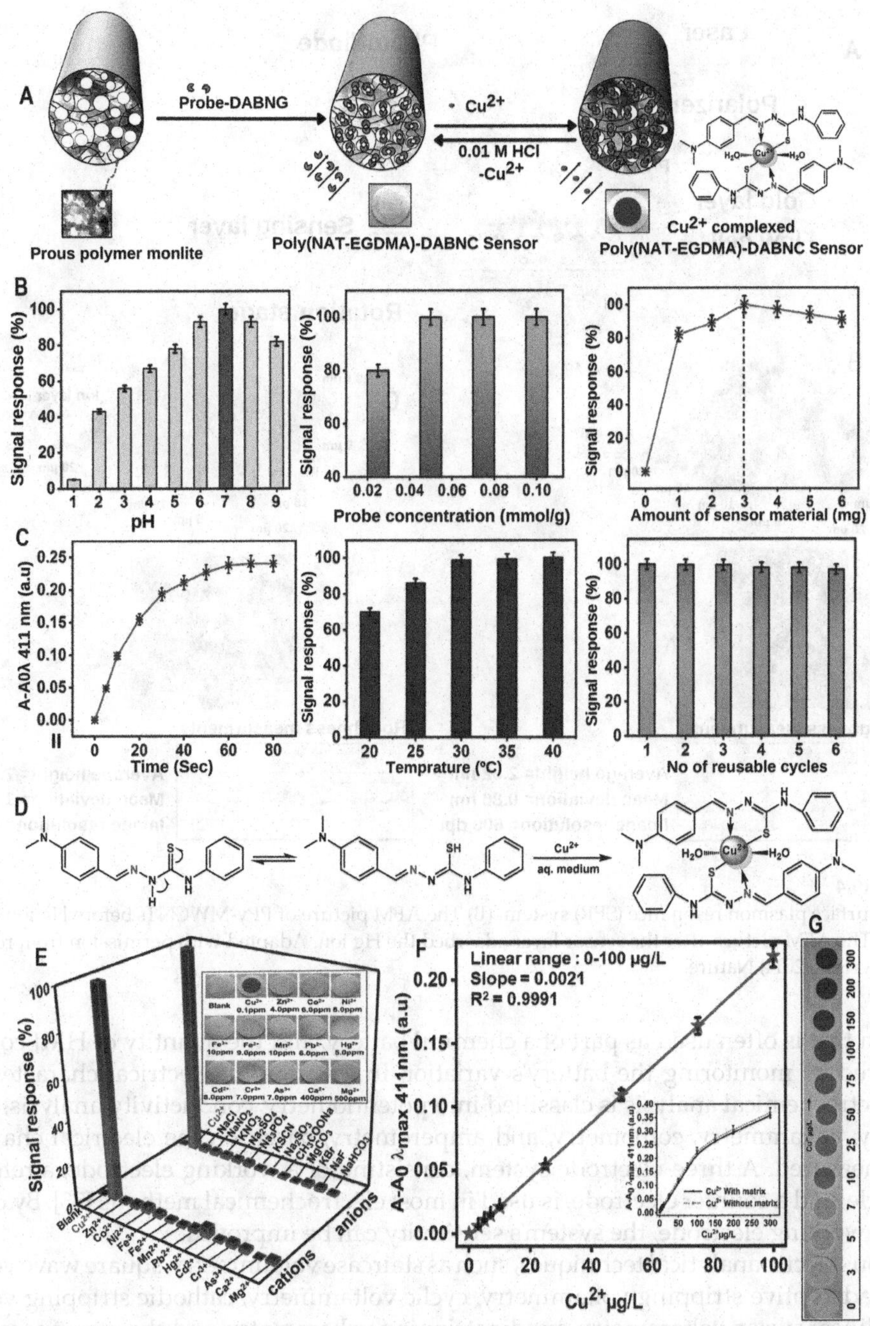

FIGURE 19.3
(A) Schematic representation of polymer-based colorimetric Cu^{2+} ion-sensor synthesis and its ion-sensing process. (B) Optimum pH, probe concentration, and amount of sensor material. (C) Effect of time, temperature, and many reusable cycles. (D) Complexation of Cu^{2+} mechanism. (E) Various colorimetric responses for Cu^{2+} and other electrolytes/cations, showing the poly(NAT-EGDMA)-DABNC sensor's ion-selectivity (and their matching naked-eye color transition (inset)). (F) The linear concentration range calibration figure and the equivalent graphs with and without matrix components. (G) Steady, visible color change shown by the sensor with incremental concentrations of Cu^{2+} ions. Adapted with permission from reference [31]. Copyright (2021) Elsevier.

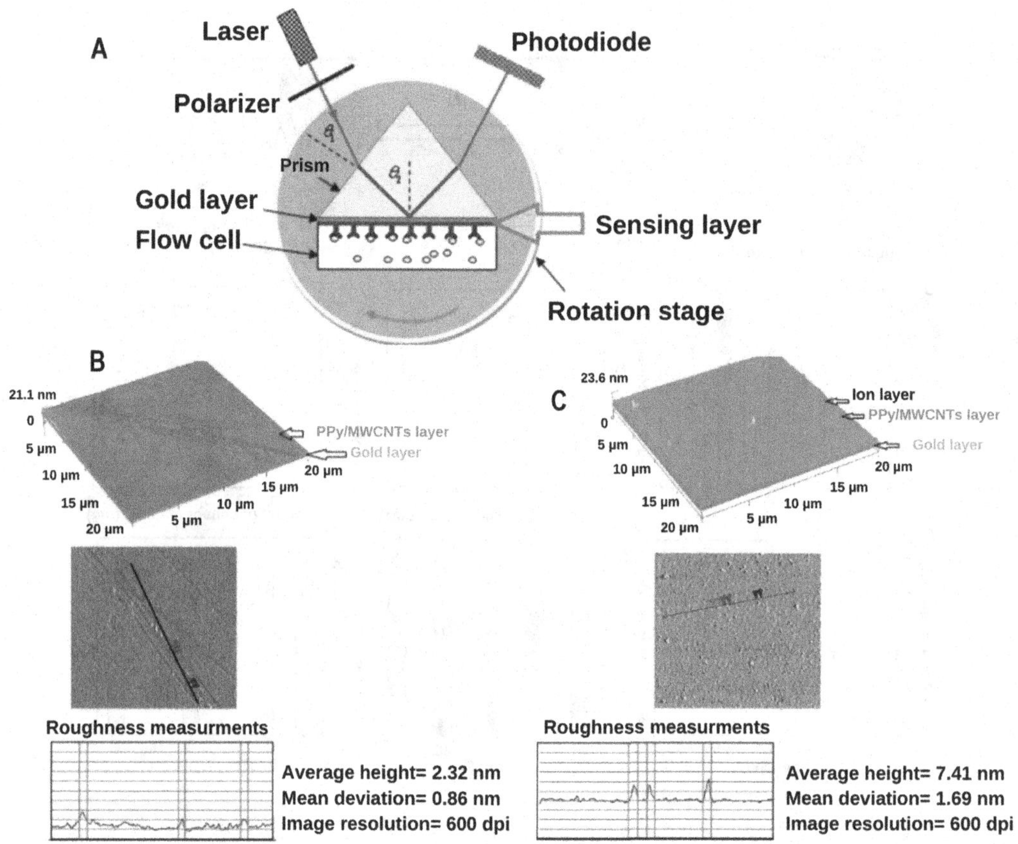

FIGURE 19.4
(A) The surface plasmon resonance (SPR) system. (B) The AFM picture of PPy-MWCNTs before Hg ion adsorption. (C) The AFM picture after the sensor layer adsorbed the Hg ion. Adapted with permission from reference [32]. Copyright (2014) Nature.

solution that is often used as part of a chemical battery, and the quantity of HMI could be measured by monitoring the battery's variation in a variety of electrical characteristics. The electrochemical analysis is classified into potentiometry, conductivity analysis, polarography, voltammetry, coulometry, and amperometry, based on the electrical characteristics measured. A three-electrode system, consisting of a working electrode, a reference electrode, and a counter electrode, is used in most electrochemical methods [33]. By changing the working electrode, the system's sensitivity can be improved.

Various electroanalytical techniques, such as staircase voltammetry, square wave voltammetry, adsorptive stripping voltammetry, cyclic voltammetry, cathodic stripping voltammetry, linear sweep voltammetry, anodic stripping voltammetry, and chronoamperometry are currently being researched and used. One drawback of electrochemical detection is the limited operational life of electrodes. For example, anodic stripping voltammetry (ASV) is a powerful electrochemical analytical tool designed to identify ionic species at the atomic level. Its high sensitivity comes mostly from a deposition phase where the concentration of target analytes on the detecting electrode's surface decreases. As a result, the electrode materials' characteristics are crucial, and materials that can enchant the HMs to be sensed are in high demand. The integration of polymer modifiers into electrochemical detection

TABLE 19.1

Electroanalytical Performances of Selected One-Dimensional Polymers in Electrochemical and Optical Techniques for Heavy Metals Detection

Nanocomposite	Metal	Detection Principle	Lod	Linear range	Reference
Ion-imprinted polymeric nanoparticles/multi-wall carbon nanotubes/polyaniline/graphite	Pb^{2+}	Potentiometry	3.4×10^{-10}	5.3×10^{-10} to 1.0×10^{-1}m	[40]
Gold/polypyrrole and polyaniline	As^{3+}	Voltammetry	2 ppb	—	[42]
Zinc oxide nanorod-doped ion-imprinted polypyrrole	Hg^{2+}	Voltammetry	1 pm	10^{-12} to 10^{-3} m	[41]
Polyaniline Sn(iv) tungstomolybdate nanocomposite	Pb^{2+}	Potentiometry	1×10^{-6} m	1×10^{-1} to 1×10^{-6}m	[43]
$Ti_3c_2t_x$/multiwalled carbon nanotubes (MWNT)/Au/polyethylene terephthalate (pet)	Zn^{2+} Cu^{2+}	Voltammetry	1.5 0.5 ppb	350–830 ppb 10–600 ppb	[39]
Imprinted polymer/multi-walled carbon nanotubes	Pb^{2+}	Voltammetry	0.0038 nm	0.01–0.50, 1–80	[44]
Ethylenediaminetetraacetic acid-polypyrrole /walled carbon nanotubes	Pb^{2+}	Voltammetry	0.07 µm	8×10^{2} µm to 0.15 µm	[45]
Fe-chitosan-coated carbon	As^{3+}	Voltammetry	1.12 ppb	2–24 ppb	[46]
Bi nanoparticles, multi-walled carbon nanotube	Cu^{2+}, Hg^{2+}	Voltammetry	0.01 nm	$0.01–5 \times 10^{6}$	[47]
Phytic acid/polypyrrole nanowires modified glassy carbon electrode	Cu^{2+}	Voltammetry	3.33 µg/l	10–60 µg/l	[48]
Functionalized polypyrrole nanotube arrays	Cu^{2+}	Voltammetry	46 nm	0.1–30 µm	[49]
Polypyrrole multi-walled carbon nanotube	Hg^{2+}, Pb^{2+}, Fe^{2+}	Surface plasmon resonance	0.1 ppm	—	[32]
pyrrole/chitosan/ITO/Ag-coated fiber	Cd^{2+}, Pb^{2+}, Hg^{2+}	Surface plasmon resonance	0.129, 0.158, 0.293 nM	0 to 200 µg/L	[50]

of HMIs has recently received a lot of interest. For example, HMI sensors based on carbon nanotubes (CNTs), conducting polymers, and conducting-polymer-based/ carbon-nanomaterial composites are drawing the attention of researchers for a variety of practical applications.

CNTs are an ideal material for the development of electro-analytical systems specialized in HMI detection because of their high surface area, excellent sorption capacity, high sorption proficiency, and regulated pore size distribution [34]. The selectivity of carbon nanotubes for HMIs was shown in Figure 19.5A. Organic conducting polymers (OCPs), on the other hand, are a desirable material for chemical sensors because of their collective redox characteristics, which make them sensitive to tiny electrochemical disturbances [35]. The sensitivity of conducting polymers for HMIs is depicted in Figure 19.5B. Low specific capacitance, hydrophobic nature, and absence of functional binding groups on virgin CNTs, instability of CNTs in hostile environments, and surface fouling are some of the constraints of CNTs for ultra-trace detection of HMIs [36]. Moreover, owing to constant charge-discharge during repetitive interpolation, and ion depletion, significant volume change occurs in the OCPs, causing mechanical instability. As a result, a combination of CNTs and OCPs is a brilliant one; OCPs have a greater affinity for metal ions while CNTs' high conductivity allows for faster electrical signal transmission (Figure 19.5C). Therefore,

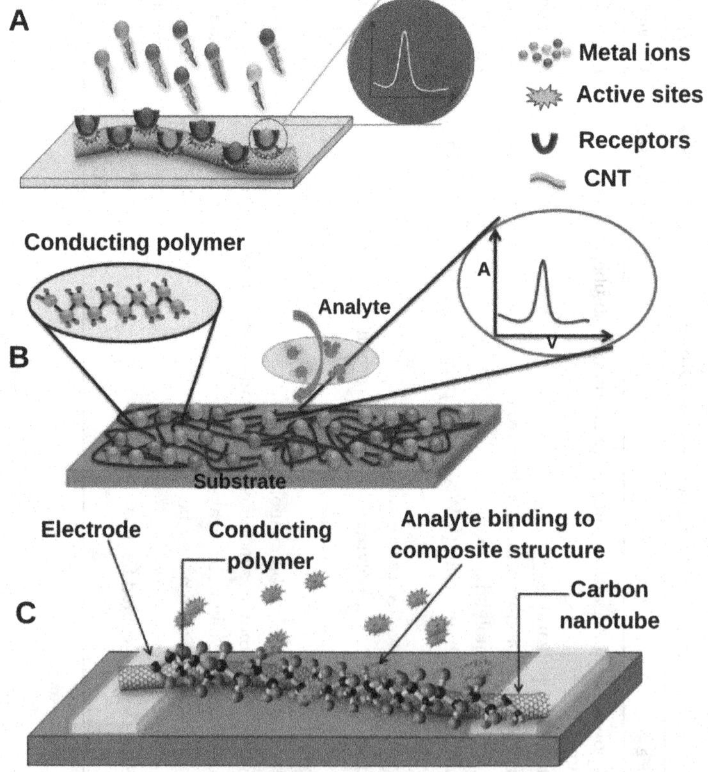

FIGURE 19.5
(A) HMI sensor based on carbon nanotubes, (B) HMI sensor based on organic conducting polymer. (C) HMI sensor with carbon nanotubes and organic conducting polymer. Adapted with permission from reference [37]. Copyright (2019) Taylor & Francis.

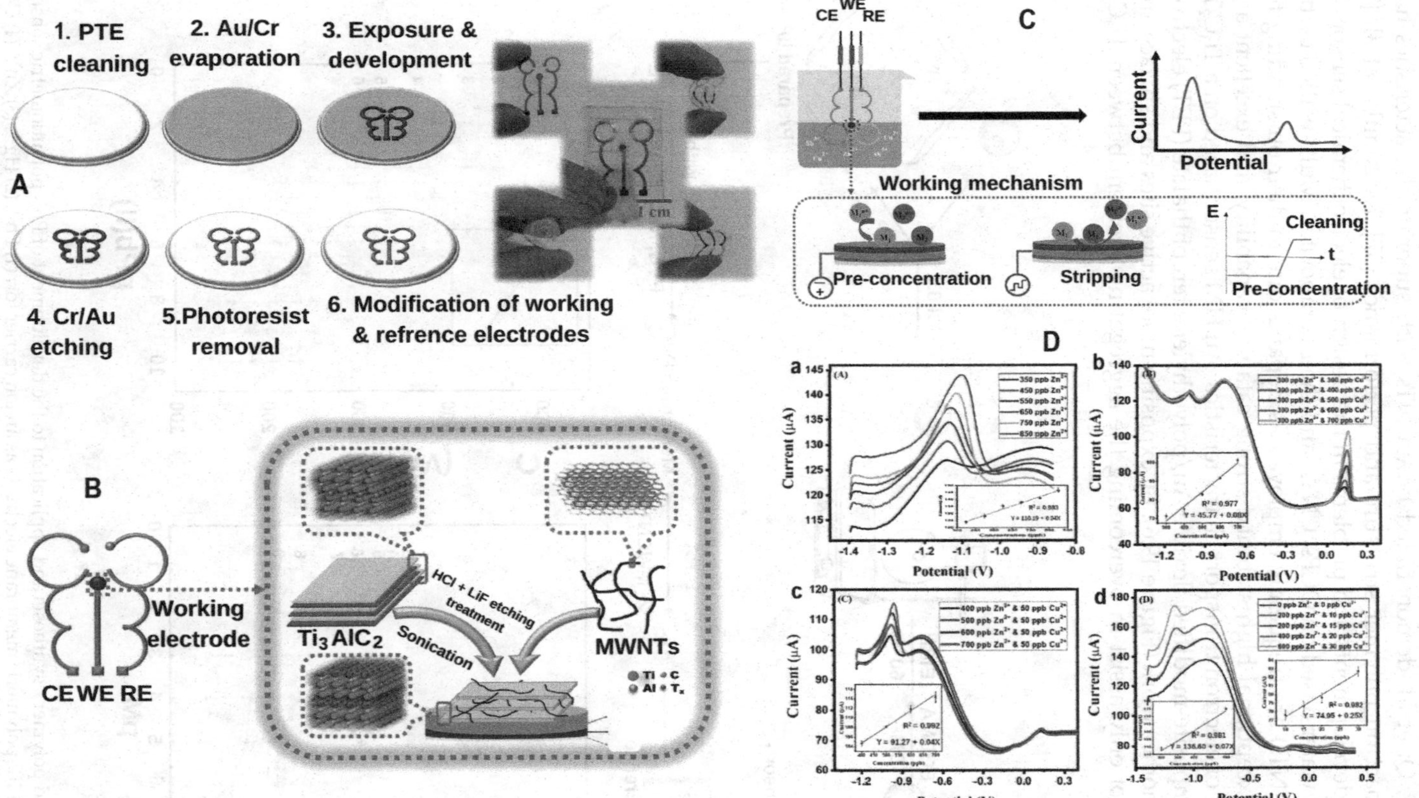

FIGURE 19.6
(A) The suggested flexible fabric synthesis processes of $Ti_3C_2T_x$/multiwalled carbon nanotube (MWNT)/Au/polyethylene terephthalate (PET) electrode. (B) The working electrode's modification process. (C) Heavy metal detecting mechanism. (D) SWASV responses of the $Ti_3C_2T_x$/MWNTs/Au/PET electrode for determination of (a) Zn(II), (b) Cu(II) having 330 ppb Zn(II), (c) Zn(II) having 50 ppb Cu(II), (d) Cu (II) and Zn (II) in acetate buffer solution (0.1 M, pH 4.6). Adapted with permission from reference [39]. Copyright (2020) American Chemical Society.

the combination of CNTs and OCPs is now appropriate for the construction of highly selective and sensitive metal ion sensors.

In their review, Qi et al. demonstrated how CNTs with superior characteristics have a substantial impact on polymer structure and performance [37]. Deshmukh et al. [38] evaluated recent developments and problems in the design of electrochemical sensors for HMI detection in water based on OCPs, CNTs, and their composites. Multi-walled carbon nanotubes (MWCNTs) are frequently employed as dopants due to their outstanding features including exceptional high stability, strong catalytic capability, and excellent anti-fouling efficiency of electron transport characteristics. An HM sensor based on a $Ti_3C_2T_x$/MWNTs nanocomposite-modified flexible Au/polyethylene terephthalate (PET) electrode is suggested in another work (Figure 19.6). By exposing more active sites and surface area, MWNTs could not only assist in overcoming the aggregating problem between $Ti_3C_2T_x$

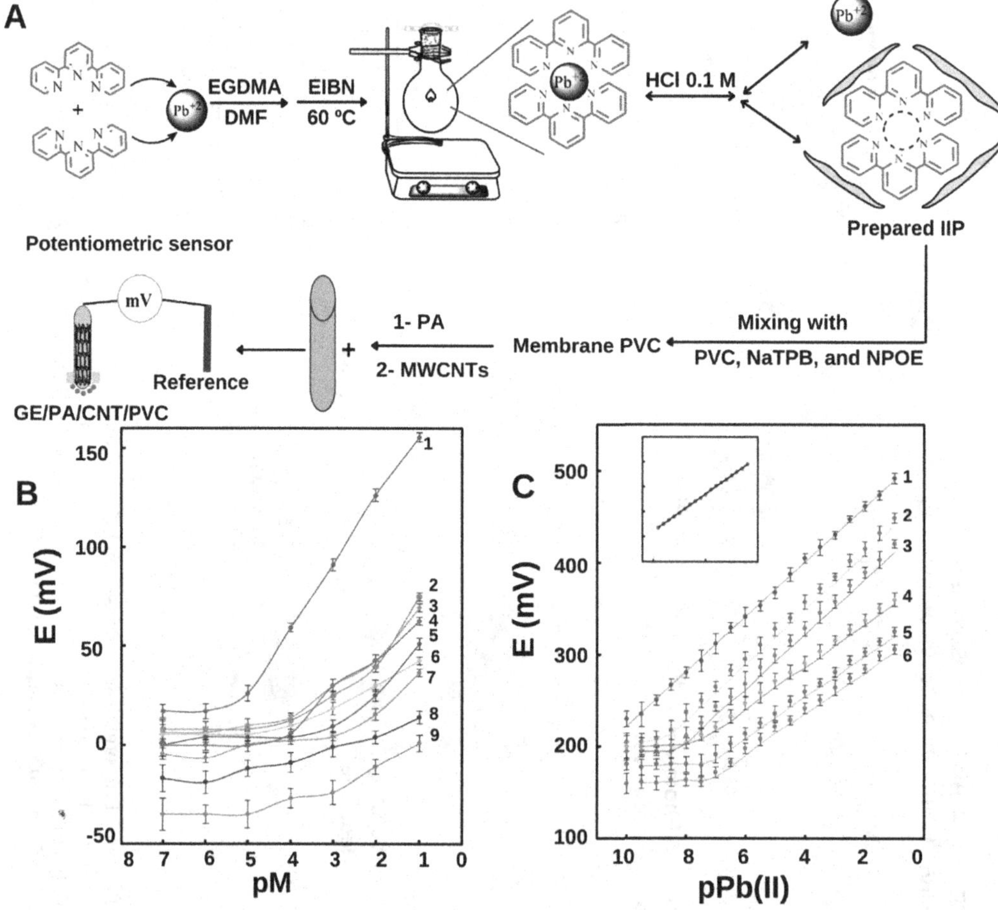

FIGURE 19.7
(A) The Pb^{2+}-imprinted polymer's synthesis and application to the development of Pb^{2+} potentiometric sensors. (B) The competence of IIP-polymeric membrane electrode as an ion carrier for: (1) Pb^{2+}, (2) Hg^{2+}, (3) Zn^{2+}, (4) Cd^{2+}, (5) Ag^{+}, (6) Cr^{3+}, (7) Cu^{2+}, (8) Ni^{2+}, and (9) Mg^{2+}. (C) All prepared electrodes' response parameters: (1) IIP-PVC/MWCNTs/PA/GE, (2) IIP-PVC/PA/GE, (3) IIP-PVC/MWCNT/GE, (4) IIP-CGE, (5) IIP-CWE, and (6) IIP-PME. Adopted with permission from reference [40]. Copyright (2020) Elsevier.

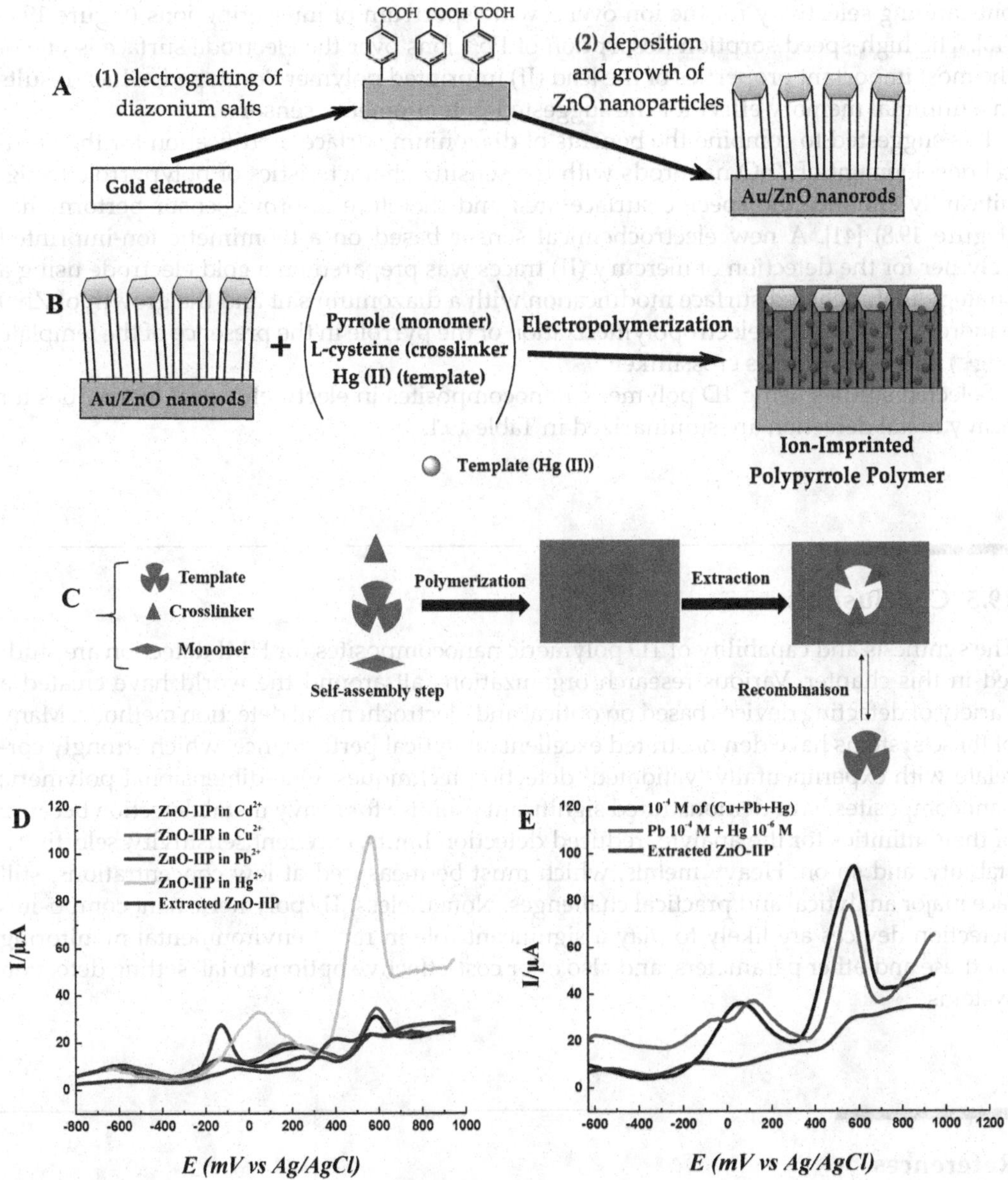

FIGURE 19.8
Experimental diagram of (A) ZnO nanorods synthesis, (B) ion-imprinted polymer synthesis, (C) the technique of forming an artificial receptor inside a polymer matrix to hold a metal ion template during the rebinding process. Sensor selectivity analysis (D) SWV graphs of ZnO/Hg(II)-IIP electrodes incubated in 10^{-4} M solutions of Hg, Cd, Pb, or Cu ions for 20 minutes in H_2O/KCl 0.1 M solution. (E) Competition tests. Conditions: E-pulse: 0.01 V, E-step: 0.01 V, frequency: 25 Hz, equilibration time: 2 s. Adapted with permission from reference [41]. Copyright (2020) Applied Sciences.

layers, but they could also allow MWNTs and $Ti_3C_2T_x$ to collaborate to increase their mutual electrochemical performances.

Ion-imprinted polymeric nanoparticles/multiwall carbon nanotubes/polyaniline/graphite electrodes are employed to detect Pb^{2+}, and the resulting sensors demonstrated outstanding selectivity for the ion over a wide spectrum of interfering ions (Figure 19.7) [40]. The high-speed sorption-desorption of Pb^{2+} ions over the electrode surface is one of the most important properties of the lead (II) imprinted polymer developed, which results in a minimal memory effect for the suggested potentiometric sensors.

It is suggested to combine the benefits of diazonium surface modification for the vertical development of ZnO nanorods with the sensing characteristics of polypyrrole to significantly enhance the specific surface area and therefore improve sensor performance (Figure 19.8) [41]. A new electrochemical sensor based on a biomimetic ion-imprinted polymer for the detection of mercury (II) traces was prepared on a gold electrode using a strategy that included surface modification with a diazonium salt and the growth of ZnO nanorods, followed by electropolymerization of the pyrrole in the presence of the template (Hg^{2+}) and L-cysteine as crosslinker.

Selected studies using 1D polymeric nanocomposites in electrochemical techniques for heavy metal detection are summarized in Table 19.1.

19.5 Conclusions

The synthesis and capability of 1D polymeric nanocomposites for HMI detection are studied in this chapter. Various research organizations all around the world have created a variety of detecting devices based on optical and electrochemical detection methods. Many of these systems have demonstrated excellent analytical performance, which strongly correlate with experimentally 'validated' detection techniques. One-dimensional polymeric nanocomposites have demonstrated significant promise for heavy metal detection because of their affinities for the analyte, reduced detection limits, excellent sensitivity, selectivity, stability, and so on. Heavy metals, which must be measured at low concentrations, still face major analytical and practical challenges. Nonetheless, 1D polymeric nanocomposites detection devices are likely to play a significant role in rapid environmental monitoring for these and other parameters, and also offer cost-effective options to lab setting detection systems.

References

1. D.J. Pain, R. Mateo, R.E. Green, Effects of lead from ammunition on birds and other wildlife: A review and update, *Ambio* 48(9) (2019) 935–953.
2. V.K. Gupta, M.R. Ganjali, P. Norouzi, H. Khani, A. Nayak, S. Agarwal, Electrochemical analysis of some toxic metals by ion-selective electrodes, *Crit. Rev. Anal. Chem.* 41(4) (2011) 282–313.
3. K. Dayanidhi, P. Vadivel, S. Jothi, N. Sheik Eusuff, White eggshells: A potential biowaste material for synergetic adsorption and naked-eye colorimetric detection of heavy metal ions from aqueous solution, *ACS Appl. Mater. Interfaces* 12(1) (2020) 1746–1756.
4. X.J. Huang, Y.K. Choi, Chemical sensors based on nanostructured materials, *Sens. Actuators, B Chem.* 122(2) (2007) 659–671.
5. B.Y. Xia, P. Yang, Y. Sun, Y. Wu, B. Mayers, B. Gates, Y. Yin, F. Kim, H. Yan, One-dimensional nanostructures: Synthesis, characterization, and applications, *Adv. Mater.* 15(5) (2003) 353–389.

6. D.W. Hatchett, M. Josowicz, Composites of intrinsically conducting polymers as sensing nanomaterials, *Chem. Rev.* 108(2) (2008) 746–769.
7. Y. Zhou, N. Shida, Y. Koizumi, K. Endo, I. Tomita, S. Inagi, Fabrication of one-dimensional polymer nanowires by templated bipolar electropolymerization promoted by electrophoretic effect, *Macromolecules* 53(18) (2020) 8123–8130.
8. J. Jang, H. Yoon, Facile fabrication of polypyrrole nanotubes using reverse microemulsion polymerization, *Chem. Commun.* 3(6) (2003) 720–721.
9. H.W. Liang, S. Liu, S.H. Yu, Controlled synthesis of one-dimensional inorganic nanostructures using pre-existing one-dimensional nanostructures as templates, *Adv. Mater.* 22(35) (2010) 3925–3937.
10. C. Li, Y. Huang, C. Chen, X. Feng, Z. Zhang, High-performance polymer electrolyte membrane modified with isocyanate-grafted Ti^{3+} doped TiO_2 nanowires for lithium batteries, *Appl. Surf. Sci.* 563 (2021) 150248.
11. M. Wan, A template-free method towards conducting polymer nanostructures, *Adv. Mater.* 20(15) (2008) 2926–2932.
12. L. Xia, Z. Wei, M. Wan, Conducting polymer nanostructures and their application in biosensors, *J. Colloid Interface Sci.* 341(1) (2010) 1–11.
13. Z. Yin, Q. Zheng, Controlled synthesis and energy applications of one-dimensional conducting polymer nanostructures: An overview, *Adv. Energy Mater.* 2(2) (2012) 179–218.
14. S.B. Kondawar, P.T. Patil, Conducting polymer nanocomposites for sensor applications. In: V. Kumar, S. Kalia, H.C. Swart, eds., *Conducting Polymer Hybrids*, Springer, 2017 223–267.
15. Z. Li, J. Peng, Z. Lin, One-dimensional hairy CNT/polymer/Au nanocomposites via ligating with amphiphilic crosslinkable block copolymers, *Giant* 5 (2021) 100048.
16. B.D. Li, Y. Xia, Electrospinning of nanofibers : Reinventing the wheel?, *Adv. Mater.* 16(14) (2004) 1151–1170.
17. A. Kausar, R. Taherian, *Electrical Conductivity in Polymer Composite Filled with Carbon Microfillers*, Elsevier Inc., 2018.
18. Y. Long, Z. Chen, N. Wang, Z. Zhang, M. Wan, Resistivity study of polyaniline doped with protonic acids, *Phys. B Condens. Matter.* 325 (2003) 208–213.
19. J. Huang, R.B. Kaner, A. General, A general chemical route to polyaniline nanofibers, *J. Am. Chem. Soc.* 126(3) (2004) 851–855.
20. J. Chen, D. Ye, X. Wu, W. Zhu, X. Wang, P. Xiao, Z. Duan, X. Yu, Large enhancement of discharge energy density of polymer nanocomposites filled with one-dimension core-shell structured $NaNbO_3@SiO_2$ nanowires, *Compos. Part A* 133 (2020) 105832.
21. X. Lu, W. Zhang, C. Wang, T.C. Wen, Y. Wei, One-dimensional conducting polymer nanocomposites: Synthesis, properties and applications, *Prog. Polym. Sci.* 36(5) (2011) 671–712.
22. X. Wang, C. Shen, C. Zhou, Y. Bu, X. Yan, Methods, principles and applications of optical detection of metal ios, *Chem. Eng. J.* 417 (2021) 129125.
23. H. Gai, Y. Li, E.S. Yeung, Optical detection systems on microfluidic chips, *Top. Curr. Chem.* 304 (2011) 171–201.
24. M. Li, H. Gou, I. Al-Ogaidi, N. Wu, Nanostructured sensors for detection of heavy metals: A review, *ACS Sustain. Chem. Eng.* 1(7) (2013) 713–723.
25. G. Li, L. Bai, F. Tao, A. Deng, L. Wang, A dual chemosensor for Cu^{2+} and Hg^{2+} based on rhodamine-terminated water-soluble polymer in 100% aqueous solution, *Analyst* 143(22) (2018) 5395–5403.
26. B. Wu, L. Xu, S. Wang, Y. Wang, W. Zhang, A PEGylated colorimetric and turn-on fluorescent sensor based on BODIPY for Hg(II) detection in water, *Polym. Chem.* 6(23) (2015) 4279–4289.
27. S. Gu, C. Hsieh, Y. Ashraf, Z. Yang, L. Li, C. Fu, R. Juang, Functionalization of activated carbons with magnetic iron oxide nanoparticles for removal of copper ions from aqueous solution, *J. Mol. Liq.* 277 (2019) 499–505.
28. H.N. Kim, Z. Guo, W. Zhu, H. Tian, Recent progress on polymer-based fluorescent and colorimetric chemosensors, *Chem. Soc. Rev.* 40(1) (2011) 79–93.

29. K. Maksymiuk, A. Michalska, Polymeric nanofiber-based ion-selective sensors, *Curr. Opin. Electrochem.* 23 (2020) 74–79.
30. W. Matysiak, T. Ta, W. Smok, K. Go, E. Schab-Balcerzak, Effect of conductive polymers on the optical properties of electrospun polyacrylonitryle nanofibers filled by polypyrrole, polythiophene and polyaniline, *Appl. Surf. Sci.* 509 (2020) 145068.
31. S.P. Sivaraman, A.M. Mohan, Chromoionophoric ligand impregnated mesoporous polymeric optical sensors for the colorimetric quantification of Cu^{2+} in water samples, *Micropor. Mesopor. Mater.* 326 (2021) 111390.
32. A.R. Sadrolhosseini, A.S.M. Noor, A. Bahrami, H.N. Lim, Z.A. Talib, M.A. Mahdi, Application of polypyrrole multi-walled carbon nanotube composite layer for detection of mercury, lead and iron ions using surface plasmon resonance technique, *PLOS ONE* 9(4) (2014) e93962.
33. L. Cui, J. Wu, H. Ju, Electrochemical sensing of heavy metal ions with inorganic, organic and bio-materials, *Biosens. Bioelectron.* 63 (2015) 276–286.
34. Y.T. Ong, A.L. Ahmad, S. Hussein, S. Zein, S.H. Tan, A review on carbon nanotubes in an environmental protection and green engineering perspective, *Braz. J. Chem. Eng.* 27(2) (2010) 227–242.
35. A. Kausaite-Minkstimiene, V. Mazeiko, A. Ramanaviciene, Physicochemical and engineering aspects evaluation of chemical synthesis of polypyrrole particles, *Colloids Surf. A Physicochem. Eng. Asp.* 483 (2015) 224–231.
36. M.M. Musameh, M. Hickey, I.L. Kyratzis, Carbon nanotube-based extraction and electrochemical detection of heavy metals, *Res. Chem. Intermed.* 37(7) (2011) 675–689.
37. X. Qi, J. Yang, N. Zhang, T. Huang, Z. Zhou, I. Kühnert, P. Pötschke, Y. Wang, Selective localization of carbon nanotubes and its effect on the structure and properties of polymer blends, *Prog. Polym. Sci.* 123 (2021) 101471.
38. M.A. Deshmukh, M.D. Shirsat, A. Ramanaviciene, A. Ramanavicius, Composites based on conducting polymers and carbon nanomaterials for heavy metal ion sensing (Review), *Crit. Rev. Anal. Chem.* 48(4) (2018) 293–304.
39. X. Hui, S. Sharma, X. Xuan, S. Zhang, S.G. Ko, S.H. Yoon, J.Y. Park, High-performance flexible electrochemical heavy metal sensor based on layer-by-layer assembly of $Ti_3C_2T_x$/MWNTs nanocomposites for noninvasive detection of copper and zinc ions in human bio fluids, *ACS Appl. Mater. Interfaces* 12(43) (2020) 48928–48937.
40. M. Ardalani, M. Shamsipur, A. Besharati-seidani, A new generation of highly sensitive potentiometric sensors based on ion imprinted polymeric nanoparticles/multiwall carbon nanotubes/polyaniline/graphite electrode for sub-nanomolar detection of lead (II) ions, *J. Electroanal. Chem.* 879 (2020) 114788.
41. Z. Ait-Touchente, H.E.E.Y. Sakhraoui, N. Fourati, C. Zerrouki, N. Maouche, N. Yaakoubi, R. Touzani, M.M. Chehimi, High performance zinc oxide nanorod-doped ion imprinted polypyrrole for the selective electrosensing of mercury II ions, *Appl. Sci.* 10(19) (2020) 7010.
42. K. Radinović, J. Milikić, U. Stamenović, V. Vodnik, M. Otoničar, S. Škapin, B. Šljukić, Tailoring gold-conducting polymer nanocomposites for sensors applications: Proof of concept for As(III) sensing in aqueous media, *Synth. Met.* 278 (2021) 116834.
43. Z.A.Al. Othman, M.M. Alam, M. Naushad, R. Bushra, Electrical conductivity and thermal stability studies on polyaniline Sn(IV)tungstomolybdate nanocomposite cation-exchange material: Application as Pb(II) ion-selective membrane electrode, *Int. J. Electrochem. Sci.* 10 (2015) 2663–2684.
44. T. Alizadeh, N. Hamidi, M.R. Ganjali, F. Rafiei, An extraordinarily sensitive voltammetric sensor with picomolar detection limit for Pb^{2+} determination based on carbon paste electrode impregnated with nano-sized imprinted polymer and multi-walled carbon nanotubes, *J. Environ. Chem. Eng.* 5(5) (2017) 4327–4336.
45. M.A. Deshmukh, G.A. Bodkhe, S. Shirsat, A. Ramanavicius, M.D. Shirsat, Nanocomposite platform based on EDTA modified Ppy/SWNTs for the sensing of Pb(II) ions by electrochemical method, *Front. Chem.* 6 (2018) 1–11.
46. J.H. Hwang, P. Pathak, X. Wang, K.L. Rodriguez, J. Park, H.J. Cho, W.H. Lee, A novel Fe-chitosan-coated carbon electrode sensor for in situ As(III) detection in mining wastewater and soil leachate, *Sens. Actuators, B Chem.* 294 (2019) 89–97.

47. Q. Bao, G. Li, Z. Yang, P. Pan, J. Liu, J. Chang, J. Wei, L. Lin, Electrochemical performance of a three-layer electrode based on Bi nanoparticles, multi-walled carbon nanotube composites for simultaneous Hg (II) and Cu (II) detection, *Chin. Chem. Lett.* 31(10) (2020) 2752–2756.
48. N. Wang, H. Dai, D. Wang, H. Ma, M. Lin, Determination of copper ions using a phytic acid/polypyrrole nanowires modified glassy carbon electrode, *Mater. Sci. Eng. C Mater. Biol. Appl.* 76 (2017) 139–143.
49. M. Lin, X. Hu, Z. Ma, L. Chen, Functionalized polypyrrole nanotube arrays as electrochemical biosensor for the determination of copper ions, *Anal. Chim. Acta* 746 (2012) 63–69.
50. R. Verma, B.D. Gupta, Detection of heavy metal ions in contaminated water by surface plasmon resonance-based optical fiber sensor using conducting polymer and chitosan, *Food Chem.* 166 (2015) 568–575.

20

One-Dimensional Polymeric Nanocomposites for Biosensors

Angelica Corpuz and Loc Thai Nguyen

CONTENTS

20.1 Introduction

In the past few decades, polymer nanocomposites (PNCs) or the radical combinations of nanoscale fillers and conventional polymers have attracted extensive academic and industrial interest due to the revolutionary performance enhancements and new prospective applications relative to traditionally filled polymers. These well-dispersed organic and/or inorganic nanoparticle fillers, in various shapes and sizes, greatly improve the mechanical

DOI: 10.1201/9781003223764-20

and physical properties of polymers. In addition, a wide array of innovative nanocomposite materials with unique properties can be developed, greatly expanding their applications and paving the way for competitive production costs [1, 2]. Among other classifications, the one-dimensional (1D) PNCs have soared in popularity due to their unique physicochemical properties. The 1D nanomaterials have a characteristic length beyond the nanoscale while two are confined to the order of nanometers (<100 nm). Its low-dimension structure and high aspect ratio enable efficient transport of electrical carriers along a defined direction, so are highly suitable for the transport of charges in integrated nanoscale systems and use in nanodevices [2] These nanocomposite structures may contain functional oxides, metals, semiconductors, and structural ceramics, along with the polymers including nanoribbons, nanorods (NRs), nanowires (NWs), nanofibers (NFs), nanobelts, hollow-structured nanotubes (NTs), and their core-shell structures [2] as shown in Figure 20.1.

Currently, PNCs play a key role in the advancement of biosensors. They serve as exceptional transducers in electroanalytical sensors owing to their morphology variants, facile synthesis, and smart responsibility, compared with their counterparts. Their inherent high electrical conductivity, fast electron-transfer kinetics, large surface area, biocompatibility, and environmental stability support the development of highly sensitive and selective biosensors with low detection limits, high reproducibility, and stability (Figure 20.2) [4]. The

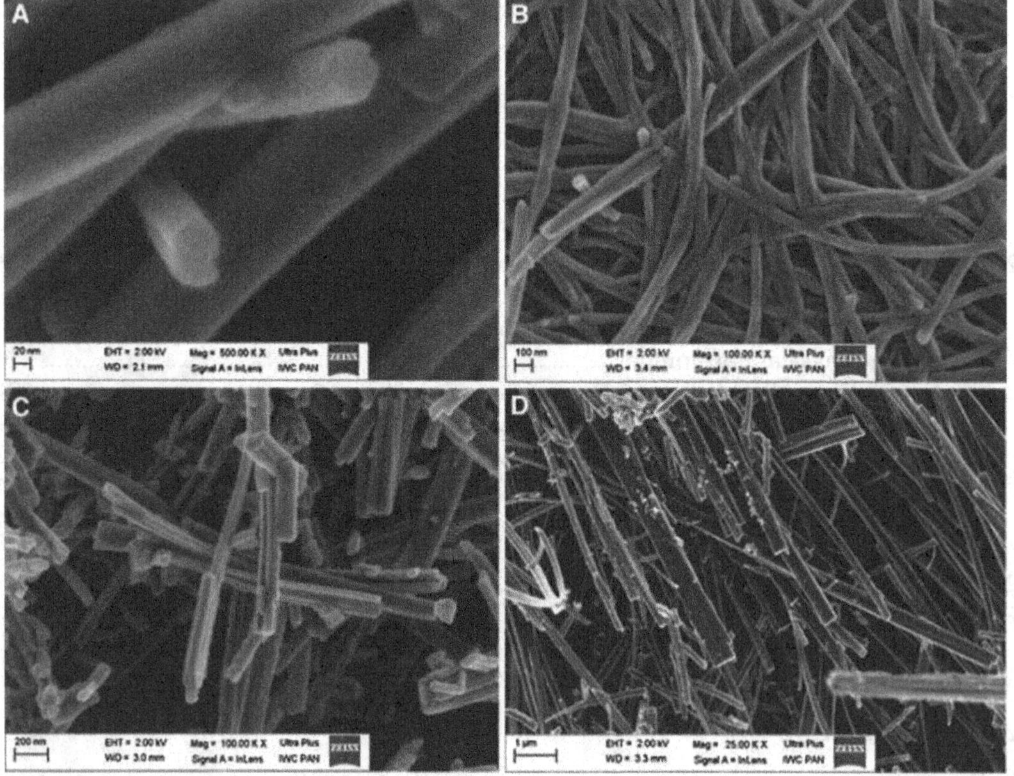

FIGURE 20.1
SEM images of 1D nanostructures: (a) nanotubes, (b) nanowires, (c) nanofibers, (d) nanoribbons. Adapted with permission from Reference [3]. Copyright B. Filipowicz, M. Pruszyn´ski, S. Krajewski, A. Bilewicz (2014), some rights reserved; exclusive licensee [Springer Nature]. Distributed under a Creative Commons Attribution License 4.0 (CC-BY) https://creativecommons.org/licenses/by/4.0/.

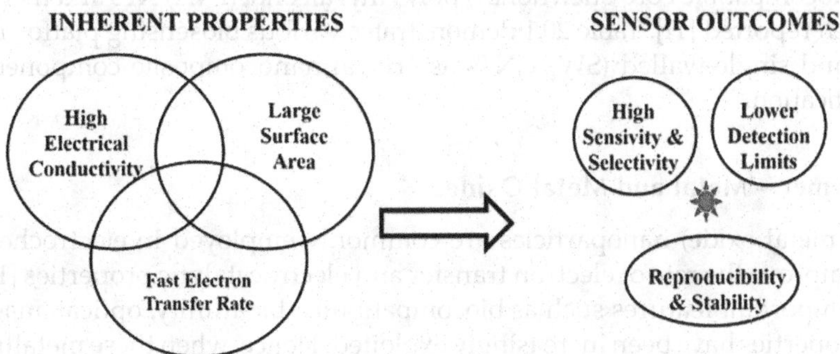

FIGURE 20.2
Key properties of PNC-based electrochemical sensors and biosensors. Adapted with permission from Reference [7].

incorporation of nanomaterials also offers functional benefits such as optical properties [5] and magnetism [6], making possible optical transduction and ease of handling.

20.2 Classifications of 1D Polymeric Nanocomposites

Nanocomposite structures manufactured through polymeric materials with different organic and inorganic nanofillers have been widely investigated. For biosensors, surface modification of the transducer using these nanostructures provides an excellent platform for the immobilization of the biorecognition molecules. Conducting or insulating polymers are commonly used as immobilization matrixes of the biological elements or other functional materials that are instrumental in improving the stability and sensitivity of the biosensing platform [8].

20.2.1 Polymer – CNTs

The outstanding properties of carbon nanotubes (CNTs) or nanofibers (CNFs) have opened up opportunities for CNT-reinforced PNCs. CNTs are thin, tubular fullerenes with hexagonal graphite structured walls [1]. CNTs exhibit a high aspect ratio, superior mechanical properties, electrical conductivity, electrocatalytic activity, and high chemical stability, with potential for use in nanodevices. Its poor processability can be complemented with polymers which are themselves limited in sensitivity and selectivity, and prone to fouling and interference [4]. In general, polymeric materials can be functionalized on CNT surfaces *via* "grafting to" and "grafting from" strategies [2]. In sensing applications, these nanocomposites offer simplicity of construction and have been shown to enhance the sensor operational characteristics in terms of selectivity, stability, or sensitivity [1, 10]. For instance, conducting multi-walled CNTs (MWCNTs) were incorporated into biopolymeric chitosan (CS) in an enzyme-modified electrode [9]. More recently, a polyaniline (PANI)/CNT nanocomposite was deposited on a screen-printed carbon electrode (SPCE) for an impedimetric aptasensor [10]. The label-by-label modification of glassy carbon electrodes (GCEs) by biocompatible core-shell honey/polyvinyl alcohol (PVA) NFs and then MWCNTs has also been reported [11]. Table 20.1 demonstrates various biosensing platforms utilizing MWCNTs and single-walled (SW) CNTs as crucial nanocomposite components in electrode modification.

20.2.2 Polymer – Metal and Metal Oxides

Metal (and metal oxide) nanoparticles are commonly employed in electrochemical (EC) sensors to improve the rate of electron transfer and electrocatalytic properties [13]. In addition, other important features such as biocompatibility, bioaffinity, optical, magnetic, and chemical properties have been increasingly exploited. Hence, when these metallic nanoparticulates or 1D nanostructures are used as nanofillers in PNCs, synergistic, unique attributes are achieved, enhancing the stability and sensitivity of the bioanalytical platform [3, 9]. There are several literature reports of biomolecular detection platforms based on these types of nanofillers. A research group has exploited the excellent electrocatalytic activity of silver nanoparticles (AgNPs) in the enhanced detection of consumption of dissolved oxygen during enzymatic oxidation of uric acid (UA) in a CS–CNF/AgNP nanocomposite-based amperometric biosensor [12]. Similarly, the high conductivity, large surface-to-volume ratio, and good biocompatibility of AgNWs in CS-based nanocomposites (Figure 20.3) were utilized in the excellent enzymatic biosensing of glucose [13]. In another study [11], MWCNTs were combined with AuNPs to enlarge the surface area of PNCs and facilitate the electron transfer kinetics for impedimetric Mucin 1 (MUC1) detection. On one hand, ZnO NPs were used to improve the conductivity of PNCs and support the immobilization of the urease enzyme [14]. On the other hand, ZnO NRs have been applied as the fluorescent and supporting material for the grafting of molecularly imprinted sol-gel polymerized 3-aminopropyl-triethoxysilane (APTES) layer for the sensing of tetracycline (TC) [5].

AuNPs have played a vital role in constructing EC biosensors such as DNA-based, enzymatic, and immunosensors ascribed to their high surface-to-volume ratio, stability, conductivity, and biocompatibility [15]. AuNPs have been dispersed onto PANI NWs for the covalent immobilization of various biomolecules including the enzyme glucose oxidase (GOx), single-stranded (ss) DNA, and Lamin A (LA) antibody for amperometric and impedimetric biosensing [16]. Similarly, the same nanofillers were dispersed in PANI/MWCNTs and polystyrene/polydopamine (PS/PDA) composites for the attachment of chlorpyrifos [15] and alpha-fetoprotein (AFP) antibodies [6], respectively, in the fabrication of voltammetric detection platforms. In addition to serving as linkers, metal and metal oxide nanoparticles act as "electronic wires" to augment the electron transfer between redox centers in biomolecules and electrode surfaces [15]. Conversely, bioaffinity of ZrO_2 was utilized to immobilize ssDNA bioreceptor onto and to improve the surface area of the electrode in CS-based PNC [17]. In another study, the magnetic properties of Fe_3O_4 in the PS/PDA-based NC facilitated its simple adsorption onto magnetic GCE [6].

20.2.3 Polymer Blend

The unique characteristics of different polymers may be combined for synergy within the hybrid nanocomposite. Conductive polymers (CPs) such as PANI and polypyrrole (PPy) have become electrode modifiers in various bioanalytical platforms [8]. Wang and Hui (2019b) combined the good conductivity of PANI NWs with the excellent antifouling capability of zwitterionic poly(carboxybetaine methacrylate) (PCBMA) in an immunosensor

TABLE 20.1

Electrochemical and Optical Biosensors Fabricated Based on 1D PNCs-Modified Electrodes

Sensor	Target Analyte	Detection Limit (Range)	Transduction	Reference
Lac–GAfCS–MWCNT/GCE	cathechol	100 nM	amperometry	[9]
NAF/HRP/PAKB NP-CNT/SPCE	H_2O_2	2.7 µM	amperometry	[35]
Au/SPAN-HRP-CS/GCE	H_2O_2	1.6 µM	amperometry	[37]
FTO/PA6/PPy/ZnO/urease	urea	0.011 mg dL^{-1}	EIS	[14]
Uricase/Chi–CNT NF/AgNP/Au	uric acid	1.0 µM	amperometry	[12]
GCE/PVA/PAMAM-Mt/PyOx	glucose	0.7 µM	amperometry	[23]
GOx@PAVE-CNT/GCE	glucose	0.36 µM	amperometry	[34]
GOD-CS/AgNW/GCE	glucose	2.83 µM	amperometry	[13]
PPyNWA–PtNP–Gox	glucose	5.6 µM	potentiometry	[39]
PANI/SnO_2-NF/GOx-HRP/CSGCE	glucose	1.8 µM	amperometry	[27]
SPCE/PANI/CNT/Apt	$VEGF_{166}$	0.4 pg/mL	DPV	[10]
Apt/f-NFPGE	NSCLC	1.2×10^3 cells mL^{-1}	EIS	[21]
Apt-MWCNT/AuNP/NF/GCE	MUC1	2.7 nM	EIS	[11]
Apt/AuNP/CNP/CeNF/GCE	*S. aureus*	1 CFU mL^{-1}	EIS	[24]
Apt/CPMCNF FET	PDGF	6 fM	FET	[30]
TTA-PANI-MWCNT/GCE	thrombin	80 fM	amperometry	[26]
SPCE-AuNP-MPA-(PDDA-oSWCNT-5-HT Apt)2- PDDA-oSWCNT-Tyr	serotonin	35 nM	DPV	[45]
SPCE/C60/MWCNT-PEI/PQdot/GLA/Apt	thrombin	2 nM	DPV	[22]
MWCNT-MIP /GCE	nolylphenol	0.02 µM	DPV	[48]
3D-CNTs@MIP /GCE	PFF	0.002 µM	amperometry, CV	[29]
MIPs-ZnO NRs	TC	1.27 µM	fluorescence	[5]
MWCNT@MIP/CCE	cholesterol	1 nM	LSV	[52]
CIP@AuNR/PGE	Ara-C	0.19 ng mL^{-1}	DPASV	[28]
MIP/PPyNW/GCE	DA	3.3×10^{-8} M	DPV	[20]
MIP/ZnO NR/GCE	4-NP	(0.5–14 µM)	fluorescence	[49]
ssDNA/HANP/PPy/MWCNT/GCE	M. tb	0.141 nM	DPV	[46]

(Continued)

TABLE 20.1 (CONTINUED)

Electrochemical and Optical Biosensors Fabricated Based on 1D PNCs-Modified Electrodes

Sensor	Target Analyte	Detection Limit (Range)	Transduction	Reference
spongy gold film/CS–MWCNT/GCE	Ara h1	0.013 fM	amperometry	[42]
GCE/Fe_3O_4/MWCNT-COOH/CS	*Bacillus cereus* DNA	2.0×10^{-15} M	DPV	[47]
ssDNA/MWNT/ZrO_2/CHIT/GCE	daunomycin	7.5×10^{-11} M	DPV	[17]
Pt/Au-PANI/GOx, ssDNA, Ab	glucose, *E. coli* DNA, Lamin A	1 μM	amperometry, DPV, EIS	[16]
DNA/PAMAM/MWNT-Chit/Au	DA, UA	0.03 μM, 0.07 μM	DPV	[33]
p-DNA/PAMAM/GaN NWs	H1N1 virus	0.1 aM	EIS	[51]
anti-PSA/AuNP–PANI/Au	PSA	0.6 pg mL^{-1}	DPV	[36]
Ab/AuNP-PANABA/SPE	2,4-D	0.3 ppb	EIS	[44]
Ab/PEG/AuNP/PANI/GCE	AFP	0.007 pg mL^{-1}	DPV	[19]
BSA/anti-chlorpyrifos/ AuNP/PANI/MWCNT/CS/GCE	chlorpyrifos	0.06×10^{-6} mg mL^{-1}	CV	[15]
PPa/1∘Ab /TiO_2-NW FET	rabbit IgG	(119 pg mL^{-1} – 5.95 ng mL^{-1})	FET	[25]
Ab/CNT-CS/AuNP/SPCE	PSA	0.6 pg mL^{-1}	ECL	[54]
FB1-BSA-SWNT/CS/GCE	FB1	2 pg mL^{-1}	DPV	[53]
AFB1-BSA-SWNT/CS/GCE	aflatoxin B1	3.5 pg mL^{-1}	DPV	[50]
IgG–AuNP/APBA/TPANI-MWCNT/GCE	IAA	0.97 pg mL^{-1}	DPV	[41]
MGCE/Fe_3O_4@PS/PDA-AuNT/Ab	AFP	0.0088 ng mL^{-1}	DPV	[6]
Ab/polyCBMA/PANI/GCE	CEA	3.05 fg mL^{-1}	DPV	[43]
Ab–P(1,5DAN)/PPyNW/SPE	CA 15-3	0.02 U mL^{-1}	DPV	[40]

Abbreviations: Lac: laccase; GafCS: GA-functionalized chitosan; NAF: Nafion; PAKB NPs: poly(AAPBA-co-VBK-co-BMA) nanoparticles; Apt: aptamer; CeNF: cellulose nanofiber; CIP: complex imprinted polymer; VEGF: Vascular endothelial growth factor; NSCLC: non-small cell lung cancer; PDGF: platelet-derived growth factor; PFF: profenofos; TC: tetracycline; Ara-C: cytosine arabinoside; DA: dopamine; 4-NP: 4-nonylphenol; M. tb: *Mycobacterium tuberculosis*; UA: uric acid; H1N1: human influenza A; PSA: prostate-specific antigen; 2,4-D: 2,4-dichlorophenoxy acetic acid; AFP: alpha-fetoprotein; FB1: fumonisin B1; IAA: indole-3-acetic acid; CEA: carcinoembryonic antigen; CA 15-3: cancer antigen 15-3; EIS: electrochemical impedance spectroscopy; DPV: differential pulse voltammetry; FET: field effect transistor; CV: cyclic voltammetry; ECL: electrochemiluminescence.

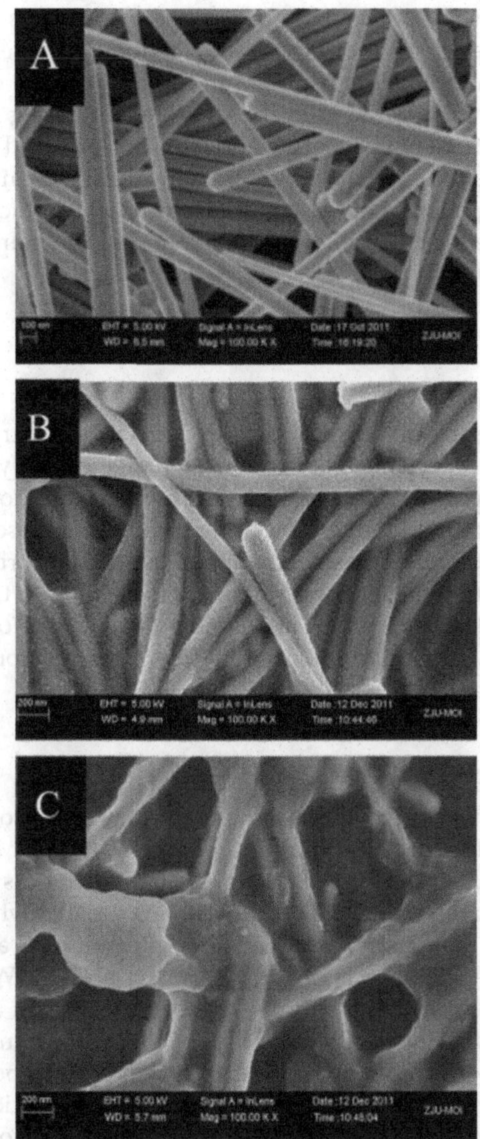

FIGURE 20.3
SEM images of (A) AgNWs, (B) CS/AgNWs nanocomposite, and (C) GOx-functionalized CS/AgNWs. Adapted with permission from Reference [13]. Copyright (2013), Elsevier.

[19]. The same researchers have applied the biocompatible and highly hydrophilic polyethylene glycol (PEG) as non-fouling material in the (PEG)/PPyNW nanocomposite. A PEG/PANI NW composite has been adapted as an immunoelectrode [19]. Nanowire structure of PPy was similarly used as a scaffold for thin and high-surface-area insulating poly(*o*-phenylenediamine) (POPD)-based molecularly imprinted film for an amperometric transducer [20]. Clearly, these studies show that aniline and pyrrole are good polymer precursors for nanowire synthesis. In other studies, PPy nanofiber was used owing to its high surface area and porous structure. For one, it was combined in an electrospun NF composite with nonconducting polyamide 6 (PA6) and ZnO for effective immobilization of the urease

enzyme [14]. In another study, polyacrylonitrile (PAN) was used as a copolymer to improve the solubility and tensile strength of PPy as an electrode modifier [21]. A nanotube configuration can provide high loading capacity and impart certain chemical properties for subsequent surface nanomaterial modifications. Non-conducting PS NTs decorated with surface-active multifunctional PDA coating and AuNPs were used in the fabrication of the immunosensor [6]. Quite recently, the superior characteristics of non-conjugated polymer quantum dots (PQDots) in terms of surface area, electrical conductivity, biocompatibility, and mechanical and chemical stability were exploited in the manufacture of aptasensor transducers [22].

20.2.4 Polymer – Nanoclay

In the past few years, there has been a growing interest in the use of clay as a promising inorganic biofunctionalization platform because of its abundance, specific surface area, higher absorption capacity, higher cation-exchange capacity, and ease of dispersal compared to its counterparts. Among the clay minerals, montmorillonite (Mt) has demonstrated utility in fabricating highly stable and reproducible biosensors. A composite NF utilizing the nonconductive PVA has been applied as the supporting polymer decorated with poly(amidoamine) (PAMAM) dendrimer-modified Mt onto the surface of GCE. The modified Mt facilitated the multi-point attachment of pyranose oxidase (PyOx) enzyme on the electrode surface, leading to the desired detection and storage stability as well as repeatability of the sensor [23].

20.2.5 Polymer – Multicomponents

As discussed, integrating two to three nanofillers may bring out synergy within the nanohybrid. New advances in CP ternary NCs have shown that blending of three components comprised of CPs, in addition to inorganic nanomaterials such as metal, metallic oxides, and carbon derivatives, exhibits excellent characteristics when compared to binary architectures [2]. This is demonstrated by the PA6/PPy/ZnO NC as previously discussed [14]. Similarly, the synergistic effects of the nanomaterials in MWCNT/AuNP/core-shell NF composite functionalized with aptamer have been reaped in an impedimetric nanoprobe design [11]. In a developed C60/MWCNTs-PEI/PQdot nanocomposite, the fullerene component was shown to enhance the electron transfer and peak current response of the developed electrode transducer [22]. Finally, carbon nanoparticles (CNPs) are another attractive carbon nanostructure providing vital features for electroanalytical applications. Along with the AuNPs, these were used to complement the poor electrical conductivity of the cellulose nanofibers (CeNFs) meant for aptamer fixation in the AuNP/CNP/CeNF composite (Figure 20.4) [24].

20.3 Strategies to Form 1D Polymeric Nanocomposites

1D PNCs can be formed by different strategies due to the varied physicochemical properties among polymers and nanofillers. Every polymer system requires distinct processing conditions built on processing efficiency, nanofiller type, and preferred features [1, 3]. In this section, recent techniques applied for the fabrication of 1D PNCs applied in bioanalytical platforms will be presented.

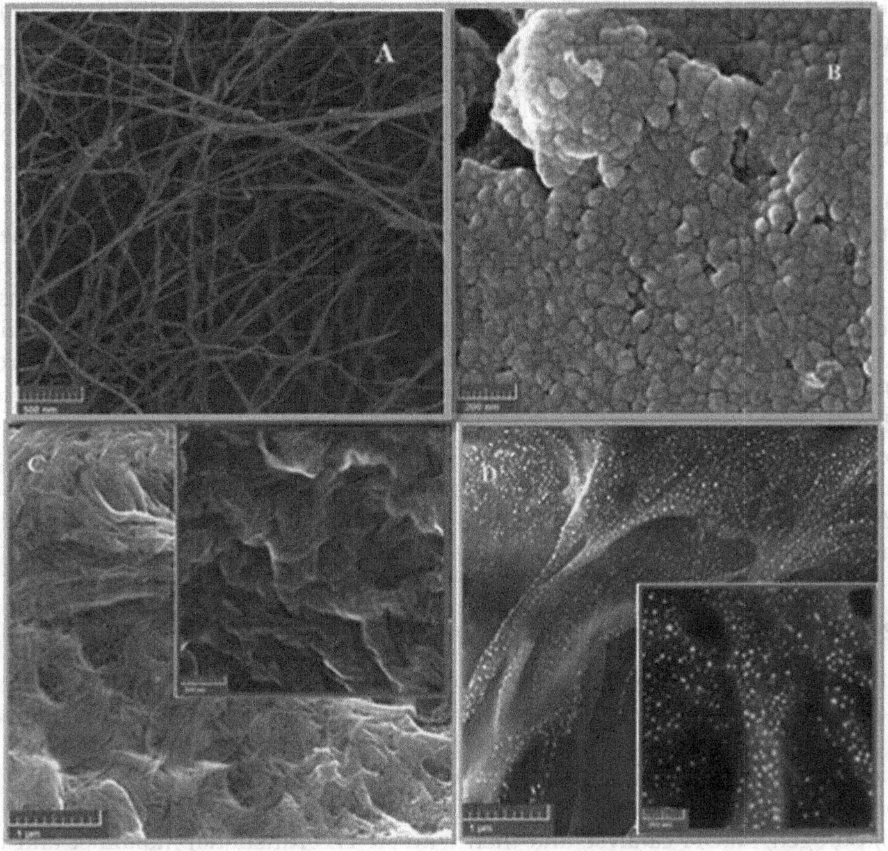

FIGURE 20.4
SEM images of (A) CNFs, (B) CNPs, (C) CNPs/CNFs and (D) AuNP/CNP/CNF nanocomposite. Adapted with permission from Reference [25].

20.3.1 *In-Situ* Polymerization of Polymer on 1D Nanofiller

The in-situ approach pertains to the synthesis of the polymer in the presence of nanofillers [2]. The synthesis route may undergo polymerization *via* chemical, electrochemical, or plasma approaches. This method follows oxidative crosslinking of monomers initiated by radical cation formation succeeded by dimer formation with another monomeric entity or radical cation. Electrochemical initiation is based on voltammetric techniques whereas the chemical approach entails the use of oxidants. As regards an electrochemical approach, core-shell nanocomposites were synthesized by thin-layer ECP on NW structures [21, 26]. In the first case, TiO_2NWs were spin-coated onto silicon substrate followed by electrochemical deposition of a thin film of PPy propylic acid (PPa). The number of immobilized biomolecules can be adjusted by controlling the scan rate and the number of scan cycles in cyclic voltammetry. In the second case, a core-shell dual-layer polymer-based transducer was fabricated by layer-by-layer ECP involving PPyNWs and POPD. The obtained 3D molecularly imprinted polymer (MIP) has high sensitivity and good selectivity for dopamine (DA) detection.

In-situ chemical synthesis is a simplified process in the formation of nano-hybrids. In this method, an oxidizing agent is mixed with the monomeric solution initiating the crosslinking. Ammonium persulphate (APS), ammonium peroxydisulfate, and APS oxidants, in the presence of the nanofillers carboxylic acid- (COOH-) functionalized CNTs, MWCNTs-COOH, and SnO_2-NFs were used to prepare PANI/CNTs [10], PANI-MWCNTs [26], and PANI/SnO_2-NF [27] nanocomposites, respectively. The surfaces of sensors may be functionalized with polymers by grafting to/from techniques. On one hand, an MIP film was grafted from PGE-attached AuNRs by linking monomeric molecules (*N*-methacryloyl-L-cysteine) and subsequently undergoing metal-ion-mediated molecularly imprinted polymerization *via* free-radical polymerization (FRP) in the presence of template (Ara-C), crosslinker (ethylene glycol dimethacrylate) (EGDMA) and initiator [2,2′-azobis(isobutyronitrile) (AIBN) [28]. On the other hand, a MIP shell has been grafted onto the CNT cores with polymerizable vinyl end groups using methacrylic acid (MAA) monomer, EGDMA cross-linker, and AIBN initiator [29]. The Ara-C and profenofos (PFF) templates were successively removed by solvent extraction to form recognition sites (Figure 20.5).

Vapor deposition polymerization is another *in-situ* chemical approaches in which the monomeric entity is in a vapor phase. Carboxylated polypyrrole (CPPy) were deposited onto ZnO-decorated ultrafine CNFs using Fe^{3+} oxidants or initiator by exposure to 3-carboxylated pyrrole (CPy) monomer vapor [30].

20.3.2 *In-Situ* Sol-Gel Processing of Polymer on 1D Nanofiller

NPs and nanocomposites can be synthesized *via* the sol-gel process which involves the chemical transformation of a colloidal suspension of nanoparticles called "sol" into a "gel" phase with subsequent conversion into solid oxide material [31]. Using this method, a mesoporous fluorescent sensor was developed for TC detection [5]. Briefly, a ZnO NR dispersion was mixed with the TC template molecules, cetyl trimethyl ammonium bromide (CTAB) porogen, NaOH catalyst, and then with APTES functional monomer and tetraethoxysilane (TEOS) crosslinker. The silica-based MIP-ZnO NRs were obtained upon hot water extraction of TC and CTAB.

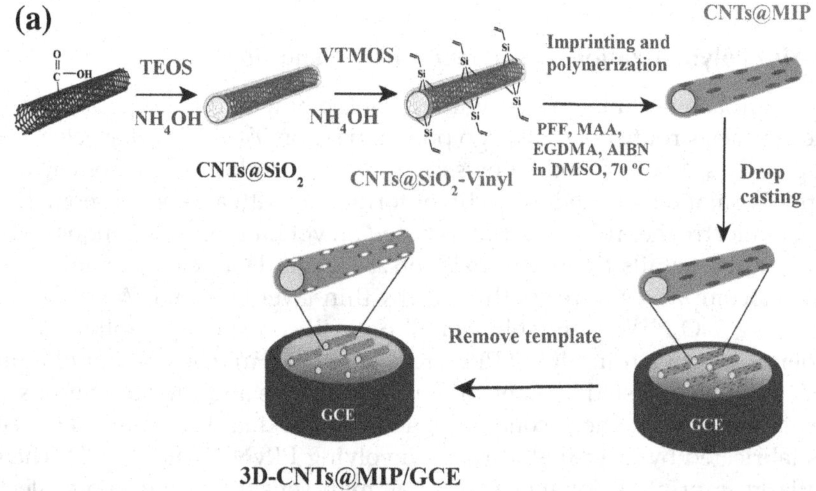

FIGURE 20.5
1D PNC-based biosensors. Fabrication of 3D-CNTs@-MIP nanocomposite based PFF biosensor. Adapted with permission from Reference [29]. Copyright (2019), Elsevier.

20.3.3 Covalent Grafting of Polymer on 1D Nanomaterial

Layer-by-layer NC fabrication may also be achieved *via* covalent grafting. It was established that achieving intimate interfacial interactions between polymers and nanofillers is difficult for the conventional physical mixing or in-situ polymerization methods, although it was suggested that covalent bonding could alleviate this condition [32] Using this concept, PEG polymer has been attached to PANI NWs *via* CV resulting in PEG/PPy NWs with the combined features of the antifouling and biocompatible PEG with that of electroactive and versatile PPy NWs [18]. Similarly, an NT nanocomposite was synthesized by grafting a polymeric material, G4 PAMAM dendrimers, onto COOH-MWCNTs by the simple spreading of PAMAM solution onto the previously drop-casted CS-dispersed MWCNTs [33].

20.3.4 Self-Assembly of 1D Polymer with Nanofiller

Co-assembly of two components aided by noncovalent interactions is another hybridization approach [2]. Strong π–π stacking interactions between aromatic rings of the functional groups of nanomaterials make way for their co-assembly. MWCNTs were co-assembled with an amphiphilic copolymer poly(acrylic acid-*r*-(7-(4-vinylbenzyloxy)-4-methyl coumarin (VMc))-*r*-ethylhexyl acrylate) (PAVE) facilitated by aromatic rings of VMc segments in PAVE polymer chains and conjugated carbon structure of MWCNTs [34]. The PAVE-CNTs were then electrodeposited onto the electrode surface, producing a robust sensing coating following photo-crosslinking of the coumarin segments. Similarly, amphiphilic copolymer poly(AAPBA-co-VBK-co-BMA) (PAKB)-based nanoparticles have been designed to synthesize PAKB NPs-CNTs aided by π–π interactions between the polymeric carbazole groups and CNTs [35]. Subsequent drop-casting of the PAKB NPs-CNTs dispersion on the SPCE surface offered a good matrix for enzyme immobilization.

Electrostatic interactions are also used in nanohybrid co-assembly. The interaction between positively charged polymeric matrixes and negatively charged gold nanoparticles is a well-established process. AuNPs were assembled from colloid solution on the surface of electrodeposited NH^+-bearing PANI molecules on the Au electrode [36]. In another study, Au NP-decorated PANI NWs were prepared *via* template-free method wherein aniline in the organic phase is mixed with a gold precursor in the aqueous phase for the interfacial polymerization process followed by thin-film drop-casting on an electrode [16]. An H_2O_2 amperometric biosensor was fabricated by synthesizing positively charged, dispersed, self-doped polyaniline (SPAN) fibers, with CTAB as a dopant, and mixed with a gold colloid solution to form an Au/SPAN NF composite [37].

20.3.5 Electrospinning Techniques

Electrospun polymeric NFs have been increasingly used in biosensors on account of their large surface area and tunable electromechanical properties. Electrospinning is accomplished by loading the polymeric solution into a syringe that is connected to a voltage source whereby 1D fibrous products may be accumulated at the collector at a certain distance. Using this technique, an electrospun nanocomposite of CS–CNTs NF has been prepared [12]. MWCNTs were dispersed in a mixture of CS and PVA solutions and electrospun to form the nanofibrous membrane on the AgNP-modified Au electrode with NaOH treatment for PVA removal. This CS–CNTs NF/AgNP/Au electrode proved to be a suitable platform for uric acid biosensors. A disposable pencil graphite electrode (PGE) was similarly modified [21] with a composite NF by electrospinning a homogenized PAN/PPy solution. Treatment with NaOH solution enabled the conversion of nitrile groups of PAN to carboxyl and amine groups for the successive aptamer functionalization.

20.3.6 Other Thin-Layer Deposition Techniques

Other core-shell layer nanocomposite structures may be fabricated using versatile thin-layer deposition techniques such as the emerging atomic layer deposition (ALD) and electrodeposition processes. ALD technique was used to coat the electrospun PAN NFs (core) with a layer of the crystalline shell of ZnO [38]. The deposition was controlled at 100 °C and included sequential exposures to diethyl zinc (DEZ) and deionized water, with dry argon purging in between. This resulted in a high surface area, solid "spaghetti-like" morphology of PAN/ZnO NFs with fine-tuned structural and optical characteristics for a photoluminescence (PL)-based immunosensor. As regards electrodeposition techniques, hard template-based ECP-deposition of well-aligned PPy nanowires array (PPyNWA) was applied on anodized aluminum oxide (AAO) template that was subsequently removed from the surface of an Au electrode and then further modified with PtNPs by the CV technique [39]. The resulting (PPyNWA–PtNP)-modified electrode found application as an enzyme biosensor transducer platform.

20.4 Biological Functionalization Techniques

The selectivity of biosensors is derived from the functionalization of a transducer with biomolecules. These bioreceptors can be immobilized on polymeric matrix of biosensors by physical adsorption, covalent bonding, or entrapment [3, 26, 41]. Proper immobilization is crucial in achieving reliability, reproducibility, and sensitivity of biosensors, and the use of biocompatible nanocomposite materials with large surface area is a major consideration [14, 41].

20.4.1 Covalent Functionalization

Among the techniques available, covalent immobilization is preferred owing to its stability and efficacy. It involves the direct chemical bonding between functional groups of aptamers, enzymes, DNA, and/or antibodies and that of nanocomposites. This may also involve surface modification to generate the functional groups and/or the use of crosslinkers to mediate the reactions [31, 41]. GOx was directly immobilized on CS/AgNW/GCE and PANI/SnO_2-NF/CS/GCE transducers using the rich amino groups of CS [15, 28]. The dense amino groups of the polymeric PAMAM dendrimer in PAMAM/MWCNT-Chit/Au electrode were used to immobilize double-stranded (ds)DNA probes for the detection of DA and uric acid (UA) [33]. The specific boronic acid-diol interaction was utilized [41] to link boronic acid-functionalized AuNPs in AuNP/PANI-MWCNT/GCE surface and the vicinal diol-functionalized AuNP-labeled immunoglobulin G (IgG–AuNPs). This boronate affinity was also applied to link horseradish peroxidase (HRP) to PAKB NPs-CNTs [35]. Thiol-ene reaction between thiolated thrombin apt (TTA) and oxidized PANI in PANI-MWCNT-modified GCE has been exploited [26]. As mentioned,

AuNP can have an additional role as an immobilization matrix in a nanocomposite. As such thiol-Au interaction has been repeatedly reported in biosensor fabrications. Firstly, thiolated aptamer was attached onto AuNP/CNP/CNF/GCE [24]. Secondly, the thiolated stem-loop Ara h1 DNA probe was connected to spongy gold film/CS–MWCNT/GCE surface using the same mechanism [42]. Lastly, thiol-ended ssDNA was connected to Au-PANI/Pt electrode aided by strong affinity of S to AuNP [17]. The use of a condensing agent facilitates amide bonding (–CONH–) between compatible molecules. For instance, the Apt/CNT/PANI/SPCE transducer has been developed by linking carboxyl of PANI/CNT and the amino group with the aid of 4-(4,6-dimethoxy-1,3,5-triazin-2-yl)-4-methyl-morpholinium chloride (DMT-MM) [10]. Using the same condensing agent, a hybrid CNF FET-type electrode was functionalized with a platelet-derived growth factor-B (PDGF-B) apt by attaching its amino group to the carboxyl group of CPPy in the functionalized nanofiber [30].

Glutaraldehyde is a preferred crosslinker for the amine coupling process over the *N*-(3- dimethylaminopropyl)-*N*-ethylcarbodiimide hydrochloride (EDC) and *N*-hydroxysuccinimide (NHS) reagents owing to its simplicity of use and stability [40]. It was used [14] to activate amino groups in CS–CNTNF/AgNP/Au electrode and subsequent crosslinking with the amine group of the uricase bioreceptor. Similarly, the amino groups in poly(1,5 diaminonaphthalene P(1,5DAN)/PPyNW/SPE were linked with Ga for coupling with capture antibodies in the immuno-platform [41]. In another study, the free amine groups in PVA/PAMAM-Mt modified transducers were crosslinked *via* Ga with PyOx. Finally, the abundant amine groups of PEI in SPCE/C60/MWCNTs-PEI/PQdot transducer were covalently attached to amino-linked thrombin apt [22] as detailed in Figure 20.6.

Carbodiimide chemistry using EDC-NHS is also commonly used. Applying this, the covalent bonding between amino groups of the aptamer and carboxyl groups of PAN in NF/PGE was achieved [21]. The same is used in the activation of carboxyl groups of functionalized MWCNTs for crosslinking with MUC1 apt in the MWCNT/AuNP/NF/GCE [11]. Using carboxylic groups as anchor sites in polyCBMA [43] and conductive poly-(aniline-co-3-aminobenzoic acid) (PANABA)[44], CEA and 2,4-D antibodies were conjugated to polyCBMA/PANI/GCE and Ab/AuNP-PANABA/SPE, respectively. Another crosslinker 2-aminoethanethiol was applied [16] for amide bonding between the $–NH_2$ moiety of GOx functionalized on AuNP of Pt/Au-PANI with that of the aldehyde group of the Lamin A antibody (LAA).

20.4.2 Noncovalent Functionalization

Noncovalent interactions between nanocomposite entities and biomolecules are also employed in biofunctionalization. Electrostatic interactions between positively charged PDDA-oxidized (o)SWCNTs nanocomposite and negatively charged 5-HT apt [45], negatively charged phosphate group of DNA with positively charged calcium groups of hydroxyapatite nanoparticles (HAPNPs) in nanocomposite electrode [46], carboxylic groups of the Liosperse 511-dispersed ZnO NPs and positively charged urease [14], negatively charged DNA and positively charged CS [47], positive charged PANI amino groups and negatively charged carboxylic groups on antibodies [36] were reported in the literature in the development of biosensor transducers. Hydrogen-bonding interactions between the PAVE copolymer in nanocomposite electrode and GOx molecules as well as the hydrophobic interactions between amphiphilic polymer and GOx are also demonstrated in the fabrication of enzyme biosensors [35].

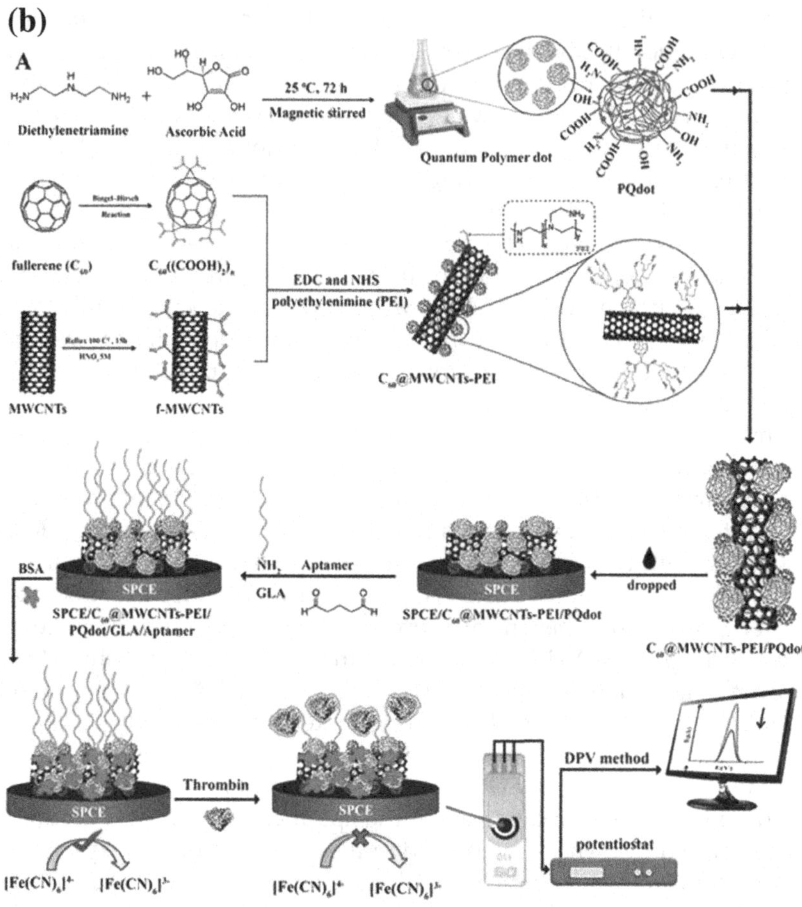

FIGURE 20.6
Fabrication of electrochemical aptasensor and C60/MWCNT-PEI/PQdot. Adapted with permission from Reference [26]. Copyright (2021), Elsevier.

20.4.3 Physical Entrapment

Physical methods such as polymer entrapment are well-established bio-immobilization protocols [26]. A classical method wherein biomolecules are encapsulated *via* the ECP of a conducting polymer on the sensing area of the sensing platform was applied [20] in embedding DA molecules for subsequent extraction in the CV-polymerized POPD matrix to generate MIP/PPyNWs/GCE biosensing platform. As mentioned, MIP generates recognition sites in synthetic polymers by embedding and then extracting the template molecules in the matrix. Other nonconducting polymers have also been used such as polyacrylamide in the entrapment of nonylphenol [48] and 4-nitrophenol [49] templates for the development of MWCNT-MIP/GCE and MIP-ZnO NRs nanocomposites, respectively, and in imprinting PFF template in CNT@SiO_2-MIP/GCE [29]. It is interesting to note that noncovalent interactions are also involved during imprinting to achieve the selective recognition of target molecules during rebinding or sensing. The latter researchers have employed π-π interactions and H-bonding between the functional monomer, template, and matrix. Silica polymer using APTES and TEOS precursors are also increasingly used in MIP such as in imprinting TC in MIP-ZnO NRs for fluorescent sensing. In a non-MIP system, electropolymerized PPa was used in immobilizing anti-rabbit IgG on a composite film coating of TiO_2-NW FETs [25].

20.5 Applications of 1D Polymeric Nanocomposites in Biosensors

20.5.1 Enzymatic Biosensors

Enzymatic biosensors have been attracting considerable attention in research owing to their large potential in various applications including bioprocess monitoring, clinical diagnostics, and food and environmental analyses [23]. Herein, enzymes are immobilized to gain superior selectivity towards target analytes as compared with their chemical counterparts [34]. The electrochemical transduction approach is preferred for greater efficiency and catalytic activity and hence greater sensitivity, simplicity, and faster response. Major design and fabrication considerations include stability of the enzyme, high enzyme loading, and catalytic efficiency in which nanomaterial functionalization plays an important role [24, 36]. As shown in Table 20.1, PNC-based enzymatic biosensors mainly utilize GOx as the bioreceptor for glucose determination. An Au/PPyNWA/PtNPs electrode was fabricated combining the biocompatible, conductive, highly stable, easily synthesized PPy with the high surface area of its nanoarray structure along with the improved EC properties and mediation of direct electron transfer (DET) by Pt nanoparticles, for GOx immobilization [39].

The analytical performance of the developed biosensor compares well with reported nanomaterial-based glucose biosensors with an option to use potentiometric mode at low (10 μM to 1 mM) and amperometric mode at high (0.1 mM to 9 mM) concentrations. Another amperometric GOx/CS/AgNW/GCE sensor was developed [13] utilizing the conductivity and catalytic activity of AgNWs that enhanced the DET between the GOx redox center and the electrode surface, and the immobilization capability of chitosan, thereby producing a highly sensitive interface with a low detection limit, highly selective, and rapid-response bioanalytical platform. Another platform employed a nanostructured GOx@PAVE-CNTs bio-nanocomposites modified on GCE that integrated the long conducting MWCNTs for enhanced DET between GOx and electrode and the fibrous, cross-linked amphiphilic PAVE copolymers that improved the stability and biological activity retention of immobilized enzymes [34] (Figure 20.7). To complement nanomaterial functionalization in enhancing DET between GOx and electrode, a redox mediator, HRP, was used to minimize interferences in sample matrixes [28]. The developed PANI/SnO_2-NF/GOx-HRP/CS/GCE biosensor also exploited the high specific surface of conductive, catalytic, biocompatible, multi-porous SnO_2 NFs, the immobilization suitability of CS along with antifouling, flexible, thermally and chemically stable, conductive PANI which contributed to good analytical performance. In addition to GOx, PyOx was also reported as an effective glucose transducer enzyme. PyOx was successfully immobilized on GCE/PVA/PAMAM-Mt electrode employing the benefits of easy-to-prepare and high-surface-area electrospun PVA NFs, the rich amine groups of PAMAM for immobilization, and the high surface area, holding capacity, and easily functionalized clay nanostructure as the immobilization support [23]. Satisfactory analytical performance was displayed with an LOD of 0.7 μM glucose and a linear range of 0.005 mM to 0.25 mM with demonstrated applicability in real sample matrixes. As shown in Table 20.1, other analytes were also

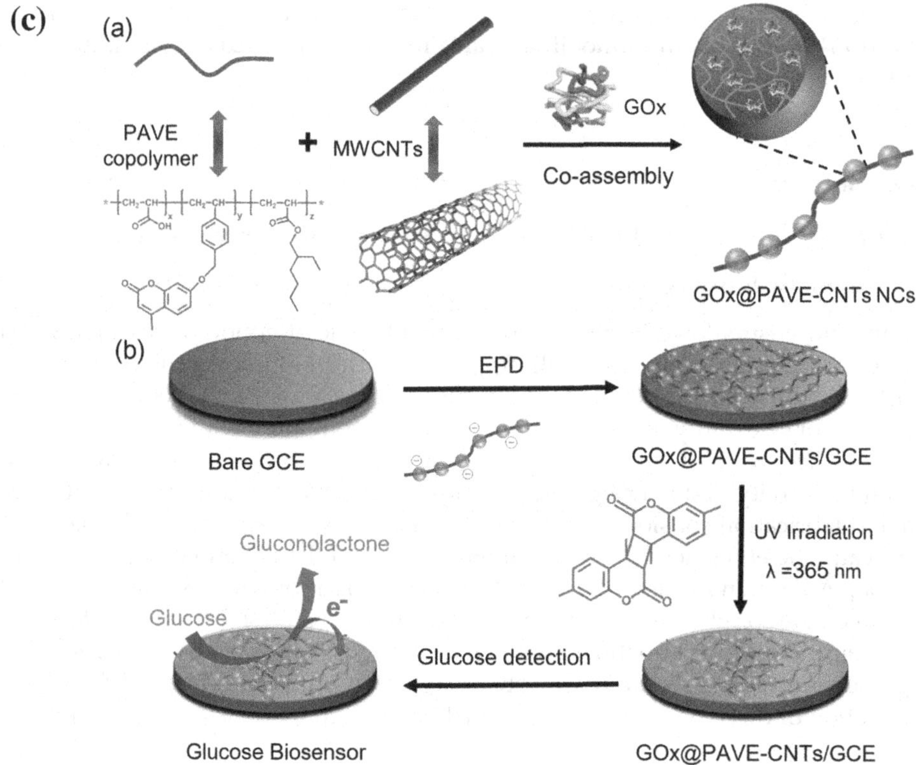

FIGURE 20.7
Fabrication of the enzymatic glucose biosensor (a) Preparation of the GOx-loaded PAVE-CNTs nanocomposite *via* one-step co-assembly. (b) Preparation process of glucose biosensor *via* direct electrophoretic deposition (EPD) of GOx@PAVE-CNTs onto GCE and subsequent photo-crosslinking. Adapted with permission from Reference [34]. Copyright (2019), Elsevier.

detected by PNC-based biosensors including phenolic compounds using laccase bioreceptor, H_2O_2 using HRP, urea using urease, and UA using uricase. A 1D-PNC-modified FTO electrode has been used for impedimetric measurements of the urea concentration in real milk samples [14].

20.5.2 Immunosensors

Immunosensors correlate transduction signals to analyte concentration based on specific antibody-antigen reactions, hence either the antigen (Ag) or the antibody (Ab) is immobilized. As with other biosensors, sensitivity, selectivity, reproducibility, and stability are among the analytical performance metrics that are optimized [3, 42].

Proper immobilization, nano-amplification, and enzyme amplification techniques, as well as assay formats, are crucial. Various analytical applications of these sensors have been reported including plant physiology [41], environmental and food safety [45, 51, 52], and disease diagnosis and management [7, 20, 37, 41, 44, 53]. Much interest has been paid to the electrochemical transduction principle although coupling it with optical techniques such as in the electro-chemiluminescence (ECL) immunoassay may also permit ultrasensitive biomolecule determination.

Label-free immunoassays employ the direct transduction method without the need for additional reagents. A dual signal amplification technique by coupling IgG–AuNPs to the immobilization matrix and AuNPs to indole-3-acetic acid (IAA) Ab receptor has helped achieve a broad linear range, low LOD, and high specificity and reproducibility of an IAA immunosensor [41]. The appropriate use of a novel Ab/AuNP-PANABA/SPE immunoassay platform gave an impedimetric response that linearly decreased with increasing 2,4-dichlorophenoxy acetic acid (2,4-D) concentration within the range of 1–100 ppb and a detection limit that is well below the regulatory limits [44]. The application of a low-fouling, polymeric immobilization matrix – PEG [19] and polyCBMA [43], respectively – greatly reduced protein fouling making possible highly sensitive DPV-based detection of AFP and a tumor marker CEA in complex serum samples. A PPa/1oAb/TiO_2-NW FET 1D PNC FET-based immunosensor that utilized a wide energy bandgap of TiO_2 gave a rapid, selective, and sensitive response to rabbit IgG with a linear range of 119 pg/mL–5.95 ng/mL [25].

To complement nanomaterial amplification, enzyme labeling in competitive or sandwich-type immunoassay formats greatly improves the analytical performance. In a DPV-based AFB1-BSA-SWNT/CS/GCE immunosensor, the secondary antibody (anti-mouse IgG Ab) was conjugated with alkaline phosphatase (ALP) which catalyzed the hydrolysis of the added alpha-naphthyl phosphate substrate, thereby giving the amplified electrochemical signal related to AFB1 concentration [50]. The same system was used by Yang et al. (2015) in analyzing the mycotoxin fumonisin B1 (FB1) from contaminated corn powder samples that gave comparable results with that of standard HPLC [53]. In a sandwich-type (Figure 20.8) technique wherein the target, an antigen, is sandwiched between capture and detection antibodies, reference [40] made use of HRP-labeled detection Ab while PtAg@

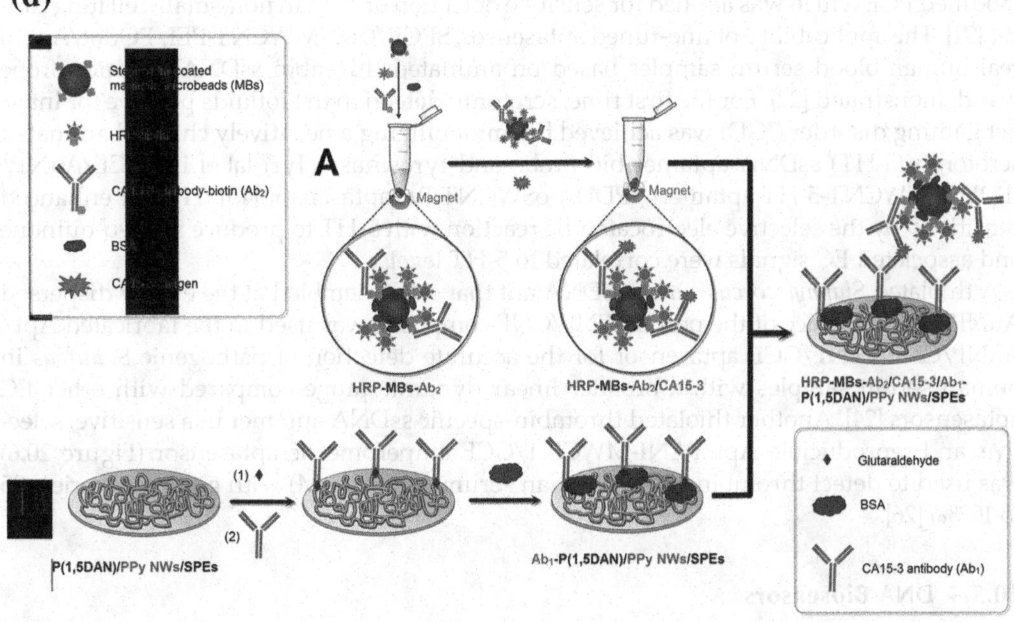

FIGURE 20.8
Fabrication of immunosensor based on P(1,5DAN)/PPy NWs bilayer and magnetic beads. Adapted with permission from Reference [40]. Copyright (2017), Elsevier.

CNC-labeling was applied [40] for an amplified EC signal from immunoreactions with CA 15-3 and PSA.

20.5.3 Aptasensors

In place of the antibodies, which suffer from poor stability, complicated synthesis, and limited shelf-life, interest in aptamers for biosensor fabrication has soared [22]. Aptamers are artificial peptides, ssDNA, or RNA molecular entities that have strong binding affinity and selectivity towards target analytes ranging from small to large molecules to even whole cells. Aptamers have a smaller size than antibodies, high stability, are easier and cheaper to synthesize, exhibit chemical tunability, and easier handling and storage [10]. Various 1D PNC-based aptasensors have been reported that can detect target molecules with high sensitivity and selectivity with applicability to real sample matrixes. They have been presented as promising alternatives to classical analytical techniques that are rather tedious, often requiring intricate sample pretreatment and costly instruments.

Aptamer immobilization has been highlighted as a critical factor in boosting sensor sensitivity, selectivity, and stability. As such, covalent immobilization is the preferred technique although a layer-by-layer electrostatic adsorption strategy has been reported. Owing to their chemical tunability, thiolated or amino-linked aptamers have been synthesized for easy functionalization on nanocomposite surfaces. A $VEGF_{165}$ RNA aptamer with linked $-NH_2$ group at 5' end was attached onto SPCE/PANI/CNT/Apt biosensor for electrochemical detection of $VEGF_{165}$ [10]. Although biomarker-related biosensors such as this are common, biomarkers are still not the recommended basis of clinical diagnostics, and hence, cell-based EC detection techniques arise. A universal impedimetric cancer cell detection platform was developed based on aptamer-cell interaction in a high-affinity 5′-aminohexyl-linked ssDNA aptamer-conjugated carboxyl-functionalized PAN/PPyNF-modified PGE which was applied for sensitive detection of human non-small-cell lung cancer [21]. The applicability of fine-tuned aptasensor, SPCE/C60/MWCNT-PEI/PQdot/Apt in real human blood serum samples based on aminated thrombin ssDNA aptamer probe was demonstrated [22]. For the first time, serotonin detection in biofluids positive for internet gaming disorder (IGD) was achieved by immobilizing a negatively charged aminated serotonin (5-HT) ssDNA aptamer bio probe and tyrosinase (Tyr) label in SPCE/AuNP/(PDDA-oSWCNT-5-HT aptamer)$_2$/PDDA-oSWCNT-Tyr aptasensor [45]. The Tyr enhanced sensitivity by the selective electrocatalytic reaction with 5-HT to produce 5-HT-o-quinone and associated EC signals were correlated to 5-HT levels.

A thiolated *Staphylococcus aureus* ssDNA apt that self-assembled at the evenly dispersed AuNPs at the surface of the porous CNP/CNF composite was used in the fabricated Apt/AuNP/CNP/CeNF/GCE aptasensor for the accurate detection of pathogenic *S. aureus* in human serum samples with a broader linear dynamic range compared with other EC aptasensors [24]. Another thiolated thrombin-specific ssDNA aptamer in a sensitive, selective, and reproducible Apt/PANI-MWCNT/GCE amperometric aptasensor (Figure 20.6) was used to detect thrombin-spiked human serum (0.2 to 4 nM) with good recoveries (95 to 102%) [26].

20.5.4 DNA Biosensors

DNA biosensors are constructed by attaching a DNA probe onto a transducer surface to recognize its complementary DNA target sequence by hybridization or to detect tiny molecules that interact with it [33]. The efficient use of nanocomposites with fine-tuned surface chemistry and suitable tethering molecules that can immobilize large quantities of probe molecules greatly enhance the sensitivity and stability of the biosensor. As such, it can achieve superior selectivity up to single-base mismatched complementary ssDNA sequence [43, 54] and carry out multiple analyte determinations in the presence of large amounts of interfering substances from biological matrixes [34]. Excellent sensitivity towards *Bacillus cereus* was achieved by GCE/Fe_3O_4/MWCNT-COOH/CS/ssDNA biosensor with a low detection limit (2.0×10^{-15} M) that surpassed that of an electrochemical sensor utilizing conventional and more sophisticated PCR technology (LOD of 9.4×10^{-12} M) [48]. Similarly superior sensitivity to peanut allergen Ara h1 in foods ($(2.88 \pm 0.05) \times 10^{-11}$ mol L^{-1}) was realized by the developed stem-loop DNA/spongy gold film/CS–MWCNT/GCE biosensor (Figure 20.9) that compares well to that of the fluorescence real-time quantitative PCR method ($2.74 \pm 0.11 \times 10^{-11}$ M). In addition, this has established the reliability of this DNA biosensor [43]. It must be noted that the establishment of the best optimum processing conditions during the development of the biosensor is also crucial and must be executed

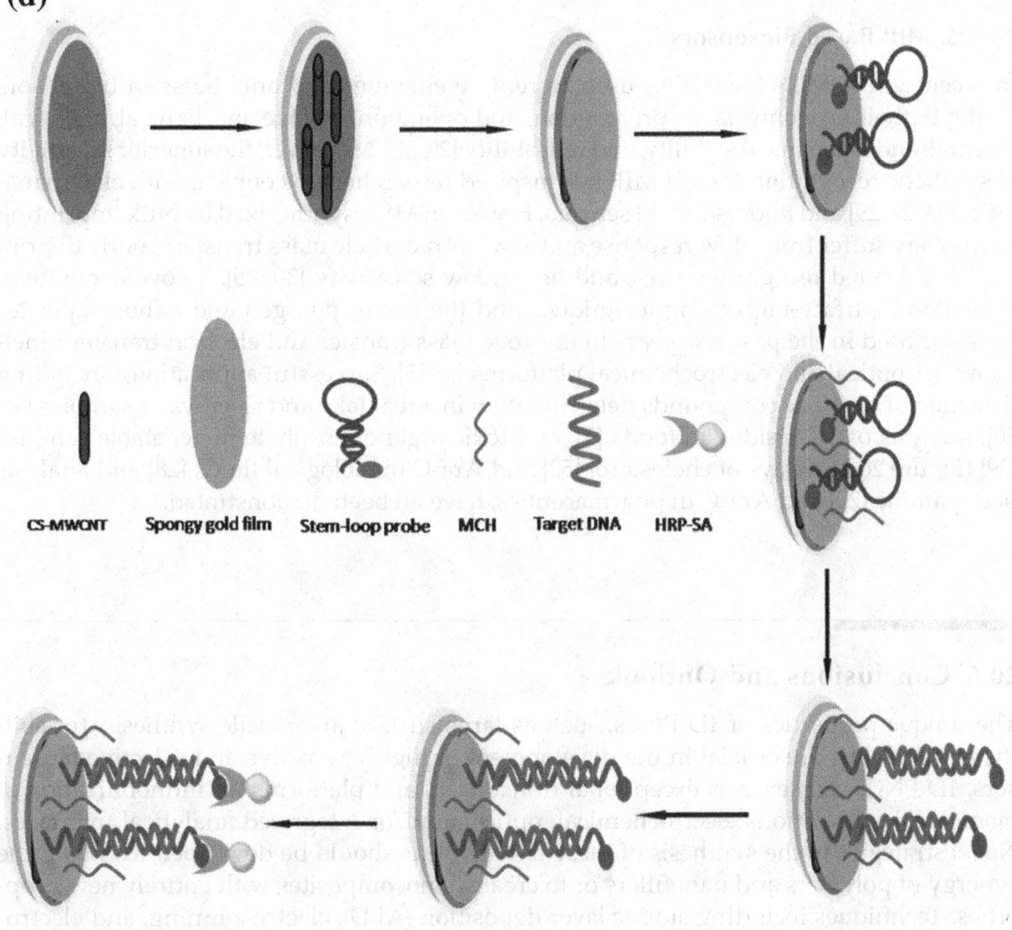

FIGURE 20.9
Fabrication of the stem-loop DNA biosensor of amplifying the signal by enzyme catalysis. Adapted with permission from Reference [42]. Copyright (2015), Elsevier.

efficiently, especially in a multivariate system. In this connection, the response surface methodology (RSM) was demonstrated [46] to be a strategic and simplified approach in optimizing *Mycobacterium tuberculosis* biosensor performance leading to a facile and fast-sensing operation.

The hybridization event or small molecule-DNA interaction may be converted to signals by the commonly used electrochemical transduction owing to its simplicity, economy, fast response, good selectivity, and high sensitivity. For electroactive analytes such as DA and UA, the electrochemical response from their oxidation (or reduction) can be correlated to their adsorbed amounts on DNA grooves. Conversely, for non-electroactive analytes, the use of redox indicators such as methylene blue (MB) [17, 47] or the redox pair $K_3(CN)_6$ / $K_4(CN)_6$ was usually preferred [47]. DNA biosensors also exploited electrochemical signals of enzymatic products when enzyme labeling is applied to adsorbed target ssDNA in the sensing electrode [43]. The use of enzyme or redox labels can be avoided altogether by monitoring the changes in the interfacial properties such as polarization resistance in impedimetric sensing of human influenza A (H1N1) virus by p-DNA/PAMAM/GaN NWs [51].

20.5.5 MIP-Based Biosensors

In recent years, MIPs have been used to replace enzymes and antibodies in biosensors owing to their economy, facile preparation, and operation, storage, mechanical, chemical, thermal, and mechanical stability, and reusability [29, 30]. Moreover, the superior selectivity of synthetic recognition sites of MIP has inspired researchers to couple it with electroanalytical [6, 21, 29] and fluorescent [5] sensors. However, MIPs synthesized by bulk imprinting technology suffer from slow response and slow intraparticle mass transfer, poorly distributed and loaded recognition sites, and hence, low sensitivity [30, 55]. To overcome these limitations, surface-imprinting techniques and the use of porogen and nanocomposites have surfaced in the past few years to improve mass transfer and electron transfer kinetics across optical and electrochemical platforms [6, 55]. Successful applications, including detection of phenolic compounds determination in a real lake and river water samples [49, 50], analysis of TC residue in foods [5], and toxic organophosphate in vegetable samples [29] (Figure 20.5), assays of cholesterol [52] and Ara-C in biological fluids [28] and analysis of dopamine [20] and Ara-C in pharmaceutics, have all been demonstrated.

20.6 Conclusions and Outlook

The unique properties of 1D PNCs, such as large surface area, facile synthesis, tunability, and stability, are crucial in the development of highly sensitive and selective biosensors. 1D PNCs can serve as exceptional transducers and platforms for immobilization of biomolecules in various electrochemical, optical, and/or integrated analytical interfaces. Novel strategies in the synthesis of these nanohybrids should be developed to exploit the synergy of polymers and nanofillers or to create nanocomposites with entirely new properties. Techniques including atomic layer deposition (ALD), electrospinning, and electrodeposition processes have been increasingly used in the preparation of 1D PNCs. With the advances in nanoscience and nanotechnology, nanofillers bearing superior characteristics, such as PAKBNPs and non-conjugated PQDots, CNPs, and nanoclays, have surfaced while the demonstrated robustness of CNTs, AuNPs, and C60 have secured their applications. Other morphological configurations, such as nanofibers and core-shell structures of various nanohybrids, have been explored due to their flexibility. In 1D PNC-based biosensors, enzymes and antibodies have been increasingly replaced with aptasensors and MIPs owing to their low cost, facile preparation and operation, stability, and reusability. Moreover, the efficient integration of 1D PNCs in DNA-based and immunoassay platforms affords direct, label-free transduction with ultrasensitive, analytical performance. The advances in PNC-based biosensors has led to promising applications in fields of environmental and biosystem monitoring, food safety, biomedical and analytical science, clinical diagnostics, and pathological research, among others.

References

1. S. Fu, Z. Sun, P. Huang, Y. Li, and N. Hu, "Some basic aspects of polymer nanocomposites: A critical review," *Nano Materials Science*, vol. 1, no. 1, pp. 2–30, 2019.
2. C. I. Idumah, "Novel trends in conductive polymeric nanocomposites, and bionanocomposites," *Synthetic Metals*, vol. 273, no. December 2020, p. 116674, 2021.
3. B. Filipowicz, M. Pruszyński, S. Krajewski, and A. Bilewicz, "Adsorption of 137Cs on titanate nanostructures," *Journal of Radioanalytical and Nuclear Chemistry*, vol. 301, no. 3, pp. 889–895, 2014.
4. S. Shrivastava, N. Jadon, and R. Jain, "Next-generation polymer nanocomposite-based electrochemical sensors and biosensors: A review," *TrAC Trends in Analytical Chemistry*, vol. 82, pp. 55–67, 2016.
5. Z. Zhou, K. Lu, X. Wei, T. Hao, Y. Xu, L. Xiaodong, and Y. Zhang, "A mesoporous fluorescent sensor based on ZnO nanorods for the fluorescent detection and selective recognition of tetracycline," *RSC Advances*, vol. 6, no. 75, pp. 71061–71069, 2016.
6. F. Peng, M. Chu, J. Sun, Y. Liu, Q. Zhang, Y. Chen, F. Wang, and W. Zhao, "Preparation of Fe_3O_4@PS/PDA-Au nanotubes for sensitive electrochemical detection of alpha-fetoprotein," *Journal of Electroanalytical Chemistry*, vol. 814, no. December 2017, pp. 52–58, 2018.
7. J. Baiju, "Polymer nanocomposite-based electrochemical sensors and biosensors," in: *Nanorods and Nanocomposites*. M. S. Ghamsari, and S. Dhara, Eds. vol. i. London: IntechOpen, pp. 159–160, 2020.
8. W. A. El-Said, M. Abdelshakour, J. H. Choi, and J. W. Choi, "Application of conducting polymer nanostructures to electrochemical biosensors," *Molecules*, vol. 25, no. 2, pp. 1–11, 2020.
9. Y. Tan, W. Deng, B. Ge, Q. Xie, J. Huang, and S. Yao, "Biofuel cell and phenolic biosensor based on acid-resistant laccase-glutaraldehyde functionalized chitosan-multiwalled carbon nanotubes nanocomposite film," *Biosensors and Bioelectronics*, vol. 24, no. 7, pp. 2225–2231, 2009.
10. Y. Park, M. S. Hong, W. H. Lee, J. G. Kim, and K. Kim, "Highly sensitive electrochemical aptasensor for detecting the VEGF165 tumor marker with PANI/CNT nanocomposites," *Biosensors*, vol. 11, no. 4, pp. 1–11, 2021.
11. G. Paimard, M. Shahlaei, P. Moradipour, V. Karamali, and E. Arkan, "Impedimetric aptamer based determination of the tumor marker MUC1 by using electrospun core-shell nanofibers," *Microchimica Acta*, vol. 187, no. 1, pp. 1–10, 2020.
12. A. Numnuam, P. Thavarungkul, and P. Kanatharana, "An amperometric uric acid biosensor based on chitosan-carbon nanotubes electrospun nanofiber on silver nanoparticles," *Analytical and Bioanalytical Chemistry*, vol. 406, no. 15, pp. 3763–3772, 2014.
13. L. Wang, X. Gao, L. Jin, Q. Wu, Z. Chen, and X. Lin, "Amperometric glucose biosensor based on silver nanowires and glucose oxidase," *Sensors and Actuators, Part B: Chemical*, vol. 176, pp. 9–14, 2013.

14. F. L. Migliorini, R. C. Sanfelice, L. A. Mercante, R. S. Andre, L. H. C. Mattoso, and D. S. Correa, "Urea impedimetric biosensing using electrospun nanofibers modified with zinc oxide nanoparticles," *Applied Surface Science*, vol. 443, pp. 18–23, 2018.
15. X. Sun, L. Qiao, and X. Wang, "A novel immunosensor based on Au nanoparticles and polyaniline/multiwall carbon nanotubes/chitosan nanocomposite film functionalized interface," *Nano-Micro Letters*, vol. 5, no. 3, pp. 191–201, 2013.
16. A. D. Chowdhury, R. Gangopadhyay, and A. De, "Highly sensitive electrochemical biosensor for glucose, DNA and protein using gold-polyaniline nanocomposites as a common matrix," *Sensors and Actuators, Part B: Chemical*, vol. 190, pp. 348–356, 2014.
17. Y. Yang, Z. Wang, M. Yang, J. Li, F. Zheng, G. Shen, and R. Yu, "Electrical detection of deoxyribonucleic acid hybridization based on carbon-nanotubes/Nano zirconium dioxide/chitosan-modified electrodes," *Analytica Chimica Acta*, vol. 584, no. 2, pp. 268–274, 2007.
18. J. Wang, and N. Hui, "Electrochemical functionalization of polypyrrole nanowires for the development of ultrasensitive biosensors for detecting microRNA," *Sensors and Actuators, Part B: Chemical*, vol. 281, no. June 2018, pp. 478–485, 2019.
19. N. Hui, X. Sun, Z. Song, S. Niu, and X. Luo, "Gold nanoparticles and polyethylene glycols functionalized conducting polyaniline nanowires for ultrasensitive and low fouling immunosensing of alpha-fetoprotein," *Biosensors and Bioelectronics*, vol. 86, pp. 143–149, 2016.
20. Y. Teng, F. Liu, and X. Kan, "Voltammetric dopamine sensor based on three-dimensional electrosynthesized molecularly imprinted polymers and polypyrrole nanowires," *Microchimica Acta*, vol. 184, no. 8, pp. 2515–2522, 2017.
21. E. Kivrak, A. Ince-Yardimci, R. Ilhan, P. B. Kirmizibayrak, S. Yilmaz, and P. Kara, "Aptamer-based electrochemical biosensing strategy toward human non-small cell lung cancer using polyacrylonitrile/polypyrrole nanofibers," *Analytical and Bioanalytical Chemistry*, vol. 412, no. 28, pp. 7851–7860, 2020.
22. H. R. Jamei, B. Rezaei, and A. A. Ensafi, "Ultra-sensitive and selective electrochemical biosensor with aptamer recognition surface based on polymer quantum dots and C60/MWCNTs-polyethylenimine nanocomposites for analysis of thrombin protein," *Bioelectrochemistry*, vol. 138, p. 107701, 2021.
23. B. Unal, E. E. Yalcinkaya, D. O. Demirkol, and S. Timur, "An electrospun nanofiber matrix based on organo-clay for biosensors: PVA/PAMAM-montmorillonite," *Applied Surface Science*, vol. 444, pp. 542–551, 2018.
24. S. Ranjbar, and S. Shahrokhian, "Design and fabrication of an electrochemical aptasensor using Au nanoparticles/carbon nanoparticles/cellulose nanofibers nanocomposite for rapid and sensitive detection of Staphylococcus aureus," *Bioelectrochemistry*, vol. 123, pp. 70–76, 2018.
25. Y. Chu, C. Lin, H. Chang, C. Li, and C. Guo, "Biosensors and bioelectronics TiO_2 nanowire FET device: Encapsulation of biomolecules by electro polymerized pyrrole propylic acid," *Biosensors and Bioelectronics*, vol. 26, no. 5, pp. 2334–2340, 2011.
26. Z. Su, X. Xu, H. Xu, Y. Zhang, C. Li, Y. Ma, D. Song, and Q. Xie, "Amperometric thrombin aptasensor using a glassy carbon electrode modified with polyaniline and multiwalled carbon nanotubes tethered with a thiolated aptamer," *Microchimica Acta*, vol. 184, no. 6, pp. 1677–1682, 2017.
27. S. Alim, A. K. M. Kafi, J. Rajan, and M. M. Yusoff, "Application of polymerized multiporous nanofiber of SnO_2 for designing a bienzyme glucose biosensor based on HRP/GOx," *International Journal of Biological Macromolecules*, vol. 123, pp. 1028–1034, 2019.
28. B. B. Prasad, R. Singh, and A. Kumar, "Gold nanorods: vs. gold nanoparticles: Application in electrochemical sensing of cytosine β-d-arabinoside using metal ion mediated molecularly imprinted polymer," *RSC Advances*, vol. 6, no. 84, pp. 1–33, 2016.
29. M. Amatatongchai, W. Sroysee, P. Sodkrathok, N. Kesangam, S. Chairam, and P. Jarujamrus, "Novel three-dimensional molecularly imprinted polymer-coated carbon nanotubes (3D-CNTs@MIP) for selective detection of profenofos in food," *Analytica Chimica Acta*, vol. 1076, pp. 64–72, 2019.

30. J. Jun, J. S. Lee, D. H. Shin, and J. Jang, "Aptamer-functionalized hybrid carbon nanofiber FET-type electrode for a highly sensitive and selective platelet-derived growth factor biosensor," *ACS Applied Materials and Interfaces*, vol. 6, no. 16, pp. 13859–13865, 2014.
31. S. Prasad, V. Kumar, S. Kirubanandam, and A. Barhoum, "Engineered nanomaterials: Nanofabrication and surface functionalization," in: *Emerging Applications of Nanoparticles and Architectural Nanostructures: Current Prospects and Future Trends, Micro & Nano Technology Series*. A. Barhoum, and A. Makhlouf, Eds. Amsterdam: Elsevier, pp. 314, 324–325, 2018.
32. K. W. Shah, S. X. Wang, D. X. Y. Soo, and J. Xu, "One-dimensional nanostructure engineering of conducting polymers for thermoelectric applications," *Applied Sciences (Switzerland)*, vol. 9, no. 7, pp. 1–22, 2019.
33. X. Liu, Y. Peng, X. Qu, S. Ai, R. Han, and X. Zhu, "Multi-walled carbon nanotube-chitosan/poly(amidoamine)/DNA nanocomposite modified gold electrode for determination of dopamine and uric acid under coexistence of ascorbic acid," *Journal of Electroanalytical Chemistry*, vol. 654, no. 1–2, pp. 72–78, 2011.
34. S. Xu, Y. Zhang, Y. Zhu, J. Wu, K. Li, G. Lin, X. Li, R. Liu, X. Liu, and C. P. Wong, "Facile one-step fabrication of glucose oxidase loaded polymeric nanoparticles decorating MWCNTs for constructing glucose biosensing platform: Structure matters," *Biosensors and Bioelectronics*, vol. 135, no. April, pp. 153–159, 2019.
35. X. Huang, S. Zu, W. Zhao, M. Xu, W. Wei, J. Luo, X. Li, and X. Liu, Screen-printed carbon electrodes modified with polymeric nanoparticle-carbon nanotube composites for enzymatic biosensing, *ACS Applied Nano Materials*, vol. 3, no. 9, pp. 1–32, 2020.
36. A. Dey, A. Kaushik, S. K. Arya, and S. Bhansali, "Mediator free highly sensitive polyaniline-gold hybrid nanocomposite based immunosensor for prostate-specific antigen (PSA) detection," *Journal of Materials Chemistry*, vol. 22, no. 29, pp. 14763–14772, 2012.
37. X. Chen, Z. Chen, J. Zhu, C. Xu, W. Yan, and C. Yao, "A novel H_2O_2 amperometric biosensor based on gold nanoparticles/self-doped polyaniline nanofibers," *Bioelectrochemistry*, vol. 82, no. 2, pp. 87–94, 2011.
38. V. Myndrul, E. Coy, M. Bechelany, and I. Iatsunskyi, "Photoluminescence label-free immunosensor for the detection of aflatoxin b1 using polyacrylonitrile/zinc oxide nanofibers," *Materials Science and Engineering* C, vol. 118, no. May 2020, p. 111401, 2021.
39. G. Xu, S. B. Adeloju, Y. Wu, and X. Zhang, "Modification of polypyrrole nanowires array with platinum nanoparticles and glucose oxidase for fabrication of a novel glucose biosensor," *Analytica Chimica Acta*, vol. 755, pp. 100–107, 2012.
40. V. A. Nguyen, H. L. Nguyen, D. T. Nguyen, Q. P. Do, and L. D. Tran, "Electrosynthesized poly(1,5-diaminonaphthalene)/polypyrrole nanowires bilayer as an immunosensor platform for breast cancer biomarker CA 15-3," *Current Applied Physics*, vol. 17, no. 11, pp. 1422–1429, 2017.
41. Z. Su, Y. Cheng, C. Li, Y. Xiong, L. Xiao, S. Chen, X. Qin, "Dispersing gold nanoparticles on thiolated polyaniline-multiwalled carbon nanotubes for development of an indole-3-acetic acid amperometric immunosensor," *Nanoscale Advances*, vol. 1, no. 9, pp. 3607–3613, 2019.
42. X. Sun, M. Jia, J. Ji, L. Guan, Y. Zhang, L. Tang, Z. Li, "Enzymatic amplification detection of peanut allergen Ara h1 using a stem-loop DNA biosensor modified with a chitosan-mutiwalled carbon nanotube nanocomposite and spongy gold film," *Talanta*, vol. 131, pp. 521–527, 2015.
43. J. Wang, and N. Hui, "Zwitterionic poly(carboxybetaine) functionalized conducting polymer polyaniline nanowires for the electrochemical detection of carcinoembryonic antigen in undiluted blood serum," *Bioelectrochemistry*, vol. 125, pp. 90–96, 2019.
44. G. Fusco, F. Gallo, C. Tortolini, P. Bollella, F. Ietto, A. De Mico, A. D'Annibale, R. Antiochia, G. Favero, F. Mazzei, "AuNPs-functionalized PANABA-MWCNTs nanocomposite-based impedimetric immunosensor for 2,4-dichlorophenoxy acetic acid detection," *Biosensors and Bioelectronics*, vol. 93, no. June 2016, pp. 52–56, 2017.
45. J. Li, Y. Si, Y. E. Park, J. S. Choi, S. M. Jung, J. E. Lee, H. J. Lee, "A serotonin voltammetric biosensor composed of carbon nanocomposites and DNA aptamer," *Microchimica Acta*, vol. 188, no. 4, pp. 1–8, 2021.

46. K. S. Rizi, B. Hatamluyi, M. Rezayi, Z. Meshkat, M. Sankian, K. Ghazvini, H. Farsiani, E. Aryan, "Response surface methodology optimized electrochemical DNA biosensor based on HAPNPTs/PPY/MWCNTs nanocomposite for detecting *Mycobacterium tuberculosis*," *Talanta*, vol. 226, no. September 2020, p. 122099, 2021.
47. L. Zuo, Q. Qu, L. Li, X. Ran, J. Gui, Q. Wang, X. Cui, C. Jiang, "Electrochemical DNA biosensor based on magnetite/multiwalled carbon nanotubes/chitosan nanocomposite for *Bacillus cereus* detection of potential marker for gold prospecting," *Electroanalysis*, vol. 30, no. 5, pp. 910–920, 2018.
48. B. Liu, J. Yan, M. Wang, and X. Wu, "Electrochemical sensor based on molecularly imprinted polymer for determination of nonylphenol," *International Journal of Electrochemical Science*, vol. 13, no. 12, pp. 11953–11960, 2018.
49. X. Wei, Z. Zhou, T. Hao, H. Li, Y. Zhu, L. Gao, Y. Yan, "A novel molecularly imprinted polymer thin film at surface of ZnO nanorods for selective fluorescence detection of para-nitrophenol," *RSC Advances*, vol. 5, no. 55, pp. 44088–44095, 2015.
50. X. Zhang, C. R Li, W. C. Wang, J. Xue, Y. L. Huang, X.X. Yang, B. Tan, X. P. Zhou, C. Shao, S. J. Ding, J. F. Qiu,., "A novel electrochemical immunosensor for highly sensitive detection of aflatoxin b1 in corn using single-walled carbon nanotubes/chitosan," *Food Chemistry*, vol. 192, pp. 197–202, 2016.
51. P. Sahoo, S. Suresh, S. Dhara, G. Saini, S. Rangarajan, and A. K. Tyagi, "Direct label free ultrasensitive impedimetric DNA biosensor using dendrimer functionalized GaN nanowires," *Biosensors and Bioelectronics*, vol. 44, no. 1, pp. 164–170, 2013.
52. Y. Tong, H. Li, H. Guan, J. Zhao, S. Majeed, S. Anjum, F. Liang, G. Xu, "Electrochemical cholesterol sensor based on carbon nanotube@molecularly imprinted polymer modified ceramic carbon electrode," *Biosensors and Bioelectronics*, vol. 47, pp. 553–558, 2013.
53. X. Yang, X. Zhou, X. Zhang, Y. Qing, M. Luo, X. Liu, C. Li, C. Li, Y. Li, H. Xia, J. Qiu, "A highly sensitive electrochemical immunosensor for fumonisin B1 detection in corn using single-walled carbon nanotubes/chitosan," *Electroanalysis*, vol. 27, no. 11, pp. 2679–2687, 2015.
54. M. Zhang, W. Dai, M. Yan, S. Ge, J. Yu, X. Song, W. Xu, "Ultrasensitive electrochemiluminescence immunosensor using PtAg@ carbon nanocrystals composites as labels and carbon nanotubes-chitosan/gold nanoparticles as enhancer," *Analyst*, vol. 137, no. 9, pp. 2112–2118, 2012.

21

One-Dimensional Polymeric Nanocomposites for Electrochemical Sensors

Rafia Nimal, Didem Nur Unal, Sevinc Kurbanoglu, Afzal Shah, Muhammad Siddiq, and Bengi Uslu

CONTENTS

21.1 Introduction

Emerging electronics aims to develop miniaturized, consistent, and exceedingly precise sensors based on functional nanomaterials [1]. Carbon nanomaterial (CNM)-based sensors have been widely used in recent years to build next-generation sensors due to their efficacy and speed [2, 3]. However, various inherent limitations of CNMs, like passive surface, chirality,

DOI: 10.1201/9781003223764-21

electrical properties being reliant on diameter, and inter-particle cluster formation, often result in non-replicable sensor response and hamper their effective utilization in electrochemical sensors. On the other hand, conducting polymers are replacing CNMs as effective sensing elements [4], primarily because they offer exceptional electrical, mechanical, and optical transduction mechanisms. The unique chemical and physical properties of conducting polymers arise from their conjugated π-electron system. Chemical and/or electrochemical doping/de-doping mechanisms readily affect the oxidation level of conducting polymers, producing precise and swift response specific to analytes [5]. In recent years, many novel conducting polymers have been designed to develop electrochemical sensors because they can offer diverse backbone compositions with discrete electrochemical behavior. This considerable chemical versatility enables us to address them with various electrochemical techniques involving voltammetry, potentiometry, and amperometry, as well as conductometry.

Additionally, conductive polymers possess functional groups that act in conjunction with organic and inorganic materials with equal competency. Thus, this greater sensitivity and selectivity make conducting polymers key substrates for detecting various types of analytes. Conductive polymer-based sensors offer various advantages, but they are often associated with the drawback of lacking long-term stability. Moreover, they have amorphous polymeric structures, which limit their efficacy in sensing applications [6].

On the other hand, one-dimensional polymeric nanocomposites have improved properties over parent conducting polymer (CPs) in bulk [4]. Versatility associated with conductive polymer and new properties arising because of nano-dimension makes them quite suitable for use in a wide range of analytical applications. They are commonly employed these days for the fabrication of disposable personal monitoring devices for various analyte diagnoses like glucose, cholesterol and the diagnosis of fertility and genetic sensors [7]. Moreover, they are employed in the care diagnostic devices for clinical use [7], i.e., for the detection of several infectious diseases, cardiovascular diseases, and carcinomas. One-dimensional polymeric nanocomposites are also used widely in the fabrication of medical monitoring devices like neural probes, tissue engineering scaffolds, and stimulus-responsive drug delivery systems [8]. They are receiving great acceptance from patients and clinicians, and they are anticipated to be a potential part of on-the-spot detection and quantification rather than limited to traditional off-site laboratory analysis. In addition, through their biomarker-based screening, conductive polymer-based biosensors would be helpful in the detection of serious illnesses such as cancer in the initial stages.

One-dimensional polymeric nanocomposites are the preferred material for electrochemical applications because of their following traits [7, 9]:

- Their preparation methods are comparatively simple, permitting production at a larger scale;
- Their surface-charge region and dimensions are compatible;
- Their greater order of crystallinity offer superior stability;
- They have reasonably high surface/volume (S/V) ratios, which are crucial to facilitate swift reaction kinetics;
- Usually, one-dimensional polymeric nanocomposite-based electrochemical sensors offer ultrahigh sensitivity, robust selectivity, and the tremendous ability to incorporate into well-defined arrays on a large scale;
- They have a higher loading density of a target analyte and permit deposition of the catalyst over the enormous surface, thus offering the advancement or inhibition of certain reactions;

- They can provide free-of-label, straightforward, real-time signal transduction at ambient temperatures.

Although one-dimensional polymeric nanocomposite-based electrochemical sensor development is still in its infancy, their promising improvements ensure that they have the potential to revolutionize the electrochemical sensing field.

21.2 Fundamentals of the Electrochemical Sensors

The type of sensors where the transducer component is the electrode are referred to as electrochemical sensors [10]. These electrochemical sensors have gained significant importance in recent years owing to the following advantages offered by them: electrons are utilized for signal obtaining, which make them a system for rational applications, with long operational life; widely used in multipurpose gadgets like micro-volume testing; speedy examination; and easy designs, allowing these procedures to be endorsed, e.g., as commercial glucose sensors [11]. Furthermore, electrochemical sensors supporting the different procedures, such as identifiers for chromatography, have expanded their effectiveness.

21.2.1 Working Principle of Electrochemical Sensors

Electrochemical sensors work by disseminating analytes of interest into the sensor, producing an electrical signal relative to the analyte (Figure 21.1). In response, the diffused analyte molecules, depending upon their nature, undergo oxidation or reduction processes at the surface of the detecting electrode. A typical electrochemical sensor contains a detecting cathode/working anode and a counter terminal electrode. The two electrodes are separated by a narrow electrolyte layer [10].

When an analyte encounters the sensor, it first passes through a small opening in the narrow structure and then diffuses through a hydrophobic medium, finally touching the

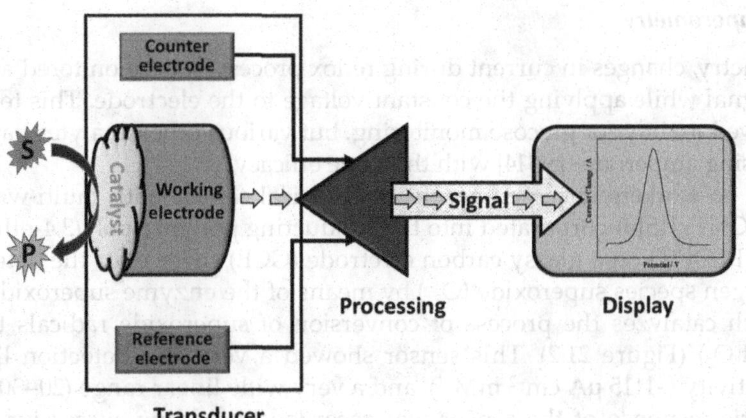

FIGURE 21.1
Electrochemical sensing principle. Adapted with permission from reference [10]. Copyright (2019) Taylor & Francis.

terminal's outer surface. This set-up results in generating an appropriate electrical sign depending upon the accurate measure of analyte that reacts with the detecting anode. It also prevents the leaking of electrolytes out of the sensor—a current stream between the electrode relative to the analyte with a resistor linked over the terminals. Current is then estimated to compute the analyte [10].

Electrochemical sensors have wide applicability and can be used solely for sensing or sometimes incorporated into chemical and biomedical sensing systems as an essential element. Blood gas sensing is an excellent example that can be commenced solely by electrochemical means, whereas various important biomedical enzymatic sensors (e.g., glucose sensors) also include an enzymatic catalyst rather than an electrochemical sensing element [12]. Electrochemical sensors are usually categorized as potentiometric, amperometric, conductivity/capacitance, or voltammetric sensors. The amperometric and voltammetric sensors are distinguished based on their current–potential relationships.

21.2.2 Characterization Methods of Electrochemical Sensors

Various electrical methods can be applied to investigate CP-based electrochemical sensors, involving potentiometry, amperometry, conductometry, or voltammetry.

21.2.2.1 Potentiometry

The potentiometric method for electrochemical characterization of CP-based electrochemical sensors includes a CP-coated electrode (sensing membrane) that can reach the equilibrium state with the targeted species in the analyte solution. The output signals result because of variations in the open-circuit potential of the CP-coated electrode measured against some appropriate reference electrode placed in the analyte solution. Ayranci et al. utilized a potentiometric approach to analyze mercury ions in an aqueous solution [13]. They used a rhodamine-functionalized, carbazole-based CP (RD-CZ), where electroactivity arises due to the carbazole (CZ) group, and the fluorescent and colorimetric changes on exposure to Hg^{2+} are provided by the rhodamine (RD) group.

21.2.2.2 Amperometry

In amperometry, changes in current during redox processes are monitored as an electrochemical signal while applying the constant voltage to the electrode. This technique has been employed mainly for glucose monitoring, but various other analytes have also been identified using amperometry [14] with the same efficacy.

Braik and co-workers designed an electrochemical sensor with multi-walled carbon nanotubes (CNT) [15] incorporated into the conducting polymer poly(3,4-ethylenedioxythiophene) (PEDOT) on a glassy carbon electrode (GCE) surface for the detection of the reactive oxygen species superoxide (O_2^-) by means of the enzyme superoxide dismutase (SOD), which catalyzes the process of conversion of superoxide radicals to hydrogen peroxide (H_2O_2) (Figure 21.2). This sensor showed a very low detection limit (1 μM), robust sensitivity (~1115 μA cm^{-2} mM^{-1}), and a very wide linear range (20–3000 μM). The amperometric response of the sensor was recorded by adding successive amounts of superoxide (KO_2) red wine in a blank solution. Thus, a comparative analysis of the slopes of the plots with and without the red wine was used to compute Relative Antioxidant Capacity (RAC).

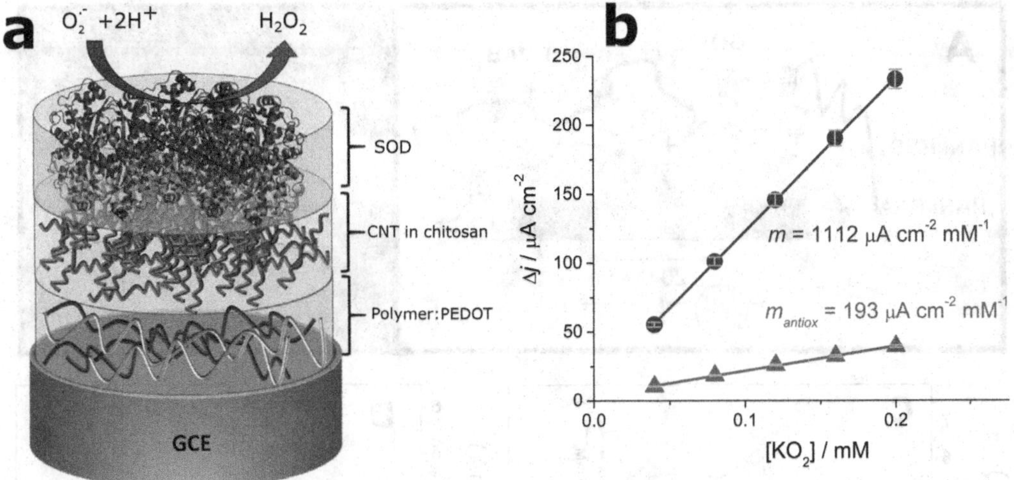

FIGURE 21.2
(a) Optimal electrode configuration, with PEDOT deposited on a glassy carbon electrode, coated with CNTs in chitosan, and then coated with superoxide dismutase (SOD). (b) Example of determination of the relative antioxidant capacity of wine; increasing amounts of KO_2 are added to the SOD/CNT/PEDOT/GCE sensor in the absence (circles) and presence (triangles) of red wine (8 μL added to 2 mL solution of 0.1 M sodium phosphate buffer containing 0.05 M NaCl (pH 7.0) at an applied potential of −0.3 V *vs.* Ag/AgCl. Adapted with permission from reference [15]. Copyright (2016) Elsevier.

21.2.2.3 Conductometry

Conductometry is based on monitoring the changes in electrical conductivity in response to an analyte contact. Conductometry has been widely used to investigate solution conductivity changes as simple electrodes encounter analytes like atmospheric CO_2 and the salt levels of tears. Nevertheless, these methods are simple and non-selective since anything that will alter solution conductivity can give a positive response. However, modified electrodes have improved the selectivity and range of analytes. Now conductometry-based electrochemical sensors have been reported to analyze NO_2, pH, NH_3 vapor, and many others. Forzani et al. developed a conductometric CP sensor for glucose detection. The group fabricated a nanosensor by depositing polyaniline (PANI) and glucose oxidase (GOx) on a pair of nanoelectrodes having a small gap (20–60 nm) in between [16]. When this system was exposed to glucose, it caused the reduction of GOx, which is then spontaneously re-oxidized in the presence of O_2, generating a molecule of hydrogen peroxide. The hydrogen peroxide produced triggers the oxidation of PANI, causing swift (<200 ms) local transformation in conductivity that could be monitored easily by observing the changes in gate current. The researchers further found that the spontaneous regeneration of GOx by dissolved O_2 molecule makes these perfect sensors candidates for *in-vivo* detection of glucose levels (Figure 21.3).

21.2.2.4 Voltammetry

The voltammetric method involves obtaining an electrical signal by electrode potential sweeping over a range related to the redox potential of the targeted analyte. The sensor's behavior depending upon the targeted redox reaction stems from the peak current shifting. Thus, the potential can be scanned using typical waveforms like cyclic, square wave,

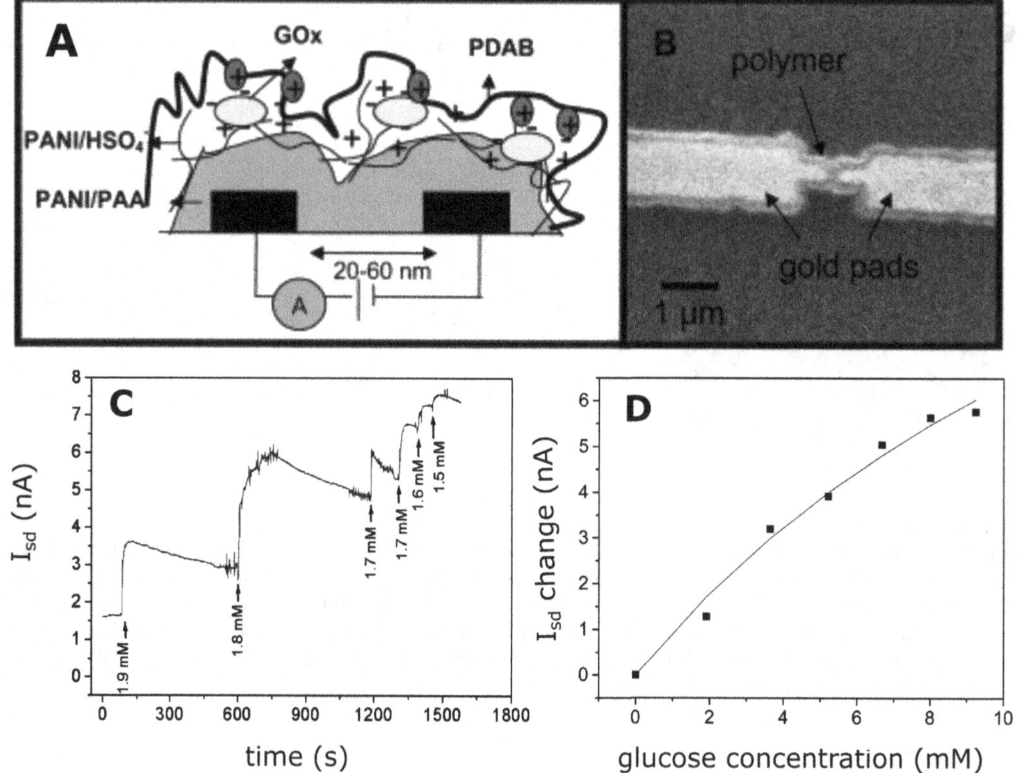

FIGURE 21.3
(a) Schematic illustration of sensor design. PANI is electrochemically deposited across the 20–60 nm gap between gold electrodes. Then, GOx is immobilized on the PANI surface. (b) Scanning electron micrograph of the sensor. (c) Drain current changes in response to successive 1 µL additions of 40 mM glucose. E_g = 35 mV *vs.* saturated calomel electrode (SCE) in 20 µL McIlvaine buffer, 0.5 M Na_2SO_4 at pH 5. (d) Corresponding calibration plot of drain current change *vs.* glucose concentration. Adapted with permission from reference [16]. Copyright (2004) American Chemical Society.

or linear sweep voltammetry [13]. This method has qualitative (in terms of the peak's location) and quantitative (majorly calculating the peak current density or area under the peak) sensing applicability.

21.3 Electrochemical Sensors Based on One-Dimensional Polymeric Nanocomposites

The basic principle of the electrochemical sensor is based on the phenomenon of recognizing the modification in charge transport properties of conducting polymers upon contacting the targeted analytes, and this change can be quantitatively interrelated to the concentration of the analytes [5]. The response mechanisms in conducting polymers could be various, including redox processes, conformational changes, swelling, charge transfer, etc. Distinctive properties of conducting polymers instead of sensor applications include [5]: i) Ease of preparation as conducting polymers can be prepared readily both by electrochemical approaches and by chemical polymerization processes; ii) Conducting polymers are exceedingly responsive to several analytes at ambient temperature; iii) Their innate transport properties, including superb electrical conductivity and high rate of energy migration, are responsible for amplified response; iv) The polymer structure is tunable so it can be easily modified or functionalized to show discerning response toward a specific analyte; v) The sensitivity can easily be modified simply by adjusting the synthetic parameters like polymerization temperature and incorporated counter ions; and vi) Sensor array fabrication method is quite simple and can take place through solution casting or electrochemical deposition method, thus facilitating the miniaturized designs and supports the mass production. These attributes make conducting polymers fit for sensor applications.

Conducting polymer composite materials of nanometer dimensions has proven their supreme transducer applications. They retain the superior properties of their heavy counterparts in addition to exceptional properties offered by small dimensions and high surface area. The large surface-to-volume ratio offered by conducting polymer nanomaterials makes them capable of exhibiting excellent sensitivity as the increased surface area facilitates the interaction between the sensor surface and analytes. The incorporated nanostructured materials in 1D polymeric nanocomposites improve sensing capability by allowing fast adsorption/desorption kinetics for analytes, causing a quick response/recovery time even at room temperature [17]. In addition, their increased signal intensity is not confined to the surface region, but they can cause signal amplification through the difference in charge transport behavior in the bulk area. 1D conducting polymer nanomaterials, composite-based electrochemical sensors have been used widely to detect various analytes like toxic gases, volatile organic compounds (VOCs), and biological species [9, 18].

21.3.1 Sensing Applications

One-dimensional polymeric composite material has gained much popularity in the recent decade owing to their increasing ability to achieve excellent catalytic activities as well as bio- and chemical sensing applications [5]. Lieber [19] is considered a pioneer in developing 1D nanostructure-based sensors that could function in an aqueous environment, which is a really crucial feature regarding the sensing application of biological or chemical species. Keeping the wide-ranging applications of one-dimensional polymeric composite materials-based sensors to various fields, it is crucial to have an essential insight into the principles upon which these devices are based, with regard to the conductance of nanostructures arising from surface charge variation specifically in metal oxide nanowire-based device configurations [20]. Nanofibers of conducting polymers are well recognized for their chemical sensing and catalytic applications. Precise electrochemical sensor device fabrication based on one-dimensional polymeric nanocomposite has matured over time and is progressing toward developing micro-electro-mechanical devices and nano-electro-mechanical devices (NEMS). The sensing domain of conducting polymeric fibers, regarding the selectivity of material and enhanced sensitivity, can be significantly improved by varying the surface electron density by integrating the functionalized nanostructures within their structures [20]. Thus, the fabrication of conducting polymeric nanofibers based on electrochemical nanosensors can substitute for the prevailing robust designs, which are sensitive down to a single molecule, and have the precise capability to recognize patterns regardless of the conditions, and requires a rational understanding of its fundamentals.

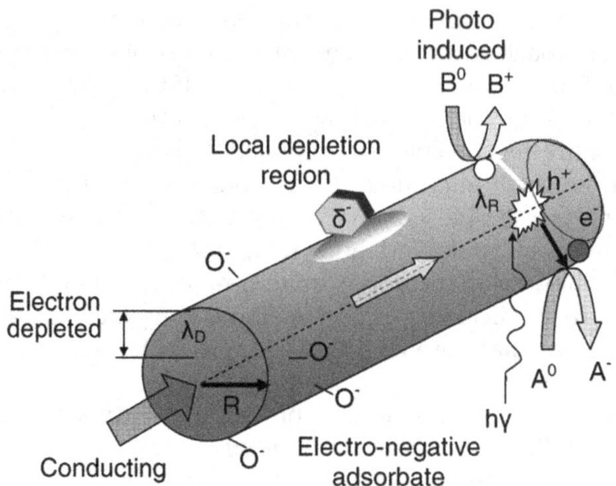

FIGURE 21.4
A summary of some of the electronic, chemical, and optical processes occurring on metal oxides that can benefit from reduction in size to the nanometer range. Adapted with permission from reference [22]. Copyright (2007) Elsevier.

There are various reasons that pique the interest of researchers to use this class for sensing applications. The foremost attribute that makes them suitable for electrochemical sensing applications is their large surface-to-volume ratio that allows more atoms from the surface to contribute to the surface reactions. The occurring surface chemistry-related sensing ability of these anisotropic nanomaterials stems from their Debye length. Debye length of 1D nanomaterials is comparable to the diameter of nanofibers [21]. The greater diffusion rate of electrons and holes to the surface accelerates the desorption process of the analyte molecule from the surface, resulting in efficient response and rapid recovery times. A greater level of crystallinity in the semiconducting nanowires could lower the volatility related to percolation or hopping conduction phenomena expected in the multigranular oxides. The surface activities in the nanomaterials are easy to control electronically simply by adjusting the Fermi level position within the bandgap of the nanowire, configured as a triply terminal field-effect transistor (FET). The decrease in nanowire dimensions causes an increase in quantum confinement effects, which are also likely to have a possible effect on the functioning of the nanowires in several applications. Figure 21.4 shows several processes that can occur at the surface [22].

21.4 Fabrication of Multi-Functionalized One-Dimensional Polymeric Nanocomposites for the Application of Electrochemical Sensors

Polymeric material-based devices are less frequent in the market due to the reproducible and reliable synthetic techniques. However, the advent of conducting-polymer nanomaterials has brought a revolution in the fabrication of miniaturized sensors. Conducting-polymer nanomaterial-based sensors are characterized by miniature sample volumes, convenience, and compactness. Furthermore, nanoparticles' incorporation within narrow-sized distribution results in more flexible sensor platforms. Nanoscience researchers have focused their attention on the development of better synthetic routes for conducting-polymeric nanomaterials based on these advantages. The conducting polymer nanomaterials' synthetic strategies can be categorized into three main groups [23]: (i) Hard-template synthesis; (ii) Soft-template synthesis; and (iii) Template-free synthesis.

Of all three methods, hard-template synthesis is deemed to be the best for controlling the dimensions. It has been extensively employed to fabricate numerous nanostructures of complex shapes, from core/shell nanoparticles to nanocapsules, nanorods, and nanotubes of desired dimensions. The conducting polymer nanomaterials have been synthesized mainly using hard templates like colloidal nanoparticles, nanofibers, and porous membranes. Palod and co-workers have recently developed a facile template-assisted electrochemical polymerization route using an alumina membrane template to synthesize polypyrrole (PPy) nanotube arrays for enzymatic glucose biosensor application [12]. They were able to control the size and diameter of the PPy nanotubes simply by regulating the polymerization time during the potentiostatic electropolymerization, thus, by physical adsorption, varying the subsequent enzyme immobilization. The resulting sensor exhibited a high sensitivity of 18.6 mA cm^{-2} M^{-1} over a wide linear range of 0.25–20 mM [12].

Vapor deposition polymerization (VDP) in conjugation with template synthesis is also a suitable method reported for the synthesis of PPy nanotubes. This method effectively controlled the thickness of the nanotube walls within the range of a few nanometers. Another study reported the fabrication of PPy-coated carbon nanofibers (CNFs) by a single-step VDP. A smooth, ultrathin layer of PPy was deposited successfully on the CNF surface [24, 25]. The thickness of the PPy layer was regulated by varying the pyrrole monomer feeding amount. The CNF/conducting polymer hybrid-based biosensors exhibit superior sensing properties owing to the synergistic activity of PPy and CNFs. Though hard-template synthesis has various pluses, some have some shortcomings associated with them, including complex synthetic procedures, limited production capacity, and relatively high cost. The soft-template methodology can effectively fabricate polymeric nanomaterials with numerous advantages. Soft templates comprise surfactants, liquid crystals, block copolymers, polyelectrolytes, and biomolecules. In a report by Yoon et al. [26] cylindrical micelle formation *via* co-operative interaction of bis(2-ethylhexyl) sulfosuccinate (AOT) molecules with iron cations in an apolar solvent took place. In this case, the iron cations were adsorbed onto the AOT cylindrical micelles' anionic headgroups, which facilitated the chemical polymerization of targeted monomers *via* oxidation of pyrrole and PEDOT (Figure 21.5).

Consequently, introducing pyrrole or PEDOT monomer into the cylindrical micelle phase results in the formation of polymer nanomaterials at the cylindrical AOT/water interface. Adjusting the input amount of oxidizing agent iron cations in the cylindrical micelle phase can result in at least three different-shaped 1D PEDOT nanomaterials. This micelle templating approach effectively regulates the 1D nanostructures' morphology by adjusting synthesis conditions. Moreover, this approach was found to be equally effective at a grander scale. The template-free synthesis approach is an upfront methodology that does not involve any specific sacrificial templates. Nevertheless, this approach is restricted to certain precursor materials. There has been a lot of research for the template-free synthesis of PANI nanofibers in recent years.

21.4.1 Transducer Incorporation to Sensing Systems

To obtain specific and sensitive responses, conducting-polymer nanomaterials are incorporated within a variety of microanalytical systems, which are united with numerous other chemicals/biological species, and they are then followed by incorporation of transducers

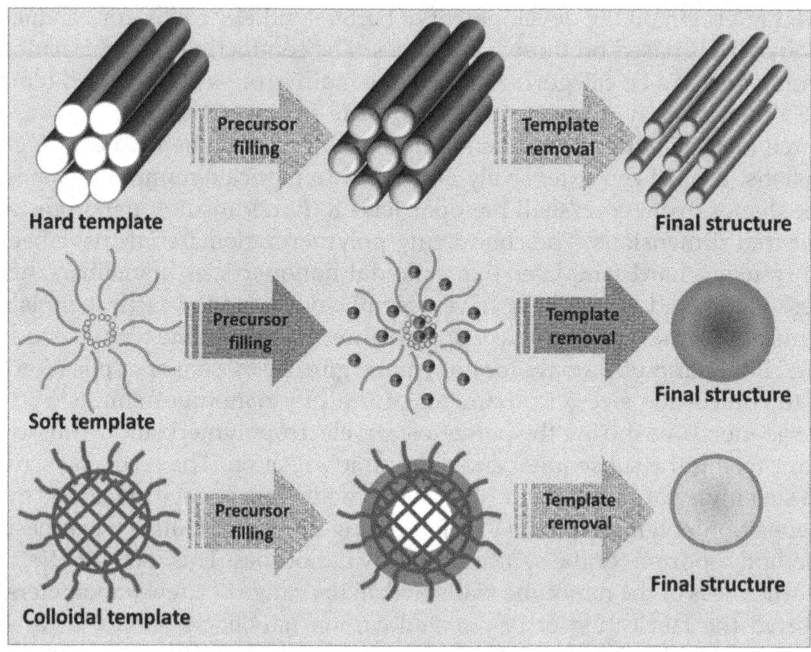

FIGURE 21.5
Schematic representation of the synthesis of materials using different types of templates. Adapted with permission from reference [26]. Copyright (2020) Royal Society of Chemistry.

into sensing systems. Suitable surface functionalization is the key to significant enhancement in the properties of the nanomaterials related to their sensor applications. Both covalent and non-covalent approaches have been widely utilized to functionalize conducting-polymer nanomaterials. Covalent approaches are associated with the advantage that conducting-polymer nanomaterials are easy to modify by simply grafting the functional groups onto the polymer backbone. They can also be modified by using intrinsically functionalized monomers during the process of polymerization [27]. At the same time, the non-covalent approaches involve the integration of suitable counterions in the polymer matrix during the synthesis process and the binding of guest molecules at the nanomaterial surface through electrostatic interaction. Compared with non-covalent approaches, covalent approaches offer additional stability and a wide range of functionalities, suitable for advanced handling and modification.

Furthermore, the control over the surface functionality from covalent approaches is both qualitative and quantitative. Several functional groups, including amino acid groups, carboxyl groups, and alkyl groups, can be incorporated successfully onto the PPy backbone for specific functionalities and defined purposes like hydrophobicity/hydrophilicity, lubrication, adhesion, and biocompatibility. For instance, the surface carboxyl groups of the nanotubes can be paired with a variety of functional materials, including organic dyes, DNA, protein, and even nanoparticles, depending upon the desired application.

21.4.2 Techniques to Enhance Adhesion of Transducers on Substrates

The most challenging step during the design and construction of a conducting-polymer nanomaterial-based electrochemical sensor is the step that involves immobilization of nanomaterial onto the surface of electrode substrate to maximize the contact between the electrodes and the nanomaterials. Carbon nanotubes and other related, inorganic nanomaterials are typically incorporated within the electrode substrate by conventional means of photolithography or e-beam lithography [3]. Nevertheless, these methods cannot work for conducting-polymer nanomaterials because chemical, thermal, and kinetic damage are possible. Several methods that involve the effective patterning of conducting polymers into the nanometer regime have been developed in recent years, of which soft lithography, electrochemical deposition, and ink-jet printing are worth mentioning. However, these techniques are limited to large-area fabrication and have restricted resolution. Additionally, insufficient adhesion to the electrode substrate further restricts their efficacy. Thus, finding efficient, scalable, and site-specific methods for incorporating nanomaterials into sensing systems is still challenging.

21.5 Role of High Aspect Ratio in Improving the Sensing Ability of 1D Nanostructures

Sensing properties of the sensing material are affected mainly by its microstructure, involving *S/V* ratio, grain size, and pore volume. The extremely high *S/V* ratio of one-dimensional polymeric nanocomposites plays a vital role in improving their electrical properties to a great extent and making them more responsive to the moieties which have been selectively adsorbed onto the electrode surface. In Figure 21.6, there is a simple representation of the estimation of the *S/V* ratio for selected nanostructures as a dimension function. It is evident from the figure that if any one of the dimensions is lower than 25 nm, then strong surface influence dominates. From the sensor perspective, *S/V* ratio strongly influences the electron-transport properties of the nanostructures [28]. When the dimensions of a material are less than that of the mean free path of electrons, then the

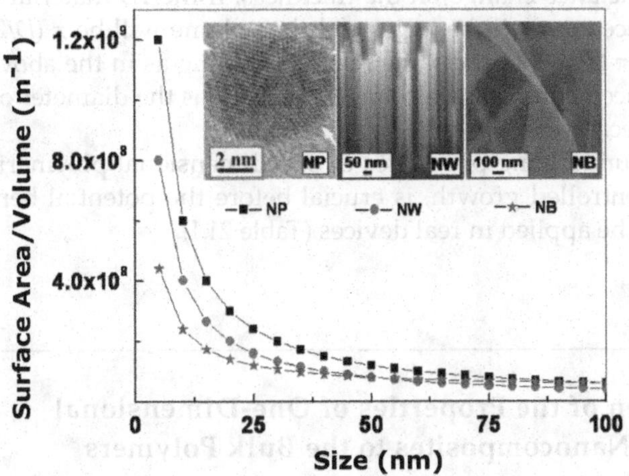

FIGURE 21.6
Surface-area-to-volume ratio *versus* size calculated for NPs, NWs, and NBs. A length *L* of 10 μm is used for estimation in the case of NWs and NBs ($D = 50$ nm). The inset shows the image of Ag NPs, ZnO NWs, and ZnO NBs. Adapted with permission from reference [28]. Copyright (2010) John Wiley and Sons.

TABLE 21.1
Sensors Based on 1D Conducting Polymer Materials

Sensor Material	Nanowire Case	Principle	Target Analyte	Lowest Concentration	Diameter of Nanowires (nm)	Reference
PANI	Multiple	Conduct	NH_3	100 ppm	—	[31]
PANI + HSCA	Multiple	Conduct	$CHCl_3$	—	20–150	[32]
Polypyrrole	Single	Conduct	NH_3	38 ppm	100	[33]
Polypyrrole	Multiple	Conduct	NH_3	1000 ppm	80–150	[34]
PEDOT: PSS	Multiple	Conduct	C_2H_5OH	325 ppm	15	[35]

chemisorption from the analyte (gas or liquid phase) can cause considerable change in the resistance of the material because the transport of electrons becomes ballistic and gives rise to the electronic coupling of adsorbates which affected the resistance [7].

Semi-classical representations have been established to rationally evaluate the change conductance as a function of adsorbate–substrate bonding length. By extending these semi-classical models, Persson and Baratoff, in 1987, developed a fundamental system to establish the relationship between the conductance change due to absorbate and the density of states of the adsorbed molecules $\rho_a(\varepsilon_F)$ at the Fermi energy of the conductor [29], so that the conductance change per adsorbate molecule will be:

$$\frac{\Delta G}{G} \propto -\frac{\Gamma \rho_a(\varepsilon_F)}{D}$$

where G represents the conductance and D represents the thickness of the material, Γ is the width of the density of states $\rho_a(\varepsilon_F)$. The above equation shows that there is an inverse relationship between the adsorbate-induced conductance change and the thickness of the material. Thus, the narrower the material is, the greater will be its sensitivity, which is quite beneficial for sensor applications. For nanofibers and related conducting material, a geometric surface-to-volume ratio argument can further simplify this dependence of the conductance change on the thickness. If the 1D material thickness is D and length L, the surface area will be πDL, and the volume will be $\pi (D/2)^2L$, which confers on $\Delta G/G \sim 4\pi DL/\pi D^2L \sim /D$, the same relation holding as in the above equation. Hence, the adsorbate-induced conductance change increases as the diameter of the 1D polymeric nanocomposites decreases [30].

Insight into the underlying properties of one-dimensional polymeric nanocomposites, as well as their controlled growth, is crucial before the potential benefits of associated nanomaterials can be applied in real devices (Table 21.1).

21.6 Comparison of the Properties of One-Dimensional Polymeric Nanocomposites to the Bulk Polymers

Conducting polymers are characterized by the π-bonds they have on their backbone and are a very important class of conjugated polymers. These π-bonds facilitate the migration of electrons all through the polymeric chain. This structure, with the presence of delocalized π electrons on the backbone of these polymers, bestows them with exceptional electrical properties and makes them remarkable candidates for electrical applications. Noteworthy conducting polymers with a wide range of applications are PANI, polyacetylene (PA), polythiophenes (PTh), PPy, polyparaphenylene (PPPh), polyparaphenylene vinylene (PPV), polyorthotoluidine (POT), etc., and their respective derivatives [17]. Conducting polymers are preferred because of their characteristic features like light weight, high electrical conductivity, corrosion resistance, photochromic properties, and minimal costs.

Additionally, they can be tailored appropriately in terms of structure and functional groups with minor modifications. However, conducting polymers are often associated with meager stability, difficult processability, and low solubility issues. Moreover, typical CPs with amorphous bulk phase have generally poor electrical, optical, and biological properties. These shortcomings can be addressed by confining their dimensions. Compared to bulk CPs, nanostructured CPs (NCPs) offer enhanced conductance, better carrier mobility, unique electrochemical activity, superior optical properties, and greater biocompatibility, etc. [5, 6] because of their more significant surface areas and well-defined nanostructures. NCPs can be assembled into varied nanostructure arrangements. These hybrid NCPs thus exhibit superior properties because they not only retain the functionalities from distinct components but are also capable of exhibiting synergistic effects that arise when they are combined with materials with other functionalities like metals and their oxides, carbonaceous materials, other non-conducting polymers, etc. Structure and morphology are of significant importance in deciding the conductivity and physicochemical efficiency of NCPs; therefore, immense efforts have been dedicated to turning their physical parameters like surface area, pore-volume, phase, lattice fringes, and patterns to make them suitable for various applications.

1D nanostructures are of prime interest regarding sensor applications because of the unique combinations they can offer in terms of anisotropic electrical properties. Unlike nanospheres, 1D NCPs and their composites offer a higher aspect ratio, which is beneficial in the efficient transportation of electrical carriers along their length in a controllable manner. Recent research focuses on the synthesis of rod-/wire-/fiber-like NCPs for their larger surface area that can provide additional electrochemically active spots and enhance electrolyte–electrode interface for ion storage, thus boosting their specific capacity and sensing capability [15].

CPs short nanorods can potentially offer superior sensing properties by regulating the morphological structure. Furthermore, their peripheral contour can be manipulated exactly by using the correct template. Hu and co-workers synthesized hexagonal spiral prismatic PPy nanorods by chemical oxidation in the hexagonal spiral prismatic micelles of IC-Fe(II) chelate. The resulting PPy nanorods exhibited conductivity of 1.6 S cm^{-1}, which is three times greater than that of PPy film [36]. Usually, pristine nanorod structures fail to develop favorable π–π stacks and therefore results in weaker chain-to-chain interactions, resulting in reduced electrical and thermoelectric properties. However, improved electrical or thermoelectric performance can be obtained from nanorods with an overall structure and thus facile inter-chain interactions, being beneficial for improved properties.

The size effect is also a key player in controlling the electrical properties of NCPs, and it is often seen in terms of dependence on the diameter. Thus, NCPs with wire or fiber morphologies are routinely developed and used in several electronic devices. Typically, nanowires can be a diameter equal to or less than 100 nm with no restriction in length [36]. Along with numerous other fascinating properties, nanowires have two prominent advantages: their superb material quality, and their ability to offer design freedom owing to their morphology. Though these properties attract the interest in CP nanowires of broad scientific communities, they are also associated with drawbacks of para-crystallinity and amorphic structure.

Moreover, they often have high defects or grain boundaries, which could act as charge carriers' trapping sites, causing decreased conductivity. However, the 1D structure of CP nanowires can support efficient doping that reinforces inter- and intra-chain interactions, improving the degree of crystallinity, and resulting in increased conductivity. This higher crystallinity order also means fewer defects or grain boundaries, leading to a lessening of the chances of charge trapping. Cho and co-workers developed single-crystal PEDOT nanowires by vapor-phase polymerization(VPP) of 3,4-ethylenedioxythiophene (EDOT) [37]. The resulting single-crystal PEDOT nanowires showed excellent conductivity up to 8.797 S cm^{-1}, which tends to be really useful in organic nanowire interconnects for cost-effective, flexible, transparent, and large-area electronic devices such as displays. The high crystallinity and a small π–π stacking distance of single-crystal PEDOT nanowires were the main factors that resulted in this remarkable improvement in the properties and ultra-high conductivity. In addition, such single-crystal PEDOT nanowires offer a great potential to be fabricated into flexible devices on a flexible substrate.

Consequently, the performance of one-dimensional polymeric nanocomposites depends on the degree of crystallinity and nanoscale of CPs. Lately, some methodologies have been presented to tailor the crystal morphological structure of NCPs. Pan et al. [38] established a practical scheme to modify the dimensions of P3HT crystals and offered an efficient methodology by which to control the polymer morphology. They tailored P3HT nanofibers into nanoribbons connected with the crystallographic transition from edge-on orientation to flat-on orientation by successively adding the alkyl thiols into P3HT nanofibers solution (Figure 21.7).

Pure CP nanofibers synthesized by traditional methods are associated with drawbacks like smaller aspect ratios, bumpy surfaces, or cluster formation. So, efforts have been devoted to improving their properties by certain modifications. Presently, there are many ways available to adjust pure NCP materials. One helpful method to enhance the electrical activity of pure CP nanofibers is by incorporating conductive filler into the polymer matrix and incorporating nanoparticles into nanowires resulting in gaps between the segments so that the surface area of the nanowires increases; consequently, they have a greater potential for electrical activity compared with continuous nanowires. General metals, metal oxides, or carbonaceous materials are employed as conductive fillers for CP nanofibers. Carbon-based materials are low cost but do not support intricate technical development for widespread application. Therefore, metals nanoparticles are preferred to transform the CP nanowires' surfaces because of their good electrical conductivity. Jang et al. conducted a study of Ag nanoparticles attached to PPy nanowires, and they found that attached Ag nanoparticles boosted the photocurrent and spectral response. The fabricated devices exhibited an unusually superior photocurrent density, up to 25.3 times under blue light illumination, and wide spectral response [15]. The authors noted that this tremendous activity was due to the mediation offered by the filler material to offset the surface defects of the polymer material, while better filler–polymer interaction enabled the material to develop new properties.

Thus, the sensitivity of one-dimensional polymeric nanocomposite-based sensors is greater than that of bulk material-based counterparts because of the high surface area they offer, as well as the briefer response time of the nanostructure-based electrochemical sensors compared with non-nanostructured sensors since the permeability offered by nanostructures allows the analyte molecules to diffuse in and out of the nanostructures channels swiftly.

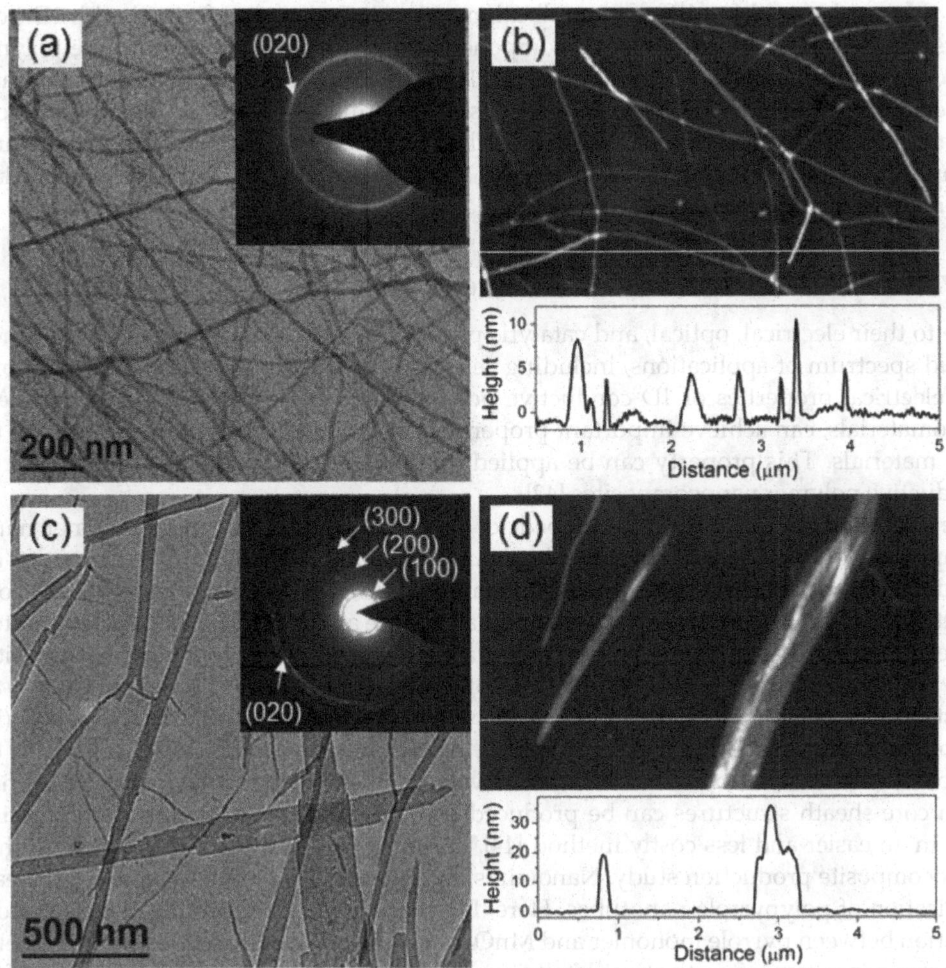

FIGURE 21.7
(a) TEM image of P3HT nanofibers drop-cast from $CH_2Cl_2/CHCl_3$ solution. (b) AFM image along with the line profile of P3HT nanofibers. (c) TEM image of P3HT nanoribbons drop-cast from $CH_2Cl_2/CHCl_3$ solution with the addition of 12-thiol at 30% by volume. (d) AFM image along with the line profile of P3HT nanoribbons. Adapted with permission from reference [29]. Copyright (2018) John Wiley and Sons.

21.7 Preparations of One-Dimensional Polymeric Nanocomposites

21.7.1 Use of Nanomaterials and Biomaterials

For the design of 1D conductive polymer nanocomposites, first of all, the production method and properties of the 1D conductive polymers should be known. In addition to the synthesis methods described in the previous sections, this section will explain how to integrate nanomaterials into polymers to be used in the production of 1D conductive polymer nanocomposites. Polymer nanocomposites are the composites of nano-sized inorganic particles dispersed into a polymer matrix [39].

These structures express the superlative properties of either component within a single structure [40]. These comprise advanced electronic, ionic, mechanical, chemical properties, etc., that can be used for different applications. In nanocomposites formed with 1D polymers, the aspect ratio of the second component and its ability to adhere to the polymer affect the properties of the nanocomposites. 1D nanostructures have also been integrated as sensing or transducing elements of a sensor, resulting in greater or sensitivity due to their high surface-area-to-volume ratio [39]. The synthesis of 1D conductive polymer nanocomposites obtained using materials such as metal nanomaterials, carbon nanomaterials, or biological materials is explained below.

21.7.2 Use of Metal-Metal Oxide Nanomaterials

Due to their electrical, optical, and catalytic properties, metal nanoparticles have found a broad spectrum of applications, including sensors [41]. Nanocomposites, which combine the electrical properties of 1D conductive polymers and the optical properties of metal nanomaterials, can achieve important properties thanks to electron transfer between the two materials. This property can be applied for the fabrication of metal-component 1D conductive polymer nanocomposites [42].

Several methods have been developed for the fabrication of 1D metal-polymer nanocomposites. 1D PPy/Au nanocomposites, Au/PANI and Pd/PANI nanocomposite are produced by methods such as template techniques, *in-situ* redox reactions. PANI/Pt composite nanofibers and many different combinations are also produced [43]. Although stencil techniques are a simple method for manufacturing 1D metal-polymer nanocomposites, they are more complex and costly because of the need to remove the stencil. One of the most important properties of conductive polymers is reversible oxidation/reduction (i.e., redox) chemistry. $HAuCl_4$, H_2PdCl_4, H_2PtCl_6, $AgNO_3$, etc., are noble metal salts with the property of polymerizing monomers oxidatively. In this way, metal/conductive polymers with core-sheath structures can be produced using a single-stage chemical polymerization in an easier and less costly method [44]. Zhang et al. used MnO_2 nanowires in their nanocomposite production study. Nanowires used as reactive templates enabled the easy fabrication of polypyrrole nanotubes. Here, PPy nanotubes were produced by the redox reaction between pyrrole monomer and MnO_2 nanowires. For the production of 1D metal-polymer nanocomposites, the template-free method is an easy and versatile technique [45].

21.7.3 Use of Carbon Nanomaterials

CNTs, which are very popular among carbon nanomaterials, are nanomaterials with high electrical conductivity because they consist of a graphene layer with a hollow tubular structure. CNTs attract attention for many applications due to their structural, thermal, mechanical, and electrical properties. On the other hand, CNTs have solubility problems, and therefore they need to be functionalized in some cases [46].

Researchers have developed a chemical modification of CNTs with organic molecules or polymers to increase the solubility of CNTs. CNTs can be functionalized with organic molecules or polymers through techniques such as esterification reaction, amidation reaction, atom transfer radical polymerization (ATRP), fragmentation chain transfer polymerization (RAFT), ring-opening metathesis polymerization (ROMP), and radical addition [47].

CNTs functionalized with 1D CNT/polymer enable the fabrication of polymeric nanocomposites in the simplest way. In these nanocomposites, it was observed that the electron transfer rate increased due to the interaction between the polymer and CNTs. In addition, studies in which 1D CNTs/conductive polymer nanocomposites are produced by a directed coating of the charged groups in surfactants on CNTs are also available in the literature [45].

21.7.4 Use of Biological Materials

Biological materials, such as nucleic acids, enzymes, and proteins, are used for the production of 1D polymer nanocomposites. The new physical and chemical properties of these composite materials have made them especially popular in the field of biosensors [48]. Nanocomposite applications using nucleic acids, such as DNA/PANI and RNA/PANI, are described in the literature [42].

Biomaterials can be immobilized to conductive polymers by covalent or physical methods. 1D biomaterial/polymer nanocomposite surfaces used in sensor applications help to create a unique sensor platform due to both catalytic activity and high selectivity. In the study of Ghanbari et al. [49], PPy nanofiber film was coated on the Pt electrode by the electropolymerization method. Then, physical immobilization of dsDNA on these films was performed. A biosensor with high electroactivity was developed with the prepared nanocomposite. DNA-spermidine interaction was investigated with a low-cost, fast-sensing biosensor with a low detection limit [49].

21.8 Conclusion

Nanotechnology is an ever-evolving field, thanks to its broad field of research, emerging technologies, and the work of scientists. Nanotechnology combined with materials science has enabled the use of new materials in various fields. 1D polymeric nanocomposites are one of these areas. It is predicted that the physical and chemical properties and developments in production methods will continue to increase the interest in these materials. The structures of 1D polymeric nanocomposites that can be integrated into different applications make them stand out compared with the other three dimensions due to their good electrical conductivity. It is expected that biocompatible, low-cost, easy-to-apply polymeric nanocomposites will be brought to science thanks to the synthesis techniques developed as well as research in this direction.

References

1. Privett BJ, Shin JH, Schoenfisch MH (2010) Electrochemical sensors. *Anal Chem* 82(12):4723–4741.
2. Bollella P, Fusco G, Tortolini C, Sanzò G, Favero G, Gorton L, Antiochia R (2017) Beyond graphene: Electrochemical sensors and biosensors for biomarkers detection. *Biosens Bioelectron* 89(1):152–166.
3. Ahammad AJS, Lee J-J, Rahman MA (2009) Electrochemical sensors based on carbon nanotubes. *Sensors (Basel)* 9(4):2289–2319.
4. Kempahanumakkagari S, Deep A, Kim KH, Kumar Kailasa S, Yoon HO (2017) Nanomaterial-based electrochemical sensors for arsenic - A review. *Biosens Bioelectron* 95:106–116.
5. Kim FS, Ren G, Jenekhe SA (2011) One-dimensional nanostructures of π-conjugated molecular systems: Assembly, properties, and applications from photovoltaics, sensors, and nanophotonics to nanoelectronics. *Chem Mater* 23(3):682–732.
6. Le T-H, Kim Y, Yoon H (2017) Electrical and electrochemical properties of conducting polymers. *Polymers (Basel)* 9(4):150.
7. Lu X, Zhang W, Wang C, Wen T-C, Wei Y (2011) One-dimensional conducting polymer nanocomposites: Synthesis, properties and applications. *Prog Polym Sci* 36(5):671–712.

8. Joulazadeh M, Navarchian AH (2015) Ammonia detection of one-dimensional nano-structured polypyrrole/metal oxide nanocomposites sensors. *Synth Met* 210:404–411.
9. Soares BG, Barra GMO, Indrusiak T (2021) Conducting polymeric composites based on intrinsically conducting polymers as electromagnetic interference shielding/microwave absorbing materials—A review. *J Compos Sci* 5(7):173.
10. Kimmel DW, LeBlanc G, Meschievitz ME, Cliffel DE (2012) Electrochemical sensors and biosensors. *Anal Chem* 84(2):685–707.
11. Stradiotto NR, Yamanaka H, Zanoni MVB (2003) Review electrochemical sensors : A powerful tool in analytical chemistry. *J Braz Chem Aoc* 14:159–173.
12. Palod PA, Pandey SS, Hayase S, Singh V (2014) Template-assisted electrochemical growth of polypyrrole nanotubes for development of high sensitivity glucose biosensor. *Appl Biochem Biotechnol* 174(3):1059–1072.
13. Ayranci R, Demirkol DO, Timur S, Ak M (2017) Rhodamine-based conjugated polymers: Potentiometric, colorimetric and voltammetric sensing of mercury ions in aqueous medium. *Analyst* 142(18):3407–3415.
14. Bilgi M, Ayranci E (2018) Development of amperometric biosensors using screen-printed carbon electrodes modified with conducting polymer and nanomaterials for the analysis of ethanol, methanol and their mixtures. *J Electroanal Chem* 823:588–592.
15. Braik M, Barsan MM, Dridi C, Ben Ali M, Brett CMA (2016) Highly sensitive amperometric enzyme biosensor for detection of superoxide based on conducting polymer/CNT modified electrodes and superoxide dismutase. *Sens Actuators B Chem* 236:574–582.
16. Forzani ES, Zhang H, Nagahara LA, Amlani I, Tsui R, Tao N (2004) A conducting polymer nanojunction sensor for glucose detection. *Nano Lett* 4(9):1785–1788.
17. Idumah CI (2021) Novel trends in conductive polymeric nanocomposites, and bionanocomposites. *Synth Met* 273:116674.
18. Jiang H, Yao X, Che J, Wang M, Kong F (2004) Preparation of ZnSe quantum dots embedded in SiO_2 thin films by sol-gel process. *Ceramics International* 30(2004:1685–1689.
19. Lieber CM (1998) One-dimensional nanostructures: Chemistry, physics & applications. *Solid State Commun* 107(11):607–616.
20. Liao Y, Duan F, Zhang H, Lu Y, Zeng Z, Liu M, Xu H, Gao C, Zhou L, Jin H, Zhang Z, Su Z (2019) Ultrafast response of spray-on nanocomposite piezoresistive sensors to broadband ultrasound. *Carbon N Y* 143:743–751.
21. Trajcheva A, Politakos N, Pérez BT, Joseph Y, Blazevska Gilev J, Tomovska R (2021) QCM nanocomposite gas sensors – Expanding the application of waterborne polymer composites based on graphene nanoribbon. *Polymer (Guildf)* 213:123335.
22. Goud KY, Satyanarayana M, Hayat A, Gobi KV, Marty JL (2019) Chapter 7 - Nanomaterial-based electrochemical sensors in pharmaceutical applications. In AM Grumezescu (ed.), *Nanoparticles in Pharmacotherapy*, pp. 195–216. Elsevier, Amsterdam.
23. Yang Z, Wang J, Hu Y, Deng C, Zhu K, Qiu J (2020) Simultaneously improved dielectric constant and breakdown strength of PVDF/Nd-$BaTiO_3$ fiber composite films via the surface modification and subtle filler content modulation. *Compos Part A Appl Sci Manuf* 128:105675.
24. Poolakkandy RR, Menamparambath MM (2020) Soft-template-assisted synthesis: A promising approach for the fabrication of transition metal oxides. *Nanoscale Adv* 2(11):5015–5045.
25. Xu Y, Wang Z, Guo Z, Huang H, Xiao Q, Zhang H, Yu XF (2016) Solvothermal synthesis and ultrafast photonics of black phosphorus quantum dots. *Adv Opt Mater* 4(8):1223–1229.
26. Yoon H, Chang M, Jang J (2007) Formation of 1D poly(3,4-ethylenedioxythiophene) nanomaterials in reverse microemulsions and their application to chemical sensors. *Adv Funct Mater* 17(3):431–436.
27. Si B, Song E (2018) Molecularly imprinted polymers for the selective detection of multi-analyte neurotransmitters. *Microelectron Eng* 187–188:58–65.
28. Ramgir NS, Yang Y, Zacharias M (2010) Nanowire-based sensors. *Small* 6(16):1705–1722.
29. Persson BNJ, Baratoff A (1987) Inelastic electron tunneling from a metal tip: The contribution from resonant processes. *Phys Rev Lett* 59(3):339–342.

30. Tikhodeev S, Natario M, Makoshi K, Mii T, Ueba H (2001) Contribution to a theory of vibrational scanning tunneling spectroscopy of adsorbates: Nonequilibrium Green's function approach. *Surf Sci* 493(1–3):63–70.
31. Wu Z, Chen X, Zhu S, Zhou Z, Yao Y, Quan W, Liu B (2013) Enhanced sensitivity of ammonia sensor using graphene/polyaniline nanocomposite. *Sens Actuators B Chem* 178:485–493.
32. Pinto NJ (2013) Electrospun conducting polymer nanofibers as the active material in sensors and diodes. *J Phys Conf Ser* 421:12004.
33. Hernandez SC, Chaudhuri D, Chen W, Myung NV, Mulchandani A (2007) Single polypyrrole nanowire ammonia gas sensor. *Electroanalysis* 19(19–20):2125–2130.
34. Lähdesmäki I, Lewenstam A, Ivaska A (1996) A polypyrrole-based amperometric ammonia sensor. *Talanta* 43(1):125–134.
35. Dos Reis MAL, Thomazi F, Del Nero J, Roman LS (2010) Development of a chemiresistor sensor based on polymers-dye blend for detection of ethanol vapor. *Sensors (Basel)* 10(4):2812–2820.
36. Chang M, Su Z, Egap E (2016) Alignment and charge transport of one-dimensional conjugated polymer nanowires in insulating polymer blends. *Macromolecules* 49(24):9449–9456.
37. Cho B, Park KS, Baek J, Oh HS, Lee YK, Sung MM (2014) Single-crystal poly(3,4-ethylenedioxythiophene) nanowires with ultrahigh conductivity. *Nano Lett* 14(6):3321–3327.
38. Pan S, Zhu M, He L, Zhang H, Qiu F, Lin Z, Peng J (2018) Transformation from nanofibers to nanoribbons in poly (3-hexylthiophene) solution by adding alkylthiols. *Macromol Rapid Commun* 39:1800048.
39. Shameem MM, Sasikanth SM, Annamalai R, Raman RG (2021) Materials today: Proceedings a brief review on polymer nanocomposites and its applications. *Mater Today Proc* 45:2536–2539.
40. Shi H, Chen J, Li G, Nie X, Zhao H, Wong P-K, An T (2013) Synthesis and characterization of novel plasmonic Ag/AgX-CNTs (X = Cl, Br, I) nanocomposite photocatalysts and synergetic degradation of organic pollutant under visible light. *ACS Appl Mater Interfaces* 5(15):6959–6967.
41. Xia Y, Xiong Y, Lim B, Skrabalak SE (2009) Shape-controlled synthesis of metal nanocrystals: Simple chemistry meets complex physics? *Angew Chem Int Ed Engl* 48(1):60–103.
42. Lu X, Zhang W, Wang C, Wen T, Wei Y (2011) Progress in polymer science one-dimensional conducting polymer nanocomposites: Synthesis, properties and applications. *Prog Polym Sci* 36(5):671–712.
43. Gallon BJ, Kojima RW, Kaner RB, Diaconescu PL (2007) Palladium nanoparticles supported on polyaniline nanofibers as a semi-heterogeneous catalyst in water. *Angew Chem Int Ed Engl* 46(38):7251–7254.
44. Liu Q, Lu X, Li JJ, Yao X, Li JJ (2007) Direct electrochemistry of glucose oxidase and electrochemical biosensing of glucose on quantum dots/carbon nanotubes electrodes. *Biosens Bioelectron* 22(12):3203–3209.
45. Zhang Z, Wan M, Wei Y (2005) Electromagnetic functionalized polyaniline nanostructures. *Nanotechnology* 16(12):2827–2832.
46. Kaur Billing B (2021) Carbon nanotubes and its potential application in sensing. *ChemistrySelect* 6(36):9571–9590.
47. Loewy M, Kirs V, Carvajal G, Venturino A, Pechen De D'Angelo AM (1999) Groundwater contamination by azinphos methyl in the Northern Patagonic Region (Argentina). *Sci Total Environ* 225(3):211–218.
48. Malhotra BD, Chaubey A, Singh SP (2006) Prospects of conducting polymers in biosensors. *Anal Chim Acta* 578(1):59–74.
49. Ghanbari K, Bathaie SZ, Mousavi MF (2008) Electrochemically fabricated polypyrrole nanofiber-modified electrode as a new electrochemical DNA biosensor. *Biosens Bioelectron* 23(12):1825–1831.

22

One-Dimensional Polymeric Nanocomposites for Biomedical Implants

Aditya Dev Rajora, Shreya Sharma, Trishna Bal, Fahad Uddin, and Itishree Jogamaya Das

CONTENTS

DOI: 10.1201/9781003223764-22

22.1 Introduction

When more than one material having a size smaller than 100 nm is evenly dispersed within a polymeric matrix, the result is a polymeric nanocomposite. The addition of these nanofillers within the matrix increases their mechanical strength, thermal properties, barrier properties, and resistance against fire but does not affect their processability [1, 2]. Based on the dimension of these incorporated nanofillers, polymeric nanocomposites are of four different dimensions: zero-, one-, two-, and three-dimensional. Of these, one-dimensional nanostructures have gained importance owing to their very small size by which they can percolate electrical carriers for an integrated nanosystem [3]. The term nanocomposites were first proposed by Theng in 1980, and later the preparation of polymeric nanocomposites started at various research and commercial organizations. The existence of the nanomaterial in the polymeric matrix changes the total structural integrity of the polymer, thereby modifying their individual properties which are far more advantageous than the general doped polymer [4]. In the case of nanosized particles, the aspect ratio of the secondary forming constituents is around 100 and the shape of the material represents the aspect ratio. Polymeric nanocomposites have diverse applications in the fields of biomedicine, textiles, construction, aerospace, automobile parts, etc. The one-dimensional nanofillers used in the polymeric composites are needle-shaped, and nanoplatelet, nanosheet, nanodisk, nanowall, nanoclay, etc. [5]. Also, the use of montmorillonite clay, nanographene platelets, zinc oxide nanosheets, Fe_3O_4 nanosheets, etc. have been explored with excellent properties for applications such as biosensors, micro- and nanoelectronic devices, and for chemical engineering aspects [4]. Nanofillers show distinct properties when compared with their bulk form, and thus it has been studied that the reinforcement capability of the cellulose nanocomposites is much better than the simple composite system. The main rationale behind the synthesis of the polymeric nanocomposites is to overcome the limitations of the individual constituents when they are used individually [5]. But the nanomaterials suffer from agglomeration because of high surface energy within the solvents. Once these nanomaterial fillers are entrapped within the polymeric matrix, however, this problem is alleviated. Polymers used for the process can be either hydrophilic or hydrophobic. The interactions of the functional group present on the polymeric backbone with the fillers can be made by grafting the nanofillers onto the polymeric matrix or by other ionic interactions. The grafting of the nanofillers with functional groups onto the polymeric matrix is a cheaper option as it involves various schemes for the synthesis of polymeric nanocomposites, which are highly biodegradable, eco-friendly, and cost effective [6]. But polymeric nanocomposites are less stable at lower pH due to the tendency toward agglomeration of the nanofillers. Despite this limitation, polymeric nanocomposites are more appreciated as a lower concentration of nanofillers is needed than in the usual composite system. Bioactive nanocomposite materials are efficient in tissue proliferation owing to their nanostructure integrity which gives cells greater ambiance and a better environment for growth [7]. By virtue of the dispersion of the nanosized fillers within the polymers, these polymeric nanocomposites are highly biocompatible, exhibit bactericidal activities, have increased modulus and strength, and possess novel optical properties [8]. The current chapter mainly deals with the fabrication strategy of one-dimensional polymeric nanocomposites and their applications in tissue regeneration.

22.2 Types of Nanocomposites for Biomedical Applications

22.2.1 Ceramic Nanocomposites

Ceramic nanocomposites (CNCs) are a unique class of materials that are used in the biomedical field due to their novel characteristics including biocompatibility, improved mechanical properties, chemical inertness, and high wear resistance. In the late eighteenth century, ceramic was introduced for biomedical applications. Porcelain was used in dentistry for the crowning of damaged teeth. CNCs are mainly used for dental-related and bone-related biomedical applications. Alumina (Al_2O_3) and zirconia (ZrO_2) are the ceramics that are used for bone tissue engineering, whereas, for dental implants, zirconia (ZrO_2) is preferred because of the aesthetic requirements. Alumina is used with other ceramics like silicon carbide, zirconia, and titania for different biomedical applications [9]. Hydroxyapatite (HA) is also a ceramic, having chemical and structural composition similar to human bones. Because of these similarities, HA found application in orthopedic implants but, in the case of other implants, the use of HA is limited because of its brittleness and inferior mechanical properties. This limitation can be avoided by reinforcing with various types of nanomaterials, namely titanium dioxide and alumina, which improve the mechanical strength of the nanocomposites. HA/graphene oxide (GO) nanocomposite was prepared using hydrothermal technique and the biocompatibility and cytotoxicity studies were performed using BSA interaction and the assay. It was found that nanocomposites do not show any cytotoxic effect so we can use HA/GO nanocomposites for various biomedical applications [10]. Karthik et al. (2018) developed HA/PVA nanocomposites for prosthetic implants for bone tissue engineering. It was found that the wear strength and the compressive strength of the nanohydroxyapatite nanocomposite increased after the infusion of PVA. The nanocomposite also showed excellent hemocompatibility with goat's blood [11].

22.2.2 Metallic Nanocomposites

The nanocomposite consisting of metal or alloy in the nanocomposite matrix is known as a metallic nanocomposite. There are two types of metal nanocomposite based on reinforced materials, continuous and discontinuous nanocomposites. The nanocomposites containing nanotubes, rod-shaped nanomaterials, are known as continuous phase nanocomposites. Metals like copper (Cu), gold (Au), silver (Ag), titanium (Ti), platinum (Pt), magnesium (Mg), zinc (Zn), and palladium (Pd) can be used for the preparation of metallic nanocomposite. Among all these metallic elements, Mg is mostly used because it presents in the body in abundant amounts and provides better biocompatibility properties and very high mechanical strength [12]. Razavi et al. (2010) fabricated magnesium-fluorapatite nanocomposite using a blending-pressing-sintering method and used those nanocomposites for bone tissue engineering. Mg increased the osteoconductivity of nanocomposites [13]. The incorporation of Mg into graphene nanoplatelets (GNP) improves the corrosion resistance, mechanical properties, and biological properties of nanocomposites in comparison to GNP. Evaluation of the *in-vitro* cytotoxicity of nanocomposites showed that the Mg-GNP nanocomposites did not show significant toxicity but showed higher cell viability values [14].

22.2.3 Polymeric Nanocomposites

The polymeric nanocomposite is produced by blending the polymer with organic/inorganic fillers. Natural polymers, as well as synthetic polymers, have been used for the preparation of nanocomposites for biomedical applications. Natural polymers include polysaccharides, such as starch, hyaluronic acid, cellulose, chitosan, and their derivatives, and proteins, including collagen, silk, fibrin, soy, etc. Synthetic polymers such as poly(Ɛ-caprolactone) (PCL), poly(lactic acid) (PLA), polypyrrole (PPy), poly(hydroxybutyrate) (PHB), polyvinyl alcohol (PVA) and polyvinyl pyrrolidine (PVP), etc., are used for the preparation of nanocomposites for biomedical applications. Baghdadi et al. (2021) prepared a multi-walled carbon nanotube/polypyrrole (MWCNT/ PPy) nanocomposite which was used for tissue engineering applications. The effect of electrical conductivity and thermal conductivity on tissue engineering was also measured and it was found that an increase in electrical conductivity led to improving the tissue engineering efficacy of nanocomposites [15]. The properties of nanocomposite are based on the interface adhesion between polymer matrix and nanoparticles. Mechanical properties of nanocomposites depend upon the functional properties of polymers and the interlinkage connections between the filler and the polymer.

22.3 Methods of Synthesis of Polymeric Nanocomposites

The fillers should be uniformly distributed within the polymeric matrix to highlight the properties of the fillers in the preparation of polymeric nanocomposites. Different polymers are used for polymeric nanocomposite preparations, like water-soluble polysaccharides like starch, alginate, and chitin/chitosan, proteins like soy, collagen, or fibrin, or synthetic polymers, e.g., poly(lactic acid), poly(ε-caprolactone), poly(glycolic acid), and poly(hydroxyl butyrate), which are used for biomedical applications. Poly(vinyl alcohol) (PVA), poly(acrylic acid) (PAA), poly(vinyl acetate) (PVAc), poly(ethylene glycol) (PEG), and guar gum are also used as water-soluble polymers. Certain conducting polymers, such as polyaniline (PANI), polyvinylpyrrolidone (PVP), polypropylene, polyethylene, polyurethane, polyacetylene, polythiophene, polyfuran, and polypyrrole (PPy), are very efficiently used for polymeric nanocomposite preparations, with wide applications in tissue engineering and as smart biosensors because of their smart response to electrical fields [16–17].

22.3.1 Generalized Methods Employed for the Preparation of Polymeric Nanocomposites

22.3.1.1 Ultrasonic-Assisted Solution Blending

Solution blending is the most commonly used method for the synthesis of nanocomposites. In this method, the material, such as ceramics, metals, polymers, or fillers, are dispersed in a suitable solvent, for instance, chloroform, tetrahydrofuran, dimethylformamide, toluene, etc. [18]. Ultrasonic waves are used for the dispersion of materials. The solvent is removed or evaporated using different methods. The nanocomposite can be collected by filtration or precipitation. Certain limitations include the selection of the solvent system and evaporation of the solvent. Graphene oxide/poly(lactic acid) (GO/PLA) nanocomposite is prepared by solution blending method. The modification of the nanocomposite with GO increases the antimicrobial activity, mechanical properties, and cytocompatibility of the system [19]. Poly(vinyl butyral) (PVB) nanocomposite is prepared by incorporation of graphene oxide using the solution blending method and it was observed that the mechanical properties were increased and the material could be used for biomedical applications (Figure 22.1) [20–22].

22.3.1.2 Melt Processing Technique

Melt processing is also known as melt blending or melt compounding. In this process, nanofillers are directly dispersed in a melted polymeric solution. In the presence of an inert gas such as nitrogen, neon, or argon, the polymer is heated to a temperature above the softening point, and the nanofillers are added to the melted polymeric solution. The melt processing method achieves superior mixing over the solution blending method. The chemical nature of the polymer, melting conditions, and feeding speed are some of the important parameters that need to be considered while preparing nanocomposites using the melt processing technique [23]. Polybutylene adipate-co-terephthalate/polymethyl methacrylate nanocomposite was prepared using melt processing technique and it was observed that the tensile strength of nanocomposite had increased over that of the pure polymer [24].

22.3.1.3 Ball Milling

This method involves the use of balls for creating collision under high shear force and compression force for mixing and grinding powders, but it can sometimes tear off two-dimensional structures. This method is mostly used in the metallurgical and mineral industries [25–30].

22.3.1.4 In-Situ *Polymerization*

In this method, the nanocomposites are prepared using the covalent and noncovalent bonds between the materials or the fillers. It yields a uniform nanocomposite in which the material or fillers are equally distributed throughout the nanocomposite. In this method, the nanoparticles are mixed in the polymeric solution and a homogeneous mixture is formed. The mixture is exposed to heat or radiation for the formation of nanocomposites. But a limitation associated with this process is that the polymeric solution must be of low viscosity. Polyacrylonitrile/graphene oxide (PAN/GO) nanocomposite was formed

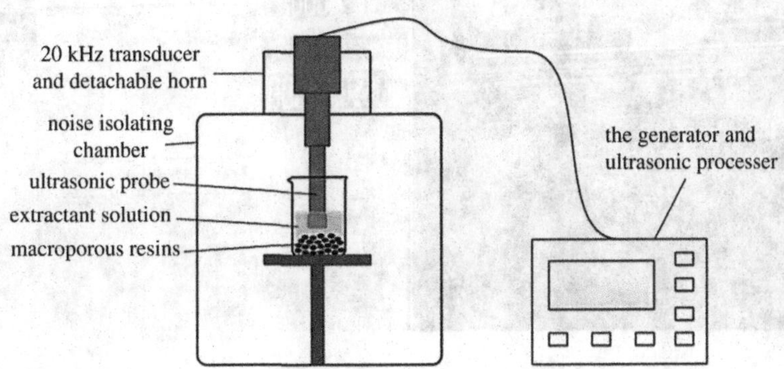

FIGURE 22.1
Schematic diagram of ultrasonic-assisted solution mixing [22].

with the help of an *in-situ* polymerization method, using dimethyl sulfoxide (DMSO) as a solvent. The drawback associated with this process is that it requires a long time for the formation of the nanocomposite. The PAN/GO nanocomposite is used as a precursor for the formation of nanofibers [31].

22.3.1.5 Electrospinning

Electrospinning is the most widely used method for the procurement of nanofiber nanocomposites. A high voltage is required to produce the nanofiber. The electrospinning apparatus consists of three parts: a high voltage supply, a syringe containing the spinning solution, and the collector to collect the nanofibers. A wide variety of natural as well as synthetic polymers are used for the fabrication of nanofibers. The prepared nanofibers can be used for various biomedical applications because they resemble the extracellular matrix. The cells are plated onto the nanofiber for recovery of the damaged tissues. Cell viability depends on numerous factors, such as the morphology and porosity of the mat that affects cell growth. Decreasing the diameter of the fiber leads to improvement of the biocompatibility of the nanofiber [32]. Zhang et al. (2021) used PAN/GO nanocomposite as a precursor for the preparation of nanofiber mats [31]. Ajmal et al. (2019) prepared PCL-gelatin nanofiber loaded with ciprofloxacin and quercetin for antibacterial and antioxidant properties, respectively, for wound healing. The attenuation of reactive oxygen species is observed during the early healing phase, while reduction of the growth of *Staphylococcus aureus* confirmed the antimicrobial as well as the antioxidant properties of the prepared nanofiber mat [33]. Dey et al. (2020) prepared gum ghatti/polyvinyl alcohol (GG/PVA) nanofibers using the electrospinning technique. Toxicity studies of the nanofiber mat were performed and the finding was that it did not show any toxic effects on local tissues of mice but promoted wound healing (Figures 22.2 and 22.3) [34].

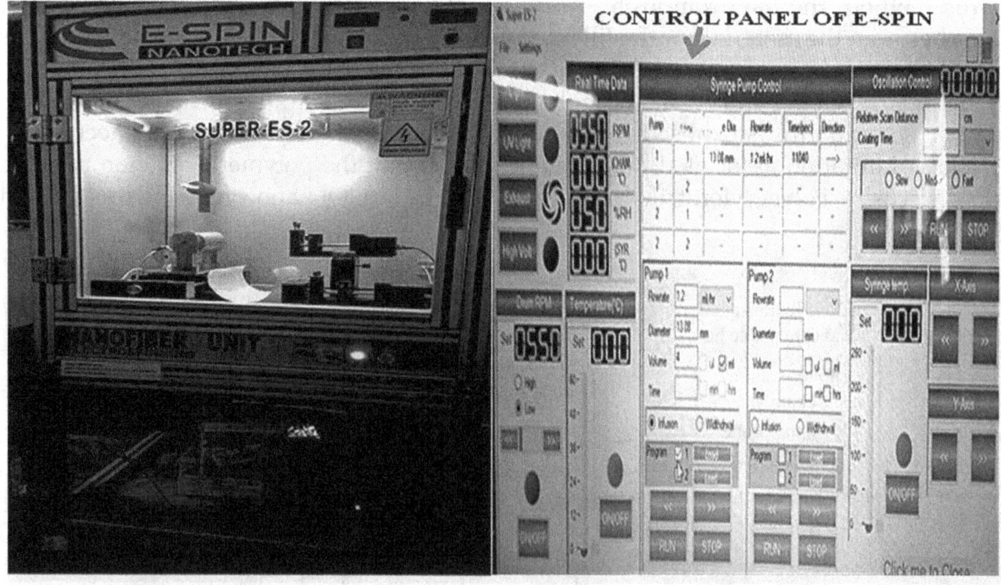

FIGURE 22.2
Espin Nanotech Super ES-2 electrospinning apparatus of Dept. of Chemical Engineering, BIT, Mesra, Ranchi, India. Adapted with permission from reference [34] Copyright © 2020 Elsevier Ltd.

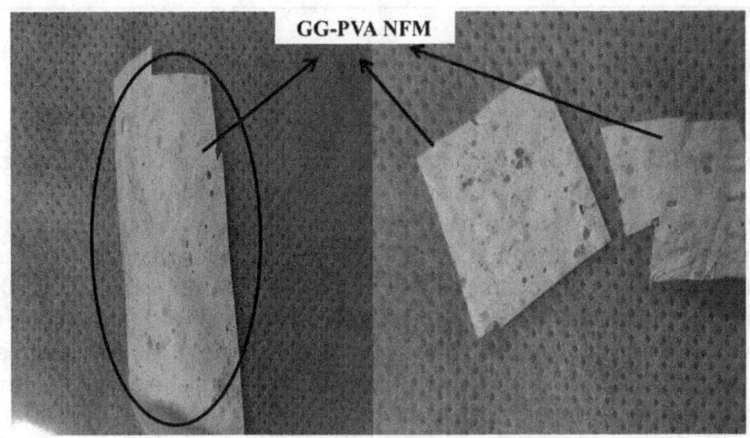

FIGURE 22.3
Electrospun GG-PVA CL-1 NFM. Adapted with permission from reference [34] Copyright © 2020 Elsevier Ltd.

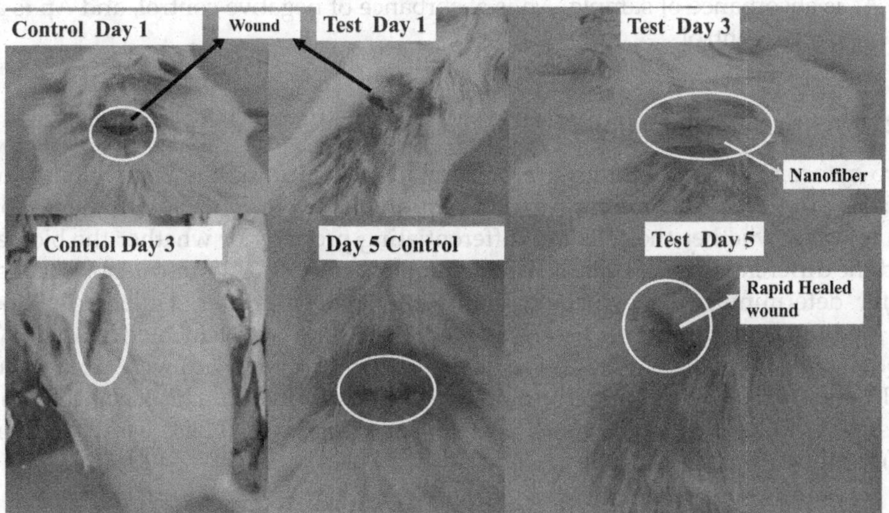

FIGURE 22.4
Progression of wound healing in test and control animals using GG/PVA nanofiber. Adapted with permission from reference [34] Copyright © 2020 Elsevier Ltd.

22.3.1.6 Microwave-Assisted Synthesis

Microwave-assisted synthesis is a green synthesis technique that utilizes microwave energy for the formation of nanocomposites. The heat is transferred through dielectric heating. The microwave-assisted technique is advantageous over conventional thermal heating process because the reaction takes place at a much faster rate. Using the microwave irradiation method, Gonzalez et al. (2012) prepared Ag-$H_2Ti_3O_7$ nanotubes having a diameter of around 8–10 nm and a few microns in length [35]. Bharath et al. (2019) prepared magnetite hydroxyapatite nanorods on a graphene oxide sheet for the controlled release of hemoglobin (Hb). The absorption capacity of the nanocomposite material was determined and, at pH 7, the maximum absorption capacity was found to be 1012 mg Hb/g (Figure 22.4) [36].

22.4 Biocompatibility Study of Polymeric Nanocomposites

22.4.1 Hemolysis Assay

The term biocompatibility indicates that the material is compatible with living tissues and does not induce any immunological responses when exposed to body fluids. If the material possesses some degree of incompatibility, it can either cause hemolysis or activate the coagulation pathway. For determination of % hemolysis, phosphate-buffered saline (PBS) extraction medium of the nanocomposite was prepared and seeded with 2% red blood cell (RBC) suspension and incubated for 2 hours. The sample was centrifuged to pellet the RBCs and the absorbance of the PBS solution was measured at 540 nm using a UV-visible spectrophotometer [37]. Blank PBS solution and Triton-X 100 is used as negative and positive s, respectively, and % hemolysis was calculated using the following equation:

$$\%\,\text{Hemolysis} = (\text{As} - \text{An}) / (\text{Ap} - \text{An}) \times 100$$

where As is absorbance of sample, An is absorbance of negative control, and Ap is absorbance of positive control.

22.4.2 In-Vitro Cytocompatibility

In-vitro cytocompatibility studies are performed to determine the compatibility of biomaterials with the cells. Different types of cells are placed in contact with the test biomaterials to check whether the cells are differentiating properly or whether the biomaterial altered the differentiation of cells. NIH3T3 fibroblast and MC3T3 pre-osteoblast cells were used for determining the cytocompatibility of nanocomposites. Dulbecco's Modified Eagle Medium and Minimum Essential Medium Alpha supplemented with 10% (v/v) fetal bovine serum and 1% (v/v) antibiotic (penicillin-streptomycin) were used to culture NIH3T3 and MC3T3 cells. Numerous assays were performed to check the cell viability. The assay includes PrestoBlue assay, lactate dehydrogenase (LDH) assay, Calcein-AM (LIVE) assay, MTT assay, etc. [38].

22.4.2.1 PrestoBlue Assay

PrestoBlue is a resazurin (7-hydroxy-3*H*-phenoxazin-3-one 10-oxide) -based solution that is blue to purple in color, weakly fluorescent, and non-toxic. PrestoBlue is used as an indicator to measure the viability of cells. When added to the cells, it is hydrolyzed by viable cells into resorufin (7-hydroxy-3*H*-phenoxazin-3-one), becoming highly fluorescent, by which we can quantitatively measure the proliferation of the cells [39].

22.4.2.2 Lactate Dehydrogenase (LDH) Assay

Lactate dehydrogenase is an enzyme that is present in almost every type of cell. Five different homologous forms of LDH are present in different tissue cells but they all catalyze the same reaction, pyruvate to L-lactate. In the LDH assay, when the plasma membrane of cells is ruptured, LDH is rapidly released outside the cell, and, in the presence of lactate, generates L-pyruvate which subsequently reacts with iodonitrotetrazolium chloride (INT) to form formazan. Formazan is water soluble and can be readily detected using colorimetry, measuring the absorbance at 490 nm. The number of cells can be determined by measuring the absorbance value of LDH at 490 nm. The final data can be plotted between the cell number and the absorbance at 490 nm [40].

22.4.2.3 Calcein-AM (LIVE) Assay

Calcein AM is a hydrophobic, non-fluorescent dye that is used to determine cell viability because it can easily penetrate inside the cells. The cells are incubated on the nanocomposite surface for 1 to 5 days, after which the calcein-AM assay is performed. The mechanism of the calcein-AM assay is that an intracellular esterase enzyme hydrolyzes the calcein-AM into the strongly fluorescent, hydrophilic calcein which is retained in the cytoplasm. The fluorescence intensity was checked at the fluorescence excitation/emission wavelength at 485 nm/530 nm, respectively, which is proportional to the number of viable cells. By using this assay, even a very small number of cells (50 viable cells) can be detected [38].

22.4.2.4 MTT Colorimetric Assay

The dye used in this assay is a tetrazolium dye, (3-(4,5-dimethylthiazol-2-yl)-2,5-diphenyltet razolium bromide. This assay works on the principle that, if a cell is alive and everything inside it is fine, mitochondria will be healthy as well. There are multiple enzymes in the mitochondria which are responsible for various metabolic reactions. NADPH-dependent oxidoreductase enzyme is only present in living cells and, in the presence of a tetrazolium dye, this enzyme tends to convert it into the visible, purple-colored compound, a formazan. When the colorless tetrazolium dye is added to a cell suspension, the cells convert the tetrazolium dye into the purple formazan, indicating that cells are viable. Qualitative and quantitative studies can be performed using different concentrations of the cell suspension, and the degree and intensity of the color indicate the density of viable cells [41].

22.5 Applications of One-Dimensional Polymeric Nanocomposites

Natural polymers like collagen, silk, hyaluronate, alginate, chitosan, fibrin, and cellulose, and synthetic polymers like polylactic acid, polylactide glycolic acid, and polycaprolactone, etc. can mimic the extracellular matrix of the cells and can also undergo enzymatic or hydrolytic degradation in response to any insult to the body, thereby making them a priority choice as one-dimensional polymeric nanocomposite systems for applications in tissue engineering.

22.5.1 As an Antibacterial

Polymeric nanocomposites can be very efficiently applied in the biomedical field as well as in drug delivery approaches, due to the anionic and cationic functional groups on their polymeric structure. This property of anion-cation arrangement also makes them appropriate for antibacterial properties due to the release of reactive oxygen species (ROS). In relation to the antibacterial properties, chitosan-based polymeric nanocomposites are found to have antibacterial properties against Gram-negative *Escherichia coli* and Gram-positive *S. aureus* [42]. Poly(*N*-isopropylacrylamide), along with carbon nanotubes, dispersed within a polyaniline matrix [(PNIPAM)/CNT/PANI], has wide applications as scaffolds and stents for cardiovascular disorders [43]. Organ or tissue replacement leads sometimes to complications and these issues can be alleviated by tissue engineering approaches. Having attained mandatory requirements of biocompatible characteristics, polymeric nanocomposites can have wide applications in cancer therapy and as an antibacterial. In a study, McGuffie et al. (2016) fabricated zinc oxide nanomaterials and coated them on polymeric implants, a treatment which not only reduced staphylococcal growth on the implants by the virtue of their surface hydrophobicity but also inhibited Gram-positive bacteria but had no impact on Gram-negative bacteria. Recently, polymeric nanocomposites are widely used for bone tissue regeneration as they have the advantage of being non-corrosive, resistant toward fatigue, with low elasticity modulus and high mechanical strength [44–46].

22.5.2 In Bone Tissue Regeneration

The polymers most commonly used as polymeric nanocomposite systems for bone tissue regeneration are polyhydroxyalkanoates (PHAs), polyether ketone, polycaprolactone (PCL), polyurethane, polyethylene (PE), and polysulfone. Porous polymeric nanocomposite structures are more efficient for the cell proliferation; such nanocomposites are fabricated using methods like thermally induced phase separation, solvent casting, emulsion-freeze-drying, supercritical gas foaming, electrospinning, leaching, and solid free-form fabrication [47]. Nanohydroxyapatite is an appropriate candidate as a one-dimensional nanomaterial for bone tissue regeneration by acting as nucleation sites for providing minerals for bone formation [48]. These nanohydroxyapatite materials are entrapped in polyethylene glycol hydrogel and formulated as hydrogel polymeric nanocomposite systems and have been used for osteoconductive applications. Also, carbon nanotubes, nanofillers and graphene oxide-based polymeric nanocomposites have been effectively used to increase electrical conductivity for tissue growth. In another study, silk fibroin polymeric nanocomposites were prepared using nanoparticles of silk (isolated from the silkworm *Bombyx mori*) followed by coating with polyethyleneimine which is further coated with calcium carbonate, encapsulating bone morphogenic protein 2 for osteogenic differentiation of mesenchymal stem cells [49]. Also, it has been studied that biodegradable polymers in combination with nanostructured materials like silica, bioglass, silicon-doped calcium phosphate, etc., are involved in the formation of cartilage and bones by enhancing cell metabolism. These materials are effective as implants in accelerating osteogenic proliferation and bone-related gene expression [50]. The inorganic bioactive nanofillers of the one-dimensional polymeric nanocomposites interact with the four proteins, namely fibronectin, vitronectin, laminin, and collagen, for osteoblast regeneration. The inorganic bioactive nanomaterials exhibit high surface reactivity for interactions by the formation of strong bonds with the bones by biological and chemical interactions [51]. Aliphatic polyesters are used as polymeric nanocomposites in the preparation of bone replacement implants. Polypeptides like poly-L-lysine embedded with montmorillonite clay as filler are efficient as tissue-engineered scaffolds.

22.5.3 In Wound Healing

The disruption of the cellular and anatomical configuration of the tissues, affecting physiological function, indicates a wound, and healing the wound with the least pain and scarless skin is the prime objective of any preparation [52]. In this aspect, polymeric nanocomposites of guar gum as the polymeric matrix embedded with biologically synthesized silver nanoparticles act effectively to achieve faster wound healing, leaving no traces of scars on the skin. In another study, mesoporous silica carboxymethyl cellulose hydrogel impregnated with zinc oxide crosslinked with citric acid acts as an effective wound dressing [53]. This nanocomposite hydrogel film, tested on adipose tissue-derived stem cells showed the material to be cytocompatible. Hyaluronic acid with zinc oxide nanocomposite hydrogels enhanced fibroblast growth on the wounded area with greater biodegradability. Wahab et al. prepared a polyurethane (PU), nanofiber-based nanocomposite, containing silver (Ag)/zinc oxide (Zn) composite nanoparticles as an antibacterial wound dressing and was evaluated against the human bacterial pathogens *S. aureus, E. coli*, and *Bacillus subtilis* [54]. These nanocomposite materials accelerate wound healing by promoting angiogenesis by achieving increased activation of vascular endothelial growth factors, angiopoietin, and fibroblast growth factors. Graphene-based three-dimensional printed polymeric nanocomposites have widespread applications in customized tissue regeneration and are also used as biosensors [55]. These 3D-printed systems can load more bio-actives, depending on their predetermined configuration, for successful wound healing and drug delivery.

22.5.4 Controlling Stem Cell Behavior

Controlling stem cell behavior artificially can be a challenging task. It has been found that graphene oxide-poly-L-lysine polymeric nanocomposites accelerated the osteogenic differentiation of mesenchymal stem cells and could be effectively used as bio scaffold coating. In addition, a polymeric nanocomposite was fabricated using poly-L-lysine impregnated with heparin and silver nanoparticles for delivering basic fibroblast growth factors and bone morphogenetic protein 2 [55, 56]. Cerium oxide nanoparticles embedded in bacterial cellulose polymeric matrix and fabricated as polymeric nanocomposites are quite promising for the proliferation of stem cells as they have high cell affinity, while gene expression in the stem cells was also improved [57].

22.5.5 Cartilage Tissue Engineering

Intra-articular fracture, osteoarthritis, ligament injury, and rheumatoid arthritis are conditions that can lead to cartilaginous tissue injury. Tissue engineering is a promising approach for treating cartilage tissue injury. Nanocomposite hydrogels consisting of starch nanocrystals (SNCs) showed around 90% cell viability of chondrogenic progenitor ATDC5 cells, which indicated the potential application of the nanocomposite for cartilage tissue engineering [58]. A dopamine-modified alginate/polydopamine nanocomposite scaffold provided a favorable environment for hydroxyapatite (HAP) so that it could accelerate the restoration of cartilage tissues [59].

22.5.6 Nerve Tissue Engineering

Better flexibility with cell adhesion, differentiation, migration, resistance to damage while implanting, and having a shape similar to the nerve are essential properties in scaffolds for nerve tissue engineering [60]. Graphene-based nanocomposites show good conductivity and high mechanical strength and it has been seen in various studies that electrical stimuli can be favorable for neuronal growth. In a study, a graphene-based nanocomposites hydrogel was fabricated consisting of polyurethane which is a thermoresponsive polymer, to improve the differentiation and growth of neural stem cells [61, 62]. Motamedi et al. in 2017 designed gold nanoparticles/polyvinylidene fluoride composites using electrospinning technique, and the gold nanoparticles were doped using the laser ablation method. Polyvinylidene fluoride expresses piezoelectric properties, and gold nanoparticles are known to be biocompatible and to enhance cell attachment. The composite showed an increase in adhesion and cell growth with no toxicity [63]. Silk fibroin-based nanocomposites have been prepared with reinforced gold nanorods which showed improved cell attachment, expression of neurogenic markers, and proliferation of cells [64].

22.5.7 Cardiac Tissue Engineering

Scaffolds with porosity, biodegradability, and mechanical strength resembling the cardiac tissue of a healthy person are utilized in cardiac tissue engineering [65]. Extensive efforts must be made to optimize the interaction between the cells and the scaffolds, starting with the selection of the best cell candidate for myocardial regeneration. There is a need to design bioactive materials with greater mechanical strength and degradation kinetics like synthetic polymers and biological specificity similar to the cardiac extracellular matrix (ECM) components, like laminin, collagen, and fibronectin [66]. Electroconductive nanofibrous patches were synthesized on polyurethane nanofibers by electrospraying of multi-walled carbon nanotubes (MWCNTs). Enhancement in cytocompatibility was observed and the scaffold's interaction with cardiomyoblasts also developed [67]. Biomaterials with poor conductivity can cause inadequate performance of cardiac constructs. Materials with good electroconductivity, like graphene oxide nanosheets or carbon nanotubes, can be utilized. Nanocomposites consisting of hydrogels of these nanomaterials can enhance the alignment of cardiomyocytes, and their electrophysiological function and maturation to some extent [68].

22.5.8 Skeletal Tissue Engineering

In the past decades, skeletal tissue engineering has proved to be a promising way of treating muscle injuries by forming functional tissues with the help of growth factors and scaffolds. Muscle tissues have the characteristic of contracting in response to an electrical signal. So, biomaterials that can conduct electrical signals are of great interest in skeletal tissue engineering [69]. In 2010, a self-contained nano-actuating scaffold was fabricated by coaxially electrospinning poly(ε -caprolactone) in which were incorporated MWCNTs to enhance conductivity and a hydrogel containing polyvinyl alcohol and polyacrylic acid for replacement of skeletal muscle tissue [70]. In 2020, Jo et al. designed nanocomposite scaffolds of polycarbonate diol and isosorbide-based polyurethane, with nanographene oxide having high flexibility, which resembled the morphology of the extracellular matrix. Nano-graphene oxide was incorporated into polyurethane to enhance the mechanical strength, elasticity, stress relaxation capacity, and hydrophilicity, and also improved the cell adhesion, proliferation, and stimulating capacity of myogenic differentiation [71].

22.6 Conclusion

The polymeric nanocomposites are multifaceted systems containing one of the phases that have one, two, or three dimensions. Nanocomposite possesses valuable tissue engineering properties, in particular larger surface area and high conductivity. Incorporation of nanofiller in nanocomposites may increase stability. Nanocomposites prepared from natural polysaccharides have the advantage of being biocompatible but show poor mechanical properties; to overcome these limitations, nanofillers, mainly one-dimensional nanofillers, may be incorporated to prepare polymeric nanocomposites which increase their mechanical properties and stability. Due to these enhanced properties, one-dimensional polymeric nanocomposites can be explored as suitable systems for different tissue engineering applications for tissue regeneration. It is expected that, in the coming years, exciting discoveries will be made in the field of polymeric nanocomposites in tissue regeneration.

References

1. Müller K, Bugnicourt E, Latorre M, Jorda M, Echegoyen Sanz Y, Lagaron JM, Miesbauer O, Bianchin A, Hankin S, Bolz U, Perez G, Jesdinszki M, Linder M, Scheuerer Z, Castello S, Schmid M (2017) Review on the processing and properties of polymer nanocomposites and nanocoatings and their applications in the packaging, automotive and solar energy fields. *Nanomaterials (Basel)* 7(4): 74.
2. Bitinis N, Hernández M, Verdejo R, Kenny JM, Lopez-Manchado MA (2011) Recent advances in clay/polymer nanocomposites. *Adv. Mater.* 23(44): 5229–5236.
3. Lu X, Zhang W, Wang C, Wen TC, Wei Y (2011) One-dimensional conducting polymer nanocomposites: Synthesis, properties and applications. *Prog. Polym. Sci.* 36(5): 671–712.
4. Li T, Ding G, Han ST, Zhou Y (2021) Introduction of polymer nanocomposites. In Y Zhou, G. Ding (eds): *Polymer Nanocomposite Materials: Applications in Integrated Electronic Devices.* Wiley Online Library, pp. 1–19.
5. Sharma AK, Kaith BS, Chandel K, Singh A (2020) Chemically modified chitosansodium alginate as chemo-sensor adsorbent for the detection of picric acid and removal of biebrich scarlet. *Int. J. Biol. Macromol.* 147: 582–594.
6. Ferreira FV, Pinheiro IF, Gouveia RF, Thim GP, Lona LMF (2018) Functionalized cellulose nanocrystals as reinforcement in biodegradable polymer nanocomposites. *Polym. Compos.* 39: E9–E29.
7. Bramhill J, Ross S, Ross G (2017) Bioactive nanocomposites for tissue repair and regeneration: A review. *Int. J. Environ. Res. Public Health* 14(1): 66.
8. Dwivedi P, Narvi SS, Tewari RP (2013) Application of polymer nanocomposites in the nanomedicine landscape: Envisaging strategies to combat implant associated infections. *J. Appl. Biomater. Funct. Mater.* 11(3): 129–142.
9. Garmendia N, Olalde B, Obieta I (2013) Biomedical applications of ceramic nanocomposites. In R Banerjee, I Manna (eds): *Ceramic Nanocomposites.* Cambridge, UK: Woodhead Publishing, pp. 530–547.
10. Ramadas M, Bharath G, Ponpandian N, Ballamurugan AM (2017) Investigation on biophysical properties of hydroxyapatite/graphene oxide (HAp/GO) based binary nanocomposites for biomedical applications. *Mater. Chem. Phys.* 199: 179–184.
11. Karthik V, Pabi SK, Chowdhury SR (2018) Development of hydroxyapatite/polyvinyl alcohol bionanocomposite for prosthesis implants. *Conf. Ser.: Mater. Sci. Eng.* 314(2018): 012031.
12. Sankar M, Vishnu J, Gupta M, Manivasagam G (2019) Magnesium-based alloys and nanocomposites for biomedical application. In AMA Inamuddin, A Mohammad (eds): *Applications of Nanocomposite Materials in Orthopedics.* Cambridge, MA: Woodhead Publishing, pp. 83–109.
13. Razavi M, Fathi MH, Meratian M (2010) Fabrication and characterization of magnesium–fluorapatite nanocomposite for biomedical applications. *Mater. Charact.* 61(12): 1363–1370.

14. Munir K, Wen C, Li Y (2020) Graphene nanoplatelets-reinforced magnesium metal matrix nanocomposites with superior mechanical and corrosion performance for biomedical applications. *J. Magnes. Alloys* 8(1): 269–290.
15. Baghdadi N, Zoromba MS, Abdel-Aziz MH, Al-Hossainy AF, Bassyouni M, Salah N (2021) One-dimensional nanocomposites based on polypyrrole-carbon nanotubes and their thermoelectric performance. *Polymers* 13(2): 278.
16. Costantini A, Luciani G, Annunziata G, Silvestri B, Branda F (2006) Swelling properties and bioactivity of silica gel/pHEMA nanocomposites. *J. Mater. Sci. Mater. Med.* 17(4): 319–325.
17. Zare EN, Makvandi P, Ashtari B, Rossi F, Motahari A, Perale G (2019) Progress in conductive polyaniline-based nanocomposites for biomedical applications: A review. *J. Med. Chem.* 63(1): 1–22.
18. Wang M, Yan C, Ma L (2012) Graphene nanocomposites. In: *Composites and Their Properties*. N Hu (Ed.). London, UK: Intech Open, p. 17.
19. Bayer IS (2017) Thermomechanical properties of polylactic acid-graphene composites: A state-of-the-art review for biomedical applications. *Materials (Basel)* 10(7): 748.
20. Hajian M, Reisi MR, Koohmareh GA, Zanjani Jam AR (2012) Preparation and characterization of Polyvinylbutyral/graphene nanocomposite. *J. Polym. Res.* 19(10): 1–7.
21. Jang J, Bae J, Yoon SH (2003) A study on the effect of surface treatment of carbon nanotubes for liquid crystalline epoxide–carbon nanotube composites. *J. Mater. Chem.* 13(4): 676–681.
22. Chen B, Bao S, Zhang Y, Zheng R (2018) Ultrasound-assisted synthesis of N235-impregnated resins for vanadium (V) adsorption. *R. Soc. Open Sci.* 5(4): 171746.
23. Guo F, Aryana S, Han Y, Jiao Y (2018) A review of the synthesis and applications of polymer-nanoclay composites. *Appl. Sci.* 8(9): 1–29.
24. Bheemaneni G, Kandaswamy R (2019) Melt processing and characterization of tricalcium phosphate filled polybutylene adipate-co-terephthalate/polymethyl methacrylate composites for biomedical applications. *Int. J. Polym. Mater. Polym. Biomater.* 68(1–3): 119–125.
25. Zhang DL (2004) Processing of advanced materials using high-energy mechanical milling. *Prog. Mater. Sci.* 49(3–4): 537–560.
26. Gupta TK, Singh BP, Mathur RB, Dhakate SR (2014) Multi-walled carbon nanotube–graphene–polyaniline multiphase nanocomposite with superior electromagnetic shielding effectiveness. *Nanoscale* 6(2): 842–851.
27. Wu H, Zhao W, Hu H, Chen G (2011) One-step in situ ball milling synthesis of polymer-functionalized graphene nanocomposites. *J. Mater. Chem.* 21(24): 8626–8632.
28. Jiang X, Drzal LT (2012) Reduction in percolation threshold of injection molded high-density polyethylene/exfoliated graphene nanoplatelets composites by solid state ball milling and solid-state shear pulverization. *J. Appl. Polym. Sci.* 124(1): 525–535.
29. Tang LC, Wan YJ, Yan D, Pei YB, Zhao L, Li YB, Wu LB, Jiang JX, Lai GQ (2013) The effect of graphene dispersion on the mechanical properties of graphene/epoxy composites. *Carbon* 60: 16–27.
30. Gu J, Li N, Tian L, Lv Z, Zhang Q (2015) High thermal conductivity graphite nanoplatelet/UHMWPE nanocomposites. *RSC Adv.* 5(46): 36334–36339.
31. Zhang Y, Zhu B, Cai X, Yuan X, Zhao S, Yu J, Qiao K, Qin R (2021) Rapid in situ polymerization of polyacrylonitrile/graphene oxide nanocomposites as precursors for high-strength carbon nanofibers. *ACS Appl. Mater. Interfaces* 13(14): 16846–16858.
32. Li Z, Wang C, Zhenyu LI (2016) *One-Dimensional Nanostructures Electrospinning Technique and Unique Nanofibers*. New York, Dordrecht, London, and Berlin Heidelberg: Springer, pp. 15–29.
33. Ajmal G, Vasant G, Mittal P, Khan G, Pandey VK, Bakade BV, Mishra B (2019) Biomimetic PCL-gelatin based Nano fi bers loaded with Cipro fl oxacin hydrochloride and quercetin: A potential antibacterial and anti-oxidant dressing material for accelerated healing of a full thickness wound. *Int. J. Pharm.* 567: 118480.
34. Dey P, Bal T, Gupta RN (2020) Fabrication and invitro evaluation of electrospun gum ghatti-polyvinyl alcohol polymeric blend green nanofibre mat (GG-PVA NFM) as a novel material for polymeric scaffolds in wound healing. *Polym. Test.* 91: 106826.

35. Rodríguez-gonzález V, Obregón-alfaro S, Lozano-sánchez LM, Lee S (2012) Chemical Rapid microwave-assisted synthesis of one-dimensional silver – $H_2Ti_3O_7$ nanotubes. *J. Mol. Catal. A Chem.* 353–354: 163–170.
36. Bharath G, Rambabu K, Hai A, Anwer S, Banat F, Ponpandian N (2020) Synthesis of one-dimensional magnetite hydroxyapatite nanorods on reduced graphene oxide sheets for selective separation and controlled delivery of hemoglobin. *Appl. Surf. Sci.* 501: 144215.
37. Roy A, Joshi M, Butola BS, Ghosh S (2020) Evaluation of biological and cytocompatible properties in nano silver-clay based polyethylene nanocomposites. *J. Hazard. Mater.* 384: 121309.
38. Farshid B, Lalwani G, Sitharaman B (2015) In vitro cytocompatibility of one-dimensional and two-dimensional nanostructure-reinforced biodegradable polymeric nanocomposites. *J. Biomed. Mater. Res. A* 103(7): 2309–2321.
39. Chen JL, Steele TW, Stuckey DC (2015) Modeling and application of a rapid fluorescence-based assay for biotoxicity in anaerobic digestion. *Environ. Sci. Technol.* 49(22): 13463–13471.
40. Kumar P, Nagarajan A, Uchil PD (2018) Analysis of cell viability by the lactate dehydrogenase assay. *Cold Spring Harb. Protoc.* 2018(6): pdb-prot095497.
41. Mokri N, Sepehri Z, Faninam F, Khaleghi S, Kazemi NM, Hashemi M (2021) Chitosan-coated Zn-metal-organic framework nanocomposites for effective targeted delivery of LNA-antisense miR-224 to colon tumor: In vitro studies. *Gene Ther.* 2021: 1–11.
42. Kaur G, Adhikari R, Cass P, Bown M, Gunatillake P (2015) Electrically conductive polymers and composites for biomedical applications. *RSC Adv.* 5(47): 37553–37567.
43. Bharatiya D, Patra S, Parhi B, Swain SK (2021) A materials science approach towards bioinspired polymeric nanocomposites: A comprehensive review. *Int. J. Polym. Mater.*: 1–16..
44. McGuffie MJ, Hong J, Bahng JH, Glynos E, Green PF, Kotov NA, Younger JG, VanEpps JS (2016) Zinc oxide nanoparticle suspensions and layer-by-layer coatings inhibit staphylococcal growth. *Nanomed. Nanotechnol.* 12(1): 33–42.
45. Dziadek M, Zagrajczuk B, Menaszek E, Cholewa-Kowalska K (2018) A new insight into in vitro behaviour of poly (ε-caprolactone)/bioactive glass composites in biologically related fluids. *J. Mater. Sci.* 53(6): 3939–3958.
46. Fouad H, AlFotawi R, Alothman OY, Alshammari BA, Alfayez M, Hashem M, Mahmood A (2018) Porous polyethylene coated with functionalized hydroxyapatite particles as a bone reconstruction material. *Materials (Basel)* 11(4): 521.
47. Brüggemann GP, Arampatzis A, Emrich F, Potthast W (2008) Biomechanics of double transtibial amputee sprinting using dedicated sprinting prostheses. *Sports Technol.* 1(4–5): 220–227.
48. Carrow JK, Gaharwar AK (2015) Bioinspired polymeric nanocomposites for regenerative medicine. *Macromol. Chem. Phys.* 216(3): 248–264.
49. Shi P, Abbah SA, Chuah YJ, Li J, Zhang Y, He P, Wong HK, Goh JC (2017) Yolk shell nanocomposite particles as bioactive bone fillers and growth factor carriers. *Nanoscale* 9(38): 14520–14532.
50. Misra SK, Valappil SP, Roy I, Boccaccini AR (2006) Polyhydroxyalkanoate (PHA)/inorganic phase composites for tissue engineering applications. *Biomacromolecules* 7(8): 2249–2258.
51. Hench LL, Polak JM (2002) Third-generation biomedical materials. *Science* 295(5557): 1014–1017.
52. Ghosh Auddy R, Abdullah MF, Das S, Roy P, Datta S, Mukherjee A (2013) New guar biopolymer silver nanocomposites for wound healing applications. *BioMed Res. Int.* 2013.
53. Rakhshaei R, Namazi H (2017) A potential bioactive wound dressing based on carboxymethyl cellulose/ZnO impregnated MCM-41 nanocomposite hydrogel. *Mater. Sci. Eng. C* 73: 456–464.
54. Jatoi AW (2020) Polyurethane nanofibers incorporated with ZnAg composite nanoparticles for antibacterial wound dressing applications. *Compos. Commun.* 19: 103–107.
55. Qi W, Yuan W, Yan J, Wang H (2014) Growth and accelerated differentiation of mesenchymal stem cells on graphene oxide/poly-L-lysine composite films. *J. Mater. Chem. B* 2(33): 5461–5467.
56. Qi W, Yan J, Sun H, Wang H (2017) Multifunctional nanocomposite films for synergistic delivery of bFGF and BMP-2. *ACS Omega* 2(3): 899–909.
57. Gofman IV, Nikolaeva AL, Khripunov AK, Ivan'kova EM, Shabunin AS, Yakimansky AV, Romanov DP, Popov AL, Ermakov AM, Solomevich SO, Bychkovsky PM, Baranchikov AE,

Ivanov VK (2021) Green polymer-based nanocomposites containing ceria and their use in the process of stem cell proliferation. *Polymers* 13(12): 1999.
58. Karthigadevi G, Malar CG, Dey N, Kumar KS, Roseline MS, Subalakshmi V (2021) Alginate-based nanocomposite hydrogels. In TK Giri, B Ghosh (eds): *Plant and Algal Hydrogels for Drug Delivery and Regenerative Medicine*. Cambridge, MA: Woodhead Publishing, pp. 395–421.
59. Shen J, Shi D, Dong L, Zhang Z, Li X, Chen M (2018) Fabrication of polydopamine nanoparticles knotted alginate scaffolds and their properties. *J. Biomed. Mater. Res. A* 106(12): 3255–3266.
60. Wahid F, Khan T, Hussain Z, Ullah H (2018) Nanocomposite scaffolds for tissue engineering; properties, preparation and applications. In AMA Inamuddin, A Mohammad (eds): *Applications of Nanocomposite Materials in Drug Delivery*. Cambridge, MA: Woodhead Publishing, pp. 701–735.
61. Bei HP, Yang Y, Zhang Q, Tian Y, Luo X, Yang M, Zhao X (2019) Graphene-based nanocomposites for neural tissue engineering. *Molecules* 24(4): 658.
62. Huang CT, Shrestha LK, Ariga K, Hsu SH (2017) A graphene–polyurethane composite hydrogel as a potential bioink for 3D bioprinting and differentiation of neural stem cells. *J. Mater. Chem. B* 5(44): 8854–8864.
63. Motamedi AS, Mirzadeh H, Hajiesmaeilbaigi F, Bagheri-Khoulenjani S, Shokrgozar MA (2017) Piezoelectric electrospun nanocomposite comprising Au NPs/PVDF for nerve tissue engineering. *J. Biomed. Mater. Res. A* 105(7): 1984–1993.
64. Afjeh-Dana E, Naserzadeh P, Nazari H, Mottaghitalab F, Shabani R, Aminii N, Mehravi B, Rostami FT, Joghataei MT, Mousavizadeh K, Ashtari K (2019) Gold nanorods reinforced silk fibroin nanocomposite for peripheral nerve tissue engineering applications. *Int. J. Biol. Macromol.* 129: 1034–1039.
65. Zhao H, Liu M, Zhang Y, Yin J, Pei R (2020) Nanocomposite hydrogels for tissue engineering applications. *Nanoscale* 12(28): 14976–14995.
66. Genovese JA, Spadaccio C, Rainer A, Covino E (2011) Electrospun nanocomposites and stem cells in cardiac tissue engineering. In AR Boccaccini, SE Harding (eds): *Myocardial Tissue Engineering*. Berlin, Heidelberg: Springer, pp. 215–242.
67. Shokraei N, Asadpour S, Shokraei S, Nasrollahzadeh Sabet M, Faridi-Majidi R, Ghanbari (2019) Development of electrically conductive hybrid nanofibers based on CNT-polyurethane nanocomposite for cardiac tissue engineering. *Microsc. Res. Tech.* 82(8): 1316–1325.
68. Paul A (2015) Nanocomposite hydrogels: An emerging biomimetic platform for myocardial therapy and tissue engineering. *Nanomedicine* 10(9): 1371–1374.
69. Dong R, Ma PX, Guo B (2020) Conductive biomaterials for muscle tissue engineering. *Biomaterials* 229: 119584.
70. McKeon-Fischer KD, Flagg DH, Freeman JW (2011) Coaxial electrospun poly (ε -caprolactone), multiwalled carbon nanotubes, and polyacrylic acid/polyvinyl alcohol scaffold for skeletal muscle tissue engineering. *J. Biomed. Mater. Res. A* 99(3): 493–499.
71. Wahid F, Khan T, Hussain Z, Ullah H (2018) Nanocomposite scaffolds for tissue engineering; properties, preparation and applications. In AMA Inamuddin, A Mohammad (eds): *Applications of Nanocomposite Materials in Drug Delivery*. Cambridge, MA: Woodhead Publishing, pp. 701–735.

23

One-Dimensional Polymeric Nanocomposites-Based Microcontainers for Biomedical Applications

Selcan Karakuş, Cemal Özeroğlu, Mizan İbrahim Kahyaoğlu, Ezgi Tan, Merve İlgar, Zeynep Orhan, and Nazlı Albayrak

CONTENTS

23.1 1D Polymeric Nanocomposite-Based Micro-/Nanocontainers

Nanobiotechnology is a branch of engineering that deals with a multidisciplinary approach to science at the nanoscale level for different applications such as pharmaceuticals, medicine, food, and agriculture. It especially focuses on physicochemical, environmental, biological, agricultural, and engineering applications to develop advanced engines, machines, and micro/nanostructures. Studies have shown that nanocarriers have distinctive physical, surface, biological, and structural properties due to their high ratio of surface area/volume, small size, biodegradability, and biocompatibility.

Polymeric nanocarriers are small particles ranging from 1 to 100 nm in size and consisting of the core loaded with active compounds (e.g., a drug) and a polymer shell. Smart polymeric nanocarriers have made a breakthrough, and the specific features of nanomaterials are made more and more demanding in a wide variety of applications, including optics, electronics, sensors, drug delivery systems, and personalized medicine. Generally, polymeric nanocarriers are classified into five different types, namely: polymeric micelles with a hydrophobic core and a hydrophilic shell; polymeric nanocapsules with a cavity-shell structure; polymeric nanogels with an average diameter ranging between 20 and 200 nm; dendrimers with a highly branched structure; and hybrid nanoparticles with porous cores [1]. In the past decade, the development and fabrication of smart polymeric nanocarriers opened up a new and effective science strategy to increase the efficiency of therapeutics by minimizing the side effects of drugs in health care applications. Accordingly, the schematic diagram of various 1D polymeric nanostructures in different biomedical applications is presented in Figure 23.1. Especially, smart drug delivery nanocarriers show

DOI: 10.1201/9781003223764-23

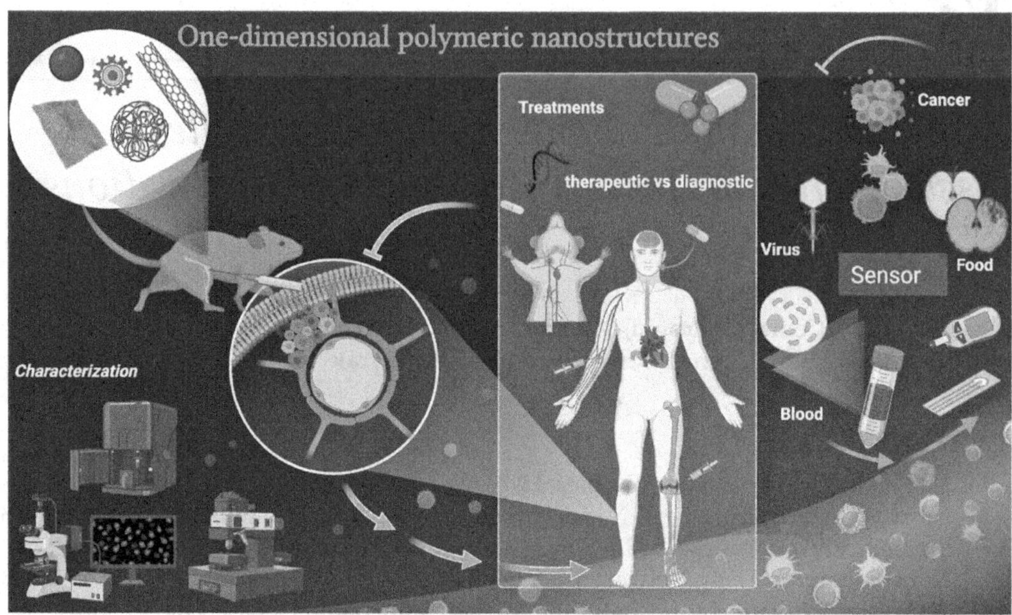

FIGURE 23.1
1D polymeric nanostructures in biomedical applications.

remarkable changes and improvements in their physical, chemical, or biological properties in response to a slight change in environmental stimulus (e.g., temperature, pH, ultrasound, enzyme, electrical, magnetic). Additionally, polymeric nanocomposites-based enzymatic or non-enzymatic sensors show extraordinary optical, electrical, and physical properties for rapid detection of the target analyte with high electrocatalytic activity, selectivity, and a lower detection limit in sensor applications.

In this chapter, the preparation, characterization, and optimization of polymeric nanocarriers with variable dimensions will be discussed, and a variety of important biotechnological and biomedical applications of nano-systems are described. It is known that polymeric nanocarriers exhibit different abilities according to their sizes, morphologies, functionalities, structures, and dimensions. Generally, polymeric nanocarriers are classified as zero-dimensional (0D), one-dimensional (1D), two-dimensional (2D), and three-dimensional (3D) based on their dimensions (x, y, or z) in the range of nanometers. In this chapter, 1D polymeric nanostructure-based microcontainers with various properties for biomedical applications are discussed.

23.2 Fabrication of 1D Polymeric Nanocomposite-Based Microcontainers

Recently, the fabrication of 1D polymeric nanocomposite-based microcontainers has attracted a great deal of attention. The most common polymers are poly(lactide), poly(acrylamide) (PAAm), polylactic acid (PLA), poly(lactide-co-glycolide) (PLGA) copolymers, poly(acrylic acid) (PAA), poly(amino acids), poly(methacrylic acid) (PMAA), poly (ε-caprolactone) (PCL), chitosan, poly(diethylaminoethyl methacrylate) (PDEAEMA), alginate, gelatin, poly(dimethylaminoethyl methacrylate) (PDMAEMA), albumin, carboxymethyl cellulose (CMC), polyvinyl alcohol (PVA), polydopamine, poly(2-methacryloyloxyethyl phosphorylcholine), and polyethylene glycol (PEG). It is important to develop biopolymer-based nanostructures that can be completely degraded in biological systems and release medicine in a control way from polymer core-shell nanocarriers under different specific conditions such as pH, temperature, concentration, light, sonication, magnetic, and electric field.

An overview of the preparation and characterization of 1D polymeric nanocomposites, such as nanotubes, nanowires, nanoparticles, nanorods, and nanofibers, are presented. These nanocomposites have been fabricated using various methods, such as solvent/solid-assisted method, solution/vapor-phase method, electrospinning, solvothermal, laser ablation, and sonication. Among 1D microcontainers, carbon nanotubes have attracted great attention for the fabrication of the drug-loaded cylindrically shaped nanostructures in smart delivery systems. Carbon nanotubes are in the form of hollow cylindrical structures with a hexagonal arrangement of hybridized carbon atoms at a nanoscale. Carbon nanotubes can be classified into the following categories: single-walled carbon nanotubes (SWCNTs) (armchair, zigzag, and chiral), double-walled carbon nanotubes (DWCNTs), and multi-walled carbon nanotubes (MWCNTs), graphenated carbon nanotubes, torus, cup-stacked carbon nanotubes, carbon nanobuds, extreme carbon nanotubes (a length of 550 nm), and carbon nano-peapods [2]. Carbon nanotubes have been reported to have been used in various biomedical fields, such as therapeutic and diagnostic agents, gene delivery to cells/organs, nervous tissue regeneration, neural scaffolds, gene therapy, myocardial conduction, tissue engineering, and cancer treatments (Figure 23.2).

Carbon nanotubes and their composites can be prepared using different techniques such as laser ablation, arc discharge, chemical vapor deposition (CVD), laser deposition method, electric arc method, and high-pressure carbon monoxide disproportionation. CVD and hydrothermal methods are widely used methods for the fabrication of the uniform diameter, controlled shape, controlled orientation, and spatial locations. The number

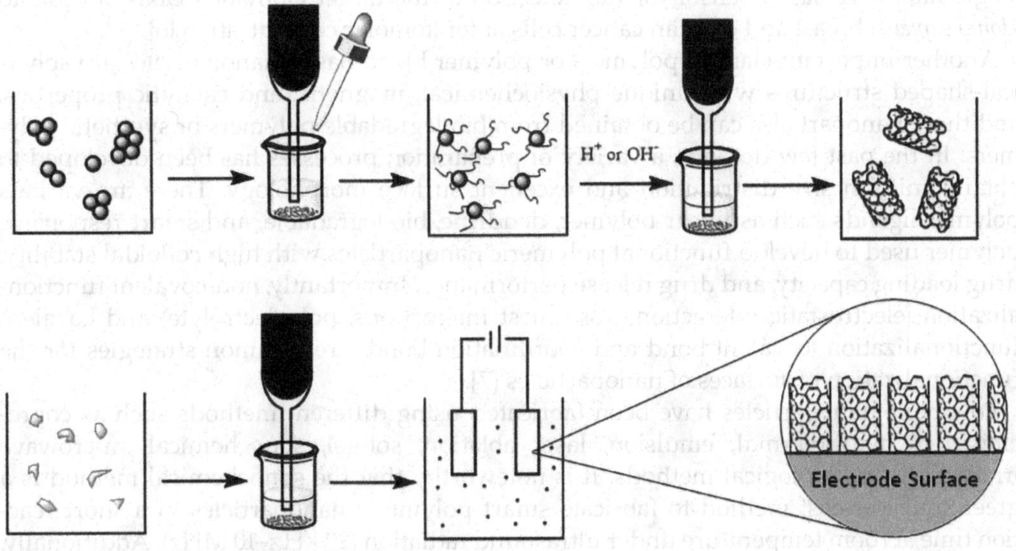

FIGURE 23.2
Different synthesis methods of nanotubes.

of biomedical applications of carbon nanotubes has increased significantly over the past years due to unique properties such as biocompatibility, light weight, high tensile strength, antimicrobial activity, and surface functionalization. Carbon nanotubes have been studied in a wide spectrum of different biomedical applications such as environmental monitoring, designing of electrochemical biosensors, sensor applications to detect pathogenic bacteria with a low detection limit, effective determination of biomolecules (glucose, bilirubin, urine, cholesterol, guanine, and adenine), clinical diagnosis, novel drug delivery systems, and cancer treatments. Yamina et al. prepared novel multi-functional poly(ethylene glycol) (PEG) – graft – halloysite nanotubes using the hydrothermal method with a shell thickness of between 20 and 26 nm. The experimental results showed that the nanostructure is a promising candidate for the controlled-release drug (quercetin) nano-system and cell imaging with the addition of a specific tumor cell targeting [3]. Ghoderao et al. reported that paclitaxel could be loaded with multi-walled carbon nanotubes equipped with iron oxide (Fe_3O_4) nanoparticles. The results proved that the nanostructure had a promising antiproliferative effect for cancer therapy [4]. The schematic diagram of different synthesis methods of nanotubes is presented in Fig. 23.2.

Nanowires have rod-shaped structures with a high aspect ratio, cross-sectional dimensions of 2–200 nm, and lengths ranging from 100 nm to several micrometers. Unique properties of nanowires support enhancing performance and chemical reactivity in the nanoscale range. Nanowires are fabricated using different techniques such as the evaporation technique, electro-deposition process, vapor–liquid–solid process, laser ablation, lithography, ultraviolet irradiation method, solvothermal method, photo-reduction technique, DNA template method, porous material template method, and solution–liquid–solid method. In recent years, different types of materials (silicon carbide, Fe_2O_3, TiO_2, and WO_3) have been used for synthesizing nanowires. Chang et al. prepared a novel magnetic polymer colloid of xanthan gum–chitosan/nickel nanowires with a variable cytocompatibility of three cell lines with N2a mouse neuroblastoma, P19 mouse teratocarcinoma, and 10T1/2 mouse multipotent mesenchymal cells [5]. In another study, Kannan et al. developed a copolymer of poly(3,4-ethylenedioxythiophene) (PEDOT) and carboxylic group-functionalized PEDOT single nanowire-based sensor for the detection of the target Oblivious DNA specific to *Homo sapiens'* breast and ovarian cancer cells at femtomolar concentration [6].

Another important class of polymer- or polymer blend-based nanoparticles are spherical-shaped structures with unique physiochemical, magnetic, and catalytic properties, and these nanoparticles can be obtained from biodegradable polymers or synthetic polymers. In the past few decades, a variety of preparation processes has been developed to obtain uniform size distribution and excellent surface morphology. There are various polymer ligands such as linear polymer, dendritic, biodegradable, and smart responsive polymer used to develop functional polymeric nanoparticles with high colloidal stability, drug loading capacity, and drug release performance. Importantly, non-covalent functionalization (electrostatic interaction, host-guest interactions, polyelectrolyte) and covalent functionalization (covalent bond and coordination bond) are common strategies for the functionalization of surfaces of nanoparticles [7].

Polymeric nanoparticles have been fabricated using different methods such as coprecipitation, hydrothermal, emulsion, laser ablation, sol-gel, sonochemical, microwave irradiation, and biological methods. It is noteworthy that the sonochemical method is a green and low-cost method to fabricate smart polymeric nanoparticles in a short reaction time at room temperature under ultrasound radiation (20 kHz–10 MHz). Additionally, green sonochemical methods lead to high yields in short reaction times and are based on the surface acoustic cavitation to reduce particle size and increase the surface area. The above-mentioned methods for synthesizing uniformly distributed polymeric nanoparticles are well known, although the control of morphology and colloidal properties of nanostructures are challenging. This situation can be attributed to the incompatibility between hydrophilic and hydrophobic materials with surface energies and wettability. To overcome these drawbacks, studies focus on a better understanding of their colloidal behavior under relevant conditions.

According to the World Health Organization (WHO), 9.6 million people died of cancer in 2018, and it is estimated that this disease will cause 11.4 million deaths in 2030. Major difficulties in cancer treatment are discriminating cancer cells from normal cells, low bioavailability of chemotherapeutic drugs, antibiotic multidrug resistance, targeting of cancer stem cells, and side effects of chemotherapeutic drugs. The extreme environments of cancer cells, such as the acidic, extracellular environment, high temperature, enzymes, surface molecules, hypoxic conditions, and reactive oxygen species offer opportunities for early detection and effective treatment. For this reason, drug-loaded polymeric nanocarriers can play a key role in the localization of drugs to the target cancer region using external conditions with minimum side effects, dosage, high bioavailability, and biocompatibility. Meng and collaborators performed a study focused on developing methotrexate- (MEX) and curcumin- (CUR) loaded triblock copolymer poly(lactic-co-glycolic acid) (PLGA)–PEG-based polymeric nanoparticles to be used for the treatment of colorectal cancer therapy [8]. Results obtained from *in-vitro* experiments showed that the novel nanostructure had a high potential for cancer treatment with low toxicity, amphiphilicity, degradation, and biocompatibility. The study conducted by Bhattacharya et al. showed recent improvements in polymeric nanoparticle-based nanocarriers composed of poly (sarcosine), PEG, and PLGA. The nanostructure was prepared using a green sonication method for the cancer hydrophobic drug (docetaxel) delivery systems [9]. More specifically, Dong et al. have developed a novel dual drug-loaded polymeric nanoparticles system based on PLGA and PEG for salinomycin (SAL) and doxorubicin (DOX) delivery systems for the glioma-specific targeting therapy [10]. Different biochemical trials such as AO-EB, Hoechst staining, and annexin V-FTIC results confirmed surface observations and apoptosis.

In recent years, rod-shaped polymeric nanomaterials have been used for controlled drug-release systems. Generally, there are three basic methods, namely template method, electrochemical method, and seed-mediated method, for the preparation of rod-shaped polymeric nanomaterials. Other approaches such as physical vapor deposition (PVD), chemical vapor deposition (CVD), and solution methods can be applied to obtain different nanorods such as carbon nanorods, ZnO nanorods, gold nanorods, NiWO nanorods, carbon nitride nanorods, Ga_2O_3 nanorods, CdS nanorods, polymer nanorods, and magnetic nanorods. In this context, Pelras et al. prepared functional polymer nanorods using atom transfer radical polymerization (ATRP) for pH-sensitive DOX release systems within MCF-7 breast cancer cells [11]. In another similar study, Bhide et al. prepared artemether-loaded PLGA nanorods using mechanical stretching of nanospheres. Results showed that there was a significant impact of the geometry of the nanocarrier on the artemether delivery system [12]. In another study, Eivazzadeh-Keihan and co-workers performed a study focused on developing alginate hydrogel–poly(vinyl alcohol) (PVA)/silk fibroin/magnesium hydroxide nanorods using a simple solution method to be used for biological and antibacterial applications [13]. The fabricated nanocomposite showed significantly greater anti-biofilm activity than pure polystyrene and cross-linked hydrogel–PVA silk fibroin ($p < 0.05$). The schematic diagram of different synthesis methods of nanorods is presented in Figure 23.3.

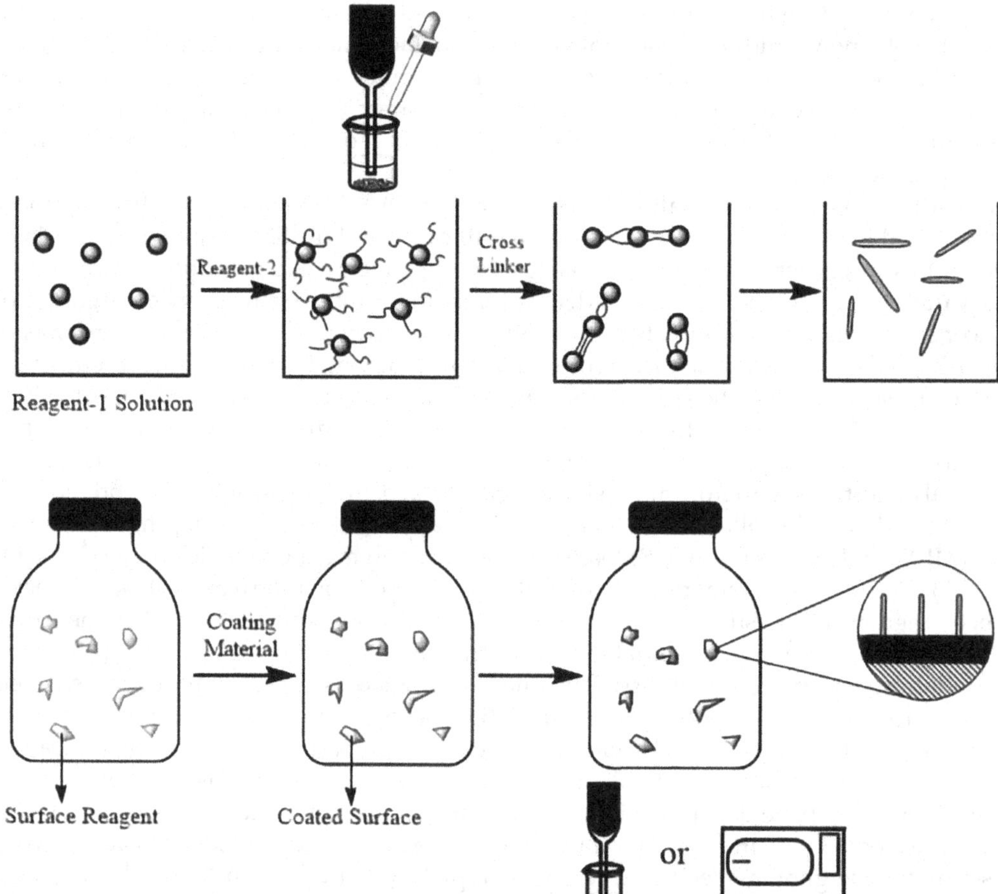

FIGURE 23.3
Different synthesis methods of nanorods.

Drug encapsulations for drug nanocarrier formulations are considered an attractive method to protect drugs and improve the solubility, bioavailability, and efficiency of bioactive compounds. There have been numerous studies to investigate the encapsulation of bioactive compounds using various techniques, such as spray-drying, lyophilization, emulsion-based systems, inclusion complexation, nanoprecipitation, liposome preparation, supercritical fluid extraction of emulsions, and electrospinning in biomedical applications. Solvent evaporation and spray-drying are the most widely used methods for industrial scale-up of production of micro/nanostructures. However, they have fundamental drawbacks owing to the high temperature involved, resulting in significant low stability, and poor encapsulation efficiency. In recent years, electrospinning is a simple and effective method used to obtain nanofibers from polymer solutions. In this method, the applied voltage, solution viscosity, flow rate, temperature, humidity, concentration, conductivity, surface tension, the distance between collector and needle, and electric field intensity play major roles in the formation and morphologies of polymer-based nanofibers and on generating a jet. A series of studies have confirmed that smart electrospun nanofibers as polymer nanocarriers can improve the drug loading efficiency with consistent

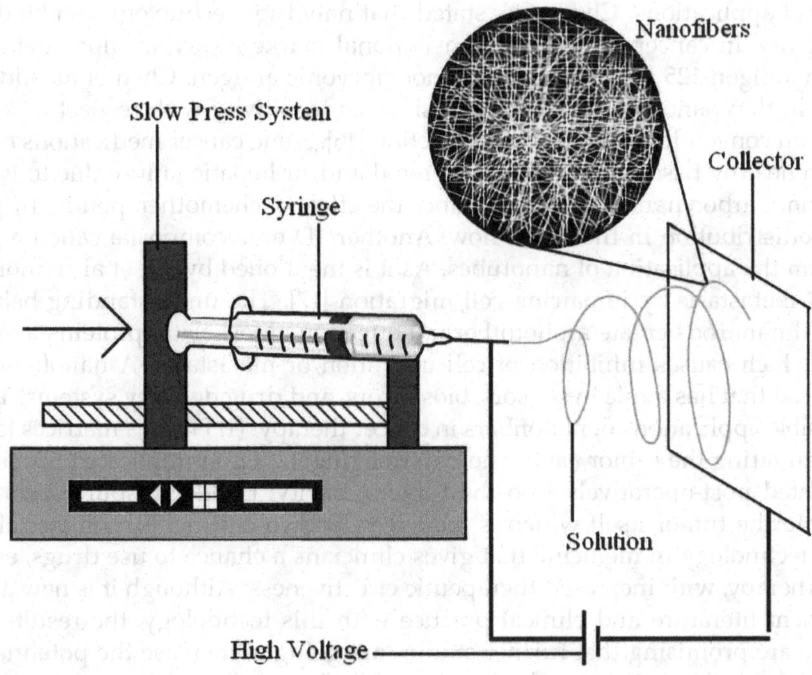

FIGURE 23.4
The electrospinning method for nanofibers.

morphology. For example, Mishra and collaborators performed a study focused on developing eucalyptol/β-cyclodextrin-based gellan/PVA nanofibers using a simple electrospinning method to be used for the antifungal drug delivery system. Results showed that the prepared nanofibers inhibited approximately 70% biofilm formation of *Candida albicans* and *Candida glabrata* [14]. In a study conducted by AnjiReddy et al., chitosan nanofilm and nanofiber have been prepared using the electrospinning method and loaded donepezil as the model drug. *In-vivo* animal model results indicated that the drug reached the maximum concentration within 3.5 h with chitosan nanoformulations to serve as an effective carrier of an anti-Alzheimer's disease drug [15]. The sample accumulates on the collector surface under a high-voltage electric field with specific structures. The schematic diagram of the electrospinning method for nanofibers is presented in Figure 23.4.

23.3 Biomedical Applications of 1D Polymeric Nanocomposite-Based Micro/Nanocontainers: Drug Delivery Systems

Since traditional chemotherapeutics require a high dose of exposure, they result in several side effects. With the help of nanotechnology, drug carriers can make it possible to expose a smaller amount of chemotherapeutics to produce the same effect, which is referred to as targeted therapy. The drug placed into the carrier molecule is not lost nor is it affected by any enzyme on its route to the tumor cell since it is protected by the carrier molecule. Nanorods, nanotubes, and nanofibers are 1D nanocomposites that are popular in biological applications. Chen et.al, stated that nanotube technology can be used as a diagnostic tool in cancer superior to conventional markers such as alpha-beta protein, carcinoma antigen 125 (CA125), and carcinoembryonic antigen. Chen et al. added to all the above in that nanotube technology, besides its sensitivity with respect to diagnosis, is faster than conventional lab tests on detection [16]. Some cancer medications have high toxicity on healthy tissues and may cause renal and/or hepatic failure due to lysosomal degradation. Carbon nanotubes can enhance the effect of chemotherapeutics by protracting the biodistribution in the blood flow. Another 1D nanocomposite called a nanorod differs from the application of nanotubes. As it is mentioned by Ali et al., nanorods aim to prevent metastasis by impairing cell migration [17]. The understanding behind this is that gold nanorods create a photothermal effect on cytoskeletal proteins and surface integrins which causes inhibition of cell migration or metastasis. A nanofiber is a 1D nanomaterial that has a role in sensors, biosensors, and drug delivery systems. There are three possible applications of nanofibers in cancer therapy: (1) Fibrous matrices loaded by the drug targeting the tumor *via* the help of imaging; (2) Drug-implicated fibrous matrices implanted post-operatively into the tumoral cavity; (3) Electrospun fibers injected directly into the tumor itself which is modeled *via* cryo-cutting. Nanomaterials are an emerging technology in medicine that gives clinicians a chance to use drugs, especially in cancer therapy, with increased therapeutic effectiveness. Although it is new and there is insufficient literature and clinical practice with this technology, the results and the knowledge are promising that further studies are going to increase the potential in this field. The schematic diagram of the nanostructure-based drug delivery systems is presented in Figure 23.5.

Nanostructure drug delivery systems of different morphologies, shapes, and sizes have been reported in many studies. Some examples and applications of therapeutic polymeric nanostructures in drug delivery systems are described in Table 23.1. Different drug release kinetic models have been used to explain the drug delivery profile of the 1D polymeric nanocomposite-based micro/nanocontainers. In Table 23.2, the equations of the drug delivery kinetic models, which were widely used in previous studies, are listed.

23.4 1D Polymeric Nanocomposite-Based Micro/Nanocontainer-Based Biosensors: Principle, Components, and Their Applications

Advances in nanoscience have led to the formation and development of biosensors, especially with the emergence of 1D nanostructures. These stable and biocompatible nanoplatforms for biosensors have sensing performances such as electrical/thermal conductivity, sensor functionalization, dispersity, wide bandgap, biosafety, high stability, and distinctive results in biosensing platforms [29]. The development of 1D biosensing nanoplatforms has contributed to the growth phase of sensor technologies. Despite common benefits supported by nanoplatforms, there are still some limitations in large-scale production, such as a lack of statistical methods to analyze complex experimental results.

Considering nonenzymatic or enzymatic electrochemical and optical nanosensors, 1D-based biosensors have been used as electrodes and photon–electron interface nanomaterials for the detection of biological molecules. This chapter broadly presents the recent advances in processing of the developed 1D polymeric micro/nanocontainer-based biosensors. Fabricating different types of functional 1D nanostructures using polymer or

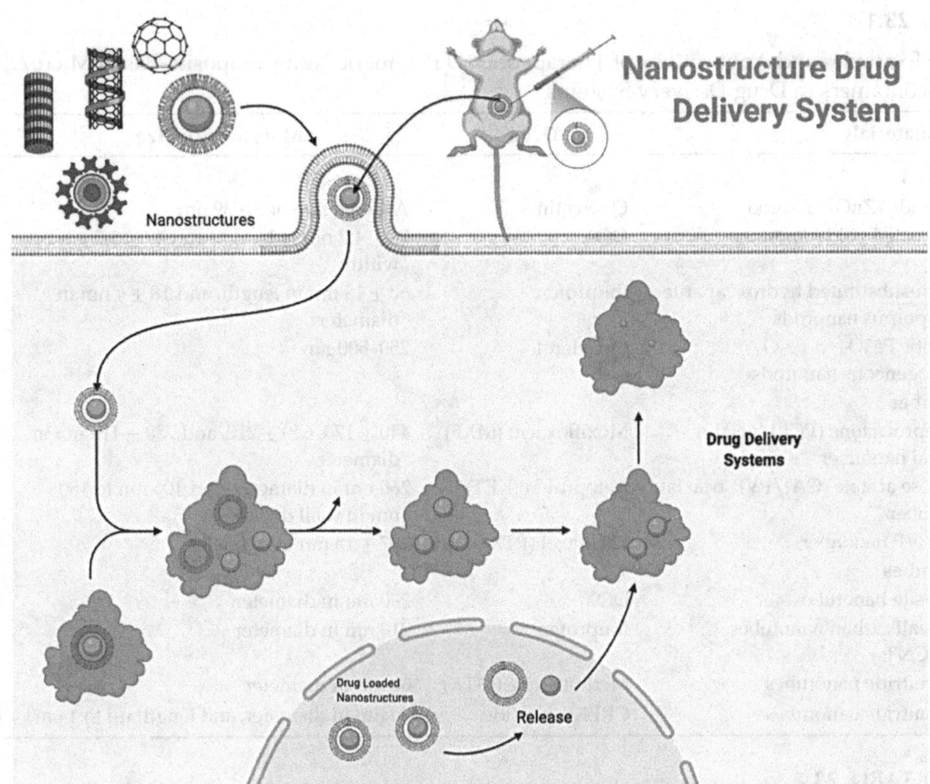

FIGURE 23.5
Drug delivery nanosystems.

polymer blends for sensor applications involves maintaining high electrochemical performance in sensitivity, specificity, selectivity, biocompatibility, and reversibility in a rapid response time. Nowadays, since we are going through the coronavirus pandemic, a wide variety of nano-based electrochemical biosensors have been developed to detect different biological or artificial fluids such as virus samples, blood, urine, saliva, semen, vaginal fluid, and hazardous materials. With the advances of nanotechnology, nanoformulation based on sensors has been fabricated. To date, a variety of nanostructures (e.g., nanorods, nanofibers, quantum dots, nanoparticles, nanowires, and nanotubes) have been utilized for sensor applications. In Table. 23.3, some examples of 1D polymeric nanocomposite-based sensors are presented.

In comparison with conventional sensors, nanostructure-based biosensors can detect the target analyte in pharmaceutical formulations or human biologic fluids at very low levels. These observations and results have demonstrated the advantageous features of nanostructures in pharmaceutical and medical applications. With the rapid advances in 1D polymeric nanostructure-based sensors, Li and collaborators reported a study focused on developing carbon quantum dot-based sensors to be used for food safety [41]. In another study, Lu et al. showed that novel CdS quantum dots-based electrochemiluminescence sensors have considerable potential for detection of DNA damage [42]. Wu et al. fabricated low-cost Fe_3O_4 nanoparticles/fluorinated multi-walled carbon nanotube electrochemical sensors to control pollution of multiple heavy metals in the environment and food [43]. With a similar approach, Wei et al. prepared multi-walled carbon nanotubes @ titanium

TABLE 23.1

Some Examples and Applications of Therapeutic 1D Polymeric Nanocomposite-Based Micro/Nanocontainers in Drug Delivery Systems

Nanomaterials	Drug	Morphology/Size	Ref
Nanorod			
Zinc oxide (ZnO) nanorods	Quercetin	Average size of 21–39 nm	[18]
Silica coated gold nanorods	DOX	78 ± 4.2 nm in length, and 49 ± 3.5 nm in width	[19]
Cerium-substituted hydroxyapatite mesoporous nanorods	Ibuprofen	60 ± 18 nm in length, and 18 ± 4 nm in diameter	[20]
Magnetic Fe3O4/α-Fe_2O_3 heterogeneous nanorods	Celasterol	250–500 nm	[21]
Nanofiber			
Polycaprolactone (PCL)/gelatin, coaxial nanofiber	Moxifloxacin (MXF)	440 ± 170, 650 ± 210, and 780 ± 110 nm in diameter	[22]
Cellulose acetate (CA)/PVP, triaxial nanofiber	Ketoprofen (KET)	260 nm in diameter, and 100 nm to 180 nm in shell thickness	[23]
PVA/PVP nanofiber	Paclitaxel (PTX)	207 ± 18 nm in diameter	[24]
Nanotubes			
Halloysite nanotubes	DOX	200 nm in diameter	[25]
Multiwall carbon nanotubes (MWCNTs)	Ibuprofen	210 nm in diameter	[26]
Boron nitride nanotubes	Methotrexate (MTX)	60 nm in diameter	[27]
Boron nitride nanotubes	CREKA peptide	30 nm in diameter, and length up to 1 μm	[28]

TABLE 23.2

Commonly Used Drug Release Models of the Nanostructures

Kinetic Model	Equation	Graph (x-y axes)
Zero-order	$M_t - M_\infty = K_0 t$	t – (M_t/M_∞)
First-order	$In\, M_t/M_\infty = -K_1 t$	t – [($\ln M_t/M_\infty$)]
Higuchi	$M_t/M_\infty = K_H t^{1/2}$	$t^{1/2}$ – (M_t/M_∞)
Hixson–Crowell	$\sqrt[3]{M_\infty} - \sqrt[3]{M_t} = K_{HC} t$	t – ($M^{1/3}_\infty - M^{1/3}_t$)
Korsmeyer—Peppas	$M_t/M_\infty = K_{KP} t^n$	ln t – [($\ln M_t/M_\infty$)]
Kopcha	$M_t = At^{1/2} + Bt$	t – A or $t^{1/2}$ – B
Bhaskar	$M_t/M_\infty = K_B t^{0.65}$	$t^{0.65}$ – [(ln(1 – M_t/M_∞))]
Elovich	$1 - M_t/M_\infty = a\ln t + b$	$\ln t^0$ – [(1 – (M_t/M_∞)]
Parabolic diffusion	$1 - M_t/M_\infty = K_{Pd} t^{-0.5} + a$	$t^{1/2}$ – (1 – (M_t/M_∞))/t

Where M_t: the amount of cumulative drug released at time t; M_∞: the total amount of loaded drug into the nanostructures; K_0: zero-order rate constant; K_1: first-order rate constant; K_H: Higuchi model rate constant; K_B: Bhaskar rate constant; K_{HC}: Hixson–Crowell rate constant; K_{KP}: Korsmeyer–Peppas rate constant; n: the release exponent; and K_{Pd}: parabolic diffusion rate constant. A and B are diffusion and erosion terms, respectively.

TABLE 23.3
Some Examples of 1D Polymeric Nanocomposite-Based Sensors

Micro/Nanocontainer	Analyte	Electrochemical Results	Morphological Results	Ref.
Nanorods				
Gold nanorod particles on conducting polymer poly(3-octylthiophene)	Glucose	Concentration range: 5–30 mM	Diameter: 24 ± 3 nm and length: 74 ± 7 nm	[30]
Conjugated polymer nanoparticles and Au nanorods	Hyaluronidase	Lower detection limit: 0.003 U/mL	Size: 20–40 nm.	[31]
Gold-conducting polymer nanocomposites	As(III)	Lower detection limit: 2 ppb	Size: 60–400 nm	[32]
Nanowires				
Conductive polymer-coated with gold nanoparticles	—	—	Size: 5 nm	[33]
Polypyrrole nanowires/reduced graphene oxide (GO) composites	Glucose	Response time: 0.5 s and a linear range: 1 nM to 5 μM	Size: 200–300 nm	[34]
Nanotubes				
Carbon nanotube	Glucose	Lower limit of detection: 28 nM, and sensitivity: 1942 nA/μM/cm^2	Length: ~40 μm	[35]
Polypyrrole nanotubes	Glyphosate	Detection range: 2.5–350 ng/mL and limit of detection: 1.94 ng/mL.	Diameter: ~120 nm and wall thickness: ~39 nm	[36]
Nanobelts				
$Ni(OH)_2$@(PEDOT)-rGO nanocomposite	Glucose	Response time: < 1 s, a wide linear range: 0.002–7.1 mM, a low detection limit: 0.6 μM and a high sensitivity: 346 μA/mM/cm^2	Size: 10 nm	[37]
Nanofibers				
Three-dimensional CuPc films decorated with well-ordered PVA parallel nanofiber	NO_2	The response time: 0.02 min and recovery time of the CuPc/PVA PNAs sensors: 25 ppm	—	[38]
CuSn bimetallic nanoparticles/carbon nanofibers	Glucose	Linear range: 0.1–9,000 μM, and low detection limit: 0.08 μM	Diameter: 395 nm	[39]
Cellulose nanofibers	Moisture	—	Thickness: 573 ± 30 nm and a sheet resistance: 458 ± 23Ω/sq	[40]

dioxides with carboxymethyl– chitosan nanocomposite electrochemical sensors for detection of pesticide residues to improve agricultural safety [44]. Hernández et al. highlighted the high potential of nanofibers for sensing cancer at the early stages [45]. To improve the detection sensitivity of sensors against anti-cancer drugs, Afzali and co-workers developed a novel sensor based on ionic liquid molecularly imprinted polymer/gold nanoparticles/graphene oxide for the selective determination of the anti-cancer drug imiquimod and investigated its electrochemical performance by square wave voltammetry and cyclic voltammetry techniques [46]. Lee et al. developed a highly selective and colorimetric urine glucose monitoring system using a paper sensor functionalized with polymer (polyaniline) nanoparticles. The electrochemical results showed that the proposed sensor had the potential for early diagnosis of diabetes with rapid screening of glucose over the range of 0–10 mg/mL with a limit of detection of 0.54 mM [47]. Jadon et al. addressed the need for a hypersensitive drug biosensor that can be used in voltammetric methods to calculate detection and validation parameters of drugs in pharmaceutical formulations and biological fluids. This research paper focused on the simultaneous quantification of amlodipine and lisinopril electrochemically with a PANI-CeO_2 modified gold electrode. The PANI-CeO_2 nanocomposite/Au sensors were obtained using a green sonication method of PANI-CeO_2 nanocomposites. The limits of detection were found to be 0.036 nmol/L for lisinopril and nmol/L for amlodipine. Results showed that the proposed sensor was more sensitive than those reported in previous studies [48]. Xu et al. presented an innovative study based on an electrochemical immune sensor with heparin-PGA-Ppy nanoparticles for the detection of α-fetoprotein in human blood, such as ascorbic acid, bovine serum albumin, uric acid, carcinoma embryonic antigen, and β-D-(+)-glucose. The high biocompatibility, uniform spherical shapes, with an average size of 50 nm for nanoparticles, and high conductivity played a role in the formation of an electrode with high electrochemical characteristics and analytical performance. Results proved that the novel immuno nanosensor developed was a promising sensor for the determination of biomarkers in whole blood [49]. Some recent studies on nanostructure-based electrochemical nanosensors for the detection of various viruses such as SARS-CoV-2, HIV-p24, Zika, HIV-1, and dengue have been reported. Very recently, Tabrizi et al. found that a prepared molecularly imprinted polymer-based electrochemical sensor had high selectivity, sensitivity, and stability for the determination of SARS-CoV-2-RBD using a macroporous gold screen-printed electrode with a low limit of detection of 0.7 pg/mL [50]. Previous studies had reported that the superior electrochemical performances of nanostructures are a rising trend topic in sensor studies with different fabrication strategies. Despite all the other properties mentioned above, nanosensors have some limitations which include the recycling of nanostructures and their association with the Internet-of-Things. Consequently, we strongly believe that the huge demand and support for ultrasensitive nanodevices for early detection of fatal, genetic, or infectious diseases can be increased in the current COVID-19 pandemic.

23.5 Future Trends

This chapter addressed the recent trends for drug delivery systems and sensors through the use of 1D polymeric nanostructure-based microcontainers. Basic emphasis was placed on the discussion of (i) the advanced strategies used to preparation methods of smart nanoformulations, (ii) different structures and morphologies (shape, size, and dispersion), (iii) the polymer matrix and additive, and (iv) the most relevant electrochemical performances of 1D nanostructure-based sensors.

Advances in new strategies focused on 1D nanostructures had led to significant advances in biomedical applications. In particular, the preparation of nanomaterials containing biodegradable and biocompatible polymers using an economic and green approach method has been made possible with a wide variety of properties. It is worth emphasizing the effects of the dual-stimulus- (pH and temperature) responsive nano-systems for drug delivery systems. 1D polymeric nanostructure-based micro/nanocontainers have emerged as promising materials for the next-generation nanostructures used for drug delivery systems with the green strategy. On the other hand, the development of nanostructure-based sensors appears to involve advanced biosensors comprising carbon dots, nanotubes, nanorods, nanoparticles, nanofibers, etc. According to these advances, experimental experience and knowledge on the drug release profile and sensor mechanism of these nanomaterials are not sufficient; therefore, further studies are required in this regard.

Acknowledgments

The authors would like to thank the Istanbul University-Cerrahpasa for their support.

References

1. S.K. Das, A. Bedar, A. Kannan, K. Jasuja, Aqueous dispersions of few-layer-thick chemically modified magnesium diboride nanosheets by ultrasonication assisted exfoliation, *Scientific Reports* 5 (2015) 10522. https://doi.org/10.1038/srep10522.
2. A. Thakur, R. Bharti, R. Sharma, Carbon nanotubes: Types, synthesis, cytotoxicity and applications in biomedical, *Materials Today: Proceedings* (2021). https://doi.org/10.1016/J.MATPR.2021.10.002.
3. A.M. Yamina, M. Fizir, A. Itatahine, H. He, P. Dramou, Preparation of multifunctional PEG-graft-halloysite nanotubes for controlled drug release, tumor cell targeting, and bio-imaging, *Colloids and Surfaces B: Biointerfaces* 170 (2018) 322–329. https://doi.org/10.1016/J.COLSURFB.2018.06.042.
4. P. Ghoderao, S. Sahare, A.A. Kulkarni, T. Bhave, *Paclitaxel Conjugated Magnetic Carbon Nanotubes Induce Apoptosis in Breast Cancer Cells and Breast Cancer Stem Cells in Vitro, Paclitaxel* (2022) 309–331. https://doi.org/10.1016/B978-0-323-90951-8.00001-1.
5. Y.T. Chang, S.H. Wang, M.S. Tsai, C.W. Hsieh, H. Bin Lee, Preparation and physicochemical and cytocompatibility analyses of a magnetic polymer colloid of xanthan gum-chitosan/nickel nanowires, *Results in Physics* 13 (2019) 102224. https://doi.org/10.1016/J.RINP.2019.102224.
6. B. Kannan, D.E. Williams, C. Laslau, J. Travas-Sejdic, A highly sensitive, label-free gene sensor based on a single conducting polymer nanowire, *Biosensors and Bioelectronics* 35(1) (2012) 258–264. https://doi.org/10.1016/J.BIOS.2012.02.058.
7. X. Wang, Q. Li, Z. Zhang, D. Liu, *Coating Polymers on Nanoparticles for Biomedical Uses, Reference Module in Materials Science and Materials Engineering* (2021). https://doi.org/10.1016/B978-0-12-822425-0.00027-0.
8. F. Meng, Z. Yun, G. Yan, G. Wang, C. Liu, Engineering of anticancer drugs entrapped polymeric nanoparticles for the treatment of colorectal cancer therapy, *Process Biochemistry* (2021). https://doi.org/10.1016/J.PROCBIO.2021.09.013.
9. S. Bhattacharya, Fabrication of poly(sarcosine), poly (ethylene glycol), and poly (lactic-co-glycolic acid) polymeric nanoparticles for cancer drug delivery, *Journal of Drug Delivery Science and Technology* 61 (2021) 102194. https://doi.org/10.1016/J.JDDST.2020.102194.
10. W. Dong, H. Liu, S. Sun, Y. Wang, J. Wang, Precise engineering of dual drug-loaded polymeric nanoparticles system to improve the treatment of glioma-specific targeting therapy, *Process Biochemistry* 102 (2021) 341–348. https://doi.org/10.1016/J.PROCBIO.2021.01.018.

11. T. Pelras, H.T.T. Duong, B.J. Kim, B.S. Hawkett, M. Müllner, A 'grafting from' approach to polymer nanorods for pH-triggered intracellular drug delivery, *Polymer* 112 (2017) 244–251. https://doi.org/10.1016/J.POLYMER.2017.02.001.
12. A.R. Bhide, A.B. Jindal, Fabrication and evaluation of artemether loaded polymeric nanorods obtained by mechanical stretching of nanospheres, *International Journal of Pharmaceutics* 605 (2021) 120820. https://doi.org/10.1016/J.IJPHARM.2021.120820.
13. R. Eivazzadeh-Keihan, F. Khalili, H.A.M. Aliabadi, A. Maleki, H. Madanchi, E.Z. Ziabari, M.S. Bani, Alginate hydrogel-polyvinyl alcohol/silk fibroin/magnesium hydroxide nanorods: A novel scaffold with biological and antibacterial activity and improved mechanical properties, *International Journal of Biological Macromolecules* 162 (2020) 1959–1971. https://doi.org/10.1016/J.IJBIOMAC.2020.08.090.
14. P. Mishra, P. Gupta, A.K. Srivastava, K.M. Poluri, R. Prasad, Eucalyptol/ β-cyclodextrin inclusion complex loaded gellan/PVA nanofibers as antifungal drug delivery system, *International Journal of Pharmaceutics* 609 (2021) 121163. https://doi.org/10.1016/J.IJPHARM.2021.121163.
15. K. AnjiReddy, S. Karpagam, Chitosan nanofilm and electrospun nanofiber for quick drug release in the treatment of Alzheimer's disease: In vitro and in vivo evaluation, *International Journal of Biological Macromolecules* 105(1) (2017) 131. https://doi.org/10.1016/J.IJBIOMAC.2017.07.021.
16. Z. Chen, A. Zhang, X. Wang, J. Zhu, Y. Fan, H. Yu, Z. Yang, The advances of carbon nanotubes in cancer diagnostics and therapeutics, *Journal of Nanomaterials* 2017 (2017). https://doi.org/10.1155/2017/3418932.
17. M.R.K. Ali, Y. Wu, Y. Tang, H. Xiao, K. Chen, T. Han, N. Fang, R. Wu, M.A. El-Sayed, Targeting cancer cell integrins using gold nanorods in photothermal therapy inhibits migration through affecting cytoskeletal proteins, *Proceedings of the National Academy of Sciences* 114(28) (2017) E5655–E5663. https://doi.org/10.1073/PNAS.1703151114.
18. P. Sathishkumar, Z. Li, R. Govindan, R. Jayakumar, C. Wang, F. Long Gu, Zinc oxide-quercetin nanocomposite as a smart nano-drug delivery system: Molecular-level interaction studies, *Applied Surface Science* 536 (2021). https://doi.org/10.1016/J.APSUSC.2020.147741.
19. Z. Song, J. Shi, Z. Zhang, Z. Qi, S. Han, S. Cao, Mesoporous silica-coated gold nanorods with a thermally responsive polymeric cap for near-infrared-activated drug delivery, *Journal of Materials Science* 53 (2018) 7165–7179. https://doi.org/10.1007/S10853-018-2117-7.
20. G. Singh, S.S. Jolly, R.P. Singh, Cerium substituted hydroxyapatite mesoporous nanorods: Synthesis and characterization for drug delivery applications, *Materials Today: Proceedings* 28 (2020) 1460–1466. https://doi.org/10.1016/J.MATPR.2020.04.821.
21. R. Liu, Z. Lv, X. Liu, W. Huang, S. Pan, R. Yin, L. Yu, Y. Li, Y. Zhang, S. Zhang, R. Lu, Y. Li, S. Li, Improved delivery system for celastrol-loaded magnetic Fe_3O_4/α-Fe_2O_3 heterogeneous nanorods: HIF-1α-related apoptotic effects on SMMC-7721 cell, *Materials Science and Engineering C* 125 (2021) 112103. https://doi.org/10.1016/J.MSEC.2021.112103.
22. X. Zhang, C. Chi, J. Chen, X. Zhang, M. Gong, X. Wang, J. Yan, R. Shi, L. Zhang, J. Xue, Electrospun quad-axial nanofibers for controlled and sustained drug delivery, *Materials and Design* 206 (2021) 109732. https://doi.org/10.1016/J.MATDES.2021.109732.
23. Y. Yang, S. Chang, Y. Bai, Y. Du, D.G. Yu, Electrospun triaxial nanofibers with middle blank cellulose acetate layers for accurate dual-stage drug release, *Carbohydrate Polymers* 243 (2020) 116477. https://doi.org/10.1016/J.CARBPOL.2020.116477.
24. B. Li, X. Xia, J. Chen, D. Xia, R. Xu, X. Zou, H. Wang, C. Liang, Paclitaxel-loaded lignin particle encapsulated into electrospun PVA/PVP composite nanofiber for effective cervical cancer cell inhibition, *Nanotechnology* 32(1) (2020) 015101. https://doi.org/10.1088/1361-6528/ABB55A.
25. K. Li, Y. Zhang, M. Chen, Y. Hu, W. Jiang, L. Zhou, S. Li, M. Xu, Q. Zhao, R. Wan, Enhanced antitumor efficacy of doxorubicin-encapsulated halloysite nanotubes, *International Journal of Nanomedicine* 13 (2017) 19–30. https://doi.org/10.2147/IJN.S143928.
26. R.H.O. Montes, A.P. Lima, R.R. Cunha, T.J. Guedes, W.T.P. dos Santos, E. Nossol, E.M. Richter, R.A.A. Munoz, Size effects of multi-walled carbon nanotubes on the electrochemical oxidation of propionic acid derivative drugs: Ibuprofen and naproxen, *Journal of Electroanalytical Chemistry* 775 (2016) 342–349. https://doi.org/10.1016/J.JELECHEM.2016.06.026.

27. Z. Mahdavifar, R. Moridzadeh, Theoretical prediction of encapsulation and adsorption of platinum-anticancer drugs into single walled boron nitride and carbon nanotubes, *Journal of Inclusion Phenomena and Macrocyclic Chemistry* 79 (2013) 443–457. https://doi.org/10.1007/S10847-013-0367-1.
28. T.H. Ferreira, L.B. de Oliveira Freitas, R.S. Fernandes, V.M. dos Santos, J.M. Resende, V.N. Cardoso, A.L.B. de Barros, E.M.B. de Sousa, Boron nitride nanotube-CREKA peptide as an effective target system to metastatic breast cancer, *Journal of Pharmaceutical Investigation* 50 (2019) 469–480. https://doi.org/10.1007/S40005-019-00467-7.
29. A. Tripathy, M.J. Nine, F.S. Silva, Biosensing platform on ferrite magnetic nanoparticles: Synthesis, functionalization, mechanism and applications, *Advances in Colloid and Interface Science* 290 (2021) 102380. https://doi.org/10.1016/J.CIS.2021.102380.
30. H. Çiftçi, U. Tamer, Functional gold nanorod particles on conducting polymer poly(3-octylthiophene) as non-enzymatic glucose sensor, *Reactive and Functional Polymers* 72(2) (2012) 127–132. https://doi.org/10.1016/J.REACTFUNCTPOLYM.2011.11.006.
31. Q. Wang, W. Fu, R. Ma, Q. Lu, Z. Liu, J. Yao, Q. Liu, X. Zhang, H. Zhao, A new fluorescence "turn on" nanoplatform based on conjugated polymer nanoparticles and Au nanorods for the detection of hyaluronidase, *Dyes and Pigments* (2021) 109844. https://doi.org/10.1016/J.DYEPIG.2021.109844.
32. K. Radinović, J. Milikić, U. Stamenović, V. Vodnik, M. Otoničar, S. Škapin, B. Šljukić, Tailoring gold-conducting polymer nanocomposites for sensors applications: Proof of concept for As(III) sensing in aqueous media, *Synthetic Metals* 278 (2021) 116834. https://doi.org/10.1016/J.SYNTHMET.2021.116834.
33. F. Zaïbi, I. Slama, C. Okolie, J. Deshmukh, L. Hawco, M. Mastouri, C. Bennett, M. Mkandawire, R. Chtourou, Electro-performance of functionalized silicon nanowires by conductive polymer-coated with gold nanoparticles, *Colloids and Surfaces. Part A: Physicochemical and Engineering Aspects* 589 (2020) 124450. https://doi.org/10.1016/J.COLSURFA.2020.124450.
34. Y. Wang, X. Qing, Q. Zhou, Y. Zhang, Q. Liu, K. Liu, W. Wang, M. Li, Z. Lu, Y. Chen, D. Wang, The woven fiber organic electrochemical transistors based on polypyrrole nanowires/reduced graphene oxide composites for glucose sensing, *Biosensors and Bioelectronics* 95 (2017) 138–145. https://doi.org/10.1016/J.BIOS.2017.04.018.
35. J. Yang, L.C. Jiang, W. de Zhang, S. Gunasekaran, A highly sensitive non-enzymatic glucose sensor based on a simple two-step electrodeposition of cupric oxide (CuO) nanoparticles onto multi-walled carbon nanotube arrays, *Talanta* 82(1) (2010) 25–33. https://doi.org/10.1016/J.TALANTA.2010.03.047.
36. S. Ding, Z. Lyu, S. Li, X. Ruan, M. Fei, Y. Zhou, X. Niu, W. Zhu, D. Du, Y. Lin, Molecularly imprinted polypyrrole nanotubes based electrochemical sensor for glyphosate detection, *Biosensors and Bioelectronics* 191 (2021) 113434. https://doi.org/10.1016/J.BIOS.2021.113434.
37. L. Sheng, Z. Li, A. Meng, Q. Xu, Ultrafast responsive and highly sensitive enzyme-free glucose sensor based on a novel Ni(OH)2@PEDOT-rGO nanocomposite, *Sensors and Actuators. Part B: Chemical* 254 (2018) 1206–1215. https://doi.org/10.1016/J.SNB.2017.08.008.
38. L. Wang, L. Wang, G. Li, Y. Zhu, C. Liu, L. Zeng, S. Zhong, L.J. Wang, Three-dimensional CuPc films decorated with well-ordered PVA parallel nanofiber arrays for low concentration detecting NO_2 sensor, *Sensors and Actuators B: Chemical* 337 (2021) 129781. https://doi.org/10.1016/J.SNB.2021.129781.
39. K. Huan, Y. Li, D. Deng, H. Wang, D. Wang, M. Li, L. Luo, Composite-controlled electrospinning of CuSn bimetallic nanoparticles/carbon nanofibers for electrochemical glucose sensor, *Applied Surface Science* 573 (2022) 151528. https://doi.org/10.1016/J.APSUSC.2021.151528.
40. A. Rivadeneyra, A. Marín-Sánchez, B. Wicklein, J.F. Salmerón, E. Castillo, M. Bobinger, A. Salinas-Castillo, Cellulose nanofibers as substrate for flexible and biodegradable moisture sensors, *Composites Science and Technology* 208 (2021) 108738. https://doi.org/10.1016/J.COMPSCITECH.2021.108738.
41. T. Li, Z. Li, T. Huang, L. Tian, Carbon quantum dot-based sensors for food safety, *Sensors and Actuators, Part A: Physical* 331 (2021) 113003. https://doi.org/10.1016/J.SNA.2021.113003.

42. L.P. Lu, L.H. Xu, T.F. Kang, S.Y. Cheng, DNA damage detection by electrochemiluminescence sensor of CdS quantum dots, *Chinese Journal of Analytical Chemistry* 41(6) (2013) 805–810. https://doi.org/10.1016/S1872-2040(13)60659-3.
43. W. Wu, M. Jia, Z. Zhang, X. Chen, Q. Zhang, W. Zhang, P. Li, L. Chen, Sensitive, selective and simultaneous electrochemical detection of multiple heavy metals in environment and food using a lowcost Fe_3O_4 nanoparticles/fluorinated multi-walled carbon nanotubes sensor, *Ecotoxicology and Environmental Safety* 175 (2019) 243–250. https://doi.org/10.1016/J.ECOENV.2019.03.037.
44. X.P. Wei, Y.L. Luo, F. Xu, Y.S. Chen, L.H. Yang, In-situ non-covalent dressing of multi-walled carbon nanotubes@titanium dioxides with carboxymethyl chitosan nanocomposite electrochemical sensors for detection of pesticide residues, *Materials and Design* 111 (2016) 445–452. https://doi.org/10.1016/J.MATDES.2016.09.023.
45. S.N.H. Hernández, G. Chauhan, Nanofibers for cancer sensing and diagnostics, *Materials Today: Proceedings* (2020). https://doi.org/10.1016/J.MATPR.2020.10.164.
46. M. Afzali, A. Mostafavi, T. Shamspur, Developing a novel sensor based on ionic liquid molecularly imprinted polymer/gold nanoparticles/graphene oxide for the selective determination of an anti-cancer drug imiquimod, *Biosensors and Bioelectronics* 143 (2019) 111620. https://doi.org/10.1016/J.BIOS.2019.111620.
47. T. Lee, I. Kim, D.Y. Cheong, S. Roh, H.G. Jung, S.W. Lee, H.S. Kim, D.S. Yoon, Y. Hong, G. Lee, Selective colorimetric urine glucose detection by paper sensor functionalized with polyaniline nanoparticles and cell membrane, *Analytica Chimica Acta* 1158 (2021) 338387. https://doi.org/10.1016/J.ACA.2021.338387.
48. N. Jadon, A. Pandey, H.K. Sharma, Selective and sensitive PANI-CeO_2 coated gold sensor for electrocatalytic sensing of hypersensitive drugs, *Sensing and Bio-Sensing Research* 22 (2019) 100256. https://doi.org/10.1016/J.SBSR.2019.100256.
49. T. Xu, B. Chi, J. Gao, M. Chu, W. Fan, M. Yi, H. Xu, C. Mao, Novel electrochemical immune sensor based on Hep-PGA-PPy nanoparticles for detection of α-fetoprotein in whole blood, *Analytica Chimica Acta* 977 (2017) 36–43. https://doi.org/10.1016/J.ACA.2017.04.045.
50. M. Amouzadeh Tabrizi, J.P. Fernández-Blázquez, D.M. Medina, P. Acedo, An ultrasensitive molecularly imprinted polymer-based electrochemical sensor for the determination of SARS-CoV-2-RBD by using macroporous gold screen-printed electrode, *Biosensors and Bioelectronics* 196 (2022) 113729. https://doi.org/10.1016/J.BIOS.2021.113729.

24

One-Dimensional Polymeric Nanocomposites for Tissue Engineering

Surabhi Sonam, Sanjay Kumar, and Shashi Singh

CONTENTS

24.1 Introduction

Since prehistoric times, materials have played a significant role in healing and have kept pace with the developments in technology. One of the significant developments in the field of materials has been nanomaterials, with dimensions in the range of 1–100 nm; their dimensionality can vary from 0D to 3D. Many kinds of nanomaterials have proven to be quite suitable for various biological applications including biomedicine and use in the human body. Polymer nanocomposites (PNC) obtained by incorporating nanomaterials into a continuous material phase without losing their identities and improved properties have also proven to be a class apart in terms of their applications. Unique properties of PNC can be attributed to the nano-size and dimension of the filler material [1].

Uniform distribution of individual nanomaterial units across the polymer phase is the hallmark of polymer nanocomposites that results in achieving more nanomaterial–polymer interactions. Polymer nanocomposites with ensured sustained segregation of

DOI: 10.1201/9781003223764-24

nanomaterial display improved mechanical properties, toughness, and conductivity, while retaining high surface-area-to-volume ratio (Figure 24.1). It is more of the interplay between the entropy of nanoparticles and the enthalpy required that describes the dispersion of nanomaterials in the composite [2]. The synthetic methods are hence designed keeping in mind that the fillers need to be dispersed to the point where each nanomaterial unit is coated with polymer. A poor dispersion of filler could lead to materials with low mechanical properties and probably increased brittleness. Alignment of the filler in the polymer matrix is not critical but there needs to be sufficient interaction between the filler and the matrix component.

Shape (aspect ratio), size, quantity, and dispersion of nanofillers affect the properties of the PNC. The region of polymer around each particle is characterized by altered properties like polymer curing or crystalline nature, thereby influencing the overall properties of PNC. The addition of different nanofillers in the composites has been reported to enhance optical properties, dielectric properties, heat resistance, stiffness, strength, wear, damage resistance, etc. In this chapter, we concentrate on the applications of polymers nanocomposites to aid the healing process using tissue engineering techniques.

24.1.1 Basic Concept of Tissue Engineering

Tissue engineering (TE) deals with the repair and replacement of damaged and dysfunctional tissues using engineered functional constructs. Tissue engineering involves the use of materials along with cells to fill in for the extracellular matrix (ECM) (Figure 24.2). ECM, composed of collagen and other proteins, glycoproteins, and proteoglycans, is not only the cementing material between cells but also provides the tissue with its characteristic mechanical properties. ECM is the creator of niches for cell function, behavior, and character status and is the hub for generating cues for the well-being of a tissue. A variety of materials with an equally wide range of techniques have been employed to develop the matrices, and developments in this area continue to meet the demands of diverse tissue types [3]. The ECM has a structure composed of a network of collagen and associated proteins with nano-architectural features, surrounding and supporting the cells. Any artificial structure should be able to hold and replace this structure and function until the

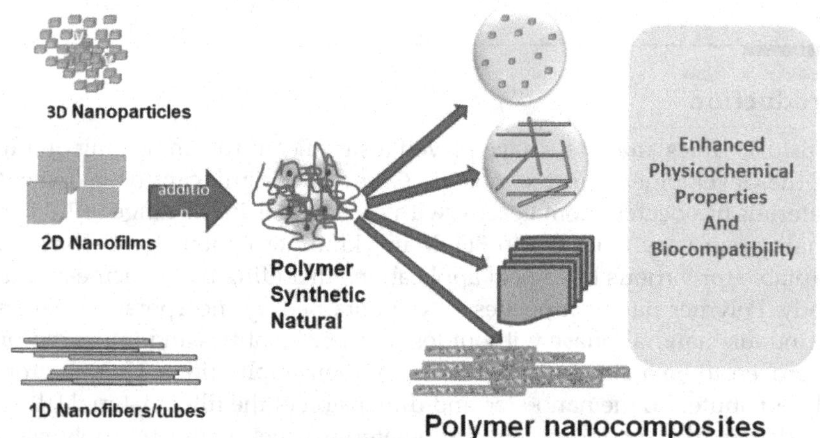

FIGURE 24.1
Concept of polymer nanoscaffolds.

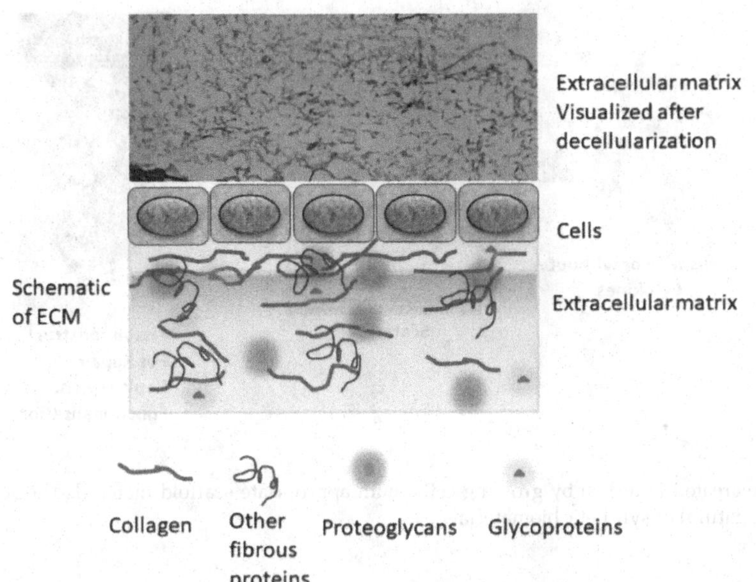

FIGURE 24.2
Extracellular matrix is the cementing material between various cells of the tissue, It can be visualized by removing cells using detergents that leave behind the empty scaffold. ECM is made up of collagen fibrils and other matrix proteins, proteoglycans, and glycoproteins, which crisscross and interact with each other to create a stable, cementing structure.

seeded cells form their own ECM. The scaffold requirement changes depending on the tissue or organ type one plans to recreate. Many specific features need to be incorporated in terms of physical, chemical, and biological requirements. The selection of material turns out to be crucial for success in tissue engineering.

There are certain requirements from a tissue engineering point of view that need to be fulfilled. The material in question has to support the cells, their growth, migration, and proliferation to qualify as being biocompatible. Biocompatibility also encompasses their non-toxic and non-immunogenic behavior in an organism. The tensile strength and mechanical properties of the material have to match the type of tissue this material is going to support. In addition, the material should be able to sustain the character of the tissue and support its function. Some tissues require a precise alignment of cells for their functionality and the scaffolds generated need to provide that guidance. The material is also required to be biodegradable and its biodegradability should be able to match the regenerative potential and time taken for the functional recovery of the damaged tissue. Furthermore, there are requirements for the fabricated scaffold to be moldable and porous to allow the exchange of metabolites for cell sustenance.

Many natural and synthetic polymers have been tested and found to be incapable of meeting all these requirements. Natural biomaterials display good biocompatibility but turn out to have poor mechanical strength. Synthetic materials fulfill the mechanical and chemical requirements but fail to elicit good biological responses. The biological performance of the biomaterials also depends on the biodegradation of the scaffold. Most of the natural biomolecules used as scaffolds are easily degradable by physiological mechanisms whereas the synthetic biomaterials are degraded by hydrolysis and can be prone to toxic accumulation.

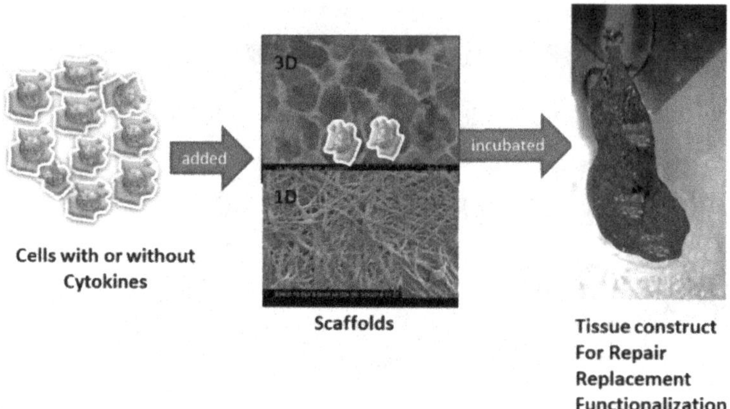

FIGURE 24.3
A tissue can be created in a dish by growing cells in an appropriate scaffold material. The scaffolds can be designed using natural or synthetic biomaterials.

The design and preparation of the multicomponent polymer systems appear to be one viable strategy to develop new, multifunctional biomaterials. The introduction of biomolecules or inorganic molecules into biodegradable polymer matrices is effective at obtaining composites with specific properties. Composite materials using synthetic and natural-based materials are being proposed for tissue engineering applications (Figure 24.3). Polymer nanocomposites may be desirable bioactive materials due to the many interfaces and structures operating at the nanoscale. It is expected that the biological performance of polymer nanocomposites is likely to be better in all terms. Ultimately, the choice of polymer and the choice of biopolymer/molecule will always depend on the kind of replaceable tissue one wants to deal with, the nature of the tissue, and the nature of regeneration time, depending on the size of the injury. Each composite has a different mechanical property, wettability, bioactivity, and degradation rate.

24.2 Applications of 1D Polymer Nanocomposites

Polymer nanocomposites can be designed with different polymers and nanofillers to suit the tissue engineering needs, though this has not been widely used as yet. Several organic nanofillers like nanocellulose, protein nanofibers, and polymer nanofibers have been used as 1D fillers. Besides, polymers with different nanofillers have also been cast in 1-Dimension by electrospinning. The polymer nanocomposites with nanofillers like carbon nanotube (CNT), graphene, carbon nanofibers, and other nanoparticles have been processed into high-performance electrospun nanofibers, many of them with applications in tissue engineering.

There are a large number of strategies by which to fabricate 1D nanocomposites. The processes employed to make polymer nanocomposites for tissue engineering are generally solution-based processes like electrospinning, freeze casting, or high-shear-rate extrusion-based, with pressurized gyration technique, melt-based blending, and *in-situ* polymerization techniques.

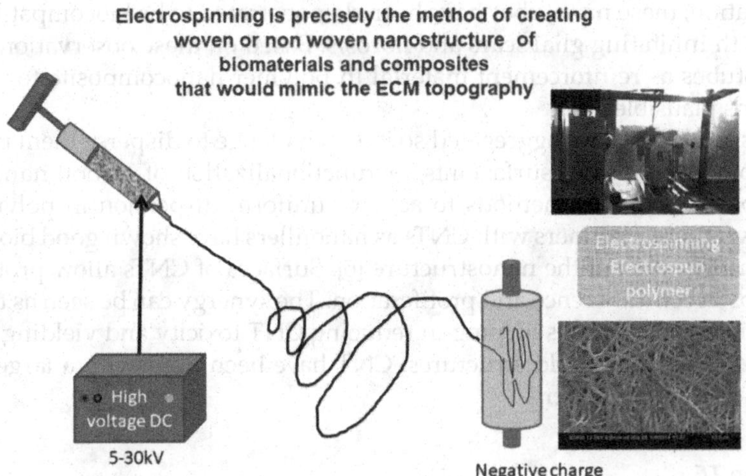

FIGURE 24.4
Electrospinning is the most favored choice of method for creating polymeric 1D structures for tissue engineering.

Because of the dimensions and the ultrastructure they possess, electrospun fibers mimic many human tissues and organs, including cartilage, bone, teeth, and skin that have ECM dimensions of this order. From the tissue engineering point of view, electrospun nanofibers using natural and synthetic polymers, such as proteins, polyurethane (PU), polyvinyl alcohol (PVA), polyethylene oxide (PEO), polycaprolactone (PCL), polylactic acid (PLA), polyglycolide (PGA), polydioxanone, and polyphosphazene derivatives, have been the materials which are quite often reported. These polymers display tunable mechanical properties, thermal stability, optical transparency, and suitable degradation rates. The process of electrospinning produces a network of fibrous mats with variable porosity that is another requirement for an ideal scaffold (Figure 24.4). Layer-by-layer deposition of fibers would result in scaffolds with appropriate microstructure, porosity, and the gradient of components sometimes desired in a scaffold [4]. Even the orientation of fibers as random or aligned makes a difference to the functioning or response of cells. Shafiee et al. [5] have demonstrated different responses of nasal septum progenitor cells on aligned and random PLLA/PCL electrospun fibers: random fibers support proliferation and aligned fibers induce differentiation as observed from expression profiles of chondrogenic markers like collagen II and aggrecan. The aligned fibers also display much better tensile strength properties.

24.2.1 Scaffolds with 1D Nanofillers in Polymer Nanocomposites

Nanotubes made of carbon and other materials, like boron nitride, zinc oxide, and silicon dioxide, show a high aspect ratio, excellent tensile strength, and a structure resembling a straw that can be filled with chemicals and macromolecules. Pristine carbon nanotubes (CNTs) display nanotoxicity towards many cell types and yet also have proven to be versatile as biomaterials. CNTs have shown some promise as a scaffolding material for bone regeneration; and have positioned themselves as biomaterial, in contact with bone-like prostheses for arthroplasty, screws, etc., for fracture fixation. There are many health risks of CNTs as inflammatory agents, once they escape from scaffolds, like becoming toxic or genotoxic. These effects can be avoided by making composites. After functionalization, these nanotubes have been demonstrated to be biocompatible, promoting cell growth-inhibiting glial scars and fibrosis. Based on these observations, the use of carbon nanotubes as reinforcement material in polymer nanocomposite for tissue engineering seems plausible.

Nanotubes tend to remain aggregated so it is a challenge to disperse them uniformly in polymers. Sonication, use of surfactants, or functionalization of carbon nanotubes have been some of the common methods to achieve uniform dispersion in polymers. Many natural and synthetic polymers with CNTs as nanofillers have shown good bioactivity due to the high surface area of the nanostructure [6]. Surfaces of CNTs allow protein adsorption that helps in cell adherence and proliferation. The synergy can be seen as the presence of polymers in the composites helping in reducing CNT toxicity and yielding a collection of mechanically stable scaffold structures. CNT have been used with a large number of natural and synthetic polymers.

24.2.1.1 Bone TE

Bone, an important skeletal element in the body protecting soft tissues, is involved in movements and has healing properties in case of limited injuries. Traumatic multiple fractures, bone loss, or tumor resection require graft interventions, which is where tissue engineering approaches come into play. Hydroxyapatite has been the favored material due to its resemblance with the natural scaffold. However, due to its non-matching mechanical properties, like tensile strength and fracture-prone properties, it cannot be used for load-bearing bone healing. The mechanical properties of CNTs can be used to optimize the bioavailability of hydroxyapatite in bone tissue engineering. CNT-based composites have shown increased potential applications in tissue engineering over the past two decades. The CNT-based composites show not only non-toxic effects but promote cell differentiation into an osteogenic lineage. Carbon nanotubes have gained prominence as reinforcement for weak polymer scaffolds in bone tissue engineering. Zhang et al. [7] demonstrated that PLGA-MWCNT (PLGA-multi-walled carbon nanotube) scaffolds had 54% more tensile strength as compared with pure PLGA scaffolds. The inclusion of MWCNTs in scaffolds increased their Young's modulus and thermal stability. Cell grown on these scaffolds showed longer filopodia, a better 3D cellular network, and 75% greater viability as compared with the control.

The inclusion of MWCNTs in chitosan, a natural polymer, with hydroxyapatite also demonstrated greater mechanical strength and good cell growth of preosteoblastic cell line. The results showed increased biocompatibility and tensile strength as MWCNT content in the composite increased [8]. Enhanced electrical conductivity along with high tensile strength, bioactivity, osteoconductivity, and mineralization were observed with polyurethane foams with CNTs deposited on the surface. The inclusion of growth factors, antibacterial agents, etc., improved the performance of the scaffolds and implants in animal studies.

24.2.1.2 Neuronal TE

Following damage, axons need a proper spatial organization in which to grow and rebuild functional connections. For neuronal regeneration with stem cells, proper material needs to be provided for aligned differentiation of cells. Aligned polymer-CNT fibers have found uses for nerve repair due to their high electrical conductivity [9]. Poly(L-lactic acid) with CNTs in a range of 0.1–5 wt% fabricated into electrospun fibrous mats showed overall higher Young's modulus and elongation break with lower electrical resistance supporting neuronal cell line proliferation [10]. In another study, a novel membrane containing CNTs and poly(dimethyldiallylammonium chloride) (PDDA) *via* layer-by-layer assembly was created. Neural stem cells (NSC) grown on such substrates performed better in terms of proliferation, differentiation, and electrophysiological maturation of neural stem cell-derived neurons. Composites favored long-term survival, better neuron-to-astrocyte ratio, and synaptic connectivity in the presence of CNTs. Initiation of a series of signaling pathways associated with NSC differentiation and cell survival was observed [11]. The use of CNTs as fillers serves two purposes; the cytotoxic effects of CNT can be reduced as that avoids direct exposure of the cells, and over-excitability of cells will be less due to reduced contact between CNTs and the membranes. Direct contact between CNTs and cell membranes causing excess electrical activity could reduce neuronal maturation.

There are very few animal studies using CNT-polymer nanocomposites, most of the studies having been performed *in vitro* under culture conditions. In one of the studies by Ahn et al [12], a 10-mm sciatic nerve defect created in a Wistar rat was implanted with CNT-interfaced glass fiber scaffold placed into three-dimensional poly(L/D-lactic acid) (PLDLA) nanotubes. In a completely transected nerve, implantation of a CNT-interlaced phosphate glass fiber (PGF) system provided a biocompatible nerve-guiding matrix. After 16 weeks, many regenerating neurons were observed with electrophysiological recovery that was aided by the presence of CNTs. The glass fibers did not provide any help in recovery mainly due to their physicochemical/morphological limitations but certainly helped in guiding the nerve growth.

24.2.1.3 Cardiac TE

There have been several attempts to use cell and material therapy for cardiac dysfunction but without much success; drawbacks such as low cell seeding efficiency and lack of implantation, poor survival rate, and low immunogenic response, as well as mechano-electrical shortcomings, mar these efforts. CNTs have also been explored for cardiac tissue engineering to provide both mechanical and electrical conductivity. Efforts are being carried out to achieve the desired mechanical strength, cell adherence, proliferation, differentiation, and remodeling for syncytial electrical conduction. Many CNT-reinforced polymers have shown moderate success with cardiomyocytes. Transfer of signals between cardiomyocytes is important for the cardiac function to maintain the rhythmicity that gets impaired with the use of biomaterials. Gelatin methacrylate (GelMA) and CNT-composite-based patches cultured with rat cardiomyocytes exhibited higher beating rates and improved cell interactions and performances. CNT-reinforced natural polymers like chitosan and collagen also improved the rhythmic activity of cardiac cells [13].

24.2.2 Non-Carbon Nanotubes

Metal/non-carbon nanotubes, such as tungsten disulfide nanotubes (WSNTs), showed greatly improved tensile strengths when electrospun with polymethyl methacrylate (PMMA) as compared with pristine fibers. Owing to better tensile strength and ease of dispersion in polymers, their use for bone polymeric implants is being explored. WS nanomaterial show optical properties that have been exploited by incorporating nanoWS in chitosan hydrogel matrix cross-linked with polyethylene glycol. WS-chitosan nanocomposite helped in the regeneration of skin lesions upon exposure to near-infrared light. The addition of ciprofloxacin into the composite helped in reducing infection [14].

Another non-carbon nanotube with promise in tissue engineering is boron nitride nanotube (BNNT), which has tensile strength similar to carbon, is chemically inert, and is more stable than carbon. BNNTs are non-cytotoxic to a variety of cells. Composites with a variety of polymers, like PMMA, PS, PLC, and PPF, show enhanced mechanical and electrical properties. Osteoblasts grown with PLC-BNNT composites show good osteoinductivity and enhanced expression of the *RUNX2* gene, a master regulator for osteogenesis [15]. By varying the amounts of nontoxic BNNTs in the reinforced polymer scaffolds, these composites can be tailored to the requirement of the bone defect and problems like stress shielding, so that refractures associated with implants can be overcome.

24.2.3 Natural Fibers

There has been considerable use of natural fibers, like jute, flax, hemp, etc., for making composites; these low-cost, renewable, and biodegradable eco-friendly fillers are also energy-economic. Natural fibers are biocompatible, having similarities to macromolecules in the living systems and elicit a better response from the body. Zero chances of leaching out, corrosion, etc., take care of toxic accumulations. Though the natural fibers are interesting candidates as fillers for tissue engineering approaches, there seems to be little effort in this direction. Many fibers like hemp and cellulose have been incorporated into the polymer composites but their use in tissue engineering has been limited. The incorporation of stiff, rod-like cellulose nanostructures has found some mention in a study by Mohan et al. [16]. Poor adhesion and interface remain key issues in the fabrication and final properties of these composites. One could still opt for natural fibers as fillers, as these offer the flexibility of changing relative content of fillers and continuous phase to suit the tissue requirements without being toxic. With better mechanical properties, these composites are less prone to fatigue failures.

Keratin, a skin, feather, hair, and epidermal appendage-dominant intermediate filament protein, can be used in a variety of biomedical applications due to its mechanical durability, biocompatibility, and biodegradability. The omnipresent keratins, with broad tunable structural diversity, lead to a wide range of characteristics features, and functions – impact resistance in hoofs, hydroresistance in whales, aerodynamic resistance in feathers, etc. Keratins have a hierarchical structure that can be a source of bioinspiration, and exploring such systems will help design new composites.

Keratin has poor solubility so keratins dissolved in any non-toxic aqueous solution containing surfactants, like SDS, could be utilized for spinning. Keratin-containing composite nanofibrous scaffolds can be obtained by electrospinning of keratin and other biodegradable polymers such as PLA in one solution [17]. The keratin-containing biodegradable composites could increase cell adhesion and accelerate cell proliferation when compared with the polymeric nanofibrous scaffolds. Osteogenesis of MG-63 cells improved on PLA/W-keratin nanofibrous scaffolds relative to pure PLA.

Composite nanofibers with a core-sheath structure using natural proteins can also be used for tissue-engineering applications. In a core-sheath system, as the name implies, one component is the core that is fully surrounded by the second component, known as the sheath; normally it is the sheath that is required for biocompatibility and biofunctionality. Core-sheath fibers are prepared by electrospinning two polymers coaxially or an emulsion system with control of the overall fiber size and controlled relative diameters of the core and sheath. Zhang et al. [18] prepared collagen-coated PCL nanofibers by a coaxial electrospinning technique. The coatings of collagen on electrospun PCL nanofibers favored cell proliferation and cell migration inside the scaffolds, indicating the excellent integration between cells and the scaffolds. The human dermal fibroblast (HDF) density on the PCL/collagen composite nanofibers with core-sheath structures increased linearly over days as compared with that of PCL nanofibers or PCL fibers randomly coated with collagen.

24.2.4 Conducting 1D Polymers

Based on the premise that many cell activities are modulated through electrical stimulation, an increasing interest in conductive polymers has been witnessed in recent years. Apart from many physiological functions like nerve conduction and muscle contractility, functions like cell attachment, migration, and differentiation are also modulated with changes in an electric field. From the plethora of conducting polymers like polyacetylene, polypyrrole(PPy), polyaniline (PANi), polythiophene, and poly(para-phenylene vinylene); PPy, and PANi have found uses in tissue engineering due to their biocompatibility [19]. 1D conducting polymers have good electrical properties, electro-activity, and large surface area, and are good candidates for use in the field of tissue engineering. In a composite form, some of these polymers can perform better in terms of porosity, or overcome the drawbacks of their crystalline nature.

Electrospinning a polymer solution containing conducting polymers and a biocompatible polymer could produce conductive composite nanofibers. Along with electrical activity, these fibers display the usual high surface area, good biocompatibility, improved tensile properties, and even appropriate biodegradability suitable for tissue engineering.

PANi/gelatin composite nanofibers fabricated by electrospinning showed inverse size relation with increasing concentration of PANi along with increased tensile strength [20]. These conductive polymer composites supported the growth of cardiomyoblast cells (H9c2) with possible uses in cardiac tissue engineering efforts [20].

PANi/PLCL composite nanofibers composed of PANi and CSA-doped PLCL exhibited higher adhesion for fibroblast cell lines and myoblasts and showed better proliferation rates. Cell growth under the electrical stimulation is also reported to be enhanced [21]. Adult porcine skeletal muscle cells attached and grew well on the synthesized PANI/collagen composite films, indicating their good support for cell growth. The conductive nature of these composites allows cells to overcome the impedance induced by injury.

Composite nanofibers with core-sheath structures are also fabricated by chemical or electrical deposition of conducting polymers on surface electrospun polymers. PCL/PPy and PLA/PPy core-sheath nanofibers, prepared using electrospun PCL and PLA nanofibers as templates and PPy coating on them, exhibited enhanced growth of neurites in orientation to the aligned fibers. Upon application of an electrical stimulus, the neurites exhibited growth with an increase in maximum length of about 83% [22].

The fabrication of extensively aligned and electrically conductive nanofibers to allow cells to mimic their natural alignment is a challenging task but important for functional accuracy. Lee et al. [23] fabricated random and aligned PLGA/PPy core-sheath nanofibers using electrospinning and *in-situ* chemical polymerization for neural tissue engineering. Electrical stimulation at a slow voltage as 10 mV/cm demonstrated enhancement for the neurite outgrowth. It was found that lower potential (10 mV/cm) was more beneficial for neurite growth than higher potential (100 mV/cm). Without any stimulation, there was no difference in the growth of cells with or without PANi on the fibers. In addition, aligned PLGA/PPy core-sheath nanofibers also exhibited the formation of more neurite-bearing cells and longer neurites compared with the random PLGA/PPy core-sheath nanofibers under electrical stimulation.

24.3 Composites with Nanoparticles to Make 1D Polymer Composites

Inorganic nano components are often incorporated into polymer nanofibers for the application of tissue engineering. Fibrous electrospun nanomaterials have been composed with a variety of natural synthetic or inorganic nanoparticles as fillers in different polymers, using a myriad of techniques.

The nanoparticles can be directly added to the polymer solution and spun as composite fibers. Marino et al. [24] have developed gelatin/cerium oxide nanoparticles electrospun into highly aligned nanocomposite fibers. Cerium oxide nanoparticles (nanoceria) effectively control ROS, inhibit cell senescence, and promote neurite development. Due to the antioxidant nature of nanoceria, an improved phenotype of neurons was seen on the nanoceria gelatin fibers. The presence of nanoceria also tends to reduce the inflammation that is a common occurrence when using any kind of conduit. Its incorporation into scaffolds not only inhibits such side effects, it also can self-regenerate [25].

Silver nanoparticles electrospun with many synthetic fibers have been able to show antibacterial activity along with cytohealing. Xing et al. [26] prepared the poly(3-hydroxyb utyrate-co-3-hydroxyvalerate)/silver (PHBV/Ag) nanocomposite fibrous scaffolds that inhibited the proliferation of bacteria; because the composites did not show *in-vitro* cell cytotoxicity, such scaffolds have the potential to heal arthro-injuries.

Bone tissue engineering is the field where composites of inorganic with polymer scaffolds have maximum potential to be utilized. The electrospun polymer scaffold is made up of either synthetic biodegradable or natural polymer with the incorporation of calcium carbonate, calcium phosphate, and hydroxyapatite (HA), Tri-calcium phosphates (TCP), with a similar crystallographic structure, were always suitable to assist in bone cell regeneration [26–28].

Fujihara et al [27] fabricated two-layered PCL fibers by electrospinning the mixture of PCL and calcium carbonate on the surface of electrospun PCL nanofibrous mats. These mats consisting of both a functional layer (PCL/calcium carbonate) and a mechanical support layer (PCL), supported bone cell growth. Similar results have also been observed for β-tricalcium phosphate (β-TCP) incorporated into electrospun PLA scaffolds and HA with PLA [28]. The use of hydrosteric acid as surfactant ensured uniform distribution of these hydrophilic nanoparticles, resulting in the fabrication of continuous and uniform nanofibers with diameters of about 1–2 μm [29]. Incorporation of HA with biomolecular 1D fibers of collagen/gelatin to create composite nanofibers always demonstrated enhanced potential bone tissue engineering as compared with nanofibers without apatite [30]. These composites always showed better growth of osteogenic cells and better expression of osteogenic markers in the cells.

Hydroxyapatite or other ceramics are quite brittle and, when combined with polymers, always improve the mechanical properties resembling bone matrix in composition, morphology, and nano-architecture. To achieve a biomimetic replica, the loading of nanoparticles has to be appropriate; apart from the ratio, the processing and dispersion of particles need attention and a suitable desirable material is one that shows the best bioactivity at the implant site. HA composed with electrospun polymer has also been successful in inducing differentiation of human mesenchymal stem cells (hMSCs) in chondrocyte-like cells that produce proteoglycan-based matrix [31].

24.4 1D Polymeric Nanocomposites for Organ-on-a-Chip

Organs-on-a-chip (OOCs) are realistic models which are designed to mimic the biological functions of tissues or organs for disease modeling and drug testing. Cells are grown inside predesigned channels to replicate the complete functionality and physiology of tissues and organs [32]. Such models aim to achieve physiological values of pressure, flow rate, pH, osmotic pressure, nutrient content, and toxin presence, among other properties (Figure 24.5). Along with the physiological condition, a mechanical condition experienced by the tissues in the body also needs to be mimicked. To imitate physiological and mechanical conditions, microfluidics, along with cells, is used. Combining microfluidics (using channels, chambers and membranes, etc) with relevant substrate architecture allows considerable control of the movement and behavior of materials and cells. These devices are generally fabricated using soft lithography, photolithography, BioMEMS (BioMicroElectroMechanical Systems), and material development techniques, which permit proper replication of the micro-scale details. These fabrication techniques also allow the use of different materials, including thermoplastics and thermoset polymers.

Consideration of substrate microarchitecture is critical for such platforms and 1D polymer nanocomposites hold a significant role in modifying the surface topography of cell attachment. Here, we discuss some of the OOC models that have been developed thus far and we then speculate the integration of various possible 1D polymer nanocomposites to make better applications of these OOCs.

24.4.1 Lung-on-a-Chip

Huh et al. [33] reconstructed a biomimetic lung-on-a-chip microsystem that reconstituted the fundamental function of the lung by recreating the functional alveolar–capillary interface of the human lung. This system also aided in the characterization of the inflammatory response to bacteria. The primary objectives of the *in-vivo* models are to aid in cell organization and to mimic the cyclic stress experienced by the diaphragm during the breathing

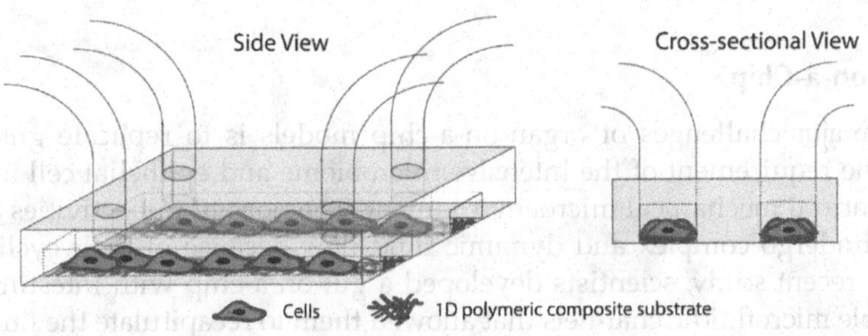

FIGURE 24.5
Organ-on-a-chip models integrated with a 1D polymeric composite substrate which allow active responses from the tissue cultures enhanced by mimicking natural tissue environment.

process [34]. As proof of principle, it is needed to engineer a multifunctional microdevice that reproduces the critical structural, functional, and mechanical properties of the human alveolar–capillary interface. Using microfluidics, a design of two closely placed micro-channels linked in parallel and separated by a thin (10 μm), porous, flexible membrane made of PDMS, many chips have been developed which allow stretching of membranes and which mimic the expansion and relaxation of alveoli during breathing. On the other side of the air exchange, the endothelial cells are exposed to blood-like fluid flow which also undergoes stretching due to blood flow. *In-vitro* 2D studies suggest that the adhesion characteristics of endothelial cells are altered due to the blood flow and the cells align along the flow direction.

The mechanically induced cellular organization can be further controlled by the architectural design of the extracellular matrix. 1D polymeric nanocomposites, like poly(vinylidene fluoride-co-trifluoroethylene) incorporating zinc oxide, have shown promising anti-inflammatory characteristics, prevention against microbial infections, sustained cell growth, and provided physiological signals, like piezoelectricity [35]. Employing these composites with a lung-on-a-chip model can add to the understanding of lung infections and respiratory disorders.

24.4.2 Liver-on-a-Chip

In the case of liver function, high metabolic activity is critical for body function. Chronic illnesses and viral infections have detrimental effects on liver tissue and it suffers beyond its regenerative capacity. To answer the questions pertaining to hepatocyte interactions, microfluidic devices with three-dimensional growth of hepatic cell cultures have been developed. Such devices can sustain monocultures and co-cultures of hepatocytes and hepatic stellate cells to study their responses in the presence or absence of the flow [36]. These devices are crucial to perform drug analysis, toxicity and screening, pathophysiology, and human physiology. Micro-engineered liver chips have also been developed for drug toxicity testing. Again, 3D cellular structures were reconstructed to represent the hepatic sinusoid [37].

Nanofibrous scaffolds of nanosilver impregnated nanoporous sodium alginate (SA)/polyvinyl alcohol (PVA) composite, made by electrospinning, allow hepatocellular regeneration [38]. Combining them in the microfluidic platforms to make 3D hepatic cell cultures will certainly allow the replication of the regeneration capacity of the liver tissue under the influence of different drugs. At the same time, a nanofibrous scaffold fabricated by combining polycaprolactone (PCL) and chitosan (CS) with better orientation and pore size for cell infiltration has enhanced mechanical properties, roughness, topology, structure, hydrophilicity, and cell growth which helps in generating better liver tissue [39].

24.4.3 Gut-on-a-Chip

One of the major challenges of organ-on-a-chip models is to replicate gut physiology because of the requirement of the intensive microbiome and epithelial cell interaction to achieve its natural mechanical microenvironment. The normal gut activities also require the cells to undergo complex and dynamic fluid flow because of their cyclic peristaltic motion. In a recent study, scientists developed a gut-on-a-chip with intestinal epithelial cells in flexible microfluidic channels that allowed them to recapitulate the fluidic dynamics by creating a peristaltic movement. The tissue formed undulating epithelium columns with polarized Caco-2 cells and multiple differentiated intestinal cell types [40, 41]. Incorporation of co-electrospun polysuccinimide/poly(vinyl alcohol) (PSI/PVA) meshes can help in strengthening the mechanical interaction of the cells with the extracellular matrix. Such meshes form 3D interpenetrating systems in which cells identify as 1D platforms and create 3D undulating columns [42]. Electrospun nanofibers incorporated with inulin provide a space for prebiotic activity associated with gut epithelium [43, 44]. The enhanced interaction between microbes and human cells will certainly allow the real assessment of absorption and assimilation of food when it passes through the intestine.

24.4.4 Skin-on-a-Chip

Skin is an important organ that provides a protective layer to the body against all environmental abrasions. Skin cells undergo several mechanical and chemical transformations to maintain homeostasis and prevent instability. A sandwich structure made by a membrane of polylactic acid (PLA)/nano-hydroxyapatite (nHAP) and polyvinyl alcohol (PVA)/sodium alginate (SA)/nHAP was fabricated for repairing a skull defect [45]. A mix of hydrophilic and hydrophobic layers allowed proper integration of membranes for tissue defect repair.

Skin models are mostly developed as grafts for wounds. A tissue-engineered scaffold made of polyurethane-gelatin composite is promising for skin regeneration and repair of complex wounds [46]. Mixing graphene in such composites will also integrate the electrical properties of the skin [47].

24.4.5 Brain-on-a-Chip

Research to replicate brain function has focused on brain formation with emphasis on its complexity and interconnectivity. Applications of brain cell functions target its interaction with drugs to cure degenerative diseases. To do so, a multilayer device was developed with pluripotent human cells to mimic the central nervous system and a functional blood-brain barrier. This model helped in understanding neurodevelopment and neuroregeneration [48]. In another work, an *in-vitro* model of Alzheimer's disease was reproduced on a chip by growing neurosteroids with flow control.

Collagen/poly(D,L-lactic-co-glycolic acid) composite fibrous scaffold has been used for dura repair [49]. In another study, fibers of cellulose acetate composites have been used for recreating the blood–brain barrier [50]. Incorporating these composites in the chip will increase the control and growth of brain cells.

24.5 Conclusion

1D polymer nanocomposites with appropriate nanofillers are promising new materials for tissue regeneration and repair strategies; with a thoughtful addition of fillers, one can think of designing a smart functional material suitable to address the individual needs of constituent cells. Polymer nanocomposites enhance properties favorable for cell growth and at the same time suppresses incompatible aspects of filler. In addition to tissue engineering, the applications of 1D polymeric composites with organ-on-a-chip can be envisaged. It will certainly strengthen the mechanical, biophysical, and biochemical responses from the tissue. Such a responsive system will help in replicating the behavior of the natural tissue which can be further used for tissue replacement. Combining all the organs to make a human-on-a-chip model will make the measurements retrieved out of such models more accurate and closer to physiological values. These multi-organ chip models are promising alternatives to animal-based testing models.

References

1. S. Fu, Z. Sun, P. Huang, Y. Li, and N. Hu, "Some basic aspects of polymer nanocomposites: A critical review," *Nano Mater. Sci.*, vol. 1, no. 1, pp. 2–30, Mar. 2019.
2. S. Porel, S. Shashi, H. S. Sree, R. D. Narayana, and R. T. P., "Nanoparticle-embedded polymer: In situ synthesis, free-standing films with highly monodisperse silver nanoparticles and optical limiting," *Chem. Mater.*, vol. 17, no. 1, pp. 9–12, Jan. 2004.
3. R. Rekulapally, K. Udayachandrika, S. Hamlipur, A. Sasidharan Nair, B. Pal, and S. Singh, "Tissue engineering of collagen scaffolds crosslinked with plant based polysaccharides," *Prog. Biomater.*, vol. 10, no. 1, pp. 29–41, Feb. 2021.
4. R. Scaffaro, A. Maio, F. Lopresti, and L. Botta, "Nanocarbons in electrospun polymeric nanomats for tissue engineering: A review," *Polym.*, vol. 9, no. 2, p. 76, Feb. 2017.
5. A. Shafiee, E. Seyedjafari, E. Sadat Taherzadeh, P. Dinarvand, M. Soleimani, and J. Ai, "Enhanced chondrogenesis of human nasal septum derived progenitors on nanofibrous scaffolds," *Mater. Sci. Eng. C Mater. Biol. Appl.*, vol. 40, pp. 445–454, Jul. 2014.
6. W. Wang, S. Liao, M. Liu, Q. Zhao, and Y. Zhu, "Polymer composites reinforced by nanotubes as scaffolds for tissue engineering," *Int. J. Polym. Sci.*, vol. 2014, 2014, pp. 805634.
7. H. Zhang, "Electrospun poly (lactic-co-glycolic acid)/ multiwalled carbon nanotubes composite scaffolds for guided bone tissue regeneration," *J. Bioact. Compat. Polym.*, vol. 26, no. 4, pp. 347–362, Jun. 2011.
8. L. Chen, J. Hu, X. Shen, and H. Tong, "Synthesis and characterization of chitosan–multiwalled carbon nanotubes/hydroxyapatite nanocomposites for bone tissue engineering," *J. Mater. Sci. Mater. Med.*, vol. 24, no. 8, pp. 1843–1851, May 2013.
9. Y. Xia, S. Li, C. Nie, J. Zhang, S. Zhou, H. Yang, M. Li, W. Li, C. Cheng, and R. Haag, "A multivalent polyanion-dispersed carbon nanotube toward highly bioactive nanostructured fibrous stem cell scaffolds," *Appl. Mater. Today*, vol. 16, pp. 518–528, Sep. 2019.
10. N. Vicentini, T. Gatti, M. Salerno, Y. S. Hernandez Gomez, M. Bellon, S. Gallio, C. Marega, F. Filippini, and E. Menna, "Effect of different functionalized carbon nanostructures as fillers on the physical properties of biocompatible poly(l-lactic acid) composites," *Mater. Chem. Phys.*, vol. 214, pp. 265–276, Aug. 2018.
11. H. Shao, T. Li, R. Zhu, X. Xu, J. Yu, S. Chen, L. Song, S. Ramakrishna, Z. Lei, Y. Ruan, and L. He, "Carbon nanotube multilayered nanocomposites as multifunctional substrates for actuating neuronal differentiation and functions of neural stem cells," *Biomaterials*, vol. 175, pp. 93–109, Aug. 2018.
12. H. S. Ahn, J. Y. Hwang, M. S. Kim, J. Y. Lee, J. W. Kim, H. S. Kim, U. S. Shin, J. C. Knowles, H. W. Kim, and J. K. Hyun, "Carbon-nanotube-interfaced glass fiber scaffold for regeneration of transected sciatic nerve," *Acta Biomater.*, vol. 13, pp. 324–334, Feb. 2015.
13. S. Mombini, J. Mohammadnejad, B. Bakhshandeh, A. Narmani, J. Nourmohammadi, S. Vahdat, and S. Zirak, "Chitosan-PVA-CNT nanofibers as electrically conductive scaffolds for cardiovascular tissue engineering," *Int. J. Biol. Macromol.*, vol. 140, pp. 278–287, Nov. 2019.
14. N. Yang, M. Zhu, G. Xu, N. Liu, and C. Yu, "A near-infrared light-responsive multifunctional nanocomposite hydrogel for efficient and synergistic antibacterial wound therapy and healing promotion," *J. Mater. Chem. B*, vol. 8, no. 17, pp. 3908–3917, May 2020.
15. D. Lahiri, V. Singh, A. Benaduce, S. Seal, L. Kos, and A. Agarwal, "Boron nitride nanotube reinforced hydroxyapatite composite: Mechanical and tribological performance and in-vitro biocompatibility to osteoblasts," *J. Mech. Behav. Biomed. Mater.*, vol. 4, no. 1, pp. 44–56, Jan. 2011.

16. T. Mohan, S. Hribernik, R. Kargl, and K. Stana-Kleinschek, "Nanocellulosic materials in tissue engineering applications," in M. Poletto, H. L. Ornaghi Jr. (eds), *Cellulose*, London, UK: IntechOpen, Dec. 2015.
17. B. Kim, K. Park, K. Park, and J. Lee, "Fabrication of nanofibrous scaffold using a PLA and hagfish thread keratin composite; its effect on cell adherence, growth, and osteoblast differentiation," *Biomed. Mater.*, vol. 8, no. 4, 2013.
18. Y. Zhang, J. Venugopal, Z. Huang, C. Lim, and S. Ramakrishna, "Characterization of the surface biocompatibility of the electrospun PCL-collagen nanofibers using fibroblasts," *Biomacromolecules*, vol. 6, no. 5, pp. 2583–2589, Sep. 2005.
19. A. Kotwal, and C. Schmidt, "Electrical stimulation alters protein adsorption and nerve cell interactions with electrically conducting biomaterials," *Biomaterials*, vol. 22, no. 10, pp. 1055–1064, 2001.
20. M. Li, Y. Guo, Y. Wei, A. G. MacDiarmid, and P. I. Lelkes, "Electrospinning polyaniline-contained gelatin nanofibers for tissue engineering applications," *Biomaterials*, vol. 27, no. 13, pp. 2705–2715, May 2006.
21. S. Jeong, I. Jun, M. Choi, Y. Nho, Y. Lee, and H. Shin, "Development of electroactive and elastic nanofibers that contain polyaniline and poly(L-lactide-co-epsilon-caprolactone) for the control of cell adhesion," *Macromol. Biosci.*, vol. 8, no. 7, pp. 627–637, Jul. 2008.
22. J. Xie, M. Macewan, S. Willerth, X. Li, D. Moran, S. Sakiyama-Elbert, and Y. Xia, "Conductive core-sheath nanofibers and their potential application in neural tissue engineering," *Adv. Funct. Mater.*, vol. 19, no. 14, pp. 2312–2318, Jul. 2009.
23. J. Lee, C. Bashur, A. Goldstein, and C. Schmidt, "Polypyrrole-coated electrospun PLGA nanofibers for neural tissue applications," *Biomaterials*, vol. 30, no. 26, pp. 4325–4335, Sep. 2009.
24. A. Marino, C. Tonda-Turo, D. De Pasquale, F. Ruini, G. Genchi, S. Nitti, V. Cappello, M. Gemmi, V. Mattoli, G. Ciardelli, and G. Ciofani, "Gelatin/nanoceria nanocomposite fibers as antioxidant scaffolds for neuronal regeneration," *Biochim. Biophys. Acta. Gen. Subj.*, vol. 1861, no. 2, pp. 386–395, Feb. 2017.
25. S. A. Gehrke, J. A. Shibli, and M. B. Salles, "Potential of the use of an antioxidant compound to promote peripheral nerve regeneration after injury," *Neural Regen. Res.*, vol. 10, no. 7, p. 1063, Jul. 2015.
26. Z.-C. Xing, W.-P. Chae, J.-Y. Baek, M.-J. Choi, Y. Jung, and I.-K. Kang, "In vitro assessment of antibacterial activity and cytocompatibility of silver-containing PHBV nanofibrous scaffolds for tissue engineering," *Biomacromolecules*, vol. 11, no. 5, pp. 1248–1253, May 2010.
27. K. Fujihara, M. Kotaki, and S. Ramakrishna, "Guided bone regeneration membrane made of polycaprolactone/calcium carbonate composite nano-fibers," *Biomaterials*, vol. 26, no. 19, pp. 4139–4147, Jul. 2005.
28. H. S. Fan, X. T. Wen, Y. F. Tan, R. Wang, H. D. Cao, and X. D. Zhang, "Compare of electrospinning PLA and PLA/β-TCP scaffold in vitro," *Mater. Sci. Forum*, vol. 475–479, pp. 2379–2382, Jan. 2005.
29. H. Kim, H. Lee, and J. Knowles, "Electrospinning biomedical nanocomposite fibers of hydroxyapatite/poly(lactic acid) for bone regeneration," *J. Biomed. Mater. Res. A*, vol. 79, no. 3, pp. 643–649, Dec. 2006.
30. J. Venugopal, S. Low, A. Choon, T. Sampath Kumar, and S. Ramakrishna, "Mineralization of osteoblasts with electrospun collagen/hydroxyapatite nanofibers," *J. Mater. Sci. Mater. Med.*, vol. 19, no. 5, pp. 2039–2046, May 2008.
31. C. Spadaccio, A. Rainer, M. Trombetta, G. Vadalá, M. Chello, E. Covino, V. Denaro, Y. Toyoda, and J. Genovese, "Poly-l-lactic acid/hydroxyapatite electrospun nanocomposites induce chondrogenic differentiation of human MSC," *Ann. Biomed. Eng.*, vol. 37, no. 7, pp. 1376–1389, Jul. 2009.
32. J. P. Wikswo, "The relevance and potential roles of microphysiological systems in biology and medicine," *Exp Biol Med (Maywood)*, vol. 239, no. 9, pp. 1061–1072, Sep. 2014. https://doi.org/10.1177/1535370214542068.
33. D. Huh, B. D. Matthews, A. Mammoto, M. Montoya-Zavala, H. Y. Hsin, and D. E. Ingber, "Reconstituting organ-level lung functions on a chip," *Science*, vol. 328, no. 5986, p. 1662, Jun. 2010.

34. A. Stucki, J. Stucki, S. Hall, M. Felder, Y. Mermoud, R. Schmid, T. Geiser, and O. T. Guenat, "A lung-on-a-chip array with an integrated bio-inspired respiration mechanism," *Lab Chip*, vol. 15, no. 5, pp. 1302–1310, Mar. 2015.
35. B. Azimi, M. S. S. Bafqi, A. Fusco, C. Ricci, G. Gallone, R. Bagherzadeh, G. Donnarumma, M. J. Uddin, M. Latifi, A. Lazzeri, and S. Danti, "Electrospun ZnO/poly(vinylidene fluoride-trifluoroethylene) scaffolds for lung tissue engineering," *Tissue Eng Part A*, vol. 26, no. 23–24, pp. 1312–1331, Dec. 2020. https://home.liebertpub.com/tea.
36. S.-A. Lee, D. Y. No, E. Kang, J. Ju, D.-S. Kim, and S.-H. Lee, "Spheroid-based three-dimensional liver-on-a-chip to investigate hepatocyte–hepatic stellate cell interactions and flow effects," *Lab Chip*, vol. 13, no. 18, pp. 3529–3537, Aug. 2013.
37. B. Delalat, C. Cozzi, S. R. Ghaemi, G. Polito, F. H. Kriel, T. D. Michl, F. J. Harding, C. Priest, G. Barillaro, and N. H. Voelcker, "Microengineered bioartificial liver chip for drug toxicity screening," *Adv. Funct. Mater.*, vol. 28, no. 28, p. 1801825, Jul. 2018.
38. G. Manikandan, M. Yuvashree, A. Sangeetha, K. P. Bhuvana, and S. K. Nayak, "Liver tissue regeneration using nano silver impregnated sodium alginate/PVA composite nanofibres," *SciMedicine J.*, vol. 2, no. 1, pp. 16–21, Mar. 2020.
39. D. Semnani, E. Naghashzargar, M. Hadjianfar, F. D. Manshadi, S. Mohammadi, S. Karbasi, and F. Effaty, "Evaluation of PCL/chitosan electrospun nanofibers for liver tissue engineering," *Int. J. Polym. Mater. Polym. Biomater.*, vol. 66, no. 3, pp. 149–157, Feb. 2016. https://doi.org/10.1080/00914037.2016.1190931.
40. H. J. Kim, D. Huh, G. Hamilton, and D. E. Ingber, "Human gut-on-a-chip inhabited by microbial flora that experiences intestinal peristalsis-like motions and flow," *Lab Chip*, vol. 12, no. 12, pp. 2165–2174, 2012.
41. H. J. Kim, and D. E. Ingber, "Gut-on-a-chip microenvironment induces human intestinal cells to undergo villus differentiation," *Integr. Biol. (Camb)*, vol. 5, no. 9, pp. 1130–1140, 2013.
42. C. Voniatis, L. Balsevicius, D. Barczikai, D. Juriga, A. Takács, L. Kőhidai, K. Nagy, and A. Jedlovszky-Hajdu, "Co-electrospun polysuccinimide/poly(vinyl alcohol) composite meshes for tissue engineering," *J. Mol. Liq.*, vol. 306, p. 112895, May 2020.
43. W. Wahbi, R. Siam, J. Kegere, W. A. El-Mehalmey, and W. Mamdouh, "Novel inulin electrospun composite nanofibers: prebiotic and antibacterial activities," *ACS Omega*, vol. 5, no. 0036, pp. 3006–3015, 2020.
44. D. Duman, and A. Karadag, "Inulin added electrospun composite nanofibres by electrospinning for the encapsulation of probiotics: Characterisation and assessment of viability during storage and simulated gastrointestinal digestion," *Int. J. Food Sci. Technol.*, vol. 56, no. 2, pp. 927–935, Feb. 2021.
45. T. Feng, B. Liang, H. Bi, and Y. Han, "A sandwich-like structure composite electrospun membrane of polylactic acid/nano-hydroxyapatite and polyvinyl alcohol/sodium alginate/nano-hydroxyapatite for skull defect repair," *Mater. Des.*, vol. 209, p. 109957, Nov. 2021.
46. M. Sheikholeslam, M. E. E. Wright, N. Cheng, H. H. Oh, Y. Wang, A. K. Datu, J. P. Santerre, S. Amini-Nik, and M. G. Jeschke, "Electrospun polyurethane–gelatin composite: A new tissue-engineered scaffold for application in skin regeneration and repair of complex wounds," *ACS Biomater. Sci. Eng.*, vol. 6, no. 1, pp. 505–516, Jan. 2019.
47. A. Sadeghianmaryan, Y. Karimi, S. Naghieh, H. Alizadeh Sardroud, M. Gorji, and X. Chen, "Electrospinning of scaffolds from the polycaprolactone/polyurethane composite with graphene oxide for skin tissue engineering," *Appl. Biochem. Biotechnol.*, vol. 191, no. 2, pp. 567–578, Dec. 2019.
48. J. Park, B. K. Lee, G. S. Jeong, J. K. Hyun, C. J. Lee, and S.-H. Lee, "Three-dimensional brain-on-a-chip with an interstitial level of flow and its application as an in vitro model of Alzheimer's disease," *Lab Chip*, vol. 15, no. 1, pp. 141–150, Dec. 2014.
49. H. K. Park, W. Joo, B. K. Gu, M. Y. Ha, S. J. You, and H. J. Chun, "Collagen/poly(d,l-lactic-co-glycolic acid) composite fibrous scaffold prepared by independent nozzle control multi-electrospinning apparatus for dura repair," *J. Ind. Eng. Chem.*, vol. 66, pp. 430–437, Oct. 2018.
50. U. Buratti, "Electrospun cellulose acetate fibers: a new approach for in vitro models of the bloodbrain barrier," p. 87316161, 2019. Master's thesis, Polytechnic University of Torino.

25

Recent Developments in One-Dimensional Polymeric Nanocomposites for Wound Healing and Infection Control

Hari Prasath Nagaiah, Priya Arumugam, Karthikeyan Kandaswamy, and Karutha Pandian Shunmugiah*

CONTENTS

25.1 Introduction

Wound healing is a complex physiological factor for repairing damaged skin tissues. It involves three overlapping biological processes: inflammation, proliferation, and remodeling. The inflammation phase occurs from the time of wound formation and ends on the sixth day. During inflammation, hemostasis occurs through the formation of blood clots, and leucocyte infiltration to the wound site acts as antimicrobials. The proliferation phase starts by covering the wound surface with granulation tissue. Under normal conditions, the proliferation phase extends from the sixth day to the denouement at two weeks and it also involves migration and activation of fibroblasts which are responsible for collagen and extracellular matrix production. In the third phase, remodeling extends to restore the function and morphology of the wounded tissue. When the chronicity of

DOI: 10.1201/9781003223764-25

wounds increases, the physiological process of wound healing has been damaged by factors such as diabetes, obesity, pressure ulcers, and burns. As the initial wound healing phase (inflammation phase) fails to be completed, the subsequent wound healing phases will fail to start. The induced stresses on the unhealed wound by external contaminations make the wound more chronic and decrease the efficacy of wound healing [1].

Wound healing chronicity increases with increased microbial pathogenesis at the wound site. Due to the disintegrated skin morphology and reduced dermal cell functions, environmental pathogens tend to colonize the wound site and cause prolonged inflammation. Multiple pathogens cause wound infection using distinct virulent pathways. A higher percentage of wound pathogens includes *Staphylococcus aureus, Pseudomonas aeruginosa, Escherichia coli, Acinetobacter baumannii, Klebsiella pneumoniae,* and *Streptococcus pyogenes* was observed on the infectious wound that results in chronicity [2]. The wound biofilms cause pathogen colonization on tissues through the production of extracellular matrix, which acts as a barrier against host immunity. Using broad-spectrum and narrow-spectrum antibiotics, pathogen colonization can be controlled, but, due to the misuse of antibiotics, resistance to these drugs has developed and it is a crucial challenge to develop a long-term drug for topically associated pathogens. The signs of infection on the wound encompass abscesses, pus discharge, delayed healing, bleeding granulation tissue, discoloration of wounded tissues, tenderness, and wound odor [3].

Drugs and dressings for wound healing have been discovered since prehistoric times, including metabolites, antibiotics, animal fats, inorganic metals (silver and zinc), and honey. The most commonly used wound healing formulations are silver sulfadiazine creams, polyamide and cellulose contour bandages, thin hydrocolloid dressings, calcium alginate fibers, polyurethane foams, and polymeric hydrogel sheets. Drugs for wound healing diverge based on the type of wounds. Recent advancements in wound therapies and dressing development enable artificial skin synthesis *via* collagen, silicon, nylon mesh, and cultured keratinocytes [3].

The advancement of nanotechnology in the past two decades is making way for targeted drug delivery for many varieties of chronic wounds. The nano delivery of drugs and antimicrobials enables wound healing without causing toxicity to neighboring cells/tissues. Among nano-delivery of drugs to wound healing, one-dimensional nanocomposites (1D-PNs) play a vital role in wound healing application. 1D-PNs include carbon nanotubes (CNTs), nanobars, nanoribbons, and nanowires. 1D-PNs were synthesized using various techniques based on their applications. Figure 25.1 shows the 1D-PNs used for the wound healing process along with phases of normal wound healing. Also, the formation of 1D-PNs from nanowires using electrospinning is illustrated. The synthesis methods used for nanocomposite synthesis were electrospinning, *in-situ* polymerization, melt mixing, sol-gel method, solution technique, and template-based method. This chapter highlights the recent findings on one-dimensional polymeric nanocomposites used as antimicrobials and wound healing agents [5].

25.2 Current Treatments for Wound Infections

Delayed wound healing is a complex burden for patients and the medical system. On a global scale, it is estimated that the wound care market would grow from $15.84 billion in 2021 to $24.01 billion in 2028 [6]. A retrospective analysis in 2018 estimated that around 8.2 million people had wounds with or without infections. The chronicity of delayed wound

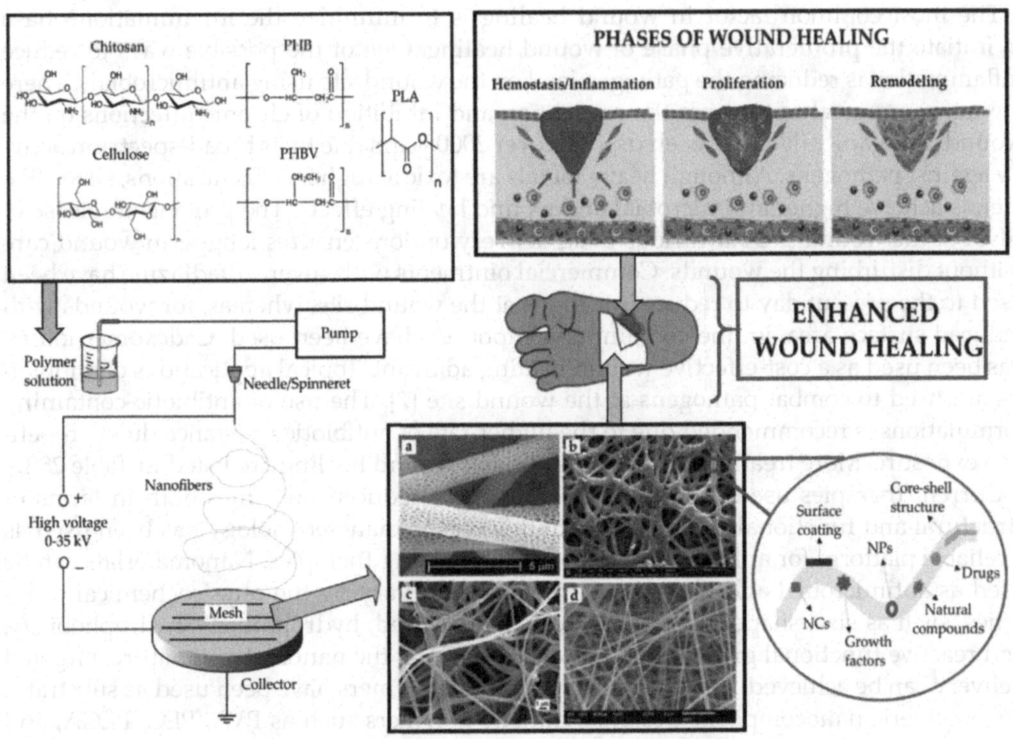

FIGURE 25.1
Enhanced wound healing by nano-dressings *via* 1D-PN synthesis. Adapted with permission from reference [4]. Copyright (2020), MDPI.

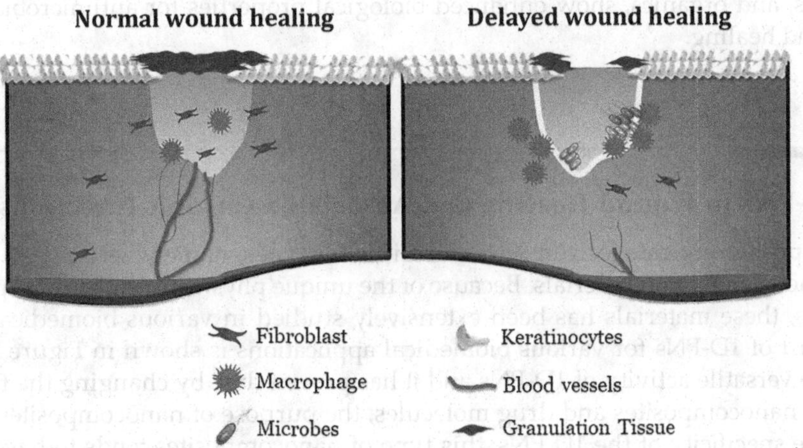

FIGURE 25.2
Comparison of normal and delayed wound healing (chronic wounds).

healing in comparison with normal wound healing is illustrated in Figure 25.2. In addition, wounds created by surgery require a great deal of attention from both a scientific and business perspective. Due to this complication in wound infections globally, the wound care product market keeps on increasing with respect to both market value and the number of drugs and wound care products.

The most common factor in wound healing is to minimize the inflammation phase to initiate the proliferative phase of wound healing. One of the possible ways to reduce inflammation is reducing the pathogen load on the wound site using antimicrobials. There are many approaches towards the prevention and inhibition of chronic infections on the wound. Inorganic silver has been used for over 2000 years due to its broad-spectrum activity against pathogens. Although heavy metals are toxic at higher concentrations, silver-like metals achieve higher antimicrobial and wound healing effects. The prolonged release of silver at the wound site *via* various drug delivery options ensures long-term wound care without disturbing the wounds. Commercial ointments with silver sulfadiazine have been used to the present day to reduce infections at the wound site, whereas, for wounds with reduced surface area, iodine-containing compounds have been used. Cadexomer iodine has been used as a cost-effective wound healing adjuvant. Topical applications of antibiotics are used to combat pathogens at the wound site [7]. The use of antibiotic-containing formulations is recommended due to the higher rate of antibiotic resistance due to repetitive exposure. More treatments for infections and wound healing are listed in Table 25.1.

Current therapies used for wound healing show reduced outcomes both in terms of structural and functional features. Due to this reason, nanotechnology has been used as a reliable platform for antimicrobial and wound healing therapies. Nanomaterials can be used as antimicrobial and wound healing agents by varying the physicochemical properties, such as size, shape, charge, type of material used, hydrophilicity/hydrophobicity, and reactive functional groups. In addition, by varying the nanomaterial nature, targeted delivery can be achieved. Synthetic and natural biopolymers have been used as substrates for polymeric nanocomposite design. Synthetic polymers such as PVA, PEG, PLGA, and polyurethane can be used for one-dimensional polymeric nanocomposite synthesis. Considering bioavailability and biodegradability, biopolymers such as chitosan, collagen, hyaluronic acid, cellulose, natural rubber, starch, and exopolysaccharides can be used. Nanocomposites prepared from the different compositions of these various polymers, inorganics, and organics, show enhanced biological properties for antimicrobial activity and wound healing.

25.3 1D-PNs in Wound Healing and Associated Chronic Infections

Nanocomposites are categorized into four different types, namely zero-, one-, two- and three-dimensional nanomaterials. Because of the unique physical and chemical properties of 1D-PNs, these materials has been extensively studied in various biomedical applications. A list of 1D-PNs for various biomedical applications is shown in Figure 25.3. This shows the versatile activity of 1D-PNs and it has proven that, by changing the functional groups of nanocomposites and drug molecules, the purpose of nanocomposites changes. Due to the specificity of the 1D-PNs, this type of nanocomposites tends to have reduced toxicity due to the stability of the drug molecule inside the 1D-PN.

There are two reasons for opting for 1D-PN over the other three types of nanocomposites. These includes:

1. Smaller dimensional structure and high aspect ratio for carrying the drug in a unidirectional way;
2. Used as device elements in many biomedical devices [19].

TABLE 25.1

Commercial Antimicrobials and Wound-Healing Agents with Their Mode of Actions

S. No	Trade name	Active Ingredients	Infection Control	Wound Healing	Mechanism of Action	References
1	**Augmentin®**	Amoxicillin, beta-lactamase inhibitor, clavulanate potassium	+	–	Inhibition of cell wall biosynthesis	[8]
2	**Thermazene®**	Silver sulfadiazine	+	+	Broad-spectrum inhibition; reduces inflammation	[9]
3	**Octenisept®**	Octenidine hydrochloride, phenoxyethanol	+	+	Depolarizes and changes the fluidity of the cell membrane; pro-inflammatory cytokines (IL-8, IL-6) and matrix-metalloproteinases (MMP1, MMP2, MMP3, MMP9) produced for remodeling tissues	[10]
4	**Prontosan® Wound Gel**	Polyhexanide	+	+	Broad-spectrum antimicrobial biocide; reduces matrix-metalloproteinase- (MMPs) induced peri wound breakdown	[11]
5	**Polysporin® Triple Antibiotic Ointment**	Polymyxin b, bacitracin, gramicidin	+	–	Broad-spectrum antimicrobial activity	[12]
6	**Betadine®**	Povidone-iodine	+	+	Protein denaturation, precipitation of bacteria; anti-inflammatory properties	[13]
7	**Rugby® Zinc Oxide Ointment**	Zinc oxide	+	+	Disruption of the cell membrane and oxidative stress; tissue re-epithelialization, angiogenesis, to inhibit fibrosis/scar formation.	[14]
8	**Soframycin ® skin cream**	Framycetin sulphate	+	+	Inhibiting translation during protein synthesis; decrease the risk of irritation (anti-inflammatory)	[15]
9	**Mupirocin ointment IP**	Mupirocin calcium	+	+	Antimicrobial activity by reversibly inhibiting isoleucyl-transfer RNA, thereby inhibiting bacterial protein and RNA synthesis; remodeling tissues	[16]
10	**Megaheal® Gel**	Nano silver colloid	+	+	Broad-spectrum antimicrobial properties; drive the differentiation of fibroblasts into myofibroblasts, and stimulate the proliferation and relocation of keratinocytes	[17]

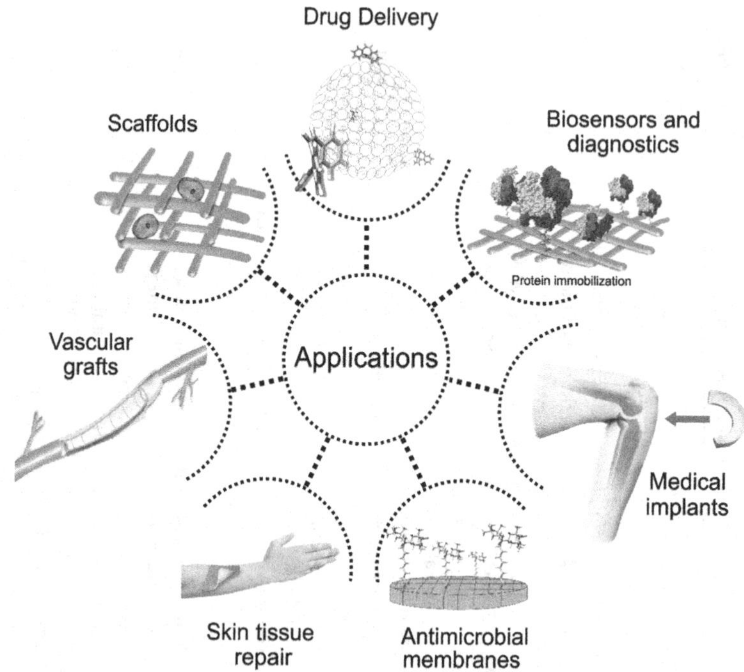

FIGURE 25.3
1D-PNs for various biomedical applications including wound dressings. Adapted with permission from reference [18]. Copyright (2013), John Wiley & Sons, Inc.

The microbial killing efficacy of the polymeric nanocomposites is better than that of the non-polymeric nanocomposites. This concept is illustrated in Figure 25.4 comparing graphene oxide/silver and graphene oxide/poly(ethylene glycol)/silver nanocomposites. Among the best killing effect is found in nanocomposites with PEG.

The synthesis methods for one-dimensional polymeric nanocomposite include template-directed methods, self-assembly methods, electrospinning techniques, solvothermal synthesis, vapor-phase approaches, etc., The 1D-PNs containing antimicrobial and wound healing properties are described in Table 25.2.

25.3.1 1D-PN as a Promising Antimicrobial Agent

25.3.1.1 Carbon Nanotube (CNT)-Loaded 1D-PNs

CNTs are the nano-cylindrical structures that contain arranged carbons from polymers. The CNTs are classified into three categories: single-walled (SWCNTs), double-walled (DWCNTs), and multi-walled (MWCNTs). CNTs enable the transfer of electrons from electro-active species and electrodes. CNTs have properties such as high absorbability, high conductivity, and high surface-to-volume ratio. These physicochemical characteristics make CNTs widely usable in drug delivery, sensors, genetic engineering, implants, and other biomedical applications. Based on the type of applications, the surface of the CNTs has been modified. The synthesis methods for CNTs are laser ablation, chemical vapor deposition, arc discharge, and ball milling [37].

CNTs act as a promising antimicrobial agent against a broad spectrum of microbes in wounds. The broad-spectrum activity of CNTs is due to the structure and size of

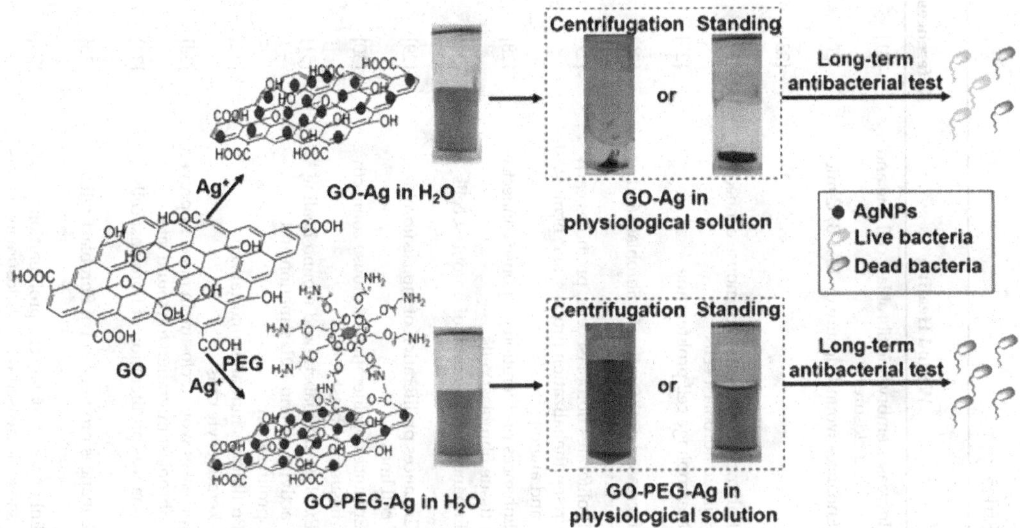

FIGURE 25.4
Greater efficacy of antimicrobial nature of 1D-PNs *vs* non-polymeric nanomaterials. Adapted with permission from reference [20]. Copyright (2017), American Chemical Society.

the nanocomposite. The first antimicrobial CNTs reported in SWCNTs and MWCNTs was against *Escherichia coli*, followed by different pathogens [38]. Varying factors such as diameter, length, residual catalyst, and surface chemistry cause a significant influence on the antimicrobial activity of CNTs. Although CNTs present good antimicrobial activity, functionalizing the CNTs with antimicrobial reactive groups enhances the antimicrobial activity even at lower concentrations. SWCNTs show higher antimicrobial activity than MWCNTs, due to the drug loading space available in SWCNTs for drug delivery. The killing mechanism of CNTs against microbes includes membrane disruption, reactive oxygen species (ROS) generation, and disruption of multiple, microbial, metabolic pathways.

For enhanced antimicrobial effect, antimicrobial compounds such as antibiotics, photosensitizers (aminopyrene trisulfonate and 3,3′-diethylthiatricarbocyanine fluorophores), antimicrobial peptides (epinecidin-1, poly(L-lysine), anionic poly(L-glutamic acid) and nisin), and enzymes (lysostaphin and laccase) can be used [39]. A study using films proved that a polymeric nanocomposite with SWCNTs can inhibit 90% of microbial cells and it is proven to be an effective antimicrobial biomaterial. Another study proves that the nanocomposite containing nisin/MWCNP/ polyethylene glycol (PEG) can effectively inhibit a broad spectrum of bacteria and fungi, with up to the 7-fold stronger antibacterial activity and 100-fold higher antibiofilm than commercially used antimicrobial agents. The significant antibacterial action of nisin in the nanocomposite interferes with cell wall production and increases cell membrane permeabilization, which may account for these remarkable outcomes.

Also, poly(ester amide)/MWCNT nanocomposites exhibited strong antibacterial activity against *Staphylococcus aureus* (2 log reduction), *Bacillus subtilis* (2–3 log reduction), and *E. coli* and *Klebsiella pneumoniae* (1 log reduction). Figure 25.5 demonstrates the synthesis of antimicrobial Tin(IV) oxide(SnO_2)/CNT nanocomposite from sunflower oil using the chemical deposition method. Due to the higher capping ability of the plant metabolites to the polymers and CNT, there is higher killing efficacy against bacterial pathogens (*B.*

TABLE 25.2

Various One-Dimensional Polymeric Nanocomposites with Antimicrobial and Wound-Healing Efficacy

S. No.	Name of One-Dimensional Polymeric Nanocomposite	Active Drug	Infection Control	Wound Healing	References
1	Arginine-chitosan nanocomposite	Antimicrobial peptides (lysozyme)	*S. aureus* and *E. coli*	Increases embryonic fibroblast cell line and wound closure	[21]
2	Graphene oxide–IPDI–Ag/ZnO@ hydroxypropyl cellulose	Silver, zinc and IDPI	*S. aureus* and *E. coli*	Enhances wound closure on BALB/C mice	[22]
3	Montmorillonite/chitosan nanocomposite	Silver sulfadiazine	Broad-spectrum antimicrobial activity	—	[23]
4	Chitosan, polyvinylpyrrolidone, and cellulose nanowhiskers nanocomposite	Curcumin	*E. coli* and *E. hirae*	Enhances proliferation of mouse connective tissue fibroblast cells	[24]
5	Cu/Zn-carbon micro/nanofiber-polymer nanocomposite	Copper, Zinc	*P. aeruginosa*	Macrophage cell proliferation	[25]
6	Non-animal mushroom carboxymethyl cellulose chitosan –ZnO nanocomposite	Chitosan, ZnO	*S. aureus*	Increased proliferation in fibroblast and keratinocyte cells	[26]
7	Halloysite and chitosan nanocomposite	Halloysite	—	Proliferates fibroblast cells, promotes early re-epithelialization process, and hemostasis and angiogenesis	[27]
8	Calendula officinalis-loaded PCL/ gum arabic nanocomposite	*Calendula officinalis*	*S. aureus* and *E. coli*	Enhances proliferation of mouse connective tissue fibroblast cells	[28]
9	Chitosan–PVP–titanium dioxide nanocomposite	TiO2	*S. aureus*, *E. coli*, *P. aeruginosa*, and *B. subtilis*	Enhances proliferation of fibroblast cells	[28]
10	Chitosan-polyvinyl alcohol-graphene oxide nanocomposite	Graphene oxide	*P. aeruginosa* and *Micrococcus luteus*	Enhances proliferation of mouse fibroblast cell lines	[29]
11	PVA/honey hybrid	Honey, PVA	—	Enhances proliferation of mouse connective tissue fibroblast cells	[30]
12	FeO Fabricated chitosan/PVA nanocomposite	FeONPs	*B. cereus*, *S. aureus*, *E. coli*, and *S. enterica*	Exhibits antidiabetic and wound healing activity *via* anti-inflammation and proliferation	[31]
13	Graphene oxide-quaternary ammonium nanocomposite	Graphene oxide, quaternary ammonium nanocomposite	*S. aureus* and *E. coli*	Proliferates human immortalized keratinocyte cells	[32]
14	QCSG/CNT nanocomposite	Carbon nano tubes	*S. aureus*, *E. coli*, and *P. aeruginosa*	Increases wound closure in mice, Improved fibroblast migration to wound site	[33]
15	ZnO-loaded chitosan/ poly (vinyl alcohol)/acacia gum nanosphere	Zinc	—	Enhances proliferation of fibroblast cells	[34]
16	Carboxymethyl starch/polyvinyl alcohol/ nanosilver nanocomposite	Nano silver	*E. coli* and *S. aureus*	Enhances proliferation of fibroblast cells	[35]
17	Keratin-chitosan/n-ZnO nanocomposite	Zinc, keratin	*E. coli* and *S. aureus*	Enhances fibroblast cell proliferations and shows increased wound closure in SD rats	[36]

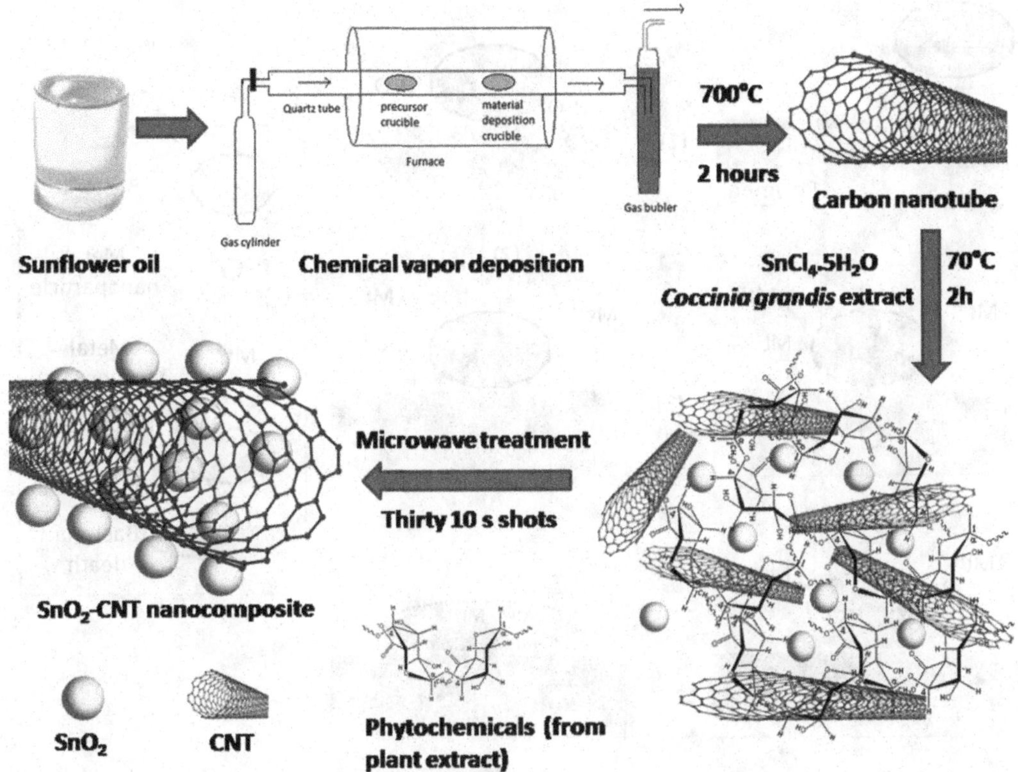

FIGURE 25.5
Synthesis of antimicrobial Tin(IV) oxide(SnO_2)/CNT nanocomposite from sunflower oil using the chemical deposition method. Adapted with permission from reference [40]. Copyright (2019), Springer Nature.

subtilis, *E. coli*, and *K. pneumoniae*) and fungal pathogens (*Aspergillus niger* and *Candida albicans*). Furthermore, the efficacy of bio-fabricated SnO_2-CNT nanocomposite paves the way to stopping the water-contaminating pathogens by increased antimicrobial activity. SnO_2 nanoparticles on the CNT surface may improve the microbicidal effect by diffusing into the cell membrane and changing cell function. Furthermore, the heterojunction structure may result in the formation of ROS that interacts with the cell wall, resulting in deadly action against microorganisms [40].

25.3.1.2 Metal/Metal Oxide/Metal Hydroxide-Loaded 1D-PNs

Polymeric nanocomposites loaded with metals, metal oxides, and metal hydroxides provide excellent biomedical applications. Based on the type of drug delivery applications; polymer-to-metal/metal oxide/metal hydroxide ratio was fixed for 1D-PN synthesis. Polymeric nanoparticles have been used as a scaffold for supporting antimicrobial metal catalysts.

Metals and metal alloys can be extremely toxic to microbes at a low concentration. It has been proved that metal and metal nanoparticles of silver, gold, copper, platinum, and palladium present an excellent broad-spectrum antimicrobial activity like Cu/Zn-Carbon-micro/nanofiber-polymer nanocomposite, which also has an excellent antimicrobial activity [25]. The biocidal activity of the metal alloy-containing nanocomposites

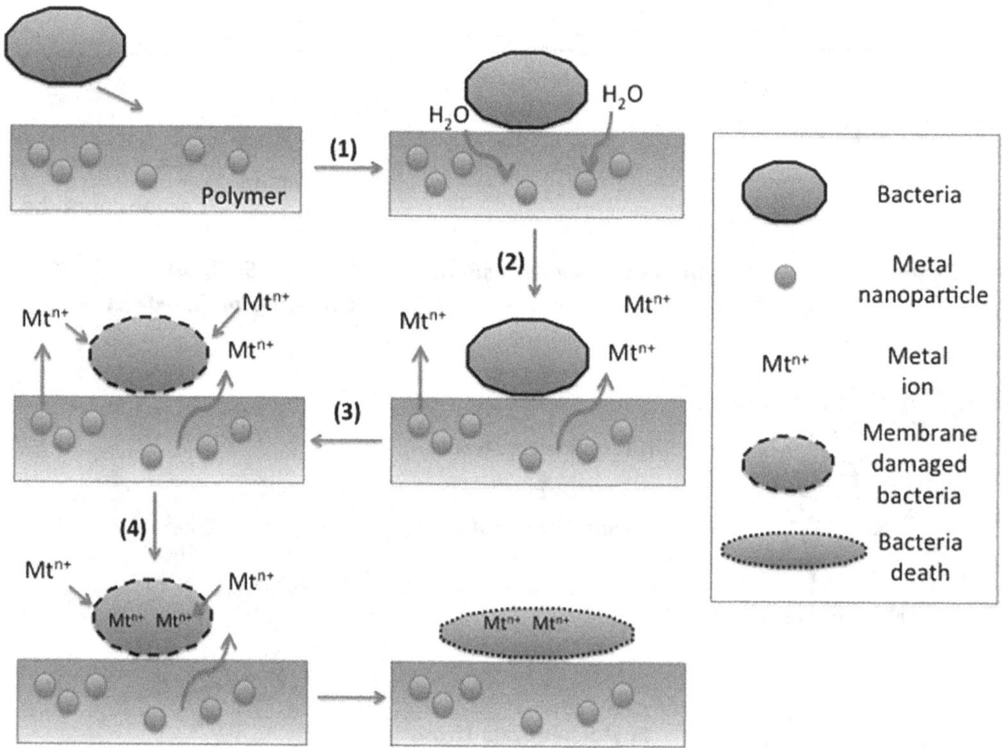

FIGURE 25.6
Antimicrobial mode of action of polymeric nanocomposites with biocidal metal nanoparticles. Adapted with permission from reference [42]. Copyright (2015), MDPI.

can be triggered by the metal donor atom selectivity; and the metal reduction potential. Polymeric nanocomposite containing metal alloys combat microbes by multiple targeting, such as cell wall inhibition, DNA/RNA damage, and by modulating the efflux pump. Due to the reduced toxicity of 1D-PNs that contain metal alloys, they have been extensively used in biomedical and food microbiology applications [41].

Numerous studies are focusing on 1D-PNs using metal, metal oxides, and metal hydroxides due to their broad-spectrum killing efficacy by multiple modes of action. The antibacterial method of action of metal nanoparticles embedded in a polymeric nanocomposite is depicted in Figure 25.6. Due to water transport, the microbial cell tends to consume heavy metals/metals after contacting the surface of the 1D-PN with metal nanoparticles. As a result, metals accumulate on metabolic enzymes, there is a lack of cell division/survival, and the pathogen is killed *via* a multi-targeted mode of action. Initially, the bacterial cells adsorb onto the polymer surface, causing water to diffuse through the polymer matrix to the medium surrounding the bacteria. Then the water containing dissolved oxygen reaches the surface of embedded metal nanoparticles, allowing dissolution or corrosion processes to occur, resulting in the formation of metal ions. After incorporation, the metal ions reach the composite surface, damaging the bacterial membrane, allowing the metal ions to diffuse, followed by killing of bacterial cells.

For example, a study formulated chitosan/AgNP/graphene oxide (GO) hybrid nanocomposite by embedding GO into room temperature/atmospheric pressure microplasma to produce chitosan/AgNPs. The resultant composite system displayed a pH-sensitive release and effective drug release profile. The nanocomposite also exhibited substantial antimicrobial and photothermal conversion capabilities against *E. coli* and *S. aureus* bacteria [43]. Another study formed 1D-PNs by covalently attaching thiol end-capped poly[2-(dimethylamino)ethyl methacrylate] to AgNPs. Antimicrobial activity was significantly increased with no bacterial resistance due to the polyvalent and synergistic antibacterial properties of these nanoparticles. The unique nanomaterials enhanced bacterial cytoplasmic membrane permeability, inducing membrane breakdown, and then infiltrated cells, inhibiting intracellular cellular functions and ultimately causing apoptosis. Furthermore, this nanocomposite combats infection caused by *Pseudomonas aeruginosa* and *S. aureus*. A diabetic rat model showed the same curative effect, demonstrating that the poly[2-(dimethylamino)ethyl methacrylate]/AgNPs nanocomposite is an effective antimicrobial material with the potential to cure bacteria-induced diseases in people with impaired immunity [44].

25.3.1.3 Chalcogenides-Loaded 1D-PNs

Chalcogenides are a prime group of glassy materials which consist of one or more chalcogen elements, such as sulfur, selenium, and tellurium. It can tune optical properties in the wide spectral range. Chalcogenides are a semi-conductor metal that delivers efficient, fast-charge transport to the conjugated polymer. They have a major role on these glasses (chalcogenides) and are involved in the development of laser power delivery systems for applications in surgery, dentistry, dermatology, and ophthalmology; they can be equally used for human tissue diagnostics [45].

Chalcogenides have an admirable result against microorganisms while being combined with copper, iron, or silver. Copper chalcogenides nanoparticles, such as copper sulfide, copper oxide, and copper selenide have superior anti-microbial activity against Gram-negative (*E. coli* and *P. aeruginosa*) and Gram-positive bacteria (*S. aureus* and *Enterococcus faecalis*) and fungi (*C. albicans*). Zinc sulfide (ZnS)/Cellulose nanocomposites exhibit good antimicrobial properties against Gram-negative bacteria (*E. coli*). In comparison with ZnS particles, ZnS/CNC demonstrated a greater degradation of up to 75.62% for phenol as they respond to visible light and can treat broad field areas containing organic dyes which potentially kill organisms **via** photodynamic therapy. For instance, because this nanocomposite has a significant number of hydroxyl groups, it interacts well with and has strong contact with nano-GO. Furthermore, GO– isophorone diisocyanate (IPDI)–Ag/ZnO@ Hydroxypropyl cellulose, silver, zinc, and IDPI nanocomposite films demonstrated significant UV resistance and the ability to efficiently inactivate both Gram-negative (*E. coli*) and Gram-positive (*S. aureus*) bacteria, as well as enhancing wound healing in BALB/C mice. Hence the multifunctional AGO@HPC nanocomposite films with mechanical, anti-UV, and antibacterial capabilities might be used in anti-UV treatment, antibacterial packaging, and biological wound-dressings [22]. Nanocomposites destabilize the outer bacterial membrane, collapse the plasma membrane potential, and deplete the levels of intracellular ATP. Due to nanocomposites' sensitivity against microorganisms, it is preferred for the diagnostics of the microbiological activity of specific organisms. Another study used microwave irradiation to create cl-Ch-pMAc@ZnO/CdSQDs by crosslinking (methacrylic acid) crosslinked chitosan with chitosan and then incorporating ZnO and cadmium sulfide quantum dots (CdS QDs) onto the crosslinked polymer surface. The composite produced is worth considering as a strong antibacterial agent against *E. coli* and *B. subtilis* [46].

25.3.1.4 Metal Phthalocyanine- and Porphyrin-Loaded 1D-PNs

Phthalocyanines are commonly associated with porphyrins due to the similarity between their structures. The cavity of porphyrins containing four pyrrolic nitrogen is well suited for the binding majority of metal ions to form metalloporphyrin. The phthalocyanines have four isoindole units linked by nitrogen atoms in an azo position, whereas the porphyrins are composed of four pyrrole units bonded by carbon atoms. Phthalocyanines are extremely resistant to chemical degradation. Recent research studies show that metalloporphyrin-based materials in drug delivery, bio-imaging determine the ferrochelatase in bone marrow, indicating their prospective clinical applications [47].

Porphyrins and phthalocyanines have been part of photodynamic therapy which can transfer electrons, catalyze peroxidase and oxidase, absorb photons, generate ROS, and split off lipids of the bacterial cell membrane. Porphyrins and phthalocyanines have an effective biocidal, antimicrobial, anti-viral capacity. Porphyrins are more toxic to Gram-positive bacteria (*S. aureus*) and Gram-negative bacteria (*E. coli*). For example, GO-quaternary ammonium nanocomposite has a higher efficacy against *S. aureus* and *E. coli* [32]. In a study, porphyrin-cellulose-based nanocrystals were developed for photodynamic antimicrobial chemotherapy against antibiotic-resistant bacteria. The photoactive porphyrin was fused on the surface of cellulose nanomaterials *via* a triazole linkage formed through an azide-alkyne Cu(I) catalyzed Huisgen 1,3-dipolar cycloaddition reaction, that occurs between the alkyne attached to the porphyrin and azide on the cellulosic surface. The surface modification of cellulose nanocrystals with the cationic porphyrin *via* covalent bonding yielded the conjugated cellulose nanocrystal (CNC)/porphyrin. The porphyrin's covalent attachment to the surface of the cellulose nanocrystals may have increased the cellulose's longer-lasting or permanent antimicrobial properties (improved durability), reduced the biocidal agent's percolation into the surrounding environment, and prevented porphyrin aggregation. By photodynamic inactivation, this 1D-PN promotes antimicrobial killing against *E. coli, Mycobacterium smegmatis*, and *S. aureus*. Due to their photodynamic inactivation in photodynamic therapy, bio-active nanoparticles become potential alternatives for the treatment and eradication of emerging bacterial resistance to antibiotics, classical antiviral, antifungal, and antiprotozoal drugs [48].

25.3.1.5 Antimicrobial Drug/Antibiotic-Loaded 1D-PNs

Even though antibiotics show a higher microbial resistance profile in a clinical environment, antibiotics have been used from the time of drug development to combat microbes. From the first-generation penicillin antibiotics to the fifth-generation cephalosporin antibiotics, resistance has been developed in clinical environments. Antibiotics are available for a broad range of applications like injectable forms of antibiotics, antibiotics for topical applications, and antibiotics for poultry/animal disease control [49]. The mechanism of action of antibiotics are listed in Figure 25.7. This illustration explains the various modes of action for various classes of antibiotics. Even though antibiotics are effective at killing microbes, there is a truism that the microbes tend to evolve resistance against each antibiotic.

A study using a PHBV/gentamicin nanocomposite reports high antimicrobial activity against *Staphylococcus haemolyticus* and *S. aureus*. The drug crystals in the PHBV/gentamicin nanocomposite are more soluble than the polymer, and their early disintegration forms porosity or channels that enable water to pass through. This speeds up the breakdown of

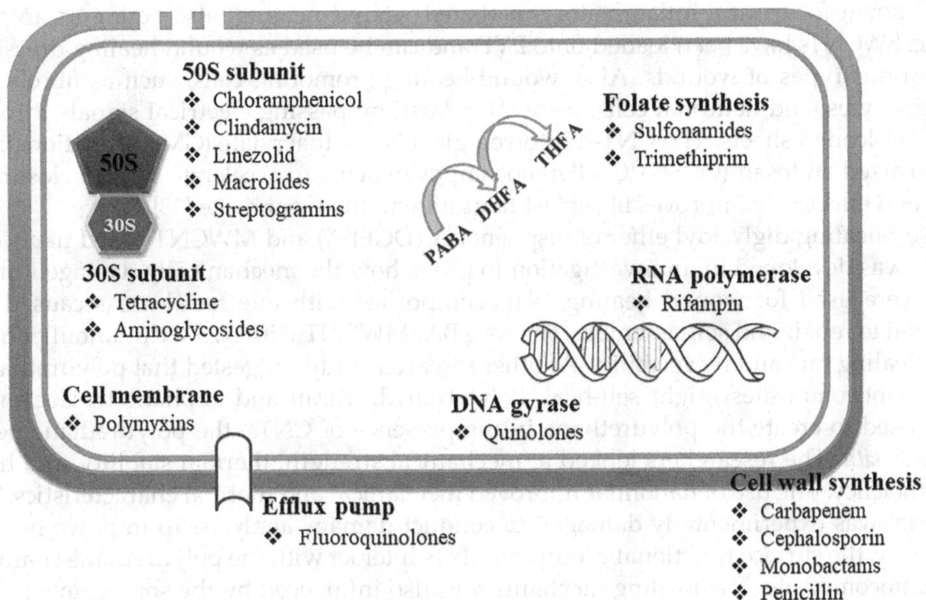

FIGURE 25.7
Targeted mechanism of action of antibiotics on microbes.

the substances within the polymeric matrix, resulting in the burst effect. The first burst effect ceases when all of the medicines on the surface that are in direct touch with each other dissolve. The elution is no longer logarithmic beyond this point, and a consistent release pattern is seen. The breakdown of the polymer now regulates the release of the antibiotic, enabling fresh drug crystals to be exposed to the medium. In reality, the quick burst release may be helpful since the high discharge kills any microorganisms introduced before or during the procedure [50].

Due to the reduced discovery of new antibiotics in recent decades, alternatives for antibiotics have been reported from various resources, such as animals, plants, yeast, algae, and bacteria. The higher eukaryotic species produce biocides against invading pathogens, which are being used as an alternative to antibiotics. Plants produce secondary metabolites, such as carvacrol, curcumin, eugenol, cinnamaldehyde, and epicatechin, which have activity on the broad-spectrum antimicrobial used in therapies. In chitosan, polyvinylpyrrolidone, and cellulose nanowhisker nanocomposite. curcumin act as an active ingredient that has antimicrobial activity against *E. coli* and *Enterococcus hirae* [24]. Both prokaryotes and eukaryotes produce short peptide sequences as defense lines, called antimicrobial peptides. These antimicrobial peptides have a broad-spectrum antimicrobial activity to kill pathogens. Some of the antimicrobial peptides include magainin (from frog), pleurocidin (flounder), beta-defensin 1 (human), and tracheal (cow) [51].

25.3.2 1D-PN as an Active Therapy for Wound Healing

25.3.2.1 CNT-Loaded 1D-PN Wound Healing

CNTs can be stable water-soluble drug/drug carriers for wound healing. SWCNTs are used along with povidone-iodine to enhance wound healing. The povidone-iodine complex is well known for its anti-inflammatory-mediated wound healing, also acting as an antiseptic. SWCNTs have been loaded onto PVP and can be used as wound healing dressings for various types of wounds. Also, wound-healing-promoting cells, such as fibroblasts, keratinocytes, and neuronal cells, were stimulated by passing electrical signals through the CNT-loaded sheets. The CNT-reinforced glycidyl methacrylate (GMA)-functionalized quaternized chitosan (QCSG)/CNT nanocomposite helps to accelerate wound closure in mice and effectively improves fibroblast migration to the wound site [33].

A self-healing diglycidyl ether of bisphenol A (DGEBA) and MWCNT-based nanocomposite was developed in an investigation to prove how the mechanically damaged materials were used for wound healing. Nanocomposites with interfacial areas caused the material to repair and self-assemble. The DGEBA/MWCNT with 0.75 wt% nanofiller has a high healing rate and recyclability. Another research study suggested that polyurethane/CNT nanocomposites might self-heal. Polytetrahydrofuran and aliphatic tri-isocyanate were used to create the polyurethane in the presence of CNTs, the polyurethane being made *in situ*. The researchers looked at mechanical strength, thermal stability, and healing efficiency. The use of nanofiller improved mechanical and thermal characteristics. The material was experimentally damaged to conduct damage analysis. To improve healing efficiency, the surface functional groups on CNTs interact with the polyurethane chains of the nanocomposite. The healing mechanism is also influenced by the soft segment composition. As a result, the self-healing efficiency of the polyurethane/CNT nanocomposite sample improved significantly up to 40% [52]. The mechanism of action of accelerated angiogenesis by europium hydroxide nanorods is illustrated in Figure 25.8. The nanorods trigger P13K/AKT, allowing for the activation of cGMP, which is essential for angiogenesis. Angiogenesis on chronic wounds enables fibroblast and keratinocyte cell migration and nutrients to the wound site. In response to electricity, the CNTs exhibit conductivity which attracts cells to the wounded site, promoting wound healing. CNTs were used in real-time monitoring of wound healing *in situ*. *In-situ* wound monitoring can be used as an effective treatment strategy for personalized wound care management. *In-situ* immunofluorescence strategies were also developed to promote wound healing by targeting growth factors and proteins in wound sites [54].

25.3.2.2 Metal/Metal Oxide/Metal Hydroxide-Loaded 1D-PNs

Currently, metals/metal oxides like silver, gold, zinc oxide, and copper oxide have been used for wound-healing purposes. A nanocomposite containing the nanometal possesses both antimicrobial and wound-healing efficiency. Metals and metal oxides in the polymeric nanocomposite enable multi-directional wound healing *via* anti-inflammation, cellular proliferation, angiogenesis, and remodeling. Due to these assorted activities and the specificity of metal functional groups, targeted wound-healing drug delivery is possible. Even today, most commercial dressings available contain silver sulfadiazine as an active drug for wound healing and antimicrobials are incorporated to combat infections and act as wound-healing agents.

Silver has wound-healing capabilities in addition to its antimicrobial characteristics and its positive influence on mechanical strength. AgNP-loaded CS/glycosaminoglycan scaffolds were shown to have considerable potential to enhance fibroblast proliferation. Matrix metalloproteinases (MMPs) that are overactive destroy fiber junctions and peptide growth factors, slowing wound healing. Silver-based materials can help in wound healing by inhibiting the development of MMPs. Another study on critical metal-binding proteins found immunohistochemical evidence that silver can cause high expression of these

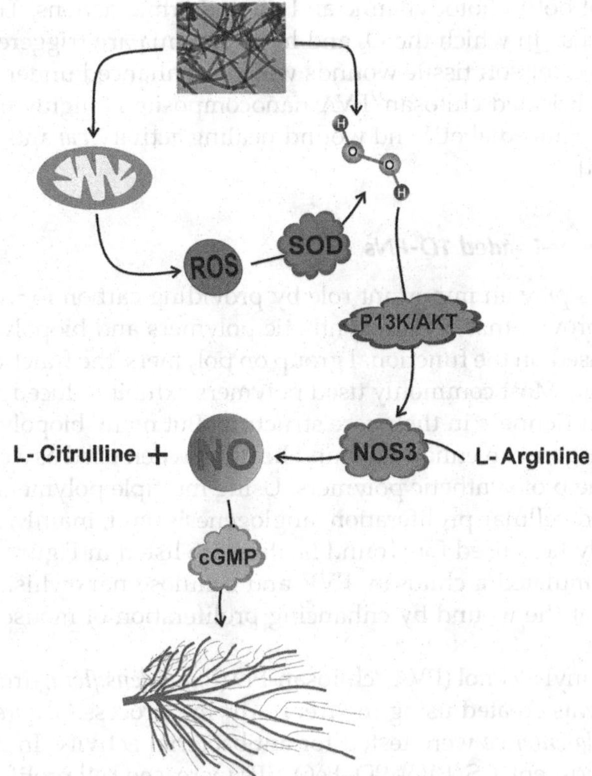

FIGURE 25.8
Nanocomposite scaffolds for boosting angiogenesis and improving chronic wound healing. Adapted with permission from reference [53]. Copyright (2021), Springer Nature.

proteins, boosting local zinc and copper concentrations. Both metals are required for the growth of epithelial cells.

With nanosheet forms, for example, ZnO-incorporated chitosan (ZnO-CS) and its antibacterial activity were discovered. MRSA and *P. aeruginosa* cells treated with ZnO showed cell membrane disruption and shrinkage, respectively. PVA was employed in other investigations to make chitosan/PVA/ZnO beads as a new antibacterial agent for wound healing. In mice skin wounds, the antibacterial activity and healing effect of Chitosan/PVA/ZnO were far superior to those of chitosan and PVA in isolation. The findings significantly confirm the nanocomposite system's suitability for wound-healing applications. However, in recent years, the impact of particle size on ZnO's antibacterial characteristics has also been studied [55].

25.3.2.3 Chalcogenides-Loaded 1D-PNs

Transition metal oxide and chalcogenide (TMO&C) nanomaterials have effective antimicrobial and wound-healing activities. TMO&C exhibits collective mechanisms of these nanoagents, including mechanical rupture, photocatalytic/photothermal activity, Fenton-type reaction, and nanozyme-assisted effect. TMO&Cs hold great promise for eradicating pathogenic bacteria and simultaneously accelerating wound healing. It has a highly efficient chitosan@MoS2 hybrid coating that presents an ability to kill bacteria due to the synergistic effects of both photodynamic and photothermal actions. There is no obvious inflammatory response in which the O_2 and hyperthermia are triggered by CS@MoS_2-Ti; it has a repair process for soft tissue wounds which is enhanced under dual lights [56]. In addition, the FeO-fabricated chitosan/PVA nanocomposite is highly effective in wound healing by exhibiting anti-diabetic and wound-healing activity *via* anti-inflammation and anti-proliferation [31].

25.3.2.4 Multi Polymer-Loaded 1D-PNs

In 1D-PNs, polymers play an important role by providing carbon for nanostructure synthesis. It has been proven that certain synthetic polymers and biopolymers are used for 1D-PN synthesis. Based on the functional group on polymers, the function and application of the product change. Most commonly used polymers exhibit reduced antibacterial activity due to the carbon richness in their core structure. But many biopolymers, such as chitin, keratin, and collagen, can enhance wound healing, when used as a co-polymer during synthesis with the help of synthetic polymers. Using multiple polymers enhances wound healing *via* increased cellular proliferation, angiogenesis, and, mainly, remodeling of the wound [57]. The polymers used for wound healing are listed in Figure 25.3. For instance, an investigation formulated a chitosan, PVP, and cellulose nanowhisker nanocomposite which helped to treat the wound by enhancing proliferation of mouse connective tissue fibroblast cells [24].

In a study, a polyvinyl alcohol (PVA)/chitosan-PD (*Pinus densiflora*)–iron oxide (FeO NPs) composite scaffold was created using the freeze-drying process. *S. aureus*, *E. coli*, *Bacillus cereus*, and *Salmonella enterica* were tested for antibacterial activity. In the *in-vitro* wound healing scratch experiment, CS/PVA–PD–FeO NPs increased cell proliferation in HEK-293 cells. Another study formed a 1D-PN using a combination of chitosan//PVP/bentonite, which is illustrated in Figure 25.9, which achieved excellent wound-healing activity in

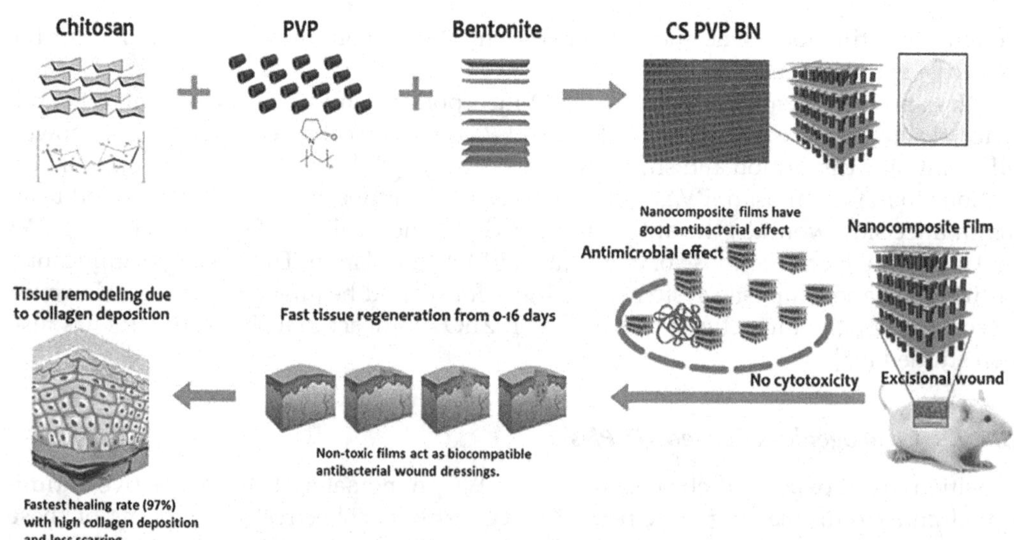

FIGURE 25.9
From 1D-PN synthesis to an active wound-healing nano-dressing formulation. Adapted with permission from reference [58]. Copyright (2018), Elsevier.

an excisional wound model in rats. The nanocomposite film exhibited better activity in 1D-PN with significantly reduced toxicity.

One more study used a solution casting approach to incorporate polyhexamethylene guanidine (PHMG)-modified graphene oxide (mGO) into the poly(vinyl alcohol) (PVA)/ chitosan (CS) films. CCK-8 tests were used to assess the biocompatibility of the PVA/CS/ mGO composite films by measuring the viability of HaCaT cells. The biocompatible surfaces disclosed by the PVA/CS/mGO composite films with low mGO concentrations in this research were appropriate for additional antibacterial investigations. As a consequence, antibacterial efficacy against *S. aureus* and *E. coli* was significantly increased, perhaps due to the synergistic bactericidal action of chitosan, GO, and polyhexamethylene guanidine. *In-vivo* tests revealed that PVA/CS mGO had the most dramatically accelerated wound-healing capabilities *via* re-epithelialization augmentation. An electrospinning-processed composite dressing, comprising CNT, PVA, and epidermal growth factor (EGF), promoted the proliferation of L929 fibroblasts, indicating that it could be used to treat wounds. The healing rate in the experimental group was up to 95% after ten days. The nanocomposite dressing effectively delivered EGF for a prolonged time, until healing was complete.

25.3.2.5 Wound-Healing Drug-Loaded 1D-PNs

Therapies for treating large wound areas include the use of wound-healing drugs responsible for reducing inflammation and enabling healing cell (keratinocytes and fibroblast) migration to the wound site. Silver-, nanosilver-, and silver sulfadiazine-containing polymeric nanocomposites were mostly preferred, due to the broad-spectrum activity of silver. For instance, in carboxymethyl starch/PVA/nanosilver nanocomposite, nanosilver acts as an active drug that helps to enhance the proliferation of fibroblast cells [35]. Silver loaded into polymers was released in a controlled manner to reduce the infection and provide the wounded site with a stressless environment for wound healing. Other metals, such as zinc oxide, cerium oxide, and gold, entrapped in polymeric nanocomposites provide increased wound-healing effects. Wound-healing drugs, such as sitagliptin, metformin, and alogliptin, were loaded into fullerene, graphene oxide, and CNTs entrapped in a polymeric film. These nanocomposites release drugs at regular intervals considering the stages of the wound-healing process [59]. For example, GO-quaternary ammonium nanocomposite, which has an active ingredient, graphene oxide and quaternary ammonium nanocomposite would help achieve a proliferation of human immortalized keratinocyte cells for wound healing [32]. The list of wound healing drugs in Table 25.2 was used in polymeric nanocomposites with other polymers.

25.3.2.6 Cell Growth Factor-Loaded 1D-PNs

Growth factors are active polypeptides that play an important role in cellular growth, differentiation, and migration. Advanced commercial dressings for wound-healing use growth factors for accelerated wound healing conditions. The US Food and Drug Administration (FDA) has approved the use of recombinant human platelet-derived growth factor (rhPDGF) in wound dressings. Also, growth factors such as recombinant human vascular endothelial growth factor (rhVEGF) and recombinant human epidermal growth factor (rhEGF) were used for various wound-healing applications. These growth factors have been effectively delivered using polymeric nanomaterials and showed excellent wound-healing effects [53]. A study has used plasma to treat the surface of CNTs. Then, to produce a unique model with controlled VEGF release, VEGF was loaded onto the nanotubes. The pig small intestine submucosa was then exposed to these modified VEGF-loaded CNTs. During the healing of the abdominal wall defect, there was a statistically significant increase in angiogenesis, according to the findings [53].

25.4 Conclusion

Nanotechnology is a rapidly growing field due to the broad range of applications in physical, chemical, and biological sciences. 1D-PN is proven to be effective against infections and for wound healing due to the activity of polymers with reduced toxicity on human cells compared with prokaryotes. Similarly, polymeric nanocomposites enable deep and targeted delivery with a broad range of drug acceptancy. Due to their one-dimensional nature, polymers in nanocomposites can be broadly modified with functional groups based on the intended applications. Additives such as metals, CNTs, antibiotics, and wound-healing drugs were used alongside nanocomposites to achieve better effects against biomedical and clinical complications. Recently, researchers have been developing 1D-PNS that are active without any drug additives, which could reduce multiple drug toxicity in drug development.

References

1. Gonzalez, A.C., et al., Wound healing - A literature review. *An Bras Dermatol*, 2016. 91(5): p. 614–620.
2. Siddiqui, A.R. and J.M. Bernstein, Chronic wound infection: Facts and controversies. *Clin Dermatol*, 2010. 28(5): p. 519–526.
3. Caldwell, M.D., Bacteria and antibiotics in wound healing. *Surg Clin North Am*, 2020. 100(4): p. 757–776.
4. Azimi, B., et al., Bio-based electrospun fibers for wound healing. *J Funct Biomater*, 2020. 11(3): p. 67.
5. Khare, H.S. and D.L. Burris, A quantitative method for measuring nanocomposite dispersion. *Polymer*, 2010. 51(3): p. 719–729.
6. Wound care market size, share & COVID-19 impact analysis. 2021. Available from: https://www.fortunebusinessinsights.com/wound-care-market-103268.
7. Sen, C.K., Human wounds and its burden: An updated compendium of estimates. *Adv Wound Care (New Rochelle)*, 2019. 8(2): p. 39–48.
8. Hu, G., et al., Bioequivalence of clavulanate potassium and amoxicillin (1:7) dispersible tablets in healthy volunteers. *J Huazhong Univ Sci Technolog (Med Sci)*, 2002. 22(3): p. 224–227.
9. Fisher, N.M., E. Marsh, and R. Lazova, Scar-localized argyria secondary to silver sulfadiazine cream. *J Am Acad Dermatol*, 2003. 49(4): p. 730–732.
10. Lademann, J., et al., Comparison of the antiseptic efficacy of tissue-tolerable plasma and an octenidine hydrochloride-based wound antiseptic on human skin. *Skin Pharmacol Physiol*, 2012. 25(2): p. 100–106.
11. Valenzuela, A.R. and N.S. Perucho, The effectiveness of a 0.1% polyhexanide gel. *Rev Enferm*, 2008. 31(4): p. 7–12.
12. Bearden, D.T., G.P. Allen, and J.M. Christensen, Comparative in vitro activities of topical wound care products against community-associated methicillin-resistant Staphylococcus aureus. *J Antimicrob Chemother*, 2008. 62(4): p. 769–772.

13. Fleischer, W. and K. Reimer, Povidone-iodine in antisepsis–State of the art. *Dermatology*, 1997. 195(Suppl 2): p. 3–9.
14. Lansdown, A.B., et al., Zinc in wound healing: Theoretical, experimental, and clinical aspects. *Wound Repair Regen*, 2007. 15(1): p. 2–16.
15. Nipanikar, S.U., et al., An open label, randomized, comparative, parallel group, multicenter, prospective, interventional, clinical study to evaluate efficacy and safety of "AHPL/AYTOP/0113" in comparison with "Framycetin Sulphate Cream" in acute wounds. *Anc Sci Life*, 2017. 36(3): p. 117–128.
16. Reagan, D.R., et al., Elimination of coincident Staphylococcus aureus nasal and hand carriage with intranasal application of Mupirocin calcium ointment. *Ann Intern Med*, 1991. 114(2): p. 101–106.
17. Ovais, M., et al., Wound healing applications of biogenic colloidal silver and gold nanoparticles: Recent trends and future prospects. *Appl Microbiol Biotechnol*, 2018. 102(10): p. 4305–4318.
18. Gaharwar, A.K., N.A. Peppas, and A. Khademhosseini, Nanocomposite hydrogels for biomedical applications. *Biotechnol Bioeng*, 2014. 111(3): p. 441–453.
19. Lu, X.F., et al., One-dimensional conducting polymer nanocomposites: Synthesis, properties and applications. *Prog Polym Sci*, 2011. 36(5): p. 671–712.
20. Zhao, R., et al., Stable nanocomposite based on PEGylated and silver nanoparticles loaded graphene oxide for long-term antibacterial activity. *ACS Appl Mater Interfaces*, 2017. 9(18): p. 15328–15341.
21. Fu, C., et al., Enhanced wound repair ability of arginine-chitosan nanocomposite membrane through the antimicrobial peptides-loaded polydopamine-modified graphene oxide. *J Biol Eng*, 2021. 15(1): p. 17.
22. Wang, Y., et al., Graphene oxide-IPDI-Ag/ZnO@Hydroxypropyl cellulose nanocomposite films for biological wound-dressing applications. *ACS Omega*, 2019. 4(13): p. 15373–15381.
23. Aguzzi, C., et al., Solid state characterisation of silver sulfadiazine loaded on montmorillonite/chitosan nanocomposite for wound healing. *Colloids Surf B Biointerfaces*, 2014. 113: p. 152–157.
24. Hasan, A., et al., Fabrication and characterization of chitosan, polyvinylpyrrolidone, and cellulose nanowhiskers nanocomposite films for wound healing drug delivery application. *J Biomed Mater Res A*, 2017. 105(9): p. 2391–2404.
25. Ashfaq, M., N. Verma, and S. Khan, Highly effective Cu/Zn-carbon micro/nanofiber-polymer nanocomposite-based wound dressing biomaterial against the P. aeruginosa multi- and extensively drug-resistant strains. *Mater Sci Eng C Mater Biol Appl*, 2017. 77: p. 630–641.
26. Rao, K.M., et al., Hemostatic, biocompatible, and antibacterial non-animal fungal mushroom-based carboxymethyl chitosan-ZnO nanocomposite for wound-healing applications. *Int J Biol Macromol*, 2020. 155: p. 71–80.
27. Sandri, G., et al., Halloysite and chitosan oligosaccharide nanocomposite for wound healing. *Acta Biomater*, 2017. 57: p. 216–224.
28. Kharat, Z., et al., Chitosan/PEO nanofibers containing Calendula officinalis extract: Preparation, characterization, in vitro and in vivo evaluation for wound healing applications. *Int J Pharm*, 2021. 609: p. 121132.
29. Samadi, A., et al., Fabrication and characterization of glycerol/chitosan/polyvinyl alcohol-based transparent hydrogel films loaded with silver nanoparticles for antibacterial wound dressing applications. *Adv Biomed Res*, 2021. 10: p. 4.
30. Kanimozhi, S., et al., Organic nanocomposite band-aid for chronic wound healing: A novel honey-based nanofibrous scaffold. *Appl Nanosci*, 2020. 10(5): p. 1639–1652.
31. Sathiyaseelan, A., et al., Antimicrobial and wound healing properties of FeO fabricated chitosan/PVA nanocomposite sponge. *Antibiotics (Basel)*, 2021. 10(5): p. 524.
32. Liu, T., et al., Synthesis of graphene oxide-quaternary ammonium nanocomposite with synergistic antibacterial activity to promote infected wound healing. *Burns Trauma*, 2018. 6: p. 16.
33. Zhao, X., et al., Injectable antibacterial conductive nanocomposite cryogels with rapid shape recovery for noncompressible hemorrhage and wound healing. *Nat Commun*, 2018. 9(1): p. 2784.

34. Guldiken, C.G., et al., ZnO microparticle-loaded chitosan/poly(vinyl alcohol)/acacia gum nanosphere-based nanocomposite thin film wound dressings for accelerated wound healing. *J Appl Polym Sci*, 2020. 137(10): p. 48445.
35. Ounkaew, A., et al., Synthesis of nanocomposite hydrogel based carboxymethyl starch/polyvinyl alcohol/nanosilver for biomedical materials. *Carbohydr Polym*, 2020. 248: p. 116767.
36. Zhai, M., et al., Keratin-chitosan/n-ZnO nanocomposite hydrogel for antimicrobial treatment of burn wound healing: Characterization and biomedical application. *J Photochem Photobiol B*, 2018. 180: p. 253–258.
37. Zhang, K. and, H.J. Choi, Smart polymer/carbon nanotube nanocomposites and their electrorheological response. *Materials (Basel)*, 2014. 7(5): p. 3399–3414.
38. Eatemadi, A., et al., Carbon nanotubes: Properties, synthesis, purification, and medical applications. *Nanoscale Res Lett*, 2014. 9(1): p. 393.
39. Elhissi, A.M., et al., Carbon nanotubes in cancer therapy and drug delivery. *J Drug Deliv*, 2012. 2012: p. 837327.
40. Ahmaruzzaman, M., D. Mohanta, and A. Nath, Environmentally benign fabrication of SnO_2-CNT nanohybrids and their multifunctional efficiency as an adsorbent, catalyst and antimicrobial agent for water decontamination. *Sci Rep*, 2019. 9(1): p. 12935.
41. Stankic, S., et al., Pure and multi metal oxide nanoparticles: Synthesis, antibacterial and cytotoxic properties. *J Nanobiotechnology*, 2016. 14(1): p. 73.
42. Palza, H., Antimicrobial polymers with metal nanoparticles. *Int J Mol Sci*, 2015. 16(1): p. 2099–2116.
43. Su, Z., et al., Chitosan/silver nanoparticle/graphene oxide nanocomposites with multi-drug release, antimicrobial, and photothermal conversion functions. *Materials (Basel)*, 2021. 14(9): p. 2351.
44. Teper, P., et al., Antimicrobial activity of hybrid nanomaterials based on star and linear polymers of N,N'-dimethylaminoethyl methacrylate with in situ produced silver nanoparticles. *Materials (Basel)*, 2020. 13(13): p. 3037.
45. Nazir, M., et al., Revealing antimicrobial and contrasting photocatalytic behavior of metal chalcogenide deposited P25-TiO_2 nanoparticles. *Photonics Nanostruct Fundam Appl*, 2019. 36: p. 100721.
46. Midya, L., et al., Novel nanocomposite derived from ZnO/CdS QDs embedded crosslinked chitosan: An efficient photocatalyst and effective antibacterial agent. *J Hazard Mater*, 2019. 369: p. 398–407.
47. Obrero, J.M., et al., Supported porous nanostructures developed by plasma processing of metal phthalocyanines and porphyrins. *Front Chem*, 2020. 8: p. 520.
48. Kou, J., D. Dou, and L. Yang, Porphyrin photosensitizers in photodynamic therapy and its applications. *Oncotarget*, 2017. 8(46): p. 81591–81603.
49. Matharu, R.K., L. Ciric, and M. Edirisinghe, Nanocomposites: Suitable alternatives as antimicrobial agents. *Nanotechnology*, 2018. 29(28): p. 282001.
50. Wang, N., et al., A novel surface plasmon resonance biosensor based on the PDA-AgNPs-PDA-Au film sensing platform for horse IgG detection. *Spectrochim Acta A Mol Biomol Spectrosc*, 2018. 191: p. 290–295.
51. Mishra, S., et al., Bioinspired nanocomposites: Applications in disease diagnosis and treatment. *Pharm Nanotechnol*, 2019. 7(3): p. 206–219.
52. Iqbal, A., et al., Synthesis and characterization of DGEBA composites reinforced with Cu/Ag modified carbon nanotubes. *Heliyon*, 2019. 5(5): p. e01733.
53. Nosrati, H., et al., Nanocomposite scaffolds for accelerating chronic wound healing by enhancing angiogenesis. *J Nanobiotechnology*, 2021. 19(1): p. 1.
54. Kittana, N., et al., Enhancement of wound healing by single-wall/multi-wall carbon nanotubes complexed with chitosan. *Int J Nanomedicine*, 2018. 13: p. 7195–7206.
55. Wang, L., C. Hu, and L. Shao, The antimicrobial activity of nanoparticles: Present situation and prospects for the future. *Int J Nanomedicine*, 2017. 12: p. 1227–1249.

56. de Fatima, A., et al., Wound healing agents: The role of natural and non-natural products in drug development. *Mini Rev Med Chem*, 2008. 8(9): p. 879–888.
57. Alven, S., X. Nqoro, and B.A. Aderibigbe, Polymer-based materials loaded with curcumin for wound healing applications. *Polymers (Basel)*, 2020. 12(10): p. 2286.
58. Shanmugapriya, K., et al., Fabrication of multifunctional chitosan-based nanocomposite film with rapid healing and antibacterial effect for wound management. *Int J Biol Macromol*, 2018. 118(Pt B): p. 1713–1725.
59. Krishnan, P.D., et al., Silver nanomaterials for wound dressing applications. *Pharmaceutics*, 2020. 12(9): p. 821.

26

Antimicrobial Activities of One-Dimensional Polymeric Nanocomposites

Parteek Prasher, Mousmee Sharma, and Pooja Chawla

CONTENTS

26.1 Introduction

The evolution of antibiotic resistance and the emergence of resistant microbial strains has manifested serious health care emergencies [1]. The unregulated use and misuse of over-the-counter antibiotics serves as the main contributing factor for the evolution of 'superbugs' [2]. Lack of sanitation and undisciplined hospital waste management further contributed to the spread of antibiotic-resistant microbes [3]. The declaration of antimicrobial resistance as one of the top ten global public health threats by the World Health Organization clearly explains the gravity of the situation [4]. The screening of huge molecular libraries for the development of novel antibiotics, their production, and public disbursement further poses a great strain on the economy [5]. Therefore, it necessitates the identification of sustainable materials that offer physiological compatibility, coupled with biodegradability, while matching the therapeutic efficacy of the customary antibiotics [6]. The 1D polymeric nanocomposites provide a robust antimicrobial profile by serving as drug carriers, microbial efflux pump inhibitors, microbial cell penetrators, and excipients for the killing of microbes [7, 8]. The 1D polymeric materials achieve a sustained release of the cargo drug molecules that further ensures the maintenance of an optimal therapeutic efficacy at the target site [9, 10]. This chapter highlights the contemporary antimicrobial applications of 1D polymeric nanocomposites. Figure 26.1 illustrates the potential applications of 1D nanofiber materials.

DOI: 10.1201/9781003223764-26

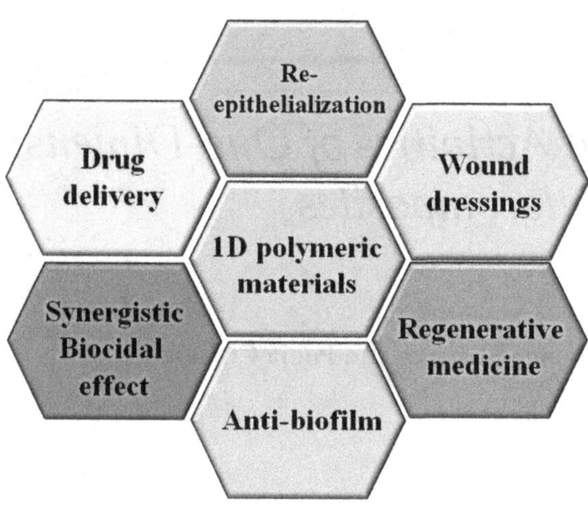

FIGURE 26.1
The antimicrobial applications of 1D polymeric nanofibers.

26.2 Polymeric Nanofiber Composites Based on Polysaccharides

The nanocomposite materials prepared from natural polysaccharide-based nanofibers, such as cellulose, chitosan, and starch, present antimicrobial applications in the form of hydrogels, packaging surfaces, anti-sweating textiles and fabrics, and antimicrobial dressing materials. The embedding of the polymeric nanofiber matrix with metal nanoparticles results in the realization of antimicrobial properties by the resultant nanocomposite material. These nanomaterials present applications as packaging materials owing to their biocidal potency. The impregnation of the polymeric nanofibers with biocidal metallic nanowires further provides a remarkable strategy for the achievement of antimicrobial effects. Li et al. prepared thiolated chitosan and impregnated it with silver nanowires *via* Ag-S covalent bonds to obtain 300-μm-thick hydrogel nanocomposite film dressings with enhanced wound-healing potency. Mainly, the thiol group of *N*-acetyl-L-cysteine forms a covalent bond with the Ag atom of the silver nanowire with an average diameter 40–80 nm, and length 5–20 μm. The thiolated chitosan serves as a soft polymeric component for the development of composite dressing *via* chemical bonding with the nanowire, whereas the antimicrobial effect arises from the Ag (I) species produced by the silver nanowire. Ag (I) species triggers a biocidal, oligodynamic effect on contacting the microbial cell and results in the production of reactive oxygen species (ROS) that result in the lysis of the microbial cell. The *in- vivo* investigation on the reported nanocomposite hydrogels further suggested cytocompatibility and a faster wound-healing effect on superficial skin wounds [11]. The impregnation of silver nanowires on cellulose nanofibrils and its subsequent deposition on the transparent polymer films strongly decreased the bacterial growth by 100% and 90% for Gram-positive *Staphylococcus aureus* and Gram-negative *Escherichia coli*, respectively. The reported nanocomposite presented applications for the development of antimicrobial packaging surfaces [12]. Further investigation of rheological properties, such as thixotropic nature and viscosity of the reported nanomaterial, led to the optimization of the conditions for processing *via* a screen-printing process. The analysis indicated that the recovery time and viscosity depended on the formulation of the nanocomposite material. Notably, the printing of the selected formulation on a flexible polyethylene terephthalate substrate exhibited 68% antimicrobial activity against *E. coli,* with improved transparency by 70%. These properties validated the application of the test nanocomposite as packaging material [13]. Furthermore, the nanocomposites synthesized from dextran and impregnated with silver nanoparticles having a hydrodynamic diameter of 26 nm, and a –32.21mV zeta potential displayed antimicrobial activity against *Bacillus subtilis,* and *E. coli* [14]. In addition to Ag-based nanomaterials, the copper nanoparticles (CuNPs) contained in starch nanocomposites presented a remarkable antimicrobial activity against *Candida albicans ATCC90028, Cryptococcus neoformance ATCC116, Escherichia coli ATCC25922, Aspergillus niger RCMB 02,724, Bacillus subtilis ATCC605, Aspergillus terreus RCMB 02,574,* and *Aspergillus fumigatus RCMB 02,568.* The cytotoxicity analysis of the copper nanoparticle-doped starch nanocomposites indicated a remarkable biological tolerance. The antimicrobial activity of the nanocomposite arises due to the generation of oxidative stress pertaining to the production of ROS [15]. The doping of CuO nanoparticles on the hydrogel based on chemically modified carboxymethyl starch resulted in the generation of nanocomposite material with remarkable antimicrobial properties and wound-healing applications. The presence of CuO nanoparticles significantly enhanced the antimicrobial potency of carboxymethyl starch against *Listeria monocytogenes, S. aureus, Pseudomonas aeruginosa, Salmonella enterica,* and *Yersinia enterocolitica,* and displayed synergistic effect for the growth inhibition of these microbes. The increased concentration of CuO nanoparticles in the nanocomposite resulted in toxicity due to the production of oxidative stress by the CuO nanoparticles. The bacterial toxicity of the nanocomposite followed dose-dependent patterns at all time intervals. As compared with carboxymethyl cellulose alone, its CuO-doped nanocomposite displayed enhanced wound-healing tendency, presenting a robust candidature for the development of wound dressings [16]. Sun et al. (2021) developed a multifunctional nanocomposite wound dressing synthesized from chitosan-gallic acid nanofibers impregnated with copper nanoparticles. The peroxidase-like activity of the reported nanocomposite led to the production of hydroxyl radicals from hydrogen peroxide that promoted antimicrobial activity and increased wound-healing effect with respect to *S. aureus*-infected wounds. Furthermore, the nanocomposite showed negligible toxicity and remarkable biocompatibility with the host cells, which supported its biological applications. The wound dressings prepared from the reported nanocomposites acted as a protective barrier between the wound and the external environment and showed excellent water-swelling properties, sterilizing the infected area efficiently [17]. Sodium alginate-based microspheres loaded with nystatin-loaded MgO/CuO nanocomposites reportedly serve as an ideal delivery system for the antifungal drug. The microspheres displayed an average diameter in the range 8–10 μm, with a highly porous surface and a semicrystalline structure. The reported microspheres downregulated the virulent traits of *Candida albicans,* such as biofilm formation at a concentration of 5 mg/mL. The nanocomposite displayed a remarkable 80% entrapment efficiency for the antifungal drug and achieved a sustained and prolonged drug release at pH 5.5 [18]. Polymer nanocomposite films based on sodium alginate and polyaniline loaded with TiO_2 nanoparticles displayed application as an antimicrobial food packaging material. The nanoparticles triggered the generation of oxidative stress in both the Gram-negative and Gram-positive bacterial species that disrupted the microbial cell wall, caused DNA damage to the microbe and denatured the essential proteins in the microbial cell [19]. The development of biodegradable material for food packaging applications requires remarkable mechanical properties for these materials. Deposition of Cu on graphitic carbon nitride nanoparticles and their reinforcement in the fabrication of starch/sodium alginate film led to the development of nanocomposites that displayed antimicrobial properties and a high tensile strength suitable for food packaging applications. Mainly, the addition of nanoparticles increased the tensile strength of the starch/sodium alginate film from 23.67 MPa to 36.57 MPa. The reported nanocomposite containing 5% Cu deposited on graphitic carbon nitride effectively hindered the growth of both the Gram-positive and Gram-negative bacteria [20]. Hydroxypropyl methylcellulose, known as hypromellose, when embedded with copper nanowire, results in the development of antibacterial nanocomposite film. The development of the nanocomposite occurs in the presence of polyethylene glycol as a plasticizer, where the copper nanowire undergoes a uniform dispersion in the polymer matrix without any apparent oxidation. The nanocomposite film displays remarkable flexibility, with antimicrobial activities against *E. coli,* and *S. aureus,* with minimum inhibitory concentration (MIC) values of 400 μg/mL, and 500 μg/mL, respectively, at 4.8% nanowire loading. The nanocomposite film exhibited cytocompatibility against normal human dermal fibroblasts due to a slower dissolution rate of the film and due to the production of a lower concentration of ROS-generating Cu^{2+} ions [21]. Arularasu et al. (2020) reported the antimicrobial potency of cellulose/TiO_2 nanocomposites with particle size in the range 20–40 nm. TiO_2 binds to the –OH group of cellulose *via* hydrogen bonding, thereby resulting in improved mechanical strength. The antimicrobial activity of the nanocomposite appears in terms of the *in-vitro* antibacterial activity against both the Gram-positive and Gram-negative bacteria *Staphylococcus aureus* and *Escherichia coli* due to the formation of reactive oxygen species that damage the microbial membrane. The reported nanocomposites presented an application as antimicrobial packaging material for food items [22].

26.3 Polymeric Nanofibercomposites Based on Synthetic Polymers

Blending of polymers, such as polyvinyl acetate and polyethylene glycol, with biocidal silver nanoparticles results in the formation of nanocomposite material with antimicrobial properties and moisture absorption efficacy. The inhibition of pathogenic *Ps. aeruginosa, E. coli,* and *S. aureus* by the reported nanocomposite, its slow degradation, and prolonged shelf life support its packaging application [23]. Multifunctional nanocomposites reported by Rehan et al. (2020) consisted of silver nanoparticle-blended polyacrylonitrile with remarkable antimicrobial efficacy. The content of AgNPs in the polymer film determined the antimicrobial activity against *E. coli, B. subtilis, S. aureus, Aspergillus niger,* and *Candida albicans* [24]. The silver ions furnished by the AgNPs interact with the negatively charged cell membrane by electrostatic interactions thereby inhibiting its growth and causing DNA damage after penetrating the microbial cell membrane. Similarly, the interaction of silver ions with thiol substituents of the enzymes causes the inhibition of microbial proteins [24]. The facile fabrication of AgI/graphitic carbon nitride nanocomposites displayed superior photocatalytic activity as compared with pure AgI and graphitic carbon nitride. The photocatalytic behavior of the nanocomposite presented applications for the photodegradation of organic dyes, such as rhodamine B, and methyl orange. Furthermore, the reported nanocomposite displayed antimicrobial activity against a wide range of pathogenic microbes, such as *Shigella dysenteriae, Aspergillus brasiliensis, Proteus vulgaris, Aspergillus niger, S. aureus, Ps. aeruginosa, Candida albicans,* and *Staphylococcus epidermidis* [25]. The composites of conducting polymers, such as polyindole, with ZrO_2 showed antimicrobial activity against the pathogenic *Salmonella typhi, E. coli, Bacillus subtilis,* and *Penicillium chrysogenum,* with a superior inhibition profile compared with amphotericin-B taken as the standard. Mainly, the presence of the biocidal indole nucleus bestowed the nanocomposite with antimicrobial properties [26]. The nanocomposites of polyvinyl alcohol fabricated with graphene oxide and nanosilver in the presence of ionic liquid displayed cytotoxicity against the MCF-7, MDA-MB-231, and MCF-10A cell lines that led to their development as antimicrobial agents. The presence of ionic liquid improved the dispersibility of the graphene oxide and nanosilver nanofillers in the matrix of polyvinyl acetate, which enhanced the thermal and mechanical properties of the composite. Furthermore, the reported nanocomposite exhibited a remarkable inhibition of *E. coli* and *S. aureus* with considerable biocompatibility, which favors their application as antimicrobial coatings [27].

26.4 Polymeric Nanocomposites Based on Polyurethane

Natural rubber-based polyurethane nanocomposites consisted of polyester-type polyurethane synthesized from –OH terminated natural rubber as soft polymer segments, and 4,4-methylene dicyclohexyl diisocyanate as hard segments, with 1,4-butan-diol as a chain extender. The impregnation of the material with TiO_2, and AgNPs led to the development of antimicrobial nanocomposites with remarkable tensile strength, improved thermal degradation, and mechanical properties. The inhibition of *S. aureus* by the reported nanocomposites supported their biomedical applications, especially as protective antimicrobial films [28]. Nanocomposite films comprising of zwitterion-impregnated rutile TiO_2 nanoparticles exhibit applications as photostable, food-packaging material owing to its antimicrobial properties. The addition of TiO_2 to polyurethane improved the antioxidant potency of the nanocomposite from 51% to 71%. The inhibition of *E. coli* by the reported nanocomposite supports its application as a food packaging material, further validated by its anti-aging effectiveness toward tomato fruits [29]. Sustainable formulations of polyurethane composites reinforced with ZnO-doped SiO_2 nanoparticles presented antimicrobial properties against fungi, Gram-positive and Gram-negative bacteria, hence serving as green packaging material. The reported nanocomposite exhibited remarkable durability due to its mechanical and thermal behavior, low flammability, hydrophobicity, water vapor, and oxygen barrier properties [30]. Polyurethane nanofibrous mats containing reduced graphene oxide/silver nanocomposites demonstrated biomedical properties, such as antimicrobial potential and cutaneous wound healing. The low diameter and porous structure of nanocomposite material containing curcumin displayed the most remarkable biological profile. As such, the reported nanocomposite material displayed 100% inactivation of *Pseudomonas*, and 95% inactivation of *S. aureus* on direct contact with the microbial cells. The presence of curcumin and reduced graphene oxide/silver nanostructures remarkably promoted the wound healing and regeneration of the epidermis layer [31]. The intercalation of drugs such as tetracycline hydrochloride in smectite clay in the presence of a nanofiber membrane containing polyurethane led to the controlled delivery of the cargo drug molecules. The drug release occurred by ion exchange and a concentration gradient mechanism *via* non-Fickian dynamics. The nanofiber composite with a web-like flat morphology promoted cell growth and proved highly beneficial for wound healing applications. *In-vitro* analysis further confirmed that the nanofiber web composites containing pure drug molecules showed a burst release, while the nanocomposites containing drug-clay conjugates achieved a sustained release of the cargo drug molecule, which proved highly desirable for the wound healing applications with prolonged activity for 10 days [32]. Polyurethane nanocomposite impregnated with graphene oxide and modified with chitosan presented applicability for the development of wound dressings. The biocidal potency of the nanocomposite against *S. aureus* and *E. coli*, in addition to its remarkable biocompatibility, highlight its wound dressing application. The chitosan fabrication of graphene oxide provides improved morphological and thermomechanical properties, and wettability of the nanocomposites [33]. Salahuddin et al. developed a polyurethane-urea nanocomposite for realizing a thermo-sensitive release of norfloxacin *in vitro*. The drug release profile of the nanocomposite was determined by the loading capacity of the drug by the nanosystem, ionic strength, pH, temperature, and composition of the carriers. At room temperature, the nanosystem showed a trivial release of the drug, which converted to controlled release at 37 °C. Furthermore, a rise in temperature to 45 °C resulted in the burst release of the cargo drug. The drug release followed Fickian diffusion mechanism at 37 °C, and first-order kinetics at 45 °C. The loading of the drug in the nanocomposite improved its antimicrobial efficacy 2-fold, as compared with the free drug [34]. Carbon quantum dot-impregnated photoactive polyurethane nanocomposites displayed light-sensitive antimicrobial activity against *S. aureus* and *E. coli* on irradiation with a dose of 200 kGγ for 15 minutes. The irradiation with gamma-radiation remarkably enhanced the production of the reactive oxygen species (ROS) that played a major role in the manifestation of the antimicrobial effect. These nanocomposites proved particularly useful for the development of antimicrobial surfaces and coatings [35].

26.5 Polymeric Nanofiber Composites Based on Polyvinyls

Nanocomposites prepared from polyvinyl alcohol-modified nanocellulose embedded with silver nanoparticles displayed antimicrobial potency against *E. coli* by inducing damage to the bacterial membrane. A dense 3D network of the films developed from these nanocomposites acted as a mechanical barrier and obstructed oxygen and water vapor permeation. These properties supported the food-packaging application of the nanocomposite films. Moreover, the biodegradability of the nanocomposite material promised an eco-friendly approach for the development of sustainable food packaging materials [36]. Fabrication of chitosan with polyvinyl alcohol to generate nanocomposite films embedded with biogenic silver nanoparticles displayed a marked inhibition of food-borne pathogens such as *Bacillus cereus, S. aureus, L. monocytogenes*, and *E. coli*. The antimicrobial effect arises from the presence of silver nanoparticles that enhanced the biocidal and anti-oxidant properties without altering the properties of the nanocomposite film. Furthermore, the biodegradability of the nanocomposite film played a significant role in extending the shelf life of the packaged material [37]. Nanocomposites constituting the blending of polyvinylchloride with polystyrene with copper chloride filler presented antimicrobial properties against *Aspergillus avenaceus*, and *Aspergillus terreus*. The Cu^{2+} ions furnished by copper chloride form the interactions with functional groups of the polyvinylchloride-polystyrene matrix. The presence of 8% wt. copper chloride in the nanocomposite blend led to a significant diminishing of the growth of fungal hyphae, hence suggesting the suitability of the composite as a base matrix. The antimicrobial effect further increased with the addition of copper chloride in the polyvinylchloride-polystyrene matrix [38]. Nanocomposites based on polyvinyl alcohol-containing NiO nanoparticles displayed antimicrobial properties against *B. cereus, E. coli, Salmonella typhimurium*, and *S. aureus* due to the presence of biocidal NiO nanoparticles. Chitosan modification of polyvinyl alcohol improved the activity of nanocomposites as compared with pure chitosan and polyvinyl alcohol [39]. Jayakumar et al. (2021) reported the blending of polyvinyl alcohol with lipopolypeptides and ZnO nanoparticles for the development of antimicrobial nanocomposites. These nanocomposite films showed skin-compatible pH and non-hemolytic properties that supported their application as an antimicrobial coating material. The reported nanocomposite films successfully inhibited the growth of *S. aureus*, *Klebsiella pneumoniae*, and *Ps. aeruginosa*, confirming their biocidal application [40].

26.6 Polymeric Nanocomposites Based on Polyacrylates

Hybrid nanocomposites prepared from polyacrylic acid and graphene oxide displayed commendable antimicrobial properties against *E. coli*, which supports its food packaging applications. The addition of 27 $\times 10^{-3}$ wt% silver nanoparticles, relative to the polyacrylic acid, to the nanocomposite matrix improved its oxygen barrier property 11-fold. The presence of silver nanoparticles enhanced the penetration of the bacterial membrane by the biogenic nanoparticles, leading to bacterial death. Furthermore, the thermal stability, chemical resistance, tensile strength, and oxygen barrier characteristics further recommend the robustness of the reported nanocomposite for packaging applications [41]. Nanocomposites consisting of a polyacrylic acid brush tethered to graphene oxide membrane coating presented an application as an anti-biofouling agent by dint of its biocidal effect. The presence of polyacrylic acid and graphene oxide coating caused a decrease in the surface charge and improved the hydrophilicity of the membrane surface used for reverse osmosis, whereas the presence of the graphene oxide coating provided antimicrobial characteristics in offering a reduction in biofouling [42]. A nanocomposite hydrogel based on poly(acrylamide-IPN-acrylic acid) displayed antimicrobial applications against *S. aureus*, *E. coli*, *Ps. aeruginosa*, *Aspergillus*, and *Penicillium*. The antimicrobial activity of the hydrogel arises due to different concentrations of peptidoglycans of the cell wall in different bacterial species. Gram-negative bacteria, having a thin cell wall, undergo a higher inhibition rate due to a loose peptidoglycan layer that allows the permeation of biocidal nanoparticles [43].

26.7 Polymeric Electrospun Nanofibers

Hybrid nanofibers based on polymeric systems offer potential antimicrobial applications due to their biocidal properties. Electrospun nanofibers obtained from hydroxyapatite nanoparticles (nHA), poly-ε-caprolactone (PCL), and gelatin (Gel) polymers loaded with doxorubicin exhibit tumor cell cytotoxicity and antimicrobial activities against Gram-positive *S. aureus* and Gram-negative *Porphyromonas gingivalis* bacteria. The co-delivery of hydroxyapatite nanoparticles and doxorubicin offers synergistic inhibition of the bacteria with greater efficacy as compared with either of the drugs alone. The nanosystem undergoes burst release to release the cargo drug molecules due to the high solubility of the nanogel in the release medium [44]. Electrospun polymeric composite nanofibers obtained from different ratios of chitosan and polyethylene oxide displayed a high loading efficacy towards moxifloxacin for inhibiting the growth of *S. aureus*, *E. coli*, and *Ps. aeruginosa*. The carrier system displayed a maximum drug release of 80% and 94% at pH 7.4 and 5.5, respectively, with a high degree of swelling (133%) at this pH. The higher drug release of the drug at pH 5.5 occurs due to the higher affinity of chitosan at lower pH, which proves highly advantageous for antimicrobial dressings as the pH of the skin occurs between 5.0 and 6.0. Interestingly, the cargo drug existed in an amorphous state in the carrier system without showing any covalent interactions with the polymeric excipients. This results in greater stability and potential toward wound-healing efficiency by the drug-loaded nanosystem [45].

Electrospun nanofibers based on polyethylene oxide-chitosan polymer loaded with ZnO nanoparticles and ciprofloxacin exhibited commendable antimicrobial activity against *E. coli* and *S. aureus* with minimal side effects against human dermal fibroblast and keratinocyte cell lines. The reported nanosystem realized a faster drug release profile at an acidic pH of 5.5, approximating to skin pH, as compared with the physiological pH of 7.4 where the controlled release of the cargo drug occurs. The rapid drug release at acidic pH occurs mainly due to the solubility of chitosan at the acidic environment, which becomes lesser at the basic pH of 7.4, where the solubility of chitosan decreases. The polyethylene oxide-chitosan nanoweb undergoes a rapid weight loss of 61% between 215–390 °C due to the thermal cleavage of inter- and intrachain linkages. The incorporation of ZnO nanoparticles resulted in lowering the degradation temperature to 160 °C due to the thermal conductivity and heat capacity of ZnO nanoparticles that favor heat transfer between polyethylene oxide-chitosan and ZnO [46].

The polyethylene oxide-chitosan electrospun nanofibers loaded with phenolic compounds extracted from *Spirulina* sp. LEB 18 exhibited improved antimicrobial activity against *E. coli* and *S. aureus* as compared with the free phenolic compounds. The high temperature of maximum degradation possessed by the reported nanosystem supports its food packaging applications. Mainly, the positive charge on chitosan results in interactions with negatively charged moieties on the surface of microbial cells, resulting in altered cell permeability, which subsequently prevents the foreign materials from entering the cells and prevents the loss of essential cell components. Furthermore, the binding of chitosan to microbial DNA results in the inhibition of its replication. The incorporation of phenolic compounds in polyethylene oxide-chitosan electrospun nanofibers retains the antimicrobial activity of the former, which further improved with the incremental addition of phenolic compounds. The incorporation of phenolic compounds to the polyethylene oxide-chitosan electrospun nanofibers plays an important role in preserving the antimicrobial potency of the former as the absence of polymeric protection results in a significant loss of their activity on coming in direct contact with the surface of products [47]. Electrospun poly(acrylic acid) and reduced graphene oxide hydrophilic nanofiber mats exhibited photothermal release of the antibiotics ampicillin and cefepime for the inhibition of *S. aureus* and *Staphylococcus epidermidis*. Irradiation of nanomats with near-infrared light at 980 nm causes an enhanced antibiotic release and presented applications for wound healing. The polymeric nanofibers displayed negligible photothermal heating after irradiation with a continuous-wave laser at 980 nm for 5 minutes, which increased the temperature to 67 °C. Furthermore, the nanofibrous mats showed no melting under these conditions, which preserved the structural integrity of the material. Importantly, the modulation of the density of laser power resulted in the realization of controlled photothermal heating of the material in a temperature range of 20–67 °C. Immersing antibiotics with nanofibers for 72 h at 4 °C resulted in their integration into the material. Mainly, the aromatic structure of antibiotic molecules and the presence of amino and carboxylic acid groups promoted their interaction with reduced graphene oxide *via* electrostatic and π-π stacking interactions. The hydrophilic nature of nanofibers resulted in their swelling in the aqueous media, with subsequent loading with antibiotics. The suitable physical, biological, and chemical properties, in addition to considerable stability over a range of temperatures, supported the application of the reported electrospun fibers in the development of wound-healing dressings [48].

26.8 Polymeric Core-Shell Nanofibers

Controlling bacterial infection at the surgical site poses a significant challenge in the present era to ensure optimal health of hospitalized patients. The susceptibility of bacteria toward antibiotics raises further concerns and necessitates the inhibition of hospital-acquired infections caused by these bacterial strains. Nylon-6/chitosan core-shell nanofibers present antimicrobial properties for the prevention of mesh-associated infection at the surgical site. The presence of nylon-6 provides mechanical stability, whereas the presence of chitosan/polyethylene oxide provides bacteriostatic activity to the nanofibers. The antimicrobial core-shell structures incorporated 5-chloro-8-quinolinol in chitosan and achieved a sustained release of poly(hexanide) from the nylon-6 fibers. The antibacterial efficacy appeared against *S. aureus* and *Ps. aeruginosa*, which are involved in causing infection at the surgical site [49].

Core-shell nanostructures comprised polyvinyl alcohol (PVA)-poly(lactic acid) (PLA) nanofibers, combining the biocompatibility of PLA and the hydrophilicity of PVA for biomedical applications. The introduction of silver nanoparticles in the PVA core resulted in the formation of nanofiber mats that exhibited antimicrobial activity against *E. coli* and *S. aureus* due to the presence of silver nanoparticles that instigate a cascade of deleterious effects and oxidative/ redox stress once they come in contact with the bacterial cell. The reported nanofibers served as a potential material for the development of antimicrobial dressings due to their considerable biocompatibility owing to the presence of PVA, although the generation of metallic silver by the reported nanofibers may prove hazardous for human use due to their oxidation to ionic silver, which is known to trigger oxidative damage to the tissues [50].

Ramalingam et al. (2021) incorporated antibiotic 'minocycline' and antimicrobial herbal extracts obtained from *Gymnema sylvestre* into core-shell structured nanofiber dressings consisting of poly-ε-caprolactone/gelatin. The resultant nanosystem promoted cutaneous wound healing and prevented the formation of antimicrobial biofilms or bacterial colonization. The most important utility of the reported nanofibers occurs in the treatment of second-degree burns by improving the healing process in patients. The presence of extract components in core-shell nanofibers improved the mechanical properties of the nanosystem, as compared to the pristine mats. The core-shell morphology of the nanosystem caused a sustained release of the antimicrobial components and antibiotics as compared with the nanofiber blends. The *in-vivo* investigations in a porcine model of cutaneous, second-degree burns indicated that the treatment of these wounds with antimicrobial dressings soaked in the reported core-shell nanofibers improved the re-epithelialization and organization of collagen at the burn site, as compared with the burn areas/ wounds which were untreated with the reported material [51].

The promotion of re-epithelialization by the core-shell nanofibers further resulted in the development of polymer-mesoporous silica nanoparticle core-shell nanofibers as dual drug delivery systems and for guiding tissue regeneration. The mesoporous silica nanoparticles containing large pores possess the ability to encapsulate the growth factors in their core and play a significant role in the preservation of their bioactivity, whereas the drug loaded in the shell nanofibers displayed a sustained release behavior. The core-shell nanofibers displayed superior ontogenetic regeneration potency towards bone marrow stromal cells, as compared with the commercial Bio-Gide membranes. Furthermore, the drug-carrying nanocomposite core-shell nanofibers displayed commendable antimicrobial properties against several species of Gram-positive and Gram-negative bacteria. The reported nano-engineered membranes displayed non-toxicity towards bone marrow stromal cells. The osteogenic tests indicated that the bone marrow stromal cells cultured with the reported nanofibrous material demonstrated better bone regeneration than was shown by the control groups. The reported nanofiber system hence provides a material for the effective delivery of antibiotics and growth factors in periodontal tissues for the better treatment of periodontitis [52].

Magnesium-mineralized nanofiber dressings containing chondroitin sulfate exhibit antimicrobial properties against *Staphylococcus epidermidis*. The composite dressings consisted of polycaprolactone and gelatin as core-shell nanofibers or blend that presented a superior tensile strength, greater stretchability, and a photoluminescence profile. The *in-vivo* analysis of core-shell nanofibers on the porcine model of cutaneous burn injury suggested improved re-epithelialization, wound closure, and better clinical outcomes as compared to untreated burn wounds. Furthermore, the histology profile of the biopsied tissues suggested a smooth regeneration and collagen organization of the burn wounds treated with core-shell nanofibers, as compared with the untreated burn wounds. The reported analysis validated the potency of core-shell nanofibrous materials in healing burn wounds by promoting tissue re-epithelialization and wound closure. Moreover, the inherent photoluminescence profile of reported core-shell nanofibers proved highly advantageous for the non-invasive monitoring of the degeneration of the wound dressing. Therefore, the reported nanofibers demonstrated multifunctional advantages that are relevant for the effective management of deep dermal wounds, in addition to protecting the underlying tissues from microbial colonization, especially by the commensal pathogens [53]. Baghali et al. further reported the fabrication of core-shell TiO_2 nanofibers containing erythromycin-loaded poly(caprolactone) (PCL/erythromycin) core, with a zein-containing titanium dioxide (zein/TiO_2) shell nanofibers. The reported material displayed a controlled release of erythromycin in 72 h, with commendable antimicrobial properties against *S. aureus, B. cereus, E. coli,* and *Salmonella* species [54].

26.9 Conclusion

The 1D polymeric materials present advanced applications for capping multidrug-resistant microbes. The nanocomposites based on these materials display extraordinary mechanical and tensile strengths and a remarkable tolerance to diverse temperatures and pH. Combined with antimicrobial properties, these physical properties validate the food packaging applications of the 1D polymeric nanocomposites. The biocidal and physicochemical characteristics of the 1D polymeric nanocomposites extend the shelf life of the packed food items. 1D polymeric nanocomposites constructed from green materials, such as natural polymers, contribute to mitigating solid waste pollution due to their biodegradability, whereas the plastic packaging materials stay in the ecosystem and elevate pollution levels. Overall, the 1D polymeric nanocomposites provide a robust candidature for the development of future packaging materials due to their extraordinary physicochemical and biological properties.

References

1. Holubar M (2020) Antimicrobial resistance: A global public health emergency further exacerbated by international travel. *J. Travel Med.* 27(1): Article 095.
2. Razzaque MS (2020) Commentary: Microbial resistant movements: An overview of global public health threats posed by antimicrobial resistance, and how best to counter. *Front. Public Health* 8: Article 629120.
3. Abushaheen MA, Muzaheed, Fatani AJ, Alosaimi M, Mansy W, George M, Acharya S, Rathod S, Divakar DD, Jhugroo C, Vellappally S, Khan AA, Shaik J, Jhugroo P (2020) Antimicrobial resistance, mechanisms and its clinical significance. *Disease-A-Month* 66(6): Article 100971.
4. https://www.who.int/news-room/fact-sheets/detail/antimicrobial-resistance.
5. https://www.downtoearth.org.in/blog/economy/antimicrobial-resistance-has-a-huge-economic-impact-too-74261.
6. Chellappan DK, Prasher P, Saravanan V, Yee VSV, Chi WCW (2021) Protein and peptide delivery to lungs by using advanced targeted drug delivery. *Chem. Biol. Interact.* https://doi.org/10.1016/j.cbi.2021.109706.
7. Cipriani CE, Ha T, Defillo OBM, Myneni M, Wang Y, Benjamin CC, Wang J, Pentzer EB, Wei P (2021) Structure-processing-property relationship of 3D printed porous polymeric materials. *ACS Mater. Au.* 1: 69–80.
8. Acquavia MA, Pascale R, Martelli G, Bondoni M, Bianco G (2021) Natural polymeric materials: A solution to plastic pollution from agro-food sector. *Polymers (MDPI)* 13: Article 158.
9. Fu Y, Yang L, Hu J, Duan G, Liu X, Li Y, Gu Z (2021) Polydopamine antibacterial materials. *Mater. Horiz.* 8(6): 1618–1633.
10. Zeng M, Xu J, Luo Q, Hou C, Qiao S, Fu S, Fan X, Liu J (2020) Constructing antibacterial polymer nanocapsules based on pyridine quaternary ammonium salt. *Mater. Sci. Eng. C Mater. Biol. Appl.* 108: Article 110383.
11. Li J, Li L, Lv J, Wang C, Liu Y (2020) Preparation of thiolated chitosan/silver nanowire composite hydrogels with antimicrobial activity for obstetric wound care. *Mater. Lett.* 280: Article 128497.
12. Spieser H, Denneulin A, Deganello D, Gethin D, Koppolu R, Bras J (2020) Cellulose nanofibrils and silver nanowires active coatings for the development of antibacterial packaging surfaces. *Carbohydr. Polym.* 240: Article 116305.
13. Spieser H, Jardin A, Deganello D, Gethin D, Bras J, Denneulin A (2021) Rheology of cellulose nanofibrils and silver nanowires for the development of screen-printed antibacterial surfaces. *J. Mater. Sci.* 56(21): 12524–12538.
14. Kokilavani S, Syed A, Thomas AM, Elgorban AM, Bahkali AH, Marraiki N, Raju LL, Das A, Khan SS (2021) Development of multifunctional Cu sensitized Ag-dextran nanocomposite for selective and sensitive detection of mercury from environmental sample and evaluation of its photocatalytic and anti-microbial applications. *J. Mol. Liq.* 321: Article 114742.
15. Hasanin M, Al Abboud MA, Alawlaqi MM, Abdelghany TM, Hashem AH (2021) Ecofriendly synthesis of biosynthesized copper nanoparticles with starch-based nanocomposite: Antimicrobial, antioxidant, and anticancer activities. *Biol. Trace Elem. Res.* https://doi.org/10.1007/s12011-021-02812-0.

16. Abdollahi Z, Zare EN, Salimi F, Goudarzi I, Tay FR, Makvandi P (2021) Bioactive carboxymethyl starch-based hydrogels decorated with CuO nanoparticles: Antioxidant and antimicrobial properties and accelerated wound healing in vivo. *Int. J. Mol. Sci.* 22(5): Article 2531.
17. Sun X, Dong M, Guo Z, Zhang H, Wang J, Jia P, Bu T, Liu Y, Li L, Wang L (2021) Multifunctional chitosan-copper-gallic acid based antibacterial nanocomposite wound dressing. *Int. J. Biol. Macromol.* 167: 10–22.
18. Abid S, Uzair B, Niazi MBK, Fasim F, Bano SA, Jamil N, Batool R, Sajjad S (2021) Bursting the virulence traits of MDR strain of *Candida albicans* using sodium alginate-based microspheres containing nystatin-loaded MgO/CuO nanocomposites. *Int. J. Nanomed.* 16: 1157–1174.
19. Abutalib MM, Rajeh A (2020) Preparation and characterization of polyaniline/sodium alginate-doped TiO_2 nanoparticles with promising mechanical and electrical properties and antimicrobial activity for food packaging applications. *J. Mater. Sci. Mater. Electron.* 31(12): 9430–9442.
20. Mousavi SN, Daneshvar H, Dorraji MSS, Ghasempour Z, Azar VP, Ehsani A (2021) Starch/alginate/Cu-g-C3N4 nanocomposite film for food packaging. *Mater. Chem. Phys.* 267: Article 124583.
21. Bagchi B, Fernandez CS, Bhatti M, Ciric L, Lovat L, Tiwari MK (2022) Copper nanowire embedded hypromellose: An antibacterial nanocomposite film. *J. Colloid Interface Sci.* 608(1): 30–39.
22. Arularasu MV, Harb M, Sundaram R (2020) Synthesis and characterization of cellulose/TiO_2 nanocomposite: Evaluation of *in vitro* antibacterial and *in silico* molecular docking studies. *Carbohyd. Polym.* 249: Article 116868.
23. Ravindran RSE, Subha V, Ilangovan R (2020) Silver nanoparticles blended PEG/PVA nanocomposites synthesis and characterization for food packaging. *Arab. J. Chem.* 13(7): 6056–6060.
24. Rehan M, Nada AA, Khattab TA, Abdelwahed NAM, El-Kheir AAA (2020) Development of multifunctional polyacrylonitrile/silver nanocomposite films: Antimicrobial activity, catalytic activity, electrical conductivity, UV protection and SERS-active sensor. *J. Mater. Res. Technol.* 9(4): 9380–9394.
25. Orooji Y, Ghanbari M, Amiri O, Niasari MS (2020) Facile fabrication of silver iodide/graphitic carbon nitride nanocomposites by notable photo-catalytic performance through sunlight and antimicrobial activity. *J. Hazard. Mater.* 389: Article 122079.
26. Anandhi S, Edward ML, Jaisankar V (2021) Synthesis, characterization and antimicrobial activity of polyindole/ZrO_2 nanocomposites. *Mater. Today Proc.* 40: 93–101.
27. Sahu G, Das M, Sethy C, Wazalwar C, Kundu CN, Raichur AM, Tripathy J (2021) Ionic liquid-assisted fabrication of poly(vinyl alcohol)/nanosilver/graphene oxide composites and their cytotoxicity/antimicrobial activity. *Mater. Chem. Phys.* 266: Article 124524.
28. Anancharoenwong E, Chueangchayaphan W, Rakkapao N, Marthosa S, Chaisrikhwun B (2021) Thermo-mechanical and antimicrobial properties of natural rubber-based polyurethane nanocomposites for biomedical applications. *Polym. Bull.* 78(2): 833–848.
29. Athir N, Shah SAA, Shehzad FK, Cheng J, Zhang J, Shi L (2020) Rutile TiO_2 integrated zwitterion polyurethane composite films as an efficient photostable food packaging material. *React. Funct. Polym.* 157: Article 104733.
30. Moustafa H, Darwish NA, Youssef AM (2022) Rational formulations of sustainable polyurethane/chitin/rosin composites reinforced with ZnO-doped-SiO_2 nanoparticles for green packaging applications. *Food Chem.* 371: Article 131193.
31. Esmaeili E, Arshaghi TE, Hosseinzadeh S, Elahirad E, Jamalpoor Z, Hatamie S, Soleimani S (2020) The biomedical potential of cellulose acetate/polyurethane nanofibrous mats containing reduced graphene oxide/silver nanocomposites and curcumin: Antimicrobial performance and cutaneous wound healing. *Int. J. Biol. Macromol.* 152: 418–427.
32. Saha K, Dutta K, Basu A, Adhikari A, Chattopadhyay D, Sarkar P (2020) Controlled delivery of tetracycline hydrochloride intercalated into smectite clay using polyurethane nanofibrous membrane for wound healing application. *Nano Struct. Nano Object* 21: Article 100418.
33. Najafabadi SAA, Mohammadi A, Kharazi AZ (2020) Polyurethane nanocomposite impregnated with chitosan-modified graphene oxide as a potential antibacterial wound dressing. *Mater. Sci. Eng. C Mater. Biol. Appl.* 115: Article 110899.

34. Salahuddin N, Rehab A, Abd-Elghany S (2020) *In vitro* thermo-triggered drug release from magnetic polyurethane-urea nanocomposite. *J. Drug Deliv. Sci. Technol.* 56: Article 101564.
35. Budimir M, Markovic Z, Vajdak J, Jovanovic S, Kubat P, Humpolicek P, Micusik M, Danko M, Barras A, Milivojevic D, Spitalsky Z, Bukherroub R, Markovic BT (2021) Enhanced visible light-triggered antibacterial activity of carbon quantum dots/polyurethane nanocomposites by gamma rays induced pre-treatment. *Radiat. Phys. Chem.* 185: Article 109499.
36. Wang W, Yu Z, Alsammarraie FK, Kong F, Lin M, Mustapha A (2020) Properties and antimicrobial activity of polyvinyl alcohol-modified bacterial nanocellulose packaging films incorporated with silver nanoparticles. *Food Hydrocoll.* 100: Article 105411.
37. Nwabor OF, Singh S, Paosen S, Vongkamjan K, Voravuthikunchai PS (2020) Enhancement of food shelf life with polyvinyl alcohol-chitosan nanocomposite films from bioactive *Eucalyptus* leaf extracts. *Food Biosci.* 36: Article 100609.
38. Nasrallah DA, Morsi MA, El-Sayed F, Metwally RA (2020) Structural, optical and electrical properties of copper chloride filled polyvinyl chloride/polystyrene blend and its antifungal properties against *Aspergillus avenaceus* and *Aspergillus terreus*. *Compos. Commun.* 22: Article 100451.
39. Ali H, Tiama TM, Ismail AM (2021) New and efficient NiO/chitosan/polyvinyl alcohol nanocomposites as antibacterial and dye adsorptive films. *Int. J. Biol. Macromol.* 186: 278–288.
40. Jayakumar A, Radoor S, Nair IC, Siengchin S, Parameswaranpillai J, Radhakrishnan EK (2021) Lipopeptide and zinc oxide nanoparticles blended polyvinyl alcohol-based nanocomposite films as antimicrobial coating for biomedical applications. *Process Biochem.* 102: 220–228.
41. Sethy PK, Mohapatra P, Patra S, Bharatiya D, Swain SK (2021) Antimicrobial and barrier properties of polyacrylic acid/GO hybrid nanocomposites for packaging applications. *Nano Struct. Nano Object* 26: Article 100747.
42. Ansari A, Bahamonde JP, Wang M, Shaffer DL, Hu Y, Rodrigues DF (2021) Polyacrylic acid-brushes tethered to graphene oxide membrane coating for scaling and biofouling mitigation on reverse osmosis membranes. *J. Membr. Sci.* 630: Article119308.
43. Sharma S, Virk K, Sharma K, Bose SK, Kumar V, Sharma V, Focarete ML, Kalia S (2020) Preparation of gum acacia-poly(acrylamide-IPN-acrylic acid) based nanocomposite hydrogels via polymerization methods for antimicrobial applications. *J. Mol. Struct.* 1215: Article 128298.
44. Agudelo AA, Scheuermann K, Garcia AG, Monteiro APF, Garcia ADP, Cortes ME, Sinisterra RD (2018) Hybrid nanofibers based on poly-caprolactone/gelatin/hydroxyapatite nanoparticles-loaded doxycycline: Effective anti-tumoral and antibacterial activity. *Mater. Sci. Eng. C* 83: 25–34.
45. Hameed M, Rasul A, Nazir A, Yousaf AM, Hussain T, Khan IU, Abbas G, Abid S, Yousafi QA, Ghori MU, Shahzad Y (2021) Moxifloxacin-loaded electrospun polymeric composite nanofibers-based wound dressing for enhanced antibacterial activity and healing efficacy. *Int. J. Polym. Mater. Polym. Biomater.* 70(17): 1271–1279.
46. Abid S, Hussain T, Nazir A, Zahir A, Ramakrishna S, Hameed M, Khenoussi N (2019) Enhanced antibacterial activity of PEO-chitosan nanofibers with potential application in burn infection management. *Int. J. Biol. Macromol.* 135: 1222–1236.
47. Kuntzler SG, Costa JAV, Morais MG (2018) Development of electrospun nanofibers containing chitosan/PEO blend and phenolic compounds with antibacterial activity. *Int. J. Biol. Macromol.* 117: 800–806.
48. Altinbasak I, Jijie R, Barras A, Golba A, Sanyal R, Bouckaert J, Drider D, Bilyy R, Dumych T, Paryzhak S, Vovk V, Boukherroub R, Sanyal A, Szuneritis S (2018) Reduced graphene-oxide-embedded polymeric nanofiber mats: An "on-demand" photothermally triggered antibiotic release platform. *ACS Appl. Mater. Interfaces* 10(48): 41098–41106.
49. Keirouz A, Radacsi N, Ren Q, Dommann A, Beldi G, Weber KM, Rossi RM, Fortunato G (2020) Nylon-6/chitosan core/shell antimicrobial nanofibers for the prevention of mesh-associated surgical site infection. *J. Nanobiotechnol.* 18(1): Article 51.
50. Maleki H, Mathur S, Klein A (2020) Antibacterial Ag containing core-shell polyvinyl alcohol-poly (lactic acid) nanofibers for biomedical applications. *Polym. Eng. Sci.* 60(6): 1221–1230.

51. Ramalingam R, Dhand C, Mayandi V, Leung CM, Ezhilarasu H, Karuppannan SK, Prasannan P, Ong ST, Sunderasan N, Kaliappan I, Kamruddin M, Barathi VA, Verma NK, Ramakrishna S, Lakshminarayanan R, Arunachalam KD (2021) Core–shell structured antimicrobial nanofiber dressings containing herbal extract and antibiotics combination for the prevention of biofilms and promotion of cutaneous wound healing. *ACS Appl. Mater. Interfaces* 13(21): 24356–24369.
52. Xu C, Cao Y, Lei C, Li Z, Kumeria T, Meka AK, Xu J, Liu J, Yan C, Luo L, Khademhosseini A, Popat A, He Y, Ye Q (2020) Polymer–mesoporous silica nanoparticle core–shell nanofibers as a dual-drug-delivery system for guided tissue regeneration. *ACS Appl. Mater. Interface* 3(2): 1457–1467.
53. Leung CM, Dhand C, Mayandi V, Ramalingam R, Lim PF, Barathi VA, Dwivedi N, Orive G, Beuerman RW, Ramakrishna S, Toh Y-C, Loh XJ, Verma NK, Chua AWC, Lakshminarayanan R (2020) Wound healing properties of magnesium mineralized antimicrobial nanofibre dressings containing chondroitin sulphate – A comparison between blend and core–shell nanofibers. *Biomater. Sci.* 8(12): 3454–3471.
54. Baghali M, Ziyadi H, Faridi-Majidi R (2021) Fabrication and characterization of core–shell TiO_2-containing nanofibers of PCL-zein by coaxial electrospinning method as an erythromycin drug carrier. *Polym. Bull.* https://doi.org/10.1007/s00289-021-03591-3.

27

One-Dimensional Polymeric Nanocomposites for Soft Electronics

Uday Shankar and Anasuya Bandyopadhyay

CONTENTS

27.1 Introduction

The ongoing research into 1D polymer nanocomposites [1] opens up new ways to build advanced soft electronics architectures [2], which are showing potential for soft robotics, wearable organic light-emitting diodes (OLEDs), bendable sensors, self-healing, biomedical, organic optoelectronic device, and energy storage devices applications, along with the advent of skin-wearable soft mechanical sensors [3]. Improvements in desirable properties, such as conductivity, toughness, optical transparency, etc., are required to facilitate the real-world uses of soft electronic devices, as illustrated in Figure 27.1 [4]. Electronic devices can be mainly divided into mechanically rigid and bendable or stretchable. Soft electronics engineering is still quite complicated because of a lack of fabrication methods and the high bulk modulus of most electronic materials. However, the advances in 1D polymeric nanocomposites and fabrication techniques has shifted from the conventional rigid form to the succeeding generation of soft electronics. This flexible architecture is also known as elastronics. The components of an elastronic system can be smoothly combined with

DOI: 10.1201/9781003223764-27

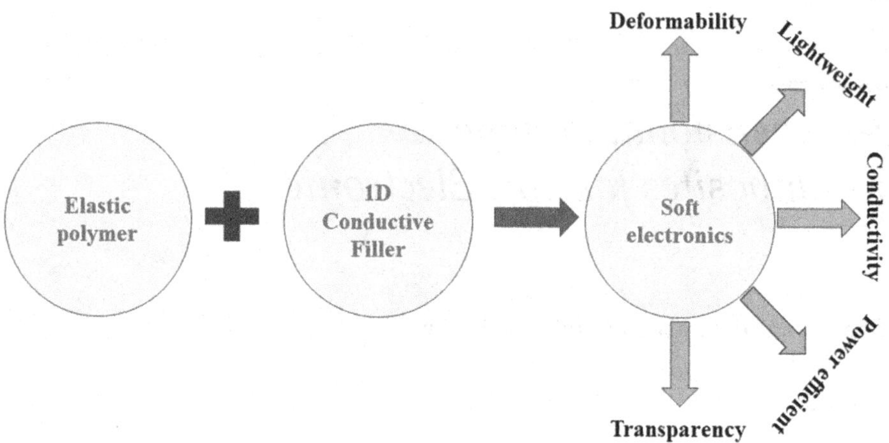

FIGURE 27.1
Schematic representation of elastomeric 1D nanocomposite for soft electronic applications with specific properties.

delicate skin to become parts of human organs, allowing for proper and *in-situ* bio-diagnostics. Elastronics is familiar for demanding a natural balance of flexibility and electrical conductivity that can be accomplished either extrinsically or intrinsically. The former is accomplished by forming stretch structures, whereas the latter is done through the development of stretchable materials. High-performance functional materials can withstand substantial deformations by modifying the characteristics of the 1D nanoconductive filler and the morphology of the composite. The interaction between the 1D nanoconductive filler and the BM is of specific curiosity for analyzing and identifying emerging alternative materials and soft electronics.

1D conducting BMs, like polypyrrole nanowires (PPy NWs) and polyaniline nanowires (PANI NWs), are used as electrode materials [5], whereas 1D insulating polymer nanomaterials, like polytetrafluoroethylene nanowires (PTFE NWs), Kapton NWs, and others [6], exhibit strong electron attraction from metal electrodes, and are therefore used in nanogenerators. Electrospinning, template-assisted, and template-free techniques have been used frequently to produce 1D BMs in recent times. Apart from these technologies, the inductively coupled plasma process has recently piqued interest in fabricating 1D BMs [7].

In microstructured and nanostructured forms, soft electronics have made promising advances in overcoming the basic mechanical incompatibility between inflexible electronics and the delicate human body. Therefore, to achieve the stretchability and conductivity of micro/nanoscale forms, a number of 1D nanofillers have emerged as viable electronic materials in upcoming soft electronics. It is well known that 1D structured materials are the versatile candidate for novel systems due to their excellent features, such as easy functionalization, size, and dimensionality. They are also helpful for energy-associated applications. They can offer various benefits like efficient electrical transport *via* direct current channels, a short ion diffusion path, and volume expansion compared to a 0D filler. The NWs, nanotubes, nanofibers (NFs), nanorods, nanofilaments, metal nanoparticles, and other conductive materials are the well-known 1D structured fillers. They offer desirable characteristics with elastomeric substrates. Usually, these materials exist in pure form; however, in some cases, they can occur in impure form due to doping in semiconductors. One option for substituting flexible for stiff electronic materials is to utilize hybrid nanomaterials. Efficiency and adaptability can be realized at a low cost by carefully considering various nanomaterial types, sizes, and concentrations. The 1D structured materials have a significant factor in terms of greater length than width. These fillers surprisingly decrease the flexural rigidity of electronic devices while acting as electrodes. The elastomeric BM has become famous for soft electronics use because of its versatility in various fabrication methods, its nontoxicity, and high moisture/gas permeability. Volume manufacturing, device performance, and construction methods mean that these nanomaterials face numerous obstacles for commercial device applications.

27.2 Why 1D Nanomaterials for Soft Electronics?

Soft nanocomposites with higher conductivities are better suited for soft electronics, whereas ones with lower conductivities are best for strain sensor applications. Rigid or brittle materials can only be used to a limited extent in the progress of soft electronics. To address this constraint, bendable conducting materials and other nanomaterials are required. These conducting fillers can be classified according to their dimension, as shown in Figure 27.2. When zero-dimensional (0D) nanoparticles and two-dimensional (2D) nanosheets are compared to one-dimensional (1D) nanomaterials, it is noticed that 1D nanomaterials have higher conductivities and lower percolation thresholds within elastic matrices due to their high aspect ratios. Thus, 1D nanomaterials can have higher conductivity at lower concentrations than can 0D and 2D nanomaterials. The π-conjugated molecules transformed into 1D nanostructure materials have exhibited novel optical, chemical, and electrical properties. The assembly of 1D nanostructured materials, including nanorods, nanotubes, nanoribbons, NWs, and nanobelts, which possess huge surface-to-volume ratio (as compared to their respective bulk materials), as well as high crystallinity of 1D nanomaterials, is an essential obligation for the development of soft electronic devices owing to efficient charge transport.

27.3 Elastic Polymer as a BM for Soft Electronics

Based on the topology, BM can be classified as 1D, 2D, or 3D; however, 3D is beyond the scope of this chapter. Linear polymers are also known as one-dimensional (1D) structures

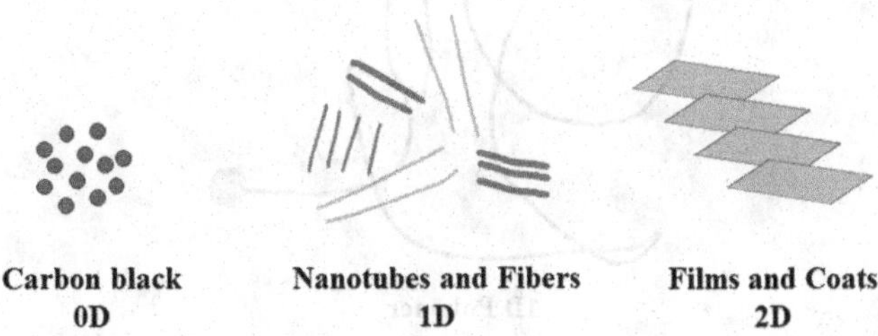

FIGURE 27.2
0D, 1D, and 2D conductive nanofillers.

(Figure 27.3) [8]. Because of having distinctive architectures, remarkable characteristics, and promising application potential, 1D polymer nanomaterials have attracted a lot more coverage than bulk polymer materials [7]. Currently, electrospinning is the utmost process used for fabricating 1D polymer nanomaterials with a wide variety of sizes for soft electronic applications [9].

BM can be stretched and released repeatedly and are used in soft electronics, where they need to last a long time without tearing. It has a significantly lower Young's modulus and higher yield strain than the other materials. Apart from conductive 1D filler, the BM also plays a pivotal role in the performance of soft electronics, as its BM mainly contributes mechanical features. To create an interconnected network in BM, elastomer chains are crosslinked with neighboring elastomer chains. As a result, this BM can be stretched, but the distortion cannot last because crosslinking between elastomer chains keeps the entire network together. The mechanical properties are obtained by the crosslinking between the backbone of BM as it is a structural characteristic. The BM becomes stiff and rigid if the crosslinking is strong and/or its density (ρ) is high. In contrast, the BM becomes soft and easily ruptured if the crosslink is weak and/or the density is low ρ.

Crosslinking can be classified into two types: physical or chemical crosslinking. Hydrogen bonds, thermally meltable ionic interactions, hydrophobic interactions, and π-π interactions can all drive physical crosslinking. These physical interactions are also recognized as noncovalent bonding, which is weaker and less stable than chemical crosslinking or covalent bonding. However, physically crosslinked BM provides various advantages, for example, superior processability and easy reprocessing. The mechanical characteristics of physically crosslinked BM can depend on the composition of blocks existing in the BM. The physically crosslinked polymer chains can be divided into physically interacted and non-interacted blocks. The physical crosslinking blocks are hard segments as the presence of strong hydrogen-bonding or polar-polar interaction within the block prevents segmental stretchability. In contrast, the other blocks are soft segments in the BM. Thus, the BM's crosslinking density ($\rho_{crosslinking}$) can be regulated by changing the ratios of these two segments in the BM.

In chemically crosslinked BM, polymer chains are strongly interconnected because of the covalent bond formation. Therefore, chemically crosslinked BM shows better stretchability than physical crosslinking. Concerning chemical inertness, the chemically crosslinked BM are not significantly influenced by polar solvents, although several organic

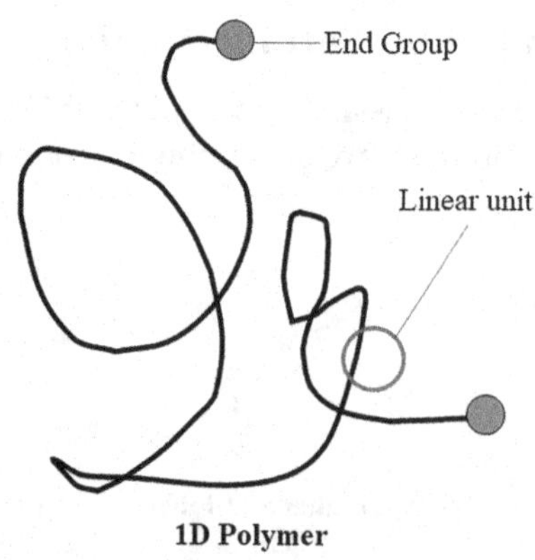

FIGURE 27.3
Schematic representation of 1D elastic polymer with their end groups and linear repeat units.

solvents can diffuse and cause swelling. The mechanical feature of chemically crosslinked elastomers can be regulated by controlling the reaction parameters, like the $\rho_{crosslinker}$, reaction temperature, etc. Furthermore, several chemical crosslinking reactions can be utilized, like enzyme- and UV-mediated reactions, based on the initiation method used to make the covalent bonds.

When fillers are added into chemically crosslinked BM to provide advanced features to the BM, high concentrations of the filler compared to the BM can reduce crosslinking of chemically crosslinked BM, resulting in inferior mechanical performance. The amount of filler in physically crosslinked BM, on the other hand, does not much affect crosslinking interactions. Therefore, physically crosslinked elastomers can have more filler concentrations than chemically crosslinked BM.

To facilitate the stretchability of nanocomposites under high strain, thermosetting elastomers, thermoplastic elastomers, and hydrogels have been frequently employed as the BM. Poly(dimethylsiloxane) (PDMS), Ecoflex, polyurethane (PU), and poly(styrene-co-ethylenebutylene-co-styrene) (SEBS) are often used for thermosetting and thermoplastic BMs (Figure 27.4) for the percolation network to offer elastic recoverability. Styrenic elastomers poly(styrene–butadiene–styrene) (SBS) and SEBS are physically crosslinked elastomers through π-interactions. Whereas the most widely used BM, PDMS, is chemically crosslinked by covalent bonds, PU contains both physical and chemical crosslinking. On the other hand, hydrogel can be either physically crosslinked (e.g., polyvinyl alcohol)

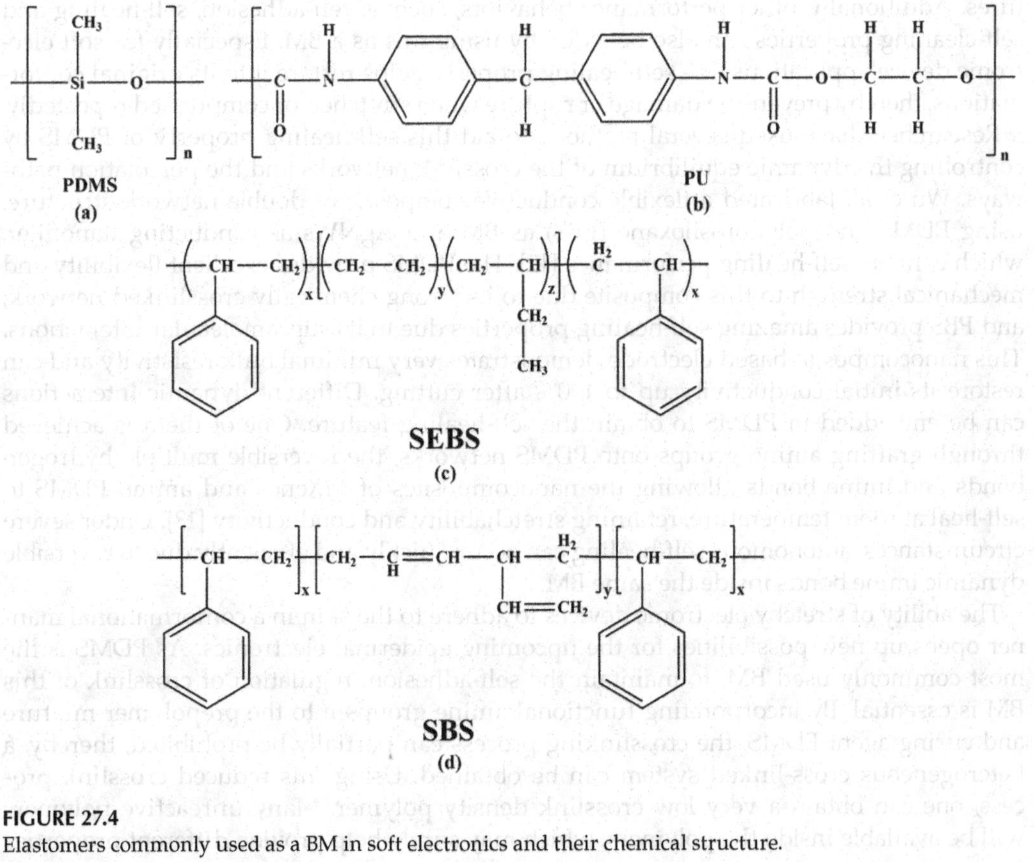

PDMS
(a)

PU
(b)

SEBS
(c)

SBS
(d)

FIGURE 27.4
Elastomers commonly used as a BM in soft electronics and their chemical structure.

or chemically crosslinked (e.g., poly(acrylamide)). The different inter- and intramolecular interactions commonly exist in these elastomers, including chemical, physical, and ionic crosslinking. Therefore, it endows the elastomers with high stretchability, up to 1000% strain, offering them as viable options for soft electronics. For bio-integrated electronics submission, the biocompatibility and nontoxicity of the BM and the conductive filler must be confirmed. Although BM is recognized to be chemically nontoxic, constituting monomers can still be poisonous. So, chemical inertness is critical for long-duration applications of bio-integrated devices. Some BM, for example silicone-based elastomers and thermoplastic PU (TPU), have been stated to be chemically compatible with the body over a long duration, making them appropriate for medical use.

27.3.1 Thermosetting Polymers

Thermoset polymers, like Ecoflex silicon rubbers and PDMS, are the most commonly used BM for fabricating flexible electronics. Such polymers display excellent flexibility, thermal stability, transparency, biocompatibility, and chemical resistance. When liquid prepolymer and curing agent are incorporated into such a polymer, a significant enhancement in Young's modulus ranging from 200 kPa to 3GPa and a simultaneous 1000% increase in failure strain makes it a potential candidate for being used in soft electronics. When the PDMS and Ecoflex are crosslinked, the formation of strong covalent bonds results in achieving excellent mechano-chemical properties [10,11]. The mechanical strength and failure strain can be tuned by adjusting prepolymer concentrations, cross-linker, and crosslink temperatures. Additionally, other performance behaviors, such as self-adhesion, self-healing and self-cleaning properties can also be aided by using this as a BM. Especially for soft electronic device applications, the self-healing property helps restore it to its original conformations, thereby preventing damage or rupture when stretched or compressed repeatedly.

Researchers have used several methods to aid this self-healing property of PDMS by controlling the dynamic equilibrium of the crosslink networks and the percolation pathways. Wu et al. fabricated a flexible conductive composite of double network structure, using PDMS and polyborosiloxane (PBS) as BM and AgNWs as conducting nanofiller, which exhibits self-healing performance [12]. The PDMS provides excellent flexibility and mechanical strength to this composite due to its strong chemically crosslinked network, and PBS provides amazing self-healing properties due to its supramolecular interactions. This nanocomposite-based electrode demonstrates very minimal bulk resistivity and can restore its initial conductivity up to 100% after cutting. Different dynamic interactions can be embedded in PDMS to obtain the self-healing feature. One of them is achieved through grafting amino groups onto PDMS networks, the reversible multiple hydrogen bonds and imine bonds allowing the nanocomposites of MXenes and amino PDMS to self-heal at room temperature, retaining stretchability and conductivity [13]. Under severe circumstances, autonomous self-healing can occur quickly and efficiently due to reversible dynamic imine bonds inside the same BM.

The ability of stretchy electronic devices to adhere to the skin in a conformational manner opens up new possibilities for the upcoming epidermal electronics. As PDMS is the most commonly used BM, to maintain the self-adhesion, regulation of crosslink of this BM is essential. By incorporating functional amine groups into the prepolymer mixture and curing agent PDMS, the crosslinking process can partially be prohibited; thereby, a heterogeneous cross-linked system can be obtained. Using this reduced crosslink process, one can obtain a very low crosslink density polymer. Many unreactive polymers will be available inside the polymers, which may surely help provide different properties like stretchability and conformability without introducing significant changes in the crosslinking protocol of PDMS.

The electrospun PVDF nanowires can be deposited on the ITO electrode surface and form a polymer composite with PDMS, which can be used for promising energy devices, i.e., a nanogenerator for mechanical energy harvesting and self-powered electronics [14]. Silicone is an extensively utilized supporting polymer owing to its superb flexibility and stretchability, along with good biocompatibility. Whereas PDMS is a transparent, flexible, and highly durable material, on the other side Ecoflex provides excellent stretchability. Researchers use Ag and graphene nanoparticles into PDMS to design different strain sensors for a broad measurement range with high sensitivity.

27.3.2 Thermoplastic Polymers

Even though thermosets are engaged in the majority of BM for soft electronics, thermoplastics such as TPU and SEBS have significant benefits, including outstanding processability and mechanical and chemical stabilities. The physical properties of soft, thermoplastic BM with comparatively weak, physical crosslinks vary from thermosetting elastomers which have strong, chemical crosslinking. TPU and SEBS BM do not need any further crosslinking procedures and can be treated directly in solution. Even though their comparatively weak, physical crosslinking within polymer chains is not always ideal for long-term thermal stability and chemical resistance, thermoplastics are progressively being used to construct various soft, electronic devices.

TPU consists of long polyol chains that form soft blocks, and hard blocks of urethane groups that form hydrogen bonds. Since the urethane groups are comprised of connected isocyanate and diol groups, they generate chemical crosslinks with other isocyanate groups. Changing the ratio of polymer chain component groups can improve the properties of these elastomers. Conductive inks, which contain conductive nanoparticles and thermoplastic BM in volatile solvents, have been extensively employed to print flexible and bendable electronics *via* a variety of printing processes. Their compositions can be tailored to fit different electromechanical specifications. A 3D printer with a 200-μm nozzle created extremely stretchable conductors with diverse designs demonstrating electrical conductivity of 1,000 S/cm and elasticity of 1,600% by incorporating conductive silver flakes to the TPU solution with acetone and DMF. Lee and co-workers produced a highly flexible and curable conductive nanocomposite centered on PUA BM, comprising conductive liquid metal and silver flakes. The resulting stretchy conductor has an elastic modulus of 2,500% and a preliminary conductivity of 6,250 S/cm, and it has been used to manufacture a triboelectric nanogenerator device. Furthermore, the reversible supramolecular hydrogen-bonding of thermoplastic BM allows them to regain their energy-harvesting capability after external mechanical forces damage them. Ma et al. have demonstrated fiber-based stretchable composites made with flower-shaped Ag nanoparticles and PU BM utilizing a wet-spinning method [15]. Lee et al. also used the wet-spinning process to make stretchy, conductive fibers out of composites of Ag NWs, Ag nanoparticles, and an SBS BM [16].

Flexible and conductive filaments made of thermoplastic BM and 1D conductive nanoparticles have been created for 3D printable, flexible electronics to meet specific needs. A 3D printed intelligent glove with elastic sensors, signal processors, and regulators has been produced by combining silicon integrated circuits into 3D printed customizable designs and using fused deposition modeling printing to manufacture a tailored wearable therapeutic device. Smart gloves with programmed heat and a photodetection technology suited to individual conditions and the body are being developed. Whereas 3D printing allows for accurate microstructure management of soft electronic devices, the existence of inexpensive, dependable 3D printing materials with high flexibility and conductivity will be crucial in exploring the benefits of tailored soft electronics.

27.3.3 Hydrogels

Hydrogels are recognized as BM for soft electronics due to their excellent characteristics, including modulated mechanical properties, optical transparency, hydrophilic nature, ion conductivity, biocompatibility, and tunable functionalization. It has highly regulated features by carefully designing the polymers and their weight ratio to water, providing a vast design space. Due to their high water content, as depicted in Figure 27.5, hydrogels are soft and stretchable like natural tissue; however, they face the problem of feebleness and fragility.

Ion or electron transport is the primary source of conductivity in hydrogels. In this BM, 1D conducting fillers, like carbon nanotubes (CNTs), metal NWs, and conductive organic materials, were employed in the electrically conductive hydrogels. Appropriate interactions between 1D conducting fillers and hydrogel BM are essential to attain homogeneous distribution of the nano conductive fillers within BM, which governs the conductivity and stretchability of the subsequent BM. In contrast, to make ion-conducting hydrogel BM, metallic salts and ionic liquids that can create free ions in water have routinely been used. This BM has a conductive route thanks to the high mobility of ions in water. Almost all documented conductive hydrogels for soft robotics are built on ion-conducting hydrogels for their great optical clarity, ease of preparation, and excellent electrochemical performance. Robust hydrogels, with better Young's modulus and fantastic failure strain up to 2,300%, have been established with the help of several interactions, including ion-polymer complexation and hydrogen bonding [17], to modify and introduce other attractive features like self-healing and anti-freeze/anti-dehydration properties. This improved hydrogel-based BM is perfect for soft electronic application submissions. The resultant flexible and elastic electronics proved a possible candidate for soft electronics submissions by directly patterning highly conductive liquid metals on hydrogels using a magnetic field [18].

Hydrogels and biological tissues have mechanical and physiological features in common, leading researchers to form more suitable and stable hydrogels as BM for soft electronics. Lin et al. have developed an alternative biodegradable and robust hydrogel BM set

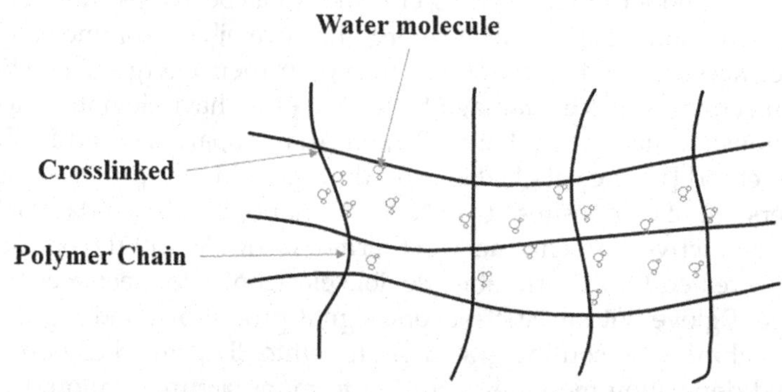

FIGURE 27.5
The water-absorbing capacity of hydrogel elastomer used as a BM in soft electronics.

that integrates stretchy electrodes with stiff electronic parts [19]. This optically clear and tough BM has encapsulated a stretchable LED array, showing stable performance under repeated stretching cycles. In addition, a smart wound dressing with temperature sensors and drug delivery channels built onto this BM and the temperatures measured at different locations can trigger the on-demand delivery of several drugs. Much research has been carried out to advance new robust hydrogels BM that has fascinating potential.

27.4 Representative 1D Nanomaterials

The advances in bendable and stretchable electrodes is a crucial component in producing soft electronics. Traditional conducting materials with outstanding electrical conductivity, including indium tin oxide (ITO), gold, copper, etc., have been used as conventional electrodes in several electronics. Nevertheless, using them in soft electronics is not viable because of their inherent brittleness and rigid structure [20, 21]. As a result, different 1D nanomaterials are being investigated as alternative materials to use in soft electronics.

NWs, NFs, and CNTs are examples of 1D nanomaterials, which are being used as an integral part of fabricated 1D nanocomposites (Figure 27.6) to improve the electrical conductivity, along with outstanding bendability because of their structural characteristic, such as high aspect ratio. The 1D configuration corroborates a direct conductive network to transport charges while reducing grain boundaries or defects responsible for its remarkable features. When deformations occur, cracks are produced, especially in the grain boundaries [22]. As such cracks substantially decrease conductivity owing to their suppression of the charge flow, it is vital to stop crack generation as far as possible [23]. Depending on these characteristics, void formation in the conducting networks at the sub-monolayer level can benefit high optical transparency. As a result, conducting networks based on 1D nanomaterials are inherently bendable, provide electrical conductivity, and have optical transparency for soft electronics applications. This section introduces the research undertaken to advance the properties of 1D polymeric nanocomposites and their applications in several areas of soft electronics.

Almost every electronic device depends on allowing electron transport to the preferred location in an electrical circuit. Therefore, stretchable and bendable conductors, which let charges flow with minimal energy losses, are vital for any soft electronics. Usually, the performance of a soft electronic device depends on the electrical conductivity of the electrodes. Therefore, 1D filler choice for the electrode is vital for efficient soft electronics to transfer electrical energy and electronic information.

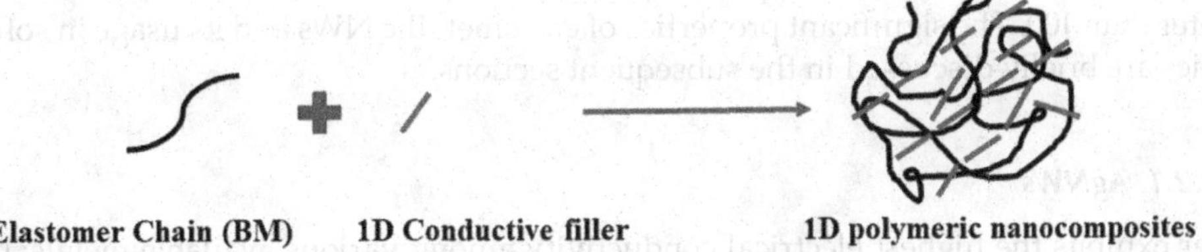

FIGURE 27.6
1D nanocomposite composed of dispersion of 1D conductive nanofiller into BM.

27.4.1 CNTs

CNTs are carbon allotropes. They are conventional 1D nanomaterials with a high aspect ratio (~333). Based on the number of layers, CNTs are categorized into single-walled CNTs (SWCNTs) or multi-walled CNTs (MWCNTs). According to their structural arrangements, they can be considered as an aligned or random network of CNTs. These CNTs have many properties that make them suitable for soft electronics, including excellent electrical conductivity (10^4 S/cm), high carrier mobility (10,000 cm^2/Vs), good mechanical properties (elastic modulus ~1 TPa), and chemical stability characteristics. As a result, various studies on CNTs have been carried out and have demonstrated promising results for the use of CNTs in soft electronics device applications.

Stretchable CNT-based electronics can be made by either dispersing CNTs into a BM to make a CNT/elastomer composite or by simply depositing CNT films onto stretchable BM. The alignment of CNTs could further optimize the device performance. Several strategies for fabricating sizable alignment of CNT film with various morphologies, such as vertically grown CNTs, buckled and aligned CNTs, or suspended CNTs, have been established. For making huge and highly conductive devices, homogeneous dispersion of CNTs is critical. It was claimed that homogeneously dispersed CNT pastes could be realized by introducing an ionic liquid to produce chemically inert dopants into a fluorinated copolymer matrix.

Strain sensors based on CNT/Ecoflex nanocomposites fabricated by Amjadi et al. [24] can detect 500% strain. Sekitani et al. [25] unfolded a stretchable printable electrode by homogeneous dispersion of SWCNTs in a fluorinated elastomer matrix with an ionic liquid. Utilizing this, they demonstrated an active, matrix-type, stretchable OLED. Wang et al. [26] successfully introduced an inherently stretchable transistor array with CNT electrodes and azide crosslinked SEBS as the BM. Hu et al. [27] also demonstrated that SWCNT ink could be used to make elastic supercapacitors, which might be used to fabricate energy gadgets.

CNTs have been studied extensively, yet there are still certain limitations. First, because CNTs are combined with metallic and semiconducting CNTs and have varied chiralities and lengths, mass manufacture of specific CNTs remains difficult. Second, because of the non-uniformity of the CNTs, steady-device performances may not be achieved, making their usage in commercial devices challenging. Third, dissolving and separating long nanotubes is difficult, and there are limited strategies for consistently building CNTs over huge surfaces.

27.4.2 Metal Nanowires

Several 1-D metal NWs, such as silver (AgNWs), copper (CuNWs), and gold (AuNWs), are widely used for designing soft electronics. The diameters of such NWs are typically in a range of 10–100 nm, and the aspect ratio (i.e., the ratio of its length to diameter (l/d)) is greater than 100. The significant properties of each metallic NWs and its usage in soft electronics are briefly discussed in the subsequent sections.

27.4.2.1 AgNWs

Silver exhibits the highest electrical conductivity among various available metals; hence, AgNWs attracted considerable attention among academics and industries. Researchers have reported several fabrication techniques to synthesize AgNWs, such as microwave-assisted, hydrothermal, polyol, electrochemical, UV irradiation, and template methods. Apart from the excellent electrical properties of AgNWs, the percolating networks display exceptional bendability, by which they demonstrate a low sheet resistance (R_S) (<20 Ω/sq) and transmittance of > 85%. Such performance behavior of AgNWs captured the overwhelming attention of scientists to fabricate flexible-cum-transparent electrode materials to replace conventionally used brittle indium tin oxides (ITOs). In discrete NWs, the electrons pass through a single crystalline NW; hence an individual AgNW has a meager resistance. However, owing to the inconsistency of crystallinity at the junctions, the electron transport is highly resistive. Consequently, researchers fabricate very long NWs to enhance their electrical properties by achieving longer percolation paths and minimizing junction resistance.

In this direction, Lee et al. [28] have fabricated long AgNWs (>100 μm) by employing sequential multistep growth techniques, using repetitive reduction of $AgNO_3$ in an ethylene glycol solution with polyvinylpyrrolidone (PVP). Another factor that lowers the performance of soft electronics is the rough surface [29] when the AgNWs are coated on the BM. For this reason, researchers have tried to embed AgNWs into BM to enhance its mechanical behavior along with improving surface smoothness. Liang et al. [30] have prepared a LED by embedding AgNWs into BM of PUA. Furthermore, Amjadi et al. [31] developed a sandwich-structured nanocomposite by incorporating AgNWs between two PDMS layers with increased stretchability, and employed is nanocomposite for sensitive strain sensors.

Large-area deposition approaches of AgNWs have been established to fulfill the criteria of large-scale, soft electronics manufacture. For example, a Meyer rod coating method was used for scalable R-2-R layer deposition of AgNWs networks, demonstrating the transparency of ~80% and R_S of 8 Ω/sq. Researchers have also used several other methods, like spray coating and ultrafiltration, for the large-scale production of AgNW film. In addition, thermal annealing, plasma treatment, and mechanical pressing are widely used for welding the junctions between NWs. But the traditional heating method during this welding process sometimes adversely affects the heat-sensitive materials. Tokuno et al. [32] demonstrated a novel approach to welding by mechanically pressing the joints tightly at 25 MPa at room temperature. Furthermore, Garnett et al. [33] reported another interesting method known as light-induced plasmonic nano-welding, wherein the authors first reduced the resistance by a factor of > 1,000. This novel, non-heating technique potentially helped not to damage the BM.

27.4.2.2 CuNWs

While the AgNWs exhibit excellent features, they also have several drawbacks: extreme roughness (as discussed in the previous section), degradation under UV and visible light, poor contact at junctions, high cost, and low Ag abundance. Such factors limit its usage from being used in real-world situations. These issues led researchers to investigate a cheap alternative with an equivalent performance behavior. CuNWs are potential candidates for conductive fillers as an alternative to AgNWs for soft electronic applications. Though the conductivity of CuNWs is marginally lower than that of AgNWs, the cost of the CuNWs is much lower. Moreover, in CuNWs, the Cu atoms are more resistant to electromigration than the Ag atoms, which helps them to be used for numerous submissions that need a comparatively higher current density.

The various techniques used to develop CuNWs include electrospinning, chemical reduction, microwave-assisted process, and self-catalytic behavior. Furthermore, distinct approaches have also been established to enhance its electrical, mechanical, and optical properties along with large-scale manufacture. CuNWs were synthesized for the first time in 2005 by reducing $Cu(NO_3)_2$ with hydrazine in an aqueous mixture of NaOH and ethylenediamine. In addition, this technique could be scaled up 200-fold. Nevertheless, the CuNWs synthesized displayed a resistance of 61 Ω/sq and an optical transparency of 67%, which is poorer than for either AgNWs or ITO.

Moreover, mass production and a simple hydrothermal process synthesized CuNWs using an electric pressure cooker. Oleylamine and oleic acid were used as dual surfactants of CuNWs, and glucose was employed as the reducing agent. A CuNW-based transparent electrode was obtained with an R_S of ~50 Ω/sq at ~90% transmittance. Almost constant optical properties over the full spectrum of light were realized, along with good conductivity. Furthermore, CuNW-based electrodes were obtained from a CuNW–poly(vinylalcohol) (PVA) aerogel. The incorporation of PVA served as a stabilizer of the CuNW scaffolds, which helped to increase the mechanical strength (an increase of up to 30 times in Young's modulus with the addition of 10 wt% of PVA) by a great extent. Furthermore, a highly porous structure was developed using the freeze-drying method and embedding it into PDMS to form an elastic amber. The material synthesized could take tensile and compressive strain up to 60% with a negligible decrease in resistance.

Several BM, for example, PU and Ecoflex, have also been stated to be used to design wearable electronic devices using CuNWs. Ding et al. [34] demonstrated a photonic sintering technique for welding CuNWs and embedding them into the PU BM, which is optically clear and has excellent mechanical stretchability, retaining its conductivity even after 1,000 cycles of stretching at 10% tensile strain, indicating its potential applicability for a soft electronics function. Even so, the performance behavior of these CuNWs is still at a lower range than AgNWs due to their lower aspect ratio, poorer solvent dispersion, and a greater tendency for oxidation.

27.4.2.3 AuNWs

AuNWs have attracted a lot of attention as vital parts of soft electronic devices as a result of their amazing electromechanical properties, chemical stability, biocompatibility, and high electrical conductivity. AuNWs exhibit low contact resistance with p-type organic semiconductors compared with AgNWs and CuNWs. Various approaches were adopted to fabricate AuNWs, including lithography, electrodeposition, chemical reduction, etc. Nonetheless, most of them produced low aspect ratios and rough surfaces. NWs, both thick and short, are stiff by nature. They have poor stretchability and conductivity, making them unsuitable for use in electronics.

In 2008, AuNWs (diameter of < 5 nm) were developed *via* oleylamine or combined chemicals as 1D growth templates and reducing agents [35]. These ultrathin AuNWs are inherently flexible, hence offering the potential for stretchable devices attributable to their high aspect ratio (>10,000) [36]. For example, a simple approach of functionalization of the BM could help direct growth of a high-aspect-ratio AuNW over a large area of a stretchable BM. This single-layered, self-assembled, thin film showed good stretchability, conductivity, and 90–97% transmittance. But as a result of fewer junctions among NWs after paralleled self-assembling, this film displayed a high R_S of ~1,142 kΩ/sq. Consequently, they planned a thin, mesh-like AuNW with a figure of merit 25 times that of a single-layer, self-assembled film, which might directly be employed for flexible transparent, stretchable touch screens, and supercapacitors [37].

Furthermore, AuNWs showed outstanding sensitivity when used in wearable sensors. For instance, a pressure sensor with a highly mechanically flexible and sensitive characteristic nature could be fabricated by sandwiching ultra-thin, AuNW-impregnated tissue paper between two thin PDMS sheets [38]. Furthermore, AuNW ink has been printed directly on soft BM to fabricate skin-attachable strain gauge sensors. The resultant AuNW thin-film stretched up to 350% with a high gauge factor of up to 60 [36]. Zhao et al. [39] developed a fiber-type seed for electroless plating by spinning a solution of AuNWs and SEBS BM. Such a material can be employed as a wearable biosensor for monitoring glucose concentration in sweat by using glucose oxidase. It can sustain the sensitivity of 11.7 μA/mM, even up to 200% strain.

27.4.2.4 Metal Nanotroughs and Metal Nanofibers

It is well known that a high aspect ratio is an essential parameter required in a 1D nanofiller for achieving good electrical conductivity. Since there is a significant limitation to increasing this aspect ratio in these nanowires, scientists are now investigating several other nanostructures such as nano-mesh, nano-troughs, and nano-fibers. Along this line, Wu et al. [40] have fabricated metal nano-trough systems with a very high aspect ratio using the electrospinning technique. Such a developed electrode is optically transparent and demonstrates no significant drop in resistance even after bending it up to a 2-mm radius of curvature. Conducting paper, produced by putting nanotrough networks onto the paper, kept its conductivity even after being crushed and dispersed. But the fabrication process for such a nanotrough involves multi-stage processing strategies, limiting its fabrication from an industrial aspect. Keeping the drawbacks of nanotrough networks in mind, Jang et al. [41] developed an ultralong AgNF by electrospinning a suspension and annealing it in a hydrogen atmosphere. The CuNF obtained possesses an ultra-high aspect ratio (~100,000). Additionally, this NF shows excellent flexibility with a marginal reduction in a radius of curvature of < 6 mm. Despite having good optical and mechanical properties, metallic 1D nanofillers require further modifications for long-term use due to their low chemical and thermal stability. In addition, one needs to think along the fabrication direction to synthesize them on an industrial scale.

27.4.3 Polymeric Conductive Nanomaterials

Conducting polymer fillers combined with other fillers, such as carbon or metal, have been at the forefront of electronics research. The fundamental advantage of these fillers is their conductivity and stretchability, which may be altered by modifying the molecular architectures. The most common conducting polymers [42,43] investigated include PPy, PANI, polyindole, polythiophene, PAN, and others that have been synthesized by template-assisted and template-free methods. However, their insolubility may constrain potential scope.

Conventional conducting polymers in soft electronics have been confined in application by their poor performance in terms of capacitance, cyclic stability, and so on. Some approaches, such as varying the morphology and structure of conducting polymers, conducting polymer composites, and electrolytes having a large electrochemical window, have been investigated to tackle these challenges. 1D conducting polymer nanoparticles, with a large surface area and a shorter, ionic diffusion path, in particular, have shown considerable promise [7]. In contrast, PEDOT: PSS has been prepared, which is water-soluble as it is made up of positively charged PEDOT, and negatively charged, water-soluble PSS. In combination with its excellent conductivity, stretchability, optical clarity, and mass production, its solution processability has proven it to be an appealing material for the soft electronics revolution.

PEDOT: PSS has a fracture strain of just 5% and a Young's modulus of only 2 GPa, despite having a high electrical (>1000 S/cm) and good optical clarity (~90%). The inclusion of plasticizers (e.g., glycerol) to reduce the elastic modulus or the formation of a polymeric blend using water-soluble polymers (e.g., poly(ethylene glycol)) are two common ways to increase the stretchability.

Buckling and Kirigami proposed a design in which a plasticizer, i.e., ionic liquid, plays the pivotal role of a dopant for PEDOT and lets PEDOT: PSS dissolve in water and improve its conductivity and elastic modulus. Excellent conductivity and stretchability (3390 S/cm at the strain of 100%) were obtained. Even its fracture strain could be up to 800% when the BM is SEBS. 1D PAN NWs fabricated by anodic aluminum oxide template-assisted synthesis also provide a porous material for a supercapacitor to improve the capacitance values [44].

Polymeric nanofiller has also been employed as an inexpensive conductive filler. On the other hand, its poor conductivity is something that needs to be resolved. These nanofillers also face problems, like large-scale manufacture, long-term stability, high prices, sophisticated microfabrication, and limited electrical performance for practical electronic applications.

27.4.4 Hybrid Structures

Every material has its advantages and disadvantages; therefore, soft electronics that rely on a single material type will unavoidably face several constraints. Combining two or more materials into hybrid nanocomposites has shown to be a promising way to make up for the shortcomings of single materials while maintaining the merits of each. For the application of flexible and bendable electronics, a binary composite of 0D/1D systems or 1D/1D systems has been thoroughly researched, such as a CNT/Ag, CNT/graphene, Ag/graphene, AuNWs/PANI, and so on. An AgNW/CNT hybrid nanocomposite for highly stretchable and bendable transparent electrodes was developed in 2014 [45]. In such hybrid systems, highly conductive AgNW percolation networks mainly serve as backbone current collector- conductors for fast electron transport, but an elastic SWCNT percolation network offers greater flexibility and bendability along with improved conductivity by filling the inter-nanowire gap of the AgNWs mesh's backbone conductor with local electron transport paths. The AgNW/CNT binary system combined stretchable (>460%) electrodes with consistency of excellent electrical conductivity and mechanical properties over up to 10,000 times in a bending cycle, exhibiting large deformation (>540°), and being fully portable, as well as highly transparency (~90%) compared with single-component system, i.e., CNTs or AgNWs [46]. 1D PEDOT/MnO_2 or RuO_2 NWs were prepared through the AAO template-assisted technique for supercapacitors. The homogeneous NWs (diameter 250 nm) developed as suitable materials for electrochemical supercapacitors, with high specific capacitance and current density, were observed [47].

Apart from the 1D/1D system, 0D structures can also be a vital addition to boosting the conductivity of 1D nanofillers. For example, MWCNTs were combined with silver flakes to construct printable and bendable hybrid electrodes. The CNTs were employed as 1D stretchable and conductive scaffolds to formulate better electrical pathways among the silver flakes. At 0% strain, this binary system had a conductivity of 5×10^3 S/cm. But, the conductivity dropped substantially due to the separation of the low-aspect-ratio Ag flakes at high strains [48]. Furthermore, we established a facile and effective method to dope PANI into AuNWs films, leading to improved conductivity and sensitivity. The binary system also exhibited a reversible and repeatable response up to 150% strain because of the comparatively low amount of 0D PANI in the binary system [36].

Due to structural defects or considerable interlayer resistance, graphene has poor conductivity and mechanical stretchability, limiting its usage in soft electronics. CNTs and metal nanowires have been adopted to enhance conductivity and stretchability. Lee et al. [49,50] deposited an AgNWs network on graphene films to effectively boost conductivity, where the AgNWs act as a bridge between graphene sheets. As a result, the low R_S of 33 Ω/sq was attained with high optical clarity of 94%. On PDMS, the deposition of graphene/AgNWs showed 100% strain tolerance and minimal deterioration in resistance. The thermal stability of this binary system is far superior to that of individual AgNWs.

27.5 Conclusion

This chapter briefly describes the different aspects of the one-dimensional nanocomposite for a soft electronic application. The advantage of 1D conducting filler over zero-dimensional is their aspect ratio, which is helpful to achieve the percolation threshold value at lower loading of the filler if the appropriate homogeneous dispersion occurs. Herein, we have summarized the different strategies and material combinations already adopted to enhance the nanocomposite's conductivity and stretchability. The judicious selection of the soft matrix and 1D conducting filler could lead to a nanocomposite matrix whose bulk conductivity would be unaffected under different types of flexural stresses.

Acknowledgment

The first author is very grateful to Dr. Sushanta K Sethi of the Indian Institute of Technology Bombay and Kumari Priya of the Indian Institute of Technology Roorkee for their support.

References

1. N. Baghdadi, M.S. Zoromba, M.H. Abdel-Aziz, A.F. Al-Hossainy, M. Bassyouni, N. Salah, One-dimensional nanocomposites based on polypyrrole-carbon nanotubes and their thermoelectric performance, *Polymers (Basel)*, 13(2), 2021, 278.
2. S. Gong, W. Cheng, One-dimensional nanomaterials for soft electronics, *Adv. Electron. Mater.*, 3(3), 2017, 1600314.
3. T.N. Do, Y. Visell, Stretchable, twisted conductive microtubules for wearable computing, robotics, electronics, and healthcare, *Sci. Rep*, 7(1), 2017, 1–12.
4. U. Shankar, D. Oberoi, S. Avasarala, S. Ali, A. Bandyopadhyay, Design and fabrication of a transparent, tough and UVC screening material as a substitute for glass substrate in display devices, *J. Mater. Sci.*, 54(8), 2019, 6684–6698.
5. N. Hui, F. Chai, P. Lin, Z. Song, X. Sun, Y. Li, S. Niu, X. Luo, Electrodeposited conducting polyaniline nanowire arrays aligned on carbon nanotubes network for high performance supercapacitors and sensors, *Electrochim. Acta*, 199, 2016, 234–241.

6. Z.-H. Lin, G. Cheng, X. Li, P.-K. Yang, X. Wen, Z.L. Wang, A multi-layered interdigitative-electrodes-based triboelectric nanogenerator for harvesting hydropower, *Nano Energy*, 15, 2015, 256–265.
7. L. Huang, W. Xu, J. Hao, Energy device applications of synthesized 1D polymer nanomaterials, *Small*, 13(43), 2017, 1701820.
8. A.D. Schlüter, Features that make macromolecules 2D polymers, *React. Funct. Polym.*, 161, 2021, 104856.
9. M. Inagaki, Y. Yang, F. Kang, Carbon nanofibers prepared via electrospinning, *Adv. Mater.*, 24(19), 2012, 2547–2566.
10. S.K. Sethi, L. Soni, U. Shankar, R.P. Chauhan, G. Manik, A molecular dynamics simulation study to investigate poly (vinyl acetate)-poly (dimethyl siloxane) based easy-clean coating: An insight into the surface behavior and substrate interaction, *J. Mol. Struct.*, 1202, 2020, 127342.
11. S.K. Sethi, U. Shankar, G. Manik, Fabrication and characterization of non-fluoro based transparent easy-clean coating formulations optimized from molecular dynamics simulation, *Prog. Org. Coat.*, 136, 2019, 105306.
12. M. Tang, P. Zheng, K. Wang, Y. Qin, Y. Jiang, Y. Cheng, Z. Li, L. Wu, Autonomous self-healing, self-adhesive, highly conductive composites based on a silver-filled polyborosiloxane/polydimethylsiloxane double-network elastomer, *J. Mater. Chem. A*, 7(48), 2019, 27278–27288.
13. K. Zhang, J. Sun, J. Song, C. Gao, Z. Wang, C. Song, Y. Wu, Y. Liu, Self-healing Ti_3C_2 MXene/PDMS supramolecular elastomers based on small biomolecules modification for wearable sensors, *ACS Appl. Mater. Interfaces*, 12(40), 2020, 45306–45314.
14. S. Wang, Z.L. Wang, Y.A. Yang, A one-structure-based hybridized nanogenerator for scavenging mechanical and thermal energies by triboelectric–piezoelectric–pyroelectric effects, *Adv. Mater.*, 28(15), 2016, 2881–2887.
15. R. Ma, B. Kang, S. Cho, M. Choi, S. Baik, Extraordinarily high conductivity of stretchable fibers of polyurethane and silver nanoflowers, *ACS Nano*, 9(11), 2015, 10876–10886.
16. S. Lee, S. Shin, S. Lee, J. Seo, J. Lee, S. Son, H.J. Cho, H. Algadi, S. Al-Sayari, D.E. Kim, T. Lee, Ag nanowire reinforced highly stretchable conductive fibers for wearable electronics, *Adv. Funct. Mater.*, 25(21), 2015, 3114–3121.
17. H. Liu, M. Li, C. Ouyang, T.J. Lu, F. Li, F. Xu, Biofriendly, stretchable, and reusable hydrogel electronics as wearable force sensors, *Small*, 14, 2018, 1801711.
18. C. Xu, B. Ma, S. Yuan, C. Zhao, H. Liu, High-resolution patterning of liquid metal on hydrogel for flexible, stretchable, and self-healing electronics, *Adv. Electron. Mater.*, 6(1), 2020, 1900721.
19. S. Lin, H. Yuk, T. Zhang, G.A. Parada, H. Koo, C. Yu, X. Zhao, Stretchable hydrogel electronics and devices, *Adv. Mater.*, 28(22), 2016, 4497–4505.
20. U. Shankar, S.K. Sethi, B.P. Singh, A. Kumar, G. Manik, A. Bandyopadhyay, Optically transparent and lightweight nanocomposite substrate of poly (methyl methacrylate-co-acrylonitrile)/MWCNT for optoelectronic applications: An experimental and theoretical insight, *J. Mater. Sci.*, 56(30), 2021, 17040–17061.
21. U. Shankar, C.R. Gupta, D. Oberoi, B.P. Singh, A. Kumar, A. Bandyopadhyay, A facile way to synthesize an intrinsically ultraviolet-C resistant tough semiconducting polymeric glass for organic optoelectronic device application, *Carbon N. Y.*, 168, 2020, 485–498.
22. H. Sun, J. Deng, L. Qiu, X. Fang, H. Peng, Recent progress in solar cells based on one-dimensional nanomaterials, *Energy Environ. Sci.*, 8(4), 2015, 1139–1159.
23. J.A. Rogers, T. Someya, Y. Huang, Materials and mechanics for stretchable electronics, *Science*, 327(5973), 2010, 1603–1607.
24. M. Amjadi, Y.J. Yoon, I. Park, Ultra-stretchable and skin-mountable strain sensors using carbon nanotubes–Ecoflex nanocomposites, *Nanotechnology*, 26(37), 2015, 375501.
25. T. Sekitani, H. Nakajima, H. Maeda, T. Fukushima, T. Aida, K. Hata, T. Someya, Stretchable active-matrix organic light-emitting diode display using printable elastic conductors, *Nat. Mater.*, 8(6), 2009, 494–499.
26. S. Wang, J. Xu, W. Wang, G.-J.N. Wang, R. Rastak, F. Molina-Lopez, J.W. Chung, S. Niu, V.R. Feig, J. Lopez, T. Lei, S.K. Kwon, Y. Kim, A.M. Foudeh, A. Ehrlich, A. Gasperini, Y. Yun, B.

Murmann, J.B. Tok, Z. Bao, Skin electronics from scalable fabrication of an intrinsically stretchable transistor array, *Nature*, 555(7694), 2018, 83–88.

27. L. Hu, M. Pasta, F. La Mantia, L. Cui, S. Jeong, H.D. Deshazer, J.W. Choi, S.M. Han, Y. Cui, Stretchable, porous, and conductive energy textiles, *Nano Lett.*, 10(2), 2010, 708–714.
28. P. Lee, J. Lee, H. Lee, J. Yeo, S. Hong, K.H. Nam, D. Lee, S.S. Lee, S.H. Ko, Highly stretchable and highly conductive metal electrode by very long metal nanowire percolation network, *Adv. Mater.*, 24(25), 2012, 3326–3332.
29. W. Gaynor, G.F. Burkhard, M.D. McGehee, P. Peumans, Smooth nanowire/polymer composite transparent electrodes, *Adv. Mater.*, 23(26), 2011, 2905–2910.
30. J. Liang, L. Li, X. Niu, Z. Yu, Q. Pei, Elastomeric polymer light-emitting devices and displays, *Nat. Photonics*, 7(10), 2013, 817–824.
31. M. Amjadi, A. Pichitpajongkit, S. Lee, S. Ryu, I. Park, Highly stretchable and sensitive strain sensor based on silver nanowire–elastomer nanocomposite, *ACS Nano*, 8(5), 2014, 5154–5163.
32. T. Tokuno, M. Nogi, M. Karakawa, J. Jiu, T.T. Nge, Y. Aso, K. Suganuma, Fabrication of silver nanowire transparent electrodes at room temperature, *Nano Res.*, 4(12), 2011, 1215–1222.
33. E.C. Garnett, W. Cai, J.J. Cha, F. Mahmood, S.T. Connor, M.G. Christoforo, Y. Cui, M.D. McGehee, M.L. Brongersma, Self-limited plasmonic welding of silver nanowire junctions, *Nat. Mater.*, 11(3), 2012, 241–249.
34. S. Ding, J. Jiu, Y. Gao, Y. Tian, T. Araki, T. Sugahara, S. Nagao, M. Nogi, H. Koga, K. Suganuma, H. Uchida, One-step fabrication of stretchable copper nanowire conductors by a fast photonic sintering technique and its application in wearable devices, *ACS Appl. Mater. Interfaces*, 8(9), 2016, 6190–6199.
35. X. Lu, M.S. Yavuz, H.-Y. Tuan, B.A. Korgel, Y. Xia, Ultrathin gold nanowires can be obtained by reducing polymeric strands of oleylamine–AuCl complexes formed via aurophilic interaction, *J. Am. Chem. Soc.*, 130(28), 2008, 8900–8901.
36. S. Gong, D.T.H. Lai, Y. Wang, L.W. Yap, K.J. Si, Q. Shi, N.N. Jason, T. Sridhar, H. Uddin, W. Cheng, Tattoolike polyaniline microparticle-doped gold nanowire patches as highly durable wearable sensors, *ACS Appl. Mater. Interfaces*, 7(35), 2015, 19700–19708.
37. S. Gong, Y. Zhao, L.W. Yap, Q. Shi, Y. Wang, J.A.P.B. Bay, D.T.H. Lai, H. Uddin, W. Cheng, Fabrication of highly transparent and flexible nanomesh electrode via self-assembly of ultrathin gold nanowires, *Adv. Electron. Mater.*, 2(7), 2016, 1600121.
38. S. Gong, W. Schwalb, Y. Wang, Y. Chen, Y. Tang, J. Si, B. Shirinzadeh, W. Cheng, A wearable and highly sensitive pressure sensor with ultrathin gold nanowires, *Nat. Commun.*, 5, 2014, 1–8.
39. Y. Zhao, Q. Zhai, D. Dong, T. An, S. Gong, Q. Shi, W. Cheng, Highly stretchable and strain-insensitive fiber-based wearable electrochemical biosensor to monitor glucose in the sweat, *Anal. Chem.*, 91(10), 2019, 6569–6576.
40. H. Wu, D. Kong, Z. Ruan, P.-C. Hsu, S. Wang, Z. Yu, T.J. Carney, L. Hu, S. Fan, Y. Cui, A transparent electrode based on a metal nanotrough network, *Nat. Nanotechnol.*, 8(6), 2013, 421–425.
41. J. Jang, B.G. Hyun, S. Ji, E. Cho, B.W. An, W.H. Cheong, J.-U. Park, Rapid production of large-area, transparent and stretchable electrodes using metal nanofibers as wirelessly operated wearable heaters, *NPG Asia Mater.*, 9(9), 2017, e432–e432.
42. U. Shankar, S. Bhandari, D. Khastgir, Carbon Black-, Carbon Black-Filled nitrile rubber composite as a flexible electrode for electrochemical synthesis of supercapacitive polyaniline, *Polym. Compos.*, 40, 2019, E1537–E1547.
43. U. Shankar, S. Bhandari, D. Khastgir, Electrodeposition of nanostructured silver particles and it's composite with polyaniline on flexible electrode, *J. Basic Appl. Res. Int.*, 2016, 267–271.
44. Y. Cao, T.E. Mallouk, Morphology of template-grown polyaniline nanowires and its effect on the electrochemical capacitance of nanowire arrays, *Chem. Mater.*, 20(16), 2008, 5260–5265.
45. P. Lee, J. Ham, J. Lee, S. Hong, S. Han, Y.D. Suh, S.E. Lee, J. Yeo, S.S. Lee, D. Lee, S.H. Ko, Highly stretchable or transparent conductor fabrication by a hierarchical multiscale hybrid nanocomposite, *Adv. Funct. Mater.*, 24(36), 2014, 5671–5678.

46. J. Liang, L. Li, D. Chen, T. Hajagos, Z. Ren, S.-Y. Chou, W. Hu, Q. Pei, Intrinsically stretchable and transparent thin-film transistors based on printable silver nanowires, carbon nanotubes and an elastomeric dielectric, *Nat. Commun.*, 6, 2015, 1–10.
47. R. Liu, S.B. Lee, MnO_2/poly (3, 4-ethylenedioxythiophene) coaxial nanowires by one-step coelectrodeposition for electrochemical energy storage, *J. Am. Chem. Soc.*, 130(10), 2008, 2942–2943.
48. K.-Y. Chun, Y. Oh, J. Rho, J.-H. Ahn, Y.-J. Kim, H.R. Choi, S. Baik, Highly conductive, printable and stretchable composite films of carbon nanotubes and silver, *Nat. Nanotechnol.*, 5(12), 2010, 853–857.
49. M.-S. Lee, K. Lee, S.-Y. Kim, H. Lee, J. Park, K.-H. Choi, H.-K. Kim, D.-G. Kim, D.-Y. Lee, S. Nam, J.U. Park, High-performance, transparent, and stretchable electrodes using graphene–metal nanowire hybrid structures, *Nano Lett.*, 13(6), 2013, 2814–2821.
50. M.-S. Lee, J. Kim, J. Park, J.-U. Park, Studies on the mechanical stretchability of transparent conductive film based on graphene-metal nanowire structures, *Nanoscale Res. Lett.*, 10, 2015, 1–9.

Index

Printed in the United States
by Baker & Taylor Publisher Services